Fenugreek

Fenugreek

Traditional and Modern Medicinal Uses

Edited by
Dilip Ghosh & Prasad Thakurdesai

CRC Press
Taylor & Francis Group
Boca Raton   London

CRC Press is an imprint of the
Taylor & Francis Group, an **informa** business

First edition published 2022
by CRC Press
6000 Broken Sound Parkway NW, Suite 300, Boca Raton, FL 33487–2742

and by CRC Press
2 Park Square, Milton Park, Abingdon, Oxon, OX14 4RN

© 2022 selection and editorial matter, Dilip Ghosh & Prasad Thakurdesai, individual chapters, the contributors

CRC Press is an imprint of Taylor & Francis Group, LLC

Reasonable efforts have been made to publish reliable data and information, but the author and publisher cannot assume responsibility for the validity of all materials or the consequences of their use. The authors and publishers have attempted to trace the copyright holders of all material reproduced in this publication and apologize to copyright holders if permission to publish in this form has not been obtained. If any copyright material has not been acknowledged please write and let us know so we may rectify in any future reprint.

Except as permitted under U.S. Copyright Law, no part of this book may be reprinted, reproduced, transmitted, or utilized in any form by any electronic, mechanical, or other means, now known or hereafter invented, including photocopying, microfilming, and recording, or in any information storage or retrieval system, without written permission from the publishers.

For permission to photocopy or use material electronically from this work, access www.copyright.com or contact the Copyright Clearance Center, Inc. (CCC), 222 Rosewood Drive, Danvers, MA 01923, 978–750–8400. For works that are not available on CCC please contact mpkbookspermissions@tandf.co.uk

Trademark notice: Product or corporate names may be trademarks or registered trademarks and are used only for identification and explanation without intent to infringe.

Library of Congress Cataloging-in-Publication Data
[Insert LoC Data here when available]

ISBN: 9780367536572 (hbk)
ISBN: 9780367566821 (pbk)
ISBN: 9781003082767 (ebk)

DOI: 10.1201/9781003082767

Typeset in Times LT Std
by Apex CoVantage, LLC

Contents

Editors ix
Contributors xi
Introduction xiii
Bhushan Patwardhan

Part I Traditional Wisdom

Chapter 1 Fenugreek (*Methika*): Traditional Wisdom and Research Evidences 3

Tanuja Manoj Nesari, Bhargav Bhide and Shivani Ghildiyal

Chapter 2 Fenugreek in Traditional Persian Medicine 15

Narges Tajik, Reza Mohammadi Nasab, Nazli Namazi and Mohammad Hossein Ayati

Chapter 3 Distribution, Biology, and Bio-Diversity of Fenugreek 21

R. S. Sharma and S. R. Maloo

Part II Chemistry

Chapter 4 Title Optimization of the Supercritical Carbon Dioxide Extraction of Phytochemicals from Fenugreek Seeds 33

Aleksandra Bogdanovic, Vanja Tadic, Slobodan Petrovic and Dejan Skala

Chapter 5 Towards the Importance of Fenugreek Proteins: Structural, Nutritional, Biological, and Functional Attributes 51

Samira Feyzi and Mehdi Varidi

Chapter 6 Extraction and Optimization of Saponin and Phenolic Compounds of Fenugreek Seed 77

Sweeta Akbari, Nour Hamid Abdurahman and Rosli Mohd Yunus

Part III Sports Supplements and Nutraceuticals

Chapter 7 Fenugreek: Nutraceutical Properties and Therapeutic Potential 91

Dilip Ghosh

Chapter 8 Fenugreek (*Trigonella foenum-graecum*): A Unique Plant with Potential Health Benefits in Sports Nutrition 103

Anand Swaroop

Chapter 9 Applications of Fenugreek in Sports Nutrition 113

Colin Wilborn and Aditya Bhaskaran

Part IV Modern Medicinal Applications

Chapter 10 The Effects of Fenugreek on Controlling Glucose in Diabetes Mellitus: An Overview of Scientific Evidence 129

Zahra Ayati, Nazli Namazi, Mohammad Hossein Ayati, Seyed Ahmad Emami and Dennis Chang

Chapter 11 Fenugreek in Management of Primary Hyperlipidaemic Conditions 149

Subhash L. Bodhankar, Amit D. Kandhare and Amol P. Muthal

Chapter 12 The Effects of Fenugreek on Cardiovascular Risk Factors: What Are Potential Mechanisms? 173

Nazli Namazi, Neda Roshanravan, Sayyedeh Fatemeh Askari, Nasrin Sayfouri and Mohammad Hossein Ayati

Chapter 13 *Trigonella foenum-graecum* L.: Therapeutic and Pharmacological Potentials— Special Emphasis on Efficacy against Inflammation and Pain 187

Sindhu G., Chithra K. Pushpan and Helen A.

Chapter 14 Medicinal Potential of Fenugreek in Neuropathy and Neuroinflammation Associated Disorders 209

Aman Upaganlawar, Chandrashekhar Upasani and Mayur B. Kale

Chapter 15 Fenugreek in Management of Neurological and Psychological Disorders 229

Rohini Pujari and Prasad Thakurdesai

Chapter 16 Fenugreek in Management of Female-Specific Health Conditions 259

Urmila Aswar and Deepti Rai

Chapter 17 Potential of Fenugreek in Management of Kidney and Lung Disorders 283

Amit D. Kandhare, Anwesha A. Mukherjee-Kandhare and Subhash L. Bodhankar

Chapter 18 Potential of Fenugreek in Management of Fibrotic Disorders 305

Amit D. Kandhare, Sunil Bhaskaran and Subhash L. Bodhankar

Contents

Chapter 19 Fenugreek in Management of Immunological, Infectious, and Malignant Disorders 319

Rohini Pujari and Prasad Thakurdesai

Part V Formulations and Food Preparations

Chapter 20 Fenugreek: Novel Delivery Technologies and Versatile Formulation Excipients 361

Ujjwala Kandekar, Sunil Ramdasi and Prasad Thakurdesai

Chapter 21 Applications of Fenugreek in Nutritional and Functional Food Preparations 401

Ujjwala Kandekar, Rohini Pujari and Prasad Thakurdesai

Part VI Regulatory Aspects

Chapter 22 Fenugreek Based Products in USA, Australia, Canada, and India 439

Savita Nimse and Sanjeevani Deshkar

Index 451

Editors

Dr. Dilip Ghosh has received his PhD in biomedical science from India & post-doc from USDA-ARS, HNRCA at Tufts University, Boston. He is an international speaker, facilitator and author and professionally associated with Nutriconnect, Australia; Adjunct-Industry Fellow, NICM Health Research Institute, Western Sydney University. He is also an Adjunct Professor, KASTURBA HEALTH SOCIETY, Medical Research Center, Mumbai, India. He is a fellow of American College of Nutrition (ACN), professional member of Australian Institute of Food Science & Technology (AIFST), an advisor and executive board member of Health Foods and Dietary Supplements Association (HADSA), & The Society for Ethnopharmacology, India (SFE-India) and also in editorial board of several journals. He is also on board of Vitafoods Content Advisory Committee. His research interest includes oxidative stress, bioactive, clinically proven functional food and natural medicine development, regulatory and scientific aspects of functional foods, nutraceuticals and herbal medicines.

Dr. Ghosh has published more than 100 papers in peer reviewed journals, numerous articles in food and nutrition magazines and books. His most recent two books, "Pharmaceutical to Nutraceutical: A Paradigm shift in disease prevention" & "Natural Medicines-Clinical efficacy, Safety and Quality" under CRC Press, USA has been published in 2017 & 2019. His latest book "Nutraceutical in Brain Health & beyond" is just published by Elsevier/Academic Press.

Dr. Prasad Thakurdesai, is Chief Scientific Officer at Indus Biotech Private Limited, Pune, India. He acquired the Bachelor of Pharmacy (B. Pharm), Master of Pharmacy (Pharmacology), and Doctor of Philosophy (PhD) in Pharmaceutical Sciences from Nagpur University, Nagpur, India. Besides, he holds many certifications on Patent drafting and IP management from World Intellectual Property Organization (WIPO), Geneva, Switzerland. He has 30 years of professional experience, including 17 years in reputed educational and research institutions of Pharmaceutical Sciences, namely Birla Institute of Science and Technology (BITS, Pilani) and Poona College of Pharmacy, Bharati Vidyapeeth University, Pune, India before joining present position in year 2008.

He is involved in wide areas of scientific research and application areas, namely efficacy studies in animal models of diseases, toxicology and safety studies, pharmacokinetics, clinical studies, standardization, authentification, formulation, stability and international regulatory affairs related to natural phytoconstituents for nutraceuticals, complementary medicines, and anti-aging segments.

He has more than 100 peer-reviewed international journal publications, 60 conference presentations, 8 awards for best research papers, one book, 2 book chapters, 11 magazine articles, and many invited lectures at international conferences and exhibitions to his credit. He has guided 4 PhD and 32 M.Pharm students for their dissertations. He is a member of many professional organizations in the area of Pharmacy, Pharmacology, Animal Welfare, Medicine writing, Medical informatics, Quality of Life, Toxicology and Community Pharmacy.

Contributors

Nour Hamid Abdurahman
Department of Chemical Engineering
Universiti Malaysia Pahang
Pahang, Malaysia

Sweeta Akbari
Department of Chemical Engineering
Universiti Malaysia Pahang
Pahang, Malaysia

Urmila Aswar
Department of Pharmacology
Bharati Vidyapeeth (Deemed to be University)
Pune, India

Zahra Ayati
NICM Heath Research Institute
Western Sydney University
Westmead, Australia

Aditya Bhaskaran
Legal Department
Indus Biotech Limited
Pune, India

Sunil Bhaskaran MD,
Indus Biotech Limited
Pune, Maharashtra, India

Bhargav Bhide
All India Institute of Ayurveda
New Delhi, India

Subhash L. Bodhankar
Bharati Vidyapeeth (Deemed to be University)
Poona College of Pharmacy
Pune, India

Aleksandra Bogdanovic
Faculty of Technology and Metallurgy
University of Belgrade
Karnegijeva, Serbia

Dennis Chang
NICM Heath Research Institute
Western Sydney University
Westmead, Australia

Sanjeevani Deshkar
Department of Pharmaceutics
Dr. D. Y. Patil Institute of Pharmaceutical Sciences & Research, Pimpri, Pune, India

Mayur B. Kale
Division of Neuroscience Smt. Kishoritai Bhoyar College of Pharmacy, Kamptee, Nagpur, India

Ujjwala Kandekar
JSPMs Rajarshi Shahu College of Pharmacy and Research
Pune, India

Amit D. Kandhare
Bharati Vidyapeeth (Deemed to be University)
Pune, India

Reza Mohammadi Nasab
Department of History of Medicine
Tabriz University of Medical Sciences
Tabriz, Iran

Anwesha A. Mukherjee-Kandhare
Bharati Vidyapeeth (Deemed to be University)
Pune, India

Amol P. Muthal
Bharati Vidyapeeth (Deemed to be University)
Pune, India

Nazli Namazi
Endocrinology and Metabolism Research Center
Tehran University of Medical Sciences,
Tehran, Iran

Tanuja Manoj Nesari
All India Institute of Ayurveda
New Delhi, India

Savita Nimse
Regulatory Affairs and Formulations
Indus Biotech Limited
Pune, India

Slobodan Petrovic
Faculty of Technology and Metallurgy, University of Belgrade, Karnegijeva 4, Belgrade, Serbia

Rohini Pujari
School of Pharmacy
Dr. Vishwanath Karad MIT World Peace University
Pune, Maharashtra, India

Deepti Rai
Department of Pharmacology
Poona College of Pharmacy
Pune, India

Sunil Ramdasi
Indus Biotech Limited
Pune, India

Vanja Tadic
Institute for Medical Plant Research
Belgrade, Serbia

Narges Tajik
Department of History of Medicine
Tehran University of Medical Sciences
Tehran, Iran

Prasad Thakurdesai
Indus Biotech Limited
Pune, India

Aman Upaganlawar
Department of Pharmacology
SNJBs Shriman Sureshdada Jain College of Pharmacy
Nashik, India

Chandrashekhar Upasani
Department of Pharmacology
SNJBs Shriman Sureshdada Jain College of Pharmacy
Nashik, India.

Colin Wilborn
Mayborn College of Health Sciences
University of Mary Hardin-Baylor,
Texas,, USA

Rosli Mohd Yunus
Faculty of Chemical & Process Engineering Technology
Universiti Malaysia Pahang
Pahang, Malaysia

Introduction

Bhushan Patwardhan

Nature has provided us most of the resources such as food, water, shelter, and many other things that human beings need to survive and thrive. Nature provided us valuable resources in the form of fruits, vegetables, grains, nutrients, and medicinal plants. One such medicinal plant is fenugreek (scientific name: *Trigonella foenum graecum*). Fenugreek is an essential ingredient documented in traditional culinary uses such as lactational aid, post-pregnancy muscle tightening, and prevention of obesity. At the same time, traditional medicinal literature of numerous countries mentions fenugreek as an essential source of treatment for many ailments such as diabetes and arthritis.

Fenugreek has been scientifically studied for chemistry, pharmacology, clinical, and health promotive benefits. Fenugreek, with its phytoconstituents such as bioactive amino acids, alkaloid, glycosides, soluble fibers, and phenolic compounds, could be a good candidate for herbal drugs. The list of potential health benefits of fenugreek is continuing to grow with new clinical evidence. It is widely used as a dietary supplement for diabetes, menstrual disorders, and as a galactagogue during breastfeeding. The search of online scientific databases results in a large number of scientific studies in publications on the efficacy and safety of fenugreek and derived phytoconstituents. For example, a PubMed search on fenugreek results in about 1,500 articles including over 50 RCTs (randomized controlled trials), systematic reviews, and meta-analysis. The National Center for Complementary and Integrative Health, National Institutes of Health describes fenugreek to be safe in the amounts commonly found in foods.

This 23-chapter book is a compilation of scientific studies, carried out by various researchers around the world on medicinal applications of fenugreek seed-based products. This book covers relevant evidence for the traditional and modern medicinal applications of fenugreek and constituents to provide good nutrients, maintain health, provide wellness, prevent diseases, and manage disorders. This book may answer a few common questions about research or evidence such as what the possible health benefits and risks are, mechanism of action, and clinical evidence from randomized controlled human studies. The book will be a helpful resource on fenugreek as an ingredient for medicinal products beyond most popular applications such as diabetes. The book is a compilation of different chapters written by highly experienced industrial and academic professionals, from broad areas of applications of fenugreek in food preparations, nutraceuticals, and complementary medicines for diverse physiological benefits.

Many Indian Ayurvedic plants have previously been used for preparation of new drugs or have shown promising results. The list of potential health benefits of fenugreek is continuing to grow with new clinical reports. Therefore, fenugreek, which possesses phenolic compounds, bioactive amino acids, glycosides, and antioxidants, might be a good candidate for natural product drug discovery.

Despite the impressive scientific and clinical profile of fenugreek, consumer understanding is still in its infancy and scientific promotion of this herb needs to focus on important commercial hurdles. This book will help to enhance the awareness about existing scientific knowledge of fenugreek and encourage future scientific research on fenugreek towards better and safer nutritional and medicinal applications.

This book is divided into broad areas of applications:

1. Introduction (Chapter 1)
2. Traditional wisdom of medicinal applications (Chapters 2, 3, 4)
3. Chemistry (Chapters 5, 6, 7)

4. Sports supplements and nutraceuticals (Chapters 8, 9, 10)
5. Modern medicinal applications (Chapters 11–20)
6. Formulation and food preparations (Chapters 21, 22)
7. Regulatory aspects (Chapter 23)

Some of the features of the book are:

- Comprehensive collection of all studies on fenugreek seeds with respect to their efficacy, safety, and mechanism (for scientific researchers, students, college and university libraries)
- Nutraceutical/dietary supplement product development of products (for manufacturers of nutraceutical and dietary supplement companies)
- Fenugreek-based products as a part of sport and excursive physiology supplement stack to satisfy the fitness objectives and achievements (for gym goers, gym trainers, fitness freaks, exercise and sports medicine practitioners)
- Fenugreek-based nutritional food products (e.g. high fiber, low fat, and ultra-low carb) (for dietitians and nutritionist)
- Fenugreek-based products as evidence-based prescription in diseases management (for general practitioners, naturopaths, Ayurvedic practitioners)
- Risk assessment and claim validation of fenugreek-based products (for regulatory bodies, regulatory professionals, toxicologists)

Dr. Bhushan Patwardhan, PhD, FNASc, FAMS
National Research Professor—AYUSH
Interdisciplinary School of Health Sciences
Savitribai Phule Pune University
Ganeshkind, Pune 411007

Part I Traditional Wisdom

1 Fenugreek (*Methika*) Traditional Wisdom and Research Evidences

Tanuja Manoj Nesari, Bhargav Bhide and Shivani Ghildiyal

CONTENTS

1.1 Introduction 3
1.2 Worldwide Ethnomedicinal Uses of Fenugreek 4
1.3 Pharmacodynamics Attributes of Fenugreek as per *Ayurveda* 5
1.4 Experiential Uses of Fenugreek as per *Ayurveda* 7
1.5 Types of Fenugreek as per *Ayurveda* 7
1.6 Research Evidence 7
1.7 Toxicology 11
1.8 Conclusion 12
References 13

1.1 INTRODUCTION

Fenugreek (*Methika*) is an ancient medicinal herb extensively used worldwide as food and medicine. It is botanically identified as *Trigonella foenum-graecum* Linn, a small herb of family Papilionaceae. The name of the genus, *Trigonella*, is derived from the old Greek word, denoting 'three-angled', referring to the triangular shape of the seeds. The first use of fenugreek was described in an ancient Egyptian papyrus dated 1500 BC.[1]

It is native to the Mediterranean, India, China, Northern Africa, and the Ukraine. Due to its utility fenugreek is extensively cultivated in various parts of the globe for its dietary, bakery, and medicinal value.[2] It is one of the oldest medicinal plants to have a food-based tradition, used for making, for example, stew with rice in Iran, flavoured cheese in Switzerland, syrup and bitter rum in Germany, and flat bread in Egypt. The young seedlings are eaten as a vegetable and its roasted grain as a coffee substitute in Africa.[3]

Its seeds are considered to be of commercial interest as a source of a steroid diosgenin, which is important to the pharmaceutical industry. The plant is known from ancient times for its medicinal value; both seeds and leaves are reported in use under ethnomedicine as abortifacients, appetite stimulants, galactagogues, and as beneficial in baldness, boils, bronchitis, cellulitis, constipation, cough, diarrhoea, eczema, flatulence, hepatic disease, hernia, indigestion, leg ulcers, myalgia, post-menopausal vaginal dryness, hyperglycaemia, tuberculosis, and wounds.[4]

In *Ayurveda*, fenugreek is mentioned as *Methika*. The leaves and seeds of the plant are used as medicine and food. However, the herb is not cited in ancient Ayurvedic classics, such as *Charakasamhita*, *Sushrutasamhita*, *Ashtangahridya*, but later it is described in various Ayurvedic lexicons (*Nighantus*) with a number of synonyms and therapeutic indications.

As per *Ayurveda*, it is bitter, pungent in taste, hot in potency, and has mild purgative action. It has *Vata kapha* pacifying potential and is beneficial for indigestion, loss of appetite, fever, and *Vata* disorders (neurological problems).[5]

DOI: 10.1201/9781003082767-2

Both single and compound formulations of the herb are enumerated in *Ayurveda* classics. Along with single use of fenugreek, compound formulations *Panchajirkapaka* and *Chaturbeeja* (combination of four seeds) are widely used in *Ayurveda*.[6,7]

The contemporary research has shown its hypoglycaemic, hypocholesterolaemic, anti-inflammatory, anti-cancer, and phyto-oestrogenic potential in preclinical and clinical studies. It is reported that fenugreek seed as a simple dietary supplement can modulate glucose, cholesterol, LDL cholesterol, and triglycerides. A decrease in serum T3 and T3/T4 ratio and an increase in T4 levels have been reported in mice and rats after use of fenugreek.[8]

Fenugreek has a long history of safety and is well tolerated by most. However, it is reported animal studies[9] that fenugreek has the potential to stimulate the uterus, thus fenugreek is contraindicated during pregnancy. For agriculture, fenugreek is a very useful legume crop for incorporation into short-term rotation and for hay and silage for livestock feed, for fixation of nitrogen in soil, and to enhance soil fertility.[10]

The biological and pharmacological actions of fenugreek are attributed to the variety of its constituents, namely steroids, N-compounds, polyphenolic substances, volatile constituents, amino acids. Fenugreek seed contains 45–60% carbohydrates, mainly mucilaginous fibre (galactomannans); 20–30% proteins high in lysine and tryptophan; 5–10% fixed oils (lipids); pyridine alkaloids – mainly trigonelline (0.2–0.38%), choline (0.5%), gentianine, and carpaine; the flavonoids apigenin, luteolin, orientin, quercetin, vitexin, and isovitexin; free amino acids, such as 4-hydroxyisoleucine (0.09%), arginine, histidine, and lysine; calcium and iron; saponins (0.6–1.7%); glycosides yielding steroidal sapogenins on hydrolysis (diosgenin, yamogenin, tigogenin, neotigogenin); cholesterol and sitosterol; vitamins A, B_1, C, and nicotinic acid; and 0.015% volatile oils (n-alkanes and sesquiterpenes).[4]

1.2　WORLDWIDE ETHNOMEDICINAL USES OF FENUGREEK

Ethnomedicine practices are the roots of present medicine. The history of fenugreek revealed that it has been in use since ancient times in various continents of the globe. Fenugreek is one of the oldest medicinal plants used in Rome and Egypt to ease childbirth and to increase milk flow. In ancient Rome, it was used to aid labour and delivery.[11] Even today, Egyptian women use this plant as Hilba tea to alleviate menstrual pains and sedate tummy problems.[12]

In China, seeds of fenugreek are used as a tonic in treatment for weakness and oedema of the legs. In the traditional medicine of China this plant has also been used to boost physique, to treat weakness of body and gout. In India, it is used as a condiment and as medicine for lactic stimulation, and as treatments of indigestion and baldness.[11]

In Iran leaves of fenugreek have been used to alleviate cold, cough, splenomegaly, hepatitis, backache, and bladder cooling reflex while seeds are used as a local emollient, a poultice for local inflammation, and as a demulcent to alleviate pain in joints (arthralgia). Infusion of fenugreek mixed with honey is recommended to treat asthma and internal oedemas. Its oil is useful for hair. This plant is used to treat skin diseases like black spots. The boiled form of fenugreek helps treat red eyes and helps soften the throat and chest and provides relief from cough. The herb is locally used as an emollient in treatment of pellagra, loss of appetite, gastrointestinal disorders, and as a general tonic. It is also useful in diabetes, eliminating mouth odour, undesired body odour and sweat, and as a laxative.[12] It has been traditionally used for gastrointestinal, pulmonary, and gynaecologic diseases.[13]

It is used for kidney problems and conditions affecting the male reproductive tract. Seeds are used as a pessary to treat cervical cancer. The aerial parts of the plant are a folk remedy for abdominal cramps associated with both menstrual pain and diarrhoea or gastroenteritis.

Fenugreek has a beneficial blood detoxification potential. It is also used for colds, influenza, catarrh, constipation, bronchial complaints, asthma, emphysema, pneumonia, pleurisy, tuberculosis, sore throat, laryngitis, hay fever, and sinusitis.

Fenugreek (Methika)

TABLE 1.1
A Glimpse of Worldwide Traditional Uses of Fenugreek

S.N.	Country	Traditional Use in Various Conditions
1.	India	Condiment and as medicine for enhancing lactation, treatments of indigestion and baldness.
2.	Iran	Cold, cough, splenomegaly, hepatitis, backache, local emollient, a poultice for local inflammation, and as a demulcent to alleviate pain of joints, etc.
3.	China	To treat weakness of body, oedema of the legs, gout, and to boost physique.
4.	Rome	Used to aid labour and delivery.
5.	Egypt	As Hilba tea to alleviate menstrual pains and sedate tummy problems.

FIGURE 1.1 Traditional uses of fenugreek.

It is often used in many teas and other products that help balance women's hormones. It is also used as a remedy to ease child birth for pregnant women and to minimize the labour pain.[14]

1.3 PHARMACODYNAMICS ATTRIBUTES OF FENUGREEK AS PER *AYURVEDA*

Ayurvedic scriptures treasure the basic principle of understanding of the Dravyas and their actions. These are known as pharmacodynamic attributes (Rasapanchaka) and include *Rasa* (taste), *Guna* (properties), *Vipaka* (biotransformation), and *Veerya* (potency).

Rasa is the object of the gustatory sense organ and is located in substance. *Rasa* is perceived through contact with the gustatory sense organ. Thus *rasa* is known from perception and also from inference on the basis of its characters such as effect on salivation, etc. *Rasa* are six in number—*madhura* (sweet), *amla* (sour), *lavana* (salty), *katu* (pungent), *tikta* (bitter), and *kashaya* (astringent). Each *Rasa* has its properties and actions.

Guna (quality or property) is defined as properties inherently existent in substance. *Guna* are 41 in number and are classified into four groups—somatic, psychic, physical, and applicative. For determination of drug action, understanding of physical properties is important.

Vipaka is the term for the final transformed state of substance after digestion. *Vipaka* determines the future course and action of the substance in body. Mainly three *Vipaka* are ascribed in *Ayurveda*, *madhura* (sweet), *amla* (sour), and *katu* (pungent). The ultimate action of substance depends on *Vipaka*.

Veerya is power or potency, which is the means of action of a substance. In common practice, grossly it is represented by the active fraction of substance. The power is located inherently in the active drug; there it is also concentrated in a particular portion. That is why specific useful parts of herbs are enumerated in *Ayurveda* classics. In *Ayurveda* mainly two *Veerya*, *Sheeta* (cold) and *Ushna* (hot) are mentioned.

Thus to understand a *Dravya* (*Ahara* or *Aushadhi*), one should understand the pharmacodynamics of that *Dravya*. Hence, rational use of drugs is mentioned on the basis of these pharmacodynamic attributes. Further, by observing the action of a drug, one can have an understanding of the pharmacodynamic attributes of drug.

The pharmacodynamics attributes of fenugreek are as follows:[5]

Rasa (taste): *Katu* (pungent)
Guna (properties): *Snigdha*, *Sara* (mild purgative)
Vipaka (biotransformation): *Katu*
Veerya (potency): *Ushna* (hot)

Thus due to its *katu rasa* and *katu vipaka* fenugreek has digestive and appetizing potential. Its *snighdha guna* (unctuous property) and *ushna veerya* (hot potency) helps to pacify *Vata* and make it a reputed drug for *Vatika* (Neurological) disorders.

Various other lexicons (*Nighantu*) have also mentioned the properties of fenugreek on the basis of three *Doshas* (humours) of the body and systems of the body. The properties and actions of fenugreek are listed in the Table 1.2.

TABLE 1.2
Properties and Uses of Fenugreek (*Methika*) from Various *Ayurveda* Lexicons

S.N.	Name of Ayurvedic Lexicon (*Nighantu*)	Properties	Uses
1.	Dhanvantari Nighantu[15]	Pungent, hot potency, produces disorders of *Rakta and Pitta*, improves taste of mouth and digestive power, and pacifies *Vatadosha*.	Indigestion, loss of appetite, *Vata* disorders (neurological disorders)
2.	Bhavaprakasha Nighantu[5]	Pacifies *Vata Kapha Dosha*.	*Vata* and *Kapha* disorders, fever
4.	Madhavadravyaguna[16]	Wholesome (*Pathya*). Pacifies three *Doshas*, especially *Vata*, and a mild purgative (*Sara*).	To balance three *Doshas*, purgative
5.	Rajanighantu[17]	Pungent, hot potency, produces disorders of *Rakta* (blood) and *Pitta*, improves taste of mouth and digestive power, and pacifies *Vatadosha*.	Indigestion, *Vata* disorders (neurological problems)
6.	Siddhamantra[18]	Pacifies *Vata* and *Kapha* and not increasing *pitta*.	*Vata* disorders (neurological disorders)
7.	Sodhal Nighantu[19]	Extremely bitter taste, hot potency, stops vomiting, pacifies *Vata* and *Kapha*, improves speech, memory, and reduces diseases by *Kapha* and *Vata*.	Vomiting, *Vata* disorders (neurological disorders), to improve memory.

Fenugreek (Methika)

1.4 EXPERIENTIAL USES OF FENUGREEK AS PER *AYURVEDA*[5]

A: EXTERNAL USES:

- The application of seeds is beneficial for good hair growth.
- The leaves paste is beneficial to cure inflammation and burning.

B: INTERNAL USES:

- **Nervous system:** The seeds of fenugreek are beneficial in nervous disorders.
- **Digestive system:** Fenugreek is an appetizer and digestive. The seed powder is beneficial to enhance the appetite. It is also beneficial in abdominal pain.
- **Reproductive system:** The seeds are galactagogue. Seeds are given to women during puerperal stage.

1.5 TYPES OF FENUGREEK AS PER *AYURVEDA*

In *Ayurveda* two types of fenugreek are mentioned, *Methika* (*Trigonella foenum-graecum* Linn.) and *Vanamethika* (*Meliotus parviflora* Desf.), a wild variety of fenugreek. Further it is stated that *Vanamethika* is inferior to *Methika*. The seeds of *Vanamethika* are said to be useful in abdominal cramps, diarrhoea, and dysmenorrhoea.[5] *Vanamethika* is more beneficial for animals as fodder.

1.6 RESEARCH EVIDENCE

1. Effective adjutant for management of type 2 diabetes mellitus

    In a randomized case-control study fenugreek seeds aqueous extract (1.32 gram) was evaluated on type 2 diabetes mellitus (T2DM) subjects. The seed extract was given as adjuvant for three months. The effects was recorded on glycaemic status, insulin resistance, homeostatic model assessment, safety parameters, and antioxidant superoxide dismutase. The study showed that glucose, HbA1C, and malondialdehyde were significantly lowered after three months in a group receiving adjunct therapy. Insulin resistance, anthropometric and antioxidant status were improved while haemoglobin, alanine transaminase, and renal functions remained unaltered. Thus fenugreek seeds aqueous extract may be used as an effective and safe adjuvant for management of T2DM.[20]

2. Effective and safe as add-on therapy in type 2 diabetes mellitus patients with insulin therapy

    A prospective, single arm, open-label, uncontrolled, multicentre trial of herbal formulation rich in standardized fenugreek seed extract (IND-2) as add-on therapy done on type 2 diabetes mellitus (T2DM) patients who were on insulin treatment to assess the safety and efficacy of add-on therapy. T2DM patients (n = 30) aged 18–80 years who were stabilized on insulin treatment with fasting blood sugar (FBS) levels between 100–140 mg/dL received IND-2 capsules (700 mg, thrice a day) for 16 weeks. The study suggested that there was no significant reduction in FBS and post-prandial blood sugar level after add-on therapy of IND-2. However, add-on therapy of IND-2 significantly reduced ($p < 0.01$) the HbA1C values, requirements of insulin and hypoglycaemic events as compared with baseline. Additionally, add-on therapy of IND-2 did not produce any serious adverse events. The results of investigation suggested that add-on therapy of IND-2 with insulin in T2DM patients improves glycaemic control through a decrease in levels of HbA1C and number of insulin doses needed per day without an increase in body weight and risk of hypoglycaemia. Thus, IND-2 may provide a safe and well-tolerated add-on therapy option for the management of T2DM.[21]

3. Effective and safe add-on therapy in type 2 diabetes mellitus patients with sulfonylurea therapy

A randomized, double-blind, placebo-controlled, multicentre study of herbal formulation rich in standardized fenugreek seed extract (IDM1) as an add on therapy was done on type 2 diabetes mellitus (T2DM) patients controlled on sulfonylurea monotherapy. At week 12, adjusted fasting plasma glucose (FPG) (20 mg%), postprandial plasma glucose (PPPG) (26 mg%), and glycated haemoglobin (HbA1C) (0.9 mg%) were reduced significantly ($p < 0.05$) from baseline as compared to placebo group (FPG: 7 mg%; PPPG: 4 mg%, and HbA1C: 0.4 mg%). These beneficial effects were seen as early as one month after consumption of IDM1 and continued until at least 15 days after withdrawal of IDM1. There were no major changes in body weight, haematology, and biochemistry at week 12 as compared to baseline. So herbal formulation rich in standardized fenugreek seed extract (IDM1) is safe, effective, and a well-tolerated add-on oral medication therapy that supports healthy blood sugar levels and glycosylated haemoglobin levels in T2DM patients inadequately controlled with a sulfonylurea.[22]

4. Therapeutic ability of fenugreek against type 2 diabetes and breast cancer

The role of fenugreek seed constituents against type 2 diabetes and breast cancer was done through molecular docking and molecular dynamics simulation-based computational drug discovery methods. The computational results revealed that the compound galactomannan is a potential drug candidate against breast cancer and type 2 diabetes rendered by higher molecular dock scores, stable molecular dynamics (MD) simulations results, and lower binding energy calculations.[23]

5. Efficacy of novel fenugreek seed (*Trigonella foenum-graecum*) extract (Fenfuro™) in patients with type 2 diabetes

A randomized, placebo-controlled, double-blind, add-on clinical study was conducted over a period of 90 consecutive days in which the efficacy of fenugreek was evaluated (daily dosage: 500 mg bid) in 154 subjects (male: 108; female: 46; age: 25–60 years) with type 2 diabetes. The study reveals that fenugreek caused significant reduction in both fasting plasma and post-prandial blood sugar levels. Approximately 83% of the subjects reported decreases in fasting plasma sugar levels in the fenugreek-treated group as compared to 62% in the placebo group, while 89% of the subjects demonstrated reduction in post-prandial plasma sugar levels in the fenugreek-treated group as compared to 72% in the placebo group. HbA1C levels were reduced in both placebo and treatment groups. The decrease in HbA1C levels was significant in both groups as compared to respective baseline values. A significant increase in fasting and post-prandial C-peptide levels compared to the respective baseline values was observed, while no significant changes in fasting and post-prandial C-peptide levels were observed between the two groups. No significant adverse effects were observed by blood chemistry analyses. Furthermore, 48.8% of the subjects reported reduced dosage of anti-diabetic therapy in the Fenugreek-treated group, whereas 18.05% reported reduced dosage of anti-diabetic therapy in the placebo group.[24]

6. Effects of fenugreek seeds on oxidative stress, pro-inflammatory cytokines and nirite as adjuncts to diet in patients with hyperlipidaemia

In a randomized, double-blind, placebo-controlled trial, the effects of fenugreek seeds on atherosclerotic risk factors, namely inflammation and nitrite deficiency, were examined. All subjects (n = 61) with hyperlipidaemia were assigned to the National Cholesterol Education Program (NCEP) step 1 diet for a period of 12 weeks and then randomized to two different test agents in identical sachets for another 12 weeks. The test agents were fenugreek seed powder rich in polyphenols (60.0 g/day) and cellulose placebo (3.0 g/day). Tumour necrosis factor (TNF)-α and interleukin (IL)-6 showed significant decline from the baseline levels in the fenugreek group, whereas these changes were non-significant in the placebo group. The incremental differences from placebo for TNF-α (22.5%) and interleukin IL-6 (27.8%) were highly significant ($p < 0.001$). The incremental differences from placebo in nitrite (indicator of NO) for fenugreek was also significant (21.4%) associated with an increase in nitrite concentrations. These findings clearly indicated anti-inflammatory activity and NO-activating effects of fenugreek, which appear to be due to significantly higher intake of polyphenol flavonoids. Blood glucose and blood lipids were significantly decreased in the fenugreek group. Thus the fenugreek seeds, which are rich sources of fibre and polyphenol antioxidants, may inhibit atherothrombosis due to their anti-inflammatory, antioxidant, and nitric oxide (NO) activating effects.[5]

7. Antidiabetic effect of fenugreek seed powder solution (*Trigonella foenum-graecum* L.) on hyperlipidaemia in diabetic patients

    This study investigated the effect of *Trigonella foenum-graecum* seed powder solution on the lipid profile of newly diagnosed type 2 diabetic patients. The study was conducted on 114 newly diagnosed type 2 diabetic patients without any significant diabetes complication. The treatment group was given 25 g *Trigonella foenum-graecum* seed powder solution orally twice a day for one month and the second group (control) (n = 57) received metformin. By the end of the intervention period, the treatment group showed significantly lower total cholesterol level by 13.6% and low-density lipoprotein cholesterol level also reduced by 23.4% but the treatment group showed significantly increased high-density lipoprotein cholesterol level by 21.7% as compared to the baseline level. However, lipid profile levels in the control group were not significantly changed. The study concluded that the administration of *Trigonella foenum-graecum* seed powder solution had pronounced effects in improving lipid metabolism in type 2 diabetic patients with no adverse effects. Therefore, *Trigonella foenum-graecum* seed may provide new alternatives for the clinical management of type 2 diabetes.[26]

8. Beneficial effects of fenugreek glycoside supplementation in male subjects during resistance training: a randomized controlled pilot study

    Glycoside fraction of fenugreek (*Trigonella foenum-graecum*) seeds (Fenu-FG) was evaluated on physiological parameters related to muscle anabolism, androgenic hormones, and body fat in healthy male subjects during an eight-week resistance training program using a prospective, randomized, double-blind, placebo-controlled design. Sixty healthy male subjects were randomized to ingest capsules of Fenu-FG (1 capsule of 300 mg, twice per day) or the matching placebo at a 1:1 ratio. The subjects participated in a supervised four-day-per-week resistance training program for eight weeks. The outcome measurements were recorded at recruitment (baseline) and at the end of the treatment (eight weeks). Fenu-FG supplementation demonstrated significant anabolic and androgenic activity as compared with the placebo. Fenu-FG treated subjects showed significant improvements in body fat without a reduction in muscle strength or repetitions to failure. The Fenu-FG supplementation was found to be safe and well-tolerated. So it was concluded that Fenu-FG supplementation has beneficial effects in male subjects during resistance training without any clinical side effects.[27]

9. Phyto-oestrogenic effect of fenugreek seed extract helps in ameliorating the leg pain and vasomotor symptoms in postmenopausal women

    The randomized, double-blinded, placebo-controlled study investigated the effect of a unique extract of fenugreek (FHE), at a dose of 250 mg × 2/day for 42 days on hormone balance and postmenopausal discomforts. Postmenopausal women having characteristic postmenopausal symptoms, as assessed by MRS questionnaire, were randomized either to FHE (n = 24) or placebo (n = 24) groups. The FHE-treated participants reported the feel of well-being within two weeks of supplementation and further improvement in somatic, psychological, and urogenital scores towards the end of the study period. There were 2.9, 4.2, and 7.2 times reduction respectively in hot flashes, night sweats, and pain in leg muscles and joints with significant improvement in irritability and vaginal dryness when compared to placebo. Hormone analysis revealed significant increase in oestradiol, free testosterone, and progesterone concentrations. So it was concluded that supplementation of FHE regulated various hormones in postmenopausal women and offered a significant reduction in vasomotor effects and leg pain without showing any significant variations in haematological and biochemical parameters.[28]

10. Fenugreek husk (FenuSMART™) alleviates postmenopausal symptoms and helps to establish the hormonal balance

    A randomized, double-blinded, placebo-controlled study investigated the effect of 90-day supplementation of a standardized extract of fenugreek (*Trigonella foenum-graecum*) (FenuSMART™), at a dose of 1000 mg/day, on plasma oestrogens and postmenopausal discomforts. Eighty-eight women having moderate to severe postmenopausal discomforts and poor quality of life were randomized either to extract-treated (n = 44) or placebo (n = 44) groups. There was a significant ($p < 0.01$) increase in plasma oestradiol (120%) and improvements on various postmenopausal discomforts and quality of life of the participants in the extract-treated group, as compared with the baseline and placebo. Further 32% of the subjects in the extract group reported no hot flushes after supplementation. Therefore despite the widespread use of hormone replacement therapy, various reports on its side effects have generated an increasing interest in the development of safe natural agents for the management of postmenopausal discomforts and fenugreek seed extract may work well in this regard.[29]

11. Efficacy of a novel fenugreek seed extract (Furocyst) in polycystic ovary syndrome (PCOS)

    An open-label, one-arm, non-randomized, post-marketing surveillance study in 50 premenopausal women (18–45 years, BMI < 42) diagnosed with PCOS using a novel *Trigonella foenum-graecum* seed extract (fenugreek seed extract, Furocyst, 2 capsules of 500 mg/day) enriched in approximately 40% furostanolicsaponins, over a period of 90 consecutive days was done. Approximately 46% of study population showed reduction in cyst size, while 36% of subjects showed complete dissolution of cyst. Further 71% of subjects reported the return of a regular menstrual cycle on completion of the treatment and 12% of subjects subsequently became pregnant. Overall, 94% of patients benefitted from the regimen. Significant increases in luteinizing hormone (LH) and follicular stimulating hormone (FSH) levels were observed compared to the baseline values. Extensive blood chemistry, haematological and biochemical assays demonstrated the broad-spectrum safety. Furocyst caused significant decrease in both ovarian volume and the number of ovarian cysts. Serum ALT, BUN, and CK were assessed to demonstrate the broad-spectrum safety of Furocyst. No significant adverse effects were observed. In summary, Furocyst was efficacious in ameliorating the symptoms of PCOS.[30]

12. Fenugreek seed extract in polycystic ovarian syndrome (PCOS) patients

    An open-labelled, single-armed, single-centric, and noncomparative study on standardized fenugreek seed extract on reduction in ovarian volume and the number of ovarian cysts on 107 female patients suffering from PCOS was used. The novel fenugreek seed extract was used for a period of 12 weeks to determine its efficacy in reduction of ovary volume and number of ovarian cysts. Results indicates significant decrease in both the ovaries' volume (p-value 0.0001). More than 65% of the patients showed reduction in cyst size in both left and right ovaries. Fifteen patients got pregnant by the end of the study and HOMA Index was reduced in 75.67% of the study population. 79.5% of the study population had regular menstrual cycles at the completion of the study and prolactin levels were significantly reduced. Hirsutism score was significantly reduced ($p = 0.002$) at the end of 12 weeks of treatment. No changes were observed on LFT, KFT, and haemogram level. The fenugreek seeds extract was proven to be safe and effective in treating PCOS in women of reproductive age by reducing the cyst volume in both ovaries as well as cyst sizes.[31]

13. Efficacy of Furosap™, a novel *Trigonella foenum-graecum* seed extract, in enhancing testosterone level and improving sperm profile in male volunteers

One-arm, open-labelled, multicentre study was conducted in 50 male volunteers (ages 35 to 65 years) over a period of 12 weeks to determine the efficacy of Furosap (FS), an innovative, patented, 20% protodioscin-enriched extract developed in the laboratory from fenugreek seeds extract (500 mg/day/subject), on free and total testosterone levels, sperm profile, sperm morphology, libido and sexual health, mood and mental alertness, and broad spectrum safety parameters.

Free testosterone levels were improved up to 46% in 90% of the study population. 85.4% of the study population showed improvements in sperm counts. Sperm morphology improved in 14.6% of volunteers. The majority of the subjects enrolled in the study demonstrated improvements in mental alertness and mood. Furthermore, cardiovascular health and libido were significantly improved. No significant changes were observed in serum lipid function, cholesterol, triglyceride, HDL and LDL levels, haemogram (CBC), hepatotoxicity, and nephrotoxicity. Overall, the results demonstrated that FS was safe and effective in attenuating testosterone levels, healthy sperm profile, mental alertness, cardiovascular health, and overall performance in human subjects.[32]

1.7 TOXICOLOGY

The preclinical studies demonstrated the acute oral LD50 was >5 g/kg in rats, and the acute dermal LD50 was >2 g/kg in rabbits.[33] In another animal study, an acute and sub-chronic regime of fenugreek powder was reported safe without any signs of toxicity or mortality in mice and rats.[34] Further, there were no significant haematological, hepatic, or histopathological changes in weanling rats fed fenugreek seeds for 90 days.

In a clinical study, no clinical, hepatic, renal toxicity, and haematological abnormalities were reported in toxicological evaluation of 60 diabetic patients who took powdered fenugreek seeds at a dose of 25 g per day for 24 weeks.[35]

TABLE 1.3
Reported Pharmacological Activities of Fenugreek

S.N.	Activities Reported	References
1.	Type 2 diabetes mellitus	Dalvi, SM 2019; Kandhare, A 2018; Rampogu, S 2018, Lakshmi, E 2017; Verma, N 2016
2.	Breast cancer	Rampogu, S 2018
3.	Hyperlipidaemia	Vargova, V 2018; Geberemeskel, GA 2019
4.	Beneficial effects in male subjects during resistance training	Wanklede, S 2016
5.	Ameliorating effect in leg pain and vasomotor symptoms in postmenopausal women	Thomas, JV 2020
6.	Alleviating postmenopausal symptoms and helping to establish the hormonal balance	Shamshad Begum, S 2016
7.	Polycystic ovary syndrome (PCOS)	Swaroop, A 2016; Sankhwar. P 2018
8.	Enhancing testosterone level and improving sperm profile in male volunteers	Maheshwari, A 2017

FIGURE 1.2 Scientific evaluation and research evidences of fenugreek.

1.8 CONCLUSION

Spices are common food adjuncts generally consumed to enhance sensory quality of foods in terms of flavour, aroma, and colour. The quantity and variety of spices consumed are extensive in India. Spices such as garlic, onion, coriander, cumin, cinnamon, clove, cardamom, asafoetida, etc. impart typical aroma (cardamom, cinnamon, bay leaves, asafoetida), provide pungency (red chilli, pepper), and impart colours (turmeric) to foods. Besides contributing to the sensory aspects of food, spices also have long been recognized to possess health beneficial physiological effects. They act as stimulus to the digestive system, relieve digestive disorders, and some spices have antiseptic value. Their attributes such as carminative, stomachic, anti-dyspeptic, diuretic, anti-spasmodic, anti-inflammatory have led to therapeutic applications in the indigenous systems of medicine.

Ayurveda is an experiential science of life in which detailed description of various *Ahara* (dietary items) and *Aushada* (drug) substances are mentioned for health promotion and disease prevention. *Ayurveda* ascribed the use of spices like pepper, cinnamon, clove, bay leaves, long pepper, turmeric, coriander, fenugreek, nutmeg, ginger, and many more as a medicine for management of various disorders.

It is observed that fenugreek is used in various part of the world as a food and also as medicine since ancient times. In *Ayurveda* the single and compound use of herbs is in practice.

In *Ayurveda* the description of fenugreek appeared firstly in *Ayurveda* lexicons. Properties and uses of fenugreek by the name of *Methika* are mentioned in detail. It is mainly advised for digestive disorders, neurological problems, and to increase lactation.

One other wild variety, *Vanamethika (Mellotus parviflora)* is also mentioned in *Ayurveda* which is beneficial for horses for increasing strength.

This is a very less-explored herb thus on the basis of *Ayurveda* wisdom it may be explored further especially in nervous disorders.

The research studies indicated that fenugreek and its compounds are beneficial for various ailments such as type 2 diabetes mellitus, breast cancer, oxidative stress, hyperlipidaemia and postmenopausal hormone regulation, polycystic ovarian syndrome in females, and in regulating testosterone level and sperm profile in males (Figure 1.2).

Most of the studies are on its antidiabetic potential. However, the potential of the herb as a hair growth promoter and galactagogue may be explored more on the line of *Ayurveda* indications. Neurological disorders (*Vata* disorders) is also an untouched area where fenugreek has promising potential.

Thus, it can be stated that fenugreek is a very potential drug which can be explored and which will be helpful to mankind in many ways in the future if researched properly and thoroughly.

REFERENCES

1. Yadav Sakshi R, Biyani Dr. Dinesh M, Umekar Dr. Milind J. Trigonella foenum-graecum: A herbal plant review. *World Journal of Pharmaceutical Research*. 2019. 8(12):402–19.
2. Petropoulos GA, editor. *Fenugreek: The genus Trigonella*. CRC Press; 11 New Fetter Lane, London. 2002 Aug 22.
3. Bitarafan Z, Asghari HR, Hasanloo T, Gholami A, Moradi F, Khakimov B, Liu F, Andreasen C. The effect of charcoal on medicinal compounds of seeds of fenugreek (Trigonella foenum-graecum L.) exposed to drought stress. *Industrial Crops and Products*. 2019 May 1;131:323–9.
4. Mehrafarin A, Ghaderi A, Rezazadeh SH, Naghdi Bh, Nourmohammadi G, Zand ES. Fenugreek. Bioengineering of important secondary metabolites and metabolic pathways in fenugreek (Trigonella foenum-graecum L.) Pg. 355–373.
5. Bhavamishra, Bhavaprakasha Nighantu. Commentary by KC Chunekar. Edited by GS Pandey. Reprint 2015. Chaukhambha Bharati Academy, Varanasi. Haritakyadi Varga, 93–5. Pg. 36.
6. Bhavamishra, Bhavaprakasha Nighantu. Commentary by Bulusu Sitaram. 1st edition 2010. reprint 2014. Chaukhambha Orientalia, Varanasi. Volume II. Chikitsasthana. 70/158–62. pg. 706.
7. Bhavamishra, Bhavaprakasha Nighantu. Commentary by KC Chunekar. Edited by GS Pandey. Reprint 2015. Chaukhambha Bharati Academy, Varanasi. Haritakyadi Varga, 98–9. Pg. 39.
8. Smith M. Therapeutic applications of fenugreek. *Alternative Medicine Review*. CRC press, Boca Raton. London, New York. 2003;8(1):20–7.
9. Verotta L, Macchi MP. *10 Healing Properties of Food. Connecting Indian Wisdom and Western Science*. Plant Usage for Nutrition and Health. 2015 Apr 24:245.
10. Ahari DS, Kashi AK, Hassandokht MR, Amri A, Alizadeh K. Assessment of drought tolerance in Iranian fenugreek landraces. *Journal of Food, Agriculture and Environment*. 2009 Jul 1;7(3–4):414–9.
11. Al-Asadi JN. Therapeutic uses of fenugreek (Trigonella foenum-graecum L.). *American Journal of Sociology*. 2014 March-April Pg. 21–36.
12. Wani SA, Kumar P. Fenugreek: A review on its nutraceutical properties and utilization in various food products. *Journal of the Saudi Society of Agricultural Sciences*. 2018 Apr 1;17(2):97–106.
13. Bahmani M, Sarrafchi A, Shirzad H, Rafieian-Kopaei M. Autism: Pathophysiology and promising herbal remedies. *Current Pharmaceutical Design*. 2016 Jan 1;22(3):277–85.
14. Moini Jazani A, Hamdi K, Tansaz M, Nazemiyeh H, Sadeghi Bazargani H, Fazljou SM, Nasimi Doost Azgomi R. Herbal medicine for oligomenorrhea and amenorrhea: A systematic review of ancient and conventional medicine. *BioMed Research International*. 2018 Mar 18;2018.
15. Dhanvantari Nighantu, Dr. Jharkhande Ojha and Umapati Mishra, Chaukhamba Surbharti Prakashan, Varanasi, Edition 2014; Suvarnadi Varga/ 104–611.
16. Madhavadravyaguna, Shaka/ Shloka 9; http://niimh.nic.in/ebooks/e-Nighantu/madhavadravyaguna/?mod=read.
17. Rajnighantu, Acharya Vishwanath Dwivedi, Chowkhamba Krishnadas Academy, Varanasi, Sixth Edition, 2016, Pippalyadi Varga/ Shloka 67–9; pg. 148.
18. Siddhamantra, Vatapittaghna/ Shloka 74; http://niimh.nic.in/ebooks/e-Nighantu/siddhamantra/?mod=read&h=methikA.
19. Shodhala Nighantu, Gyanendra Pandey, Chaukhamba Krishnadas Academy Varanasi, First Edition 2009, Shatapushpadi Varga/ Shloka 260; pg. 56.
20. Dalvi SM, Patwardhan MS, Yeram N, Patil VW, Patwardhan SY. Evaluation of biochemical markers in type 2 diabetes mellitus with adjunct therapy of fenugreek seed aqueous extract. *Journal of Medicinal Plants*. 2019;7(1):109–13.
21. Kandhare A, Phadke U, Mane A, Thakurdesai P, Bhaskaran S. Add-on therapy of herbal formulation rich in standardized fenugreek seed extract in type 2 diabetes mellitus patients with insulin therapy: An efficacy and safety study. *Asian Pacific Journal of Tropical Biomedicine*. 2018 Sep 1;8(9):446.

22. Kandhare, A.D., Rais, N., Moulick, N., Deshpande, A., Thakurdesai, P. and Bhaskaran, S. Efficacy and safety of herbal formulation rich in standardized fenugreek seed extract as add-on supplementation in patients with type 2 diabetes mellitus on sulfonylurea therapy: A 12-week, randomized, double-blind, placebo-controlled, multi-center study. *Pharmacognosy Magazine.* 2018; 14(57): 393.
23. Rampogu S, Baek A, Zeb A, Lee KW. Exploration for novel inhibitors showing back-to-front approach against VEGFR-2 kinase domain (4AG8) employing molecular docking mechanism and molecular dynamics simulations. *BMC Cancer.* 2018 Dec 1;18(1):264.
24. Verma N, Usman K, Patel N, Jain A, Dhakre S, Swaroop A, Bagchi M, Kumar P, Preuss HG, Bagchi D. A multicenter clinical study to determine the efficacy of a novel fenugreek seed (Trigonella foenum-graecum) extract (Fenfuro™) in patients with type 2 diabetes. *Food & Nutrition Research.* 2016 Jan 1;60(1):32382.
25. Vargova V, Fedacko J, Singh RB, Kartikey K, Hristova K, Maheshwari A, Saxena M, Isaza A. Effects of fenugreek seeds on oxidative stress, pro-inflammatory cytokines and nirite as adjuncts to diet in patients with hyperlipidemia. *World Heart Journal.* 2018 Oct 1;10(4):265–75.
26. Geberemeskel GA, Debebe YG, Nguse NA. Antidiabetic effect of fenugreek seed powder solution (Trigonella foenum-graecum L.) on hyperlipidemia in diabetic patients. *Journal of Diabetes Research.* 2019 Sep 5;2019.
27. Wankhede S, Mohan V, Thakurdesai P. Beneficial effects of fenugreek glycoside supplementation in male subjects during resistance training: A randomized controlled pilot study. *Journal of Sport and Health Science.* 2016 Jun 1;5(2):176–82.
28. Thomas JV, Rao J, John F, Begum S, Maliakel B, Krishnakumar IM, Khanna A. Phytoestrogenic effect of fenugreek seed extract helps in ameliorating the leg pain and vasomotor symptoms in postmenopausal women: A randomized, double-blinded, placebo-controlled study. *Pharma Nutrition.* 2020 Dec 1;14:100209.
29. Shamshad Begum S, Jayalakshmi HK, Vidyavathi HG, Gopakumar G, Abin I, Balu M, Geetha K, Suresha SV, Vasundhara M, Krishnakumar IM. A novel extract of fenugreek husk (FenuSMART™) alleviates postmenopausal symptoms and helps to establish the hormonal balance: A randomized, double-blind, placebo-controlled study. *Phytotherapy Research.* 2016 Nov;30(11):1775–84.
30. Swaroop A, Jaipuriar AS, Kumar P, Bagchi D. Efficacy of a novel fenugreek seed extract (Furocyst) in polycystic ovary syndrome (PCOS). *Planta Medica.* 2016 Dec;82(S 01):P1097.
31. Sankhwar P, J SP GA, Tiwari K. Clinical evaluation of furostanolic saponins and flavanoids in polycystic ovarian syndrome (PCOS) Patients. *Journal of Endocrinology and Diabetes: JEAD-102.* 2018;(1):1–9.
32. Maheshwari A, Verma N, Swaroop A, Bagchi M, Preuss HG, Tiwari K, Bagchi D. Efficacy of Furosap™, a novel Trigonella foenum-graecum seed extract, in enhancing testosterone level and improving sperm profile in male volunteers. *International Journal of Medical Sciences.* 2017;14(1):58.
33. Opdyke DL. Fenugreek absolute. *Food Cosmetics Toxicology.* 1978;16:S755–S756.
34. Muralidhara, Narasimhamurthy K, Viswanatha S, Ramesh BS. Acute and subchronic toxicity assessment of debitterized fenugreek powder in the mouse and rat. *Food Chemical Toxicology.* 1999;37:831–8.40. Rao PU, Sesikeran B, Rao PS, et al. Short term nutritional and safety evaluation of fenugreek. *Nutrition Research.* 1996;16:1495–505.
35. Sharma RD, Sarkar A, Hazra DK, et al. Toxicological evaluation of fenugreek seeds: A long term feeding experiment in diabetic patients. *Phytotherapy Research.* 1996;10:519–20.

2 Fenugreek in Traditional Persian Medicine

Narges Tajik, Reza Mohammadi Nasab, Nazli Namazi and Mohammad Hossein Ayati

CONTENTS
2.1 Fenugreek 16
References 18

The earliest evidence of civilization in Iran goes back to over 10,000 years ago (1–3). Iranians have traditionally been well acquainted with medicinal herbs and their uses, and have played a crucial role in discovering their properties and their application in treating diseases, in pharmacy and in medical knowledge (4, 5). They had even developed a special script for medical writings, as is, for example, evidenced by the Ibn al-Nadim reference in his *Kitab al-Fihrist* to a specific script, namely *Nim kashtaj* (fragrant plant), solely reserved for philosophical and medical texts in ancient Iran (6, 7).

The history of Traditional Persian Medicine is traceable in Zoroaster's teachings. What is known as Avestan or Zoroastrian medicine has its roots in sacred scripture (8, 9).

Regarding the use of medication in ancient Iran, the foremost and major consultable sources on the Achaemenid and earlier medical traditions include the Avesta, the holy book of Zoroastrianism, and other contemporaneous religious texts (10–13).

References in the Vandidad, the Yasna, and the Yashts as well as in other ancient texts make it clear that physicians in ancient Iran were familiar with many diseases and their treatments, and had already fostered proper drugs to deal with them (14–16).

According to the Avesta, the first physician in the pre-Zoroastrian times was Thrita, who lived, as suggested by Vedic narrations, in the period that coincided with the coexistence of Indians and Iranians. According to Vandidad, Fargard 2, Thrita was sent by Ahuramazda to alleviate malaises. Fargard 7, 44 informs us that Ahuramazda endowed Thrita with 10,000 plants to fight diseases (17–19).

Hamza al-Isfahani, a historian living in the 3rd century AH/9th century AD, credits Thrita as the founder of medical science and pharmacology, who managed to dispatch evil creatures and cure diseases by distinct techniques (20).

Herbal medicine was of particular prominence in ancient Iran, so that the Avesta lists herbalists among the different classes of physicians; and according to the Ordibehesht Yasht, a group called urvarō baešazau (herbal therapy) consisted of physicians who practiced treatment, prescription of medicine, and removal of harm using herbs. The herbalist's job was to recognize herbaceous plants and use them in the treatment of various diseases, whereby they were typically administered orally, topically, or through inhalation. He treated his patients with herbal extracts and medicines. Herbal medicine is praised several times in the Vandidad, and its practitioners knew hundreds of healing plants and herbs and regarded them as sacred (19, 21).

These physicians always carried a drug box, known as *tabanguk* (hutch), laden with such herbal sources as roots, bark, leaves, flowers, buds, fruits, seeds, kernels, and sap. They had to collect their medicine in different seasons from mountains, plains, deserts, and forests, or obtain them from such countries as India (22, 23).

In addition to oral and topical administration, herbalists made use of the healing properties of herbs through distillation, oil extraction, and burning (24).

Medicinal plants remained objects of veneration throughout Iranian history. In the Achaemenid period, such plants were particularly popular, so that about 60 different herbal species are claimed to have been known at the time (25, 26). Denkard III, Fargard 308 says, "They believed that one should strive to be acquainted with every medication and medicinal plant, because by so doing one can cure diseases and pains that are otherwise incurable" (27). Further, in his Cyropaedia, the Greek philosopher and historian Xenophon states that in ancient times the Iranians taught children the benefits of plants and introduced the poisonous plants so as to exploit the beneficial species and evade the poisons (28–30).

With the onset of the Islamic era, Iranian and non-Iranian physicians thrived – thanks to the focus of Islam on learning and knowledge and the transition of hubs and the transference of the Jundishapur model through the massive migration of physicians, various disciplines, among them being the *materia medica* and pharmacology. And, the translation movement made available the Indian, Greek, and Iranian pharmaceutical legacies in Arabic to the Muslims, who would soon begin to scrutinize them (31–35).

Following the Arab conquest of Iran and the demise of the Sassanian rule and the spread of the Arabic language, the Muslims inherited the Sassanian as well as the great Greek legacy (36–38).

Medicine and medicinal plants have been principal subjects in Iranian medical texts, and the resources with references to pharmacy constitute a considerable part of Islamic medical literature. One can rarely find medical texts without pharmacological discussions (19, 34, 39).

In the traditional Iranian medicine doctrine, the foremost step is to maintain health and prevent disease, and in case a disease arises, the use of food is preferred over medication (40, 41). Treatment with drugs is also contingent on certain rules. Hence, a proper dose should be decided on in view of the conditions specified by Iranian medicine for the disease, and the physician should appraise the temper of the patient and prescribe the medication according to their sex, age, lifestyle, and habits, time of year, the city in which they live, and their profession, vigor, and appearance (42, 43). Further, this traditional medicine makes use of all herbal components such as roots, stems, etc. in curing different diseases, and each component has to be used in a certain phase of the plant's lifetime. For example, the best time for the stem is when the plant is fully grown and has not yet started withering (44–47).

2.1 FENUGREEK

Fenugreek is a medicinal plant with a long tradition in Iranian medicine and other regions, and possesses significant healing properties. Its high therapeutic power has placed it among the most important medicinal plants in the world (48, 49).

The plant is indigenous to North Africa and India, and was medically used in both extract and powder forms (50–52).

Evidence on its application comes from ancient Egypt, where it was used as an inhalant and also in mummification (53, 54). It remains in common use in modern Egypt in the bread industry as a flour complement. In ancient Rome, it was applied to facilitate childbirth, while the ancient Chinese used its seeds to treat weakness and inflammations. It served as a spice and breast milk booster in ancient India (55–59).

Similar to other regions, in Iranian medicine fenugreek has been used in treating a wide range of maladies (60–62).

Iranian scholars and physicians have authored numerous works in Persian and Arabic (31, 32, 63). Regarding their opinion on the therapeutic use of fenugreek, one can consult a wide array of medical texts written in the medieval period (64, 65). In addition to the works dealing specifically with the identification of drugs or their administration methods, in almost the entire medical literature we find pages set apart for introducing and describing the importance, properties, and administration modes of fenugreek (66–74).

Fenugreek in Traditional Persian Medicine

In technical texts dealing with Iranian medicine and pharmacy, the plant is occasionally referred to as *holba*. On the authority of al-Biruni in his book *al-Saydana*, it was called *falilasa* in Syriac, *shalmit* in Persian, and *Fariqa* in the Syrian dialect (69–75).

Regarding fenugreek, Iranian medical texts generally state that its oil fosters hair growth; its seed mucilage, especially if mixed with rose oil, can heal cracked skin caused by cold weather; it is used in medicines that remove facial black spots and alter bad breath and unpleasant body odor; if used in hair shampooing, it removes dandruff; if boiled in water, it can clear up a red spot on the eye; if applied to the eye, it will remove the thick substances that cause eye inflammation; it clears the voice; it provides a little nourishment for the lungs; it softens the chest and throat; it relieves cough. The use of the plant as powder, decoction, and ointment has been very popular in traditional medicine since ancient times (45, 71, 76–78). Table 2.1 gives various forms that fenugreek has been administered in Iranian medicine.

Apart from its benefits, one also comes across in Iranian medical literature references to side effects prompted by the use of the plant, among them being headache, nausea, and harm for onthayan (testis and ovary). And to reduce its side effects it is recommended to be used together with sour oxymel and anise, and sucking the juice of sweet-and-sour pomegranate; and in the case of those with warm temper, its leaf should never be used without chicory, and its leaf decoction should be accompanied by spinach leaf, purslane leaf, or parsnip (69–74).

TABLE 2.1
Various Forms of Fenugreek Administration in Iranian Medicine

Administration Form	Description	Disease	Other *Materia Medica*	Resources
la'uq (lincut)	preparations meant to be licked	useful for cleaning up lungs from thick *balgham*		al-Abnyia'an Haqa'iq al-Adwiya
tila (embrocation)	thin liquid preparation for embrocation	severe inflations	mixed with flaxseed, improves dissolution	al-Abnyia'an Haqa'iq al-Adwiya
matbukh (tabikh)	decoction	chronic chest pain and its ulcers	tamarind, fig and raisin, which bind its juice with honey	Tohfat al-Mo'menin
natul (medicinal rinse)	water with boiled medicine, poured on an area	its boiled form for easing labor, abortion of *mashimah* (a layer of amniotic membrane), and womb cleanse		Tohfat al-Mo'menin
zimad (poultice)	inedible preparation as thick or thin liquid	for overt or covert rigid tumefactions, splenitis, pelvic pain, hair loss, nail distortion, cold induced fissures, burns from fire, ulcers, dandruff, and kalaf		Tohfat al-Mo'menin
hoqnah (enema)	liquid preparation injected through anus	for proctitis		Tohfat al-Mo'menin
naqi' (*naqu'*)	a beverage made by steeping in water	epiphora, solaq (thickened eyelids), red spot, and remains of eye inflation	in rosewater	Tohfat al-Mo'menin
zimad (poultice)		dissolution of the spleen	its flour with borax	Makhzan al-Adwiya

In addition, there is a strictly stipulated dosage for each plant, which is, for example, 10 dirhams for fenugreek, and a maximum of 5 dirhams for its seed (73, 79).

Iranian scholars have even heedfully ascertained apposite alternatives in case a plant is unavailable. Thus, flax seed represents the alternative acclaimed in the related texts for fenugreek seed (45, 77).

REFERENCES

1. Daryaee T. *Sasanian Persia: The rise and fall of an empire*. Bloomsbury Publishing; 2014.
2. Daryaee T, Daryāyī T. *The Oxford handbook of Iranian history*. Oxford University Press; 2012.
3. Frye RN. *The history of ancient Iran*. CH Beck; 1984.
4. Elgood C. *A medical history of Persia and the eastern caliphate: From the earliest times until the year A.D. 1932*. Cambridge University Press; 2010.
5. Joshee N, Dhekney SA, Parajuli P. *Medicinal plants: From farm to pharmacy*. Springer International Publishing; 2019.
6. al-Nadim IAY, Flügel G, Roediger J, Mueller A. *Kitab al-Fihrist mit Anmerkungen*. F.C.W. Vogel; 1872.
7. اسحاق، المب, Ramadan I. *al-Fihrist li-Ibn al-Nadim*. Dar al-Ma'rifah; 1994.
8. Bodeker G, Organization WH, Ong CK. *WHO global atlas of traditional, complementary and alternative medicine*. WHO Centre for Health Development; 2005.
9. Boyce M. *Zoroastrians: Their religious beliefs and practices*. Routledge; 2001.
10. Hastings J, Selbie JA. *Encyclopædia of religion and ethics: Hymns-liberty*. T. & T. Clark; 1915.
11. *Chambers's Encyclopedia*. United Kingdom: Pergamon Press; 1967.
12. Waterhouse JW. *Zoroastrianism*. Book Tree; 2006.
13. Jackson AW. *Zoroaster, the prophet of ancient Iran*. Franklin Classics Trade Press; 2018.
14. Sayce AH, Gibbon E, Johnston C. *Ancient empires of the east*. Collier; 1913.
15. Olmstead AT, Olmstead ATE, Olmstead AT. *History of the Persian empire*. University of Chicago Press; 1948.
16. Malandra WW. *An introduction to ancient Iranian religion: Readings from the Avesta and Achaemenid inscriptions*. University of Minnesota Press; 1983.
17. Stein A. *Innermost Asia: Text*. Cosmo; 1981.
18. Alamdari A, Zarifi A, Mohammad Hossaini S. Medicine in Zoroastrian school. *Iranian Journal of Pharmaceutical Research*. 2010(Supplement 2):33.
19. Elgood C. *A medical history of Persia and the eastern caliphate: From the earliest times until the year AD 1932*. Cambridge University Press; 2010.
20. al-Isfahānī H. K. *ta'rīkh sinī mulūk al-ard wa l-anbiyad*, Ed. JME Gottwaldt; 902 AD.
21. Cantera Glera A. *Medical fees and compositional principles in Avestan*. Videvdad; 2004.
22. Levey M. *Early Arabic pharmacology: An introduction based on ancient and medieval sources*. Brill Archive; 1973.
23. Hobhouse P. *The gardens of Persia*. Kales Press; 2004.
24. Kapoor S, Lata Saraf S. Topical herbal therapies an alternative and complementary choice. *Research Journal of Medicinal Plant*. 2011;5(3):650–69.
25. Griggs B, Van der Zee B. *Green pharmacy: The history and evolution of western herbal medicine*. Inner Traditions/Bear & Co; 1997.
26. Sahranavard S, Ghafari S, Mosaddegh M. Medicinal plants used in Iranian traditional medicine to treat epilepsy. *Seizure*. 2014;23(5):328–32.
27. Arberry AJ. *Classical Persian literature*. Taylor & Francis; 2006.
28. Flower MA. *The Cambridge companion to Xenophon*. Cambridge University Press; 2017.
29. Dillery J. *Xenophon and the history of his times*. Taylor & Francis; 2002.
30. Miller W, Brownson CL, Todd OJ, Marchant EC. *Xenophon in seven volumes*. W. Heinemann; 1968.
31. Gutas D. *Greek thought, Arabic culture: The Graeco-Arabic translation movement in Baghdad and Early'Abbasaid society* (2nd-4th/5th-10th c.). Routledge; 2012.
32. Pormann PE, Joosse NP. Commentaries on the Hippocratic aphorisms in the Arabic tradition: The example of melancholy. Epidemics in context Greek commentaries on Hippocrates in the Arabic tradition. de Gruyter. 2012:211–50.
33. Nasr SH. *Traditional Islam in the modern world: K. Paul International*. Distributed by Routledge; 1987.
34. Nasr SH, De Santillana G. *Science and civilization in Islam*. Harvard University Press; 1968.
35. Sarton G. *Introduction to the history of science*. Harvard University Press; 1962.

36. Kennedy H. *The great Arab conquests: How the spread of Islam changed the world we live in:* Da Capo Press. Incorporated; 2007.
37. Nicolle D. *The great Islamic conquests AD 632–750.* Bloomsbury Publishing; 2012.
38. McDonough S. The legs of the throne: Kings, elites, and subjects in Sasanian Iran. In *The roman empire in context: Historical and comparative perspectives.* Wiley-Blackwell; 2011:290–321.
39. Farzaei MH, Farzaei F, Abdollahi M, Abbasabadi Z, Abdolghaffari AH, Mehraban B. A mechanistic review on medicinal plants used for rheumatoid arthritis in traditional Persian medicine. *Journal of Pharmacy and Pharmacology.* 2016;68(10):1233–48.
40. Saad B, Said O. *Greco-Arab and Islamic herbal medicine: Traditional system, ethics, safety, efficacy, and regulatory issues.* John Wiley & Sons; 2011.
41. Fieldhouse P. *Food and nutrition: Customs and culture.* Springer; 2013.
42. Hamedi A, Zarshenas MM, Sohrabpour M, Zargaran A. Herbal medicinal oils in traditional Persian medicine. *Pharmaceutical Biology.* 2013;51(9):1208–18.
43. Ghadiri MK, Gorji A. Natural remedies for impotence in medieval Persia. *International Journal of Impotence Research.* 2004;16(1):80–3.
44. Sharafzadeh S, Alizadeh O. Some medicinal plants cultivated in Iran. *Journal of Applied Pharmaceutical Science.* 2012;2(1):134–7.
45. Avicenna. *Al-Qanun Fi'l-tibb [by] Al-Shaikh Al-Ra'is Abu 'Ali Al-Husain Bin 'Abdullah Bin Sina (980–1037 AD): Book I. Critical edition prepared under the auspices of institute of history of medicine and medical research.* Vikas Publishing House; 1982.
46. Hosseinzadeh H, Nassiri-Asl M. Avicenna's (Ibn Sina) the canon of medicine and saffron (Crocus sativus): A review. *Phytotherapy Research.* 2013;27(4):475–83.
47. Paavilainen H. *Medieval pharmacotherapy-continuity and change: Case studies from Ibn Sīnā and some of his late Medieval commentators.* Brill; 2009.
48. Gurib-Fakim A. Medicinal plants: Traditions of yesterday and drugs of tomorrow. *Molecular aspects of Medicine.* 2006;27(1):1–93.
49. Morton J. Mucilaginous plants and their uses in medicine. *Journal of Ethnopharmacology.* 1990;29(3):245–66.
50. Bahmani M, Shirzad H, Mirhoseini M, Mesripour A, Rafieian-Kopaei M. A review on ethnobotanical and therapeutic uses of fenugreek (Trigonella foenum-graecum L). *Journal of Evidence-Based Complementary & Alternative Medicine.* 2016;21(1):53–62.
51. Maier H, Anderson M, Karl C, Magnuson K, Whistler RL. *Guar, locust bean, Tara, and fenugreek gums. Industrial gums.* Elsevier; 1993:181–226.
52. Acharya S, Thomas J, Basu S. Fenugreek, an alternative crop for semiarid regions of North America. *Crop Science.* 2008;48(3):841–53.
53. Campbell JM. Pharmacy in ancient Egypt. *Egyptian Mummies and Modern Science.* 2008:216–33.
54. Boi M. *The ethnocultural significance for the use of plants in ancient funerary rituals and its possible implications with pollens found on the Shroud of Turin.* Valencia; 2012.
55. Du Toit L. *Celling properties of cactus pear mucilage-hydrocolloid combinations in a sugar-based confectionery.* University of the Free State; 2018.
56. French V. Midwives and maternity care in the Roman world. Midwifery and the medicalization of childbirth. *Comparative Perspectives.* 2004;13(2):53.
57. Ögenler O, Ün İ, Uzel İ. Medical plants used for treatment of gynecological disorders in Ottomans in the 15th century. *Journal of Complementary Medicine.* 2018;7(2):171–7.
58. Khoja KK, Shaf G, Hasan TN, Syed NA, Al-Khalifa AS, Al-Assaf AH, et al. Fenugreek, a naturally occurring edible spice, kills MCF-7 human breast cancer cells via an apoptotic pathway. *Asian Pacific Journal of Cancer Prevention.* 2011;12(12):3299–304.
59. Xie W, Du L. Diabetes is an inflammatory disease: Evidence from traditional Chinese medicines. *Diabetes, Obesity and Metabolism.* 2011;13(4):289–301.
60. Thomas JE, Bandara M, Lee EL, Driedger D, Acharya S. Biochemical monitoring in fenugreek to develop functional food and medicinal plant variants. *New Biotechnology.* 2011;28(2):110–7.
61. Zareian MA, Yargholi A, Khalilzadeh S, Shirbeigi L. Etiology and treatment of dandruff according to Persian medicine. *Dermatologic Therapy.* 2019;32(6):e13102.
62. Leppik EE. Cercospora Traversiana and some other pathogens of fenugreek new to North America. *The Plant Disease Reporter.* 1960;44:40.
63. Four words involving Sogdian in the (guidance of the learned in medicine).
64. Riddle JM. *Dioscorides on pharmacy and medicine.* University of Texas Press; 1986.

65. Ardalan MR, Shoja MM, Tubbs RS, Eknoyan G. Diseases of the kidney in medieval Persia—the Hidayat of Al-Akawayni. *Nephrology Dialysis Transplantation*. 2007;22(12):3413–21.
66. Al-Baytar I. Al-Jami'Li-Mufradat al-Adwiya wa-'l-Aghdiya. *Cairo: Al-Matba'a al-Amiriyya al-Masriyya*. 1874;1291:178.
67. Ghāfiqī AJAiM, Gacek A, Miller P, Ragep FJ, Wallis F. *The herbal of Al-Ghāfiqī: A facsimile edition of MS 7508 in the Osler library of the history of medicine, McGill University, with critical essays*. Osler Library of McGill University; 2014.
68. Hindu I. *Miftah al-tibb wa-minhaj al-tullab. [The key to the science of medicine and the students' guide]*. Institute of Islamic Studies, McGill University Tehran Branch, in collaboration with Tehran University; 1989.
69. Akhvayni Bokhari A. *Hedayat al-Motalemin fi-Teb*. Translated by-Matini j. Mashhad University; 1992.
70. Ardakani MRS, Farjadmand F, Rahimi R. Makhzan al Adviyeh and pointing to the scientific names of medicinal plants for the first time in a Persian book. *Traditional and Integrative Medicine*. 2018:186–95.
71. Bakhtiar L, Gruner OC. *The canon of medicine (al-Qānūn Fī'l-ṭibb)*. Abjad Book Designers & Builders; 1999.
72. Mansour Ibn Mohammad EE S. *Kefaye Mansoori. Iran university of medical sciences*. Iran University of Medical Sciences; 2003.
73. Heravi AMM. *Al-Abniye an Haghayegh al-Adviye*. Corrected by Bahmanyar A. Tehran University Publisher; 1967:1371.
74. Momen H. *Tohfe almomenin*. The Parliament Library; 2009:116.
75. Biruni A. *Kitab-al Saydana fi-Tibb*. Academy of Persian Language and Literature; 2004.
76. Akhaveini Bokhari. *Hedayat al-mota'allemin fi al-tibb (An educational guide for medical students)*. Ferdowsi University of Mashhad Publication; 1992.
77. Razi M. Al-Hawi fi'l-tibb (Comprehensive book of medicine). *Hyderabad: Osmania Oriental Publications Bureau*. 1968;20:548–53.
78. Hossaini-Tabib M. *Tohfe of Hakim Momen*. Mostafavi Press, Tehran and Qum (In Persian); 1959.
79. Jorjani E. *Zakhireye Khwaram Shahi (Treasure of Khwarazm Shah)*. Entesharat-e Bonyade Farhang-e Iran; 1976.

3 Distribution, Biology, and Bio-Diversity of Fenugreek

R. S. Sharma and S. R. Maloo

CONTENTS

3.1 Introduction 21
3.2 Species, Names, Origin, and Distribution 22
3.3 Botanical Perspective 26
3.4 Chromosome Number 26
3.5 Biodiversity of Fenugreek 27
    4.5.1 Molecular Genetic Diversity 27
3.6 Conclusion 28
References 28

3.1 INTRODUCTION

Fenugreek (*Trigonella foenum-graecum* L.) is an annual, herbaceous, dicotyledonous, diploid, self-pollinated and the oldest spice crop grown worldwide (Amiriyan et al., 2019). It is a multipurpose plant where the seeds have been traditionally used for medicinal and culinary purposes since ancient times and the leaves used as animal forage. Fenugreek is considered as a high-value, low-volume crop with enormous health benefits. Nowadays every part of the fenugreek plant, seeds, tender shoots, roots, and leaves, are commercially being used for preparation of various medicinal and Ayurvedic herbs. Fenugreek roots have a symbiotic association with various soil microflora and beneficial bacteria which act as nitrogen fixation and soil stabilizers. Therefore in India fenugreek is grown as a short-term crop rotation as well as an intercrop to replenish nitrogen in the soil. The seed and leaf of fenugreek contain a high amount of animal growth-promoting substance diosgenin, which is a potential source for growth and development of the animal. Leaves and tender stems are also directly consumed as a salad to improve the taste of various dishes (Malhotra, 2011). The crop generally matures in 100–120 days after sowing and different growth and developmental stages of the fenugreek are shown in Figure 3.1.

Fenugreek crops have been recommended and presently are grown at large scale in the arid and semiarid regions of Asia, Africa, and Latin America (Zandi et al., 2015). It is widely distributed throughout the world in wild and cultivated form and can be grown in diverse climatic conditions. The plant is highly sensitive to frost and high temperature; therefore it is not suitable in high or low temperature regions. In India it is mostly grown in Rajasthan, Madhya Pradesh, Gujarat, Bihar, and Karnataka. Recently fenugreek cultivation has been started within a protected environment to reduce the temperature and frost losses. India is the major producer and consumer of fenugreek but due to high internal demand and consumption does not have a major share of the global fenugreek trade. Presently India is contributing more than 68% of the global fenugreek production (Basu, 2006; Malhotra, 2011). Several breeding efforts have been carried out in India for fenugreek improvement and as a recent result, new cultivars and varieties are available for year-round crop cultivation.

Fenugreek has been used traditionally in Tibetan and Chinese medication, Indian Ayurvedic medicines, and traditional Tibetan medicines for the curing of several human and animal diseases. Ancient Islamic scholars and physicians have also reported the use of fenugreek in traditional Islamic

FIGURE 3.1 Various stages of the fenugreek crops.

medicinal practices in ancient texts and scriptures (Acharya et al., 2007). Modern scientific research has shown that all parts of the fenugreek plant are useful for the treatment of various curative diseases including reducing blood sugar and blood cholesterol levels in both animals and humans (Acharya et al., 2006). The crop has a good potential to be considered as a panacea for the treatment of diabetic, microbial, and cancer diseases. The reason behind the rich medicinal properties of fenugreek is due to the presence of a wide diversity of important phytochemicals (diosgenin, trigonelline, fenugreekine, galactomannan, and 4-hydroxy isoleucine (Zandi et al., 2015)). Recently new research is being carried out to identify the potential fenugreek lines with the highest amount of diverse phytochemicals. Several plant chemotypes have been reported for specific purposes such as high trigonelline content. Hence, the crop has huge international demand in the associated pharmaceutical, nutraceutical, and functional food industries. Being known as a chemurgic crop, fenugreek has a widespread adoption in industrial sectors. Its seeds contain substantial amount of fiber and a reliable source of steroid diosgenin, which acts as a supplement in the pharmaceutical industry (Mehrafarin et al., 2010). Fenugreek is also well known as a global spice crop grown in all the major continents (depending on soil and climatic conditions) across the globe including parts of North Africa, Mediterranean Europe, Russia, the Middle East, China, India, Pakistan, Iran, Afghanistan, parts of the Far East and Southeast Asia, Australia, the USA, Canada, and Argentina (Acharya et al., 2006, 2007).

3.2 SPECIES, NAMES, ORIGIN, AND DISTRIBUTION

The genus name, *Trigonella*, means 'little triangle' because of the triangular shape of its small yellowish-white flowers. The species name *foenum-graecum* means 'Greek hay' in reference to its primary introgression from Greece (Basu, 2006; Petropoulos, 2002). To date, several local traditional names have been ascribed to the plant depending on the nation's local language and culture in which

Distribution, Biology, and Bio-Diversity

the crop is grown and/or consumed. For instance, fenugreek in Arabic is called Hulba; in Persian called Shanbalilae; in Greek called Tili, Tipilina, Trigoniskos, Tintelis, Tsimeni, and Moschositaro; in Uzbekistani called Boidana, Ul'ba, and Khul'ba; in Armenian called Shambala; in Chinese called K'u-Tou; in Ethiopian called Abish; in Japanese called Koroba; in English called fenugreek or Fenigrec; in Pakistani and Indian called Methi; in Italian called Fieno Greco; in Russian called Pazhitnik; and in French called Senegre (Petropoulos, 2002; Mehrafarin et al., 2011).

Saraswat (1984) recovered carbonized fenugreek seed from a Rohira village in the Sangrur district of Punjab, India indicating its use in trade by people of the Harappan civilization as far back as 2000–1700 BC (Lust, 1986) and said it was the oldest medicinal plant documented in history. Usually, *Trigonella* has been grown for centuries across the Indian subcontinent, the Mediterranean region, North Africa, and Yemen. Vavilov (1951) has suggested that fenugreek originated from the Mediterranean region, while De Candolle (1964) and Fazli and Hardman (1968) proposed it as an Asian origin. Apart from this, fenugreek is also grown in a few parts of North Africa, the Middle East, Mediterranean Europe, China, Southeast Asia, Australia, the USA, Argentina, and Canada. A lot of discrepancies have been reported about the fenugreek species (around 70–97) in the literature; however, older taxonomies like Linnaeus have clearly accentuated on the existence of ~260 species (Basu, 2006). Vasil'chencko (1953) recognized 128 species, Hector (1936). Hutchinson (1964) recorded about 70 species and Fazli (1967) considered 97 species. Across the reported species of fenugreek, the following are mostly prominent for their medicinal and pharmaceutical properties (Basu, 2006): *T. foenumgraecum, T. balansae, T. corniculata, T. maritima, T. spicata, T. occulta, T. polycerata, T. calliceras, T. cretica, T. caerulea, T. lilacina, T. radiata, T. spinosa*. Among which *T. foenum-graecum* is widely cultivated and distributed throughout the world (Petropoulos, 2002) (Table 3.1).

Fenugreek is ancient and the oldest spice crop found in diverse geographical latitudes. Several wild species of fenugreek have been reported in the five continents of Africa, Asia, Australia, and Europa; being cultivated mostly in North America, Australia, West and South Asia, the Middle East, Russia, North West of Africa. Potential new areas for fenugreek cultivation are some parts of Japan, Southeast Asia, Central Asia (Mongolia), wide parts of Africa, and South America (Figure 3.2).

India is the major fenugreek producer and holds the highest cultivated area in the world. In India fenugreek is grown in an area of 21.97 million hectares with an annual production of 31.12 metric tons (www.nrcss.res.in). A wide range of diversity is available in wild and cultivated forms in

TABLE 3.1
Major *Trigonella* Species and Their Distribution

Trigonella Species	Distribution
T. arabica Delile (syn. *T. pectin* Schenk)	N. Africa, especially in Arabia; Syria to NE Egypt
T. caerulea (L.) Ser. (syn. *Melilotus caeruleus* Desr., *Trifolium caeruleum* Moench., *Trigonella melilotus caoerulea* L. Aschers. et Graebn)	E. Mediterranean region; SE Europe, origin in Mediterranean region
T. caerulea (L.) Ser. ssp. *Caerulea*	C., W., and S. Europe; N. Africa, widely cultivated in gardens
T. corniculata (L.) L. (syn. *Medicago corniculata* L. Trautv., *Trifolium corniculata* L., *Trigonella esculenta* L.)	Mediterranean region; Near East countries
T. stellata Forsk.	N. Africa, Arabia, Egypt, Tunisia, Algeria, Morocco, Canary Islands, W. Asia, Iran, Iraq, Middle East, Israel, Lebanon, Kuwait
T. foenum-graecum L. (syn. *Foenum graecum officinale* Moench, *T. graeca* St. Lag.)	Caucasus, ex-Soviet Union, C. Asia, E. Europe

Source: Table developed from the text of Seidemann, J. 2005. *World spice plants, economic usages, botany and taxonomy*, 372–374. Berlin: Springer Verlag.

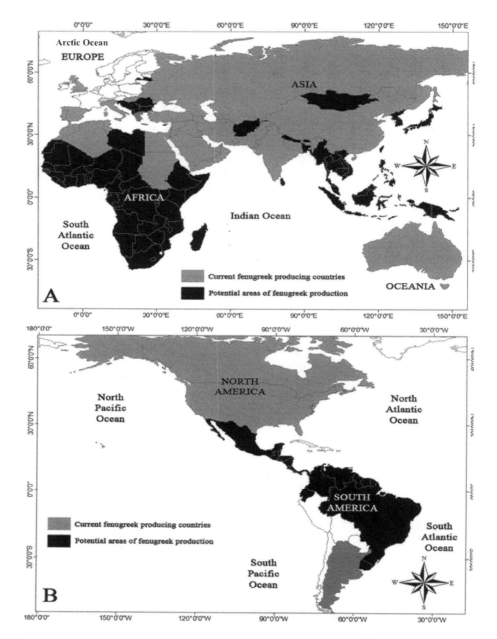

FIGURE 3.2 World map (A, B) showing major growing and potential area of fenugreek production.

different parts of the country. Substantial efforts have been done by Indian Council of Agriculture Research to collect and conserve the germplasm of fenugreek for the crop improvement program. Recently some new plant types have been developed through mutation breeding with improved stature of plants, high trigonelline content, and better yield. Similarly a large number of fenugreek wild accessions is reported with distribution and their relative frequencies in the other parts of the globe listed in Table 3.2.

TABLE 3.2
Origin, Number of Reported Fenugreek Accessions with Relative Frequencies of Fenugreek Available across the Given Countries

Origin	Number of Reported Fenugreek Accessions	Relative Frequency (%)	Average Forage Production 1961–2019 (Tons/Acre) per Country
Afghanistan	27	2.48	>550,682
Algeria	2	0.18	3,835,860
Australia	7	0.64	2,868,759
Austria	1	0.09	1,539,297
Azerbaijan	1	0.09	624,045
Canada	54	4.95	>550,682
China	44	4.04	51,489,483
Egypt	16	1.47	>550,682
England	17	1.56	>550,682
Eritrea	1	0.09	>550,682
Ethiopia	152	13.94	>550,682
France	3	0.28	>550,682
Germany	4	0.37	7,445,345
Greece	6	0.55	>550,682
Hungary	3	0.28	>550,682
India	401	36.79	91,881,132
Iran	46	4.22	2,653,431
Iraq	7	0.64	230,685
Israel	3	0.28	>550,682
Italy	4	0.37	9,906,660
Jordan	5	0.46	30,439
Kenya	1	0.09	>550,682
Libya	5	0.46	573,678
Morocco	5	0.46	1,694,057
Nepal	2	0.18	>550,682
Oman	77	7.06	>550,682
Pakistan	41	3.76	13,819
Poland	2	0.18	7,142,332
Portugal	1	0.09	6,894,828
Romania	15	1.38	694,961
Russia	3	0.28	10,701,727
Slovenia	4	0.37	86,966
South Africa	1	0.09	>550,682
Spain	8	0.73	163,784
Sudan	10	0.92	1,187,407
Sweden	3	0.28	>550,682
Switzerland	2	0.18	3,218,208
Syria	15	1.38	33,113
Taiwan	1	0.09	>550,682
Tunisia	42	3.85	>550,682
Turkey	40	3.67	>550,682
Turkmenistan	1	0.09	>550,682
USA	3	0.28	>550,682
Ukraine	1	0.09	475,316
Yemen	3	0.28	92,026

3.3 BOTANICAL PERSPECTIVE

Fenugreek is the oldest medicinal plant recognized in ancient history. Linnaeus described the species *Trigonella foenum-graecum* first as an annual dicotyledonous plant belonging to the subfamily Papilionaceae, family Leguminacae (=Fabacecae).

TAXONOMY:

Kingdom: Plantae
Division: Magnoliophyta
Class: Magnoliopsida
Order: Fabales
Family: Fabaceae
Subfamily: Papilionaceae
Genus: Trigonella
Species: foenum-graecum

COMMON NAMES:

General name: Fenugreek
English name: Fenugreek
Latin: *Trigonella foenum-graecum* L.; *Foenugraeci semen* (for the seed)
Arabic name: Hhulbah, Hhelbah
French name: Trigonelle, Senegrain, Foingrec
German name: Gemeiner, Hornklee, Bockshornklee
Indian name: Sagmethi, Methi, Kasuri methi
Italian name: Fienogreco, Erbamedica
Persian name (Irani): Shanbelileh, Schemlit
American name: Shambala
Chinese name: K'u-Tou
Dutch name: Fenegriek
Ethiopian name: Abish
Greek name: Trigoniskos, Tsimeni, Tintelis, Moschositaro, Tili, Tipilina
Russian name: Pazhitnik grecheski, Shambala, Pazhitnik cennoj

3.4 CHROMOSOME NUMBER

In the genus *Trigonella* many diploid species are present with varying in the chromosome numbers. According to Darlington and Wylie's (1955) reports, haploid chromosome numbers of most species in the *Trigonella* genus is 8, 9, 11, or 14 and diploid chromosome number is $2n = 16$. For example, *T. balansae, T. foenum-graecum, T. sprunerana* Boiss., *T. monspeliaca* L., *T. uncata, T. anguina, T. cassia, T. macrorrhyncha, T. stellata, T. corniculata, T. astroites, T. cariensis, T. gladiata, T. berythea, T. homosa,* and *T. foenum-graecum* have $2n = 16$ chromosomes (Darlington and Wylie, 1955; Singh and Roy, 1970; Ladizinsky and Vosa, 1968; Srivastava and Raghuvanshi, 1987; Reasat et al., 2002; Dundas et al., 2006) but, some exceptions are also present in *T. polycerata* from the Mediterranean and Asia, which has 28, 30, and 32 chromosomes; *T. geminiflora* from Iran and Asia Minor, *T. grandiflora* from Turkey, and *T. hamosa* from Egypt have 44 chromosomes; and *T. ornithopodioides* from Europe was reported to have 18 chromosomes within its genome (Petropoulos, 2002). Although generally most of the fenugreek species in the *Trigonella* genus contain $2n = 16$ chromosomes. Some studies show that chromosome duplication and mutation in the genome of some species have led to changes in the chromosome numbers. Singh and Singh (1976) get a number of trisomics along with five double trisomics in an auto-tetraploid population which had 18 ($2n + 1 + 1$)

chromosomes and thought that among 13 species of *Trigonella*, only *T. neoana* has $2n = 30$ chromosomes. In a another study carried out by Marc and Capraru (2008) to assess the cytogenetic effects of sodium phosphate on meristematic cells of root tips of fenugreek, they observed the negative effect on the mitotic index. Basu (2006) produced fenugreek ($2n + 2n = 32$) by treating the seeds with colchicines and Roy and Singh (1968) also obtained tetraploid fenugreek by treating shoot apices with colchicine.

A study about the variation in chromosome number in fenugreek was also carried out by Joshi and Raghuvanshi (1968) and reported an extra B chromosome in some genotypes of the fenugreek. The presence of the extra B chromosome in the genotypes affected growth and development with varying behaviors (Petropoulos, 2002). The chromosomes are large and varying sizes (longest = 26.28 µm and shortest = 13.52 µm). Reasat et al. (2002) offer a good scope for cytogenetic and molecular studies.

3.5 BIODIVERSITY OF FENUGREEK

3.5.1 Molecular Genetic Diversity

Knowledge of genetic diversity in the genetic material can help to provide useful information in the selection of breeding materials for hybridization programs and mapping quantitative trait loci (Sharma et al., 2012). A review of the literature shows that using DNA markers for investigating the genetic diversity of fenugreek does not have a long history in the world.

Dangi et al. (2004) assess genetic diversity using RAPD and ISSR markers among the two different fenugreek species (*T. caerulea* and *T. foenum-graecum*) and observed that *T. caerulea* species are more diverse than *T. foenum-graecum*. Further, authors suggested the use of these two methods for grouping the genotypes and determining the genetic relationship among them.

Sundaram and Purwar (2011) estimated the genetic diversity using 18 RAPD markers among two *Trigonella* species and 61 accessions and established the phylogenic relationship. A total of 141 bands were produced using 18 RAPD markers among which 74 were polymorphic. Genetic similarity in the accessions ranged between 0.66 and 0.90, showing a moderate to high genetic diversity among the studied fenugreek populations. The dendrogram obtained using RAPD markers formed two major clusters, with each cluster having two separate subgroups. Based on the study, the authors suggested that the RAPD marker is a useful tool for the assessment of genetic diversity and relationship among different *Trigonella* species.

Kumar et al. (2012) evaluated the genetic diversity among five Indian fenugreek varieties through 9 RAPD and 7 AFLP primers. The RAPD primers produced a total of 47 bands with a size range of 200–5000 bp with an average polymorphism of 62.4% while the AFLP marker amplified a total of 669 bands with a size range of 50–538 bp. Further, the study authors suggested that RPAD markers are more polymorphic and AFLP markers are more reproducible.

Ahari et al. (2014) used 20 Iranian fenugreek landraces to assess the genetic diversity through AFLP markers. Five different AFLP primers combinations were used for the study and a total of 147 bands were obtained, ranging from 50 to 500 base pairs with a polymorphism of 87%. The polymorphism information content (PIC) value showed that a high polymorphism existed among Kashan (0.79), Broojerd and Kashan (up to 0.93) landraces, which shows the moderate and high genetic diversity among these populations. Results based on the study showed that AFLP markers are more useful for investigating the genetic diversity among Iranian fenugreek populations.

Haliem and Al-Huqail (2014) established the association among biochemical characteristics such as glutamate-oxaloacetate, and acid phosphatase transaminase isozymes, and amino acid composition and molecular variations of seven wild *T. foenum-graecum* L. Two hundred fourteen active ingredients from aromatic and medicinal plants accessions were found using RAPD markers. RAPD marker-based analysis revealed the high polymorphism (94.12%) which can be used in the differentiation of the genotypes effectively.

Al-Maamari et al. (2014) assessed the genetic relationship in the 20 Omani fenugreek accessions and evaluated their relationship with four accessions from Iraq and Pakistan with six AFLP primer combinations. A total of 1,852 polymorphic loci were produced using various combinations with a high level of genetic diversity (H) among Omani accessions (0.2146) compared to Pakistani (0.0844) and Iraqi (0.1620) accessions. Further, the authors suggested that the average level of genetic variation among fenugreek accessions shows their long history of cultivation and frequent exchange of fenugreek genetic material among regions in Oman.

Choudhary et al. (2013) evaluated the genetic variability in 17 Indian fenugreek varieties through morphological and molecular markers. Fifteen Random Amplified Polymorphic DNA (RAPD) markers were used and showed a 57.66% polymorphism. Further, based on the results all the varieties were classified into two major clusters, cluster-I and cluster-II, cluster-I irrespective of their geographical distribution.

Hora et al. (2016) studied the genetic diversity and phylogenetic relationships among diverse fenugreek collections of northern India using RAPD and ISSR markers. They obtained the high similarity coefficient values and suggested that the huge genetic diversity is present in the fenugreek populations in India. Further they concluded that ISSR and RAPD markers are potential markers to evaluate the genetic diversity and assess genetic relationship.

Miraj et al. (2017) assessed the genetic diversity among 21 fenugreek germplasms of different geographical origins using SDS-PAGE. Results based on the SDS analysis of 21 fenugreek accessions indicated that two of them were monomorphic and 14 accessions were polymorphic. Further, on the basis of cluster analysis, four clusters were identified at a similarity level of 1.8. Overall on the basis of total seed protein, the grouping pattern of *Trigonella foenum-graecum* was mostly compatible with their species status.

Amiriyan et al. (2019) evaluated the genetic diversity and population structure of 88 individuals of eight landraces of Iranian fenugreek using SRAP markers. A total 72 amplicon were generated using six primers with 80.11% polymorphism. Authors further suggested that SRAP is an efficient technique to reveal genetic diversity and population structure of Iranian fenugreek landrace.

3.6 CONCLUSION

The fenugreek plant becomes a great interest for researchers due to the presence of a wide range of pharmaceutical and commercial food value. It is widely grown throughout the globe with wider adaptability. Huge morphological and molecular diversity were present in the core collection of the fenugreek germplasm. Recently, some conventional and non-conventional breeding approaches have been used for the improvement of the fenugreek accessions. Some biotechnological tools such as mutation breeding will be helpful for creating variation in the germplasm and for providing the focus needed in such research areas.

REFERENCES

Acharya, S.N., Srichamroen, A., Basu, S., Ooraikul, B., and Basu, T. 2006. Improvement in the nutraceutical properties of fenugreek (*Trigonella foenum-graecum* L.). *Songklanakarin Journal of Science and Technology* 28(1): 1–9.

Acharya, S.N., Thomas, J.E., and Basu, S.K. 2007. Fenugreek (*Trigonella foenum-graecum* L.) an alternative crop for semiarid regions of North America. *Crop Science* 48: 841–853. DOI: 10.2135/cropsci2007.09.0519.

Ahari, S.D., Hassandokht, M., Kashi, A., and Amri, A. 2014. Evaluation of genetic diversity in Iranian fenugreek (*Trigonella foenum-graecum* L) landraces using AFLP markers. *Seed Science Technology Improv Journal* 30: 155–171.

Al-Maamari, I.T., Al-Sadi, A.M., and Al-Saady, N.A., 2014. Assessment of genetic diversity in Fenugreek (*Trigonella foenumgraecum*) in Oman. *International Journal of Agriculture & Biology* 13: 560–8530.

Amiriyan, M., Abdolali, S., Abbas, Y., Masoud, M., and Zeinab, B. 2019. Genetic diversity analysis and population structure of some Iranian Fenugreek (*Trigonella foenum-graecum* L.) landraces using SRAP markers. *Molecular Biology Research Communications* 8(4): 181–190. DOI: 10.22099/mbrc.2019.34952.1440.

Basu, S.K. 2006. *Seed production technology for fenugreek (Trigonella foenum-graecum L.) in the Canadian*. Master of Science thesis. Department of Biological Sciences University, Lethbridge, Alberta, Canada, 2006.

Choudhary, S., Meena, R.S., Singh, R., Vishal, M.K., Choudhary, V., and Panwar, A. 2013. Assessment of genetic diversity among Indian fenugreek (*Trigoi ella foenum-graecum* L.) varieties using morphological and RAPD markers. *Legume Research-an International Journal* (36): 289–298.

Dangi, R.S., Lagu, M.D., Choudhary, L.B., Ranjekar, P.K., and Gupta, V.S. 2004. Assessment of genetic diversity in *Trigonella foenum-graecum* and *Trigonella caerulea* using ISSR and RAPD markers. *BMC Plant Biology* 4: 13.

Darlington, C.D., and Wylie, A.P. 1955. *Chromosome atlas of flowering plants*. 2nd ed. George Allen & Unwin, London, p. 519.

De Candolle, A. 1964. *Origin of cultivated plants*. Hafner, New York.

Dundas, I.S., Nair, R.M., and Verlin, D.C. 2006. First report of meiotic chromosome number and karyotype analysis of an accession of *Trigonella balansae*. *New Zealand Journal of Agricultural Research* 49: 55–58.

Fazli, F.R.Y. 1967. *Studies in steroid-yielding plants of the genus Trigonella*. Ph.D. Thesis. University of Nottingham, England, Thesis, 1967.

Fazli, F.R.Y., and Hardman, R. 1968. The spice fenugreek *(Trigonella foenum-graecum* L.). Its commercial varieties of seed as a source of diosgenin. *Tropical Science* 10: 66–78.

Haliem, E.A., and Al-Huqail, A.A. 2014. Correlation of genetic variation among wild *Trigonella foenum-graecum* L. accessions with their antioxidant potential status. *Genetics and Molecular Research: GMR* 13: 10464–10481. DOI: 10.4238/2014.December.12.8.

Hector, J.N. 1936. *Introduction to the botany of fields crops (non cereals)*. Central News Agency Ltd., Johannesburg, 1936.

Hora, A., Malik, C., and Kumari, B. 2016. Assessment of genetic diversity of *Trigonella foenum-graecum* L. in northern India using RAPD and ISSR markers. *International Journal of Pharmacy and Pharmaceutical Sciences* 8: 179–183.

Hutchinson, J. 1964. *The genera of flowering plants*, Vol. 1. Clarendon Press, Oxford, 1964.

Joshi, S., and Raghuvanshi, S.S. 1968. B-chromosome, pollen germination *in situ* and connected grains in *Trigonella foenum-graecum* L. *Beitrage zur Biologie der Pflanzen* 44: 161–166.

Kumar, V., Srivastava, N., Singh, A., Vyas, M.K., Gupta, S., Katudia, K., Vaidya, K., Chaudhary, S., Ghosh, A., and Chikara, S.K. 2012. Genetic diversity and identification of variety-specific AFLP markers in fenugreek (*Trigonella foenum-graecum*). *African Journal of Biotechnology* 11: 4323–4329.

Ladizinsky, G., and Vosa, C.G. 1968. Karyotype and C-banding in *Trigonella* section *foenum-graecum* (Fabaceae). *Plant Systematics and Evolutions* 153(1–2): 1–5.

Lust, J.B. 1986. *The herb book*. Bantam Books Inc, New York, pp. 1–55.

Malhotra, S.K. 2011. Breeding potential of indigenous germplasm of seed spices. In: Singh, D.K., and Chowdhary, H, editors. *Vegetable crops: Genetic resources and improvement*. New India Publishing House, New Delhi, 2011, pp. 477–497.

Marc, R.C., and Capraru, G. (2008) Influence of sodium phosphate (E 339) on mitotic division in *Trigonella foenum-graecum* L. Scientific Annals of the "Alexandru Ioan Cuza" University of Iasi, Section II a. *Genetic and Molecular Biology* 9(1): 67–70.

Mehrafarin, A., Qaderi, A., Rezazadeh, Sh., Naghdibadi H., Noormohammadi Gh., and Zakati, E. 2010. Bioengineering of important secondary metabolites and metabolic pathways in fenugreek (*Trigonella foenum-graecum* L.). *Journal of Medicinal Plants* 9(35): 1–18.

Mehrafarin, A., Rezazadeh, S., Naghdibadi, H., Noormohammadi Gh., and Qaderi, A. 2011. A review on biology, cultivation and biotechnology of fenugreek (*Trigonella foenum- graecum* L.) as a valuable medicinal plant and multipurpose. *Journal of Medicinal Plants* 10(37): 6–24.

Miraj, M., Durrishahwar, S., Muhammad, M. A., Nawab, A., and Sajid, I. 2017. Assessment of genetic diversity in exotic Fenugreek (Trigonella foenum-*graecum*) germplasm using SDS-Page analysis. *Middle East Journal of Agriculture Research* 6(4): 1285–1294.

Petropoulos, G.A. 2002. *Fenugreek -The genus Trigonella*. 1st ed. Taylor and Francis, London and New York, pp. 1–127.

Reasat, M., Karapetyam, J., and Nasirzadeh, A. 2002. Karyotypic analysis of *Trigonella* genus of Fars province. *Iranian Journal of Rangelands and Forests Plant Breeding and Genetic Research* 11(1): 127–145.

Roy, R.P., and Singh, A. 1968. Cytomorphological studies of the colchicines-induced tetraploids *Trigonella foenum-graecum* L. *Genetica Iberian* 20(1–2): 37–54.

Saraswat, K.S. 1984. Discovery of Emmer wheat and fenugreek from India. *Current Science* 53(17): 925.

Sharma, R., Joshi, A., Maloo, S.R., and Rajamani, G. 2012. Assessment of genetic finger printing using molecular marker in plants: A review. *Scientific Research Impact* 1: 29–36.

Singh, A., and Roy, R.P. 1970. Karyological studies in *Trigonella, Indigofera* and *Phaseolus*. *The Nucleus (Calcutta)* 13: 41–54.

Singh, A., and Singh, D. 1976. Karyotype studies in Trigonella. *The Nucleus (Calcutta)* 19: 13–16.

Srivastava, A., and Raghuvanshi, S.S. 1987. Buffering effect of B-chromosome system of *Trigonella foenum-graecum* against different soil types. *Theoretical and Applied Genetics* 75: 807–810. DOI: 10.1007/BF00265609.

Sundaram, S., and Purwar, S. 2011. Assessment of genetic diversity among fenugreek *(Trigonella foenum-graecum* L.), using RAPD molecular markers. *Journal of Medicinal Plants Research* 5(9): 1543–1548.

Vasil'chencko IT Berich uber die Arten der Gattung (SSSR.1: 10). 1953. *Trigonella* Trudy Bot. *Inst. Akad. Nauk* 17: 331–333.

Vavilov, N.I. 1951. The origin, variation, immunity and breeding of cultivated plants. *Chronica Botanica* 1: 6.

Zandi, P., Basu, S.K., Bazrkar, Khatibani L., Balogun, M., Aremu, M.O., Sharma, M., Kumar, A., Sengupta, R., Li, X., Li, Y., Tashi, S., Hedi, A., and Cetzal-Ix, W. 2015. Fenugreek (*Trigonella foenum-graecum* L.) seed: A review of physiological and biochemical properties and their genetic improvement. *Acta Physiologiae Plantarum* 37: 1714. DOI:10.1007/s11738-014-1714-6.

Part II Chemistry

4 Title Optimization of the Supercritical Carbon Dioxide Extraction of Phytochemicals from Fenugreek Seeds

Aleksandra Bogdanovic, Vanja Tadic, Slobodan Petrovic and Dejan Skala

CONTENTS

4.1 Introduction	33
4.2 Supercritical Extraction of Phytochemicals from Fenugreek	35
4.2.1 Obtaining Steroidal Sapogenins Rich Extracts from Fenugreek Seeds	36
4.2.1.1 Pre-treatment of Plant Seeds	37
4.2.1.2 Supercritical CO_2 Extraction-SFE of Steroidal Sapogenins and Triterpene Saponins	37
4.2.2 Total Sterols, Vitamin E and Vitamin D in Fenugreek Extracts	37
4.2.2.1 SC CO_2 Extraction of Total Phytosterols, Vitamins E and D	37
4.2.3 Content of Unsaturated Fatty Acids (UFA) Obtained by SCE of Fenugreek	38
4.2.3.1 SC CO_2 Extraction of UFA from Fenugreek	38
4.2.4 Analysis of Phytochemicals from Fenugreek	38
4.2.4.1 Analysis of Steroidal and Triterpenoid Sapogenins, Phytosterols, Vitamin E and Vitamin D and Unsaturated Fatty Acids (UFA)	38
4.3 Supercritical Extraction of Phytochemicals from Fenugreek Seeds	39
4.3.1 Supercritical Extraction of Sapogenins	39
4.3.2 Supercritical Extraction of Sterols, Vitamin E and Vitamin D	40
4.3.3 Extraction of Unsaturated Fatty Acids (UFA) from Fenugreek by Using CO_2	40
4.3.4 Optimization of Supercritical Extraction of Sapogenins from Fenugreek	43
4.3.4.1 The RSM-CCRD Analysis	43
4.3.4.2 Analysis of Response Surface	43
4.3.5 RSM-CCRD Analysis	45
4.3.6 Optimization of SCE Yields of Important Phytochemicals from Fenugreek	47
4.4 Summary	47
References	48

4.1 INTRODUCTION

Various types of secondary metabolites—phytochemicals, present in medicinal plants and spices—have been gaining increased interest because of their proven health benefits. They exhibit numerous advantages in comparison to synthetic drugs due to their safe application, with no toxic or side effects, and having better bioavailability. The application of phytochemicals as active compounds in extracts obtained from medical plants in the treatment of a variety of disorders and diseases has been known since ancient times. But, because of a lack of more comprehensive and available studies of their characterization, optimization, and pharmacological activity and mechanism of action, phytochemicals were taken with caution due to unreliability, while preference has been given to synthetic drugs. Recently, due to the proven disadvantages and side effects of synthetic drugs, the increased use of plants and their active compounds as phytochemicals in medicine has been analyzed. The various extraction methods, characterization of extracts, mechanism of their action with aim to enable their most effective application in health care were analyzed [Djurdjevic et al. 2017; Zugic et al. 2016; Stamenic et al 2014; Maksimovic et al. 2017, 2018; Bogdanovic et al. 2016a].

The classical example when the medicinal plants might be used with significant health improvement is in a case of metabolite disorders, namely in a case of hyperlipidemia.

Fenugreek (*Trigonella foenum-graecum* L., Fabaceae), abundant in bioactive phytochemicals of significant importance, has been applied in traditional medicine for its beneficial health effects for centuries. Application of fenugreek in traditional medicine during its long history has been widely beneficial for health disorders, with a general impact on immune system strengthening. Since its wide traditional application in ancient Egypt, Rome, and Greece, it has been used in resolving health issues in metabolic disorders, kidney and liver issues, as well as in managing hormone imbalance.

Most known applications of fenugreek were for the semi-synthesis of steroid hormones as well as its beneficial role against metabolic diseases such as hypercholesterolemia, dyslipidemia, diabetes, and obesity [Mehrafarin et al. 2010]. Ability of fenugreek to enhance quality and productivity of milk in breastfeeding has increased the use as a lactation stimulant. Fenugreek is widely known for its use as a nutraceutical, especially as an ingredient of curry powder. In traditional medicine use of fenugreek has been known in decreasing infective skin processes including dandruff treatment [Czygan et al. 2001]. Fenugreek is composed of a high amount of fibers, carbohydrates, proteins, vitamins; contains a low content of essential oil; and possesses significantly valuable secondary metabolites as protoalkaloids (trigonelline and choline), steroidal saponins (diosgenin, yamogenin, sarsapogenin, tigogenin, protodioscin, etc.); phytosterols (β-sistosterol, cholesterol, etc.); and flavonoids (orientin, isoorientin, isovitexin) with amino acids as 4-hydroxyisoleucine, arginine, histidine, and lysine [Czygan et al. 2001; Hagers Handbuch Der 1969; Dab 10 1991; European Pharmacopoeia 6.0; WHO 2007].

Diosgenin and phytosterols take a significant place as very important raw materials for the semi-synthesis of steroid hormones, and their application and use in the pharmaceutical industry. Diosgenin, a naturally occurring steroid saponin, abundantly present in fenugreek, is a precursor of various synthetic steroidal drugs that are extensively used therapeutically and commercially. Recently, many conducted studies have reported the beneficial role of diosgenin in metabolic disorders such as hypercholesterolemia, dyslipidemia, diabetes, and obesity, as well anti-inflammation and anticancer activity [Fernandes et al. 2003; Peng et al. 2011]. Diosgenin has been extensively proven to contribute to the treatment of several diseases; therefore its usefulness in maintenance of health is widespread, leading to increasing demand for diosgenin extraction from fenugreek [Lepage et al. 2011; Sauvaire et al. 1991; Jung et al. 2010].

In addition to diosgenin, a main abundant representative within steroidal sapogenins, other steroidal sapogenins detected in fenugreek, namely sarsapogenin, yamogenin, protodioscin, tigogenin, etc., revealed significant anticancer, hypocholesterolemic, and hypoglycemic activity, protective effects on the cardiovascular system with regulation of high blood pressure and antidepressant activities [Bhatia et al. 2006; Tong et al. 2012; Francis et al. 2002; Swaroop et al. 2017; Wang et al. 2017; Moon et al. 2012; Hostettmann and Marston 2005]. It has been revealed that fenugreek seed lipids represent one of the richest phytosterols sources compared to commodity oils and has higher content of cholesterol in total sterols [Ciftci et al. 2011; Philips et al. 2002]. Positive effects of phytosterols have been maintained through their ability to reduce plasma cholesterol levels in humans, concurrently exhibiting anticancer, anti-atherosclerotic, anti-inflammatory, and anti-oxidative properties [Ostlund et al. 2002; Jong et al. 2003; Sharma et al. 1990; Petit et al. 1995; Awad et al. 2003; Bouic et al. 2001]. The main reason in the increased number of studies and development of functional food with phytosterols is because of the growing interest in their beneficial effects on the decrease of serum low-density lipoprotein cholesterol (LDL).

Fenugreek seed lipids are also an excellent source of vitamin E and vitamin D. Vitamin E has been established as an efficient antioxidant that protects the cell membranes from oxidation, while vitamin D exhibits beneficial activity in the treatment of osteoporosis [Seppanen et al. 2010; Holick et al. 2003].

Besides, some compounds of interest within the triterpene saponins group of compounds, ursolic acid and oleanolic acid, have been studied according to their different biological activities, like their ability to reduce blood pressure, to regulate cholesterol level, to induce the apoptosis of cancer cells [Sharma et al. 1990; Jong et al. 2003; Sierksma et al. 1999].

The focus of researchers' interest in the extraction of bioactive compounds from fenugreek has been related to significant phytochemicals bioactivity, biocompatibility, and wide application, highlighting their use in the synthesis of steroid drugs in pharmaceutical industries, as well as their application in the treatment of hypercholesterolemia, hyperglycemia, hypertension and cancer.

Because of the wide pharmacological properties useful in the treatment of many health disorders, the fenugreek seed extracts rich in steroidal saponins, sterols, essential amino acids, triterpene saponins, vitamin E, and vitamin D attracted our attention and we set the goal of our research studies.

Widely applied extraction with conventional solvents is characterized by poor selectivity and additionally usually requires high temperature for solvent recovery, which could result in degradation of the desired compounds. The obtained extracts usually have solvent residues which are not convenient in medicinal use. Contrarily, supercritical extraction (SCE) shows higher selectivity compared to conventional techniques of extraction. Furthermore, the selectivity of SCE towards some groups of compounds might be tuned by changing extraction conditions (pressure and temperature). The separation of supercritical fluid from the extract is achieved by lowering the pressure, which allows obtaining extracts without residual solvents, useful for application in pharmaceutical, food, and cosmetics industry.

In our research work, the wide range of working conditions of SCE were examined in order to exhibit the optimal process parameters of pressure, temperature, and consumption of CO_2 (different time of process) which would enable extracts rich in steroidal sapogenins, sterols, essential amino acids, triterpene saponins, vitamin E, and vitamin D, as phytochemicals of significant importance. The aim of these studies was to establish the comprehensive mathematical model, defining the influence of SCE process parameters and the best operating conditions to achieve extract abundant in the phytochemicals recognized as carriers of significant bioactivity. Based on applied wide range of pressure, temperature, and consumption of SC solvent, an optimal working condition of process extraction was determined by response surface methodology (RSM) and central composite rotatable design (CCRD). RSM and CCRD gave the information related to influence of process parameters on total bioactive compounds yields, their interactive effect, and signification on maximal achieved content of these compounds in the obtained extracts.

4.2 SUPERCRITICAL EXTRACTION OF PHYTOCHEMICALS FROM FENUGREEK

Hypercholesterolemia has been taking a dominant role in the medical disorders of human health, requiring efficient and effective solutions instead of globally used synthetic drugs. Plants are abundant with various phytochemicals of important bioactivity such as sapogenins, fibers, amino acids, phytosterols, flavonoids, triterpenes, proteins, and others, therefore having great potential in lowering content of cholesterol [Jungmin et al. 2013]. Therefore, requests for safe and efficient phytopreparations with no side effects are of great importance in the pharmaceutical industry, as well as much simpler and cheaper production processes. Fenugreek seed revealed a high potential of bioactive compounds regarding their potential hypolipidemic, hypocholesterolemic, and hypoglycemic activity. Based on the identified phytochemicals in fenugreek, further investigations were undertaken in order to find the adequate methods for obtaining the extracts with defined chemical composition, namely rich in the components/secondary metabolites known for their effect on normalization of cholesterol level.

Often, conventional methods of extraction, comprising organic solvents usage, enable high yields of extraction of the different types of secondary metabolites. Although, the quantitative and qualitative composition of extracts might be satisfactory, the serious drawbacks of this type of extraction include the presence of toxic and harmful organic solvents and need for further purification manifesting in more complex, prolonged, and more expensive processes. Revealed in recent decades, supercritical fluid extraction of biologically active extracts has surpassed conventional methods as extracts contain no toxic residues of solvents and have no further need for purification. The supercritical extraction, realized at low temperature, prevents thermal degradation of thermolabile compounds, as carriers of antibacterial and antioxidant activity [Bakkali et al. 2008; Tajkarimi et al. 2010].

Taking into account that fenugreek seeds contain the significant content of non-polar phytochemicals responsible for hypolipemic properties, extraction using supercritical carbon dioxide was performed.

The supercritical CO_2 (SC CO_2) was chosen as most the convenient and adequate solvent, environmentally suitable, easily removable from extracts by decreasing temperature and pressure below its critical values (31.1°C, 7.38MPa). Solvent power of SC CO_2 can be controlled by different pressure and temperature of SCE which induce different solvent densities, directly correlating and affecting achieved extraction yield, specifically influencing and determining characteristics and composition of bioactive compounds in obtained extracts. Therefore, optimizing and fitting pressure, temperature, and amount of used CO_2 in SCE can affect achieved extracts yields with specific bioactive ingredients content in obtained extracts [Stamenic et al. 2014; Bogdanovic et al. 2016a].

The extraction conditions (pressure, temperature, and CO_2 flow rate) as well as the pre-treatment (defatting and hydrolysis) were systematically varied in the performed experiments. Our experiments were designed in order to enable the extraction of specified and selected phytochemicals from fenugreek seed by varying the wide range of SCE process parameters. Modifying and changing process conditions by applying a wide range of design experiments, fitting pressure, and temperature induced different densities of SC CO_2. Thus, affecting the process extraction conditions allows modified SC CO2 diffusivity through the plant matrix and achieves more efficient extraction of phytochemicals from fenugreek. The impact of the varying process parameters on the extraction yield and content of specific phytochemicals in the obtained extracts were observed with the aim to establish comprehensive correlation models.

4.2.1 Obtaining Steroidal Sapogenins Rich Extracts from Fenugreek Seeds

Fenugreek seeds are known as an important source of steroidal saponins, with several of them present as sapogenins, but most of them are present in the form of glycosides, meaning that prior to the extraction process, the removal of sugar moiety by applying a hydrolysis process is necessary. In addition, having in mind that the fat oil content in seeds is more than 6–10%, it does not exhibit any effects on cholesterol level decrease, so a pre-treatment including the defatting of the fenugreek seeds was required [Czygan et al. 2001]. Hence, methods for steroidal sapogenins extraction from fenugreek require pre-treatment of the plant material comprising the defatting as the first step, followed by acid hydrolysis and supercritical extraction. The dominant interest, while developing the method of extraction from fenugreek seeds, was to obtain the extracts abundant in diosgenin as well as other steroidal sapogenins that play a significant role in cholesterol metabolism, lowering its level in blood by inhibiting its absorption, as well as reducing liver cholesterol levels by promotion of bile excretion [Sauvaire et al. 1991].

The most common source of diosgenin within plants is usually plants from genus *Dioscorea*, followed by *T. foenum-graecum* L., *Tribulus terrestris* L. Zygophyllaceae, *Smilax china* L. Smilacaceae, which are rich in other steroidal sapogenins as well [Ghoreishi et al. 2012; Shu et al.

2004; Yang et al. 2012]. Although fenugreek as a source has a lower level of steroidal sapogenins compared to the *Dioscorea* species, the easier cultivation of fenugreek and its faster growth make fenugreek competitive and efficient for steroidal sapogenin extraction.

4.2.1.1 Pre-treatment of Plant Seeds

Diosgenin and other steroidal compounds isolation could be performed by the SCE process, with the previously applied pre-treatment of fenugreek seeds, as mentioned earlier. Pre-treatment includes the defatting process as the first step, followed by a second step of hydro-isolate preparation from the fenugreek seeds. The acid hydrolysis of steroidal saponins was necessary to employ in diosgenin extraction from seeds, which led to the release of diosgenin from the sugar chain, while the defatting step enabled the acquisition of SC CO_2 extracts without dispensable fats, maintaining the high efficiency in application as hypocholesterolemic actives.

4.2.1.2 Supercritical CO_2 Extraction-SFE of Steroidal Sapogenins and Triterpene Saponins

Steroidal sapogenins SCE as a process of extraction from fenugreek took place in recently reported studies, and was revealed to be an efficient and convenient method compared to previous traditionally used techniques of extraction. The numerous studies were performed in order to define process extraction in order to obtain the extracts rich in steroidal sapogenins estimating the best operating conditions [Merkli et al. 1997; Petit et al. 1995; Yang et al. 2012; Kang et al. 2013]. Supercritical extraction was efficient in gaining the high-quality fenugreek extracts rich in steroidal sapogenin. Recently reported studies revealed information regarding the process conditions of SCE that was followed by microwave-assisted extraction, but none gave enough data regarding the wide range of applied conditions to set the best relevant operating conditions. SCE was carried out in a semi-batch Autoclave Engineers Screening System. The reactor system was filled with previously pre-treated seeds by defatting and acid hydrolysis, dried to constant moisture content less than 5%. SC CO_2 extraction of pre-treated defatted hydro-isolate was exhibited at the wide range of the designed operating conditions of pressure, temperature, and consumption of CO_2. Applied process conditions in supercritical extraction were based on preliminary research which formed a range of working conditions, applying the appropriate set of experiments by RSM and CCRD design. The effects of applied conditions of pressure, temperature, and amount of SC CO_2 on the yield of total extract and phytochemicals content in them were investigated.

4.2.2 Total Sterols, Vitamin E and Vitamin D in Fenugreek Extracts

Besides extracts rich in steroidal sapogenins, the main goal was to obtain fenugreek extracts abundant in phytosterols as well, as they exhibit significant hypocholesterolemic and hypolipidemic effects because of their ability to reduce blood cholesterol levels, probably due to structural similarity with cholesterol. In some previously reported investigations, the phytosterols in high amounts in fenugreek extracts were determined, mostly obtained by conventional extraction methods [Ciftci et al. 2011; Philips et al. 2002; Lagarda et al. 2006; Brenac and Sauvaire 1996]. The fenugreek seeds represent a significant source of lipophilic vitamins, tocopherols, with vitamin E as the main constituent, known as a powerful antioxidant whose absence in organisms can lead to chronic diseases [Ciftci et al. 2011]. Moreover, the presence of vitamin D in the fenugreek extracts has a significant and important role as a lipophilic vitamin used in protection from osteoporosis, as well as in treatment of hypertension [Holick et al. 2003].

4.2.2.1 SC CO_2 Extraction of Total Phytosterols, Vitamins E and D

Extraction was carried out at the selected operating conditions of pressure, temperature, and for different time of extraction, that is, for various consumptions of SC CO_2. Extract was collected at the outlet line from the extractor into the small glass vial and after measuring its weight, the yield of total extract was determined as well as solubility of total extract on the basis of the rate of extraction in the initial period of the process and the amount of used SC CO_2. Collected extracts were used for further determination of the total sterols, vitamin E and vitamin D content in extracts using gas chromatography and mass spectroscopy (GC/MS). An experimental design matrix of experiments was followed; by combining the pressures and temperatures from the set of experiments of the formed matrix, such operating conditions of SCE defined the range of SC CO_2 densities between 800–930 kg/m³.

4.2.3 Content of Unsaturated Fatty Acids (UFA) Obtained by SCE of Fenugreek

Extracts abundant in unsaturated fatty acids (UFA) revealed potential hypolipemic action, leading to wide application for lowering the levels of cholesterol [Kahn et al. 1963; Yu et al. 1995]. Recent research data indicated that fenugreek seeds were a good source of unsaturated fatty acids. They have significant commercial use, taking into account that the unsaturated fatty acids are important as a nutritional supplement in diets. If a deficiency of UFA is detected, some chronic diseases might be revealed. Recent research has shown that fenugreek seed contains about 7% lipids consisting mostly of UFA [Yang et al. 2012].

The aim of our study was to define optimal supercritical extraction conditions for UFA extraction from fenugreek seeds, as well as to determine a unique class of chemical constituents with a wide range of biological activities.

4.2.3.1 SC CO_2 Extraction of UFA from Fenugreek

Extraction was performed using various operating conditions of pressure and temperature in a wide range. The extract was discharged from the extractor into the sample container and the yield of extraction was measured. Achieved extracts were derivatized by refluxing with methanol and H_2SO_4 for 3 hours, then solutions were cooled, filtered, and neutralized by adding petroleum ether and distilled water. A fraction of petroleum ether was separated and evaporated to dryness and analyzed using gas chromatography (GC/FID and GC/MS). The yield of fatty acids was varied in function of used supercritical extraction, the pressure, and temperature, in order to optimize the extraction efficiency.

4.2.4 Analysis of Phytochemicals from Fenugreek

Based on the chemical profile of the obtained extracts from fenugreek, substances that were characterized in the literature as carriers of hypolipidemic activity were specific targets of the obtained study and their contents were determined and analyzed.

4.2.4.1 Analysis of Steroidal and Triterpenoid Sapogenins, Phytosterols, Vitamin E and Vitamin D and Unsaturated Fatty Acids (UFA)

Quantitative and qualitative analysis of steroidal and triterpenoid sapogenins in obtained extracts was performed by high-performance liquid chromatography (HPLC), according to the standards revealed in details in [Bogdanovic et al. 2016b, 2020]. Total sterols, vitamin E and vitamin D, as well as unsaturated fatty acids in extracts were analyzed using gas chromatography and mass spectroscopy (GC/MS) [Bogdanovic et al. 2016c].

Central composite design was applied to study the SCE process exhibited to obtain fenugreek seeds extracts rich in steroidal sapogenins, triterpenoid saponins (ursolic and oleanolic acids), phytosterols, and vitamin E and vitamin D. RSM and CCRD were used to study effects of pressure, temperature, and duration of extraction (consumption of SC CO_2) as well as to define interactions among these working conditions. A similar procedure was applied as those described elsewhere [Yang et al. 2012; Ivanov et al. 2012; Ghoreishi et al. 2012; Gelmez et al. 2009; Xu et al. 2011]. The design of the experiment was based on a three-factor central composite with 20 individual experimental points. Three variables were provided, and a total of 20 experiments included 9 factorial, 5 star, and 6 central points for RSM-CCRD analysis. The investigated parameters were pressure (determined as X_1, MPa), temperature (X_2, °C), and amount of consumed CO_2 (X_3, m_{co2}/m_{dm}). The actual and coded variables used in the experimental design were defined on the basis of preliminary tests realized at different SC CO_2 density. A second-order polynomial equation was considered for prediction of target phytochemicals yield.

Second-order polynomial equations were used to express effects of independent coded variables of pressure (X_1), temperature (X_2), and consumption of CO_2 (X_3) and to represent the extraction yield (Y_1) and content of steroidal sapogenins (Y_2), triteprensaponins (Y_3), total sterols (Y_4), vitamin E (Y_5), and vitamin D (Y_6) as have been shown by general equation:

$$Y_i = \beta_{0,i} + \beta_{1,i}X_1 + \beta_{2,i}X_2 + \beta_{3,i}X_3 - \beta_{11,i}X_1^2 + \beta_{22,i}X_2^2 - \beta_{33,i}X_3^2 + \beta_{12,i}X_1X_2 + \beta_{13,i}X_1X_3 + \beta_{23,i}X_2X_3 \tag{1}$$

MATLAB software was used to perform multiple regression coefficients to fit the second-order polynomial equations to dependent variables affected by independent variables, to plot response 3D surfaces and to perform optimization.

4.3 SUPERCRITICAL EXTRACTION OF PHYTOCHEMICALS FROM FENUGREEK SEEDS

4.3.1 Supercritical Extraction of Sapogenins

Fenugreek extracts abundant in steroidal and triterpenoid sapogenins were obtained applying the supercritical extraction. The goal was to optimize process parameters and to find the best operating conditions for achieving the maximal yield of target compounds. Supercritical extraction of sapogenins from fenugreek seeds was performed in a wide range of selected process parameters as pressure and temperature, defining the various range of SC CO_2 density, in the range 790–930 kg/m³. The content of steroidal and triterpenoid sapogenins, diosgenin, sarsapogenin, protodioscin, oleanolic acid, and ursolic acid were determined in all performed SCE runs with wide range of SC CO_2 density and presented in Figure 4.1.

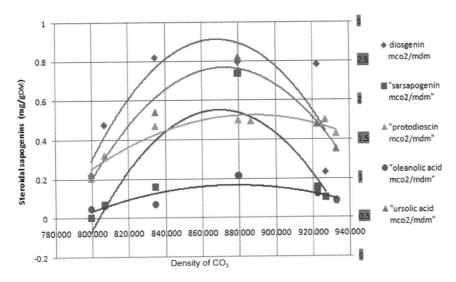

FIGURE 4.1 Steroidal sapogenins content in obtained extracts of fenugreek determined on a wide range of SC CO_2 density.

Source: By permission of Bogdanovic et al. (2020).

From data shown in Figure 4.1, it might be concluded the same tendency of increased yield of steroidal sapogenins content is in accordance with the rising SC CO_2 density. The highest content of steroidal sapogenins was detected when SC CO_2 density ranged from 840 kg/m^3 to 900 kg/m^3, with maximum of steroidal content occurring at SC CO_2 density of 880 kg/m^3. The achieved content of triterpenoid saponins, ursolic and oleanolic acid, in accordance to rising values of SC CO_2 density was not significant as in the case of steroidal sapogenins.

4.3.2 SUPERCRITICAL EXTRACTION OF STEROLS, VITAMIN E AND VITAMIN D

Determined content of total sterols and vitamins (D + E) in obtained total extract related to applied of SC CO_2 density is presented in Figure 4.2.

The maximal yield of total extract from the fenugreek seeds was almost four times smaller compared to the yield of defatted and hydrolyzed fenugreek seed used for steroidal sapogenin extraction. Determined content of different sterols in extract obtained from fenugreek seed was between 8.6% and 13.6%. These values are in concordance with literature data as shown in Table 4.1.

Influence of time of extraction or consumed amount of SC CO_2 with density of 880 kg/m^3 was shown in Figure 4.3. Obviously, total yield of isolated extract was significantly increased until some value of consumed CO_2, and after that a very small yield of extract was determined.

4.3.3 EXTRACTION OF UNSATURATED FATTY ACIDS (UFA) FROM FENUGREEK BY USING CO_2

The highest yield of obtained extracts was achieved at 30MPa and 50°C, as shown in Figure 4.4. These experiments were performed to analyze the kinetic of extraction process. The optimal conditions for supercritical extraction of UFA from derivatized fenugreek seed were detected. It was found that the total composition of UFA in extracts slightly varies with changing the conditions of pressure and temperature. It was also found that maximal yield could be obtained at the highest value of pressure and temperature (30MPa and 50°C).

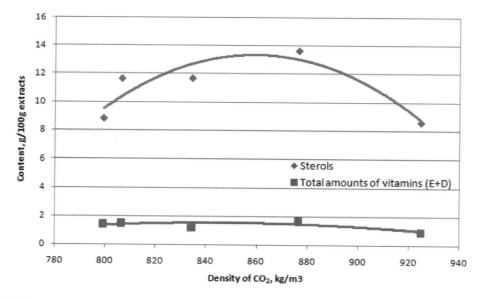

FIGURE 4.2 Content of sterols and total amount of vitamins (D + E) in the fenugreek seeds extracts.

Source: By permission of Bogdanovic et al. (2016c).

Supercritical Carbon Dioxide Extraction

TABLE 4.1
Data of Sterols, Vitamin E, Triacylglycerol, and Fatty Acids Content in Fenugreek Extracts

	Content g/100 g extracts	Mw (g/mol)
Total sterols	6–14%	
Campesterol	0.19–1.17	400.7
Sitosterol	2.48–6.47	414.7
Fucosterol	0.06–0.27	412.7
Cycloartenol hanolianol	1.87–2.17	426.7
Stigmastan-3,5 dien	1.01–2.17	396.7
Others	1.00–3.79	
Tocoferol (vitamin E)	0.2–0.6%	430.71
Triacylglycerols[a]	89.1–90%	793–845
Free fatty acids (Ln; L; O; P, S)	<5%	256–284

Source: By permission of Bogdanovic et al. (2016c).
[a]Type of triacylglycerols: LnLnLn, LnLnL, LLLn, LnLnO, LnLnP, LLL, LnLO, PLLn, OLL, PLL, OOL, PLO, PLP, OOO.
Abbreviations: Ln—linolenic, L—linoleic, O—oleic, P—palmitic, S—stearic.

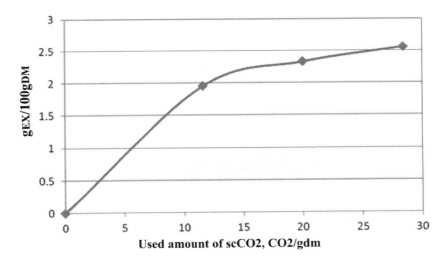

FIGURE 4.3 Extraction yield determined using SC CO2 density of 880 kg/m³ as function of consumption of CO_2 (time of extraction).

Source: By permission of Bogdanovic et al. (2016c).

Analyzed extracts contained 10 mostly present UFA. Extract obtained at 50°C and 20MPa contained the highest content of UFA (99.77%), according to the extracts obtained at other SFE conditions. The GC/MS analyses of obtained extracts have shown the highest percent of linoleic acid (~40%), linolenic acid (~40%), palmitic acid (~10%), and stearic acid (~4%), determined as methyl esters of UFA. Composition profile of UFA in obtained extracts was similar with the highest percent of methyl linoleate, methyl linolenate, and methyl palmitate, determined by GC/MS shown in Figure 4.5.

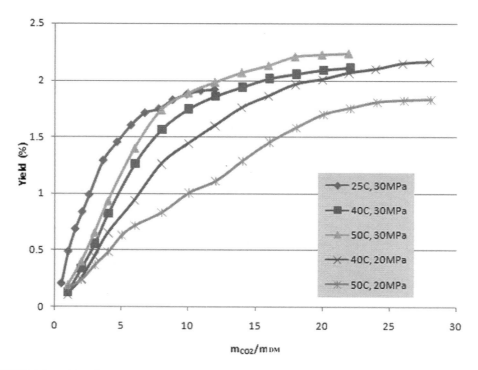

FIGURE 4.4 Yield of the SFE of fenugreek seeds as a function of specific amount of solvent (kg CO_2/kg herbaceous material).

Source: By permission of Bogdanovic et al. (2016c).

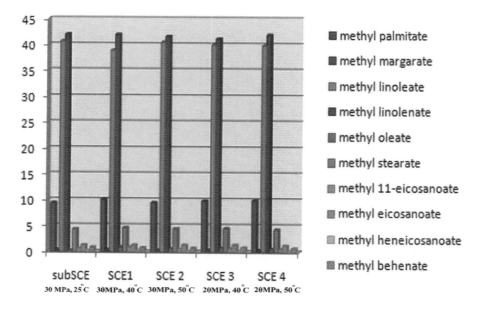

FIGURE 4.5 Composition of UFA in fenugreek extracts obtained by subcritical and supercritical conditions.

4.3.4 Optimization of Supercritical Extraction of Sapogenins from Fenugreek

4.3.4.1 The RSM-CCRD Analysis

The application of supercritical extraction enables easy separation of the steroidal sapogenins from the seed matrix, as well as the selective extraction according to applied density of SC CO_2. Selectivity of the supercritical process obtained by regulation of the solvent power of SC CO_2 density might be changed and thus the yield and composition of steroidal sapogenins could be regulated. Therefore, this extraction process is of special interest to find the optimal conditions for achieving maximal content of steroidal sapogenins. The mathematical model has been developed to describe the SCE processes.

Response surface methodology and central composite design was applied with the goal to optimize the extraction process with maximal achieved sapogenins content in extract. The solvent power under determined conditions induces the solubility of steroidal compounds. The supercritical extraction process was investigated according to the influence through three independent process parameters: pressure, temperature, and consumption of SC CO_2. The variation of applied process conditions induced different yields of obtained extracts followed by individual and different content of sapogenins in them. Yield of extract, content of steroidal and triterpenoid sapogenins, in correlation to applied condition were analyzed and optimized by the chosen RSM and CCRD model. The RSM model examined the correlation to fit the quadratic equation, proven and estimated through the analysis of variance (ANOVA) test.

4.3.4.2 Analysis of Response Surface

3D response of the applied model was constructed to give visual insights into the influence of process parameters on the yield and content of steroidal sapogenins and triterpenoid saponins. Interaction of the process conditions and summarized effect on obtaining maximum yield of target compounds were observed and estimated on 3D surfaces.

RSM analysis of process parameters on steroidal sapogenins content was presented through 3D plotted surfaces in Figure 4.6.

Adjusting pressure and temperature as process conditions induces evident differences in density of SC CO_2 and dissolution power of solvent, followed by different selectivity. Selectivity of SC CO_2 can be adjusted with insights into the basic behavior of the process parameters, as the temperature rising will lead to decrease of solvent density, while rising pressure will increase solvent power by increasing density. Temperature effect on solvent power can be the opposite: the increase will lead to decrease of solvent density, while an increase influences the vapor pressure of the solution. A greater effect on solubility is increase of vapor pressure than a decrease in the SC CO_2 density. Finally it is followed by enhanced SC CO_2 extraction leading to greater solubility of solutes.

RSM analysis proved that diosgenin yield was significantly affected by the influence of pressure and consumption of CO_2 in linear and quadratic term, while interaction between process conditions and diosgenin content was not observed. Increased solvent density at higher pressure resulted in higher extraction potential of SC CO_2 (Figure 4.6). In addition, temperature had significant effect on protodioscin content in obtained extracts. In this case temperature has more significant influence on the vapor pressure than the reduced effect of solvent power, resulting in a higher yield of protodioscin. Sarsapogenin yield was significantly influenced by solvent power, its larger density, while the quadratic term indicated significance of temperature to vapor pressure of sarsapogenin until some value. However, further increase of temperature decreased solvent power and consequently the content of sarsapogenin in obtained extracts was smaller, wherefrom solvent reducing power overwhelmed influence on decreasing yield of sarsapogenin content in obtained extracts. According to the results of performed investigation and on the basis of derived correlation of applied process parameters, estimated density and selectivity as well as solubility of steroidal compounds, the optimal conditions for achieving the maximum yields for all of them were determined.

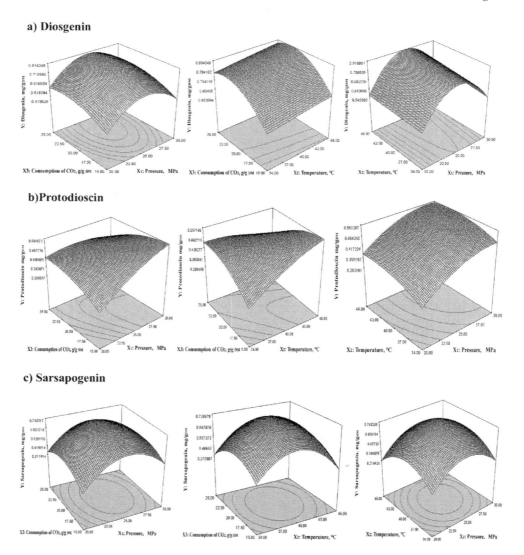

FIGURE 4.6 3D surface plot of influence of process parameters on extraction yield on steroidal sapogenins.

Source: By permission of Bogdanovic et al. (2020).

3D surface plots (Figure 4.7) of the RSM model show the influence of process parameters on extract yield and triterpenoid saponins—ursolic and oleanolic acid content in them.

Extract yield was augmented by increasing pressure and temperature, although became almost constant at 25MPa. A higher temperature followed by increased pressure gave a higher yield of extract, confirming positive effect of increased vapor pressure which overwhelmed decrease of solvent density.

In the case of ursolic and oleanolic acid yield, influence of process parameters was similar to those estimated for extraction yield. The yields of these compounds were increased at higher temperature, pressure, and amount of used solvent for SCE until some value. Further increase of these process parameters led to decrease of their yields.

Supercritical Carbon Dioxide Extraction

a) Extract yield

b) Ursolic acid

c) Oleanolic acid

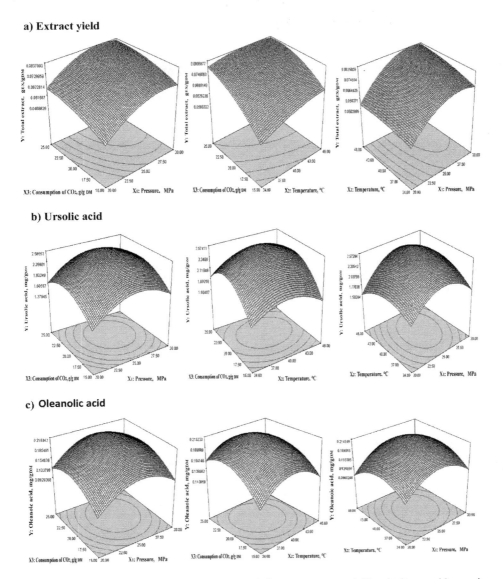

FIGURE 4.7 3D surface plots of process parameters influence on extract yield and triterpenoid saponins.

Source: By permission of Bogdanovic et al. (2020).

4.3.5 RSM-CCRD Analysis

The wide ranges of selected operating conditions of SCE (pressure and temperature) enable the extractions of various natural active compounds which might be correlated to their estimated solubility. The use of SC CO_2 in extraction of active ingredients from fenugreek seeds gave extracts of different quality and quantity with active compounds in obtained extracts depending on applied density of SC CO_2.

To estimate the optimal process conditions for extraction of phytochemicals from fenugreek seeds (phytosterols vitamin E and vitamin D), it was necessary to determine their solubility in the SC CO_2.

The solubility of compounds was a result of vapor stress and solvent-solution interaction, which could be understood and explained by RSM-CCRD models and illustrated in 3D surface plots.

A 3D plot illustration showed influence of pressure, temperature, and amount of used CO_2 on extraction of total sterols, vitamin E and vitamin D content in obtained extracts from fenugreek. The yield of total sterols was significantly influenced with increase of SC CO_2 consumption and temperature, as shown in Figure 4.8a. The similar tendency of process parameters on yields of vitamin E and vitamin D was observed, estimating that optimal SC CO_2 density is 880 kg/m^3 which gave the maximal yield as illustrated by surface plots shown in Figure 4.8b and Figure 4.8c.

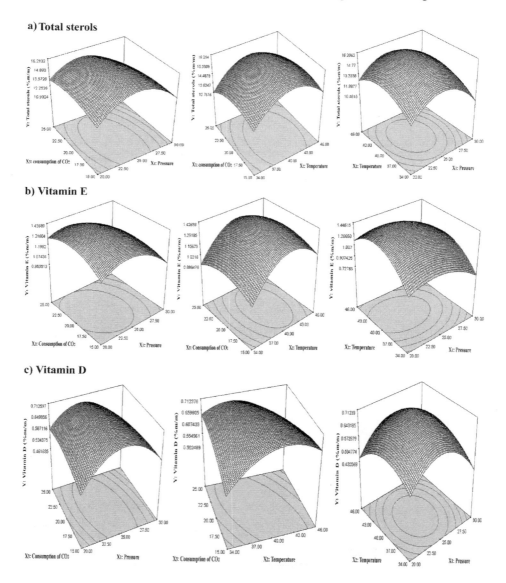

FIGURE 4.8 3D surface plots of process parameters influence on total sterols yield, vitamin E and vitamin D yield.

Source: By permission of Bogdanovic et al. (2016c).

TABLE 4.2
RSM Response Optimization (Global Optimum and Sub-Optimal Condition for Individual Maximal Yield of Target Compounds as well as Global Maximum Yield of All of Them)

Extraction conditions of SCE for maximum gained yields of phytochemicals (steroidal sapogenins, triterpene saponins, sterols, vitamin E and D) according to extraction yield

SCE Parameters for Maximal Response			Maximal Response of Total Extraction Yield and Some Specific Solutes						
Pressure (MPa)	Temperature (°C)	Amount of CO_2 (gCO_2/g)	Total extract, g/100g$_{dm}$	total sterols, g/100g$_{dm}$	vitamin E, g/100g$_{dm}$	vitamin D, g/100g$_{dm}$	Total steroidal sapogenins mg/g$_{dm}$	Ursolic acid mg/g$_{dm}$	Oleanolic acid mg/g$_{dm}$
24.48	40.12	18.63	2.285	0.372	0.032	0.016			
25.87	40.51	20.62	0.079				2.055	0.216	2.587
SCE parameters for obtaining individual maximal responses of total extraction yield and corresponding maximal yields of total sterols, or vitamins (E or D), total steroidal sapogenins, ursolic and oleanolic acid									
26.25	36.34	15.35		0.454					
23.05	40.68	18.70			0.037				
25.82	39.64	15.00				0.019			
25.96	41.00	20.26					2.055		
26.61	41.77	21.38						2.568	
25.73	39.88	21.11							0.216

Source: By permission of Bogdanovic et al. (2016c).

4.3.6 Optimization of SCE Yields of Important Phytochemicals from Fenugreek

The derived mathematical model based on experimental results of SCE enabled estimation of the correlation between process parameters of supercritical extraction of phytochemicals. It was used for better understanding and monitoring of selectivity, solubility, and efficiency of phytochemicals extraction by SC CO_2. The model determined the correlation and might be used for selection of process parameters in order to obtain the largest amount of desired bioactive compounds. The optimal SCE conditions could ensure complete extraction of target compounds, achieved in the shortest possible time with highest possible yield, which directly define the cost of their isolation.

It was determined that 25.87MPa, 40.51°C, and 20.62g/gdm were optimal conditions for obtaining extract with highest yield of sapogenins such as diosgenin, sarsapogenin, protodioscin, oleanolic and ursolic acid, while 24.48MPa, 40.12°C, and 18.63g/gdm were determined optimal conditions to obtain maximal yield of total sterols, vitamin E and vitamin D.

4.4 SUMMARY

Fenugreek provides a valuable source of various compounds applied in the food and pharmaceutical industries, regarding their potential bioactivity. Supercritical carbon dioxide extraction might ensure high-quality and safe fenugreek extracts, preserving the presence of compounds known for their significant bioactivity. At specific supercritical extraction conditions the costs and time of extraction could be reduced and desired extracts with specific activity could be obtained. Thus, the aim of this contribution was to present results of supercritical extraction of fenugreek, as advanced technology, focused to optimize the extraction conditions for specific compounds isolation from seeds, which enabled the maximal yield of specific composition of active components.

An RSM and CCRD analysis model was applied to optimize and study the influence of process parameters on yield of obtained extracts and content of target steroidal sapogenins as diosgenin, sarsapogenin, and protodioscin, as well as oleanolic and ursolic acid as triterpenoid sapogenins, total sterols content with liposoluble vitamins E and D. The optimal conditions for obtaining maximal extract yield as well as content of phytochemicals of importance was defined: pressure of 24.7MPa, temperature of 38.2°C, and consumption of SC CO_2 of 19.2 g/g_{dm} [Bogdanovic et al. 2016b, 2016c, 2020]. The best conditions to achieve the maximal yield of extract with highest content of steroidal and triterpenoid sapogenins was obtained under SC CO_2 density of 885 kg/m^3.

After applying defatting and acid hydrolysis of fenugreek seeds followed by supercritical extraction, the extract abundant in diosgenin, as main steroidal sapogenin and hormone precursor, could be obtained as a purified extract with higher content of diosgenin. It might be used for hypolipemic preparation and application, in lowering cholesterol activity, regulation of blood pressure, and steroidal hormone precursor. With aim to obtain unsaturated fatty acids rich extracts, subcritical and supercritical extractions were applied. Obtained extracts contained 10 important unsaturated fatty acids, determined by GC/MS analysis in the form of methyl esters. The highest content of methyl linoleat, methyl linolenat, methyl palmitat among other estimated methyl esters of UFA in obtained extracts from fenugreek were also achieved.

The efficiency of supercritical fluid extraction revealed dependence on the estimated process parameters of pressure and temperature which had induced density of SC CO_2, followed by data of solubility of compounds which could be extracted from the plant material.

The solubility of the solutes in the SC CO_2, selectivity and solvent power were estimated with aim to accomplish efficient and effective isolation of active compounds by supercritical fluid extraction. Optimization of SCE estimated by mathematical correlations and analysis has a goal to define conditions for achieving the highest possible content of desired bioactive phytochemicals in obtained extracts from fenugreek.

REFERENCES

Awad, A. B. Roy, R. Fink, C. S. 2003. Beta-sitosterol, a plant sterol, induces apoptosis and activates key caspases in MDA-MB-231 human breast cancer cells. *Oncology Reports* 10: 497–500.

Bakkali, F. Averbeck, S. Averbeck, D. Idaomar, M. 2008. Biological effects of essential oils-a review. *Food and Chemical Toxicology* 46: 446–475.

Bhatia, K. Kaur, M. F. Atif, M. et al. 2006. Aqueous extract of *Trigonella foenum-graecum L.* ameliorates additive urotoxicity of buthionine sulfoximine and cyclophosphamide in mice. *Food and Chemical Toxicology* 44: 1744–1750.

Bogdanovic, A. Tadic, V. Petrovic, S. Skala, D. 2020. Supercritical CO2 extraction of steroidal sapogenins from Fenugreek (trigonella foenum-graecum l.) Seed. *Chemical Industry & Chemical Engineering Quarterly* 26: 171–182.

Bogdanovic, A. Tadic, V. Skala, D. et al. 2016a. Supercritical and high pressure subcritical fluid extraction from Lemon balm (Melissa officinalis L., Lamiaceae). *Journal of Supercritical Fluids* 107:234–242.

Bogdanovic, A. Tadic, V. Skala, D. et al. 2016b. Supercritical carbon dioxide extraction of *Trigonella foenum-graecum* L. seeds: Process optimization using response surface methodology. *Journal of Supercritical Fluids* 107: 44–50.

Bogdanovic, A. Tadic, V. Skala, D. et al. 2016c. Optimization of supercritical CO_2 extraction of fenugreek seed (*Trigonella foenum-graecum* L.) and calculating of extracts solubility. *Journal of Supercritical Fluids* 117: 297–307.

Bouic P. J. 2001. The role of phytosterols and phytosterolins in immune modulation: A review of the past 10 years. *Current Opinion in Clinical Nutrition & Metabolic Care* 4: 471–475.

Brenac, P. Sauvaire, Y. 1996. Accumulation of sterols and steroidal sapogenins in developing fenugreek pods: Possible biosynthesis in situ. *Phytochemistry* 41: 415–422.

Ciftci, O. N. Przybylski, R. Rudzinska, M. et al. 2011. Characterization of Fenugreek (Trigonella foenum-graecum) seed lipids. *Journal of the American Oil Chemists' Society* 88: 1603–1610.

Czygan, F.-C. Frohne, D. Holtzel, C. 2001. *Herbal drugs and phytopharmaceuticals*, 203–205, 329–332.

Dab 10. 1991. *Deutsches Arzneibuch*, Band 3. Stuttgart: Monographien M-Z.

Đurđević, S. Milovanović, S. Šavikin, S. et al. 2017. Improvement of supercritical CO2 and n-hexane extraction of wild growing pomegranate seed oil by microwave pretreatment. *Industrial Crops & Products* 104:21–27.

Fernandes, P. Cruz, A. Angelova, B. et al. 2003. Microbial conversion of steroid compounds: Recent developments. *Enzyme and Microbial Technology* 32: 688–705.

Francis, G. Kerem, Z. Makkar, H. P. et al. 2002. The biological action of saponins in animal systems: A review. *British Journal of Nutrition* 88:587–605.

Gelmez, N. Suzan Kıncal, N. Esra Yener, M. 2009. Optimization of supercritical carbon dioxide extraction of antioxidants from roasted wheat germ based on yield, total phenolic and tocopherol contents, and antioxidant activities of the extracts. *The Journal of Supercritical Fluids* 48: 217–224.

Ghoreishi, S. M. Bataghva, E. Dadkhah, A. A. 2012. Response surface optimization of essential oil and diosgenin extraction from *Tribulus terrestris* via supercritical fluid technology. *Chemical Engineering & Technology* 35: 133–141.

Hagers Handbuch Der. 1969. *Pharmazeutischen praxis*. Berlin, 268–272, 759–762.

Holick, M. F. 2003. Evolution and function of vitamin D. *Recent Results in Cancer Research* 164: 3–28.

Hostettmann, K. Marston, A. 2005. *Saponins*. Cambridge, New York: Cambridge University Press, 310–325.

Ivanov, D. S. Colovic, R. J. Levic, D. S. et al. 2012. Optimization of supercritical fluid extraction of linseed oil using RSM. *European Journal of Lipid Science and Technology* 114: 807–815.

Jong, A. Plat, J. Mensink, R. P. 2003. Metabolic effects of plant sterols and stanols (review). *The Journal of Nutritional Biochemistry* 14: 362–369.

Jong, D. Plat, J. Mensink, R. P. 2003. Metabolic effects of plant sterols and stanols (review). *The Journal of Nutritional Biochemistry* 14: 362–369.

Jung, D. H. Park, H. J. Byun, H. E. et al. 2010. Diosgenin inhibits macrophage-derived inflammatory mediators through downregulation of CK2, JNK, NF-kappaB and AP-1 activation. *International Immunopharmacology* 10: 1047–1054.

Jungmin, O. Heonjoo, J. Reum, C. et al. 2013. Antioxidant and antimicrobial activities of various leafy herbal teas. *Food Control* 31: 403–409.

Kahn, S. G. Wind, S. Slocum, A. et al. 1963. A study of the hypocholesterolemic activity of the ethyl esters of the ethyl esters of the polyunsaturated fatty acids of cod liver oil in the chicken. II Effect on serum and tissue cholesterol and aortic and coronary atherosclerosis. *Journal of Nutrition* 80: 414–424.

Kang, L.-P. Zhao, Y. Pang, X. et al. 2013. Characterization and identification of steroidal saponins from the seeds of *Trigonella foenum-graecum* by ultra high-performance liquid chromatography and hybrid time-of-flight mass spectrometry. *Journal of Pharmaceutical and Biomedical Analysis* 74: 257–267.

Lagarda, M. Garcıa-Llatas, G. Farre, R. 2006. Analysis of phytosterols in foods. *Journal of Pharmaceutical and Biomedical Analysis* 41: 1486–1496.

Lepage, C. Léger, D. Y. Bertrand, J. et al. 2011. Diosgenin induces death receptor-5 through activation of p38 pathway and promotes trail induced apoptosis in colon cancer cells. *Cancer Letter* 301: 193–202.

Liu, L. Dong, Y.-S. Xiu, Z.-L. 2010. Three-liquid-phase extraction of diosgenin and steroidal saponins from fermentation of *Dioscorea zingibernsis*. *Process Biochemistry* 45: 752–756.

Maksimovic, S. Tadić, V. Ivanović, J. et al. 2018. Utilization of the integrated process of supercritical extraction and impregnation for incorporation of Helichrysum italicum extract into corn starch xerogel. *Chemical Industry and Chemical Engineering Quarterly* 24:191–200.

Maksimovic, S. Tadic, V. Skala, D. et al. 2017. Separation of phytochemicals from Helichrysum italicum: An analysis of different isolation techniques and biological activity of prepared extracts. *Phytochemistry* 138:9–28.

Mehrafarin, A. Qaderi, A. Rezazadeh, Sh. et al. 2010. Bioengineering of important secondary metabolites and metabolic pathways in fenugreek (Trigonella foenum-graecum L.). *Journal of Medicinal Plants* 9: 1–18.

Merkli, A. Christen, P. Kapetanidis, I. 1997. Production of diosgenin by hairy root cultures of *Trigonella foenum-graecum* L. *Plant Cell Reports* 16: 632–636.

Moon, E. Kim, A. Kim, S. Y. 2012. Sarsasapogenin increases melanin synthesis via induction of tyrosinase and microphthalmia-associated transcription factor expression in Melan-a cells. *Biomolecules & Therapeutics (Seoul)* 20: 340–345.

Ostlund, R. E. Racette, S. B. Okeke, A. et al. 2002. Phytosterols that are naturally present in commercial corn oil significantly reduce cholesterol absorption in humans. *The American Journal of Clinical Nutrition* 75: 1000–1004.

Peng, Y. Yang, Z. Wang, Y. 2011. Pathways for the steroidal saponins conversion to diosgenin during acid hydrolysis of Dioscorea zingiberensis. *Chemical Engineering Research and Design* 89: 2620–2625.

Petit, P. R. Sauvaire, Y. D. Hillaire-Buys, D. M. et al. 1995. Steroid saponins from fenugreek seeds: Extraction, purification, and pharmacological investigation on feeding behavior and plasma cholesterol. *Steroids* 60: 674–680.

Phillips, K. M. Ruggio, D. M. Toivo, J. I. et al. 2002. Free and esterified sterol composition of edible oils and fats. *Journal of Food Composition and Analysis* 15: 123–142.

Sauvaire, Y. Ribes, G. Baccou, J.-C. et al. 1991. Implication of steroid saponins and sapogenins in the hypocholesteolemic effect of fenugreek. *Lipids* 26: 191–197.

Seppanen, C. M. Song, Q. Csallany, A. S. 2010. The antioxidant functions of tocopherol and tocotrienol homologues in oils, fats, and food systems. *Journal of the American Oil Chemists' Society* 87: 469–481.

Sharma, R. D. Raghuram, T. C. Sudhakar, R. N. 1990. Effect of fenugreek seeds on blood glucose and serum lipids in type I diabetes. *European Journal of Clinical Nutrition* 44: 301–306.

Shu, X.-S. Gao, Z.-H. Yang, X.-L. 2004. Supercritical fluid extraction of sapogenins from tubers of Smilax China. *Fitoterapia* 75: 656–661.

Sierksma, A. Sierksma, J. A. Weststrate, G. W. 1999. Spreads enriched with plant sterols, either esterified 4,4-dimethylsterols or free 4-desmethylsterols, and plasma total- and LDL-cholesterol concentrations. *British Journal of Nutrition* 82: 273–282.

Stamenic, M. Vulic, J. Djilas, S. et al. 2014. Free-radical scavenging activity and antibacterial impact of Greek oregano isolates obtained by SFE. *Food Chemistry* 165:307–315.

Swaroop, A. Maheshwari, A. Verma, N. et al. 2017. A novel protodioscin-enriched fenugreek seed extract (*Trigonell foenum-graecum*, family Fabaceae) improves free testosterone level and sperm profile in healthy volunteers. *Functional Foods in Health and Disease* 7: 235–245.

Tajkarimi, M. Ibrahim, S. Cliver, D. 2010. Antimicrobial herb and spice compounds in food. *Food Control* 21: 1199–1218.

Tong, Q.-Y. He, Y. Zhao, Q.-B. et al. 2012. Cytotoxicity and apoptosis-inducing effect of steroidal saponins from Dioscorea zingiberensis Wright against cancer cells. *Steroids* 77: 1219–1227.

Wang, W. D. Wang, Z. Yao, G. et al. 2017. Synthesis of new sarsasapogenin derivatives with cytotoxicity and apoptosis-inducing activities in human breast cancer MCF-7 cells. *European Journal of Medicinal Chemistry* 15: 62–71.

World Health Organization, 2007. *WHO monographs on selected medicinal plants*, Vol. 3. Geneva: World Health Organization, 338–348.

Xu, X. Dong, X. Mu, X. et al. 2011. Supercritical CO_2 extraction of oil, carotenoids, squalene and sterols from lotus (Nelumbo nucifera Gaertn) bee pollen. *Food and Bioproducts Processing* 8: 47–52.

Yang, R. Wang, H. Jing, N. et al. 2012. Trigonella foenum-graecum L. Seed oil obtained by supercritical CO2 extraction. *Journal of the American Oil Chemists' Society* 89: 2269–2278.

Yu, S. Derr, J. Etherton, T. D. et al. 1995. Plasma cholesterol-predictive equations demonstrate that stearic acid is neutral and monounsaturated fatty acids are hypocholesterolemic. *American Journal of Clinical Nutrition* 61: 1129–1139.

Zugic, A. Jeremic, I. Isakovic, A. et al. 2016. Evaluation of anticancer and antioxidant activity of a commercially available CO2 supercritical extract of old man's beard (Usnea barbata). *PLoS One* 11(1): e0146342. doi:10.1371/journal.pone.0146342.

5 Towards the Importance of Fenugreek Proteins

Structural, Nutritional, Biological, and Functional Attributes

Samira Feyzi and Mehdi Varidi

CONTENTS

5.1	Introduction	51
5.2	Chemistry of Fenugreek Proteins	52
	5.2.1 Amino Acids Profile and Structures of Fenugreek Proteins	52
	5.2.2 Effect of Processing Conditions on Amino Acids Profile and Structures of Fenugreek Proteins	56
5.3	Nutritional Value, Anti-Nutritive Compounds, Digestibility, and Allergenicity of Fenugreek Proteins	58
5.4	Biological Effects of Fenugreek Proteins, Peptides, and Amino Acids	61
5.5	Functional Properties of Fenugreek Proteins in Food Systems	63
5.6	Thermal Properties	67
5.7	Methods of Protein Extraction	68
5.8	Application of Fenugreek in Food Systems	69
5.9	Future Challenges and Trends in Using Fenugreek Proteins	70
5.10	Conclusion	71
References		71

5.1 INTRODUCTION

Food proteins are consumed to provide a wide range of functional, biological, and nutritional properties in human beings' diets. Proteins impart the synthesis of muscles and tissues for mechanical support of the body; mediate the cholesterol and insulin levels; stimulate excretion of an enzyme or substance in the body; interact reversibly with other molecules known as ligands which is the basis of oxygen transport, muscle contraction, immune function, etc.; catalyze reactions of biological systems such as enzymes; act as hormones; or may result in toxic and allergenic effects which altogether indicate their biological importance. Amino acids composition, on top of other factors, indicates nutritional value of proteins. Moreover, functional properties of proteins including solubility, gelation, interfacial, texture and flavor refinement within food matrices rely on proteins' structural attributes and their modifications during processing. These benefits are expected to result in 6.8 million tons of global demand for protein ingredients by 2025 (Grand View Research, 2016). Consequently, various sources of proteins are required to fully address the aforementioned requirements owing to some insufficiencies of each specific protein source from animals or plants.

Fenugreek (*Trigonella foenum graecum*), a plant from the Leguminosae family, is widely used in Iran (seeds and leaves in foods and salads or as a herbal medicine), Turkey (seeds as fenugreek paste or Çemen), Sudan and Egypt (seeds in beverages containing milk or water), India (seeds or leaves in salads or as a herbal medicine), Saudi Arabia (as a herbal medicine), owing to its profound

TABLE 5.1
Proximate Chemical Composition of Fenugreek Seeds and Leaves (%)

	Moisture	Carbohydrates	Proteins	Lipids	Saponins	Alkaloids	Polyphenols	β-carotene	Minerals
Seeds	5.0–8.0	40.0–55.0	20.0–35.0	5.0–8.0	1.0–5.5	0.2–0.4	8.5–12.0	-	-
Leaves	86.0	6.0	4.50	1.0	-	-	-	<0.1	1.50

nutritional, biological, and functional properties (Isıklı & Karababa, 2005; El-Nasri & El-Tinay, 2007; Khorshidian et al., 2016). The rewarding contribution of fenugreek to food quality and health promotion is attributed to its constituents. Fenugreek seeds and leaves contain carbohydrates including mainly mucilaginous fiber (galactomannans), proteins, lipids including glycerides and free fatty acids, as well as saponins like fenugrin B, graecunins, and fenugreekine. Seeds also have some other minor compounds such as alkaloids (e.g., trigonelline, gentianine, and carpaine), β-carotene, polyphenols, especially flavonoids including quercetin, rutin, vitexin, isovitexin, minerals like Ca, Zn, Na, K, Mg, and Fe, as well as vitamins including A, B1, and C (Abd- El-Aal & Rahma, 1986; Feyzi et al., 2015; Zandi et al., 2017; Wani & Kumar, 2018; Mahmood & Yahya, 2017). Table 5.1 indicates the approximate chemical composition of fenugreek leaves and seeds. Fenugreek leaves have less amounts of total solid compounds compared with seeds, owing to high moisture content thereof (about 86%) (Wani & Kumar, 2018). Considering the growing attentions of health-conscious consumers towards the importance and consumption of plant proteins worldwide nowadays, more specifically in developing countries where the average protein consumption is less than the recommended allowance, along with less environmental impacts of various plant proteins compared with those of animal sources in terms of time, land, fuel, water, and pesticides, researchers are driven to conduct studies on novel plant proteins like fenugreek proteins and their utilization. Consequently, there are some reports about rewarding properties and drawbacks of fenugreek proteins which shed light on how fenugreek proteins could be used in food or pharmaceutical industries.

Therefore, the focus of this chapter is on the chemistry of fenugreek proteins, their extraction and functional properties, nutritional and biological values, toxicity, along with their application in foodstuffs.

5.2 CHEMISTRY OF FENUGREEK PROTEINS

5.2.1 Amino Acids Profile and Structures of Fenugreek Proteins

Almost all research on fenugreek proteins has been done on the seeds, owing to the dominant distribution of fenugreek proteins in seeds compared with leaves. As an endospermic legume, fenugreek endosperm contains higher protein content (43.8%) compared with its husk (7.9%) (Naidu et al., 2011). However, the seed's endosperm is the exclusive source of gum galactomannan, and cotyledon is mainly occupied with proteins (Bewley et al., 2013). Figure 5.1 shows the fenugreek seed and its structure. Generally, the total protein content of fenugreek seeds (29.0–30%) is reported to be more than lentil (28.0%), chickpea (22.9–25.8%), pea (22.5%), and faba bean (23.4%). However, this amount is reported less than that of soybean (38.0%) and lupin (31.2%) (Abdel-Aal et al., 1986; Torki et al., 1987; Sauvaire et al., 1998).

Legume proteins seemingly do not have well-defined solubility fractions. However, the Osborne (1924) method is widely used for classification of plant proteins. Accordingly, albumins (43–48%) as heat liable and water soluble proteins at neutral pH followed by globulins (20–27%) which are salt-soluble proteins at neutral pH are the most dominant fenugreek proteins. Fenugreek is considered as a legume due to dominant contribution of these two storage proteins. Glutelins (14–18%), partially soluble fractions in diluted NaOH, followed by prolamins (less than 7%), soluble proteins in 60–70% v/v alcohol, are the next dominant proteins of fenugreek seed. Generally, legumes contain no prolamins or very little amounts (Sauvaire et al., 1984; Abd El-Aal & Rahma, 1986;

Importance of Fenugreek Proteins

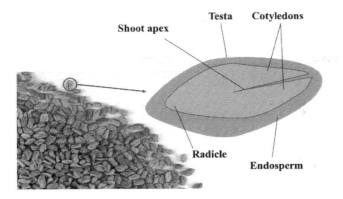

FIGURE 5.1 Fenugreek seed and its structure.

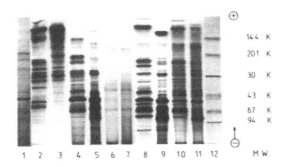

Vertical PAGE–SDS of reduced and nonreduced proteins from fenugreek, cv. Gouka. 1, protein soluble with SDS; 2, reduced albumins; 3, nonreduced albumins; 4, reduced globulins; 5, nonreduced globulins; 6, reduced glutelins; 7, nonreduced glutelins;

FIGURE 5.2 SDS-PAGE pattern of albumins, globulins, and glutelins of fenugreek proteins. *Source*: Adapted and with permission from Sauvaire et al. (1984).

Mansour & El-Adawy, 1994; Lafarga, 2019). Results of research by El Mahdy and El Sebay (1985) showed a concentration ratio of 4:3.5:2.8:1 among albumins, globulins, glutelins, and prolamins in fenugreek proteins seed.

Fenugreek albumins with isoelectric point (pI) of 4.0 encompass different subunits with molecular weights of 15–100 kDa or higher according to the SDS-PAGE electrophoretic pattern (Torki et al., 1987). Such finding coincided with results about albumins of six legumes including mung bean, chickpea, lentil, faba bean, pea, vetch, lathyrus, and dry bean seeds (Bhatty, 1982). Moreover, differences in electrophoretic pattern of fenugreek albumins under reducing and non-reducing conditions indicated cleavage of S-S bonds under reducing conditions which led to less mobility of the subunits and their emerging at higher M_r values. This phenomenon confirms considerable portions of Cys in albumins (Sauvaire et al., 1984). The Cys-rich fraction of legumes'—including lupin, pea, and peanut—albumins is known to be the 2S albumin, based on the sedimentation coefficient (SC) (S values) (Boulter & Croy, 1997). Fæste et al. (2010) confirmed the 2S albumin-like fenugreek seed proteins at M_r 80 kDa of SDS-PAGE pattern using MOLDI-ToF MS. Figure 5.2 indicates the

SDS-PAGE pattern of various fenugreek proteins. The contribution of higher molecular weight fractions in fenugreek albumins is obvious.

Globulins, although known as the main storage protein of many legumes like soybean, lentil, peas, etc., encompass the second major fenugreek seeds' proteins (Torki et al., 1987; Fukushima, 2011). The pI of fenugreek globulins equals to 4.3. Moreover, globulins are built of several subunits with MW of about 9–100 kDa or even higher MW (Torki et al., 1987). Similar to albumins, higher MW bands emerged under reducing conditions of electrophoresis confirm S-S bonds in globulins (Figure 5.2) (Sauvaire et al., 1984). Generally, globulins are divided into two main fractions being known as vicilin or β-conglycinin with average SC of 7S, and legumin or glycinin with average SC of 11S (Boulter & Croy, 1997; Fukushima, 2011). Vicilins (β-conglycinins) are usually glycoprotein with trimeric structure and M_r of 150–200 kDa in various legumes. The major subunits are α, α′, and β, from which the first two subunits contain Cys but it is suggested that vicilin has no S-S. Legumins or glycinins, on the other hand, usually have non-glycosylated subunits and assemble to a hexameric structure with about 11S value. The subunits of this hexameric structure usually are composed of acidic (35–40 kDa) and basic (~ 20 kDa) polypeptides linked by S-S, and are dissociated under reducing conditions (Boulter & Croy, 1997; Fukushima, 2011).

To the best of our knowledge, no study about the secondary, tertiary, and crystallographic structures of the fenugreek globulins or other protein fractions is available. Having said that, there are reports about the total amino acids profile, FTIR, and electrophoresis pattern of the whole fenugreek seed proteins.

Numerous literature data are published about amino acids composition of fenugreek seeds' proteins; altogether compromised about profound amounts of Glu, Asp, Leu, Arg, and Lys (Sauvaire et al., 1984; Mansour & El-Adawy, 1994; Feyzi et al., 2015). Similarly, Glu, Asp, and Arg have been known as the representative amino acids of legumes including soybean, pea, jack bean, and grass pea (Boulter & Croy, 1997; Feyzi et al., 2018a). Similar to other legumes including soybean, lupin, pea, faba bean, and grass pea, fenugreek contains low amounts of sulfur-containing amino acids of Met and Cys (Sauvaire et al., 1998; Feyzi et al., 2018b). Protein isolation from fenugreek seeds decreases the amounts of sulfur-containing amino acids compared with those of the whole seeds' proteins (Osman & Simon, 1991), which could be as a consequence of their sensitivity to alkaline conditions, oxidation, and heat they undergo. Table 5.2 summarizes essential and non-essential amino acid compositions of fenugreek protein isolate compared with commercial soy protein isolate (Feyzi et al., 2015).

Fenugreek contains a non-proteinogenic branched-chain amino acid, 4-hydroxyisoleucine, contributing mainly to the free amino acid of seed (80%). The major isomer is (2S,3R,4S)-4 hydroxyisoleucine (~90%), illustrated in Figure 5.3 (Fowden et al., 1973; Alcock et al., 1989; Avalos-Soriano et al., 2016).

Secondary structures of proteins are mainly reflected by FTIR bands within the wave number of 1600–1700 cm^{-1} (amide I) owing to its sensitivity towards conformational changes during folding, unfolding, and aggregation (Carbonaro et al., 2012). Such vibrations occur at 1610–1640 cm^{-1} for β-sheet structures, 1640–1650 cm^{-1} for random coil structures, 1650–1658cm^{-1} for α-helix structures, and at 1660–1700 cm^{-1} for β-turn structures (Wang et al., 2011). FTIR results about fenugreek protein isolate reported by Feyzi et al. (2017) and Feyzi et al. (2018b) were in compromise with those of El-Bahy (2005) on fenugreek seed powder about various amide regions of I, II, and III. Figure 5.4 indicates the FTIR spectrum of fenugreek protein isolate. Contribution of α-helix and β-sheet structures in amide I region was confirmed through relatively strong bands at 1650–1657 cm^{-1} owing to C=O stretching and 1612–1640 cm^{-1}, respectively (Feyzi et al., 2017, 2018b). Moreover, anti-parallel β-sheet structures are usually accompanied with a relative weak band at 1680–1700 cm^{-1} (Guerrero et al., 2014). Similarly, El-Bahy (2005) reported a Raman line at 1661 cm^{-1} as an indication of the amide I band of fenugreek proteins. The amide II bands of fenugreek proteins were observed at 1517 and 1550 cm^{-1} owing to N-H bending and related to β-sheet structures. Also, amide III bands appeared at the region between 1240 and 1472 cm^{-1} owing to N-H bending (El-Bahy, 2005; Feyzi

TABLE 5.2
Amino Acids Compositions of Fenugreek Protein Isolate (FPI) and Soy Protein Isolate (SPI) (g Kg protein^{-1})

Amino Acid	FPI	SPI	FAO/WHO/UNU* Child	Adult
Essential Amino Acids				
Histidine	16.00	20.40	19.00	16.00
Threonine	80.01	28.30	34.00	9.00
Valine	54.70	44.10	35.00	13.00
Methionine	7.70	14.20	27.00	17.00
Phenylalanine	23.90	41.60	63.00	19.00
Isoleucine	59.70	56.00	28.00	13.00
Leucine	93.70	58.40	66.00	19.00
Lysine	51.70	63.50	58.00	5.00
Total essential amino acids	387.41^{a**}	326.50^b		
Non-Essential Amino Acids				
Tyrosine	37.00	31.80		
Cysteine	7.40	17.40		
Aspartic acid	116.80	112.10		
Glutamic acid	199.80	166.70		
Serine	27.50	51.40		
Glycine	46.00	39.10		
Arginine	75.70	76.50		
Alanine	35.70	40.30		
Proline	9.70	59.50		
Percentage of Amino Acids with Different Characterizations				
Basic [1]	143.40^b	160.40^a		
Acidic [2]	316.60^a	278.80^b		
Hydrophobic [3]	285.10^b	314.10^a		
Uncharged polar [4]	197.91^a	168.00^b		
Charged polar [5]	460.00^a	439.20^b		
Sulfur-containing [6]	15.10^b	31.60^a		

[1]Basic: Lysine, arginine, histidine. [2]Acidic: Aspartic acid, glutamic acid. [3]Hydrophobic: Alanine, isoleucine, leucine, methionine, phenylalanine, valine, proline. [4]Uncharged polar: Glycine, serine, tyrosine, cysteine, threonine. [5]Charged polar: Basic and acidic amino acids. [6]Sulfur-containing: Cysteine, methionine.

*In each row different superscript letters (a–b) indicate significant difference ($p < 0.05$).
**Table is adapted with permission from Feyzi et al. (2015).

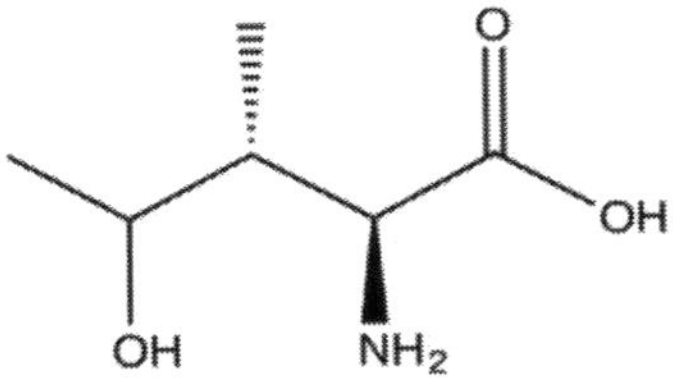

FIGURE 5.3 Free amino acid (2S,3R,4S)-4-hydroxyisoleucine.

et al., 2017, 2018b). These observations are in accordance with findings about grass pea protein isolate (Feyzi et al., 2018a). To the best of our knowledge there is no quantitative report about the secondary structures of fenugreek proteins. However, results about other legumes including soybean, lentil, and pea confirmed the dominant contribution of β-sheet and β-turn structures at the expense of α-helix structures in legume proteins (Tang & Ma, 2009; Carbonaro et al., 2014).

Several literature data are available about SDS-PAGE profile of fenugreek protein isolate. Figure 5.5 indicates the SDS-PAGE profile of fenugreek protein isolate and soy protein isolate, under both reducing and non-reducing conditions.

Similar to findings about fractions of fenugreek protein, soybean, and other legumes, reducing conditions lead to dissociation of subunits with S-S and band emerging at higher MW owing to less mobility. This could be mainly attributed to the albumins followed by globulins which contain higher MW and free sulfhydryl groups (Sauvaire et al., 1984; Feyzi et al., 2015).

5.2.2 Effect of Processing Conditions on Amino Acids Profile and Structures of Fenugreek Proteins

Germination is of the primary seeds' processes whose effect on fenugreek proteins is studied, owing to the consumption of germinated seeds as snacks in some countries like Egypt (El-Mansour & El-Adawy, 1994). Results showed that while having no effect on the total crude protein content, short-time germination (about 12 h) would decrease the amount of albumins and globulins in spite of the increase of glutelins and non-protein nitrogen. Moreover, it lowered the amount of essential amino acids (El-Mansour & El-Adawy, 1994). However, similar to the short-time germination, the prolonged process (5 days) decreased the amount of albumins, globulins, as well as prolamins, while it increased the glutelins and non-protein nitrogen levels (Abd El-Aal & Rahma, 1986). These

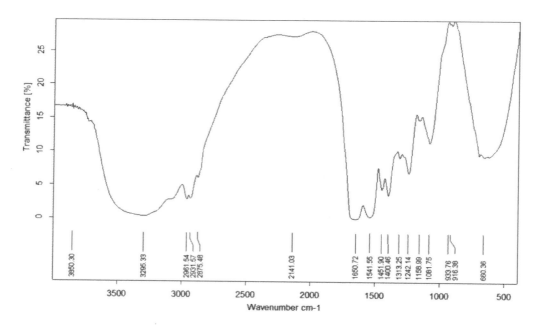

FIGURE 5.4 FTIR spectrum of fenugreek protein isolate.

Source: Adapted with permission from Feyzi et al. (2017).

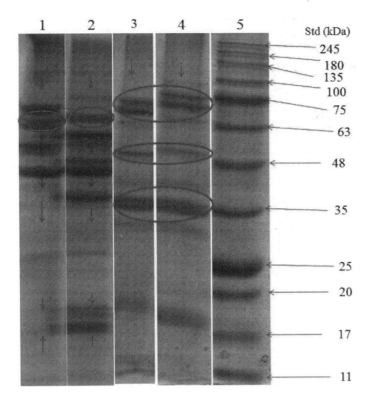

FIGURE 5.5 Electrophoretograms of fenugreek protein isolate (FPI) and soy protein isolate (SPI) using reducing and non-reducing buffers. (1) FPI under non-reducing condition; (2) FPI under reducing condition; (3) SPI under non-reducing condition; (4) SPI under reducing condition; and (5) standard ladder with molecular size ranging from 11 to 245 kDa.

Source: Adapted with permission from Feyzi et al. (2015).

trends in findings could be attributed to the various degrees of hydrolysis in proteins by proteolytic enzymes during germination. This conclusion is in agreement with observation of various dissociated proteins bands with slightly higher mobility by Abd El-Aal and Rahma (1986).

Heating and boiling are common treatments used in food industry. To give illustrations, fenugreek flour is traditionally used as a supplement for bread-making in Egypt (El-Mansour & El-Adawy, 1994), and the flour is recommended to be used in combination with wheat flour in biscuits (Hooda & Jood, 2005; Hussein et al., 2011). Also, the boiled fenugreek extract is commonly consumed as a soft drink in Egypt (El-Mansour & El-Adawy, 1994). El-Mansour and El-Adawy (1994) observed an increase in total protein content of fenugreek seed upon boiling for 20 min. In addition, no substantial changes were found in the total essential amino acids. However, boiling caused a dramatic reduction of albumins, globulins, prolamins, and glutelins in spite of a considerable higher residual protein compared with raw seeds. In contrast, Abdel-Nabey and Damir (1990) observed reduction of total protein, as well as total amino acids content (10%), especially for threonine, proline, glycine, isoleucine, tyrosine, histidine, and arginine, in fenugreek seeds which underwent 15 min boiling treatment. Variation of findings on total protein content between the two aforementioned research studies could be related to differences in the methods applied. Addition of seeds to water at the high temperature of boiling would result in proteins denaturation which reduces their leakage from seeds. However, seeds' gums and other compounds with high solubility leak to water. Therefore, the ratio of proteins to the seeds' total solid compounds increases.

Drying using spray, freeze, and vacuum oven methods, etc., as a tool for preservation, is the common step of protein isolates' processing. Although accompanied with more limited changes in proteins' attributes, freeze drying is less used on an industrial scale owing to the time and energy required. Spray drying, in contrast, is more favored in the industry because of its speed and some characteristics in its final products, including smaller and more uniform particles. Feyzi et al. (2018b) found that while vacuum oven drying resulted in less Lys, Met, and total amino acids content in fenugreek protein isolate, no considerable difference would occur between freeze and spray dried samples. Such phenomena occur as a result of Maillard and oxidation during prolonged vacuum oven procedure. Moreover, results of FTIR proved that freeze and spray drying methods resulted in more retention of α-helix structures compared with vacuum oven procedure. However, bands related to β-sheet structures were more profound in spray dried samples than freeze dried ones, probably due to conversion of α-helix structures to β-sheet ones during the cooling step after heating. Finally, no protein aggregation was detected for three dried fenugreek protein samples based on SDS-PAGE and FTIR results. These findings coincided with those reported about various grass pea protein isolates dried using vacuum oven and freeze drying methods (Feyzi et al., 2018b).

Defatting is an important procedure in plant protein isolation. Hexane is the most common solvent used for this purpose. However, researchers are interested in application of other solvents owing to some valuable benefits in terms of structural and functional properties each solvent might cause. While comparable total amino acids content was observed, fenugreek protein isolate from hexane defatted flour had higher hydrophobic amino acids and denaturation temperature (T_d) than other samples defatted with more polar solvents (diethyl ether, chloroform, ethanol, and acetone) (Feyzi et al., 2017). These could be attributed to the fact that, first, hydrophobic solvents have a protective effect on native structure of globular proteins owing to the more rigid structure in the absence of polar solvents; and second, hydrophobic interactions are responsible for sustaining the interior core of globular proteins. Chloroform impaired the β-sheet structures of fenugreek proteins, as confirmed by the FTIR results and low T_d and enthalpy of the fenugreek protein isolate. SDS-PAGE profiles of various defatted fenugreek proteins showed no severe aggregations, and were comparable except about samples produced by chloroform and ethanol.

5.3 NUTRITIONAL VALUE, ANTI-NUTRITIVE COMPOUNDS, DIGESTIBILITY, AND ALLERGENICITY OF FENUGREEK PROTEINS

Nutritional or nutritive value of proteins is an indicative of their quality and bioavailability, and relies on amino acid composition, essential amino acids, proteins' susceptibility to hydrolysis during digestion, anti-nutrients, and the effects of processing. One important aspect of protein quality interacts with essential amino acids. Protein quality refers to the balance of the amino acids that are absorbed and utilized for growth and other purposes (Friedman, 1996).

Amino acids provide essential nitrogen for the synthesis of protein and other biological molecules, and are known as a common tool to measure protein quality and nutritional value. Accordingly, three categories were developed based on their absolute or relative rates of protein synthesis *in vivo* including (I) *indispensables* (*essential amino acids*): histidine, isoleucine, leucine, lysine, methionine, phenylalanine, threonine, tryptophan, valine; (II) *conditionally indispensables*: arginine, cysteine, tyrosine; and (III) *dispensables* (*non-essential amino acids*): alanine, asparagine, aspartic acid, glutamine, glutamic acid, glycine, proline, serine (Mercer et al., 1989). Generally, the higher ratio of indispensable amino acids of a protein is, the more qualified that protein source is. Legumes like soybean, grass pea, lupin, and fenugreek are poor in sulfur-containing amino acids like Met + Cys (Sauvaire et al., 1998; Feyzi et al., 2015, 2018b). Feyzi et al. (2015) and Mansour and El-Adawy (1994) studied the amino acid composition of fenugreek protein isolate and fenugreek seed, respectively, and compared it with that FAO/WHO recommended for the adult and child (Table 5.1). Results of both research studies coincided and showed that except about Met and, to some extent, Phe, fenugreek protein could meet the recommended allowance of FAO/WHO for essential amino acids. However, reports claim a low level of Trp in fenugreek seed, as well. Considering the high amount of Lys, as the limiting amino acid of cereals, and low Met, the limiting amino acid of legumes, in fenugreek proteins, cereals and fenugreek proteins could act as nutritional complements (Duranti & Gius, 1997).

Protein efficiency ratio (PER), ratio of weight gain of a test group to total protein consumed, and corrected PER (i.e., PER of a test protein to PER of the casein control) of fenugreek seeds as indicators of protein quality and its bioavailability were studied by Rao and Sharma (1987). They found PER of 2, 1.86, and 1.4 for diets with 35% raw defatted fenugreek seed, 35% cooked defatted fenugreek seed, and 10% fenugreek protein, respectively, in rats. PER below 1.5, 1.5–2.0, and more than 2.0 approximately describes a protein of poor, intermediate, good to high quality (Friedman, 1996). However, considering the fact that FAO/WHO proposed amino acids scores as an accurate method, the nutritional value of protein isolates are mainly reported based on comparison between amino acid content of desired protein isolate with those required according to WHO/FAO.

Anti-nutritive compounds like hydrolase inhibitors affect digestive enzymes irreversibly. Trypsin and chymotrypsin inhibitors, as well as α-amylase inhibitors are known as the most important anti-nutritive compounds affecting protein digestibility (Duranti & Gius, 1997). Digestibility is a measure of protein hydrolysis and absorption of the liberated amino acids. Although an indication, digestibility could not thoroughly imply protein quality (Friedman, 1996). However, due to perquisite role of protein digestibility (i.e. susceptibility of peptide bonds to enzyme hydrolysis) on availability of amino acids and their integrity to oxidation, heat, etc., along with their absorption, it is worthy to measure protein digestibility under various conditions.

There is evidence that consumption of dietary fenugreek seeds stimulate digestive actions of rats through stimulation of liver for production and secretion of more bile enriched in bile acids, along with activities of pancreatic lipase and chymotrypsin (Platel & Srinivasan, 2001). However, it is reported that pancreatic amylase and trypsin were decreased by fenugreek diet in rats (Platel & Srinivasan, 2000). This could be as a result of trypsin and amylase inhibitors activity in fenugreek. Weder and Haufiner (1991) isolated three inhibitors from fenugreek proteins using ion-exchange chromatography, two of which indicated high trypsin inhibitor activity and low chymotrypsin inhibition activity. These inhibitors were more affective on human enzymes compared with bovine ones. Moreover, their MWs were less than 6 kDa, and they were rich in cystine, Asp, Ser, Pro, and Thr, while lacked Met and Val. These inhibitors were linked to the Bowman-Birk soybean proteinase inhibitor family owing to high disulfide bridges, comparable MWs. and aggregate formation in SDS-PAGE and IEF-IPG. Also, Oddepally et al. (2013) extracted and purified Kunitz trypsin inhibitor, a serine protease inhibitor, from fenugreek seed. It had a MW of about 20 kDa, and mainly composed of β-sheet and random structures. Interestingly, the Kunitz trypsin inhibitor retained the inhibitory activity over a broad range of pH (pH 3–10), temperature (37–100 °C), and salt concentration (up to 3.5%), and had high binding affinity toward trypsin ($K_i = 3.01 \times 10^{-9}$ M) and chymotrypsin ($K_i = 0.52 \times 10^{-9}$ M). El-Mahdy and El-Sebaiy (1982) proved a positive correlation between protein content of fenugreek extract and its inhibitory effects on the tryptic digestion of casein. The trypsin inhibitor unit (TIU) of fenugreek extract was 27.860 per gram on the wet basis. However, germination process intensified the trypsin inhibitor activity by 66%. In contrast, Mansour and El-Adawy (1994) observed that processes including boiling, toasting followed by boiling, and 12 h germination lowered the trypsin inhibitor activity of fenugreek extract from 6.02 TIU/mg to 0.93, 0.84, and 2.80 TIU/mg, respectively. This finding coincided with declined trypsin inhibitor activity, in terms of Kunitz inhibitor, of soybean up to 144 h germinations at temperatures of 25 and 35 °C.

Protein digestibility could be determined by pepsin, pancreatin, and pepsin followed by pancreatin, under in-vitro conditions. However, PDCAAS (Protein Digestibility-Corrected Amino Acid Score), an in-vivo approach using rats, is the recommended protein digestibility method by FAO/

WHO. The method was endorsed by FAO/WHO in 2001. The importance of protein digestibility arises from its effect on protein quality. However, legumes proteins showed lower digestible proteins compared with animal ones owing to the existence of various anti-nutritional compounds like trypsin and chymotrypsin inhibitors, tannins, phytates, oxidation of sulfur amino acids (such as methionine sulfoxide, methionine sulfone, and cysteic acid), Maillard products, etc. Consequently, while protein digestibility of milk, milk casein, egg, egg albumin, and various meats are in the range of 94 to 100%, its amount for soybean, soy flour, and soy protein isolate was 78, 86, and 95%, respectively (Sarwar Gilani et al., 2005). Singh and Jambunathan (1981) observed higher in-vitro digestible protein for Kabuli chickpea (72.7–79.1% for dhal and 52.4–69% for whole seed) than Desi chickpea (63.7–76% for dhal and 52.4–69.0% for whole seed). In agreement with this finding, trypsin and chymotrypsin inhibitors were also in higher amounts for Desi chickpea than Kabuli. In-vitro protein digestibility using pepsin followed by pancreatin gives a realistic illustration of protein digestibility in the gastrointestinal tract. Generally, pancreatic digestibility of proteins is lower compared with other enzymes. Protein digestibility of fenugreek seeds was in the order of 55.0%, 61.2%, 66.7%, and 74.9% by trypsin, pepsin, pancreatin, and trypsin-pancreatin, respectively. However, results showed that in-vitro protein digestibility through each of these enzymes improves upon germination, boiling, and toasting followed by boiling processes (Mansour & El-Adawy, 1994). Such refinements as a consequence of heat treatments could be attributed to protein denaturation and partially destruction of enzyme inhibitors. Moreover, proteases activities increase during germination. In contrast, El-Mahdy and El-Sebaiy (1982) observed that in-vitro protein digestibility of germinated fenugreek decreased by pepsin followed by pancreatin compared with un-germinated samples, while pancreatic digestibility improved after germination. Venkataraman et al. (1976), also, observed that in-vivo protein digestibility of chickpea and cowpea decreased after germination. Hooda and Jood (2005) studied the in-vitro protein digestibility, employing pepsin and pancreatin, of biscuits containing different amounts of raw and germinated fenugreek flours. Their results showed that in spite of the unchanged protein digestibility of biscuits containing raw fenugreek flour, application of germinated fenugreek flours in biscuits formulation improved their digestibility slightly. Such finding could occur as a consequence of hydrolytic enzymes effects in combination with baking heat on phytic acid and polyphenols, and less insoluble complexes of proteins with these compounds. Formation of insoluble complexes or aggregation of proteins, which are usually accompanied with higher amounts of intermolecular β-sheet structures in proteins like soy protein isolate, reduces protein susceptibility to proteases and its digestibility (Carbonaro et al., 2014). Rao and Sharma (1987) observed that application of diets containing 10% legumes proteins to rats concluded the in-vivo protein digestibility of 60–90% based on the nitrogen content. Fenugreek seed resulted the lowest protein digestibility (60%) compared with other legumes like chickpea, red gram, green gram, and black gram. This might have occurred owing to high amounts of fibers and other compounds in fenugreek seeds. In-vivo protein digestibility of soy protein isolate (92–98%) was comparable to that of pea concentrate (92%) (Sarwar Gilani et al., 2005). Taking all this to account, further investigations are required on in-vivo digestibility of fenugreek proteins and its improvement.

Allergenicity risks of legumes might include mild skin reaction, oral allergy or extreme anaphylactic reactions. Allergenic legumes are reported in the order of peanut > soybean > lentil > chickpea > pea > mung bean. Allergenic proteins of peanut, as the most serious potential allergen among legumes, are reported as peanut profilin (Ara h 5), pathogenesis-related (PR-10), pollen protein (Ara h 8), prolamins (Ara h 2, Ara h 6, Ara h 7, and Ara h 9), cupins (Ara h 1, Ara h 3, and Ara h 4) and oleosins (Ara h 10 and Ara h 11) (Fæste et al., 2010). Allergenic legumes proteins usually show high resistance to prolonged heat treatments or extensive proteolysis (Carbonaro et al., 2014). Research studies indicate that sensitization might happen in peanut allergic patients by consumption of fenugreek-containing foods, probably owing to extensive cross-reactivity between these two legumes. Cross-reactivity occurs when one antibody binds to different allergens due to highly similar epitopes, homologous proteins containing conserved sequence motifs (Vinge et al., 2012). Such cross-reactivity between other members of the Leguminosae family such as peanut, soy, and lupin has previously been documented (Lallès & Peltre, 1996; Jensen et al., 2008; Fæste et al., 2010).

IgE-binding epitopes with considerable cross-reactivity with peanut are known as allergens of fenugreek proteins. Patil et al. (1997) proved the IgE-binding of fenugreek and chickpea proteins through the immunoblotting method and SDS-PAGE. Vinge et al. (2012) proved IgE-binding of fenugreek proteins and anaphylactic reaction in treated mice with a fenugreek diet. Further investigations through immunoblotting followed by MS-based proteomic analysis using patient sera showed that fenugreek 7S vicilin-like protein designated "Tri f 1" with main IgE-binding band at 50.1 kDa, which is in analogy to Ara h 1 of peanut protein. Fenugreek 11S legumin-like protein was designated "Tri f 3" with bands at 20.1, 50, and 80 kDa, in analogy to Ara h 3 from peanut. However, fenugreek 2S albumin, preliminarily called "Tri f 2" showed IgE-binding affinity, as well. In conclusion, results confirmed that 7S vicilin-like protein with MW of 50 kDa–110 kDa and mainly including 50.1 kDa and 66.1 kDa bands is responsible for the strong IgE-binding signals of fenugreek (Fæste et al., 2010).

5.4 BIOLOGICAL EFFECTS OF FENUGREEK PROTEINS, PEPTIDES, AND AMINO ACIDS

Diets containing legumes and fenugreek seeds or extract are beneficial for well-being owing to their biological activities including anti-hypercholesterolemic, antidiabetic, anti-obesity, anticancer, anti-inflammatory, as well as antioxidant effects (Al-Habori & Raman, 1998; Venkata et al., 2017). Some of these biological activities have been attributed to fenugreek proteins, peptides, and amino acids.

Antioxidant properties of fenugreek proteins, peptides, and amino acids could be studied under in-vitro and in-vivo conditions, and are important in both food and biological systems including living animals. Oxidation of food lipids during processing and storage impairs the quality of foods as well as their nutritive values. Moreover, normal reactions within the body during respiration in aerobic organisms generate free radicals and reactive species like reactive oxygen species (ROS) and reactive nitrogen species (RNS), particularly in humans. Consequently, such phenomena give rise to accumulation of free radicals in cells, damage proteins, DNA, and modification of LDL to oxidized-LDL, followed by some diseases including atherosclerosis, arthritis, diabetes, and cancer (Sarmadi & Ismail, 2010).

Although cost-effective and efficient, synthetic antioxidants might be accompanied with toxic and hazardous effects. Consequently, natural antioxidants like polyphenols, proteins, and peptides have been drawing attention toward their potential health benefits. It has been proved that legumes proteins like soy, lentil, fenugreek, and their hydrolysates and peptides, 2–20 amino acids with molecular weight of less than 6 kDa, are potential natural antioxidants (Sarmadi & Ismail, 2010; Allaoui et al., 2018; Setti et al., 2018).

Setti et al. (2018) investigated the antioxidant activity of fenugreek protein isolate by means of DPPH, ABTS radical scavenging tests, along with β-carotene bleaching test. They observed IC_{50} values of 35 ± 0.99 and 250 ± 9.00 mg/ml for DPPH• and ABTS•, respectively. Moreover, β-carotene bleaching test resulted in value of 485 ± 19. However, fermentation of fenugreek protein isolate through *Lc. Lactis* concluded 42.9%, 40%, and 23.7% increase in DPPH•, ABTS•, and antioxidant activity coefficient of β-carotene bleaching, respectively. Such observations could be attributed to the modification effects of microbial proteases on molecular mass and molecular structures of fenugreek proteins. That is, microbial proteases caused hydrolysis of high MW proteins (>35 kDa) and emerging of new hydrolysates with MW less than 25 kDa, as well as free amino acids. On the molecular level, efficient antioxidant ability under such conditions might arise from aromatic amino acids which donate protons to electron-deficient radicals, and better interaction of hydrophobic amino acids with lipid targets through hydrophobic associations. Moreover, hydrolysates with MW of 15–50 kDa were accompanied with the highest antioxidant activity compared with fractions with either lower (<15 kDa) or higher (>50 kDa) MWs. The dominant amino acids of the desired hydrolysate fraction were Glu, Ala, Tyr, Ile, and Lys. Allaoui et al. (2018) found that fenugreek protein isolate and in particular its Esperase or Purafect hydrolysates refined the anti-oxidative action of HDL on VLDL-LDL oxidation level. However, identification of the peptide sequence and main mechanism contributing to such antioxidant activity of fenugreek protein hydrolysates has remained elusive. Martínez-Villaluenga et al. (2008) detected an active peptide in β-conglycinin (7S globulin) of soy protein isolate with antioxidant activity. Chen et al. (1998), also, detected a Leu-Leu-Pro-His-His peptide sequence with antioxidant activity from soybean β-conglycinin. Moreover, they observed that, first, addition of a Leu or Pro to the N-terminus of a His-His dipeptide improved its antioxidant activity and interaction with other antioxidant molecules (e.g., butylated hydroxytoluene); second, removing a His residue from the C-terminus reduces its antioxidant activity.

Antidiabetic effect of fenugreek proteins is mainly referred to its 4-hydroxyisoleucine amino acid through glucose concentration-dependent stimulation of insulin secretion (Sauvaire et al., 1998; Rangari et al., 2015). This unique characteristic as an advantage avoids an undesirable hypoglycemia side effect, in contrast to other drugs for type 2 diabetes like sulfonylureas (Petit et al., 1995; Rangari et al., 2015; Venkata et al., 2017). Type 2 diabetes, noninsulin-dependent disorder, could be diagnosed by urine, plasma glucose, and glycated hemoglobin levels (HbA1C) (Rangari et al., 2015).

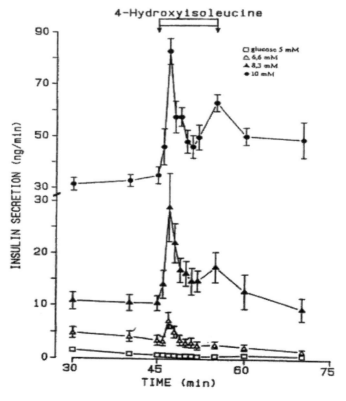

FIGURE 5.6 Kinetics of insulin secretion effect of 4-hydroxyisoleucine (200 μM) in the presence of different glucose concentrations (5, 6.6, 8.3, and 10 mM).

Source: Adapted and with permission from Sauvaire et al. (1998).

The underlying mechanism of 4-hydroxyisoleucine antidiabetic property is glucose-induced insulin release through a direct effect on isolated pancreatic islets β-cells in both rats and humans. That is, 4-hydroxyisoleucine, at concentration of 200 µM, stimulates insulin secretion under high blood glucose conditions (8.3–16.7 mM) (Petit et al., 1995; Rangari et al., 2015; Venkata et al., 2017). Figure 5.6 indicates kinetics of insulin secretion effect of 4-hydroxyisoleucine as a function of glucose concentration. Sauvaire et al. (1998) confirmed the antidiabetic effect of 4-hydroxyisoleucine on humans through insulinotropic activity and its direct effect on β-cells of the pancreas. There is other research which confirms the antidiabetic effect of 4-hydroxyisoleucine on both type 1 and 2 diabetes (Haeri et al., 2012). In conclusion, the well-demonstrated insulinotropic properties of 4-hydroxyisoleucine may suggest it as a potential anti-diabetic nutraceutical and pharmacologic compound, especially at elevated glucose concentrations (≥ 8.3 mM), resulting in successful hyperglycemia reduction.

Anti-hypercholesterolemic and anti-hypertriglyceridemic activities of proteins and peptides may prevent from atherosclerosis and cardiovascular diseases (CVD). Generally, anti-hypercholesterolemic activity is linked with the bile acid binding affinity followed by disruption of cholesterol micelles in the gastrointestinal tract or through changes in hepatic and adipocytic enzymes activities and gene expression of lipogenic proteins (Allaoui et al., 2018). Fenugreek protein isolate and its Purafect hydrolysate considerably reduced total liver cholesterol and triglyceride of rats (up to 30%) (Allaoui et al., 2018). This cholesterol-lowering effect in rats was aligned with that of soy protein isolate and its hydrolysate (Fassini et al., 2011). Moreover, fenugreek protein isolate and its Esperase hydrolysate decreased the amounts of serum total cholesterol and triglycerides of rats (up to 60%).

Such findings may indicate that fenugreek and soy protein isolates and their hydrolysates inhibit cholesterol and lipid absorption. The more hydrolyzed proteins are, the better they pass the intestinal barrier and reach the blood stream followed by the liver, where they could exert their bioactivities. Such a decrease coincided with a high fecal cholesterol excretion in rats (Cho et al., 2007). Moreover, both the low-density lipoproteins (LDL) and high density lipoproteins (HDL) of rats were reduced upon treatment with fenugreek protein isolate and its Esperase hydrolysate (Allaoui et al., 2018). However, to the best of our knowledge, there is no more report about the main mechanisms of hypocholesterolemic and hypotriglyceridemic effects of fenugreek protein isolate and its hydrolysates.

Moreover, the dyslipidemic effect of 4-hydroxyisoleucine was studied by Narender et al. (2006) whose research proved the lowering effect of this amino acid on plasma total triglycerides, cholesterol, and free fatty acids.

5.5 FUNCTIONAL PROPERTIES OF FENUGREEK PROTEINS IN FOOD SYSTEMS

There is limited research on functional properties of fenugreek proteins in spite of their importance in generation of novel and valuable raw materials and development of non-animal-based products. Generally, plant proteins which are globular, usually with high MWs, and low flexibility might expose weaker functional properties compared with some animal proteins (Boyle et al., 2018). However, in order to produce high-quality, low-cost, and appealing plant-based products, food industries need to improve the techno-functional properties of plant proteins like fenugreek proteins. These properties affect proteins' behavior within a food matrix through changes in their conformation and interaction with other components. Numerous parameters, including defatting solvent, extraction method, matrix conditions like pH, ionic strength, temperature, matrix constitutes like sugar, oil, etc., would alter proteins' structures and functionality. Consequently, this section is dedicated to functional properties of fenugreek proteins.

Solubility is known as the main perquisite to get insights into the majority of functional and biological properties of proteins. Low solubility of some plant proteins, even at neutral pH, necessitates pH adjustment prior to their application in food formulations (Lafarga, 2019). However, Feyzi et al. (2015) reported significantly higher amounts ($p < 0.05$) of average protein solubility for fenugreek protein isolate than commercial soy protein isolate over the pH range of 2–10. This finding may eliminate the need to adjust the pH of solution containing fenugreek protein isolate for food applications. The least protein solubility of albumins, globulins, and whole protein isolate of fenugreek seeds was found at pH 4, 4.3, and 4.5, respectively; indication their pI (Torki et al., 1987; El-Nasri & El-Tinay, 2007; Feyzi et al., 2015). In addition to pH, heat, ionic strength, salt type, defatting process, type of solvent, drying method, and novel processing and modification technologies like ultrasound, microwave, enzymatic hydrolysis, etc. influence protein solubility, owing to their effects on proteins' structures and side-chains orientations which leads to various degrees of charged groups hydration. L'hocine et al. (2006) observed no difference between protein solubility of soy protein isolates prepared from defatted flour using hexane, ethanol, and methanol at specific pH 7. However, Feyzi et al. (2017) observed that fenugreek protein isolates from hexane or acetone defatted flour had higher protein solubility compared with those from diethyl ether, ethanol, and chloroform solvents over the wide pH range of 2–10. Moreover, applying freeze and spray drying methods led to higher protein solubility of fenugreek protein isolate compared to vacuum oven drying (Feyzi et al., 2018b). This could be attributed to the lower outlet temperature of the spray dryer (about 60 °C) compared to the denaturation temperature of fenugreek proteins, its fast speed, along with the small particle size of the final powder. In contrast, vacuum oven drying led to formation of Maillard products with higher MW. Hu et al. observed better protein solubility of spray dried soy protein isolate compared to the freeze dried sample.

Surface hydrophobicity (H_0) of proteins shows the available hydrophobic sites for probes—like 1-anilinonaphthalene-8-sulfonicacid (ANS), 6-propionyl-2-(dimethylamino)-naphthalene (PRODAN), and *cis*-parinaric acid (CPA)—binding and does not necessarily show the structure flexibility. The binding sites of CPA, as an anionic aliphatic probe, on protein molecules might be different from those of ANS, an aromatic probe containing a sulfonic acid group, owing to the existence of aromatic rings in ANS in spite of the aliphatic hydrocarbon chain of CPA (Alizadeh-Pasdar & Li-Chan, 2000). Wang et al. (2011) observed a positive correlation between H_0 and amounts of β-sheet and random coil structures in spite of its negative correlation with amount of α-helix structures of various soy protein isolates, using an ANS probe. Besides, it has been proved that high ionic strength would increase the H_0 of soy protein isolate accompanying random coil structures at the expense of α-helices, altogether leading to looser tertiary structures (Jiang et al., 2015). To the best of our knowledge there are limited studies about H_0 of fenugreek proteins. However, Feyzi et al. (2018b) observed higher H_0 in various dried fenugreek protein isolates compared with soy protein isolates using an ANS probe. Meanwhile, vacuum oven dried fenugreek protein isolate had higher H_0 compared with the freeze and spray dried samples, which was in accordance with its lower protein solubility. This indicates that vacuum oven drying resulted in conformational changes which were accompanied with higher hydrophobic and lower charged amino acids on proteins' surface and led to less protein hydration. Defatting solvents, also, altered H_0 of the fenugreek protein isolates such that samples produced from acetone and chloroform defatted flours had the lowest and the highest H_0, respectively (Feyzi et al., 2017).

Interfacial properties including dynamic surface and interfacial tensions (σ and Υ, respectively) are important parameters due to their link with proteins' conformations and other functional properties like foaming and emulsifying. σ refers to the migration ability of proteins into the water-air interface, while Υ indicates the migration ability of proteins into the water-oil interface, altogether leading to tension reduction and preventing from phase separation. The better surface active a protein is, the more effectively it reduces water-air surface or water-oil interfacial tensions (σ_0 and Υ_0, respectively). Generally, σ and Υ of fenugreek protein isolate (46.69 and 8.96 mN/m, respectively) were less than those of soy protein isolate (49.88 and 14.00 mN/m) at pH 7, indicating superior function of fenugreek proteins in reduction of interfacial tension (Feyzi et al., 2018b). This might be as a consequence of higher albumins contents of fenugreek proteins which have more α-helix structures and water solubility at neutral pH compared with soy proteins which are rich in globulins. Interestingly, while affecting σ, drying methods showed no influence on Υ of fenugreek proteins

(Feyzi et al., 2018b). Moreover, various defatting solvents showed significant contribution to both σ and Y of fenugreek proteins, owing to their influence on proteins' structures and solubility, along with removal of other surface active agents. Fenugreek protein isolate from chloroform defatted flour exhibited the maximum σ and Y among five samples, indicating weak surface and interfacial behaviors. In contrast, defatting with ethanol resulted in the lowest σ and appropriate Y among samples, which could be as a consequence of its ability in elimination of phospholipids as low-molecular weight active agents and creating enough space for proteins' migration into the surface and/or interface (Feyzi et al., 2017). However, this phenomenon would not necessarily overcome the key role of protein solubility in various interfacial properties.

Foaming properties include foaming capacity and stability. Creation of stable foams, consisting of gases, liquids, and solid components whose gas phases are separated and stabled by a thin layer called the lamellar phase, is crucial in many food systems such as whipped toppings, chiffon mixes, confections, and cakes. Egg white, milk and soy proteins are usually used in food industries to produce foam, while there are some reports about promising function of other plant proteins as foaming agents (Lafarga, 2019). Pea proteins including pea, chickpea, and lentil protein isolates, for instance, are reported as rewarding foaming agents with foaming capacity of about 95–105% (Boye et al., 2010b). Similar to other functional properties, foaming capacity and stability vary upon changes in pH, salt concentration, temperature, protein concentration, drying method, etc. El-Nasri and El-Tinaly (2007), and Feyzi et al. (2015) studied the dependency of foaming properties of fenugreek protein isolate on pH shift. Both research groups found that foaming capacity followed a similar trend to changes in protein solubility against pH. That is, the lowest foaming capacity occurred at pI (about 30%), while it increased under acidic and especially basic conditions. Such a behavior was observed about soy protein isolate, as well. However, the average foaming capacity of fenugreek protein isolate was considerably higher than that of soy protein isolate over pH range of 2–10. The highest foaming capacity of fenugreek protein isolate reported by Feyzi et al. (2015) was observed at pH 8 (136.67%), showing the most appropriate balance between protein solubility and H_0 which led to low σ and profound entrapment of air particles. Acetone followed by hexane and ethanol defatted fenugreek flours resulted in protein isolates with high foaming capacity (31.2, 18.7, and 17.0%, respectively) (Feyzi et al., 2017). This phenomenon coincided with superior foam expansion of soy protein isolate produced from ethanol and hexane defatted flour compared with that from methanol defatted flour (L'hocine et al., 2006). Drying, also, affected foaming capacity of fenugreek protein isolate such that spray dried powder had higher foaming capacity, followed by freeze and vacuum oven dried powders (Feyzi et al., 2018b). These observations confirm the rewarding agreement between foaming ability of fenugreek protein isolates and high protein solubility, along with low σ thereof.

Appropriate viscosity is necessary for stable protein film around air bubbles to prevent from rupture and film distortion. While El-Nasri and El-Tinaly (2007) reported the best foam stability of fenugreek protein concentrate at acidic pH 2 at times up to 105 min, Feyzi et al. (2015) observed considerable foam stability at pH 3 to 10 at prolonged times up to 120 min. It is worth to mention that in the latter research foam stability of fenugreek protein isolate was more than soy protein isolate at all pH at 30–120 min. Such variations in findings of two research groups could be related to the differences in protein concentrations and extraction methods employed. Various defatting solvents could neither improve nor distort foam stability of the fenugreek protein isolates at pH 7 (Feyzi et al., 2017). In contrast, applying vacuum oven drying was accompanied with weak protein film stability of fenugreek protein isolate (45% foam stability) compared with freeze and spray drying methods (86 and 76%, respectively) (Feyzi et al., 2018b).

Emulsifying properties importance arises from contribution of the emulsions to many food systems like bakery, dairy, and meat products. Such comprehensive use raises researchers' interests to employ various emulsifiers or their mixture to meet food industries' requirements. On top of this, plant-derived emulsifiers (e.g., plant proteins) would draw more attention owing to their rewarding health benefits. Animal-derived proteins including milk and egg proteins are commonly used to stabilize emulsions (Lynch et al., 2017). However, the solubility and amphiphilic nature required for migration into oil/water interfacial layer, and adequate reduction of Υ are characteristics of legume proteins, as well. The more flexibility and lower molecular weight proteins have, the superior emulsion activity they show.

There was no significant ($p < 0.05$) difference between emulsion capacity of fenugreek protein isolate and soy protein isolate over pH range of 2–10. In addition, both protein isolates showed their lowest emulsion capacity at their pI (4.5) (Feyzi et al., 2015). Similarly, El-Nasri and El-Tinaly (2007) found the lowest emulsion capacity of fenugreek protein isolate at pH 4.5. Feyzi et al. (2018b) observed that various drying methods did not make any differences among emulsion activity indices (80–170 m^2/g) and emulsion capacities (28–31%) of fenugreek protein isolates. However, defatting solvents varied these two properties of fenugreek protein isolate considerably, such that sample from chloroform defatted flour resulted in the least amount (Feyzi et al., 2017). Various protein fractions have different emulsification properties due to their differences in conformation, structures, and MWs. β-Conglycinin (vicilin or 7S globulin) fraction of soy protein had better emulsifying properties compared with glycinin (legumin or 11S globulin) (Keerati-u-rai & Corredig, 2010). However, glycinin showed better emulsion stability owing to its higher molecular weight and thicker interfacial layer it creates around oil droplets (Keerati-u-rai & Corredig, 2009).

Emulsion stability is crucial in food products to prevent coalescence of oil droplets. It is proved that higher molecular weight of the legumes proteins compared with dairy proteins led to a thicker interfacial layer and better emulsion stability, especially upon heating or salt addition treatments (Keerati-u-rai & Corredig, 2009; Day, 2013). Emulsion stability of fenugreek protein isolate was better than soy protein isolate over pH range of 2–10. The lowest emulsion stability was found at pI for both protein isolates, while the highest amounts were measured at pH 8 and 6 for fenugreek and soy protein isolates, respectively. Interestingly, vacuum oven dried fenugreek protein isolate resulted in the best emulsion stability compared with freeze and spray dried samples which could be as a consequence of Maillard products formation with higher MWs (Feyzi et al., 2018b). Also, defatting with chloroform resulted in the lowest emulsion stability of the fenugreek protein isolate compared with other samples (Feyzi et al., 2017). Taking all these findings to account, results show that obtaining desirable emulsion stability and activity require relatively high protein solubility along with low Υ.

Gelation generally refers to phase transition of sol state (liquid solution of proteins) into a gel-like structure through physical or chemical approaches. Gelation of plant proteins, as globular proteins, usually occurs through heat-induced reaction. High temperatures of about 100 °C result in proteins dissociation, denaturation, aggregation, sometimes followed by arrangement of aggregates, altogether creating a developed three-dimensional network "gel". Strong gel is a function of balance between protein-protein and protein-water interactions. The predominance of protein-protein interactions owing to excess hydrophobic interactions at the expense of electrostatic interactions would result in coagulum encompassing rough structures with low water content. However, the more exposed charged groups are, the higher possibility of weak transparent gel is owing to the repulsive electrostatic interactions (Arntfield & Maskus, 2011; Phillips & Williams, 2011; Feyzi et al., 2015). Moreover, the kinetics of denaturation and aggregation processes is important in final gel structure and attributes. That is, aggregating slower than unfolding provides more time for ordered intra-molecular interactions and ordered assembled aggregate structures. Such types of developed structures create fine and transparent networks. In contrast, a faster aggregation compared with unfolding would result in random aggregate clusters and finally a turbid gel. Many forces are involved in gel formation and strength including mainly hydrophobic interactions, as well as electrostatic interactions, hydrogen and disulfide bonds (Hettiarachchy et al., 2013). These facts bring to mind that gel formation is a multi-step process, such that even the heating process must be considered as more than one phase in order to produce the desired gel.

One well-known and simple approach to evaluate gelation ability of plant proteins is least gelation concentration (LGC), based on thermal-induced gel formation, indicating the lowest protein concentration required to form a gel. The better gelling agent a protein is, the lower protein concentration is required to form a gel. This amount is usually between 4.0–20.0% for legume protein isolates. The LGC of commercial soy protein isolate (4.0%) was considerably less than that of fenugreek protein isolate (8.0%), reported by Feyzi et al. (2015). Such finding was in agreement with superior water binding ability, as well as higher hydrophobic amino acids content of soy protein isolate compared with fenugreek protein isolate. Withana-Gamage et al. (2011) found the LGC of 10–14% for soy protein and chickpea protein isolates. Sun and Arntfield (2010) reported LGC of 5.5–14.5% for various pea proteins.

LGC and other gel characteristics could vary as a result of many factors including pH, ionic strength, temperature, and the rate it increases various agents like urea, 2-mercaptoethanol, etc. For example, gel strength which could be modified by pH and salt confirm the stable role of electrostatic interactions in gel formation (Arntfield & Maskus, 2011; Phillips & Williams, 2011; Hettiarachchy et al., 2013). There is, to the best of our knowledge, no literature data about gelation properties of fenugreek proteins or its characteristics, which could be studied by various rheology methods.

5.6 THERMAL PROPERTIES

Various inter-molecular interactions in proteins result in their low molecular mobility, as well as high softening or melting temperature. In other words, once the protein is folded to its final native form, it is stabilized through hydrophobic and electrostatic interactions, hydrogen bonds, along with further strong covalent crosslinks. Softening of proteins requires their denaturation, meaning partial unfolding of structured native protein into an unstructured state with no or little fixed residual structures. Consequently, melting temperature of proteins could be considered as their denaturation temperature (T_d). However, a complete unfolding into a fully amorphous structure may not occur as true melting means (Ricci et al., 2018). Considering that any changes in secondary, tertiary, or quaternary structures of proteins may refer to proteins denaturation, DSC not only gives insights into the differences between thermal characteristics of various proteins like legumes, but also could help to study the effects of various parameters on those attributes. Generally, denaturation of proteins is an endothermic process, owing to the heat they absorb to thermally unfold over a temperature range.

Some thermal characteristics of fenugreek protein isolate have been studied in its dry state or in solutions using differential scanning calorimetry (DSC). Feyzi et al. (2015) found two peaks in the thermograph of 10% solution of fenugreek protein isolate, indicating two transition temperatures. The first peak had onset temperature (T_o), T_d, and end set temperature (T_e) of 100.00, 104.96, and 120.00 °C and ΔH 2.69 J/g, while the second peak was bigger with T_o, T_d, and T_e, and ΔH of 120.00, 121.40 °C, and 130.00 °C, and 5.66 J/g, respectively. Under the similar conditions, soy protein isolate also had two peaks with T_o, T_d, and T_e, and ΔH of 111.30, 117.58, 119.50 °C, and 3.70 J/g and 119.50, 128.50, 130.040 °C, and 9.72 J/g for the first and second peaks, respectively. The researchers contributed the higher T_d of soy protein isolate in comparison with fenugreek protein isolate to high hydrophobic vs. low polar amino acids contents in soy proteins isolate (Feyzi et al., 2015). Existence of more than one peak in thermograph of protein isolates could be related to the various fractions with different thermal attributes. In agreement with this, L'hocine et al. (2006) observed two different peaks for soy protein isolate. Further studies revealed that the first one with lower T_d and ΔH was referred to β-conglycinin (7S globulin) and the second one was referred to glycinin (legumin or 11S globulin) fractions (L'hocine et al., 2006). Such phenomenon could be as a consequence of detected disulfide bonds in legumin fraction.

Various parameters in extraction procedure including pH, temperature and ionic strength, defatting solvents, drying methods, etc. affect the thermal properties of proteins owing to their impact on proteins structures and their denaturation. Feyzi et al. (2017) observed the lowest T_d of fenugreek protein isolates for chloroform followed by ethanol defatted samples. However, fenugreek protein isolates from hexane, diethyl ether, and acetone defatted flour showed no considerable difference. The structural stability of sample from chloroform defatted flour was too low that ΔH could not be distinguished and calculated. L'hocine et al. (2006) observed no considerable differences among T_d and ΔH of both vicilin 7S and legumin 11S factions of soy protein isolates prepared from hexane, ethanol, and methanol defatted flours. Moreover, lower ΔH of vacuum oven dried fenugreek protein isolate compared with those of freeze and spray dried samples indicated less stability of secondary, tertiary, and/or quaternary structures. This was in agreement with lower α-helix structures of the first sample. Also, higher T_d of vacuum oven dried sample compared with the other two ones could be as a result of covalent bonds formation through Maillard interactions (Feyzi et al., 2018b). Such observation coincided with findings of Feyzi et al. (2018b) about vacuum oven dried grass pea protein isolate. On the other hand, L'hocine et al. (2006) observed that extraction procedure, in terms of heat precipitation or acidic pre-treatments, had no effect on T_d and ΔH of both vicilin 7S fraction and legumin 11S fractions of soy protein isolate.

5.7 METHODS OF PROTEIN EXTRACTION

Various approaches could be employed to extract legumes proteins. These include air classification, alkaline extraction followed by isoelectric precipitation, acid extraction, salt extraction (micellization), water extraction, enzyme hydrolysis, ultrafiltration, as well as combination of two or more of these methods. Moreover, extraction procedure not only affects protein content of final concentrate/isolate, but also results in varied nutritional, biological, or functional properties (Boye et al., 2010a). Some of these methods have been used for extraction of fenugreek proteins, explained in the following.

Alkaline extraction/isoelectric precipitation is the most widespread protein extraction approach, and was applied for extraction of fenugreek proteins by some researchers. The technique takes advantage of high solubility of legume proteins at alkaline pH followed by their precipitation at pI (pH 4–5). This requires flour dispersion and stirring in water at appropriate ratio (usually 1:5 to 1:20) for a definite time (usually 30 to 180 min), during which the pH is adjusted at a certain amount above the isoelectric point (usually 7–11), and temperature could be elevated below the denaturation temperature (25–60 °C). Distancing from the isoelectric point increases proteins solubility. This provides higher protein concentrations in supernatant phase for precipitation at pI, which is equal to 4.5 for fenugreek proteins. This procedure could be followed by re-solubilizing of the precipitated protein at pH 7 and drying to produce final protein isolate (Boye et al., 2010a; Feyzi et al., 2015). Abdel-Aal et al (1986) extracted fenugreek proteins of Egyptian fenugreek seed, with 30.6% protein; using alkaline extraction (pH 11) followed by isoelectric precipitation the extraction reached to 60.37% protein content. Feyzi et al. (2015) applied this procedure on Iranian fenugreek seed with 34.7% protein content at various alkaline pH (6, 9, and 12), and produced fenugreek protein isolates with 90.8, 84, and 82.1% protein content, respectively. Higher protein content at lower pH might be achieved as a result of dominant contribution of albumins to fenugreek proteins (Feyzi et al., 2015).

Salt extraction (micellization) is another frequent approach in protein extraction from legumes. The procedure lies in the salting-in and salting-out phenomena of food proteins. Accordingly, proteins solubility increases at appropriate ionic strength (salting in). This is usually followed by protein precipitation through centrifugation or filtration. Similar to other extraction methods, the procedure could be accomplished by employment of a drying method. El-Nasri and El-Tinay (2007) used the micellization method for extraction of Sudanese fenugreek seed with 28.4% protein using 1 M NaCl and flour to solvent ratio of 1:10 for 30 min. The extraction was accomplished by further precipitation of protein at pI (4.5) and drying in open air. Final protein concentrate contained 73.9% protein. Abdel-Aal et al. (1986) applied a similar procedure, using 0.5 M NaCl and stirring for 1 h, for protein extraction from Egyptian fenugreek seeds, and produced protein isolate with 94.7% protein. The influential role of salt on protein extraction from legumes like fenugreek is possibly attributed to their high amount of globulins, salt soluble proteins. Globulins are the second major fraction of fenugreek and the most dominant one in others including soybean, peas, etc. (Feyzi et al., 2015).

Combination of alkaline extraction/isoelectric precipitation and salt extraction is another extraction method which takes advantage of high proteins solubility using both appropriate ionic strength and alkaline medium. This method could be considered as an optimized technique for extraction of legumes proteins since albumins and globulins are soluble at mild alkaline and slat solutions, respectively. Feyzi et al. (2015) applied this procedure for optimization of protein extraction from Iranian fenugreek seed with 34.7% protein content based on the dry basis. They dissolved defatted flour in NaCl solutions (0, 0.5, and 1 M) with flour to solvent ratio of 1:20, while the solution pH was adjusted to 6, 9, and 12. After blending for 2 h, aliquots were centrifuged followed by precipitation at pI (4.5). The precipitated proteins were separated through centrifugation, and were solubilized in deionized water at pH 7. Final product was freeze dried. The optimization results through response surface methodology (RSM) revealed that pH 9.25 and NaCl concentration of 0.33 M conclude fenugreek protein isolate with 89.1% protein content and 2.1 g protein weight (from 10 g initial defatted flour). Such extraction optimization process under both alkaline and salt media has also been used for bitter melon seed by Horax et al. (2011).

Water (aqueous) extraction method is a general and simple approach for protein extraction. This method could be simply done by solubilization of flour in water followed by acid precipitation at pI. However, such a simple method may not result in high extraction yield or protein content. Osman and Simon (1991) used this method to extract proteins of Egyptian fenugreek seed, and only found 25.5% protein in final product.

Aqueous enzymatic extraction is another protein extraction method used by researchers, although it is not frequent. In brief, aqueous or saline solutions containing pepsin or pancreatin are added to the flour at desired ratio, and the mixture is blended for the appropriate time, followed by enzyme inactivation. The solution is centrifuged and the supernatant phase containing proteins are lyophilized. Abdel-Aal et al. (1986) applied this technique for protein extraction from fenugreek seed, using 1–3 mg enzyme/1 g flour. Final product had 62.4% protein content.

There are other protein extraction methods like acid precipitation, ultrafiltration, etc. which have not been used for fenugreek proteins. However, their application and comparison of their functions with other methods could be recommended for fenugreek proteins. More information about other protein extraction procedures is available elsewhere (Boye et al., 2010a; Arntfield & Maskus, 2011; Singhal et al., 2016).

5.8 APPLICATION OF FENUGREEK IN FOOD SYSTEMS

Increasing demands for animal proteins' analogues raise the necessity in searching for new types of proteins as raw materials. Considering the chemical composition, nutritional value, bioactive compounds, ease of cultivation, environmentally friendly nature, and affordable price, plant-based analogues like proteins can be effectively produced using legumes seeds, which also contain a number of substances giving them a specific taste and smell.

Hegazy and Ibrahium (2009) investigated the effect of partial replacement (5, 10, 15, and 20%) of wheat flour by soaked or germinated fenugreek flour on amino acids and chemical compositions, protein quality, and sensory evaluation of biscuits. Protein content of biscuits increased by higher substitution levels of wheat flour by fenugreek flour. Flour substitution up to 10% significantly ($p < 0.05$) improved essential amino acids of biscuits compared with control biscuits (100% wheat flour). Also, the essential amino acids of biscuits met the FAO/WHO recommended allowance, except about Lys and Met. Such substitution refined the PER of mice under study from 0.99 in mice with diet of control biscuits to 1.49 and 1.60 in mice who consumed biscuits with 10% soaked and germinated fenugreek flour, respectively. The sensory evaluation, flavor, taste, color, texture, and appearance of the latter two biscuit samples were also better than the control sample. Consequently, biscuits with 10% substitution of wheat flour with fenugreek flour were recommended in terms of better protein quality. These findings coincided with those previously Hooda and Jood (2005) reported about wheat biscuits containing raw, soaked, and germinated fenugreek flours (0–20%).

They stated that, first, biscuits containing various types of fenugreek flours had higher protein content, and, second, biscuits with germinated flour were accompanied with better in-vitro protein digestibility (>70.0%) compared with control biscuits (69.4%). However, overall acceptability of biscuits was reduced after substitution of wheat flour with fenugreek flours.

Recently, Motuzka and Koshelnyk (2020) suggested fenugreek seeds for production of fenugreek-based milk analogue, owing to its considerable amino acids content. However, a reduction in Arg, with explicitly bitter taste, occurred upon frying for 15 min which would change the organoleptic characteristics of the final product. Also, Met, His, and Lys which contribute to bitter aftertaste in food products were decreased. However, Asp and Pro, responsible for sweet aftertaste, and Glu, a factor of pleasant taste, were reduced. Considering that reduction of the three latter amino acids was more profound than those contributing to bitter taste, it is expected that a more bitter taste is perceived upon heating.

Although fenugreek seeds and flour are traditionally used in Çemen (fenugreek paste) as a local food in Turkey, a kind of soft drink, bread, and *hulba makoda* (a local food) in Egypt, and etc., the importance of functional, nutritional, and biological effects of fenugreek proteins, peptides, and amino acids remained elusive. Moreover, to the best of our knowledge there is no scientific report about application of fenugreek proteins (protein concentrate or isolate) in real food systems. In conclusion, more practical studies are required to give more insights on wide range of proteins' characteristics in foods containing fenugreek.

5.9 FUTURE CHALLENGES AND TRENDS IN USING FENUGREEK PROTEINS

Considering the novel nature of legumes proteins and especially fenugreek proteins, consumers are unaware of their beneficial nutritional, biological, and functional characteristics. Consequently, food researchers and industries are in charge of this task to unveil the benefits of fenugreek proteins to the consumers along with removing the barriers of their use for consumers, owing to the key role of consumer awareness, perception, and acceptance followed by their demand on product development.

Similar to other legumes, one dominant barrier in development of fenugreek products like its proteins is seemingly their organoleptic properties, anti-nutritional compounds, and relatively low digestibility. Seeds' bitterness is mainly as a consequence of the oil, steroidal saponins, and alkaloids, while de-bitterized seeds are rich in protein and Lys. In compromise with this, Shirani and Ganesharanee (2009) observed that incorporation of more than 2% fenugreek flour in extruded products formulation resulted in lower sensory scores as a result of bitter taste, flavor, and overall acceptability. This may necessitate removal of the bitter compounds from flour prior its use in food formulations or extraction of fenugreek proteins.

Moreover, limiting amino acids of fenugreek protein including Met and Trp/Phe may necessitate its incorporation with cereals which are rich enough in those amino acids. Allergenicity, which is evidenced by cross-reactivity among legumes, anti-nutritional compounds like trypsin and chymotrypsin inhibitors, and low protein digestibility of fenugreek proteins may be eliminated through germination, protein modifications, or hydrolysis.

Some other limitations might occur in use of fenugreek proteins from a techno-functional point of view. That is, similar to other legumes, incorporation of fenugreek flour or its protein products might result in considerable changes in texture, color, flavor, rheological properties, and as a consequence the palatability and consumer acceptance of the final product. In agreement with this statement, Hooda and Jood (2005) observed that although germination enhances various attributes of fenugreek flour in biscuits compared with raw flour, its incorporation of more than 10% in biscuits formulation resulted in lower texture, flavor, color, and overall acceptability. Similarly, Shirani and Ganesharanee (2009) found that incorporation of fenugreek flour of more than 2% in extruded products led to lower moisture retention and water solubility index, lower radial expansion, and higher hardness compared with control product containing 70:30 chickpea–rice blend. However, incorporation of de-bitterized fenugreek polysaccharides into the formulations improved the aforementioned functional properties in extruded products.

Although combination of alkaline extraction/isoelectric precipitation and salt extraction procedures were considerably effective in producing protein isolate with high protein content, extraction yield, and promising thermo and functional properties, various extraction methods coupled with chemical and physical modifications including ultrasound, high hydrostatic pressure, protein hydrolysis, etc. could refine the aforementioned techno-functional attributes of fenugreek protein-based products, along with nutritional and biological values.

5.10 CONCLUSION

Fenugreek seed could be recommended as a promising legume protein source owing to its superior or comparable nutritional value, functional properties, and biological effects with other legumes proteins like soy protein, as the commercial and worldwide plant protein. The nutritive value of fenugreek protein isolate in terms of indispensable (essential) amino acids is more than commercial soy protein isolate and meets the recommended allowance of FAO/WHO, except about Met, Phe, and/or Trp (for children). However, fenugreek proteins with MWs less than 20 kDa demonstrated trypsin and chymotrypsin inhibitors activity, similar to soy proteins. Limiting amino acids, antinutritive compounds, oxidative products of sulfur amino acids, Maillard products, phytates, along with tannins which could interact with proteins and create insoluble aggregates, altogether may impair digestibility of fenugreek proteins. Moreover, IgE-binding signals of fenugreek relating to 7S vicilin-like protein may result in allergenic potential. From the biological point of view, fenugreek proteins and peptides are accompanied with anti-oxidative activity, which could also improve the in-vivo anti-oxidative activity of HDL on VLDL-LDL oxidation, as well as in-vivo hypocholesterolemic and hypotriglyceridemic effects. Moreover, 4-hydroxyisoleucine has anti-diabetic (Type 2) activity through glucose-induced insulin release mechanism. In addition to these attributes, the techno-functional properties of fenugreek proteins are confirmed. Fenugreek protein isolate has lower surface and interfacial tensions, higher protein solubility, foaming capacity and stability, as well as emulsifying activity and stability compared with soy protein isolate, in model food systems. Considering the fact that preparation processes including defatting, extraction method, modifications, and drying procedures, along with treatments like soaking, germination, toasting, and boiling vary all these properties, it is expected that fenugreek proteins with more profound characteristics could be produced in the future. In conclusion, authors believe in light of this chapter information about both the potentials and drawbacks of fenugreek proteins, researchers could design further studies to take advantage of the rewarding characteristics of fenugreek proteins in foods and medicines, as well as to eliminate their weaknesses, and finally extend their use.

REFERENCES

Abd El-Aal, M.H., & Rahma, E.H. (1986). Changes in gross chemical composition with emphasis on lipid and protein fractions during germination of fenugreek seeds. *Food Chemistry*, 22, 193–207.

Abdel-Aal, E.M., Shehata, A.A., El-Mahdy, A.R., & Youssef, M.M. (1986). Extractability and functional properties of some legume proteins isolated by three different methods. *Journal of the Science of Food and Agriculture*, 37, 553–559.

Abdel-Nabey, A.A., & Damir, A.A. (1990). Changes in some nutrients of fenugreek (Trigonella Foenumgraecum L.) seeds during water boiling. *Plant Foods for Human Nutrition*, 40, 267–274.

Alcock, N.W., Crout, D.H., Gregorio, M.V.M., Pike, E.L.G., & Samuel, C.J. (1989). Stereochemistry of 4-Hydroxyisoleucine from Trigonella foenum graecum. *Phytochemistry*, 28, 1835–1841.

Al-Habori, M., & Raman, A. (1998). Antidiabetic and hypocholesterolaemic effects of fenugreek. *Phototherapy Research*, 12, 233–242.

Alizadeh-Pasdar, N., & Li-Chan, E.C.Y. (2000). Comparison of protein surface hydrophobicity measured at various pH values using three different fluorescent probes. *Journal of Agricultural and Food Chemistry*, 48, 328–334.

Allaoui, A., Barranquero, C., Yahia, S., Herrera-Marcos, L.V., Benomar, S., Jridi, M., Navarro, M.A., Rodriguez-Yoldi, M.J., Nasri, M., Osada, J., & Boualga, A. (2018). Fenugreek proteins and their hydrolysates prevent hypercholesterolemia and enhance the HDL antioxidant properties in rats. *Nutrition & Food Science*, 48, 973–989.

Arntfield, S.D., & Maskus, H.D. (2011). Peas and other legume proteins. In *Handbook of Food Proteins*, eds. Phillips, G.O., & Williams, P.A. Woodhead Publishing Series in Food Science, Technology and Nutrition, pp. 233–266.

Avalos-Soriano, A., Cruz-Cordero, R.D., Rosado, J.L., & Garcia-Gasca, T. (2016). 4-Hydroxyisoleucine from Fenugreek (Trigonella foenum-graecum): Effects on insulin resistance associated with obesity. *Molecules*, 21, 1596–1608.

Bewley, J.D. (2013). Structure and composition, In *Seeds: Physiology of Development, Germination and Dormancy*, eds. Bewley, J.D., Bradford, K., Hilhorst, H., & Nonogaki, H. 3rd Edition. Springer, New Delhi, India, pp. 1–25.

Bhatty, R.S. (1982). Albumin proteins of eight edible grain legume species: Electrophoretic patterns and amino acid composition. *Journal of Agricultural and Food Chemistry*, 30, 620–621.

Boulter, D., & Croy, R.R.D. (1997). The structure and biosynthesis of legume seed storage proteins: A biological solution to the storage of nitrogen in seeds. *Advances in Botanical Research*, 27, 1–92.

Boye, J.I., Aksay, S., Roufik, S. et al. (2010a). Comparison of the functional properties of pea, chickpea and lentil protein concentrates processed using ultrafiltration and isoelectric precipitation techniques. *Food Research International*, 43, 537–546.

Boye, J.I., Zare, F., & Pletch, A. (2010b). Pulse proteins: Processing, characterization, functional properties and applications in food and feed. *Food Research International*, 43, 414–431.

Boyle, C., Hanse, L., Hinnenkamp, C., & Ismail, B.P. (2018). Emerging Camelina protein: Extraction, modification, and structural/functional characterization. *Journal of American Oil Chemical Society*, 95, 1049–1062.

Carbonaro, M., Maselli, P., & Nucara, A. (2012). Relationship between digestibility and secondary structure of raw and thermally treated legume proteins: A Fourier transform infrared (FT-IR) spectroscopic study. *Amino Acids*, 38, 679–690.

Carbonaro, M., Maselli, P., & Nucara, A. (2014). Structural aspects of legume proteins and nutraceutical properties. *Food Research International*, 76, 19–30.

Chen, H. M., Muramoto, K., Yamauchi, F., Fujimoto, T., & Nokihara, K. (1998). Antioxidant properties of histidine-containing peptides designed from peptide fragments found in the digests of a soybean protein. *Journal of Agricultural and Food Chemistry*, 46, 49–53.

Cho, S.J., Juillerat, M.A., & Lee, C.H. (2007). Cholesterol lowering mechanism of soybean protein hydrolysate. *Journal of Agricultural and Food Chemistry*, 55, 10599–10604.

Day, L. (2013). Proteins from land plants—Potential resources for human nutrition and food security. *Trends in Food Science & Technology*, 32, 25–42.

Duranti, M., & Gius, C. (1997). Legume seeds: Protein content and nutritional value. *Field Crops Research*, 53, 31–45.

El-Bahy, G.M.C. (2005). FTIR and Raman spectroscopic study of fenugreek (Trigonella foenum-graecum L.) seeds. *Journal of Applied Spectroscopy*, 72, 111–116.

El-Mahdy, A.R., & El-Sebaiy, L.A. (1982). Effect of germination on the nitrogenous constituents, protein fractions, in vitro digestibility and antinutritional factors of fenugreek seeds (trigonella foenum-graecum L.). *Food Chemistry*, 8, 253–262.

El-Mahdy, A.R., & El-Sebaiy, L.A. (1985). Proteolytic activity, amino acid composition and protein quality of germinating fenugreek seeds (Trigonella foenum-graecum L.). *Food Chemistry*, 18, 19–33.

El-Nasri, N.A., & El-Tinay, A.H. (2007). Functional properties of fenugreek (Trigonella foenum graecum) protein concentrate. *Food Chemistry*, 103, 582–589.

Fæste, C.K., Christians, U., Egaas, E., & Jonscher, K.R. (2010). Characterization of potential allergens in fenugreek (Trigonella foenum-graecum) using patient sera and MS-based proteomic analysis. *Journal of Proteomics*, 73, 1321–1333.

Fassini, P.G., Noda, R.W., Ferreira, E.S., Silva, M.A., Neves, V.A., & Demonte, A. (2011). Soybean glycinin improves HDL-C and suppresses the effects of rosuvastatin on hypercholesterolemic rats. *Lipids in Health and Disease*, 10, 165–171.

Feyzi, S., Milani, E., & Golimovahhed, Q.A. (2018a). Grass pea (Lathyrus sativus L.) protein isolate: Study of extraction optimization, protein characterizations, structure and functional properties. *Food Hydrocolloids*, 74, 187–196.

Feyzi, S., Varidi, M., Zare, F., & Varidi, M. J. (2015). Fenugreek (Trigonella foenum graecum) seed protein isolate: Extraction optimization, amino acid composition, thermo and functional properties. *Journal of the Science of Food and Agriculture*, 95, 3165–3176.

Feyzi, S., Varidi, M., Zare, F., & Varidi, M. J. (2017). A comparison of chemical, structural and functional properties of fenugreek (Trigonella foenum graecum) protein isolates produced using different defatting solvents. *International Journal of Biological Macromolecules*, 105, 27–35.

Feyzi, S., Varidi, M., Zare, F., & Varidi, M.J. (2018b). Effect of drying methods on the structure, thermo and functional properties of fenugreek (Trigonella foenum graecum) protein isolate. *Journal of the Science of Food and Agriculture*, 98, 1880–1888.

Fowden, L., Pratt, H.M., & Smith, A. (1973). 4-Hydroxyisoleucine from seed of Trigonella foenum graecum. *Phytochemistry*, 12, 1701–1712.

Friedman, M. (1996). Nutritional value of proteins from different food sources. A Review. *Journal of Agricultural and Food Chemistry*, 44, 6–29.

Fukushima, D. (2011). Soy proteins, In *Handbook of Food Proteins*, eds. Phillips, G.O., & Williams, P.A. Woodhead Publishing Series in Food Science, Technology and Nutrition, pp. 210–232.

Grand View Research. (2016). Protein ingredients market size worth USD 48.77 billion by 2025. Retrieved from www.grandviewr esearch.com/press-release/global-protein-ingredients-market-analysis.

Guerrero, P., Kerry, J.P., & de la Caba, K. (2014). FTIR characterization of protein-polysaccharide interactions in extruded blends. *Carbohydrate Polymer*, 111, 598–605.

Haeri, M.R., Limaki, H.K., White, C.J., & White, K.N. (2012). Non-insulin dependent anti-diabetic activity of (2S, 3R, 4S) 4-hydroxyisoleucine of fenugreek (Trigonella foenum graecum) in streptozotocin-induced type I diabetic rats. *Phytomedicine*, 19, 571–574.

Hegazy, A.I., & Ibrahium, M.I. (2009). Evaluation of the nutritional protein quality of wheat biscuit supplemented by fenugreek flour. *World Journal of Dairy and Food Science*, 4, 129–135.

Hettiarachchy, N., Kannan, A., Schafer, C., & Wagner, G. (2013). Gelling of plant based proteins. In *Product Design and Engineering: Formulation of Gels and Pastes,* eds. Brockel, U., Meier, W., & Wagner, G, 1st ed. Wiley-VCH Verlag GmbH & Co. KGaA.

Hooda, S., & Jood, S. (2005). Effect of fenugreek flour blending on physical, organoleptic and chemical characteristics of wheat bread. *International Journal of Food Sciences and Nutrition*, 35, 229–242.

Horax, R., Hettiarachchy, N., Kannan, A., & Chen, P. (2011). Protein extraction optimization, characterization, and functionalities of protein isolate from bitter melon (Momordica charantia) seed. *Food Chemistry*, 124, 545–550.

Hussein, A.M.S., Amal, S., Amany, M., & Abeer, A. (2011). Physiochemical, sensory and nutritional properties of corn-fenugreek flour composite biscuits. *Australian Journal of Basic and Applied Sciences*, 5, 84–95.

Isıklı, N.D., & Karababa, E. (2005). Rheological characterization of fenugreek paste (Çemen). *Journal of Food Engineering*, 69, 185–190.

Jensen, L.B., Pedersen, M.H., Skov P.S., Poulsen, L.K., Bindslev-Jensen, C, Andersen, S.B., et al. (2008). Peanut cross-reacting allergens in seeds and sprouts of a range of legumes. *Clinical & Experimental Allergy*, 38, 1969–1977.

Jianga, L., Wanga, Z., Lia, Y., Mengb, X., Suia, X., Qia, B., & Zhou, L. (2015). Relationship between surface hydrophobicity and structure of soy protein isolate subjected to different ionic strength. *International Journal of Food Properties*, 18, 1059–1074.

Keerati-u-rai, M., & Corredig, M. (2009). Heat-induced changes in oil-in-water emulsions stabilized with soy protein isolate. *Food Hydrocolloids*, 23, 2141–2148.

Keerati-u-rai, M., & Corredig, M. (2010). Heat-induced changes occurring in oil/water emulsions stabilized by soy glycinin and beta-Conglycinin. *Journal of Agricultural and Food Chemistry*, 58, 9171–9180.

Khorshidian, N., Yousefi Asli, M., Arab, M., Adeli Mirzaie, A., & Mortazavian, A.M. (2016). Fenugreek: Potential applications as a functional food and nutraceutical. *Nutrition and Food Sciences Research*, 3, 5–16.

L'hocine, L., Boye, J.I., & Arcand, Y. (2006). Composition and functional properties of soy protein isolates prepared using alternative defatting and extraction procedures. *Journal of Food Science*, 71, 137–145.

Lafarga, T. (2019). Potential applications of plant-derived proteins in the food industry. In *Novel Proteins for Food, Pharmaceuticals, and Agriculture: Sources, Applications, and Advances*, ed. Hayes, Maria, 1st ed. Wiley & Sons Ltd, pp. 117–138.

Lallès, J.P., & Peltre, G. (1996). Biochemical features of grain legume allergens in humans and animals. *Nutrition Reviews*, 54, 101–107.

Lynch, S.A., Mullen, A.M., O'Neill, E.E., & García, C.Á. (2017). Harnessing the potential of blood proteins as functional ingredients: A review of the state of the art in blood processing. *Comprehensive Reviews in Food Science and Food Safety*, 16, 330–344.

Mahmood, M.N., & Yahya, I.K. (2017). Nutrient and phytochemical of fenugreek (Trigonella foenum graecum) seeds. *International Journal of Sciences: Basic and Applied Research*, 36, 203–213.

Mansour, E.H., & El-Adawy, T.A. (1994). Nutritional potential and functional properties of heat-treated and germinated fenugreek seeds. *LWT—Food Science and Technology*, 27, 568–572.

Martínez-Villaluenga, C., Bringe, N.A., Berhow, M.A., & González de Mejía, E. (2008). Beta-conglycinin embeds active peptides that inhibit lipid accumulation in 3T3-L1 adipocytes in vitro. *Journal of Agricultural and Food Chemistry*, 56, 10533–10543.

Mercer, L.P., Dodds, S.J., & Smith, S.L. (1989). Dispensable, indispensable, and conditionally indispensable amino acid ratios in the diet. In *Absorption and Utilization of Amino Acids*, ed. Friedman, M. CRC Press, Boca Raton, FL, pp. 1–13.

Motuzka, Y., & Koshelnyk, A. (2020). Formation of quality of milk analogue from Greek fenugreek seeds. *International Scientific Journal Commodities and Markets*, 34, 110–118.

Naidu, M.M., Shyamala, B.N., Naik, J.P., Sulochanamma, G., & Srinivas, P. (2011). Chemical composition and antioxidant activity of the husk and endosperm of fenugreek seeds. *LWT—Food Science and Technology*, 44, 451–456.

Narender, T., Puri, A., Khaliq, T., Saxena, R., Bhatia, G., & Chandra, R. (2006). 4-Hydroxyisoleucine an unusual amino acid as antidyslipidemic and antihyperglycemic agent. *Bioorganic & Medicinal Chemistry Letters*, 16, 293–296.

Oddepally, R., Sriram, G., & Guruprasad, L. (2013). Purification and characterization of a stable Kunitz trypsin inhibitor from Trigonella foenum-graecum (fenugreek) seeds. *Phytochemistry*, 96, 26–36.

Osborne, T.B. (1924). *The Vegetable Proteins*. Longman Green and Co., London.

Osman, M.K., & Simon, L.S. (1991). Biochemical studies of some non-conventional sources of protein. Part 5. Extraction and characterization of protein from fenugreek seed (Trigonella foenum L.). *Die Nahrung*, 35, 303–308.

Patil, S.P., Niphadkar, P.V., & Bapat, M.M. (1997). Allergy to fenugreek (Trigonella foenum graecum). *Annals of Allergy, Asthma & Immunology*, 78, 297–300.

Petit, P., Sauvaire, Y., Hillaire-Buys, D., Manteghetti, M., Baissac, Y., Gross, R., & Ribes, G. (1995). Insulin stimulating effect of an original amino acid, 4-hydroxyisoleucine, purified from fenugreek seeds. *Diabetologia*, 38.

Phillips, G.O., & Williams, P.A. (2011). Introduction to food proteins. In *Handbook of Food Proteins*, eds. Phillips, G.O., & Williams, P.A. Woodhead Publishing Series in Food Science, Technology and Nutrition, pp. 1–12.

Platel, K., & Srinivasan, K. (2000). Influence of dietary spices and their active principles on pancreatic digestive enzymes in albino rats. *Nahrung*, 44, 42–46.

Platel, K., & Srinivasan, K. (2001). Studies on the influence of dietary spices on food transit time in experimental rats. *Nutrition Research*, 21, 1309–1314.

Rangari, V.D., Shukla, P., & Badole, S.L. (2015). 4-Hydroxyisoleucine: A potential antidiabetic agent from Trigonella foenum-graecum. Glucose intake and utilization in pre-diabetes and diabetes implications for cardiovascular disease, Part I, 191–198.

Rao, P.U., & Sharma, R.D. (1987). An evaluation of protein quality of fenugreek seeds (Trigonella foenum graecum) and their supplementary effects. *Food Chemistry*, 24, 1–9.

Ricci, L., Umiltà, E., Righetti, M.C., Messina, T., Zurlini, C., Montanari, A., Broncoa, S., & Bertoldo, M. (2018). On the thermal behavior of protein isolated from different legumes investigated by DSC and TGA. *Journal of the Science of Food and Agriculture*, 98, 5368–5377.

Sarmadi, B.H., & Ismail, A. (2010). Antioxidative peptides from food proteins: A review. *Peptides*, 31, 1949–1956.

Sarwar Gilani, G., Cockell, K., & Sepehr, E. (2005). Effects of antinutritional factors on protein digestibility and amino acid availability in foods. *Journal of AOAC International*, 88, 967–986.

Sauvaire Y.D., Baccou, J.F., & Kobrehel, K. (1984). Solubilization and characterization of fenugreek seed proteins. *Journal of Agricultural and Food Chemistry*, 32, 34–47.

Sauvaire Y.D., Petit, P., Broca, C., Manteghetti, M., Baissac, Y., Fernandez-Alvarez, J., Gross, R., Roye, M., Leconte, A., Gomis, R., & Ribes, G. (1998). 4-Hydroxyisoleucine A novel amino acid potentiator of insulin secretion. *Diabetes*, 47, 206–210.

Setti, K., Kachouri, F., & Hamdi, M. (2018). Improvement of the antioxidant activity of fenugreek protein isolates by Lactococcus lactis fermentation. *International Journal of Peptide Research and Therapeutics*, 24, 499–509.

Shirani, G., & Ganesharanee, R. (2009). Extruded products with Fenugreek (Trigonella foenum-graecium) chickpea and rice: Physical properties, sensory acceptability and glycaemic index. *Journal of Food Engineering*, 90, 44–52.

Singh, U., & Jambunathan, R. (1981). Studies on desi and kabuli chickpea (Cicer arietinum L.) cultivars: Levels of protease inhibitors, levels of polyphenolic compounds and in vitro protein digestibility. *Journal of Food Science*, 46, 1364–1367.

Singhal, A., Karaca, A.C., Tyler, R., & Nickerson, M. (2016). Pulse proteins: From processing to structure-function relationships. In *Grain Legumes, Aakash Kumar Goyal*. IntechOpen, p. 55.

Sun, X., & Arntfield, S.D. (2010). Gelation properties of salt-extracted pea protein induced by heat treatment. *Food Research International*, 43, 509–515.

Tang, C.H., & Ma, C.Y. (2009). Effect of high pressure treatment on aggregation and structural properties of soy protein isolate. *LWT—Food Science and Technology*, 42, 606–611.

Torki, M.A., Shabana, M.K.S., Nadia, A., & Abd El-Alim, I.M. (1987). Protein fractionation and characterization of some legumenous seeds. *Annals of Agricultural Science*, 25, 278–290.

Venkata, K.C.N., Swaroop, A., Bagchi, D., & Bishayee, A. (2017). A small plant with big benefits: Fenugreek (Trigonella foenum-graecum Linn.) for disease prevention and health promotion. *Molecular Nutrition & Food Research*, 61, 1–26.

Venkataraman, L.V., Geneshkumar, K., & Jaya, T.V. (1976). Studies on the changes of proteins, carbohydrates, trypsin inhibitor, amylase inhibitor and haemagglutinins of some legumes during germination. *Proceedings of the Society Biological Chemistry*, 34.

Vinje, N.E., Namork, E., & Løvik, M. (2012). Cross-allergic reactions to legumes in lupin and Fenugreek-sensitized mice. *Scandinavian Journal of Immunology*, 76, 387–397.

Wang, C., Jianga, L., Weia, D., Li, Y., Suia, S., Wang, Z., & Li, D. (2011). Effect of secondary structure determined by FTIR spectra on surface hydrophobicity of soybean protein isolate. *Procedia Engineering*, 15, 4819–4827.

Wani, S.A., & Kumar, P. (2018). Fenugreek: A review on its nutraceutical properties and utilization in various food products. *Journal of the Saudi Society of Agricultural Sciences*, 17, 97–106.

Weder, J.K.P., & HaufIner, K. (1991). Inhibitors of human and bovine trypsin and ehymotrypsin in fenugreek (Trigonella foenum-graecum L.) seeds. *International Journal of Food Research and Technology*, 192, 535–540.

Withana-Gamage, T., Wanasundara, J.P.D., Pietrasika, Z., & Shand, P.J. (2011). Physicochemical, thermal and functional characterisation of protein isolates from Kabuli and Desi chickpea (Cicer arietinum L.): A comparative study with soy (Glycine max) and pea (Pisum sativum L.). *Journal of the Science of Food and Agriculture*, 91, 1022–1031.

Zandi, P., Basu, S.K., Cetzal-Ix, W., Kordrostami, M., Khademi Chalaras, S., & Bazrkar Khatibai, L. (2017). In *Active Ingredients from Aromatic and Medicinal Plants*, ed. El-Shemy, H.A., 1st ed. InTech, pp. 207–224.

6 Extraction and Optimization of Saponin and Phenolic Compounds of Fenugreek Seed

Sweeta Akbari, Nour Hamid Abdurahman and Rosli Mohd Yunus

CONTENTS

6.1 Introduction 77
6.2 Materials and Methods 78
    6.2.1 Fenugreek Seed Extraction 78
    6.2.2 Optimization of the Process 78
    6.2.3 Model Validation and Statistical Analysis 79
    6.2.4 Determination of Total Saponin Content in the Extracts 80
    6.2.5 Determination of Total Phenolic Content in the Extract 80
    6.2.6 Antioxidant Activities 80
        6.2.6.1 DPPH Radical Scavenging Activity 80
        6.2.6.2 ABTS Radical Scavenging Activity of the Extract 81
    6.2.7 Analysis of LC-QTOF-Mass Spectrometer 81
6.3 Results and Discussion 82
    6.3.1 Optimization and Model Fitting Using RSM 82
    6.3.2 Validation of the Model 82
    6.3.3 DPPH and ABTS Antioxidant Activities 82
    6.3.4 LC-QTOF-MS Analysis of Extracted Compounds 83
6.4 Conclusion 87
References 87

6.1 INTRODUCTION

Fenugreek (*Trigonella foenum-graecum*) is a medicinal plant with a long history of application in Asia, Africa, Egypt, the Middle East, and European countries. The drugs obtained from plants have played an important role in treatment of many diseases in humans. The use of synthetic medicines or drugs involves adverse or side effects on humans. Reports revealed that yearly around 100,000 people die due to the side effects or sensitivity to certain medicines (Karimi et al., 2015). According to the World Health Organization (WHO), since the early civilizations more than 80% of people all around the world consume herbs as traditional medicines to cure and treat various ailments (Baruah et al., 2016). The side effects and negative impacts of herbal remedies are very rare and even hard to find. Therefore, in developed countries, such as the United Status (US) and Europe, people turn to herbal medicine as they believe that medicines produced from herbal remedies are free of risk and other side effects to the health (Karimi et al., 2015). Fenugreek is also one of these plants with a wide range of health benefits. The seed of fenugreek is reported to contain many phytochemical

DOI: 10.1201/9781003082767-8

compounds including alkaloids, phenols, tannins, steroids, flavonoids, saponins, which have potential health benefits for humans due to their antioxidant, anticancer, antitumor, anti-inflammation, and antidiabetic properties (Akbari et al., 2019a, 2019b; Maan et al., 2018; Yadav & Baquer, 2014). The aim of this chapter is to assess the total saponin and phenolic bioactive compounds of fenugreek seed extracted via microwave-assisted extraction (MAE) in the optimum conditions.

6.2 MATERIALS AND METHODS

Seeds of fenugreek were collected from a local retail market located in Kuantan, Pahang, Malaysia. Before, drying the seeds at 50 °C in an air-oven for 24 h, the seeds were manually cleaned to separate all the unwanted species and foreign matters. The seeds had moisture content of 5.51 ± 0.14% (d.w. basis). Further, the dried seeds were ground using an ultra-centrifugal grinder (Retsch ZM-200, Germany) equipped with a ring sieve at size of 0.5 mm trapezoid holes. Finally, the powdered seeds were stored at 4 °C in a dark airtight plastic container before extraction.

Ethanol (99.5%), methanol (99.9), vanillin, diosgenin, Folin-Ciocalteu reagent, gallic acid (GA), 2,2-diphenyl-1-picrylhydrazyl (DPPH), 2,20-azino-bis (3-ethylbenzothiazoline-6-sulfonic acid) (ABTS), sodium carbonate, sulphuric acid (H_2SO_4), ascorbic acid, and potassium persulfate ($K_2S_2O_8$) were obtained from Sigma Aldrich Sdn. Bhd, Selangore, Malaysia. All reagents and chemicals used in this research were of analytical grade with high purity.

6.2.1 Fenugreek Seed Extraction

Microwave-assisted extraction was performed using the closed system of an ethos microwave extractor (Frequency 2450, 1000 W; Milestone, Italy) as shown in Figure 6.1. As it can be seen, the whole microwave extractor system includes a microwave oven, a condenser connected to the extraction flask, and a water-cooling system connected to the inlet and outlet of the condenser. The ethos microwave system was connected to a parameter control device which was used for setting up the experimental parameters. This device was applied to set up the extraction time at three different levels, preheating for 2 min, the desired extraction time, and 2 min for cooling the extract inside the system. The purposes of pre- and post-treatment were to heat the samples efficiently while absorbing the heat; however, post-treatment was applied to cool down the samples at room temperature. Fenugreek seed (10g) was extracted using parameters of extraction time, microwave power, ethanol concentration, feed-to-solvent ratio, and temperature. Ethanol was used as extraction solvent due to the environmental concern and high efficiency.

After removing the sample from the oven, a qualitative filter paper No.1 (Advantec®) was used to separate the particles from the mixture and a vacuum filtration system was used to accelerate the process. Afterwards, the solvents (ethanol and water) were removed at 50 °C from the extract via a rotary evaporation system (Büchi, R-200, Germany). Then, based on Equation (1) the yield of extraction was calculated. The extracts were stored at 4 °C to avoid the biodegradation of compounds. Then, a UV-Vis spectrophotometer (Hitachi, U-1800; Japan) was used for the measurement of TPC and TSC in the extract. The experiment was repeated three times.

$$Extraction\ yield\ (\%) = \frac{Weight\ of\ extracted\ yield\ (w)}{Weight\ of\ dry\ sample\ used\ (w)} \times 100 \qquad 1$$

6.2.2 Optimization of the Process

The optimization process was accomplished via RSM by applying the face centred composite design (FCCCD). From the experiment, three responses were recorded including yield of extraction (Y_{Ex}), total content of saponin (Y_{TSC}), and phenolic (Y_{TPC}). The four variables (A, B, C, and D)

Saponin and Phenolic Compounds

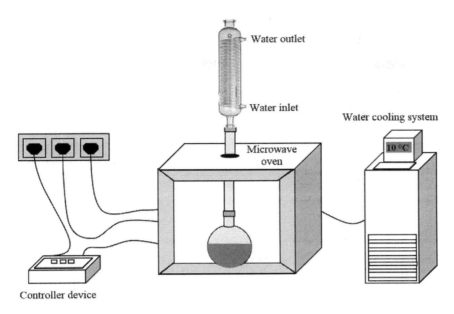

FIGURE 6.1 Schematic diagram of microwave-assisted extraction method.

TABLE 6.1
Independent Variables and Factor Levels Used in the Experiment

Independent Variables	Factor Levels		
	-1	0	+1
A: Extraction time (min)	2	3	4
B: Microwave power (W)	500	600	700
C: Ethanol concentration (%)	40	60	80
D: Feed-to-solvent-ratio (g/mL)	1:8	1:10	1:12

are displaced in Table 6.1, respectively. Basically, a total of 30 runs including 6 centre points with a quadratic polynomial equation were generated as shown in Equation (2).

$$Y = \beta_0 + \sum_{i=1}^{n} \beta_i x_i + \sum_{i=1}^{n} \beta_{ii} x_i^2 + \sum_{i=1}^{n-1} \sum_{j=i+1}^{n} \beta_{ij} x_i x_j \qquad 2$$

Where, the response, the coefficients of regression for the intercept, linear, square, and interaction, the independent variables and number of factors were indicated as Y, $\beta 0$, βi, βii, βij, Xi, Xj, and n, respectively.

The model terms, adequacy based on R-squares, lack-of-fit, F-value, interaction of factors, and adequate precision were evaluated from the analysis of variance (ANOVA).

6.2.3 Model Validation and Statistical Analysis

The validation and adequacy of the model were carried out by conducting verification experiments in the optimum conditions of extraction yield, TPC and TSC. To validate the model, the obtained results were compared with the actual and predicted values. The model accuracy was evaluated by triplicating the experiment in the optimal condition and R-squares were assessed. The results were reported as mean ± SD. A paired samples t-test was performed to evaluate the difference between actual and predicted values.

6.2.4 Determination of Total Saponin Content in the Extracts

The TSC in the extracts of fenugreek seed was measured according to the methods described by Hu et al. (2012), Moyo et al. (2013), and Venegas-Calerón et al. (2017). Initially, 0.2 mL of the extract with 0.35 mL of 8% vanillin in absolute ethyl alcohol and 0.80 mL of methanol were mixed in a test tube. Then, from a 72% sulphuric acid about 1.25 mL was added and shaken vigorously. The mixture was then transferred into a water bath to heat for 10 min at 60 °C. Then the sample was cooled in an ice crystals bath for 300 sec. The TSC in the extracts of fenugreek seed was then determined using a UV-Vis spectrophotometer device (Hitachi, U-1800, Japan) at a wavelength of 544 nm. The standard curves were prepared using diosgenin at concentrations of 100–600 mg/L, respectively. The TSC in the extract of fenugreek seed was presented as milligram of diosgenin equivalent (DE) per gram of dry weight (mg DE/g d.w.). The absorbance was read against methanol (blank) and the measurements were repeated thrice. Then, the total content of saponin was calculated using Equation (3) showing milligrams of standard equivalent (StE) per gram of dry weight sample.

$$TSC\left(mg\,DE\,/\,g\,d.w\right) = \frac{V \times C}{m} \qquad 3$$

In Equation (3), V, C, and m indicate the volume of solvent (mL), sample concentration from the standard curve (mg/L), and weight of dry sample used for extraction (g).

6.2.5 Determination of Total Phenolic Content in the Extract

The determination of TPC in the extracts of fenugreek seed was done by applying the methods used by Sookjitsumran et al. (2016) and Nickel et al. (2016) with minor changes. First, the equal amount (0.2 mL) of extract and reagent of Folin-Ciocalteu were added together in a test tube and placed in the dark for around 5 min at 30°C. Afterwards, 0.6 mL of 20% sodium carbonate (Na_2CO_3) solution was added to the mixture and stored for another 2 h. The wavelength of 765 nm was selected in a UV-Vis spectrophotometer to take the absorbance of the sample using methanol as blank. A standard curve at concentrations of (100–500 mg/L) was prepared from gallic acid (GA) to measure the sample concentration. The result was expressed as milligram of GA equivalent per g of the extract (mg GAE/g d.w.). Equation (4) was used to calculate the amount of TPC in the extract.

$$TPC\left(mgGAE\,/\,g\,d.w\right) = \frac{V \times C}{m} \qquad 4$$

Where V is the volume of solvent (mL), C is the sample concentration taken from standard curve (mg/L) and m is the weight of sample (g) used for extraction.

6.2.6 Antioxidant Activities

6.2.6.1 DPPH Radical Scavenging Activity

The DPPH antioxidant activity of the extract was performed using the methods described by Alara et al. (2018) with some changes. To prepare a 1 mM of DPPH stock solution, 0.004 g of DPPH solid particles were dissolved in 100 mL of pure methanol and the mixture was stored at 4 °C. Next, 10 mL of DPPH stock solution was mixed with 90 mL of pure methyl alcohol to make a 0.1 mM of fresh DPPH working solution. The dried sample (extract) or ascorbic acid was dissolved in methanol as 10 mg of dried extract in 10 mL (1:1 g/mL) of methyl alcohol to make the stock

Saponin and Phenolic Compounds

solution of (1 mg/mL). Then, different concentrations of 100–500 µg/mL were prepared from the stock extraction solution of plant extract of Soxhlet extraction and optimized MAE. Afterwards, the same amount (0.2 mL) of the extract or the standard ascorbic acid and DPPH working solution were mixed together and kept in dark for half an hour. The mixture was then stored at room temperature in the dark for 30 min. The antioxidant properties of extracts were presented as IC_{50} which indicates the concentration of µg/mL of the extract to inhibit 50% of the DPPH radical. Finally, the absorbance recorded at 517 nm using a UV visible Spectrophotometer and the analysis were done three times. The DPPH inhibition was calculated using Equation (5) against methanol as blank.

$$DPPH\ radical\ scavenging\ activity\ (\%) = \frac{A_1 - A_0}{A_1} \times 100 \qquad 5$$

Where A_1 and A_0 indicate the absorbance of DPPH-methanol mixture, and the mixture of extract-DPPH prepared solution, respectively.

6.2.6.2 ABTS Radical Scavenging Activity of the Extract

The ABTS assay of the extract in the optimized condition of MAE was determined according to Zielinski et al. (2014) and Cheng et al. (2013) with little modification. The assay was started by preparing 7 mM and 2.45 mM stock solution of ABTS and $K_2S_2O_8$, respectively. To prepare the solutions, 7.6 mg of ABTS and 1.32 mg of $K_2S_2O_8$ were separately dissolved in 2 mL of water (solution a and b), respectively. Then, the solutions of a and b mixed together in a same ratio (1:1 v/v) and incubated at room temperature for 12–16 h in the dark. The dried sample (extract) or ascorbic acid were dissolved in methanol as 10 mg of dried extract in 10 mL (1:1 g/mL) of methanol to prepare the stock solution of (1 mg/mL). Then, different concentrations of 100–500 µg/mL were prepared from the stock extraction solution of plant extract of Soxhlet extraction and optimized MAE. Next, to get a constant absorbance of 1.1 ± 0.02 at 734 nm, 1 mL of the solution was diluted with 60 mL of methanol. Afterwards, 0.15 mL of the extracts and 2.85 mL of ABTS solution were added together. The sample mixture was stored at 26 ± 2 °C for 2 h and the absorbance obtained via UV-Vis spectrophotometer at 734 nm. The antioxidant properties of the extract were represented as IC_{50} which indicates the concentration of µg/mL of the extract to inhibit 50% of the ABTS radical. The blank was ethanol and the ABTS scavenging activity calculated as shown in Equation (6).

$$ABTS\ radical\ scavenging\ activity\ (\%) = \left(1 - \frac{A_{sample}}{A_{control}}\right) \times 100 \qquad 6$$

where $A_{control}$ and A_{sample} represent the mixture of ABTS and methanol absorbance and the blend of extract and ABTS solution, respectively.

6.2.7 Analysis of LC-QTOF-Mass Spectrometer

The phenolic and saponin compounds in extract of fenugreek seed obtained at the optimal condition were identified using LC-QTOF-MS analysis (Waters® Vion IMS, USA). The LC-QTOF-MS was equipped with a PDA detector and a symmetry C18 column (100 mm x 2.1 mm, 1.8 µm particle size) was used for the tentative identification. Ionization was performed in positive and negative electrospray (ESI) modes. The extract of fenugreek seed was prepared at 20 ppm in LCMS grade methanol. The identification of saponins and phenolic compounds were performed at negative and positive ion modes and different concentrations of acetonitrile and water were used as mobile phase. The condition of MS was selected between 100–1000 m/z. The temperature of desolvation, column, and sample were set to 550, 40, and 15 °C, respectively. The flow rate (0.5 mL/min) and injection volume (20 µL) were set to operate the system. The capillary voltage and desolvation gas flow rate were set at 1.50 kV and 800 L/h, respectively.

The positive ion mode was operated with the mobile phase of A (water + 0.1% formic Acid) and B (acetonitrile). The initial conditions were set at 90% phase A and 10% phase B and the time was changed from 0.00 to 8.34 min during the entire operation. Then, the operation conditions were changed based on the operational time of 4.17 min A 45% and B 55%, at 6.25 min the mobile phases were A 10% and B 90%, while at 8.34 min, A 90% and B 10%. The operation was carried out at constant flow rate of 0.5 mL/min. The low collision energy was at 4.00 eV and the high collision energy started at 10.00 eV and ended at 45.00 eV. The same operation method was applied for the negative ion mode, but no formic acid was used in the mobile phase. UNIFI software with a scientific library was used to identify the bioactive compounds.

6.3 RESULTS AND DISCUSSION

6.3.1 Optimization and Model Fitting Using RSM

TSC and TPC were irradiation time A (2–4 min), microwave oven power B (500–700 W), ethanol concentration C (40–80%), and feed-to-solvent ratio D (1:8–1:12 g/mL as seen in Table 6.1). The temperature during the extraction process did not indicate a considerable contribution; therefore, it was fixed at 70 °C based on best points of OFAT. The second-order polynomial equations of optimized conditions for extraction yield, TSC and TPC of fenugreek are shown in Equations (7 and 8), respectively.

$$Y_{TSC(F.S)} = +192.80 - 4.49A - 4.49B + 6.52C - 8.97D - 4.01AB - 3.64AC - 3.89AD \\ + 1.70BC + 4.67BD - 5.43CD + 1.62A^2 - 11.18B^2 - 14.38C^2 - 14.23D^2 \qquad 7$$

$$Y_{TPC(F.S)} = +81.01 - 0.95A - 0.95B + 1.04C - 0.90D - 1.26AB + 0.14AC + 0.81AD \\ - 1.47BC + 0.31BD + 0.25CD - 5.29A^2 - 2.24B^2 - 5.16C^2 - 3.84D^2 \qquad 8$$

Where Y is the response, A, B, C, and D indicate the variables, respectively.

6.3.2 Validation of the Model

The model validation was performed based on the suggested optimal condition of RSM which were 2.84 min, 572.50 W, 63.68%, and 1:9 g/mL. Where, based on these conditions, the extraction yield, TSC and TPC of fenugreek seed extract were 27.14%, 196.49 mg DE/g d.w. and 81.01 mg GAE/g d.w., respectively. To evaluate the validity of the predicted model, experimental runs were performed in suggested optimal conditions and the responses were 26.04 ± 0.88%, 195.89 ± 1.07 mg DE/g d.w., 81.85 ± 0.61 mg GAE/g d.w., respectively. By applying the paired t-tests for each optimization processes it is found that there was no statistically significant difference observed between predicted and actual values ($p > 0.05$) observed. Thus, it is suggested that the obtained model is suitable for the study.

6.3.3 DPPH and ABTS Antioxidant Activities

The antioxidant activities of fenugreek seed were evaluated using DPPH and ABTS free radical scavenging assays. The half maximal inhibitory concentration (IC50) of the samples was also determined. Table 6.2 shows the free radical scavenging activities of fenugreek seed against ascorbic acid. It is also seen that, the IC50 values of DPPH in fenugreek seed extract (195.27 ± 0.56) are higher when compared to IC50 values of ABTS assay (157.92 ± 1.11), respectively. The smaller the IC50 value the better the antioxidant activity (Lee et al., 2015). The proton free radicals of ABTS and DPPH compounds can be reduced when exposed to proton radical

TABLE 6.2
TSC, TPC, DPPH, and ABTS Antioxidant Activities of Fenugreek Seed

Extraction Technique	TSC (mg DE/g d.w.)/ (mg OAE/g d.w.)	TPC (mg GAE/g d.w.)	DPPH IC_{50} (µg/mL)	ABTS IC_{50} (µg/mL)
Fenugreek seed extract	195.89 ± 1.07	81.85 ± 0.61	195.27 ± 0.56	157.92 ± 1.11
Ascorbic acid			129.89 ± 1.33	70.57 ± 0.78

scavengers (K. J. Lee et al., 2015). Previous findings suggested that the capacity of antioxidant activities in most of the plants was better with ABTS rather than with DPPH, and the reason is the sensitivity of ABTS assay which increases the rate of kinetic reaction. As reported, phenolic and flavonoids also act as antioxidants or hydrogen donors, which is a chemical reaction inside the human body, specifically, phenolic compounds due to the presence of hydroxyl (-OH) group. The free -OH group as an antioxidant is responsible for controlling the oxidative damage by inhibiting the oxidation reaction caused by reactive oxygen species (ROS) in foods (Altemimi et al., 2017; Paj et al., 2019).

6.3.4 LC-QTOF-MS Analysis of Extracted Compounds

Saponin and phenolic bioactive compounds are among the secondary metabolites that are known to be useful for the treatment of many diseases and are commonly found in many plants. Saponins are found in either steroid or triterpenoid glycosides. In terms of health benefits, both saponins and phenolic compounds existing in plants have demonstrated potential medicinal values for the treatment of numerous diseases such as cancers, heart-related illnesses, tumour, infections, and diabetes (Chan et al., 2014).

The identification of saponin and phenolic bioactive compounds of fenugreek seed extract were performed using LC-QTOF-MS analysis. A total of 58 and 27 saponin and phenolic compounds were identified in optimized fenugreek seed extracted via MAE in both positive and negative ion modes, respectively (Table 6.3). The obtained biologically active compounds are responsible for many activities in the human body such as antioxidant, anti-inflammatory, antidiabetic, and anticancer properties (Lidia et al., 2017). As seen in Table 6.3, the saponin components belong to steroid and terpenoid saponins. It is also seen that different types of phenolic compounds such as simple phenols, phenolic aldehydes, phenolic acids, and polyphenols were presented in the extract of fenugreek seed. Saponins such as terrestrosin, timosaponin, markogenin, protodiosgenin, yamogenin, and epi-Smilagenin and phenolic compounds such as protocatechuic, cistanoside C, campneoside, forsythoside E, have been reported to possess anticancer, antidiabetic, anti-inflammatory, antibiotic, antioxidant, hormone balancing, and antidepressant properties (B. Lee et al., 2009; Lidia et al., 2017; Wei et al., 2014). Cistanoside C is used to repair DNA damage (Sperandio et al., 2002). Figure 6.2 (a and b) shows the identified compounds of fenugreek seed through LC-QTOF-MS analysis based on retention time and observed intensity of positive and negative ion modes, respectively. It is seen in Table 6.3 that most of the saponins are observed in positive ion modes, while phenolic compounds tend to be visible in negative ion modes. This may be due to the high affinity of alkali cations in saponin compounds; therefore, most of the saponins are detected in the positive ion mode and charged hydrogen, sodium, and even potassium adducts of the molecules [+H, +Na, and +K]. Relatively, most of the phenolic compounds are detected in the negative ion mode and charged hydrogen adducts of the molecules [-H] (Bahrami et al., 2014).

TABLE 6.3
Saponin and Phenolic Compounds of Fenugreek Seed Extract Identified by LC-QTOF-MS in Positive and Negative Ion Modes

	Component Name	Formula	Observed Neutral Mass (Da)	Observed m/z	Observed RT (min)	Response	Adducts	Total Fragment
Saponin Compounds								
1	Gentiopicroside	$C_{16}H_{20}O_9$	356.1127	395.0759	2.43	22086	+K	10
2	Periplocoside L	$C_{28}H_{46}O_7$	494.3232	517.3125	2.93	1716	+Na	12
3	Timosaponin D	$C_{45}H_{74}O_{19}$	918.482	919.4892	2.99	20828	+H	66
4	25(S)-Ruscogenin	$C_{27}H_{42}O_4$	430.309	431.3163	3	6952	+H	12
5	Prosapogenin 2	$C_{32}H_{48}O_8$	560.3351	599.2983	3	1192	+H	19
6	Zingiberogenin	$C_{27}H_{42}O_4$	430.3092	431.3165	3.08	7355	+H	17
7	Markogenin	$C_{27}H_{44}O_4$	432.3237	433.331	3.12	62957	+H	28
8	Protodiosgenin	$C_{33}H_{54}O_9$	594.3764	595.3837	3.12	552505	+H	42
9	Tigogenin-3-O-β-Dglucopyranosyl (1 4)-β-D-galactopyranoside	$C_{39}H_{64}O_{13}$	740.4347	741.442	3.1	113363	+H	18
10	Terrestrosin A	$C_{45}H_{74}O_{18}$	902.4878	903.4951	3.24	537223	+H	69
11	Timosaponin B2	$C_{45}H_{76}O_{19}$	920.498	943.4872	3.24	16023	+Na	68
12	Ophiopogonin C'	$C_{39}H_{62}O_{12}$	722.424	723.4313	3.25	57789	+H	25
13	Yamogenin	$C_{27}H_{42}O_3$	414.3131	415.3204	3.25	11482	+H	5
14	epi-Smilagenin	$C_{27}H_{44}O_3$	416.3277	417.335	3.25	16045	+H	25
15	Trigoneoside IIa	$C_{44}H_{74}O_{18}$	890.4865	913.4757	3.31	8245	+Na	37
16	Quinatoside D	$C_{39}H_{60}O_{11}$	704.4144	705.4217	3.43	10962	+H	6
17	Timosaponin A2	$C_{39}H_{64}O_{14}$	756.4311	779.4203	3.44	10412	+Na	16
18	Cimifoetiside VII	$C_{43}H_{70}O_{16}$	842.4668	843.474	3.48	1218	+H	45
19	Timosaponin G	$C_{39}H_{64}O_{14}$	756.4304	757.4377	3.53	14169	+H	40
20	Abrisaponin I	$C_{48}H_{74}O_{20}$	970.4773	971.4845	3.53	260947	+H	45
21	Diosgenone	$C_{27}H_{40}O_3$	412.2978	413.305	3.53	4485	+H	7
22	Macrostemonoside F	$C_{45}H_{74}O_{18}$	902.4882	903.4954	3.59	62506	+H, Na	66
23	Terrestroside F	$C_{33}H_{54}O_7$	562.3877	563.395	3.6	2686	+H	15
24	Soyasaponin βg	$C_{47}H_{74}O_{17}$	910.4909	911.4982	3.67	24485	+H	56
25	3-O-β-D-Galactopyranosyl-(1 2)-β-Dglucuronopyranosyl gypsogenin	$C_{42}H_{64}O_{16}$	824.4205	825.4277	3.68	5161	+H	53
26	Ophiogenin	$C_{27}H_{42}O_5$	446.3032	447.3105	3.69	1702	+H	11
27	Trigoneoside IIIb	$C_{45}H_{76}O_{18}$	904.5044	927.4936	3.7	27683	+K, +Na	61
28	Neohecogenin-3-O-β-D-glucopyranoside	$C_{33}H_{52}O_9$	592.3605	615.3498	3.71	1351	+Na	12
29	Atroposide D	$C_{39}H_{62}O_{13}$	738.4185	739.4258	3.72	147331	+H	38
30	Δ3,5-Deoxytigogenin	$C_{27}H_{40}O_2$	396.3033	397.3106	3.79	4616	+H	13
31	Gracillin	$C_{45}H_{72}O_{17}$	884.4773	885.4846	3.92	378625	+H, + Na	63
32	Toosendanin	$C_{30}H_{38}O_{11}$	574.243	575.2503	3.95	2266	+H	16
33	Kingianoside D	$C_{39}H_{60}O_{14}$	916.4653	917.4725	4.07	38900	+H	49
34	Atroposide B	$C_{33}H_{52}O_8$	576.3661	577.3733	4.14	42670	+H	17
35	Hecogenone	$C_{27}H_{40}O_4$	428.2938	429.301	4.19	3752	+H	3
36	Picfeltarraenin IV	$C_{47}H_{72}O_{18}$	924.4726	925.4798	4.19	5948	+H	31
37	Ophiopogonin E	$C_{38}H_{60}O_{13}$	724.4046	725.4119	4.24	18991	+H, Na	25

	Component Name	Formula	Observed Neutral Mass (Da)	Observed m/z	Observed RT (min)	Response	Adducts	Total Fragment
38	Hookeroside C	$C_{38}H_{62}O_{15}$	758.4093	781.3985	4.29	19378	+Na	32
39	Timosaponin A1	$C_{33}H_{54}O_{8}$	578.3821	579.3894	4.3	17525	+H	17
40	Kingianoside B	$C_{39}H_{60}O_{13}$	736.4019	737.4092	4.38	21728	+H	33
41	Azukisaponin II	$C_{42}H_{68}O_{14}$	796.4612	797.4685	4.4	66788	+H	146
42	Azukisaponin I	$C_{42}H_{68}O_{13}$	780.4663	781.4736	4.45	26138	+H	143
43	Soyasaponin I	$C_{48}H_{78}O_{18}$	942.5201	943.5273	4.45	1202823	+H, +Na, +K	160
44	3-O-(β-DGlucuronopyranosyl)-soyasapogenol B	$C_{36}H_{58}O_{9}$	634.4083	635.4156	4.45	26648	+H	129
45	Achyranthoside II	$C_{42}H_{68}O_{13}$	780.4663	781.4736	4.45	3564	+H	29
46	(25R)-Ruscogenin-1-O-β-D-xylopyranosyl(1 3)-β-Dfucopyranoside	$C_{38}H_{60}O_{12}$	708.4102	709.4175	4.49	7368	+H	13
47	Kaikasapomin III	$C_{48}H_{78}O_{17}$	926.524	949.5132	4.58	23277	+Na	12
48	Esculentoside A	$C_{42}H_{66}O_{16}$	826.4348	827.442	4.71	5466	+H	23
49	Su-diosgenin-3-O-α-Lrhamnosus(1 2)-O-[α-L-rhamnopyranosyl(1 4)]-β-Dglucopyranoside	$C_{45}H_{72}O_{16}$	868.4814	869.4887	4.72	15860	+H	4
50	Celosin C	$C_{42}H_{66}O_{13}$	778.4518	779.4591	4.74	1293	+H	37
51	Esculentoside J	$C_{47}H_{76}O_{20}$	960.4934	983.4826	4.78	2535	+Na	94
52	Eleutheroside K	$C_{41}H_{66}O_{11}$	734.4592	735.4664	4.87	10546	+H	16
53	Picfeltarraenin IA	$C_{41}H_{62}O_{13}$	762.4208	763.4281	5.19	9037	+H	4
54	Esculentoside B	$C_{36}H_{56}O_{11}$	664.383	687.3722	5.53	2049	+H	15
55	23,27-Dihydroxypennogenin	$C_{27}H_{42}O_{6}$	462.2969	463.3042	6.29	5218	+H	5
56	Pulchinenoside A	$C_{41}H_{66}O_{12}$	750.4535	751.4607	6.71	14882	+H	17
57	Eclalbasaponin XIII	$C_{37}H_{58}O_{10}$	662.4012	663.4084	6.73	29031	+H	8
58	Smilaxin	$C_{17}H_{16}O_{6}$	316.0948	315.0875	2.86	144596	-H	10
Phenolic Compounds								
1	Cimicifugic acid B	$C_{21}H_{20}O_{11}$	448.1003	449.1076	1.79	21004	+Na	14
2	E-p-Coumatic acid	$C_{9}H_{8}O_{3}$	164.0473	165.0545	0.68	7808	+H	3
3	Auraptenol	$C_{19}H_{22}O_{4}$	314.1535	337.1427	1.06	3237	+Na	1
4	Actinidioionoside	$C_{19}H_{34}O_{9}$	406.2198	429.209	1.89	65198	+Na	6
5	Cichorioside B	$C_{21}H_{28}O_{10}$	440.1681	463.1573	1.41	5223	+Na	0
6	Quercetin-3-Oneohesperidoside	$C_{27}H_{30}O_{16}$	610.153	611.1603	1.79	16702	+H,+Na	10
7	Sweroside	$C_{16}H_{22}O_{9}$	358.1263	381.1155	1.57	6491	+Na	0
8	Onitin-2'-O-β-Dallopyranoside	$C_{21}H_{30}O_{8}$	410.1951	433.1843	3.79	8673	+Na	3
9	Estrone	$C_{18}H_{22}O_{2}$	270.16	293.1492	2.09	13866	+Na	1
10	Ulmoside	$C_{21}H_{32}O_{14}$	508.1812	547.1443	2.37	339575	+Na	47
11	5,7,2',5'-Tetrahydroxyflavone	$C_{15}H_{10}O_{6}$	286.0479	287.0551	2.74	7172	+H	1
12	Oleuropein	$C_{25}H_{32}O_{13}$	540.1862	579.1494	3.28	23590	+H	12
13	6'-O-Galloylhomoarbutin	$C_{20}H_{22}O_{11}$	438.1162	437.109	0.51	2365	-H	1
14	Meliadanoside B	$C_{15}H_{20}O_{8}$	328.1155	327.1082	0.59	4384	-H	2
15	Protocatechuic aldehyde	$C_{7}H_{6}O_{3}$	138.0322	137.0249	0.59	2326	-H	0
16	2,4,6Trihydroxyacetophenone-2,4-di-O-β-Dglucopyranoside	$C_{20}H_{28}O_{14}$	492.1487	491.1415	0.76	80020	-H	23
17	Cistanoside C	$C_{30}H_{38}O_{15}$	638.222	637.2147	0.78	137579	-H	15

(Continued)

TABLE 6.3 (Continued)

	Component Name	Formula	Observed Neutral Mass (Da)	Observed m/z	Observed RT (min)	Response	Adducts	Total Fragment
18	(−)-Suspensaside B	$C_{33}H_{44}O_{16}$	696.2635	695.2563	0.87	217570	-H	29
19	Campneoside I	$C_{30}H_{38}O_{16}$	654.2163	653.209	0.89	16268	-H	34
20	Osmanthuside H	$C_{19}H_{28}O_{11}$	432.164	431.1567	1.3	14481	-H	0
21	Decaffeoylacteoside	$C_{20}H_{30}O_{12}$	462.1744	461.1671	1.5	6002	-H	8
22	Forsythoside E	$C_{20}H_{30}O_{12}$	462.1727	461.1654	1.64	11463	-H	1
23	(+)-Suspensaside A	$C_{29}H_{34}O_{15}$	622.1905	621.1833	2.16	22531	-H	6
24	Erigoster A	$C_{27}H_{26}O_{13}$	558.1377	557.1304	2.7	15171	-H	45
25	Kuzubutenolide A	$C_{23}H_{24}O_{10}$	460.1376	459.1304	2.8	128449	-H	25
26	Dihydroresveratrol	$C_{14}H_{14}O_3$	230.0945	229.0872	2.82	3188	-H	1
27	2,3,5,4'-Tetrahydroxystilbene-2,3-O-β-Dglucopyranoside	$C_{26}H_{32}O_{14}$	568.1801	567.1728	2.92	33603	-H	32

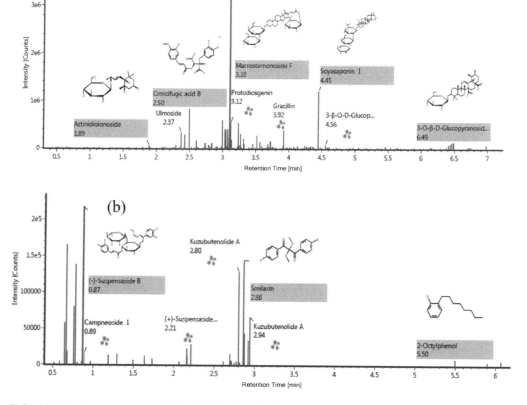

FIGURE 6.2 Chromatogram of LC-QTOF-MS analysis of fenugreek seed in positive (a) and negative (b) ion modes.

6.4 CONCLUSION

Bioactive compounds extracted from plants have achieved potential interest as an ingredient in food, cosmetic and pharmaceutical products. The optimization of MAE parameter for fenugreek seed showed that the optimal conditions of parameters were at 2.84 min, 572.50 W, 63.68%, and 1:9 g/mL. Based on the optimum condition, the responses of TSC and TPC of fenugreek seed were 26.04%, 195.89 mg DE/g d.w., 81.85 mg GAE/g d.w., respectively. The LC-QTOF-MS analysis of the extracts confirmed the presence of saponin and phenolic compounds in fenugreek seed. A total of 58 and 27 saponin and phenolic compounds were identified in optimized fenugreek seed extracted via MAE in both positive and negative ion modes. Considering the results of this study, it can be suggested that fenugreek seed extract obtained via MAE can be used as a good source of saponin, phenolic, and antioxidants in food, and in the cosmetic and pharmaceutical industries.

REFERENCES

Akbari, S., Abdurahman, N. H., & Yunus, R. M. 2019. Optimization of saponins, phenolics, and antioxidants extracted from fenugreek seeds using microwave-assisted extraction and response surface methodology as an optimizing tool. *Comptes Rendus Chimie*, 22(11–12), 714–727. https://doi.org/10.1016/j.crci.2019.07.007.

Akbari, S., Abdurahman, N. H., Yunus, R. M., & Fayaz, F. 2019. Microwave-assisted extraction of saponin, phenolic and flavonoid compounds from Trigonella foenum-graecum seed based on two level factorial design. *Journal of Applied Research on Medicinal and Aromatic Plants*, 14(March), 100212. https://doi.org/10.1016/j.jarmap.2019.100212.

Alara, O. R., Abdurahman, N. H., Ukaegbu, C. I., & Azhari, N. H. 2018. Vernonia cinerea leaves as the source of phenolic compounds, antioxidants, and anti-diabetic activity using microwave-assisted extraction technique. *Industrial Crops and Products*, 122, 533–544. https://doi.org/10.1016/j.indcrop.2018.06.034.

Altemimi, A., Lakhssassi, N., Baharlouei, A., Watson, D., & Lightfoot, D. 2017. Phytochemicals: Extraction, isolation, and identification of bioactive compounds from plant extracts. *Plants*, 6(4), 42. https://doi.org/10.3390/plants6040042.

Bahrami, Y., Zhang, W., Chataway, T., & Franco, C. 2014. Structural elucidation of novel saponins in the sea cucumber Holothuria lessoni. *Marine Drugs*, 12(8), 4439–4473. https://doi.org/10.3390/md12084439.

Baruah, A., Bordoloi, M., & Deka Baruah, H. P. 2016. Aloe vera: A multipurpose industrial crop. *Industrial Crops and Products*, 94, 951–963. https://doi.org/10.1016/j.indcrop.2016.08.034.

Chan, K. W., Iqbal, S., Khong, N. M. H., Ooi, D. J., & Ismail, M. 2014. Antioxidant activity of phenolics-saponins rich fraction prepared from defatted kenaf seed meal. *LWT—Food Science and Technology*, 56(1), 181–186. https://doi.org/10.1016/j.lwt.2013.10.028.

Cheng, H., Feng, S., Jia, X., Li, Q., Zhou, Y., & Ding, C. 2013. Structural characterization and antioxidant activities of polysaccharides extracted from Epimedium acuminatum. *Carbohydrate Polymers*, 92(1), 63–68. https://doi.org/10.1016/j.carbpol.2012.09.051.

Hu, T., Guo, Y. Y., Zhou, Q. F., Zhong, X. K., Zhu, L., Piao, J. H., Chen, J., & Jiang, J. G. 2012. Optimization of ultrasonic-assisted extraction of total saponins from Eclipta prostrasta L. using response surface methodology. *Journal of Food Science*, 77(9). https://doi.org/10.1111/j.1750-3841.2012.02869.x.

Karimi, A., Majlesi, M., & Rafieian-Kopaei, M. 2015. Herbal versus synthetic drugs; beliefs and facts. *Journal of Nephropharmacology*, 4(1), 27–30. www.ncbi.nlm.nih.gov/pubmed/28197471%0Awww.pubmedcentral.nih.gov/articlerender.fcgi?artid=PMC5297475.

Lee, B., Jung, K., & Kim, D. H. 2009. Timosaponin AIII, a saponin isolated from Anemarrhena asphodeloides, ameliorates learning and memory deficits in mice. *Pharmacology Biochemistry and Behavior*, 93(2), 121–127. https://doi.org/10.1016/j.pbb.2009.04.021.

Lee, K. J., Oh, Y. C., Cho, W. K., & Ma, J. Y. 2015. Antioxidant and anti-inflammatory activity determination of one hundred kinds of pure chemical compounds using offline and online screening HPLC assay. *Evidence-Based Complementary and Alternative Medicine*, 2015, 1–13. https://doi.org/10.1155/2015/165457.

Lidia, D.-A., Delia, M.-H., Nancy, O.-T., & Rosa Isela, G.-G. 2017. Microwave-Assisted Extraction of Phytochemicals and Other Bioactive Compounds. *Reference Module in Food Science*, 2016, 1–10. https://doi.org/10.1016/b978-0-08-100596-5.21437-6.

Maan, A. A., Nazir, A., Khan, M. K. I., Ahmad, T., Zia, R., Murid, M., & Abrar, M. 2018. The therapeutic properties and applications of Aloe vera: A review. *Journal of Herbal Medicine*, 1–10. https://doi.org/10.1016/j.hermed.2018.01.002.

Moyo, M., Amoo, S. O., Ncube, B., Ndhlala, A. R., Finnie, J. F., & Staden, J. Van. 2013. Phytochemical and antioxidant properties of unconventional leafy vegetables consumed in southern Africa. *South African Journal of Botany*, 84, 65–71. https://doi.org/10.1016/j.sajb.2012.09.010.

Nickel, J., Spanier, L. P., Botelho, F. T., Gularte, M. A., & Helbig, E. 2016. Effect of different types of processing on the total phenolic compound content, antioxidant capacity, and saponin content of Chenopodium quinoa Willd grains. *Food Chemistry*, 209, 139–143. https://doi.org/10.1016/j.foodchem.2016.04.031.

Paj, P., Socha, R., Broniek, J., Królikowska, K., & Fortuna, T. 2019. Antioxidant properties, phenolic and mineral composition of germinated chia, golden flax, evening primrose, phacelia and fenugreek. *Food Chemistry*, 275(September 2018), 69–76. https://doi.org/10.1016/j.foodchem.2018.09.081.

Sookjitsumran, W., Devahastin, S., Mujumdar, A. S., & Chiewchan, N. 2016. Comparative evaluation of microwave-assisted extraction and preheated solvent extraction of bioactive compounds from a plant material: A case study with cabbages. *International Journal of Food Science and Technology*, 51(11), 2440–2449. https://doi.org/10.1111/ijfs.13225.

Sperandio, O., Fan, B. T., Panaye, A., Doucet, J. P., El Fassi, N., Zakrzewska, K., Jia, Z. J., & Zheng, R. L. 2002. Theoretical study of fast repair of DNA damage by cistanoside C and analogs: Mechanism and docking. *SAR and QSAR in Environmental Research*, 13(2), 243–260. https://doi.org/10.1080/10629360290002749.

Venegas-Calerón, M., Ruíz-Méndez, M. V., Martínez-Force, E., Garcés, R., & Salas, J. J. 2017. Characterization of Xanthoceras sorbifolium Bunge seeds: Lipids, proteins and saponins content. *Industrial Crops and Products*, 109, 192–198. https://doi.org/10.1016/j.indcrop.2017.08.022.

Wei, S., Fukuhara, H., Chen, G., Kawada, C., Kurabayashi, A., Furihata, M., Inoue, K., & Shuin, T. 2014. Terrestrosin D, a steroidal saponin from tribulus terrestris L., Inhibits growth and angiogenesis of human prostate cancer in vitro and in vivo. *Pathobiology*, 81(3), 123–132. https://doi.org/10.1159/000357622.

Yadav, U. C. S., & Baquer, N. Z. 2014. Pharmacological effects of Trigonella foenum-graecum L. in health and disease. *Pharmaceutical Biology*, 52(2), 243–254. https://doi.org/10.3109/13880209.2013.826247.

Zielinski, A. F., Silva, M. V, Pontes, P. V. D. A., Iora, S. R. F., Maciel, G. M., Haminiuk, C. W. I., & Granato, D. 2014. Original article evaluation of the bioactive compounds and the antioxidant capacity of grape pomace. *International Journal of Food Science and Technology*, 50(1), 62–69. https://doi.org/10.1111/ijfs.12583.

Part III Sports Supplements and Nutraceuticals

7 Fenugreek
Nutraceutical Properties and Therapeutic Potential

Dilip Ghosh

CONTENTS

7.1 Introduction 91
7.2 History of Fenugreek 92
7.3 Taxonomy 92
7.4 Biology and Cultivation 93
7.5 Plant Chemistry 93
7.6 Nutritional Constituents 94
      7.6.1 Active Compounds 94
7.7 Traditional Use 95
7.8 Therapeutic Claims 95
      7.8.1 Hypoglycaemic and Hypocholesterolaemic Effect 96
      7.8.2 Free Testosterone Boosting 97
7.9 Application in Food and Functional Food Products 97
7.10 Fenugreek on Gut Microbiota: Links to Metabolic Syndrome 98
7.11 Recent Scientific Development 99
7.12 Conclusion 99
7.13 Future 99
References 99

ABBREVIATIONS

WHO—World Health Organization; 4-HI—4-hydroxyisoleucine; FBG—Fasting Blood Glucose; 2hBG—2-h Blood Glucose; HbA1c—Glycated Haemoglobin; TC—Total Count; PCA—Principal Component Analysis

---

7.1 INTRODUCTION

For the first time, the World Health Organization is recognizing traditional medicine as an influential global medical compendium (WHO, 2013). Recently, there is an increasing tendency towards traditional medicine, motivated by the occurrence of harmful effects of chemical drugs on human health as well as the various deficits of modern medicine in treating some diseases.

"There has been a definite change in the conversation around Ayurveda, and an increase in interest." Herbal supplement sales are growing at an average of over 20%, with some products like turmeric, moringa and ashwagandha seeing 40% to 50% growth (ET HealthWorld November 19,

2018). Brahmi (Bacopa), ashwagandha, turmeric and tulsi are the forerunner of the single component products which dominate the global market. The coronavirus disease 19 (COVID-19) pandemic has challenged health care systems across the globe. The coronavirus pandemic has turned the world's attention to the immune system against disease-causing bacteria, viruses and other organisms that we touch, ingest and inhale every day. COVID-19 has brought about a shift in the complementary medicine industry (Lakshmanan & Aggarwal, 2020). Two big Indian Ayurvedic medicine companies, The Dabur and Himalaya have both witnessed huge spikes in sales during this pandemic.

Apart from these Ayurvedic herbs, I strongly believe that **fenugreek** will be the next blockbuster Indian herb, with substantial and convincing scientific and clinical support building around it.

7.2 HISTORY OF FENUGREEK

Fenugreek belongs to the Fabaceae family. The name, Trigonella, comes from the Latin language and means "little triangle," due to its yellowish-white triangular flowers. It is also named methi (Hindi, Bengali, Urdu, Punjabi and Marathi), Hulba (Arabic), Moshoseitaro (Greek), Uluva (Malayalam), Shoot (Hebrew), Dari (Persian) and hayseed in English. Fenugreek (*Trigonella foenum-graecum* L.) is one of the oldest medicinal plants from the Fabaceae family, originating in central Asia around 4000 BC (Acharya et al., 2006). Its description and benefits had been reported in the Ebers Papyrus (one of the oldest maintained medicinal documents) in 1500 BC in Egypt. It is being commercially grown in India, Pakistan, Afghanistan, Iran, Nepal, Egypt, France, Spain, Turkey, Morocco, North Africa, the Middle East and Argentina (Mabberley, 1997). Due to the presence of a substantial amount of fibre, phospholipids, glycolipids, oleic acid, linolenic acid, linoleic acid, choline, vitamin A, B1, B2, C, nicotinic acid, niacin, and many other functional elements, fenugreek seed is one of the hot ingredients in the health and wellness domain (Ahmed et al., 2016).

7.3 TAXONOMY

The genus Trigonella L. belongs to the subtribe Trifolieae (Fabaceae), together with the genera Medicago Inc., Trifolium L., and Melilotus L. This subtribe is monophyletic within "vicioid clade," and their taxa are morphologically characterized by trifoliate leaves with stipules adnate to the stem that do not cover it entirely (Banerjee and Kole, 2004; Dangi et al., 2016). Trigonella is the genus within this subtribe with more taxa and high economic importance (Table 7.1); this genus is monophyletic and includes annual or perennial aromatic herbs that are characterized by a campanulate or tubular sepals with two (large) and three (small) equal lobes, diadelphous stamens, uniform anthers, terminal stigma, and ovary with numerous ovules (Banerjee and Kole, 2004). Taxa are widely distributed in the dry regions around Mediterranean, West of Asia, Europe, North and South Africa, North America and South Australia (Townsend et al., 1974). The exact number of taxa accepted in the Trigonella depends on unresolved names, accepted synonymies and authors; for example, Linnaeus listed 260 species, and other authors have recognized between 62 and 128 species.

The most recent study based on Small (1987) indicates that the genus currently includes 62 species. However, The Plant List (www.theplantlist.org/) lists 98 taxa accepted, 97 unresolved names, three misapplied names and 27 accepted synonymies. Trigonella has a wide diversity of species with economic potential (Table 7.1), including among these *T. foenum-graecum*, which is the most popular species in the genus by its countless uses and properties. Also, different species of fenugreek were assigned several potential uses by different authors; however, some currently are synonymous (Table 7.1) or belong to other genera. For example, *T. monspeliaca* is a synonym for *Medicago monspeliaca* (L.) Trautv., *T. polycerata* for *Medicago orthoceras* Trautv., and *T. radiata* Boiss, for *Medicago radiata* L.

TABLE 7.1
Diversity of Species of Fenugreek with Several Potential Economic Uses

Taxa	Uses
T. anguina Delile	FO, IR, NU
T. arabica Delile	FO, IR, NU
T. baccarinii Chiov. (Syn. *T. marginata* Baker)	FO, IR, UN
T. balansae Boiss. and Reut.	FO, IR, NU, PH
T. caerulea (L.) Ser.	IR, ME, PH, SP
T. calliceras Fischer ex M. Bieb.	PH
T. coerulescens Halacsy	FO, IR, UN
T. corniculata Sibth. and Sm.	CO, FO, IR, NU, PH
T. cretica (L.) Boiss.	PH
T. foenum-graecum L.	FO, IR, ME, NU, PH
T. glabra subsp. *uncata* (Boiss. and Noe) Lassen (Syn. *T. uncata* Boiss. and Noe)	ME
T. glabra Thunb.	FO, IR, NU
T. hamosa Del. ex Smith	FO, IR, UN
T. laciniata L.	FO, IR, UN
T. lilacina Boiss.	PH
T. maritima Poir	PH
T. occulta Ser.	FO, IR, ME, NU, PH
T. spicata Sm.	FO, IR, NU, PH
T. spinosa L.	FO, IR, NU, PH
T. spruneriana Boiss (Syn. *T. sibthorpii* Boiss.)	FO, IR, NU
T. stellata Forssk.	FO, IR, NU
T. suavissima Lindl.	IR

CO, comestible; **IR**, insect repellent; **FO**, forage; **ME**, medicinal; **NU**, nutraceutics; **PH**, pharmaceutical; **SP**, species.

Source: Adopted from Basu et al. (2019).

7.4 BIOLOGY AND CULTIVATION

Fenugreek is an annual legume, diploid (2n = 16) plant with no aneuploidy. Morphologically, it is an erect, aromatic annual closely resembling large clover. The stem is long, cylindrical (30–60 cm long) and pinkish in colour, whereas its roots are massive finger-like structures. Fenugreek blooms with white to yellowish white, axillary and sessile flowers that are hermaphrodite and insect pollinated. Fenugreek flowers produce brownish to yellowish brown 15 cm long 2–8 pods. Each pod contains 10–20 seeds; seeds are small (5 mm long), hard, smooth, dull yellow to brownish yellow in colour. It is a fast-growing plant, which may grow on dry grasslands, cultivated or uncultivated lands, hillsides, plains as well as field edges but it requires a fair amount of sunlight. Fenugreek needs four to seven months to reach maturity, the flowering period is midsummer (June to August) and seeds ripen during late summer. It is a drought-tolerant plant and grows well in tropical climates with mild winters and cool summers; however, its leaf and flower development are temperature dependent. Please refer to these review articles for details (Basu, 2006; Moradi-Kor and Moradi, 2013; Mehrafarin et al., 2011).

7.5 PLANT CHEMISTRY

Till now, more than 100 phytochemicals have been isolated and identified from fenugreek seeds, mainly including polysaccharides, saponins, alkaloids, polyphenols and flavonoids (Acharya et al.,

TABLE 7.2
Chemical Constituents of Fenugreek

Major Groups	Chemical Constituents
Alkaloids	Trimethylamine, neurin, trigonelline, choline, gentianine, carpaine and betain
Amino acids	Isoleucine, 4-hydroxyisoleucine, histidine, leucine, lysine, L-tryptophan, arginine
Saponins	Graecunins, fenugrin B, fenugreekine, trigofoenosides A–G
Steroidal sapinogens	Yamogenin, diosgenin, smilagenin, sarsasapogenin, tigogenin, neotigogenin, gitogenin, neogitogenin, yuccagenin, saponaretin
Flavonoids	Quercetin, rutin, vitexin, isovitexin
Fibres	Gum, neutral detergent fibre
Lipids	Triacylglycerols, diacylglycerols, monoacylglycerols, phosphatidylcholine phosphatidylethanolamine, phosphatidylinositol, free fatty acids.
Other	Coumarin, lipids, vitamins, minerals. 28% mucilage; 22% proteins; 5% of a stronger-swelling, bitter fixed oil.

Source: Yadav et al. (2011); Wani and Kumar (2018).

2006; Basu et al., 2014; Mandegary et al., 2012; Zandi et al., 2015). The main ingredients of the seed contain steroidal saponins, alkaloids, mucilage and fibres (Table 7.2). The most important steroidal saponins are diosgenin and yamogenin. The seeds also contain a sapogenin peptide ester named fenugreekine. Trigonelline is the alkaloid of this plant that has been extracted at up to 36% concentration. The amount of protein in this plant is high (22–25%), and its protein is rich in lysine, arginine, tryptophan, and to some extent, histidine. It contains low levels of sulfur-containing amino acids, threonine, valine, methionine and high levels of lysine, arginine and gelicin. 4-hydroxyisoleucine (4-HI) constitutes for about 80% of the total content of free amino acids in fenugreek seeds and are exclusively found in this plant. The amount of carbohydrates of this plant is about 8%. The seeds of fenugreek also contain proteinase-inhibiting compounds. They are also reported to contain minerals such as iron, phosphate, calcium and vitamins such as nicotinic acid, B1, C, A and D. Fenugreek also contains galactomannan, a highly bioactive molecule. Fenugreek seeds contain fixed oil comprising golden yellow and odourless unsaturated fatty acids (6% to 10%). Oil is easily dissolved in ether, benzene, sulfur and petroleum ether. Fenugreek contains wide varieties of flavonoids and other phenolics compounds. The main flavonoids identified in this plant include glycoside, orientin, isoorientin, vitexin, epigenin and quercetin.

7.6 NUTRITIONAL CONSTITUENTS

Fenugreek green leaves are one of the most ancient medicinal herbs containing β-carotene, ascorbate, fibre, iron, calcium and zinc, even more than the regular food items (Basu et al., 2019). Its seeds, biologically endosperm, are the most valuable plant part. Raw seeds are golden in colour with maple flavour but bitter in taste. However, this bitterness may be reduced by roasting. The seeds are fibrous, sticky and gummy in nature. Saponins and alkaloids are considered to be bioactive (and antinutritional) factors in seeds. However, defatted seeds are free from these compounds and may be consumed by people who have issues with fat.

7.6.1 Active Compounds

Fenugreek has powerful antioxidant properties, more particularly from the aqueous fraction than the flavonoids and phenolics linked to its health benefits. Fenugreek contains a fairly high amount

TABLE 7.3
Pharmacological and Therapeutic Actions of Fenugreek

Disease/Disorders	Active Constituents	Mechanism of Action
Diabetes	4-hydroxyisoleucine; polyphenolics	Stimulation of insulin production and antidiabetic effects
Cancer	Flavonoids; saponins, alkaloids, galactomannans, fibres	Inhibition of hyperplasia and carcinogenesis; decrease of incidence decrease
Hypercholesterolemia	Flavonoids and polyphenols, saponins	Controls high blood cholesterol and triglycerides; reduction in blood lipid levels
Inflammation	Various	Stimulatory effect on immune system; reduction of swelling and pain
Indigestion and flatulence	Fibres	Stimulates appetite and acts as laxative
Aging/Kidney disorders/Bacterial infection	Polyphenols, flavonoids, alkaloids	Antioxidants reduce cell death and aging; Protects functional and histopathologic abnormalities of diabetics; inhibition of *E. coli, S. typhi, S. aureas*

Source: Adapted and modified from Goyal et al., 2016. Saudi J Biol Sci.

of flavonoids, alkaloids, saponins, amino acids and other antioxidants. It contains a major class of phenolics like gallic acid, protocatechuic acid, catechin, gentisic acid, chlorogenic acid, vanillic acid and syringic acid of the seed extract. Fenugreek endosperm contains 35% alkaloids, primarily trigonelline. All these compounds are classified as biologically active, as these have pharmacological effects on the human body when ingested (Table 7.3). Their use should, therefore, be promoted in daily diet or through supplementation to manage hypercholesterolemia, cancer and diabetes mellitus as they possess hypoglycaemic, antilipidemic, anticarcinogenic and cholagogic properties.

7.7 TRADITIONAL USE

The medicinal value of fenugreek seeds is mentioned in Ayurvedic texts as well as in Greek and Latin pharmacopoeia. Since ancient times, fenugreek has played an extensive role in treating and preventing diseases (Roberts et al., 2011; Acharya et al., 2006, 2010). The Ayurvedic texts praise this herb for its power as an aphrodisiac, but modern Vaidyas (registered Ayurvedic doctors) seem to be using it more for digestive and respiratory problems stemming from an excess of kaph (phlegm) and vat (wind). In ancient Egypt, methi was used to ease childbirth and to increase milk flow, and modern Egyptian women are still using it today to relieve menstrual cramps, as well as making hilba tea out of it to ease other kinds of abdominal pain. The Chinese call it hu lu ba, and also use it for treating abdominal pain. In India, fresh methi ka saag (the stems and leaves of the plant) is very commonly cooked as a winter vegetable to control cough and cold, and the seeds are also eaten raw as sprouts and used medicinally. Fenugreek has been used to relieve colds, bronchial complaints, influenza, asthma, catarrh, constipation, sinusitis, pleurisy, pneumonia, sore throat, laryngitis, hay fever, tuberculosis and emphysema.

7.8 THERAPEUTIC CLAIMS

Numerous studies conducted so far have confirmed many of these traditional applications and have shown the clear therapeutic value of this plant (Askarpoura et al., 2020; Friedewald et al., 1972; Fedacko et al., 2016; Geberemeskel et al., 2019). Due to the presence of high levels of antioxidant compounds, bioactive compounds, phenols, flavonoids and anthocyanin, various experimental and clinical trials have shown that fenugreek has the ability to combat pathologic conditions, especially for the treatment and prevention of life-threatening diseases such as diabetes, some cancers, infections and gastrointestinal disorders. 4-hydroxyisoleucine (4-HI) constitutes about 80% of the total content of free amino acids in *Trigonella foenum-graecum* seeds that are exclusively found in this plant (Table 7.4). Based on convincing evidence on increased serotonin turnover in the brain, several studies demonstrated anti-anxiety and antidepressant-like effects of 4-HI. Fenugreek has a considerable antidiabetic effect. It (galactomannan component) can slow down the absorption of sugar in the gastrointestinal tract, whereas the 4-HI component stimulates insulin release, lowering the blood sugar in diabetic patients. Low molecular weight galactomannan-based fenugreek seed extract is reported to prevent fat accumulation. It is used for kidney complications due to its high antioxidant property. The antioxidant property of the plant has been attributed to the presence of many active phytochemicals, including flavonoids, plant sterols, vitamins, cumarins, terpenoids, carotenoids, curcumins, lignin and saponin.

7.8.1 Hypoglycaemic and Hypocholesterolaemic Effect

Fenugreek is a widely used herb for the treatment of diabetes mellitus (DM) but the effects in randomized controlled trials (RCTs) were controversial (Gong et al., 2016). Few meta-analyses had been conducted to evaluate the hypoglycaemic effect of fenugreek (Bordia etal.,1997; Suksomboon et al., 2011; Neelakantan et al., 2014; Gong et al., 2016). The pooled results showed that fenugreek could decrease FBG, 2hBG, HbA1c and TC. In prediabetes or DM, high blood glucose and blood lipids interacted to generate a vicious circle and are associated with macrovascular and microvascular disease including a recent human trial (Geberemeskel et al., 2019). The results suggest fenugreek has the hypoglycaemic and TC-lowering efficacy; however, the effects on TG, LDL-c and HDL-c need further confirmations.

TABLE 7.4
Potential Nutraceutical Applications of Fenugreek

Beneficial Effects	Components Used	References
Hypoglycaemic effect	Seeds	Roberts, 2011
Hypocholesterolaemic effect	Seeds	Srivastava et al., 2012
Antioxidant	Seed, leaves	Naidu et al., 2010
Lactation aid	Seeds	Snehlata and Payal, 2012
Immunomodulatory effect	Seeds	Meghwal and Goswami, 2012
Digestive effect and appetizer	Seed, leaves	Platel and Srinivasan, 2000
Decreases blood pressure	Seed, leaves	Sowmya and Rajyalakshmi, 1999
Wounds and sore muscles treatment	Seed, leaves	Mathern et al., 2009
Anticancer agent	Seed, leaves	Mathern et al., 2009
Asthma, emphysema, pneumonia	Seeds	Emtiazy et al. 2018
Anti-ulcer agent	Seed, leaves	Figer et al., 2017
Induces growth and reproduction hormones	Seeds	Blank, 1996
Gastro- and hepatoprotective	Seed, leaves	Blank, 1996
Healthy heart	Seeds	Blank, 1996
Prevents constipation	Seeds	Sowmya and Rajyalakshmi, 1999
Testosterone (free) booster	Seeds	Wankhede et al., 2016; Wilborn et al., 2010
Muscle cells and lean fat	Seeds	Dandona & Rosenberg, 2010; Kaufman & Vermeulen, 2005

Fenugreek comprises saponins, flavonoids, polysaccharides and other active ingredients. Diosgenin, 4-hydroxyisoleucine, trigonelline and soluble dietary fibre have received the most widespread concern (Fuller and Stephens, 2015). The hypoglycaemic and lipid-lowering mechanisms of fenugreek were varied. The possible mechanisms included delayed gastric emptying, selective reduction of glucose and fat absorption, inhibition of glucose transport, the direct stimulation of insulin secretion of islet β cells, increased insulin sensitivity, improved oxidative stress and the modulation of glucagon-like peptide-1 (Goyal et al., 2016; King et al., 2015; Mathern et al., 2009). Another recent metabolomics study (Jiang et al., 2018) using the principal component analysis (PCA) model revealed significant differences among the animal groups, and identified fenugreek flavonoids-induced changes of 11 potential biomarkers involved in lipid metabolism, amino acid metabolism and kidney function-related metabolism. They demonstrated that flavonoids are bioactive components of fenugreek with potent antidiabetic activity, which exert their therapeutic effects by multiple mechanisms, including reducing insulin resistance, improving gluconeogenesis and protecting islet cells and kidneys from damage.

7.8.2 Free Testosterone Boosting

Testosterone as the main male sex hormone is responsible for the development of male reproductive tissues and anabolic functional in men. Evidence have well indicated that serum total testosterone declines gradually with age (Fabbri et al., 2016). The primary clinical manifestations of low serum total testosterone include decreased libido, erectile dysfunction, fatigue and negative mood states. Secondary outcomes are changes in body composition, including increasing fat mass, decreasing lean body mass and bone mineral density as well as loss of muscle mass and strength (Dandona & Rosenberg, 2010; Kaufman & Vermeulen, 2005), which have significant impact on athletic performance (Wankhede et al., 2016; Wilborn et al., 2010).

Animal and human studies indicated that fenugreek extract through several mechanisms improves serum testosterone levels. The glycoside-rich fraction of fenugreek seed such as saponins and sapogenins has shown androgenic and anabolic activity (Wankhede et al., 2016). Evidence suggested the efficacy of saponins, especially protodioscin-like components and diosgenin, on testosterone and anabolic status (Salgado et al., 2017; Rao &Kale, 1992).

7.9 APPLICATION IN FOOD AND FUNCTIONAL FOOD PRODUCTS

Fenugreek can modify food texture owing to the high content of proteins and fibres, especially a soluble dietary fibre called gum (about 20.9 g/100 g in the seed), as well as neutral detergent. This fibre content, in addition to the flavour components modulates the organoleptic properties of foods. Soluble fibres can be utilized in nutrition and cereal bars, yogurts, dairy products and nutritional beverages. Plain powders of soluble fibre or total dietary fibre can be mixed with fruit juices, other spice mixes and seasonings. It can also be formulated as tablets or capsules along with the other vitamins and nutrients for direct supplements. It might further be applied to milk shakes, soups, dressings, sweets and candies or to fortify bakery flour for pizza, bread, bagels, muffins, cake mix, noodles, tortillas, flat bread, and fried and baked corn chips (Im et al., 2008; Srinivasan, 2006; Baba et al., 2018; Khorshidian et al., 2016). In general, fenugreek is beneficial to food processing as a food stabilizer, food adhesive, food emulsifier and gum (Jani et al., 2009, Table 7.5). Hooda and Jood (2004) noted that the addition of 10% of fenugreek flour to wheat flour increased protein content, fibre, total calcium and total iron; this indicates that fenugreek can be incorporated to prepare acceptable biscuits, and may also be mixed with cereals as a supplement for some limiting amino acids, and hence, for improving their protein quality through amino acid balance (Hooda and Jood, 2004). The muffin volumes increased up to 10% by addition of fenugreek seed husk which led to a decrease in texture hardness. Mixtures of fenugreek flour and de-bittered fenugreek polysaccharide were investigated on the physical and sensory quality characteristics and glycaemic index (GI) of chickpea–rice-based extruded products (Shirani et al., 2009).

TABLE 7.5
Potential Application of Fenugreek in Food and Functional Food Products

Component Used	Area of Utilization
Seeds	Bread (Wani and Kumar, 2018)
Fenugreek seeds, leaves	Biscuits (Wani and Kumar, 2018)
Seeds and gum	Extruded product (Wani and Kumar, 2018)
Seed, leaves	Culinary use as spice and seasoning (colour, flavour, aroma) (Wani and Kumar, 2018)
Seed, leaves	Organoleptic character improver (Wani and Kumar, 2018)
Seed	Maple syrup and artificial flavouring (Wani and Kumar, 2018
Seed	Dietary fibre, galactomannan (Wani and Kumar, 2018)
Seed	Food stabilizer, adhesive and emulsifying agent (Jani et al., 2009; Sowmya and Rajyalakshmi, 1999)

7.10 FENUGREEK ON GUT MICROBIOTA: LINKS TO METABOLIC SYNDROME

While beneficial effects of fenugreek on hyperlipidaemia and hyperglycaemia have been widely reported, the mechanisms of fenugreek-mediated actions on metabolic function are unknown. Here Bruce-Keller et al. (2020) recently demonstrated the robust effects of fenugreek on gut microbiota and describe a novel combination of statistical analyses to generate insight into the role of intestinal microbial changes in the protective effects of fenugreek. Specifically, sequencing analyses reveal that fenugreek significantly increased overall microbiome diversity in mice, and specifically reversed the actions of high dietary fat on key intestinal taxa. Furthermore, the representation of fenugreek-corrected taxa significantly correlated with metabolic function, including changes in body weight and composition, glucose regulation and hyperlipidaemia. These findings are in agreement with the extensive body of literature documenting the ability of a high fat diet to reduce bacterial diversity and disrupt the balance of pathogenic/commensal bacteria within the intestine (Kim et al., 2012; Everard et al., 2011). Data from many laboratories demonstrated that this pattern of gut dysbiosis is sufficient to impair both metabolic and neurologic function within the context of unhealthy, Western-style diets (Bruce-Keller et al., 2020; Shtriker et al., 2018; Anwar et al., 2018).

These pooled data indicate that fenugreek is particularly effective against hyperlipidaemia (Chaturvedi et al., 2013; Roberts, 2011). While there are several mechanisms whereby changes in gut microbiota could mediate the effects of fenugreek on serum lipids, most ultimately impact the absorption of dietary fat. They proposed several mechanisms, for example, intestinal microbiota alter the metabolism of diet-derived long-chain fatty acids such as conjugated linoleic acid, modulating absorption (Kishino et al., 2013). Published data suggest that fenugreek supplementation can dose-dependently increase faecal excretion of cholesterol from rats given high fat/high calorie diets (Muraki et al., 2011), mostly through members of Lachnospiraceae and Runinococcacea families (Kriaa et al., 2019).

These data indicate that fenugreek supplementation can stabilize metabolic function within the context of high fat consumption. Thus, use of fenugreek to maintain a healthy population of intestinal microbes in the context of a high fat diet could foster metabolic resilience even when diets/lifestyles are not optimal. Overall, data in these recent studies strongly suggest that the development of microbially targeted therapies—both primary and adjunctive—that are built upon safe, natural, plant-based products like fenugreek could be used to attain significant advancements in public health within the context of contemporary dietary environments.

7.11 RECENT SCIENTIFIC DEVELOPMENT

Several well-designed animal studies have demonstrated that combined therapy with 4-HI and pioglitazone is more beneficial than pioglitazone alone and also more beneficial than the combination of 4-HI and glyburide in the treatment of diabetes (Kulkarni et al., 2012; Kamble et al., 2013). The Herbalgram (2014, Vol.103, page 33) also concluded that IBHB, a standardized fenugreek seed extract (Indus Biotech Pvt Ltd., Pune, India), slows the progression of Parkinson's disease when taken as an adjunct to L-Dopa therapy and has a good safety profile (Nathan et al., 2014). The glycoside-based fenugreek seed extract is reported to have beneficial effects in animal models of pulmonary fibrosis. Several recent clinical trials demonstrated the efficacy of the special extract of fenugreek in conditions like hay fever and sinusitis. The seeds are considered useful in heart disease prevention and aphrodisiac, and as a galactagogue promoting lactation. A few recent randomized placebo controls in clinical trials demonstrated that standardized extracts (Libifem and Testofen) of *T. foenum-graecum* seed may be a useful treatment for increasing sexual arousal and desire in women and middle-aged and older men by increasing serum testosterone level (Rao et al., 2015, 2016).

Despite the fact that fenugreek is well explored by medical science for its potential contribution to human health, the average production in India remains low (1,200 kg/ha). The major factors responsible for low production are low level of awareness among the farming community, less availability of high yielding and resistant seed varieties, traditional farm mechanization and plant protection. An extensive genomic and agronomic characterization/clustering is needed to identify the potential genes which could further help in breeding programs followed by targeted mutation and genetic improvement for abiotic stress tolerance.

7.12 CONCLUSION

Fenugreek having antidiabetic, antifertility, anticancer, antimicrobial, antiparasitic, lactation stimulant and hypocholesterolaemic effects has been discussed in this review. Fenugreek has been found to have important bioactive compounds. From this review it was observed that fenugreek has been used as a food stabilizer, food adhesive, food emulsifier and gum. Fenugreek has been used to produce various types of bakery products and extruded products. Based on this health usefulness as discussed in this review, based on various past reported scientific findings, fenugreek can be recommended and must be taken as a part of our daily diet as its liberal use is safe and various health benefits can be drawn from this natural herb. The previously mentioned studies on Fenugreek suggest that the functional, nutritional and therapeutic characteristics of fenugreek can be exploited further in the development of healthy products (Pal and Mukherjee, 2020).

7.13 FUTURE

The list of potential health benefits of Fenugreek is continuing to grow with new clinical evidence. Therefore, fenugreek, which possesses phenolic compounds, bioactive amino acids, glycosides and antioxidant activity, could be a good candidate for an herbal drug. Despite the impressive scientific and clinical profile of fenugreek, consumer understanding is still in its infancy and promoters of this herb need to focus on this important commercial hurdle.

REFERENCES

Acharya, S.N., J. Thomas, and S. Basu. 2006. Fenugreek: An "old world" crop for the "new world". *Biodiversity* 3: 27–30.

Acharya, S.N., S.K. Basu, B.S. Datta, and R. Prasad. 2010. Genotype X environment interactions and its impact on use of medicinal plants. *The Open Nutrition Journal* 3: 47–54.

Ahmad, A., S. Salem, A.K. Mahmood, et al. 2016. Fenugreek a multipurpose crop: Potentialities and improvements. *Saudi Journal Biological Sciences* 23: 300–310.

Anwar, S., U. Bhandari, B.P. Panda, et al. 2018. Trigonelline inhibits intestinal microbial metabolism of choline and its associated cardiovascular risk. *Journal of Pharmaceutical and Biomedical* 159: 100–112.

Askarpoura, M., F. Alami, M.S. Campbell, K. Venkatakrishnan, A. Hadi, and E. Ghaedi. 2020. Effect of fenugreek supplementation on blood lipids and body weight: A systematic review and meta-analysis of randomized controlled trials. *Journal of Ethnopharmacology* 253: 112538.

Baba, W. N., Q. Tabasum, S. Muzzaffar, et al. 2018. Some nutraceutical properties of fenugreek seeds and shoots (Trigonella foenum-graecum L.) from the high Himalayan region. *Food Bioscience* 23: 31–37.

Banerjee, A., and P.C. Kole. 2004. Analysis of genetic divergence in fenugreek (Trigonella foenum-graecum L.). *Journal of Spices and Aromatic Crops* 13: 49–51.

Basu, S.K. 2006. Seed production technology for fenugreek (Trigonella foenum-graecum L.) in the Canadian Prairies (thesis). University of Lethbridge, Faculty of Arts Sci., Lethbridge, Alberta, Canada.

Basu, S.K., and G. Agoramoorthy. 2014. Fenugreek (Trigonella foenum-graecum L): Production challenges and opportunities for Asia, Africa and Latin America. *American Society of Human Genetics Social Issues* 12 Fenugreek Special Issue (March/April).

Basu, S.K., P. Zandi, and W. Cetzal-Ix. 2019. Fenugreek (*Trigonella foenum-graecum* L.): Distribution, Genetic diversity, and potential to serve as an industrial crop for the global pharmaceutical, nutraceuticals, and functional food industry. In: *The Role of Functional Food Security in: Global Health*. Eds. R.B. Singh, R.R. Watson, and T. Takahashi, Pages 471–497. https://doi.org/10.1016/B978-0-12-813148-0.00028-1.

Blank, I. 1996. The flavor principle of fenugreek. Nestlé research center. 211th ACS Symposium. New Orleans, 24–28.

Bordia, A., S.K. Verma, and K.C. Srivastava. 1997. Effect of ginger (Zingiberofficinale Rosc.) and fenugreek (Trigonella foenum-graecum L.) on blood lipids, blood sugar and platelet aggregation in patients with coronary artery disease. *Prostaglandins Leukot Essent Fat Acids* 56(5): 379–384.

Bruce-Keller, A.J., A.J. Richard, S-O. Fernandez-Kim, et al. 2020. Fenugreek counters the effects of high fat diet on gut microbiota in mice: Links to metabolic benefit. *Scientific Reports*, 10: 1245. https://doi.org/10.1038/s41598-020-58005-7.

Chaturvedi, U., A. Shrivastava, S. Bhadauria, J.K. Saxena, and G. Bhatia. 2013. A mechanism-based pharmacological evaluation of efficacy of Trigonella foenum-graecum (fenugreek) seeds in regulation of dyslipidemia and oxidative stress in hyperlipidemic rats. *Journal of Cardiovascular Pharmacology* 61: 505–512.

Dandona, P., and M. Rosenberg. 2010. A practical guide to male hypogonadism in the primary care setting. *International Journal of Clinical Practice* 64(6): 682–696.

Dangi, R., S. Tamhankar, R.K. Choudhary, and S. Rao. 2016. Molecular phylogenetics and systematics of Trigonella L. (Fabaceae) based on nuclear ribosomal ITS and chloroplast trnL intron sequences. *Genetic Resources and Crop Evolution* 63: 79–96.

Emtiazy, M., L. Oveidzadeh, M. Habibi, et al. 2018. Investigating the effectiveness of the Trigonella foenum-graecum L. (fenugreek) seeds in mild asthma: A randomized controlled trial. *Allergy Asthma Clinical Immunology* 14: 19. https://doi.org/10.1186/s13223-018-0238-9.

Everard, A., V. Lazarevic, M. Derrien, et al. 2011. Responses of gut microbiota and glucose and lipid metabolism to prebiotics in genetic obese and diet-induced leptin-resistant mice. *Diabetes* 60: 2775–2786.

Fabbri, E., Y. An, M. Gonzalez-Freire, et al. 2016. Bioavailable testosterone linearly declines over a wide age spectrum in men and women from the Baltimore longitudinal study of aging. *Journals of Gerontology Series A: Biomedical Sciences and Medical Sciences* 71(9): 1202–1209.

Fedacko, J., R.B. Singh, M.A. Niaz, et al. 2016. Fenugreek seeds decrease blood cholesterol and blood glucose as adjunct to diet therapy in patients with hypercholesterolemia. *World Heart Journal* 8: 239–249.

Figer, B., R. Pissurlenkar, P. Ambre, et al. 2017. Treatment of gastric ulcers with fenugreek seed extract; In vitro, in vivo and in silico approaches. *Indian Journal of Pharmaceutical Sciences* 79(5): 724–730.

Friedewald, W.T., R.I. Levy, and D.S. Fredrickson. 1972. Estimation of low density lipoprotein cholesterol concentration in plasma without use of the preparative ultracentrifuge. *Clinical Chemistry* 18: 499–502.

Fuller, S., and J.M. Stephens. 2015. Diosgenin, 4-hydroxyisoleucine, and fiber from fenugreek: Mechanisms of actions and potential effects on metabolic syndrome. *Advances in Nutrition* 6(2): 189–197.

Geberemeskel, G.A., Y.G. Debebe, and N.A. Nguse. 2019. Antidiabetic effect of Fenugreek seed powder solution (Trigonella foenum-graecum L.) on hyperlipidemia in diabetic patients. *Journal of Diabetes Research*, Article ID 8507453, 8 pages. https://doi.org/10.1155/2019/8507453.

Gong, J., K. Fang, H. Dong, et al. 2016. Effect of fenugreek on hyperglycaemia and hyperlipidemia in diabetes and pre-diabetes: A meta-analysis. *Journal of Ethnopharmacology* 194: 260–268.

Goyal, S., N. Gupta, and S. Chatterjee. 2016. Investigating therapeutic potential of Trigonella foenum-graecum L. as our defence mechanism against several human diseases. *Journal of Toxicology* 1250387. doi: 10.1155/2016/1250387.

Herbalgram 2014, Vol 103, Page 33, https://www.herbalgram.org/media/11928/issue103.pdf

Hooda, S., and S. Jood. 2004. Nutritional evaluation of wheat—fenugreek blends for product making. *Plant Foods for Human Nutrition* 59(4): 149–154.

Im, K.K., and B.P. Maliakel. 2008. Fenugreek dietary fibre a novel class of functional food ingredient. *Agro Food Industry Hi-Tech* 19(2): 18–21.

Jani, R., S. Udipi, and P. Ghugre. 2009. Mineral content of complementary foods. *The Indian Journal of Pediatrics* 76(1): 37–44.

Jiang, W., L. Si, P. Li, et al. 2018. Serum metabonomics study on antidiabetic effects of fenugreek flavonoids in streptozotocin-induced rats. *Journal of Chromatography B* 1092: 466–472.

Kamble, H., D. Amit.. K. Subhash et al. 2013. Effect of low molecular weight galactomannans from fenugreek seeds on animal models of diabetes mellitus. *Biomedicine & Aging Pathology* 3: 45–151.

Kaufman, J.M., and A. Vermeulen. 2005. The decline of androgen levels in elderly men and its clinical and therapeutic implications. *Endocrine Reviews* 26(6): 833–876.

Khorshidian, N., M-Y. Asli, M. Arab, A.M. Mortazavian, et al. 2016. Fenugreek: Potential applications as a functional food and nutraceutical. *Nutrition and Food Sciences Research* 3(1): 5–16.

Kim, K.A., W. Gu, I.A. Lee, E.H. Joh, and D.H. Kim. 2012. High fat diet-induced gut microbiota exacerbates inflammation and obesity in mice via the TLR4 signaling pathway. *PLoS One* 7(10): e47713. doi: 10.1371/journal.pone.0047713.

King, K., N.P. Lin, Y.H. Cheng, G.H. Chen, and R.J. Chein. 2015. Isolation of positive modulator of glucagon-like peptide-1 signaling from Trigonella foenum-graecum (Fenugreek) seed. *Journal of Biological Chemistry* 290(43): 26235–26248.

Kishino, S., M. Takeuchi, S-B. Park, et al. 2013. Polyunsaturated fatty acid saturation by gut lactic acid bacteria affecting host lipid composition. *Proceedings of the National Academy of Sciences of the United States of America* 110: 17808–17813.

Kriaa, A., M. Bourgin, A. Potironet, et al. 2019. Microbial impact on cholesterol and bile acid metabolism: Current status and future prospects. *Journal of Lipid Research* 60: 323–332.

Kulkarni, C., S. Bodhankar., A.E. Ghule et al. 2012. Antidiabetic activity of Trigonella foenumgraecum L. seeds extract (IND01) in neonatal streptozotocin-induced (n-STZ) rats. *Diabetologia Croatica* 41(1).

Lakshmanan, R., and A. Aggarwal. 2020. Invigorating Ayurveda. In: *National Investment Promotion and Facilitation Agency*.

Mabberley, D.J. 1997. *The Plant-Book: A Portable Dictionary of the Vascular Plants*. Cambridge, UK: Cambridge University Press.

Mandegary, A., M. Pournamdari, F. Sharififar, et al. 2012. Alkaloid and flavonoid rich fractions of fenugreek seeds (Trigonella foenum-graecum L.) with antinociceptive and anti-inflammatory effects. *Food and Chemical Toxicology* 50(7): 2503–2507. doi: 10.1016/j.fct.2012.04.020.

Mansoori, A., S. Hosseini, M. Zilaee, R. Hormoznejad, and M. Fathi. 2020. Effect of fenugreek extract supplement on testosterone levels in male: A meta-analysis of clinical trials. *Phytotherapy Research* 34(8). doi: 10.1002/ptr.6627.

Mathern, J.R., S.K. Raatz, W. Thomas, and J.L. Slavin. 2009. Effect of fenugreek fiber on satiety, blood glucose and insulin response and energy intake in obese subjects. *Phytotherapy Research* 23: 1543–1548.

Meghwal, M., and T.K. Goswami. 2012. A review on the functional properties, nutritional content, medicinal utilization and potential application of fenugreek. *Journal of Food Process Technology* 3: 9.

Mehrafarin, A., S.H. Rezazadeh, B.H. Naghdi, et al. 2011. A review on biology, cultivation and biotechnology of fenugreek (Trigonella foenum-graecum L.) as a valuable medicinal plant and multipurpose. *Journal of Medicinal Plants* 10: 6–24.

Moradi-Kor, N., and K. Moradi. 2013. Physiological and pharmaceutical effects of fenugreek (Trigonella foenum-graecum L.) as a multipurpose and valuable medicinal plant. *Global Journal of Medicinal Plants Research* 1: 199–206.

Muraki, E., Y. Hayashi, H. Chiba, N. Tsunoda, and K. Kasono. 2011. Dose-dependent effects, safety and tolerability of fenugreek in diet induced metabolic disorders in rats. *Lipids in Health and Disease* 10: 240.

Naidu, M.M., B.N. Shyamala, P.J. Naik, G. Sulochanamma, and P. Srinivas. 2010. Chemical composition and antioxidant activity of the husk and endosperm of fenugreek seeds. *Food Science and Technology* 44: 451–456.

Nathan, J., S. Panjwani., V. Mohan et al. 2014. Efficacy and safety of standardized extract of Trigonella foenum-graecum L seeds as an adjuvant to L-Dopa in the management of patients with Parkinson's disease. *Phytother Researh*. 28: 172–178.

Neelakantan, N., M. Narayanan, R.J. deSouza, and R.M. vanDam. 2014. Effect of fenugreek (Trigonella foenum-graecum L.) intake on glycemia: A meta-analysis of clinical trials. *Nutrition Journal* 13: 7. doi: 10.1186/1475-2891-13-7.

Pal, D.K., and S. Mukherjee. 2020. Fenugreek (Trigonella foenum) seeds in health and nutrition. In: *Nuts and Seeds in Health and Disease Prevention*, 2nd ed., Elsevier Inc., 161–170.

The Plant List. 2016. A working list of all plants, Version 1.1. www.theplantlist.org. Accessed June 18, 2021.

Platel, K., and K. Srinivasan. 2000. Influence of dietary spices and their active principles on pancreatic digestive enzymes in albino rats. *Nahrung* 44: 42–46.

Rao, A., and R. Kale. 1992. Diosgenin—a growth stimulator of mammary gland of ovariectomized mouse. *Indian Journal of Experimental Biology* 30(5): 367–370.

Rao, A., E. Steels., G. Beccaria et al. 2015. Influence of a specialized Trigonella foenum-graecum seed extract (Libifem), on testosterone, estradiol and sexual function in healthy menstruating women, a randomised placebo controlled study. *Phytotherapy Research* 29:1123–1130.

Rao, A., E. Steels, W.J. Inder et al. 2016. Testofen, a specialised Trigonella foenum-graecum seed extract reduces age-related symptoms of androgen decrease, increases testosterone levels and improves sexual function in healthy aging males in a double-blind randomised clinical study. *Aging Male* 19: 134–42.

Roberts, K.T. 2011. The potential of fenugreek (Trigonella foenum-graecum) as a functional food and nutraceutical and its effects on glycemia and lipidemia. *Journal of Medicinal Food* 14: 1485–1489.

Salgado, R.M., M.H. Marques-Silva, E. Goncalves, A.C. Mathias, J.G. Aguiar, and P. Wolff. 2017. Effect of oral administration of *Tribulus terrestris* extract on semen quality and body fat index of infertile men. *Andrologia* 49(5). doi: 10.1111/and.12655.

Shirani, G., and R. Ganesharanee. 2009. Extruded products with Fenugreek (Trigonella foenum-graecium) chickpea and rice: Physical properties, sensory acceptability and glycaemic index. *Journal of Food Engineering* 90(1): 44–52.

Shtriker, M.G., M. Hahn, E. Taiebet, et al. 2018. Fenugreek galactomannan and citrus pectin improve several parameters associated with glucose metabolism and modulate gut microbiota in mice. *Nutrition* 46: 134–142.e3. doi: 10.1016/j.nut.2017.07.012.

Small, E. 1987. A taxonomical study of the Medicagoid Trigonella leguminosae. *Canadian Journal Botany* 65: 1199–1211.

Snehlata, H.S., and D.R. Payal. 2012. Fenugreek (Trigonella foenumgraecum L.): An overview. *International Journal of Current Pharmaceutical Research* 2(4): 169–187.

Sowmya, P., and P. Rajyalakshmi. 1999. Hypocholesterolemic effect of germinated fenugreek seeds in human subjects. *Plant Foods for Human Nutrition* 53: 359–365.

Srinivasan, K. 2006. Fenugreek (Trigonella foenum-graecum): A review of health beneficial physiological effects. *Food Reviews International* 22(2): 203–224.

Srivastava, D., R.J. Naidu, et al. 2012. Effect of fenugreek seed husk on the rheology and quality characteristics of muffins. *Food Science & Nutrition* 3: 1473–1479.

Suksomboon, N., N. Poolsup, S. Boonkaew, and C.C. Suthisisang. 2011. Meta-analysis of the effect of herbal supplement on glycemic control in type 2 diabetes. *Journal of Ethnopharmacology* 137(3): 1328–1333.

Townsend, C.C., E. Guest, and A. Al-Rawi. 1974. Trigonella L. In: C.C. Townsend and E. Guest, editors. *Flora of Iraq*. Vol. 3. Baghdad, Iraq: Ministry of Agriculture and Agrarian Reform.

Wani, S.A., and P. Kumar. 2018. Fenugreek: A review on its nutraceutical properties and utilization in various food products. *Journal of the Saudi Society of Agricultural Sciences* 17: 97–106.

Wankhede, S., V. Mohan, and P. Thakurdesai. 2016. Beneficial effects of fenugreek glycoside supplementation in male subjects during resistance training: A randomized controlled pilot study. *Journal of Sport and Health Science* 5(2): 176–182.

WHO traditional medicine strategy: 2014–2023, 15 May 2013, ISBN: 978 92 4 150609 0, www.who.int/publications/i/item/9789241506096.

Wilborn, C., L. Taylor, C. Poole, C. Foster, D. Willoughby, and R. Kreider. 2010. Effects of a purported aromatase and 5 α-reductase inhibitor on hormone profiles in college-age men. *International Journal Sport Nutrition and Exercise Metabolism* 20(6): 457–465.

Yadav, R., R. Kaushik, and D. Gupta. 2011. The health benefits of Trigonella foenum-graecum: A review. *International Journal of Engineering Research and Applications* 1(1): 32–35.

Yao, D., B-Z. Zhang, J. Zhu, et al. 2020. Advances on application of Fenugreek seeds as functional foods: Pharmacology, clinical application, products, patents and market. *Critical Reviews in Food Science and Nutrition* 60(14): 2342–2352. doi: 10.1080/10408398.2019.1635567.

Zandi, P., S.K. Basu, L. Bazrkar-Khatibani, et al. 2015. Fenugreek (Trigonella foenum-graecum L.) seed: A review of physiological and biochemical properties and their genetic improvement. *Acta Physiologiae Plantarum* 37: 1714. https://doi.org/10.1007/s11738-014-1714-6.

8 Fenugreek (*Trigonella foenum-graecum*)
A Unique Plant with Potential Health Benefits in Sports Nutrition

Anand Swaroop

CONTENTS

8.1 Introduction 103
8.2 Ethnobotany: The Origin of the Name 103
8.3 Ayurvedic Therapeutics 104
8.4 Chemical Constituents 105
8.5 Diverse Health and Sports Nutrition Benefits 105
8.6 Novel Fenugreek Seed Extracts 107
8.7 Fenfuro® and Enhanced Insulin Sensitivity 107
8.8 Furocyst® and Polycystic Ovary Syndrome (PCOS) 108
8.9 Furosap®, a Novel Fenugreek Seed Extract enriched in 20% Protodioscin, and Sports Nutrition 108
8.10 Summary and Conclusions 109
References 110

8.1 INTRODUCTION

Fenugreek (*Trigonella foenum-graecum*, family Fabaceae) is an ancient Indian medicinal plant with a myriad of health benefits [1–9]. This novel plant is native to central and southeast Asia, predominantly in India [1–5]. Several other countries including Iran, Egypt, Turkey, China, France, Spain, Morocco, and Argentina are now cultivating this popular leguminous annual plant, however, the largest producer is India [1–11]. In India, Fenugreek seeds are termed as "*methi*" seeds in Hindi [1].

8.2 ETHNOBOTANY: THE ORIGIN OF THE NAME

In Latin little triangle is termed as "Trigonella", thus, triangular-shaped flowers are the origin of the Genus "Trigonella", while the Species term "foenum-graecum" originated from the historical perspective of the Romans [1–5]. This terminology is acquired from Greek-hay, illustrating this as the common crop used as fodder for animals in Greece [1–3].

Typically, fenugreek plants attain a height of 1–2 feet and bear green trifoliate leaves (Figure 8.1) and its flowers are white to yellow in color, while the plant carries thin pods (approximately 15 cm in length) and an average pod contains 10–20 seeds [1–5]. Fenugreek seeds are golden yellow in color and their average height, width, and thickness are 4.01–4.19 mm, 2.35–2.60 mm, and 2.40–2.66 mm, respectively [1–3]. The dried fenugreek seeds are grounded to obtain fenugreek seed powder which is used as a condiment. Fenugreek gum is obtained from the endosperm of the seeds [4, 5].

FIGURE 8.1 Fenugreek plant and seeds.

8.3 AYURVEDIC THERAPEUTICS

In Ayurveda, fenugreek has been demonstrated for ameliorating *Kapha*, and in low quantities for *Vata*; however, fenugreek boosts *Pitta* [1, 2]. Ayurvedic therapeutics mainly focuses on maintaining ideal health besides the treatment of a disease and with the belief that a human body is composed of five elements: ether, water, air, fire, and earth. Ayurvedic therapeutics greatly depends on these three doshas, namely, (1) *Kapha*, constituted of air and water; (2) *Vata*, dependent on the elements of fire and ether; while (3) *Pitta* is composed of water and fire. Ayurvedic therapeutics largely depends on these three doshas for maintaining ideal health and curing disease [1, 2, 4].

Ayurvedic practitioners greatly believe that fenugreek seeds and the plant exhibit versatile health benefits including (1) glucose homeostasis by balancing insulin and glucagon to maintain a healthy blood glucose; (2) attenuating digestive and stomach disorders; (3) boosting energy, endurance,

Fenugreek (*Trigonella foenum-graecum*)

and vitality; (4) supporting dermal health; (5) assisting in women's health, reproductive health, and fertility; and (6) providing nephroprotective benefits [1, 9–14]. In fact, both Ayurveda and Unani therapeutics have been using fenugreek for versatile anti-inflammatory, anti-diabetic, antiseptic, aphrodisiac, and women's health benefits for several thousands of years. Fresh and dried leaves, twigs, roots, sprouts, microgreens, and the cuboid-shaped, yellow- to amber-colored seeds of fenugreek are extensively used in diverse culinary dishes, salads, pickles, brewed in tea, lentil soups, and baked into bread, while the roasted seeds are employed as a flavor-enhancer in the Indian subcontinent. Later, researchers emphasized that soluble dietary fiber "galactomannan" in the endosperm of fenugreek seeds, potentiates the nutritional and physicochemical benefits of the bread [1, 7–11]. In fact, Fenugreek seeds have been extensively studied by the researchers around the globe. Fenugreek leaves and seeds have long been demonstrated in Ayurvedic and Chinese medicine for their extensive use in muscle building, sports nutrition, and wrestling [1, 11–17].

8.4 CHEMICAL CONSTITUENTS

Fenugreek plants, especially fenugreek seeds, are rich in soluble dietary fibers, furostanolic saponins, 28% mucilage, 5% stronger-smelling and bitter oil, phosphates, lecithin, 4-hydroxyisoleucine, lysine, niacin, and numerous amino acids, L-tryptophan, nucleoalbumin, readily absorbable iron in an organic form, trigonelline, trimethylamine, choline, biotin, inositol, vitamins A, B1, B2, B3, B5, B6, B9, B12, C, and D, diosgenin, diosgenin-β–D-glucoside, protein, potassium, neurin, betaine, vitexin, vitexin-7-glucoside, yamogenin, vicenin, saponaretin, tigogenin, neotigogenin, and isoorientin. It is important to emphasize that fenugreek seeds don't contain an essential oil, while its characteristic flavor is due to trace amounts of a powerful odorant 4,5-dimethyl-3-hydroxy-2[5H]-furanone (termed as fenugreek lactone) [1–4]. These novel chemical constituents in fenugreek seeds exhibit an array of beneficial effects including cholesterol absorption in the intestine as well as inhibition of hepatic tissue-induced production of cholesterol, lower triglyceride, and LDL levels [1–4]. It is worthwhile to emphasize that fenugreek is rich in furostanolic steroidal saponin or the spirostanol saponin class. Especially the broad-spectrum safety, pharmacokinetic profile, and efficacy.

The efficacy, pharmacokinetic profile, and broad-spectrum safety of furostanolic saponins exhibit potent cardiometabolic endurance, sports performance, and anabolic potential in boosting testosterone [1, 5–8, 11–19].

8.5 DIVERSE HEALTH AND SPORTS NUTRITION BENEFITS

Research studies exhibited that Fenugreek has diverse health benefits. As indicated earlier, fenugreek is a unique medicinal plant, and the therapeutic benefits are categorically emphasized in ancient medical literatures [1, 11–20]. In the traditional Indian medical system, Ayurveda has employed fenugreek in diverse disease maladies as well as for enhancing vitality, vigor, and endurance, while ancient Egyptians used it to incense and embalm mummies and as a lactation aid [1, 11–20]. In ancient Chinese therapeutics, fenugreek was used to treat edema in the legs. Several folkloric uses of fenugreek, including the treatment of lung and sinus congestion, indigestion, baldness in men and use as a hair tonic and conditioner and as a galactagogue, have been demonstrated [2–4, 16, 20, 21].

A great body of evidence exhibits that fenugreek has a broad spectrum of health benefits: it is an antioxidant, improves cardiovascular functions, is anti-inflammatory, regulates glucose-insulin, improves glucose tolerance and insulin sensitivity; it exhibits potent therapeutic benefits against hypercholesterolemia [3–8], polycystic ovary syndrome, gastric ulcer and hyperthyroidism, and offers significant benefits in boosting endurance, serum testosterone level, vitality, vigor, and sports performance [14–17]. Diverse beneficial medicinal properties of fenugreek have been repeatedly demonstrated, including free radical scavenging and antioxidant benefits [1, 2]; anti-inflammatory efficacy [4, 5], anti-diabetic efficacy [18, 22]; uses in sports nutrition; anti-obesity [2–4], anti-cancer [2,4], hepatoprotective [4], and antihyperlipidemic [8] properties; and it is supportive of women's health [23] and sexual function [14–16]. The presence of diverse phytopharmaceuticals in fenugreek and their diverse pharmacological and therapeutic benefits are attributed to numerous health benefits, disease prevention, and sports performance [1].

It is important to emphasize that most of the testosterone is metabolized to androsterone, which ultimately conjugates with glucuronic acid and is secreted as a 17-keto-steroid metabolite [14]. Testosterone in most target cells is enzymatically transformed by the microsomal enzyme 5-alpha-reductase into dihydrotestosterone (DHT) and, in turn, binds to the intracytoplasmic receptor protein to form a DHT-protein complex and then is transported into the nucleus [15–17]. This protein complex undergoes a conformational transformation, which is involved in chromatin binding. This transformation synthesizes mRNA and cytoplasmic proteins, which induces cellular growth and sequential effects mediated by androgens [14–16]. As it is well documented, testosterone and anabolic androgenic steroids enhance athletic performance not only through its long-term anabolic effects but also through enhanced effects on behavior [16, 17]. A positive testosterone booster can enhance athletic performance.

Research studies have exhibited that protodioscin induces a dose-dependent relaxation effect on the cavernous smooth muscle by attenuating the NO/NOS pathway in the corpus cavernosum endothelium. In a male reproductive organ, NO is constitutively produced from the autonomic nerve terminals and endothelial cells in corporal tissue. NO is disseminated locally into the adjacent smooth muscle cells and, in turn, attaches with the intracellular guanylate cyclase, which functions as a physiological "receptor" [1, 14–16]. This causes a conformational change in guanylate cyclase, stimulating the enzyme which initiates the conversion of guanosine triphosphate to cGMP, which in turn mediates via a cGMP-dependent protein kinase to optimize the contractile state of the corporal smooth muscle. Administration of protodioscin-enriched standardized *Tribulus terrestris* extract has been exhibited to enhance lean body mass and fat free mass. Hwang et al. [24] exhibited that L-Citrulline, an amino acid which can be readily converted to L-Arginine, in situ transformed into nitric oxide, which promotes lean body mass.

Tomcik et al. [25] conducted an in vitro study and exhibited that fenugreek seeds in conjunction with insulin potently produced creatine, which is independent of the activity of sodium- and chloride-dependent creatine transporter, SLC6A8. The efficacy of fenugreek seed extract (10 mg/kg s.c. bi-weekly or 10 and 35 mg/kg body weight p.o.) was investigated on immature castrated male Wistar rats by Aswar et al. [26], and some anabolic effect was exhibited in these rats without androgenic activity. Effect of fenugreek seed extract (500 mg/day) was assessed in a double-blind placebo-controlled study in 49 resistance-trained male subjects, which demonstrated a pronounced effect on both upper- and lower-body strength and body composition.

Three independent clinical investigations [27–29] on a proprietary fenugreek extract were conducted by Rao et al. [27, 28] and Steels et al. [29] in both male and female volunteers, which demonstrated enhanced serum testosterone levels. This randomized, double blind and placebo-controlled investigation was conducted by Steels et al. on this standardized fenugreek seed extract in 60 healthy males (age: 25–52 years; dose: 600 mg/day) over a period of 6 consecutive weeks. This study exhibited a positive effect on enhanced libido, muscle strength, and energy. The second independent clinical investigation was conducted in 80 healthy menstruating women (age: 20–49 years) who complained of compromised low sexual drive. Supplementation of fenugreek seed extract (600 mg/day) to these women boosted sexual arousal and desire. The third independent placebo-controlled, randomized, and double-blind investigation was conducted over a period of 12 weeks in 120 men (age: 43–70 years; dose: 600 mg/day). Following 12 weeks of supplementation, both free and total testosterone levels, as well as sexual function, increased significantly [29].

Ikeuchi et al. (2006) conducted a dose-dependent efficacy study on fenugreek seed extract (0, 150, or 300 mg/kg) in male mice over a period of 4 weeks on endurance capacity in a swimming model [15]. At 300 mg/kg body weight dose, the swimming endurance was significantly increased. Researchers emphasized that the utilization of fatty acids was also significantly enhanced as an energy source at this dose. Another swimming exercise investigation by Arshadi et al. (2015)

assessed the efficacy of fenugreek seed extract (0, 0.8 or 1.6 g/kg) and compared to glibenclamide in type 2 diabetic male rats [16, 17]. The researchers concluded that fenugreek, in conjunction with swimming exercise, induced a therapeutic benefit on the improvement of diverse anti-diabetic indices including blood insulin level, plasma leptin, HOMA-IR, and adiponectin [16, 17].

Poole et al. (2010) conducted a placebo-controlled, double-blind investigation over a period of 8 weeks in 49 resistance-trained male volunteers to assess the efficacy of fenugreek (500 mg/day) on body strength, body composition, muscle endurance, power output, and hormonal profiles in a structured resistance training program [30]. Fenugreek potently enhanced upper- and lower-body strength, reduced body fat, and enhanced overall body composition, as well as non-significantly impacted muscular endurance, hormonal concentrations, and hematological variables. Broad spectrum safety was also confirmed by these clinical researchers [30].

Fenugreek glycosides have exhibited potent androgenic and anabolic effects in males [31]. Randomized clinical studies from four trials exhibited the benefits of fenugreek extract in boosting serum total testosterone in males [31].

Another study ascertained the benefits of fenugreek compared to glibenclamide on regulating blood glucose in 12 patients with uncontrolled type II diabetes on metformin treatment. Six subjects took 2 g fenugreek/day, while the remaining six subjects received glibenclamide (5 mg/day) and the glycemic control and lipid profile were assessed at the beginning and completion of 12 weeks of treatment [32]. Fenugreek supplementation demonstrated an insignificant reduction in fasting blood glucose, while the fasting insulin level increased significantly. HDL:LDL ratio also significantly reduced during the treatment. No adverse effects were observed. The researchers concluded that fenugreek could be employed in conjunction to anti-diabetic drugs to regulate blood glucose [32].

8.6 NOVEL FENUGREEK SEED EXTRACTS

Our research team developed a strategic US patented (Patent# 8,754,205B2 17 June 2014; #8,217,165B2 10 July 2012) water-ethanol extraction and fractionation technique of standardized fenugreek seeds in a cGMP manufacturing facility to obtain three distinct fractions (1) Fenfuro®, enriched in 40% furostanolic saponins, (2) Furocyst® complemented with 70% saponins, and (3) Furosap® containing 20% protodioscin. These standardized ingredients are >95% water soluble and contain <5% moisture content; both residual solvent and pesticide content comply with USP38 requirements and exhibits a shelf-life of 2+ years.

8.7 FENFURO® AND ENHANCED INSULIN SENSITIVITY

A scientific team in our laboratories conducted a multicenter, randomized, placebo-controlled, double-blind, add-on 90-day clinical investigation to determine the efficacy of Fenfuro® (daily dosage: 500 mg b.i.d.) in 154 male and female subjects (age: 25–60 years) suffering from type 2 diabetes. All subjects were on standard metformin therapy. Subjects took a dietician-affirmed vegetarian or non-vegetarian diet of approximately 2,000 kcal/day (protein 18–22%, carbohydrate 52–56%, and fat 22–26%). The clinical team meticulously assessed body weight, blood pressure, and pulse rate, fasting and post-prandial plasma sugar (mg/dL), glycosylated hemoglobin (HbA1c), and fasting and post-prandial C-peptide levels. The Fenfuro® group exhibited significant reduction in both fasting plasma and post-prandial blood sugar levels. In the Fenfuro®-treated group, 83% subjects reported a decrease in fasting plasma sugar levels as compared to 62% in the placebo group [20]. Simultaneously, approximately 89% subjects exhibited a decrease in post-prandial plasma sugar levels in the Fenfuro® group as compared to 72% in the placebo. Significant reduction in glycosylated hemoglobin (HbA1c) levels was observed in both placebo and treatment groups. In both fasting and post-prandial C-peptide levels, a significant increase was observed, while no significant changes were observed in fasting and post-prandial C-peptide levels between the placebo- and Fenfuro®- groups [20]. In addition, it was interesting to note that 48.8% of the subjects exhibited reduced dosage of metformin therapy in the Fenfuro®-treated group, whereas 18.05% reported reduced dosage of anti-diabetic therapy in the placebo group. No adverse effects were observed. Thus, Fenfuro® demonstrated broad spectrum safety and efficacy in diabetic subjects [20].

It is important to emphasize that diabetics have a strong correlation with testosterone deficiency and other intricate aspects of compromised metabolic syndrome [33]; therefore, anti-diabetic efficacy of fenugreek may play a pioneering role in boosting serum testosterone level and enhanced sports performance.

8.8 FUROCYST® AND POLYCYSTIC OVARY SYNDROME (PCOS)

The Chereso and Cepham research teams conducted an open-label, one-arm, non-randomized study in 50 premenopausal women (age = 18–45 years, BMI <42) suffering from PCOS. It is important to emphasize that PCOS, the most prevalent hormonal disorder among women of reproductive age, leads to irregular menstrual cycles, excessive body or facial hair, miscarriage, and infertility. All recruited subjects took Furocyst® (2 x 500 mg capsules/day) over a period of 90 consecutive days. This study focused on determining two important aspects: (1) reduction of ovarian volume, and (2) the number of ovarian cysts. Institutional review board (IRB) approvals were obtained [21, 22].

Furocyst® intake significantly reduced the ovary volume. Interestingly, 46% of subjects exhibited a significant reduction in cyst size, and 36% demonstrated complete dissolution of cysts. Furocyst® also exhibited a significant decrease in both ovarian volume and the number of ovarian cysts. Following completion of the treatment, approximately 71% of the subjects reported the return of a regular menstrual cycle, while 12% subjects subsequently became pregnant. A total of 94% participants benefited from the supplementation of Furocyst®. Luteinizing hormone (LH) and follicular stimulating hormone (FSH) levels significantly increased. No adverse events were reported [21, 22].

Moreover, McCartney and Marshall (2016) indicated that PCOS has an intricate connection with low testosterone level and compromised glucose-insulin sensitivity [34]. Thus, protective ability of fenugreek against PCOS may play a vital role in boosting serum testosterone level.

8.9 FUROSAP®, A NOVEL FENUGREEK SEED EXTRACT ENRICHED IN 20% PROTODIOSCIN, AND SPORTS NUTRITION

Our scientific team assessed the efficacy of a novel, patented fenugreek seed extract enriched in 20% protodioscin (Furosap®, US Patents# US 8,217,165 B2; US 8,754,205 B2) to boost free and total testosterone levels, exercise tolerance and allied cardiometabolic parameters, sperm profile and morphology, sexual health, mood and mental alertness, and broad-spectrum safety in human volunteers [19, 23].

The first one-arm, open-labelled, multicenter study assessed the broad-spectrum safety and efficacy of Furosap® (500 mg/day) over a period of 12 weeks in 50 male volunteers (age: 35–65 years). This study was registered at the clinicaltrials.gov (NCT02702882), and Institutional Review Board (IRB) approval was obtained [23]. Multiple relevant parameters including testosterone levels (both free and total testosterone), cardiovascular health, sperm profile and morphology, mood, mental alertness, libido, reproductive function, and broad-spectrum safety were assessed. Interestingly, 90% of the enrolled subjects exhibited an increase of 46% in free testosterone level, while 85.4% demonstrated an improvement in sperm counts [23]. Improvement in sperm morphology was reported in 14.6% study participants. Remarkable improvements in mental alertness and mood were observed, while significant improvement was observed in cardiovascular health and libido. Blood chemistry analysis exhibited broad spectrum safety and no adverse events were reported. Taken together, Furosap® significantly improved cardiovascular health, free testosterone level, sperm profile, mental acuity, and overall performance in the study participants [23] (Table 8.1).

TABLE 8.1
Furosap®-Induced Changes in Free and Total Testosterone Levels at the Beginning and Completion of 12 Weeks of Treatment

Groups	Time Point	Free Testosterone (pg/dL)	Total Testosterone (ng/dL)
Placebo	0 Day	8.17 ± 5.04	405.9 ± 156.95
Furosap®-Treated	12 Weeks post-treatment	11.97 ± 5.65* (*$p < 0.05$)	436.34 ± 189.94

TABLE 8.2
Furosap®-Induced Changes in Serum Testosterone Level, Lean Body Mass, and Fat Free Mass at the Initiation and Completion of 12 Weeks of Treatment

Groups	Time Point	Serum Testosterone (ng/dL)	Lean Body Mass (g)	Fat Free Mass (g)
Placebo	0 Day	608 ± 50	59604 ± 3107	62829 ± 1650
	12 Weeks post-treatment	545.6 ± 59	59858 ± 1573	63161 ± 3246
Furosap®-Treated	0 Day	596 ± 45	59058 ± 1521	62297 ± 1599
	12 Weeks post-treatment	669.6 ± 54*	60278 ± 3141*	63646 ± 3279*

Effect of Furosap® on the placebo and treatment groups at the beginning and completion of 12 weeks of treatment. *$p < 0.05$

Another randomized, double-blind, placebo-controlled, clinical investigation was conducted in 40 healthy male athletes [19]. Subjects took either a placebo or Furosap® capsules (250 mg/day b.i.d.) over a period of 12 consecutive weeks. Body fat mass, lean mass, fat mass, fat-free mass, grip strength, upper and lower body strength, maximal graded exercise stress using a digital hand dynamometer, dual-energy X-ray absorptiometry (DEXA), force plate, and treadmill with open-circuit spirometry were assessed at the baseline and at the end of 12 weeks of treatment [19]. Total testosterone level and C-reactive proteins (CRP) were assessed in the serum at baseline and at the end of 12 weeks of treatment. Furosap® supplementation significantly increased mean lean body mass, fat-free mass, and elevated serum testosterone levels (Table 8.2). Furosap® demonstrated a tendency towards reducing blood pressure during exhaustion [19]. Overall, Furosap® significantly enhanced exercise endurance and sports performance. No adverse effects were observed.

Effect of 20% protodioscin-enriched Furosap® on the placebo and treatment groups at the beginning and completion of 12 weeks of treatment. *$p < 0.05$

8.10 SUMMARY AND CONCLUSIONS

Taken together, fenugreek plants and seeds exhibit significant promise for ameliorating and treating numerous diseases and dysfunctions; however, additional studies are essential to ascertain the versatile potential of fenugreek-derived standardized extracts as novel and efficacious nutraceutical supplements, medical foods, botanical drugs, and OTC (over the counter) drugs [1]. The studies presented and analyzed in this review highly suggest that fenugreek is a unique medicinal plant with versatile health benefits including enhanced endurance and sports performance. A great body of evidence demonstrates the efficacy in enhancing endurance, vigor, vitality, sports nutrition, insulin sensitivity, and diverse health benefits. In two independent clinical studies [19, 23], our research team affirmed that supplementation (500 mg/day) of standardized, dietary fiber rich, patented 20% protodioscin-enriched *Trigonella foenum-graecum* extract (Furosap®) over a period of 12 consecutive weeks significantly boosted serum testosterone, lean body mass, fat free mass, vitality, vigor, and endurance without causing any adverse events. Moreover, sperm count and sperm motility increased significantly at 4, 8, and 12 weeks of Furosap® treatment, while abnormal sperm morphology significantly decreased at both 8 and 12 weeks of treatment. Significant alleviation in mental alertness, mood alleviation, reflex erection, cardiovascular health, and overall performance was also observed [19, 23]. Multiple safety evaluation and toxicological studies in conjunction with extensive blood chemistry and lipid profile data affirmed the broad-spectrum safety of standardized fenugreek seed extract [19, 23]. Furthermore, a convincing link exists between diabetes and low testosterone level [33], and thus, potent anti-diabetic efficacy [33] of fenugreek seed extract as well as its efficacy against diabetes, which in turn boosts serum testosterone level, may provide encouraging evidence in boosting endurance and sports performance. Also, women with PCOS exhibit compromised testosterone level [34]; again our investigation exhibited that fenugreek has potential protective abilities against PCOS [21, 22]. Moreover, our studies [19, 23] as well as several independent research studies [14–19, 26–33, 35, 36] further strengthen our findings about the potential use of standardized fenugreek seed extract in enhancing sports performance and endurance. Future investigations are in progress to exhibit its extensive application in muscle building, sports nutrition, and endurance exercise, as well as to unveil the diverse molecular mechanism of action.

REFERENCES

[1] Venkata, K.C., Swaroop, A., Bagchi, D., Bishayee, A. 2017. A small plant with big benefits: Fenugreek (Trigonella foenum-graecum Linn.) for disease prevention and health promotion. *Mol. Nutr. Food Res.* Mar 7. doi: 10.1002/mnfr.201600950.

[2] Basch, E., Ulbricht, C., Kuo, G., Szapary, P., Smith, M. 2003. Therapeutic applications of fenugreek. *Altern Med Rev.* 8(1): 20–7.

[3] Bahmani, M., Shirzad, H., Mirhosseini, M., Mesripour, A., Refieian-Kopaei, M. 2016. A review of ethnobotanical and therapeutic uses of fenugreek (Trigonella foenum-graecum L.). *J. Evid. Based Complementary Altern. Med.* 21(1): 53–62.

[4] Yadav, U.C., Baquer, N.Z. 2014. Pharmacological effects of Trigonella foenum-graecum L. in health and disease. *Pharm. Biol.* 2(2): 243–54.

[5] Roberts, K.T. 2011. The potential of fenugreek (Trigonella foenum-graecum) as a functional food and nutraceutical and its effects on glycemia and lipidemia. *J Med Food.* 14(12): 1485–9.

[6] Smith, J.D., Clinard, V.B. 2014. Natural products for the management of type 2 diabetes mellitus and comorbid conditions. *J Am Pharm Assoc.* 54(5): e304–e18.

[7] Baquer, N.Z., Kumar, P., Taha, A., Kale, R.K., Cowsik, S.M., McLean, P. 2011. Metabolic and molecular action of Trigonella foenum-graecum (fenugreek) and trace metals in experimental diabetic tissues. *J. Biosci.* 36(2): 383–96.

[8] Haber, S.L., Keonavong, J. 2013. Fenugreek use in patients with diabetes mellitus. *Am. J. Health Syst Pharm.* 70(14): 1196, 1198, 1200, 1202–3.

[9] Swaroop, A., Bagchi, M., Kumar, P., Preuss, H.G., Tiwari, K., Marone, P.A., Bagchi, D. 2014. Safety, efficacy and toxicological evaluation of a novel, patented anti-diabetic extract of Trigonella Foenum-Graecum Seed Extract (Fenfuro®). *Toxicol. Mech. Meth.* 24(7): 495–503. doi: 10.3109/15376516.2014.943443.

[10] Hua, Y., Ren, S., Guo, R., Rogers, O., Bagchi, D., Swaroop, A., Nair, N. 2015. Trigonella Foenum-Graecum seed extract (Fenfuro®) inhibits diet-induced insulin resistance and hepatic fat accumulation. *Mol. Nutr. Food Res.* 59(10): 2094–100.

[11] Fuller, S., Stephens, J.M. 2015. Diosgenin, 4-hydroxyisoleucine, and fiber from fenugreek: Mechanisms of actions and potential effects on metabolic syndrome. *Adv. Nutr.* 6(2): 189–97.

[12] Hibasami, H., Moteki, H., Ishikawa, K., Katsuzaki, H., Imai, K., Yoshioka, K., Ishii, Y., Komiya, T. 2003. Protodioscin isolated from fenugreek (Trigonella foenum-graecum L.) induces cell death and morphological change indicative of apoptosis in leukemic cell line HL-60, but not in gastric cancer cell line Kato III. *Int. J. Mol. Med.* 11(1): 23–6.

[13] Krishnaswamy, K. Traditional Indian spices and their health significance. 2008. *Asia. Pac. J. Clin. Nutr.* 17(Suppl 1): 265–68.

[14] Gupta, R.K., Jain, D.C., Thakur, R.S. 1986. Minor steroidal sapogenins from fenugreek seeds, Trigonella foenum-graecum *J. Nat. Prod.* 49(6): 1153–3.

[15] Ikeuchi, M., Yamaguchi, K., Koyama, T., Sono, Y., Yazawa, K. 2006. Effects of fenugreek seeds (Trigonella foenum-graecum) extract on endurance capacity in mice. *J. Nutr. Sci. Vitaminol. (Tokyo).* 52(4): 287–92.
[16] Arshadi, S., Bakhtiyari, S., Haghani, K., Valizadeh, A. 2015a. Effects of fenugreek seed extract and swimming endurance training on plasma glucose and cardiac antioxidant enzymes activity in streptozotocin-induced diabetic rats. *Osong. Public Health Res. Perspect.* 6(2): 87–93.
[17] Arshadi, S., Azarbayjani, M.A., Hajaghaalipor, F., Yusof, A., Peeri, M., Bakhtiyari, S., Stannard, R.S., Osman, N.A., Dehghan, F. 2015b. Evaluation of Trigonella foenum-graecum extract in combination with swimming exercise compared to glibenclamide consumption on type 2 diabetic rodents. *Food Nutr. Res.* 59: 29717. doi: 10.3402/fnr.v59.29717. eCollection2015.
[18] Swaroop, A., Maheshwari, A., Verma, N., Tiwari, K., Bagchi, M., Preuss, H.G., Bagchi, D. 2017. A novel protodioscin-enriched fenugreek seed extract (Trigonella foenum-graecum, family Fabaceae) improves free testosterone level and sperm profile in healthy volunteers. *Functional Foods in Health and Disease* 7(4): 235–45.
[19] Guo, R., Wang, Q., Nair, R.P., Barnes, S.L., Smith, D.T., Dai, B., Robinson, T.J., Nair, S. 2018. Furosap®, a novel fenugreek seed extract improves lean body mass and serum testosterone in a randomized, placebo-controlled, double-blind clinical investigation. *Functional Foods in Health and Disease.* 8(11): 508–19.
[20] Verma, N., Usman, K., Patel, N., Jain, A., Dhakre, S., Swaroop, A., Bagchi, M., Preuss, H.G., Bagchi, D. 2016. A multi-center clinical study to determine the efficacy of a novel Fenugreek seed (Trigonella foenum-graecum) extract (Fenfuro®) in patients with type-2 diabetes. *Food Nutr. Res.* Oct 11;60:32382. doi: 10.3402/fnr.v60.32382.
[21] Swaroop, A., Jaipuriar, A.S., Gupta, S.K., Bagchi, M., Kumar, P., Preuss, H.G., Bagchi, D. 2015. Efficacy of a novel Fenugreek seed extract (*Trigonella foenum-graecum*, Furocyst®) in polycystic ovary syndrome (PCOS). *Int. J. Med. Sci.* 12: 825–31.
[22] Swaroop, A., Jaipuriar, A.S., Gupta, Kumar, P., Bagchi, D. 2016. Efficacy of a novel fenugreek seed extract (Furocyst®) in polycystic ovary syndrome (PCOS). *Planta Med.* 81: S1–S381.
[23] Maheshwari, A., Verma, N., Swaroop, A., Bagchi, M., Preuss, H.G., Bagchi, D. 2017. Efficacy of efficacy of Furosap®, a novel Trigonella foenum-graecum seed extract, in enhancing testosterone level and improving sperm profile in male volunteers. *Int. J. Med. Sci.* 14: 58–66.
[24] Tomcik, K.A., Smiles, W.J., Camera, D.M., Hügel, H.M., Hawley, J.A., Watts, R. 2017. Fenugreek increases insulin-stimulated creatine content in L6C11 muscle myotubes. *Eur. J. Nutr.* 56(3): 973–9. doi: 10.1007/s00394-015-1145-1.
Eight weeks of resistance training in conjunction with glutathione and L-Citrulline supplementation increases lean mass and has no adverse effects on blood clinical safety markers in resistance-trained males.
[25] Hwang, P., Morales Marroquín, F.E., Gann, J., Andre, T., McKinley-Barnard, S., Kim, C., Morita, M., Willoughby, D.S. 2018. *J. Int. Soc. Sports Nutr* Jun 27;15(1):30. doi: 10.1186/s12970-018-0235-x.
[26] Aswar, U., Bodhankar, S.L., Mohan, V., Thakurdesai, P.A. 2010. Effect of furostanol glycosides from Trigonella foenum-graecum on the reproductive system of male albino rats. *Phytother. Res.* 24(10): 1482–8. doi: 10.1002/ptr.3129.
[27] Rao, A., Steels, E., Inder, W.J., Abraham, S., Vietta, L. 2016. Testofen, a specialized Trigonella foenum-graecum seed extract reduces age-related symptoms of androgen decrease, increases testosterone levels and improves sexual function in healthy aging males in a double-blind randomized clinical study. *Aging Male.* 19(2): 134–42.
[28] Rao, A., Steels, E., Beccaria, G., Inder, W.J., Vietta, L. 2015. Influence of a specialized Trigonella foenum-graecum seed extract (libifem), on testosterone, estradiol and sexual function in healthy menstruating women, a randomised placebo-controlled study. *Phytother Res.* 29(8): 1123–30.
[29] Steels, E., Rao, A., Vietta, L. 2011. Physiological aspects of male libido enhanced by standardized Trigonella foenum-graecum extract and mineral formulation. *Phytother. Res.* 25(9): 1294–300.
[30] Poole, C., Bushey, B., Foster, C., Campbell, B., Willoughby, D., Kreider, R., Taylor, L., Wilborn, C. 2010. The effects of a commercially available botanical supplement on strength, body composition, power output, and hormonal profiles in resistance-trained males. *J. Int. Soc. Sports Nutr.* Oct 27;7: 34. doi: 10.1186/1550-2783-7-34.
[31] Mansoori, A., Hosseini, S., Zilaee, M., Hormoznejad, R., Fathi, M. 2020. Effect of fenugreek extract supplement on testosterone levels in male: A meta-analysis of clinical trials. *Phytother. Res.* 34(7): 1550–5. doi: 10.1002/ptr.6627.

[32] Najdi, R.A., Hagras, M.M., Kamel, F.O., Magadmi, R.M. A randomized controlled clinical trial evaluating the effect of Trigonella foenum-graecum (fenugreek) versus glibenclamide in patients with diabetes. *Afr Health Sci.* 19(1): 1594–601. doi: 10.4314/ahs.v19i1.34.

[33] Fernández-Miró, M., Chillarón, J.J., Pedro-Botet, J. 2016. Testosterone deficiency, metabolic syndrome and diabetes mellitus. *Med. Clin. (Barc).* 146(2): 69–73. doi: 10.1016/j.medcli.2015.06.020.

[34] McCartney, C.R., Marshall, J.C. 2016. Polycystic ovary syndrome. *N Engl J Med.* 375(1): 54–64. doi: 10.1056/NEJMcp1514916.

[35] Hosseini, S.A., Hamzavi, K., Safarzadeh, H., Salehi, O. 2020. Interactive effect of swimming training and fenugreek (Trigonella foenum-graecum L.) extract on glycemic indices and lipid profile in diabetic rats. *Arch. Physiol. Biochem.* Oct 5:1–5. doi: 10.1080/13813455.2020.1826529.

[36] Wankhede, S., Mohan, V., Thakurdesai, P. 2016. Beneficial effects of fenugreek glycoside supplementation in male subjects during resistance training: A randomized controlled pilot study. *J. Sport Health Sci.* Jun;5(2):176–82. doi: 10.1016/j.jshs.2014.09.005.

9 Applications of Fenugreek in Sports Nutrition

Colin Wilborn and Aditya Bhaskaran

CONTENTS

9.1   Introduction 113
9.2   Fenugreek for Body Composition 113
9.3   Fenugreek for Strength and Endurance 115
9.4   Fenugreek for Testosterone 116
9.5   Fenugreek for Post-Exercise Muscle Recovery 118
9.6   Fenugreek for Creatine Delivery 119
9.7   Conclusion 120
References 121

9.1 INTRODUCTION

Fenugreek (*Trigonella foenum-graecum* L.) is a leguminous annual plant, used worldwide as a food. In some geographies it is also used as a traditional medicine for nutritional and health benefits. In recent years, the extracts or powder of fenugreek leaves and seeds have been explored for preventive and therapeutic applications. The newest fenugreek application is an ingredient for sports and exercise physiology applications such as testosterone boosters, improving body composition, strength, endurance, muscle recovery and creatine utilization.

9.2 FENUGREEK FOR BODY COMPOSITION

All performance outcomes depend on body composition to some extent. A controlled resistance training program is known to influence body composition across multiple populations (American College of Sports Medicine 2009; Kraemer et al. 2002; Steele et al. 2017). At the same time, scientific evidence on fenugreek extract supplementation suggests the beneficial effects during resistance training programs (Aswar et al. 2008; Poole et al. 2010; Wilborn et al. 2010; Woodgate and Conquer 2003). In one of these studies (500 mg, once a day, 8 weeks), the group of resistant-trained men supplemented with fenugreek seed extract experienced a significant reduction in body fat percentage and a significant increase in lean body mass as compared to the placebo group (Poole et al. 2010). This finding suggested that supplementing with 500 mg of the galactomannan-based commercially available supplement of fenugreek extract with resistance training can alter body composition to a greater extent than resistance training alone (Poole et al. 2010).

Besides, the fenugreek extract as an effective stimulus for decreasing fat mass by 1.77% compared to 0.55% in the placebo group ($p < 0.05$) was clinically proven during the double-blind placebo-controlled study in college-age men (Wilborn et al. 2010). Increased serum testosterone levels are reported to accelerate lipolysis via hormone-sensitive lipase activation (Hossain and Hornick 1994). In contrast, diminished fat-oxidation efficiency and resting energy expenditure are liked with hypogonadism (Hayes 2000). Researchers observed significant increases in total and bioavailable testosterone levels (purported aromatase and 5-α reductase inhibitor effects), without any noticeable change in estradiol between fenugreek extract and placebo. Even though lipolysis markers were not assessed in the study, a possible connection between increased androgen levels and decreased fat mass can be suggested (Wilborn et al. 2010).

The preclinical evidence suggests that the galactomannans from fenugreek seeds might be the responsible phytoconstituent for significant anabolic potential without any androgenic effects as demonstrated in a study using castrated rats (Aswar et al. 2008). During the subacute oral administration (10 and 35 mg/kg, once a day, 4 weeks), fenugreek derived galactomannans significantly increased the weight of the levator ani, a skeletal muscle, suggesting anabolic effects without a change in testosterone levels (Aswar et al. 2008). An increase in a skeletal muscle's weight reflects muscle hypertrophy or an increase in the cross-sectional area of the muscle. The direct and positive correlation between a muscle's cross-sectional area and the overall strength of that particular muscle has been reported previously (Jones et al. 2008; Maughan, Watson, and Weir 1983). Therefore, the levator ani's increased weight suggested fenugreek extract's potential for strength increase through a non-hormone mediated channel, even though strength measurements were not assessed in this study (Aswar et al. 2008).

However, the evidence from the multi-ingredient supplementation (with fenugreek seed extract as one of the ingredients) is not clear about its effect on body composition (Tinsley et al. 2019; Woodgate and Conquer 2003). A multi-ingredient supplement containing a proprietary blend of papaya leaf, cascara sagrada bark, slippery elm bark, peppermint leaf, red raspberry leaf, fenugreek seed, ginger root, and senna leaf did not affect the body composition or waist circumference during a double-blind, randomized, placebo-controlled trial in healthy young adult females (Tinsley et al. 2019). Twenty-two participants were randomly assigned to consume a multi-ingredient supplement containing 1,350 mg/serving or placebo, daily, for 4 weeks. Although the multi-ingredient supplement was found safe during the study, it did not affect body composition, waist circumference, gastrointestinal symptoms or blood markers (Tinsley et al. 2019). However, another stimulant-free multi-ingredient supplement, containing fenugreek extract, improved the body composition in obese adults with a normal diet and exercise practices for 6 weeks during a single-center, prospective, randomized, double-blind, placebo-controlled study (Woodgate and Conquer 2003). The 6 weeks of supplementation of the proprietary blend containing glucomannan, fenugreek, chitosan, *G sylvestre* and vitamin C significantly reduced body fat percentage and absolute fat mass (v/s placebo), suggesting a greater effect on the body composition of the group supplemented with the proprietary blend as compared to the placebo group. Although these effects cannot be attributed to fenugreek alone, fenugreek might have assisted in the reported changes (Woodgate and Conquer 2003).

An increase in lean body mass is known to contribute to strength and power development in athletes (Barbieri et al. 2017; Kavvoura et al. 2018; Zaras et al. 2020). The relationship between muscle architectural characteristics, such as muscle thickness and strength, power and athletic performance is known (Stasinaki et al. 2019; Zaras et al. 2016). Furthermore, the body fat percentage was strongly correlated with power output in Division I male hockey players (Potteiger et al. 2010), elite male and female elite wrestlers (Vardar et al. 2007) and Division III women's volleyball players (Boldt et al. 2011). The correlation between power output and performance of athletes is also reported (Abe et al. 2011). Therefore, the increase in lean muscle mass and decrease in body fat percentage is highly desirable to increase the performance. In addition, the training method implemented by an athlete has a great impact on the amount of lean muscle mass and amount of body fat of the athlete. An increase in body fat, without a concomitant increase in lean muscle mass, may decrease acceleration, power and jumping ability in athletes (Jeukendrup and Gleeson 2004, 2010).

Taken together, the available evidence demonstrated the potential efficacy of fenugreek supplementation over a chronic period to improve body composition and reducing body fat resulted in higher strength and performance (Aswar et al. 2008; Poole et al. 2010; Wilborn et al. 2010; Woodgate and Conquer 2003).

Based on the mentioned studies, it can be seen that highly purified fractions/extracts of fenugreek and its phytoconstituents, especially galactomannan, have shown significant effects on body composition and lean muscle mass parameters and hence could be used as a supplementation regimen for individuals/athletes who are looking at increasing strength and improvement of body composition.

9.3 FENUGREEK FOR STRENGTH AND ENDURANCE

Recently, many studies have reported on fenugreek seed extract as a naturally derived testosterone booster that can increase strength and improve overall men's health in terms of muscle mass, body composition and libido (Rao et al. 2016; Rao, Mallard, and Grant 2020; Wankhede, Mohan, and Thakurdesai 2016; Wilborn et al. 2010). Strength increase resulting from a resistance training regimen are well established (Anderson and Kearney 1982; Brown and Wilmore 1974; Chilibeck et al. 1998; Faigenbaum et al. 1999; Hagerman et al. 2000; Morganti et al. 1995; Starkey et al. 1996). The strength changes in initial stages in untrained populations can be attributable to neural adaptations (Aagaard et al. 2002; Sale 1987), while neutrally adapted individuals experience hypertrophic changes in weeks to months after the onset of resistance training (Staron et al. 1994).

The commercially available galactomannans-rich fenugreek extract supplement (Torabolic), demonstrated a significant impact on both upper- and lower-body strength and body composition in resistance-trained men during a randomized, double-blind study (Poole et al. 2010). During the study, 49 resistance-trained men, matched according to body weight, were randomly assigned to ingest, in a double-blind manner, capsules containing either 500 mg of fenugreek extract supplementation or placebo. Subjects participated in a supervised resistance-training program of 4 days per week for a total of 8 weeks. Each program was split into two upper and two lower extremity workouts per week. After 8 weeks of supplementing fenugreek extract (500 mg, once daily), significant improvements in strength and anaerobic exercise performance, body fat, lower body strength, and upper body strength were observed (Poole et al. 2010). In this study, the researchers employed an eight-week linear resistance training program, an efficient stimulus that increases muscular strength and lean muscle mass (Kerksick et al. 2009). The 8 weeks of fenugreek extract supplementation showed a 9.19% increase in the bench press, 1-RM. Additionally, significant differences were observed between the fenugreek extract and placebo groups for leg press 1-RM (Poole et al. 2010). Further, non-pairwise comparisons, no changes between the groups were noticed during hormonal (DHT, leptin, free testosterone) and clinical safety evaluations (lipid panel, liver function, kidney function or CBC panel) (Poole et al. 2010).

These results were supported by the reports of strength increase by fenugreek extract (saponin glycoside rich) supplementation (Rao, Mallard, and Grant 2020; Wankhede, Mohan, and Thakurdesai 2016). In the first report, fenugreek supplementation (with furostenol glycoside as the main phytoconstituent) (300 mg, twice a day, 8 weeks) demonstrated significant improvements in muscle strength and repetitions to failure (bench press) and a reduction in body fat during the resistance training program (Wankhede, Mohan, and Thakurdesai 2016). The second study examined changes in muscular strength and endurance, body composition, functional threshold power, and sex hormones in response to an 8-week exercise program with daily supplementation of 600 mg of a fenugreek extract, standardized to a 50% saponin glycosides or a placebo (Rao, Mallard, and Grant 2020). Participants completed a whole-body workout program three times a week and observed for changes in muscle strength, endurance, functional threshold power, body composition, and sex hormones for 8 weeks at baseline, weeks 4 and 8 (Rao, Mallard, and Grant 2020). All groups improved their maximal leg press within the group at 8 weeks, whereas the fenugreek extract-treated group showed improvement more than placebo (Rao, Mallard, and Grant 2020). The fenugreek extract group showed body mass decrease (1.2 kg), body fat decrease (1.4%) with lean mass increase (1.8%) at 8 weeks.

Based on the previously mentioned studies, it can be seen that highly purified fractions/extracts of fenugreek and its phytoconstituents, especially galactomannan and glycosides, have shown significant effects on strength parameters.

9.4 FENUGREEK FOR TESTOSTERONE

Athletes are continuously searching for novel ways to enhance performance, including the use of anabolic steroids (i.e., the testosterone derivatives). Anabolic steroids are capable of inducing a positive nitrogen balance, increasing fat-free mass, stimulating protein synthesis, and minimizing protein breakdown. Several publications reported the increase in muscle size and strength on administration of testosterone derivatives to younger (Bhasin et al. 1996) and older (Ferrando et al. 2002; Schroeder, Terk, and Sattler 2003; Snyder et al. 1999), and hypogonadal (Bhasin et al. 1997, 2000) men. This contrasts with exercise-induced testosterone changes that do not appear to have such a profound response to muscle protein synthesis (West et al. 2009).

Fenugreek supplementation has demonstrated increased levels of free and bioavailable testosterone, presumably through inhibition of aromatase and 5-α reductase (Wilborn et al. 2010). Endogenous (within the human body) testosterone in the blood is found in three forms: (1) free, unbound testosterone (2–3%), (2) testosterone weakly bound to albumin (a protein) (37–38%), or (3) testosterone tightly bound to another protein, sex hormone-binding globulin (SHBG) (60%) (Dunn, Nisula, and Rodbard 1981; Li et al. 2016). The first two forms of the hormone collectively are known as "bioavailable testosterone" (Figure 9.1). The bioavailable testosterone, roughly 40% of the endogenous testosterone, is vital for protein synthesis and muscle hypertrophy and hence critical in any supplementation scheme. The third form, SHBG bound testosterone, is tightly bound to SHBG and constitutes about 60% of total testosterone. The free endogenous or exogenously administered testosterone (which does not bind to SHBG) can bind to androgen receptors. This binding promotes intracellular transcriptional and translational events to increase the fat-free mass and induce muscle hypertrophy. Hence, for increased performance and muscle hypertrophy, the availability of free testosterone is critical. Once bound to its receptor, testosterone can convert to dihydrotestosterone (DHT) and estradiol through enzymatic action of 5-α reductase and aromatase respectively.

Because of the legal and ethical repercussions surrounding anabolic steroid use, nutritional supplement companies have designed prohormones or testosterone precursors that are marketed to

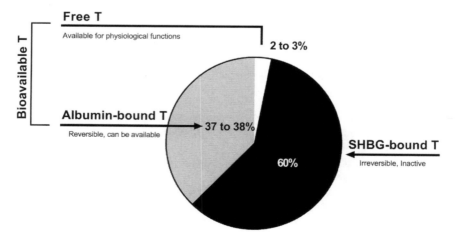

FIGURE 9.1 Distribution of endogenous testosterone (T) in the human body.

increase testosterone production similar to that of anabolic steroids. Nutritional supplement companies are continuously attempting the development of natural products with ergogenic potential comparable to anabolic steroids. The latest line of nutritional supplements fitting this persona is aromatase inhibitors (AIs). The AIs are proposed to suppress estrogen levels and subsequently increase parameters such as the endogenous free testosterone levels, free testosterone: estrogen (Test:Est) ratio, fat-free mass and strength. Synthetic AIs are classified as drugs, as they have been utilized as a medicinal preventative and treatment of breast cancer, and the effects of pharmacologic AIs, such as anastrozole and exemestane, on the Test:Est are well substantiated in both young and old men (Hayes 2000; Leder et al. 2004; Mauras et al. 2003; Taxel et al. 2001). Testosterone deficiency in males results in a considerable decrease in protein synthesis, decreased strength, decreased fat oxidation, increased adiposity (Mauras et al. 1998), and are all regarded as negative physiological conditions for any individual.

The effects of a commercially available glycoside-based fenugreek extract, purported to inhibit aromatase and 5-α reductase activity on strength, body composition, and hormonal profiles in resistance-trained males during an eight-week resistance training program was reported (Wilborn et al. 2010). The results of this study illustrate that, over the allotted 8-week supplementation time frame, fenugreek extract increases total testosterone and bioavailable testosterone by 6.57% and 12.26% respectively ($p < 0.05$). These results demonstrated that the fenugreek supplement increased endogenous testosterone levels, without a change in serum dihydrotestosterone or estradiol, probably due to the blocking aromatase and 5-α reductase. These results on estradiol in this study are in concurrence with other reports on an aromatase-inhibiting supplement (Rohle et al. 2007), which increases estradiol marginally after 8-weeks of supplementation. Aromatase inhibits the conversion of testosterone to estradiol, sending feedback to the hypothalamus and pituitary glands to enhance testosterone production (Hayes 2000). Therefore, estradiol levels would likely decrease while testosterone levels inversely elevate. This data is in agreement with the study (Wilborn et al. 2010), as estradiol decreased 9.64% from week 0 to week 4 before rising above baseline values by the conclusion of the 8-week study period. Due to a significant increase in total and bioavailable testosterone without a corresponding increase in estradiol and DHT, the authors suggested that the fenugreek supplement successfully inhibited aromatase and 5-α reductase activity (Wilborn et al. 2010).

The androgenic and anabolic effects of 600 mg of furostenol saponin glycoside based standardized fenugreek extract (Fenu-FG) in healthy male subjects was analyzed during an 8-week resistance training program during a randomized, double-blind, placebo-controlled study (Wankhede, Mohan, and Thakurdesai 2016). Sixty healthy male subjects were randomized and supplemented with capsules of Fenu-FG or placebo. During the supervised 4 day per week resistance-training program for 8 weeks, subjects were assessed for the androgenic hormones (total and free testosterone levels in serum, muscle strength and repetitions to failure, metabolic markers anabolic activity and body fat percentage. The authors found that Fenu-FG supplementation demonstrated significant anabolic and androgenic activity as compared with the placebo. In the present study, Fenu-FG supplementation was found to increase free testosterone without reducing total testosterone. These results are in agreement with the study of fenugreek seeds extract supplementation to increase serum testosterone levels in resistance-trained men resulting in increased free (i.e., bioavailable and unbound) testosterone by potentially decreasing the metabolism of serum testosterone (Wilborn et al. 2010).

In another randomized, double-blind, placebo-controlled, clinical study, elevated serum testosterone levels were reported by a protodioscin rich fenugreek seed extract, Furosap®, in 40 healthy male athletes for 12 consecutive weeks (Guo et al. 2018). Subjects were given either placebo or fenugreek seed extract (Furosap®) and the serum samples were used to assess serum total testosterone levels at baseline and at the end of 12 weeks of treatment. Lean body mass (body fat mass, lean mass, fat mass, fat-free mass), strength (grip strength, upper and lower body strength, force plate), maximal graded exercise stress (a digital hand dynamometer), dual-energy X-ray absorptiometry and treadmill with open-circuit spirometry were assessed at the baseline and at the end of 12 weeks of treatment. The fenugreek supplementation significantly increased mean lean body mass, fat-free mass, and serum testosterone compared to subjects receiving placebo. There were no changes in strength or endurance (Guo et al. 2018).

Finally, a recent meta-analysis reported the effects of fenugreek extract on total testosterone levels in males (Mansoori et al. 2020). The randomized clinical trials comparing intake of fenugreek extract with a control group, published up to November 2018, were searched in Medline via PubMed, Scopus databases, Cochrane Library, Web of Science and Google Scholar. Data of change in serum total testosterone (but not free testosterone) were pooled from four trials using random-effects for effects on total testosterone (Mansoori et al. 2020). However, such meta-analysis is not available on free or bioavailable testosterone.

Fenugreek's testosterone-enhancing effect was supported by evidence from published studies in non-resistant trained, healthy men (Mokashi et al. 2014; Rao et al. 2016). During a double-blind, randomized, crossover study, a single dose of 600 mg of glycoside-based standardized fenugreek extract (two capsules of 300 mg) was reported to increase free, bioavailable and total testosterone within physiological limits as compared with placebo (change from baseline, at 10 h) in healthy sedentary male subjects (Mokashi et al. 2014). A subsequent randomized, double-blind, parallel-group study, 600 mg/day supplementation for 12 weeks of a standardized specialized fenugreek seed extract, Testofen®, reported an increase in serum testosterone in healthy middle-aged and older men (Rao et al. 2016).

Given the significant reports on the benefits of fenugreek extract supplementation on free and bioavailable testosterone, it can be seen that highly purified fractions/extracts of fenugreek and its phytoconstituents, especially glycosides, can be incorporated in a supplementation regime for resistance-trained athletes or even sedentary males. The reports point to fenugreek extract as an alternative AI to synthetic pharmacologics. The reports also lends credence for using fenugreek extract to improve free and bioavailable testosterone levels in both resistance-trained males as well as sedentary males. The safety profile couple with the significant efficacy in increasing free testosterone levels makes fenugreek extract supplementation a good option for men looking at increasing performance in sports as well as aging men looking at improving their hormonal profile for healthy living.

9.5 FENUGREEK FOR POST-EXERCISE MUSCLE RECOVERY

Eccentric exercise is known to have several health-promoting benefits. For example, chronic eccentric exercise is reported for favorable changes in blood lipid profile (Nikolaidis et al. 2008; Panayiotou et al. 2013), insulin sensitivity, oxidative status (Paschalis et al. 2011) and inflammatory status (Paulsen et al. 2012). The acute eccentric exercise induces muscle microdamage and upregulates proinflammatory/pro-oxidant agents (e.g., increased IL-6), but down-regulates proinflammatory/pro-oxidant agents and upregulates anti-inflammatory/anti-oxidant agents (e.g., increased IL-10), chronically over a longer term (Paulsen et al. 2012). However, the increase in muscle strength due to eccentric exercise-induced muscle damage is greater than that of concentric exercise, due to greater muscle hypertrophy associated with eccentric training than concentric training (Farthing and Chilibeck 2003; Hortobágyi et al. 1996).

The earliest evidence of the efficacy of fenugreek seed extract on endurance capacity was reported in male mice (Ikeuchi et al. 2006). During the study, oral supplementation of fenugreek seed extract (300 mg/kg, once a day, 4 weeks) showed a significant increase in swimming time with decreased fat accumulation, perhaps due to increased fatty acid utilization during exercise (Ikeuchi et al. 2006). Subsequent preclinical studies showed evidence of anti-inflammatory activity in decreasing inflammatory cytokine levels associated with fenugreek seed extract (Hassan et al. 2015; Suresh et al. 2012).

Recently, clinical evidence of a decrease in eccentric exercise-induced inflammatory markers in non-resistance-trained male and female subjects by a proprietary blend rich in glycoside-based standardized fenugreek seed extract (IBPR) was shown (Wilborn et al. 2017). During this double-blind placebo-controlled study, 14-days of supplement dosage of IBPR (400 mg of glycoside-based standardized fenugreek seed + 25 mg each of curcumin and cinnamon each) with acute physical activity (24 sets of 10 eccentric knee extensor repetitions, with one leg at 30 °/s on an isokinetic device) in non-resistance-trained male and female subjects on inflammatory markers was assessed after an exercise bout. Significant time-dependent effects were observed in IL1b, IL6 and creatinine values from baseline, whereas significant treatment-dependent effect was seen in IL-1ra. Following 14 days of supplement dosage, the serum level of the inflammatory marker, IL6 in the IBPR-supplemented group was significantly less than placebo at 1 hr and 3 hrs post-exercise which supports anti-inflammatory properties of IBPR. The absence of classical proinflammatory cytokines (TNF-α and IL-1β) in the exercise-induced cytokine cascade caused us to increase the levels of IL-6, IL-1ra, IL-19 and sTNF-R (Ostrowski, Schjerling, and Pedersen 2000).

The responses of IBPR were consistent with past reports of reduction in the acute-phase inflammatory response, after daily intake of fenugreek seed extract, during animal studies in arthritic conditions through decreased cytokine gene expression in Freund's complete adjuvant-induced arthritis in rats (Pundarikakshudu et al. 2016; Sindhu et al. 2012). In addition, fenugreek seed extract is also reported to cause significant reduction in proinflammatory cytokines *in vitro* (Hassan et al. 2015). Therefore, the anti-inflammatory efficacy of IBPR in the study (Wilborn et al. 2017) can be attributed to its major component, a glycoside-based standardized fenugreek seed extract, which constitutes 88.89% and suggests IBPR as the potential option for post-exercise inflammation and muscle soreness.

9.6 FENUGREEK FOR CREATINE DELIVERY

Creatine, in the form of monohydrate, is one of the most popular supplements for sports nutrition and ergogenic aids. Creatine helps to supply additional substrates for the phosphocreatine system in high-intensity exercise and allows for the energy restoration and maintenance of high-power output for the skeletal muscles (Cooper et al. 2012; Greenhaff et al. 1994). Therefore, creatine is included in the supplementation regimen of many power athletes, including weightlifters and football players. Today, creatine supplementation has evolved to find various ways to maximize absorption rates in the body and in the muscle in an attempt to maximize the ergogenic effect of creatine in sport-specific applications. Higher amounts of creatine in skeletal muscle fulfil high-energy-demanding endeavors during exercise. Therefore, increasing the level of creatine absorption and uptake into the skeletal muscle is an important issue when supplementing creatine. The increased amount of intracellular creatine in skeletal muscle translates into more substantial effects on performance in anaerobic settings.

Creatine is absorbed in the skeletal muscle in a similar manner to that of glucose, whose uptake is highly regulated by the hormone insulin (Cooper et al. 2012; Pereira et al. 2017). The concurrent ingestion of creatine and carbohydrates are known to increase the creatine absorption, which does seem to translate into performance adaptations when combined with resistance training and produces greater increases in strength and body composition (Becque, Lochmann, and Melrose 2000; Bemben, Bemben et al. 2001; Bemben, Tuttle et al. 2001; Kreider et al. 1998).

In the last few years, insulin and insulinotropic agents gained interest in promoting creatine uptake due to insulin's stimulatory actions on GLUT-4 transporters (Pereira et al. 2017). Although fenugreek's actions on the glucose, insulin and GLUT-4 pathways are well established in animal and human studies, the use of fenugreek extract for improving creatine uptake was not known until the clinical evidence in this regard was reported (Taylor et al. 2011). In this study, the significant impact of the addition of fenugreek extract with creatine monohydrate on upper body strength and body composition on resistance-training adaptations during a double-blind, placebo-controlled study is reported (Taylor et al. 2011). Forty-seven resistance-trained men orally ingested either 70 g of a dextrose (placebo), 5 g creatine + 70 g of dextrose, or 3.5 g creatine + 900 mg fenugreek extract and participated during a periodized resistance-training program of 4 days per week for 8-weeks. With a structured resistance training program for eight weeks, both creatine supplementation combination groups showed significant and similar benefits on strength and body composition in resistance-trained males. Both creatine supplementation groups demonstrated significant increases in bench press 1-RM, leg press 1-RM and lean mass with minimal differences in body composition and performance parameters. These results suggested that fenugreek supplementation might have increased creatine absorption and retention during the initial weeks of the study, more than creatine + carbohydrates ingestion, thus showing the earlier onset of training adaptations (Taylor et al. 2011). This notion is supported by existing scientific evidence of fenugreek's prominent antihyperglycemic effects (Gupta, Gupta, and Lal 2001; Madar et al. 1988; Neeraja and Rajyalakshmi 1996) by mimicking insulin's action to increase glucose sensitivity (Madar et al. 1988; Neeraja and Rajyalakshmi 1996; Sharma and Raghuram 1990; Sharma et al. 1996; Vijayakumar et al. 2005). The defatted fenugreek seed extract was reported to decrease 24-hour urinary excretion of glucose by 64% in non-insulin-dependent diabetes mellitus subjects, suggesting an increased glucose absorption rate by the tissues (Sharma and Raghuram 1990).

Ingesting a large amount of simple carbohydrates over an extended period may prove harmful to an athlete's performance and overall health, especially when caloric restriction is commonly practiced (Taylor et al. 2011). Therefore, the addition of fenugreek extract to creatine can be a safer option for comparable creatine delivery that may prove beneficial to specific populations that are concerned with the negative implications of consuming large quantities of simple carbohydrates.

9.7 CONCLUSION

While the research on fenugreek's application for health is extensive and there is a body of evidence that substantially supports its claims, the effects of fenugreek supplementation in sports performance is limited. However, over the last decade, evidence on the varied physiological benefits of fenugreek extract supplementation on sports performance (by improving strength, free testosterone, reducing inflammation, improving body composition, etc.) is growing (Figure 9.2). Multiple studies and a meta-analysis demonstrated fenugreek's ability to increase levels of free or bioavailable testosterone, strength, endurance, with decrease in body fat, and improvement of body composition. Although more research in this area is recommended, fenugreek has emerged as a promising ingredient for sports nutrition and exercise performance applications.

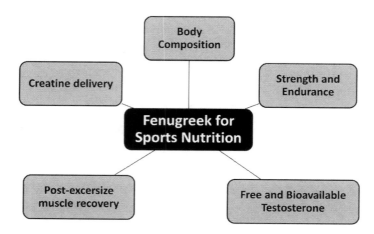

FIGURE 9.2 Fenugreek for sports nutrition.

REFERENCES

Aagaard, P., E. B. Simonsen, J. L. Andersen, P. Magnusson, and P. Dyhre-Poulsen. 2002. "Increased rate of force development and neural drive of human skeletal muscle following resistance training." *J Appl Physiol* 93 (4):1318–26. doi: 10.1152/japplphysiol.00283.2002.

Abe, T., Y. Harada, K. Kawamoto, and S. Fukashiro. 2011. "Relationship between body composition and 100-M running time in an elite female sprinter: A 7-year retrospective study." *Medicina Sportiva* 15:227–9. doi: 10.2478/v10036-011-0028-7.

American College of Sports Medicine. 2009. "American college of sports medicine position stand. Progression models in resistance training for healthy adults." *Med Sci Sports Exerc* 41 (3):687–708. doi: 10.1249/MSS.0b013e3181915670.

Anderson, T., and J. T. Kearney. 1982. "Effects of three resistance training programs on muscular strength and absolute and relative endurance." *Res Q Exerc Sport* 53 (1):1–7. doi: 10.1080/02701367.1982.10605218.

Aswar, U., V. Mohan, S. Bhaskaran, and S. L. Bodhankar. 2008. "Study of galactomannan on androgenic and anabolic activity in male rats." *Pharmacology online* 2:56–65.

Barbieri, D., L. Zaccagni, V. Babić, M. Rakovac, M. Mišigoj-Duraković, and E. Gualdi-Russo. 2017. "Body composition and size in sprint athletes." *J Sports Med Phys Fitness* 57 (9):1142–6. doi: 10.23736/s0022-4707.17.06925-0.

Becque, M. D., J. D. Lochmann, and D. R. Melrose. 2000. "Effects of oral creatine supplementation on muscular strength and body composition." *Med Sci Sports Exerc* 32 (3):654–8. doi: 10.1097/00005768-200003000-00016.

Bemben, M. G., D. A. Bemben, D. D. Loftiss, and A. W. Knehans. 2001. "Creatine supplementation during resistance training in college football athletes." *Med Sci Sports Exerc* 33 (10):1667–73. doi: 10.1097/00005768-200110000-00009.

Bemben, M. G., T. D. Tuttle, D. A. Bemben, and A. W. Knehans. 2001. "Effects of creatine supplementation on isometric force-time curve characteristics." *Med Sci Sports Exerc* 33 (11):1876–81. doi: 10.1097/00005768-200111000-00012.

Bhasin, S., T. W. Storer, N. Berman, C. Callegari, B. Clevenger, J. Phillips, T. J. Bunnell, R. Tricker, A. Shirazi, and R. Casaburi. 1996. "The effects of supraphysiologic doses of testosterone on muscle size and strength in normal men." *N Engl J Med* 335 (1):1–7.

Bhasin, S., T. W. Storer, N. Berman, K. E. Yarasheski, B. Clevenger, J. Phillips, W. P. Lee, T. J. Bunnell, and R. Casaburi. 1997. "Testosterone replacement increases fat-free mass and muscle size in hypogonadal men." *J Clin Endocrinol Metab* 82 (2):407–13. doi: 10.1210/jcem.82.2.3733.

Bhasin, S., T. W. Storer, M. Javanbakht, N. Berman, K. E. Yarasheski, J. Phillips, M. Dike, I. Sinha-Hikim, R. Shen, R. D. Hays, and G. Beall. 2000. "Testosterone replacement and resistance exercise in HIV-infected men with weight loss and low testosterone levels." *JAMA* 283 (6):763–70. doi: joc90874 [pii].

Boldt, M., D. Gregory, D. Jaffe, T. Dodge, and M. Jones. 2011. "Relationship between body composition and performance measures in NCAA division III women's volleyball players." *J Strength and Conditioning Res* 25. doi: 10.1097/01.JSC.0000395706.44342.e2.

Brown, C. H., and J. H. Wilmore. 1974. "The effects of maximal resistance training on the strength and body composition of women athletes." *Med Sci Sports* 6 (3):174–7.

Chilibeck, P. D., A. W. Calder, D. G. Sale, and C. E. Webber. 1998. "A comparison of strength and muscle mass increases during resistance training in young women." *Eur J Appl Physiol Occup Physiol* 77 (1–2):170–5. doi: 10.1007/s004210050316.

Cooper, R., F. Naclerio, J. Allgrove, and A. Jimenez. 2012. "Creatine supplementation with specific view to exercise/sports performance: An update." *J Int Soc Sports Nutr* 9 (1):33. doi: 10.1186/1550-2783-9-33.

Dunn, J. F., B. C. Nisula, and D. Rodbard. 1981. "Transport of steroid hormones: Binding of 21 endogenous steroids to both testosterone-binding globulin and corticosteroid-binding globulin in human plasma." *J Clin Endocrinol Metab* 53 (1):58–68. doi: 10.1210/jcem-53-1-58.

Faigenbaum, A. D., W. L. Westcott, R. L. Loud, and C. Long. 1999. "The effects of different resistance training protocols on muscular strength and endurance development in children." *Pediatrics* 104 (1):e5. doi: 10.1542/peds.104.1.e5.

Farthing, J. P., and P. D. Chilibeck. 2003. "The effects of eccentric and concentric training at different velocities on muscle hypertrophy." *European Journal of Applied Physiology* 89 (6):578–86. doi: 10.1007/s00421-003-0842-2.

Ferrando, A. A., M. Sheffield-Moore, C. W. Yeckel, C. Gilkison, J. Jiang, A. Achacosa, S. A. Lieberman, K. Tipton, R. R. Wolfe, and R. J. Urban. 2002. "Testosterone administration to older men improves muscle function: Molecular and physiological mechanisms." *Am J Physiol Endocrinol Metab* 282 (3):E601–E7. doi: 10.1152/ajpendo.00362.2001.

Greenhaff, P. L., K. Bodin, K. Soderlund, and E. Hultman. 1994. "Effect of oral creatine supplementation on skeletal muscle phosphocreatine resynthesis." *Am J Physiol* 266 (5 Pt 1):E725–E30. doi: 10.1152/ajpendo.1994.266.5.E725.

Guo, R., Q. Wang, R. P. Nair, S. L. Barnes, D. T. Smith, B. Dai, T. J. Robinson, and S. Nair. 2018. "Furosap, a novel Fenugreek seed extract improves lean body mass and serum testosterone in a randomized, placebo-controlled, double-blind clinical investigation." *Functional Foods in Health and Disease* 8 (11):519. doi: 10.31989/ffhd.v8i11.565.

Gupta, A., R. Gupta, and B. Lal. 2001. "Effect of *Trigonella foenum-graecum* (fenugreek) seeds on glycaemic control and insulin resistance in type 2 diabetes mellitus: A double blind placebo controlled study." *J Assoc Physicians India* 49:1057–61.

Hagerman, F. C., S. J. Walsh, R. S. Staron, R. S. Hikida, R. M. Gilders, T. F. Murray, K. Toma, and K. E. Ragg. 2000. "Effects of high-intensity resistance training on untrained older men. I. Strength, cardiovascular, and metabolic responses." *J Gerontol A Biol Sci Med Sci* 55 (7):B336–B46. doi: 10.1093/gerona/55.7.b336.

Hassan, N., C. Withycombe, M. Ahluwalia, A. Thomas, and K. Morris. 2015. "A methanolic extract of *Trigonella foenum-graecum* (fenugreek) seeds regulates markers of macrophage polarization." *Functional Foods in Health and Disease* 5 (12):417–26. doi: 10.31989/ffhd.v5i12.216.

Hayes, F. J. 2000. "Testosterone—fountain of youth or drug of abuse?" *J Clin Endocrinol Metab* 85 (9):3020–3. doi: 10.1210/jcem.85.9.6868.

Hortobágyi, T., J. Barrier, D. Beard, J. Braspennincx, P. Koens, P. Devita, L. Dempsey, and J. Lambert. 1996. "Greater initial adaptations to submaximal muscle lengthening than maximal shortening." *J Applied Physiology* 81 (4):1677–82. doi: 10.1152/jappl.1996.81.4.1677.

Hossain, A., and C. A. Hornick. 1994. "Androgenic modulation of lipid metabolism at subcellular sites in cholestatic rats." *Horm Metab Res* 26 (1):19–25. doi: 10.1055/s-2007-1000766.

Ikeuchi, M., K. Yamaguchi, T. Koyama, Y. Sono, and K. Yazawa. 2006. "Effects of fenugreek seeds (*Trigonella foenum graecum*) extract on endurance capacity in mice." *J Nutr Sci Vitaminol (Tokyo)* 52 (4):287–92. doi: 10.3177/jnsv.52.287.

Jeukendrup, A. E., and M. Gleeson. 2004. *Sport nutrition: An introduction to energy production and performance*. Champaign, IL; Leeds: Human Kinetics.

Jeukendrup, A. E., and M. Gleeson. 2010. *Sport nutrition: An introduction to energy production and performance*. 2nd ed. ed. Leeds: Human Kinetics.

Jones, E. J., P. A. Bishop, A. K. Woods, and J. M. Green. 2008. "Cross-sectional area and muscular strength: A brief review." *Sports Med* 38 (12):987–94. doi: 10.2165/00007256-200838120-00003.

Kavvoura, A., N. Zaras, A.-N. Stasinaki, G. Arnaoutis, S. Methenitis, and G. Terzis. 2018. "The importance of lean body mass for the rate of force development in taekwondo athletes and track and field throwers." *J Funct Morphol Kinesiol* 3 (3):43. doi: 10.3390/jfmk3030043.

Kerksick, C. M., C. D. Wilborn, B. I. Campbell, M. D. Roberts, C. J. Rasmussen, M. Greenwood, and R. B. Kreider. 2009. "Early-phase adaptations to a split-body, linear periodization resistance training program in college-aged and middle-aged men." *J Strength Cond Res* 23 (3):962–71. doi: 10.1519/JSC.0b013e3181a00baf.

Kraemer, W. J., K. Adams, E. Cafarelli, G. A. Dudley, C. Dooly, M. S. Feigenbaum, S. J. Fleck, B. Franklin, A. C. Fry, J. R. Hoffman, R. U. Newton, J. Potteiger, M. H. Stone, N. A. Ratamess, T. Triplett-McBride, and M. American College of Sports. 2002. "American college of sports medicine position stand. Progression models in resistance training for healthy adults." *Med Sci Sports Exerc* 34 (2):364–80. doi: 10.1097/00005768-200202000-00027.

Kreider, R. B., M. Ferreira, M. Wilson, P. Grindstaff, S. Plisk, J. Reinardy, E. Cantler, and A. L. Almada. 1998. "Effects of creatine supplementation on body composition, strength, and sprint performance." *Med Sci Sports Exerc* 30 (1):73–82. doi: 10.1097/00005768-199801000-00011.

Leder, B. Z., J. L. Rohrer, S. D. Rubin, J. Gallo, and C. Longcope. 2004. "Effects of aromatase inhibition in elderly men with low or borderline-low serum testosterone levels." *J Clin Endocrinol Metab* 89 (3):1174–80. doi: 10.1210/jc.2003-031467.

Li, H., T. Pham, B. C. McWhinney, J. P. Ungerer, C. J. Pretorius, D. J. Richard, R. H. Mortimer, M. C. d'Emden, and K. Richard. 2016. "Sex hormone binding globulin modifies testosterone action and metabolism in prostate cancer cells." *Int J Endocrinol* 2016:6437585. doi: 10.1155/2016/6437585.

Madar, Z., R. Abel, S. Samish, and J. Arad. 1988. "Glucose-lowering effect of fenugreek in non-insulin dependent diabetics." *Eur J Clin Nutr* 42 (1):51–4.

Mansoori, A., S. Hosseini, M. Zilaee, R. Hormoznejad, and M. Fathi. 2020. "Effect of fenugreek extract supplement on testosterone levels in male: A meta-analysis of clinical trials." *Phytother Res* 34 (7):1550–5. doi: 10.1002/ptr.6627.

Maughan, R. J., J. S. Watson, and J. Weir. 1983. "Strength and cross-sectional area of human skeletal muscle." *J Physiol* 338:37–49. doi: 10.1113/jphysiol.1983.sp014658.

Mauras, N., J. Lima, D. Patel, A. Rini, E. di Salle, A. Kwok, and B. Lippe. 2003. "Pharmacokinetics and dose finding of a potent aromatase inhibitor, aromasin (exemestane), in young males." *J Clin Endocrinol Metab* 88 (12):5951–6. doi: 10.1210/jc.2003-031279.

Mauras, N., V. Hayes, S. Welch, A. Rini, K. Helgeson, M. Dokler, J. D. Veldhuis, and R. J. Urban. 1998. "Testosterone deficiency in young men: Marked alterations in whole body protein kinetics, strength, and adiposity." *J Clin Endocrinol Metab* 83 (6):1886–92. doi: 10.1210/jcem.83.6.4892.

Mokashi, M., R. Singh-Mokashi, V. Mohan, and P. A. Thakurdesai. 2014. "Effects of glycosides based fenugreek seed extract on serum testosterone levels of healthy sedentary male subjects: A exploratory double blind, placebo controlled, crossover study." *Asian J Pharmaceutical and Clin Res* 7 (2):177–81.

Morganti, C. M., M. E. Nelson, M. A. Fiatarone, G. E. Dallal, C. D. Economos, B. M. Crawford, and W. J. Evans. 1995. "Strength improvements with 1 year of progressive resistance training in older women." *Med Sci Sports Exerc* 27 (6):906–12.

Neeraja, A., and P. Rajyalakshmi. 1996. "Hypoglycemic effect of processed fenugreek seeds in humans." *J Food Sci Technol* 33 (5):427–30.

Nikolaidis, M., V. Paschalis, G. Giakas, I. Fatouros, G. Sakellariou, A. Theodorou, Y. Koutedakis, and A. Jamurtas. 2008. "Favorable and prolonged changes in blood lipid profile after muscle-damaging exercise." *Med Sci Sports+ Exercise* 40 (8):1483. doi: 10.1249/MSS.0b013e31817356f2.

Ostrowski, K., P. Schjerling, and B. K. Pedersen. 2000. "Physical activity and plasma interleukin-6 in humans—effect of intensity of exercise." *Euro J Applied Physiology* 83 (6):512–15. doi: 10.1007/s004210000312.

Panayiotou, G., V. Paschalis, M. Nikolaidis, A. Theodorou, C. Deli, N. Fotopoulou, I. Fatouros, Y. Koutedakis, M. Sampanis, and A. Jamurtas. 2013. "No adverse effects of statins on muscle function and health-related parameters in the elderly: An exercise study." *Scand J Med Sci Sports* 23 (5):556–67. doi: 10.1111/j.1600-0838.2011.01437.x.

Paschalis, V., M. G. Nikolaidis, A. A. Theodorou, G. Panayiotou, I. G. Fatouros, Y. Koutedakis, and A. Z. Jamurtas. 2011. "A weekly bout of eccentric exercise is sufficient to induce health-promoting effects." *Med Sci Sports Exerc* 43 (1):64–73. doi: 10.1249/MSS.0b013e3181e91d90.

Paulsen, G., U. R. Mikkelsen, T. Raastad, and J. M. Peake. 2012. "Leucocytes, cytokines and satellite cells: What role do they play in muscle damage and regeneration following eccentric exercise." *Exerc Immunol Rev* 18 (1):42–97.

Pereira, R. M., L. P. D. Moura, V. R. Muñoz, A. S. R. d. Silva, R. S. Gaspar, E. R. Ropelle, and J. R. Pauli. 2017. "Molecular mechanisms of glucose uptake in skeletal muscle at rest and in response to exercise." *Motriz: Revista de Educação Física* 23 (SPE). doi: 10.1590/s1980-6574201700si0004.

Poole, C., B. Bushey, C. Foster, B. Campbell, D. Willoughby, R. Kreider, L. Taylor, and C. Wilborn. 2010. "The effects of a commercially available botanical supplement on strength, body composition, power output, and hormonal profiles in resistance-trained males." *J Int Soc Sports Nutr* 7:34. doi: 10.1186/1550-2783-7-34.

Potteiger, J., D. L. Smith, M. L. Maier, and T. S. Foster. 2010. "Relationship between body composition, leg strength, anaerobic power, and on-ice skating performance in division I men's hockey athletes." *J Strength Cond Res* 24:1755–62. doi: 10.1519/JSC.0b013e3181e06cfb.

Pundarikakshudu, K., D. H. Shah, A. H. Panchal, and G. C. Bhavsar. 2016. "Anti-inflammatory activity of fenugreek (*Trigonella foenum-graecum* Linn) seed petroleum ether extract." *Indian J Pharmacol* 48 (4):441. doi: 10.4103/0253-7613.186195.

Rao, A. J., A. R. Mallard, and R. Grant. 2020. "Testofen (Fenugreek extract) increases strength and muscle mass compared to placebo in response to calisthenics. A randomized control trial." *Transl Sports Med* 3 (4):374–80. doi: 10.1002/tsm2.153.

Rao, A. J., E. Steels, W. J. Inder, S. Abraham, and L. Vitetta. 2016. "Testofen, a specialised *Trigonella foenum-graecum* seed extract reduces age-related symptoms of androgen decrease, increases testosterone levels and improves sexual function in healthy aging males in a double-blind randomised clinical study." *Aging Male* Jan (2):1–9. doi: 10.3109/13685538.2015.1135323.

Rohle, D., C. Wilborn, L. Taylor, C. Mulligan, R. Kreider, and D. Willoughby. 2007. "Effects of eight weeks of an alleged aromatase inhibiting nutritional supplement 6-OXO (androst-4-ene-3,6,17-trione) on serum hormone profiles and clinical safety markers in resistance-trained, eugonadal males." *J Int Soc Sports Nutr* 4:13. doi: 10.1186/1550-2783-4-13.

Sale, D. G. 1987. "Influence of exercise and training on motor unit activation." *Exerc Sport Sci Rev* 15:95–151.

Schroeder, E. T., M. Terk, and F. R. Sattler. 2003. "Androgen therapy improves muscle mass and strength but not muscle quality: Results from two studies." *Am J Physiol Endocrinol Metab* 285 (1):E16–E24. doi: 10.1152/ajpendo.00032.2003.

Sharma, R. D., and T. Raghuram. 1990. "Hypoglycaemic effect of fenugreek seeds in non-insulin dependent diabetic subjects." *Nutr Res* 10 (7):731–9. doi: 10.1016/S0271-5317(05)80822-X.

Sharma, R. D., A. Sarkar, D. K. Hazra, B. Mishra, J. B. Singh, S. K. Sharma, B. B. Maheshwari, and P. K. Maheshwari. 1996. "Use of fenugreek seed powder in the management dependent diabetes mellitus." *Nutr Res* 16 (8):1331–9. doi: 10.1016/0271-5317(96)00141-8.

Sindhu, G., M. Ratheesh, G. L. Shyni, B. Nambisan, and A. Helen. 2012. "Anti-inflammatory and antioxidative effects of mucilage of Trigonella foenum-graecum (Fenugreek) on adjuvant induced arthritic rats." *Int Immunopharmacol* 12 (1):205–11. doi: 10.1016/j.intimp.2011.11.012.

Snyder, P. J., H. Peachey, P. Hannoush, J. A. Berlin, L. Loh, D. A. Lenrow, J. H. Holmes, A. Dlewati, J. Santanna, C. J. Rosen, and B. L. Strom. 1999. "Effect of testosterone treatment on body composition and muscle strength in men over 65 years of age." *J Clin Endocrinol Metab* 84 (8):2647–53. doi: 10.1210/jcem.84.8.5885.

Starkey, D. B., M. L. Pollock, Y. Ishida, M. A. Welsch, W. F. Brechue, J. E. Graves, and M. S. Feigenbaum. 1996. "Effect of resistance training volume on strength and muscle thickness." *Med Sci Sports Exerc* 28 (10):1311–20. doi: 10.1097/00005768-199610000-00016.

Staron, R. S., D. L. Karapondo, W. J. Kraemer, A. C. Fry, S. E. Gordon, J. E. Falkel, F. C. Hagerman, and R. S. Hikida. 1994. "Skeletal muscle adaptations during early phase of heavy-resistance training in men and women." *J Appl Physiol (1985)* 76 (3):1247–55. doi: 10.1152/jappl.1994.76.3.1247.

Stasinaki, A. N., N. Zaras, S. Methenitis, G. Bogdanis, and G. Terzis. 2019. "Rate of force development and muscle architecture after fast and slow velocity eccentric training." *Sports (Basel)* 7 (2). doi: 10.3390/sports7020041.

Steele, J., J. Fisher, M. Skivington, C. Dunn, J. Arnold, G. Tew, A. M. Batterham, D. Nunan, J. M. O'Driscoll, S. Mann, C. Beedie, S. Jobson, D. Smith, A. Vigotsky, S. Phillips, P. Estabrooks, and R. Winett. 2017. "A higher effort-based paradigm in physical activity and exercise for public health: Making the case for a greater emphasis on resistance training." *BMC Public Health* 17 (1):300. doi: 10.1186/s12889-017-4209-8.

Suresh, P., C. N. Kavitha, S. M. Babu, V. P. Reddy, and A. K. Latha. 2012. "Effect of ethanol extract of *Trigonella foenum-graecum* (Fenugreek) seeds on Freund's adjuvant-induced arthritis in albino rats." *Inflammation* 35 (4):1314–21. doi: 10.1007/s10753-012-9444-7.

Taxel, P., D. G. Kennedy, P. M. Fall, A. K. Willard, J. M. Clive, and L. G. Raisz. 2001. "The effect of aromatase inhibition on sex steroids, gonadotropins, and markers of bone turnover in older men." *J Clin Endocrinol Metab* 86 (6):2869–74. doi: 10.1210/jcem.86.6.7541.

Taylor, L., C. Poole, E. Pena, M. Lewing, R. Kreider, C. Foster, and C. Wilborn. 2011. "Effects of combined creatine plus fenugreek extract vs. Creatine plus carbohydrate supplementation on resistance training adaptations." *J Sports Sci Med* 10:254–60.

Tinsley, G., S. Urbina, E. Santos, K. Villa, C. Foster, C. Wilborn, and L. Taylor. 2019. "A purported detoxification supplement does not improve body composition, waist circumference, blood markers, or gastrointestinal symptoms in healthy adult females." *J Diet Suppl* 16 (6):649–58. doi: 10.1080/19390211.2018.1472713.

Vardar, S. A., S. Tezel, L. Oztürk, and O. Kaya. 2007. "The relationship between body composition and anaerobic performance of elite young wrestlers." *J Sports Sci Med* 6 (2):34–8.

Vijayakumar, M. V., S. Singh, R. R. Chhipa, and M. K. Bhat. 2005. "The hypoglycaemic activity of fenugreek seed extract is mediated through the stimulation of an insulin signalling pathway." *Br J Pharmacol* 146 (1):41–8. doi: 10.1038/sj.bjp.0706312.

Wankhede, S., V. Mohan, and P. Thakurdesai. 2016. "Beneficial effects of fenugreek glycoside supplementation in male subjects during resistance training: A randomized controlled pilot study." *J Sport and Health Sci* 5 (2):176–82. doi: 10.1016/j.jshs.2014.09.005.

West, D. W., G. W. Kujbida, D. R. Moore, P. Atherton, N. A. Burd, J. P. Padzik, M. De Lisio, J. E. Tang, G. Parise, M. J. Rennie, S. K. Baker, and S. M. Phillips. 2009. "Resistance exercise-induced increases in putative anabolic hormones do not enhance muscle protein synthesis or intracellular signalling in young men." *J Physiol* 587 (Pt 21):5239–47. doi: 10.1113/jphysiol.2009.177220.

Wilborn, C., L. Taylor, C. Poole, C. Foster, D. Willoughby, and R. Kreider. 2010. "Effects of a Purported Aromatase and 5 α-Reductase Inhibitor on Hormone Profiles in College-Age Men." *International journal of sport nutrition* 20 (6):457. doi: 10.1123/ijsnem.20.6.457.

Wilborn, C., S. Hayward, L. Taylor, S. Urbina, C. Foster, P. Deshpande, V. Mohan, and P. Thakurdesai. 2017. "Effects of IBPR, a proprietary blend rich in glycoside based standardized fenugreek seed extract on inflammatory markers during acute eccentric resistance exercise in young subjects." *Asian J Pharmaceutical and Clin Res* 10 (10):99–104. doi: 10.22159/ajpcr.2017.v10i10.18811.

Woodgate, D. E., and J. A. Conquer. 2003. "Effects of a stimulant-free dietary supplement on body weight and fat loss in obese adults: A six-week exploratory study." *Curr Ther Res Clin Exp* 64 (4):248–62. doi: 10.1016/S0011-393X(03)00058-4.

Zaras, N. D., A. N. Stasinaki, P. Spiliopoulou, M. Hadjicharalambous, and G. Terzis. 2020. "Lean body mass, muscle architecture, and performance in well-trained female weightlifters." *Sports (Basel)* 8 (5). doi: 10.3390/sports8050067.

Zaras, N. D., A. N. Stasinaki, S. K. Methenitis, A. A. Krase, G. P. Karampatsos, G. V. Georgiadis, K. M. Spengos, and G. D. Terzis. 2016. "Rate of force development, muscle architecture, and performance in young competitive track and field throwers." *J Strength Cond Res* 30 (1):81–92. doi: 10.1519/jsc.0000000000001048.

Part IV Modern Medicinal Applications

10 The Effects of Fenugreek on Controlling Glucose in Diabetes Mellitus

An Overview of Scientific Evidence

Zahra Ayati, Nazli Namazi, Mohammad Hossein Ayati, Seyed Ahmad Emami and Dennis Chang

CONTENTS

10.1 Introduction 129
10.2 Hyperglycaemia, Diabetes Mellitus and the Use of Complementary Medicine 130
10.3 Antihyperglycaemic Effects of Fenugreek 131
    10.3.1 *In Vitro* Observations in Cell Culture Studies 131
    10.3.2 *In Vivo* Observations in Animal Studies 131
10.4 Clinical Studies of Fenugreek 133
10.5 Conclusion 137
References 141

10.1 INTRODUCTION

Diabetes mellitus is a group of metabolic disorders which arise from complex interactions between several genetic, environmental or lifestyle factors. Diabetes has reached epidemic proportions in the past few decades with worldwide prevalence approaching 400 million people. It is characterised by a chronic hyperglycaemic condition resulting from impaired insulin secretion, insulin action or both, and is classified into two major categories, type 1 and type 2. Type 1 diabetes accounts for 5–10% of those with diabetes and results from cellular-mediated autoimmune destruction of the beta-cells of the pancreas. Type 2 diabetes accounts for more than 90% of diabetes and is characterised by impaired insulin secretion and insulin resistance (1–3). Long-term diabetes is associated with several co-morbidities, such as heart disease and kidney failure. Effective control of hyperglycaemia in diabetic patients is critical for reducing the risk of micro- and macrovascular diseases (3). At present, the best course of action to reduce higher than normal blood sugar is to adhere to healthy dietary patterns and increase physical activity. For some, rigorous and sustained behavioural change is not enough. In these cases, pharmaceutical interventions such as metformin may be needed to delay or suppress the onset of diabetes. However, this is not always an adequate long-term solution. A Kaplan-Meier analysis showed a cumulative incidence of monotherapy failure at 5 years of 21% with metformin and 34% with glybenclamide (4). Low-risk low-cost alternatives to conventional pharmaceutical interventions are clearly needed where lifestyle modifications have failed to adequately improve glucose tolerance.

Herbal medicines are relatively safe, accessible and affordable alternatives which have been used for the management of diabetes over centuries and some of them are reported to be useful in the management of hyperglycaemia in diabetes.

Fenugreek (*Trigonella foenum-graecum* L.) is an angiosperm plant from the Fabaceae family. It is also known as bird's foot and Greek hay and is originated to the Mediterranean, southern Europe, Asia and North Africa; however, it is now grown in many parts of the world including northern Africa, India and the United States (5–7). Fenugreek seed has a long history of use for nutritional value and for a broad range of therapeutic effects, including labour induction, indigestion and as a general tonic to improve metabolism and health (8). It has been suggested for the management of hyperglycaemia by several traditional and folklore medicines around the world, especially in Asian traditional medicines including Ayurvedic, Chinese and Persian medicines (9–11).

Fenugreek seed contains various categories of compounds including dietary fibre, steroidal saponins, galactomannan rich polysaccharides, alkaloids (e.g. trigonelline), oil, mucilage, amino acid derivatives (e.g. 4-hydroxyisoleucine), carbohydrates, phenolic compounds of both flavonoids (e.g. apigenin, luteolin, arabinose and orientin) and aromatic acids, and coumarins. The seeds contain 50% fibre (30% soluble fibre and 20% insoluble fibre) (6, 12, 13).

Several preclinical and clinical studies have investigated the antihyperglycaemic effects of fenugreek and its bioactive constituents (14, 15). Recent evidences from meta-analysis of clinical trials support the beneficial effects of fenugreek seeds on glycaemic control in patients with diabetes (9, 16).

The antidiabetic activity of fenugreek can be mediated through different pathways including stimulation of the pancreas to produce and secrete insulin, regulation of glucose absorption in the intestine, promotion of glycogen synthesis and antioxidant and anti-inflammatory activities in the liver (17–20).

Several research data support that besides the direct effects of fenugreek and its bioactive constituents on insulin regulation and glucose control, they can also influence the production and activity of some other metabolic hormones, such as insulin-like growth factor 1 (IGF-1), growth hormone (GH) and thyroid hormones (T3 and T4), which all play indirect but important roles in the regulation of glucose metabolism (21–25).

The hypoglycaemic effect of fenugreek can be attributed to some of its bioactive constituents such as 4-hydroxysoleucine, trigonelline and orientin (8, 26, 27).

In this chapter, the potential effects of fenugreek in management of glucose in diabetes mellitus will be reviewed and an insight into its active constituents and mechanisms of action will be provided. In addition, the results of related clinical trials will be discussed.

10.2 HYPERGLYCAEMIA, DIABETES MELLITUS AND THE USE OF COMPLEMENTARY MEDICINE

Diabetes is a complex condition which is highly prevalent worldwide. It is generally characterised by hyperglycaemia, which can lead to multiple body system damage and dysfunction. There are two main types of diabetes; type 1, also referred as insulin-dependent diabetes mellitus, is caused by impaired insulin production. Type 2, however, is commonly associated with the inability of cells to respond to insulin due to insulin resistance and therefore is referred to as non-insulin dependent diabetes mellitus. Type 2 diabetes is the predominant form of diabetes and accounts for at least 90% of all cases of diabetes mellitus (2). Hyperglycaemia is the hallmark metabolic abnormality associated with diabetes. Diabetes is a leading cause of coronary heart disease, renal failure, visual impairment and non-traumatic lower limb amputations (28). The risk of morbidity and mortality of diabetes can be considerably reduced by achieving specific glycaemic goals (29). Complementary and alternative medicines (CAMs), especially herbal medicines, are emerging aspects of identifying new pharmacological interventions for the management of glucose in diabetes and can effectively reduce the risk of morbidity and mortality.

10.3 ANTIHYPERGLYCAEMIC EFFECTS OF FENUGREEK

10.3.1 IN VITRO OBSERVATIONS IN CELL CULTURE STUDIES

Evidence from several *in vitro* studies indicate on the antihyperglycaemic effect of fenugreek and its active compounds such as 4-hydroxyisoleucine.

4-hydroxyisoleucine, a peculiar nonprotein amino acid isolated from the seeds, exhibits considerable effects on insulin resistance in obesity and diabetes. It improves liver function and enhances glucose-induced insulin release (30–33). In an *in vitro* study, the effect of 4-hydroxyisoleucine and aqueous extract of fenugreek seed was evaluated in human liver cells; insulin was used as a positive control. Cells were treated under normoglycaemic and hyperglycaemic conditions. Under both conditions, fenugreek seed extract and 4-hydroxyisoleucine caused a significant increase in glycogen synthase kinase-3a/b, insulin receptor-b and glucose transporter 2, comparable to insulin and metformin. In addition, fenugreek seed extract caused a significant glucose uptake in both conditions. These findings can show that the mechanism in which fenugreek exerts antihyperglycaemic properties is comparable to that of metformin and insulin. This can be explained by the effect of fenugreek through improving insulin-signalling pathways and gene expression and enhancing glucose uptake (34).

Skeletal muscle is the main insulin target tissue which is responsible for whole body glucose homeostasis. In some *in vitro* studies, 4-hydroxyisoleucine and its derivatives showed a stimulatory effect on glucose reuptake in L-6 skeletal muscle cells. This effect is mainly via enhancing the translocation of glucose transporter 4 (GLUT4) to the plasma membrane through the involvement of the phosphatidylinositol-3-kinase pathway (35, 36).

Fenugreek contains a high amount of saponins. In an *in vitro* study, diosgenin, a major aglycone saponin in fenugreek, improved glucose uptake in 3T3-L1 cells by enhancing adipocyte differentiation and inhibiting the expression of inflammatory factors. It also decreased the m-RNA expression level and protein secretion of pro-inflammatory cytokine, monocyte-chemoattractant protein-1 (MCP-1), which can result in the suppression of macrophage infiltration (37). Another study on 3T3-L1 adipocytes and human hepatoma cells (HepG2) revealed that fenugreek seed extract, like insulin, can induce tyrosine phosphorylation in a range of proteins including insulin receptors (38). Polyphenol stilbenes (rhaponticin, desoxyrhaponticin, rhapontigenin) from fenugreek seed extract also showed to improve insulin sensitivity and mitochondrial function in 3T3-L1 adipocytes (39). Steroidal saponins and sapogenins of fenugreek have shown inhibitory activity against α-glucosidase in *in vitro* model (40).

Fenugreek has shown a considerable antioxidant effect in cellular studies which can be one of the possible mechanisms in controlling diabetes mellitus (41). *In vitro* study on the radical scavenging activity of fenugreek in isolated mouse pancreatic islets and mitochondrial of rat liver showed that fenugreek has considerable antioxidant capacity which can be due to the high amount of phenolic and flavonoid contents (42). Another study evaluated the effect of fenugreek seed extract on lipid peroxidation which was induced by ferrous sulphate, hydrogen peroxide and carbon tetrachloride in mice liver. The result showed that fenugreek seed extract inhibited hepatic lipid peroxidation in a concentration-dependent manner. Moreover, when both fenugreek seed extract and glibenclamide were added to the incubation mixture, the inhibition was further improved (43).

10.3.2 IN VIVO OBSERVATIONS IN ANIMAL STUDIES

Several animal studies have examined the efficacy of fenugreek in controlling hyperglycaemia in the chemical- or diet-induced models of diabetes mellitus.

It has been shown that fenugreek seed extract has the potential to slow enzymatic digestion of carbohydrates, reduce gastrointestinal absorption of glucose, and thus reduce postprandial glucose levels. Daily oral administration of soluble dietary fibre fraction of fenugreek for 28 days significantly improved oral glucose tolerance in normal, type 1 and type 2 diabetic rats (44). In addition, *in vivo* study of fenugreek seed extract in alloxan-induced diabetic mice showed a comparable hypoglycaemic activity of fenugreek seed extract with that of insulin. It also improved intraperitoneal glucose tolerance in normal mice. This can be explained, at least partly, by the effect of fenugreek in activating an insulin-signalling pathway in liver and adipocytes cells (38). In another study on normal and streptozotocin-induced diabetic rats, oral administration of fenugreek extract for 14 days reduced serum glucose in diabetic rats which was comparable to that observed for glibenclamide. The insulin level was increased in diabetic rats in the treatment group when compared with diabetic control rats. However, administration of the extract did not change glucose and insulin levels in non-diabetic rats (45).

Some studies suggest that fenugreek and diosgenin can modify insulin sensitivity. In a rat model of insulin resistance induced by a high-fat diet, diosgenin and fenugreek seed were administered orally for 6 weeks. Diosgenin and fenugreek seed in some doses improved peripheral insulin sensitivity (46, 47). Study on diabetic mice evaluated the mechanism underlying the protective effect of fenugreek against insulin resistance. The results showed that it can modulate the expression of Glut-2, Glut-4, SREBP-1c and IRS-2, which are involved in the normal response to insulin (48).

Trigonelline, a major alkaloid compound of fenugreek, is suggested to be responsible for some of the antihyperglycaemic activities of fenugreek. In an *in vivo* study on high-fat-fed/streptozotocin-induced type 2 diabetic rats, oral administration of trigonelline significantly attenuated the elevated levels of glucose and improved the insulin level. It also inhibited β cells apoptosis in streptozotocin-induced type 1 diabetic mice (49, 50). Likewise, administration of whole fenugreek seed extract in diabetic rats increased the number and size of pancreatic islet β-cells (51).

Flavonoids are also among the bioactive compounds of fenugreek which have shown a considerable antihyperglycaemic effect. In a rat model of streptozotocin-induced diabetes, administration of the extracted flavonoids of fenugreek seeds resulted in a considerable reduction in serum glucose compared to control (52).

It is well known that adipocyte hypertrophy and chronic inflammation in adipose tissues may cause insulin resistance and type 2 diabetes (53). In an *in vivo* study, obese mice were fed with a high-fat diet supplemented with 2% fenugreek. The results showed that fenugreek not only improved insulin resistance, but also miniaturised the adipocytes and inhibited obesity-related inflammation by decreasing the expression of the inflammatory mRNA genes and inhibiting macrophage infiltration into adipose tissues (37). Trigonelline also showed to reduce some inflammatory factors such as interleukin-6, tumour necrosis factor-α and interleukin-1β in streptozotocin-induced diabetic mice (50).

It is suggested that fenugreek modulates the activity of key enzymes involved in glucose metabolism in both kidney and liver. In an *in vivo* study on alloxan-induced diabetic rats, oral administration of fenugreek seed powder (5% in the diet) significantly reduced fasting blood glucose levels and total liver weight when compared with control. Moreover, the altered activities of glycolytic, gluconeogenic and lipogenic enzymes were reversed to normal in the fenugreek-treated group compare to control (54). The results of a study on largemouth bass showed that the inclusion of fenugreek seed extract elevated hepatic glycogen content and decreased glycolysis through reducing the expression of genes involved in glycolysis and insulin pathway (55).

It is well established that abnormal gut microbiota composition is linked with metabolic disorders, including obesity and diabetes. Galactomannan, the water-soluble polysaccharides in fenugreek gum, has been considered to have the potential to regulate gut microbiota and improve glucose metabolism. In an *in vivo* study on mice, administration of galactomannan in the diet altered microbiota population in the gut. It also increased adenosine monophosphate-activated protein kinase (AMPK) activation in the liver. AMPK is considered as a regulator of glucose metabolism in the liver, which inhibits the activity of acetyl-coenzyme A carboxylase (ACC) and thus leads to a decrease of lipogenesis (56). A metabonomic study evaluated the underlying antidiabetic mechanism of fenugreek galactomannan in streptozotocin-induced diabetic rats. The results of urine and serum analysis of treated rats revealed that administration of galactomannan modifies a range of biomarkers which are involved in energy metabolism and several other metabolisms, such as phenylalanine, histidine, tryptophan, sphingolipid, glycerophospholipid and arachidonic acid (57). Galactomannan is also suggested to improve and protect the pancreas function. In an *in vivo* study on diabetic mice, administration of galactomannan protected the pancreas from histological changes (58).

Glucose overload can damage the cells through oxidative stress and antioxidant treatment may attenuate hyperglycaemia and diabetic complications (59). The results of a study on the effect of fenugreek extract and trigonelline on diabetic rabbits revealed that fenugreek extract and trigonelline decreased the plasma malondialdehyde (MDA) and increased the level of plasma glutathione (GSH) after four weeks of oral administration; however, trigonelline was less effective than the fenugreek extract (58). In some other studies on streptozotocin or alloxan-induced diabetic rats, fenugreek significantly improved several cardiac, liver and pancreas antioxidants, such as superoxide dismutase, compared to control (60–64). Fenugreek could increase antioxidant enzymes and reverse the altered peroxidase damage of the heart and kidney in diabetic rats (61, 65). Trigonelline also showed to reduce oxidative stress parameters in the pancreas (50). Results from animal studies are summarised in Table 10.1.

10.4 CLINICAL STUDIES OF FENUGREEK

Considering the strong preclinical evidence of the effect of fenugreek on glycaemic control, several randomised clinical trials have explored the use of fenugreek as a potential management option for diabetes mellitus.

The first recorded clinical trials on antihyperglycaemic activity of fenugreek date back to the late 1900s. One of those was a clinical trial that evaluated the effect of fenugreek on postprandial glucose and insulin levels following the meal tolerance test (MTT) on non-insulin-dependent diabetic patients. The results showed that fenugreek reduced the subsequent postprandial glucose levels significantly (96).

A recent randomised clinical trial evaluated the effect of fenugreek versus glibenclamide in 12 patients with diabetes. While fenugreek caused insignificant reduction in fasting blood glucose compared to that of glibenclamide, however, it caused significant increase in fasting insulin level (97). In another recent RCT involving 72 patients, administration of an herbal formulation containing 150 mg of fenugreek and some other herbal medicines for 3 months effectively reduced the fasting and 2-hour postprandial glucose level in diabetic participants who were resistant to two conventional medicines (98). Administration of fenugreek powder to patients with type 2 diabetes that were receiving antidiabetic conventional medications caused significant reduction in fasting plasma glucose and improved liver and kidney function in the treatment group compared to the placebo group (99). A multicentre double-blind RCT compared fenugreek seed extract with placebo on 154 diabetic patients. Over a period of 3 months, fenugreek caused significant reduction in fasting blood sugar and non-significant reduction in HbA1c (100).

It is suggested that fenugreek dietary supplement can modulate glucose homeostasis and prevent diabetes mellitus in prediabetic individuals. A long-term randomised, controlled parallel clinical trial evaluated the effect of fenugreek on glucose control in 140 prediabetic individuals for 3 years. Participants were randomly divided to receive fenugreek powder, 5 g twice daily or matched control. By the end of the trial, the incidence rate of diabetes reduced considerably in the

TABLE 10.1
Animal Studies on the Effects of Fenugreek on Glycaemic Status and Other Related Metabolic Factors

Part/Extract/Phytochemical	Model	Animal	Dosage/Rout	Duration	Results	References
seed/soluble dietary fibre fraction	sucrose-induced type 2 diabetes/non-diabetic	rat	0.5 g/kg, twice daily, oral	28 d[1]	↓serum glucose, ↓hepatic glycogen, no change in serum and pancreatic insulin	(44)
seed/dialysed aqueous extract	alloxan-induced diabetes	Swiss albino mice	(15 mg/kg), ip[2]	single dose	↓serum glucose	(38)
seed/ethanol extract	normal and STZ-induced diabetes	rat	0.1, 0.25 and 0.5 g/kg, oral	14 d	↓serum glucose, ↓triacylglycerol, ↑serum insulin	(45)
seed/aqueous extract	obesity	mice	2% of diet, oral	28 d	↓macrophage infiltration to white adipose tissue, ↓insulin resistance, ↓obesity-related inflammation	(37)
seed/hydroalcoholic extract	alloxan-induced diabetes	rat	500, 1000 and 2000 mg/kg, oral	30 d	↑(GSH[3] and SOD[4] in liver and pancreas), inhibition of histopathological change in pancreas and liver	(66)
seed/aqueous and methanolic extract	normal	mice	1 g/kg		↓serum glucose (with both extracts)	(67)
seed/flavonoids	STZ-induced diabetes	rat	0.5 g daily, intragastrical	28 d	↑serum insulin, ↓serum glucose, ↑liver glycogen contents	(52)
seed/powder	alloxan-induced diabetes	rat	5% of diet, oral	21 d	↓serum glucose, ↓body weights, ↑glycolytic enzymes	(54)
seed/oil	alloxan-induced diabetes	rat	10% of diet, oral	4 w	restoring IL-6 rate, ↑serum insulin, ↓serum glucose	(68)
seed/extract	HFD[5]-induced insulin resistance	mice	30, 60 and 100 mg/kg, twice daily, oral	12 w	↓insulin resistance, inhibition of histopathological change in pancreas	(48)
seed/trigonelline	HFD/STZ[6]-induced type 2 diabetes	rat	150 mg/kg twice daily, oral	30 d	↑serum insulin, ↓serum glucose, ↓HbA1c%	(49)
seed/powder	STZ-induced diabetes	rat	1g/kg, oral	8 w	↓serum glucose, alteration in histopathology of pancreas	(69)
seed/powder	STZ-induced diabetes	rat	1g/kg, oral		↑(insulin, GSH, GST, CAT, liver and muscle glycogen content)	(70)
seed/trigonelline and ethanol extract	alloxan-induced diabetes	rabbit	10 mg of trigonelline and 7.7 ml ethane extract twice daily, oral	28 d	↓plasma malondialdehyde (MDA), ↑plasma glutathione (GSH), ↓serum glucose	(71)
seed/aqueous extract	STZ-induced diabetes	rat	1.74 g/kg, oral	56 d	↓serum glucose, ↑glutathione peroxidase, ↑catalase, ↑superoxide dismutase	(60)
seed/powder	alloxan-induced diabetes	rat	(5% w/w), oral	21 d	↑glutathione peroxidase, ↑catalase, ↑superoxide dismutase, ↓peroxidative damage, ↓serum glucose	(61)

Fenugreek on Controlling Glucose

Form	Model	Animal	Dose	Duration	Effects	Ref
seed/powder	STZ-induced diabetes, high-fat diet	rat	not found	not found	↓serum glucose, ↑catalase, ↑superoxide dismutase, ↑glutathione peroxidase, ↑glutathione reductase	(63)
seed/dialysed aqueous extract	alloxan-induced diabetes	mice	150 mg/kg twice daily, ip	15 d	↓serum glucose	(72)
seed/aqueous extract	STZ-induced diabetes	rat	0.44, 0.87, 1.74 g/kg, oral intragastric intubation	6 w	↓serum glucose, ↓HbA1c%	(73)
seed/extract	STZ-induced diabetes	rat	0.8 and 1.6 g/kg, oral	6 w	↓insulin resistance, ↓plasma leptin, ↑adiponectin	(74)
leaf/powder	STZ-induced diabetes	rat	1 g/kg, oral	45 d	↑antioxidants in liver, kidney and pancreas	(75)
leaf/aqueous and alcohol extract	Normal and alloxan-induced diabetes	rat	0.06, 0.2, 0.5, 1 g/kg, ip	1 d	ethanolic extract: ↓serum glucose in diabetic rats/no significant effects in normal rats aqueous extract: ↓serum glucose (in both normal and diabetic rats)	(76)
seed/powder	alloxan-induced diabetes	rat	2 g/kg, oral	30 d	↑antioxidants in liver, kidney and pancreas	(62)
seed/powder	alloxan-induced diabetes	rat	15 mg/100 g, oral	21 d	↓serum glucose, ↓oxidative stress	(77)
seed/water extract	alloxan-induced diabetes	rabbit	50 mg/kg twice daily, oral	29 d	↓FBS, ↓HbA1c%	(78)
seed/extract	—	largemouth bass	0.05–0.3% of diet twice daily, oral	14 w	↓expression of insulin receptor IRb, ↓expression of glycolysis-related genes, ADP dependent glucokinase (ADPGK) and pyruvate kinase (PK), ↑hepatic glycogen accumulation	(55)
seed/powder	STZ-induced diabetes	rat	1g/kg, oral	8 w	↓oxidative stress, ↑liver function	(79)
seed	STZ-induced diabetes	rat	not reported	13 d	↓FBS	(80)
seed/oil	STZ-induced diabetes	mice	60 mg/kg twice daily, oral	28 d	↓serum glucose	(81)
seed/ethanol extract	alloxan-induced diabetes	rat	100 mg/kg, oral	21d	↓serum glucose, ↑pancreas recovery	(82)
seed/hydroalcoholic extract	HFD-induced diabetes	mice	2 g/kg, oral	20 w	↓FBS, ↓plasma insulin, ↓insulin resistance	(83)
seed/hydroalcoholic extract	STZ-induced diabetes	rat	100 mg/kg, oral	8 w	↑number and size of pancreatic islet β-cells, ↓HbA1c%	(51)
seed/powder	nitrate-induced diabetes	rat	5% (w/w) of the diet, oral	16 w	↑glycogen content, inhibiting weight changes, alleviating hyperglycaemia	(84)
seed/4-hydroxyisoleucine	STZ-induced diabetes	rat	50 mg/kg, oral	4 w	↓serum glucose	(85)
seed/4-hydroxyisoleucine	C57BL/KsJ-db/db	mice	50 mg/kg, oral	10 d	↓serum glucose, ↓plasma insulin	(86)
seed/4-hydroxyisoleucine	HFD-induced dyslipidaemia	hamster	50 mg/kg, oral	7 d	↓glycerol, ↓body weight	(87)
seed/4-hydroxyisoleucine	sucrose-lipid-fed-induced diabetes	rat	20 mg/kg/hr, intravenous	—	activation of the PI 3-kinase activity in liver and muscle, ↓insulin resistance	(88)

(Continued)

TABLE 10.1 (Continued)

Part/Extract/Phytochemical	Model	Animal	Dosage/Rout	Duration	Results	References
diosgenin and whole seed	HFD-induced insulin resistance	rat	diosgenin:1 mg/kg, 10 mg/kg, and 50 mg/kg/seed:0.2 g/kg, oral	6 w	↑serum insulin, ↑insulin sensitivity index, ↑metabolic clearance rate of insulin, ↑fasting T4 levels	(46)
seed/diosgenin	HFD and STZ-induced diabetes	rat	60 mg/kg, oral	30 d	↓body weight gain, ↓blood glucose, ↓insulin, ↓insulin resistance	(47)
seed/galactomannan	normal	mice	15% (w/w) of the fibre-enriched diet, oral	3 w	↑AMPK,[8] ↑microbiota populations, ↑adiponectin receptor expression, ↓weight gain, ↓serum glucose	(56)
seed/galactomannan	STZ-induced diabetes	rat	50, 100 and 200 mg/kg, oral	28 d	↓serum glucose, ↓serum arachidonic acid, ↓urocanic acid	(57)
seed/galactomannan	alloxan-induced diabetes	mice	50, 100 and 200 mg/kg, oral	28 d	↓serum glucose (dose-dependently), pancreas protection	(58)
seed/trigonelline	STZ-induced type 1 diabetes	mice	70 mg/kg, oral	4 w	↓serum glucose, ↓serum tumour necrosis factor-α, ↓interleukin-6, ↓interleukin-1β, ↑serum insulin, ↑glutathione concentration, serum activities of superoxide dismutase and catalase in pancreas)	(50)
seed		rat	0.056% of diet, oral	43 d	↑CAT, ↓GPX[9] activities in erythrocyte and liver	(89)
seed/furostanolic saponins	HFD-induced diabetes	mice	not reported	24 w	↓serum glucose, no significant effect on energy expenditure	(90)
seed/ethanol extract	HFD and STZ-induced type 2 diabetes	mice	20 and 80 mg/kg	4 w	↓serum glucose, ↑serum superoxide dismutase, ↑catalase, ↓malondialdehyde	(65)
seed	STZ-induced type 2 diabetes	rat	10% of diet, oral	6 w	↓activity of SOD, ↓activity of CAT, ↑ascorbic acid, ↓glutathione level	(64)
seed (plus mulberry leaf and ginseng root)	HFD and alloxan-induced diabetes	rat	unknown dosage of fenugreek, oral	7 d	↓α-amylase, ↓α-glucosidase, ↑insulin sensitivity, ↓serum glucose, ↓fasting insulin resistance	(91)
seed/saponin fraction	alloxan-induced diabetes	rat	100mg/kg, oral	15 d	↓serum glucose, ↑liver glycogen	(92)
seed/mucilage	STZ-induced type 2 diabetes	rat	25 g, oral	4 w	↓serum glucose, ↓urine glucose	(93)
seed	alloxan-induced diabetes	rat	2 and 8 g/kg	2 w	↓serum glucose	(94)
seed extract/subfractions (a: rich in saponin, b: rich in fibres and protein)	alloxan-induced diabetes	dog			subfraction a: ↓serum glucose, ↓urine glucose; subfraction b: no significant effect on glucose parameters	(95)

fenugreek group when compared to the control group (101). In an uncontrolled clinical trial on diabetic patients who were receiving insulin, add-on therapy with fenugreek seed extract not only reduced the number of insulin doses needed per day but also reduced the hypoglycaemic events in patients (102).

Galactomannans are suggested to have glycaemic control activities. In a recent RCT involving 64 patients with type 2 diabetes, galactomannan was compared with placebo. The results revealed that galactomannan significantly reduced fasting blood glucose, HbA1c, total cholesterol and triglyceride compared to placebo (103).

Saponins are also among active constituents of fenugreek and are known to have antihyperglycaemic properties. In a double-blind and controlled clinical trial, total saponins of fenugreek in combination with sulfonylureas could significantly reduce glucose parameters in diabetic patients compared to sulfonylureas alone over 12 weeks (104).

As discussed, some preclinical studies indicated on the effect of fenugreek on oxidative and inflammatory parameters. However, there is inconsistent evidence from clinical trials to approve its antioxidant and anti-inflammatory activities. In a parallel control clinical trial, 48 patients with type 2 diabetes were divided to intake fenugreek seed powder (15 g/day) or no treatment. After 8 weeks, fenugreek consumption resulted in a significant reduction in high-sensitive protein (hs-CRP) and significant increase in superoxide dismutase (SOD) compared to control; however, there was no significant effect of fenugreek on total antioxidant capacity (TAC), glutathione peroxidase (GPX) activity, tumour necrosis factor (TNF)-α and serum interleukin (IL)-6 (105).

A randomised-controlled crossover clinical trial was carried out to evaluate the metabolic mechanism involved in the glycaemic control activity of fenugreek. A glucose tolerance test (GTT) was performed at the end of each study period. The results of the study indicated that fenugreek significantly reduced the area under the plasma glucose curve (AUC) and increased the metabolic clearance rate. Moreover, fenugreek increased erythrocyte insulin receptors. The outcomes suggest that fenugreek improves peripheral glucose utilisation which can result in an improvement in glucose tolerance. Fenugreek can exert its hypoglycaemic activities not only through insulin receptors, but also at the gastrointestinal level (106).

More details about the related clinical trials are summarised in Table 10.2.

A recent meta-analysis evaluated the effects of fenugreek on cardiometabolic risk factors (including glucose parameters) in adults, using 12 studies which were published between 2001 and 2018. The results of the meta-analysis showed that fenugreek seed compared to placebo can reduce fasting blood glucose by 12.94 mg/dL (95% CI: -21.39 mg/dL, -4.49; I2: 85.0%, p heterogeneity = 0.0001), and can reduce HbA1c by -0.58% (95% CI: -0.99, -0.17%; I2:0%, p heterogeneity = 0.61) (107). Some other previous meta-analysis also confirmed the efficacy of fenugreek in glucose management (9, 108, 109). Considering the high heterogeneity in glycaemic status and low quality of most of the included clinical trials in the mentioned meta-analysis, the findings from meta-analysis need to be interpreted with caution.

10.5 CONCLUSION

Fenugreek is a unique herbal medicine with considerable potential to be used as an effective complementary medicine for the management of hyperglycaemia. Some bioactive components of fenugreek may also serve as potential leads for pharmaceutical drug development as novel antidiabetic agents. Improvement of insulin secretion, regulation of glucose absorption, elevation of glycogen synthesis and anti-inflammatory and antioxidant activities in the liver are among the most important mechanisms of actions underlying the pharmacological effects of fenugreek. However, more RCTs with larger sample sizes and longer durations are warranted to further evaluate the clinical effectiveness of fenugreek as a complementary therapy for prediabetic and diabetic patients.

TABLE 10.2
Clinical Studies on the Effects of Fenugreek on Controlling Glycaemic Status

Fenugreek Part/ Extract/Phytochemical	Type of Study	Intervention	Participants (Number/ Disorder)	Concomitant Drugs	Duration	Results	References
seed	open-label RCT[10]	fenugreek 2 g/day or glibenclamide 5 mg/day	type 2 diabetes	metformin	12 w	↓serum glucose ($p = 0.115$), ↓HbA1c[11] ($p = 0.04$), ↓insulin resistance ($p = 0.007$)	(97)
seed	double-blind RCT	Capsule containing fenugreek (150 mg) plus nettle leaf, berry leaf, onion and cinnamon bark	type 2 diabetes, resistant to conventional medicines	—	12 w	↓FBS[12] ($p = 0.0001$), ↓HbA1c ($p = 0.0001$), 2hppBG[13] ($p = 0.002$)	(98)
seed	parallel RCT	5 g 3 times a day mixed with water	type 2 diabetes	conventional medication	8 w	↓FBS ($p = 0.024$), ↓AST ($p = 0.014$), irisin ($p = 0.001$)	(99)
galactomannan of the seed	randomised, single-blind RCT	1 g/day	type 2 diabetes (newly diagnosed)	—	12 w	↓FBS ($p = 0.024$), ↓HbA1c ($p = 0.0001$)	(103)
seed	parallel RCT	5 g 3 times a day	type 2 diabetes	—	8 w	↓hs-CRP[14] ($p = 0.012$), ↑SOD[15] ($p = 0.001$), ↓IL-6 ($p = 0.69$), ↓TNF-α ($p = 0.84$), ↑GPX ($p = 0.47$)	(105)
seed extract	multicentre RCT	500 mg twice daily	154/type 2 diabetes	—	12 w	↓C-peptide ($p = 0.000$), ↓FBS ($p = 0.000$)	(100)
total saponins	double-blind RCT	2.1 g 3 times daily	69/type 2 diabetes	sulfonylureas	12 w	↓FBS ($p < 0.05$), 2 h PBG ($p < 0.01$), ↓HbA1c ($P < 0.05$)	(104)
seed extract	randomised, cross-over	12.5 g twice daily	5/type 2 diabetes	—	3 w	↓serum glucose ($P < 0.05$), ↓24-hour urinary glucose output ($P < 0.05$)	(110)
seed	randomised, cross-over	25 g daily	5/type 2 diabetes	—	2 w	↓serum glucose, ↑glucose metabolic clearance rate, ↑erythrocyte insulin receptors	(106)
defatted seed	randomised, cross-over	50 g twice daily	15/type 2 diabetes	conventional medication	10 d	↓FBS ($p = 0.05$), ↓serum insulin ($p = 0.01$), ↑glucose tolerance after GTT[16] ($p = 0.01$)	(111)
seed	randomised, controlled, parallel	5 g twice daily	140/prediabetes	—	144 w	↓incidence rate of diabetes ($P < 0.01$), ↓FBS ($P < 0.05$), ↓PPPG[17] ($P < 0.01$), ↑serum insulin ($P < 0.01$)	(101)

Form	Study design	Dose	Subjects	Co-treatment	Duration	Outcomes	Ref
seed extract	double-blind RCT	1 g daily	25/type 2 diabetes (newly diagnosed)	—	8 w	no changes in FBS and 2 h PBG ↓AUC of blood glucose and serum insulin ($P < 0.05$)	(112)
seed flour (incorporated into bread (9%))	double-blind RCT	5 g daily	8/type 2 diabetes	—	evaluation just after consumption	↓AUC of insulin ($P < 0.05$), ↓AUC of blood glucose ($P > 0.05$)	(113)
seed flour (incorporated into bread (10%))	randomised, crossover	20 g daily	10/healthy subjects	—	12 w	↓AUC of blood glucose ($P < 0.05$)	(114)
seed (plus olive leaf polyphenols and bergamot extract)	double-blind RCT	not reported	100/prediabetes	—	24 w	no significant change in FBS, HbA$_{1c}$ or insulin	(115)
seed extract	randomised, double-blind, placebo-controlled, multicentre	700 mg 3 times daily	119/type 2 diabetes	sulfonylurea	12 w	↓HbA1c ($P < 0.001$), ↓PPPG ($P < 0.05$), ↓FBS ($P < 0.001$)	(116)
seed (plus aloe vera leaf gel, black seed, garlic, fenugreek seed, psyllium seed, and milk thistle seed)	RCT	2.5 g twice daily	50/type 2 diabetes	statin	12 w	↓HbA1c ($P < 0.05$)	(117)
seed	randomised, double-blind, placebo-controlled	1000 mg, 3 times daily	13/healthy volunteers	—	10 d	↑glucose tolerance, ↑insulin sensitivity	(118)
seed extract	prospective, open-label, uncontrolled, single-arm, multicentre	700 mg, 3 times daily	30/type 2 diabetes	insulin	16 w	↓HbA1c ($P < 0.01$), ↓number of insulin doses needed per day ($P < 0.01$), ↓hypoglycaemic events ($P < 0.05$)↑	(102)
seed	parallel group, randomised single-blind	10 g daily	60/type 2 diabetes	conventional medicines/insulin	24 w	↓FBS ($p = 0.042$), ↓HbA1c ($p = 0.02$)	(119)
seed	triple-blind randomised controlled clinical trial	5 g daily	88/type 2 diabetes	—	8 w	↓FBS ($p = 0.007$), ↓HbA1c ($p = 0.0001$), ↓serum insulin ($p = 0.03$), ↓homeostatic model assessment for insulin resistance ($p = 0.004$), ↑serum levels of adiponectin ($p = 0.001$)	(120)

(Continued)

TABLE 10.2 (Continued)

Fenugreek Part/Extract/Phytochemical	Type of Study	Intervention	Participants (Number/Disorder)	Concomitant Drugs	Duration	Results	References
leaves, aqueous extract powder	RCT	40 mg/kg	20/healthy volunteers	—	single dose	↓FBS ($P \leq 0.05$)	(121)
seed powder	controlled trial	5 g daily	60/non-insulin-dependent diabetes mellitus	—	24 w	↓24-hour urinary glucose excretion ($P < 0.001$), ↓FBS	(122)
seed powder	crossover trial	100 g daily	/insulin-dependent diabetes mellitus	—	20 d	↓FBS, ↓24-hour urinary sugar excretion	(123)
seed powder (plus jambu seed and bitter ground fruit)	non-controlled trial	3 g daily	60/non-insulin-dependent diabetes mellitus	conventional medicines	12 w	↓FBS, ↓24-hour urinary glucose excretion	(124)
seed powder (in yogurt or hot water)	non-controlled	10 g daily	24/type 2 diabetes	—	8 w	fenugreek seed soaked in hot water: ↓FBS; fenugreek seed mixed with yoghurt: no significant changes	(125)
seed extract (4-hydroxyisoleucine)	randomised, crossover, double-blind trial	148 mg	8/normoglycaemic male endurance athletes	—	immediately after and 2 h after exercise	no difference in muscle glycogen	(126)
seed powder (soaked in dextrose solution)	RCT	2.5 and 5 g	166/type 2 diabetes		single dose	↓postprandial glucose, ↓2 h PBG	(127)

NOTES

1. day(s)
2. intraperitoneal injection
3. glutathione
4. superoxide dismutase
5. high-fat diet
6. streptozotocin
7. glutathione-transferase
8. adenosine monophosphate-activated protein kinase
9. glutathione peroxidase
10. randomised controlled trial
11. glycated haemoglobin A1c
12. fasting blood sugar
13. 2-h postprandial blood glucose
14. high-sensitivity C-reactive protein
15. superoxide dismutase
16. glucose tolerance test
17. postprandial plasma glucose

REFERENCES

1. Nathan DM. Diabetes: Advances in diagnosis and treatment. *Jama*. 2015;314(10):1052–62.
2. Ozougwu J, Obimba K, Belonwu C, Unakalamba C. The pathogenesis and pathophysiology of type 1 and type 2 diabetes mellitus. *Journal of Physiology and Pathophysiology*. 2013;4(4):46–57.
3. Mellitus D. Diagnosis and classification of diabetes mellitus. *Diabetes Care*. 2006;29:S43.
4. Garber A, Henry R, Ratner R, Garcia-Hernandez PA, Rodriguez-Pattzi H, Olvera-Alvarez I, et al. Liraglutide versus glimepiride monotherapy for type 2 diabetes (LEAD-3 Mono): A randomised, 52-week, phase III, double-blind, parallel-treatment trial. *The Lancet*. 2009;373(9662):473–81.
5. Altuntaş E, Özgöz E, Taşer ÖF. Some physical properties of fenugreek (*Trigonella foenum-graceum* L.) seeds. *Journal of Food Engineering*. 2005;71(1):37–43.
6. Bahmani M, Shirzad H, Mirhosseini M, Mesripour A, Rafieian-Kopaei M. A review on ethnobotanical and therapeutic uses of fenugreek (*Trigonella foenum-graecum* L). *Journal of Evidence-Based Complementary & Alternative Medicine*. 2016;21(1):53–62.
7. Necyk C, Zubach-Cassano L. Natural health products and diabetes: A practical review. *Canadian Journal of Diabetes*. 2017;41(6):642–7.
8. Smith M. Therapeutic applications of fenugreek. *Alternative Medicine Review*. 2003;8(1):20–7.
9. Neelakantan N, Narayanan M, de Souza RJ, van Dam RM. Effect of fenugreek (*Trigonella foenum-graecum* L.) intake on glycemia: A meta-analysis of clinical trials. *Nutrition Journal*. 2014;13(1):7.
10. Medagama AB, Senadhira D. Use of household ingredients as complementary medicines for perceived hypoglycemic benefit among Sri Lankan diabetic patients; a cross-sectional survey. *Journal of Intercultural Ethnopharmacology*. 2015;4(2):138.
11. Zarshenas MM, Khademian S, Moein M. Diabetes and related remedies in medieval Persian medicine. *Indian Journal of Endocrinology and Metabolism*. 2014;18(2):142.
12. Habtemariam S. The chemical and pharmacological basis of fenugreek (*Trigonella foenum-graecum* L.) as potential therapy for type 2 diabetes and associated diseases. *Medicinal Foods as Potential Therapies for Type-2 Diabetes and Associated Diseases*. 2019: 579–637.
13. Khorshidian N, Yousefi Asli M, Arab M, Adeli Mirzaie A, Mortazavian AM. Fenugreek: Potential applications as a functional food and nutraceutical. *Nutrition and Food Sciences Research*. 2016;3(1):5–16.
14. Kouzi SA, Yang S, Nuzum DS, Dirks-Naylor AJ. Natural supplements for improving insulin sensitivity and glucose uptake in skeletal muscle. *Front Biosci (Elite Ed)*. 2015;7(1):94–106.
15. Nagulapalli Venkata KC, Swaroop A, Bagchi D, Bishayee A. A small plant with big benefits: Fenugreek (*Trigonella foenum-graecum* Linn.) for disease prevention and health promotion. *Molecular Nutrition & Food Research*. 2017;61(6):1600950.
16. Demmers A, Korthout H, van Etten-Jamaludin FS, Kortekaas F, Maaskant JM. Effects of medicinal food plants on impaired glucose tolerance: A systematic review of randomized controlled trials. *Diabetes Research and Clinical Practice*. 2017;131:91–106.

17. Pradeep SR, Srinivasan K. Chapter 28 — synergy among dietary spices in exerting antidiabetic influences. In: Watson RR, Preedy VR, editors. *Bioactive Food as Dietary Interventions for Diabetes* (Second Edition). Academic Press; 2019, pp. 407–24.
18. Sanlier N, Gencer F. Role of spices in the treatment of diabetes mellitus: A minireview. *Trends in Food Science & Technology*. 2020;99:441–9.
19. Koupý D, Kotolova H, Rudá JK. Effectiveness of phytotherapy in supportive treatment of type 2 diabetes mellitus II. Fenugreek (*Trigonella foenum-graecum*). *Ceska a Slovenska farmacie: Casopis Ceske farmaceuticke spolecnosti a Slovenske farmaceuticke spolecnosti.* 2015;64(3):67–71.
20. Madhava Naidu M, Shyamala BN, Pura Naik J, Sulochanamma G, Srinivas P. Chemical composition and antioxidant activity of the husk and endosperm of fenugreek seeds. *LWT—Food Science and Technology*. 2011;44(2):451–6.
21. Shim SH, Lee EJ, Kim JS, Kang SS, Ha H, Lee HY, et al. Rat growth-hormone release stimulators from fenugreek seeds. *Chemistry & Biodiversity*. 2008;5(9):1753–61.
22. Panda S, Tahiliani P, Kar A. Inhibition of triiodothyronine production by fenugreek seed extract in mice and rats. *Pharmacological Research*. 1999;40(5):405–9.
23. Tahiliani P, Kar A. The combined effects of Trigonella and Allium extracts in the regulation of hyperthyroidism in rats. *Phytomedicine*. 2003;10(8):665–8.
24. Bian D, Li Z, Ma H, Mu S, Ma C, Cui B, et al. Effects of diosgenin on cell proliferation induced by IGF-1 in primary human thyrocytes. *Archives of Pharmacal Research*. 2011;34(6):997–1005.
25. Garg R. *Chapter 12-Nutraceuticals in Glucose Balance and Diabetes. Nutraceuticals*. Boston: Academic Press; 2016.
26. Kong Z-L, Che K, Hu J-X, Chen Y, Wang Y-Y, Wang X, et al. Orientin protects podocytes from high glucose induced apoptosis through mitophagy. *Chemistry & Biodiversity*. 2020;17(3):e1900647.
27. Zhou J, Chan L, Zhou S. Trigonelline: A plant alkaloid with therapeutic potential for diabetes and central nervous system disease. *Current Medicinal Chemistry*. 2012;19(21):3523–31.
28. Luitse MJA, Biessels GJ, Rutten GEHM, Kappelle LJ. Diabetes, hyperglycaemia, and acute ischaemic stroke. *The Lancet Neurology*. 2012;11(3):261–71.
29. Nathan DM, Buse JB, Davidson MB, Heine RJ, Holman RR, Sherwin R, et al. Management of hyperglycemia in type 2 diabetes: A consensus algorithm for the initiation and adjustment of therapy: A consensus statement from the American diabetes association and the European association for the study of diabetes. *Diabetes Care*. 2006;29(8):1963–72.
30. Avalos-Soriano A, la Cruz-Cordero D, Rosado JL, Garcia-Gasca T. 4-Hydroxyisoleucine from fenugreek (*Trigonella foenum-graecum*): Effects on insulin resistance associated with obesity. *Molecules*. 2016;21(11):1596.
31. Jain D, Bains K, Singla N. Mode of action of anti-diabetic Phyto-compounds present in traditional Indian plants: A review. *Current Journal of Applied Science and Technology*. 2020:19–38.
32. Rangari VD, Shukla P, Badole SL. Chapter 15–4-hydroxyisoleucine: A potential antidiabetic agent from *Trigonella foenum-graecum*. In: Watson RR, Dokken BB, editors. *Glucose Intake and Utilization in Pre-Diabetes and Diabetes*. Boston: Academic Press; 2015, pp. 191–8.
33. Lu F, Cai Q, Zafar MI, Cai L, Du W, Jian L, et al. 4-Hydroxyisoleucine improves hepatic insulin resistance by restoring glycogen synthesis *in vitro*. *International Journal of Clinical and Experimental Medicine*. 2015;8(6):8626.
34. Naicker N, Nagiah S, Phulukdaree A, Chuturgoon A. *Trigonella foenum-graecum* seed extract, 4-hydroxyisoleucine, and metformin stimulate proximal insulin signaling and increase expression of glycogenic enzymes and GLUT2 in HepG2 cells. *Metabolic Syndrome and Related Disorders*. 2016;14(2):114–20.
35. Korthikunta V, Pandey J, Singh R, Srivastava R, Srivastava AK, Tamrakar AK, et al. In vitro antihyperglycemic activity of 4-hydroxyisoleucine derivatives. *Phytomedicine*. 2015;22(1):66–70.
36. Jaiswal N, Maurya CK, Venkateswarlu K, Sukanya P, Srivastava AK, Narender T, et al. 4-Hydroxyisoleucine stimulates glucose uptake by increasing surface GLUT4 level in skeletal muscle cells via phosphatidylinositol-3-kinase-dependent pathway. *European Journal of Nutrition*. 2012;51(7):893–8.
37. Uemura T, Hirai S, Mizoguchi N, Goto T, Lee JY, Taketani K, et al. Diosgenin present in fenugreek improves glucose metabolism by promoting adipocyte differentiation and inhibiting inflammation in adipose tissues. *Molecular Nutrition & Food Research*. 2010;54(11):1596–608.
38. Vijayakumar MV, Singh S, Chhipa RR, Bhat MK. The hypoglycaemic activity of fenugreek seed extract is mediated through the stimulation of an insulin signalling pathway. *British Journal of Pharmacology*. 2005;146(1):41–8.

39. Li G, Luan G, He Y, Tie F, Wang Z, Suo Y, et al. Polyphenol stilbenes from Fenugreek (*Trigonella foenum-graecum* L.) seeds improve insulin sensitivity and mitochondrial function in 3T3-L1 adipocytes. *Oxidative Medicine and Cellular Longevity.* 2018;2018:9.
40. Zhang H, Xu J, Wang M, Xia X, Dai R, Zhao Y. Steroidal saponins and sapogenins from fenugreek and their inhibitory activity against α-glucosidase. *Steroids.* 2020;161:108690.
41. Anusha MB, Shivanna N, Kumar GP, Anilakumar KR. Efficiency of selected food ingredients on protein efficiency ratio, glycemic index and *in vitro* digestive properties. *Journal of Food Science and Technology.* 2018;55(5):1913–21.
42. Dixit PP, Devasagayam TPA, Ghaskadbi S. Formulated antidiabetic preparation Syndrex® has a strong antioxidant activity. *European Journal of Pharmacology.* 2008;581(1):216–25.
43. Neha S, Anand K, Sunanda P. Administration of Fenugreek seed extract produces better effects in glibenclamide-induced inhibition in hepatic lipid peroxidation: An *in vitro* study. *Chinese Journal of Integrative Medicine.* 2019;25(4):278–84.
44. Hannan J, Ali L, Rokeya B, Khaleque J, Akhter M, Flatt P, et al. Soluble dietary fibre fraction of *Trigonella foenum-graecum* (fenugreek) seed improves glucose homeostasis in animal models of type 1 and type 2 diabetes by delaying carbohydrate digestion and absorption, and enhancing insulin action. *British Journal of Nutrition.* 2007;97(3):514–21.
45. Eidi A, Eidi M, Sokhteh M. Effect of fenugreek (*Trigonella foenum-graecum* L) seeds on serum parameters in normal and streptozotocin-induced diabetic rats. *Nutrition Research.* 2007;27(11):728–33.
46. Kiss R, Pesti-Asbóth G, Szarvas MM, Stündl L, Cziáky Z, Hegedűs C, et al. Diosgenin and its fenugreek based biological matrix affect insulin resistance and anabolic hormones in a rat based insulin resistance model. *BioMed Research International.* 2019;2019.
47. Naidu PB, Ponmurugan P, Begum MS, Mohan K, Meriga B, RavindarNaik R, et al. Diosgenin reorganises hyperglycaemia and distorted tissue lipid profile in high-fat diet—streptozotocin-induced diabetic rats. *Journal of the Science of Food and Agriculture.* 2015;95(15):3177–82.
48. Kandhare AD, Bodhankar SL, Mohan V, Thakurdesai PA. Prophylactic efficacy and possible mechanisms of oligosaccharides based standardized fenugreek seed extract on high-fat diet-induced insulin resistance in C57BL/6 mice. *Journal of Applied Pharmaceutical Science.* 2015;5:35–45.
49. Subramanian SP, Prasath GS. Antidiabetic and antidyslipidemic nature of trigonelline, a major alkaloid of fenugreek seeds studied in high-fat-fed and low-dose streptozotocin-induced experimental diabetic rats. *Biomedicine & Preventive Nutrition.* 2014;4(4):475–80.
50. Liu L, Du X, Zhang Z, Zhou J. Trigonelline inhibits caspase 3 to protect β cells apoptosis in streptozotocin-induced type 1 diabetic mice. *European Journal of Pharmacology.* 2018;836:115–21.
51. Kulkarni CP, Bodhankar SL, Ghule AE, Mohan V, Thakurdesai PA. Antidiabetic activity of *Trigonella foenum-graecum* L. seeds extract (IND01) in neonatal streptozotocin-induced (n-STZ) rats. *Diabetologia Croatica.* 2012;41(1).
52. Jiang W, Si L, Li P, Bai B, Qu J, Hou B, et al. Serum metabonomics study on antidiabetic effects of fenugreek flavonoids in streptozotocin-induced rats. *Journal of Chromatography B.* 2018;1092:466–72.
53. Gustafson B, Hedjazifar S, Gogg S, Hammarstedt A, Smith U. Insulin resistance and impaired adipogenesis. *Trends in Endocrinology & Metabolism.* 2015;26(4):193–200.
54. Raju J, Gupta D, Rao AR, Yadava PK, Baquer NZ. *Trigonella foenum-graecum* (fenugreek) seed powder improves glucose homeostasis in alloxan diabetic rat tissues by reversing the altered glycolytic, gluconeogenic and lipogenic enzymes. *Molecular and Cellular Biochemistry.* 2001;224(1–2):45–51.
55. Wang A, Li S, Liu Y, Han Z, Chen N. The inclusion of fenugreek seed extract aggravated hepatic glycogen accumulation through reducing the expression of genes involved in insulin pathway and glycolysis in largemouth bass, Micropterus salmoides. *Aquaculture.* 2020;528:735567.
56. Shtriker MG, Hahn M, Taieb E, Nyska A, Moallem U, Tirosh O, et al. Fenugreek galactomannan and citrus pectin improve several parameters associated with glucose metabolism and modulate gut microbiota in mice. *Nutrition.* 2018;46:134–42.
57. Jiang W, Gao L, Li P, Kan H, Qu J, Men L, et al. Metabonomics study of the therapeutic mechanism of fenugreek galactomannan on diabetic hyperglycemia in rats, by ultra-performance liquid chromatography coupled with quadrupole time-of-flight mass spectrometry. *Journal of Chromatography B.* 2017;1044–5:8–16.
58. Kamble H, Kandhare AD, Bodhankar S, Mohan V, Thakurdesai P. Effect of low molecular weight galactomannans from fenugreek seeds on animal models of diabetes mellitus. *Biomedicine & Aging Pathology.* 2013;3(3):145–51.
59. Ceriello A, Testa R. Antioxidant anti-inflammatory treatment in type 2 diabetes. *Diabetes Care.* 2009;32(Suppl 2):S232–S6.

60. Haghani K, Bakhtiyari S, Doost Mohammadpour J. Alterations in Plasma glucose and cardiac antioxidant enzymes activity in streptozotocin-induced diabetic rats: Effects of *Trigonella foenum-graecum* extract and swimming training. *Canadian Journal of Diabetes*. 2016;40(2):135–42.
61. Genet S, Kale RK, Baquer NZ. Alterations in antioxidant enzymes and oxidative damage in experimental diabetic rat tissues: Effect of vanadate and fenugreek (*Trigonella foenum graecum*). *Molecular and Cellular Biochemistry*. 2002;236(1–2):7–12.
62. Anuradha C, Ravikumar P. Restoration on tissue antioxidants by fenugreek seeds (*Trigonella Foenum graecum*) in alloxan-diabetic rats. *Indian Journal of Physiology and Pharmacology*. 2001;45(4):408–20.
63. Sankar P, Subhashree S, Sudharani S. Effect of *Trigonella foenum-graecum* seed powder on the antioxidant levels of high fat diet and low dose streptozotocin induced type II diabetic rats. *European Review for Medical and Pharmacological Sciences*. 2012;16:10–7.
64. Pradeep SR, Srinivasan K. Amelioration of oxidative stress by dietary fenugreek (*Trigonella foenum-graecum* L.) seeds is potentiated by onion (*Allium cepa* L.) in streptozotocin-induced diabetic rats. *Applied Physiology, Nutrition, and Metabolism*. 2017;42(8):816–28.
65. Xiao-yan L, Shuang-shuang L, Hong-lun W, Li G, Yan-feng H, Xiao-yu L, et al. Effects of the fenugreek extracts on high-fat diet-fed and streptozotocin-induced type 2 diabetic mice. *Animal Models and Experimental Medicine*. 2018;1(1):68–73.
66. Yella SST, Kumar RN, Ayyanna C, Varghese AM, Amaravathi P, Vangoori Y. The combined effect of Trigonella foenum seeds and Coriandrum sativum leaf extracts in alloxan-induced diabetes mellitus wistar albino rats. *Bioinformation*. 2019;15(10):716.
67. Hamza N, Berke B, Cheze C, Le Garrec R, Umar A, Agli A-N, et al. Preventive and curative effect of Trigonella foenum-graecum L. seeds in C57BL/6J models of type 2 diabetes induced by high-fat diet. *Journal of Ethnopharmacology*. 2012;142(2):516–22.
68. El-Wakf AM, Hassan HA, Mahmoud AZ, Habza MN. Fenugreek potent activity against nitrate-induced diabetes in young and adult male rats. *Cytotechnology*. 2015;67(3):437–47.
69. Haeri MR, Limaki HK, White CJB, White KN. Non-insulin dependent anti-diabetic activity of (2S, 3R, 4S) 4-hydroxyisoleucine of fenugreek (*Trigonella foenum graecum*) in streptozotocin-induced type I diabetic rats. *Phytomedicine*. 2012;19(7):571–4.
70. Singh A, Tamarkar A, Shweta, Narender T, Srivastava AK. Antihyperglycaemic effect of an unusual amino acid (4-hydroxyisoleucine) in C57BL/KsJ-db/db mice. *Natural Product Research*. 2010;24(3):258–65.
71. Narender T, Puri A, Khaliq T, Saxena R, Bhatia G, Chandra R. 4-Hydroxyisoleucine an unusual amino acid as antidyslipidemic and antihyperglycemic agent. *Bioorganic & Medicinal Chemistry Letters*. 2006;16(2):293–6.
72. Broca C, Breil V, Cruciani-Guglielmacci C, Manteghetti M, Rouault C, Derouet M, et al. Insulinotropic agent ID-1101 (4-hydroxyisoleucine) activates insulin signaling in rat. *American Journal of Physiology-Endocrinology and Metabolism*. 2004;287(3):E463–E71.
73. Yoshinari O, Takenake A, Igarashi K. Trigonelline ameliorates oxidative stress in type 2 diabetic Goto-Kakizaki rats. *Journal of Medicinal Food*. 2013;16(1):34–41.
74. Hua Y, Ren SY, Guo R, Rogers O, Nair RP, Bagchi D, et al. Furostanolic saponins from Trigonella foenum-graecum alleviate diet-induced glucose intolerance and hepatic fat accumulation. *Molecular Nutrition & Food Research*. 2015;59(10):2094–100.
75. Kan J, Velliquette RA, Grann K, Burns CR, Scholten J, Tian F, et al. A novel botanical formula prevents diabetes by improving insulin resistance. *BMC Complementary and Alternative Medicine*. 2017;17(1):352.
76. Bansode TS, Salalkar B, Dighe P, Nirmal S, Dighe S. Comparative evaluation of antidiabetic potential of partially purified bioactive fractions from four medicinal plants in alloxan-induced diabetic rats. *Ayu*. 2017;38(3–4):165.
77. Kumar GS, Shetty A, Sambaiah K, Salimath P. Antidiabetic property of fenugreek seed mucilage and spent turmeric in streptozotocin-induced diabetic rats. *Nutrition Research*. 2005;25(11):1021–8.
78. Khosla P, Gupta D, Nagpal R. Effect of *Trigonella foenum-graecum* (Fenugreek) on blood glucose in normal and diabetic rats. *Indian Journal of Physiology and Pharmacology*. 1995;39:173.
79. Ribes G, Sauvaire Y, Costa CD, Baccou J, Loubatieres-Mariani M. Antidiabetic effects of subtractions from fenugreek seeds in diabetic dogs. *Proceedings of the Society for Experimental Biology and Medicine*. 1986;182(2):159–66.
80. Madar Z, Abel R, Samish S, Arad J. Glucose-lowering effect of fenugreek in non-insulin dependent diabetics. *European Journal of Clinical Nutrition*. 1988;42(1):51–4.

81. Najdi RA, Hagras MM, Kamel FO, Magadmi RM. A randomized controlled clinical trial evaluating the effect of *Trigonella foenum-graecum* (fenugreek) versus glibenclamide in patients with diabetes. *African Health Sciences.* 2019;19(1):1594–601.
82. Parham M, Bagherzadeh M, Asghari M, Akbari H, Hosseini Z, Rafiee M, et al. Evaluating the effect of a herb on the control of blood glucose and insulin-resistance in patients with advanced type 2 diabetes (a double-blind clinical trial). *Caspian Journal of Internal Medicine.* 2020;11(1):12.
83. Hadi A, Arab A, Hajianfar H, Talaei B, Miraghajani M, Babajafari S, et al. The effect of fenugreek seed supplementation on serum irisin levels, blood pressure, and liver and kidney function in patients with type 2 diabetes mellitus: A parallel randomized clinical trial. *Complementary Therapies in Medicine.* 2020;49:102315.
84. Verma N, Usman K, Patel N, Jain A, Dhakre S, Swaroop A, et al. A multicenter clinical study to determine the efficacy of a novel fenugreek seed (*Trigonella foenum-graecum*) extract (Fenfuro™) in patients with type 2 diabetes. *Food & Nutrition Research.* 2016;60(1):32382.
85. Gaddam A, Galla C, Thummisetti S, Marikanty RK, Palanisamy UD, Rao PV. Role of Fenugreek in the prevention of type 2 diabetes mellitus in prediabetes. *Journal of Diabetes & Metabolic Disorders.* 2015;14(1):74.
86. Kandhare A, Phadke U, Mane A, Thakurdesai P, Bhaskaran S. Add-on therapy of herbal formulation rich in standardized fenugreek seed extract in type 2 diabetes mellitus patients with insulin therapy: An efficacy and safety study. *Asian Pacific Journal of Tropical Biomedicine.* 2018;8(9):446–55.
87. Rashid R, Ahmad H, Ahmed Z, Rashid F, Khalid N. Clinical investigation to modulate the effect of fenugreek polysaccharides on type-2 diabetes. *Bioactive Carbohydrates and Dietary Fibre.* 2019;19:100194.
88. Lu F-r, Shen L, Qin Y, Gao L, Li H, Dai Y. Clinical observation on *Trigonella foenum-graecum* L. total saponins in combination with sulfonylureas in the treatment of type 2 diabetes mellitus. *Chinese Journal of Integrative Medicine.* 2008;14(1):56–60.
89. Tavakoly R, Maracy MR, Karimifar M, Entezari MH. Does fenugreek (*Trigonella foenum-graecum*) seed improve inflammation, and oxidative stress in patients with type 2 diabetes mellitus? A parallel group randomized clinical trial. *European Journal of Integrative Medicine.* 2018;18:13–7.
90. Raghuram T, Sharma R, Sivakumar B, Sahay B. Effect of fenugreek seeds on intravenous glucose disposition in non-insulin dependent diabetic patients. *Phytotherapy Research.* 1994;8(2):83–6.
91. Khodamoradi K, Khosropanah MH, Ayati Z, Chang D, Nasli-Esfahani E, Ayati MH, et al. The effects of fenugreek on cardiometabolic risk factors in adults: A systematic review and meta-analysis. *Complementary Therapies in Medicine.* 2020;52:102416.
92. Gong J, Fang K, Dong H, Wang D, Hu M, Lu F. Effect of fenugreek on hyperglycaemia and hyperlipidemia in diabetes and prediabetes: A meta-analysis. *Journal of Ethnopharmacology.* 2016;194:260–8.
93. Suksomboon N, Poolsup N, Boonkaew S, Suthisisang CC. Meta-analysis of the effect of herbal supplement on glycemic control in type 2 diabetes. *Journal of Ethnopharmacology.* 2011;137(3):1328–33.
94. Joshi DV, Patil RR, Naik SR. Hydroalcohol extract of *Trigonella foenum-graecum* seed attenuates markers of inflammation and oxidative stress while improving exocrine function in diabetic rats. *Pharmaceutical Biology.* 2015;53(2):201–11.
95. Zia T, Hasnain SN, Hasan SK. Evaluation of the oral hypoglycaemic effect of *Trigonella foenum-graecum* L. (methi) in normal mice. *Journal of Ethnopharmacology.* 2001;75(2):191–5.
96. Hamden K, Masmoudi H, Carreau S, Elfeki A. Immunomodulatory, β-cell, and neuroprotective actions of fenugreek oil from alloxan-induced diabetes. *Immunopharmacology and Immunotoxicology.* 2010;32(3):437–45.
97. Haritha C, Reddy AG, Reddy YR, Anilkumar B. Pharmacodynamic interaction of fenugreek, insulin and glimepiride on Sero-biochemical parameters in diabetic Sprague-Dawley rats. *Veterinary World.* 2015;8(5):656.
98. Marzouk M, Soliman A, Omar T. Hypoglycemic and antioxidative effects of fenugreek and termis seeds powder in streptozotocin-diabetic rats. *European Review for Medical and Pharmacological Sciences.* 2013;17(4):559–65.
99. Hamadi SA. Effect of trigonelline and ethanol extract of Iraqi Fenugreek seeds on oxidative stress in alloxan diabetic rabbits. *Journal of the Association of Arab Universities for Basic and Applied Sciences.* 2012;12(1):23–6.
100. Vijayakumar MV, Bhat MK. Hypoglycemic effect of a novel dialysed fenugreek seeds extract is sustainable and is mediated, in part, by the activation of hepatic enzymes. *Phytotherapy Research.* 2008;22(4):500–5.

101. Xue W-L, Li X-S, Zhang J, Liu Y-H, Wang Z-L, Zhang R-J. Effect of *Trigonella foenum-graecum* (fenugreek) extract on blood glucose, blood lipid and hemorheological properties in streptozotocin-induced diabetic rats. *Asia Pacific Clinical Nutrition.* 2007;16(Suppl 1):422–6.
102. Arshadi S, Ali Azarbayjani M, Hajiaghaalipour F, Yusof A, Peeri M, Bakhtiyari S, et al. Evaluation of Trigonella foenum-graecum extract in combination with swimming exercise compared to glibenclamide consumption on type 2 Diabetic rodents. *Food & Nutrition Research.* 2015;59(1):29717.
103. Annida B, Prince PSM. Supplementation of fenugreek leaves reduces oxidative stress in streptozotocin-induced diabetic rats. *Journal of Medicinal Food.* 2005;8(3):382–5.
104. Abdel-Barry JA, Abdel-Hassan IA, Al-Hakiem MH. Hypoglycaemic and antihyperglycaemic effects of Trigonella foenum-graecum leaf in normal and alloxan induced diabetic rats. *Journal of Ethnopharmacology.* 1997;58(3):149–55.
105. Kumar P, Kale RK, Baquer NZ. Antihyperglycemic and protective effects of *Trigonella foenum-graecum* seed powder on biochemical alterations in alloxan diabetic rats. *European Review for Medical and Pharmacological Sciences.* 2012;16(Suppl 3):18–27.
106. Puri D, Prabhu K, Murthy P. Antidiabetic effect of GII compound purified from fenugreek (Trigonella foenum-graecum Linn) seeds in diabetic rabbits. *Indian Journal of Clinical Biochemistry.* 2012;27(1):21–7.
107. Haritha C, Reddy AG, Reddy YR, Anjaneyulu Y, Rao TM, Kumar BA, et al. Evaluation of protective action of fenugreek, insulin and glimepiride and their combination in diabetic Sprague Dawley rats. *Journal of Natural Science, Biology, and Medicine.* 2013;4(1):207.
108. Mondal D, Yousuf B, Banu L, Ferdousi R, Khalil M, Shamim K. Effect of fenugreek seeds on the fasting blood glucose level in the streptozotocin induced diabetic rats. *Mymensingh Medical Journal: MMJ.* 2004;13(2):161–4.
109. Al-Amoudi NS, Araki HAA. Evaluation of vegetable and fish oils diets for the amelioration of diabetes side effects. *Journal of Diabetes & Metabolic Disorders.* 2013;12(1):13.
110. Sharma RD. Effect of fenugreek seeds and leaves on blood glucose and serum insulin responses in human subjects. *Nutrition Research.* 1986;6(12):1353–64.
111. Sharma R, Raghuram T. Hypoglycaemic effect of fenugreek seeds in non-insulin dependent diabetic subjects. *Nutrition Research.* 1990;10(7):731–9.
112. Gupta A, Gupta R, Lal B. Effect of *Trigonella foenum-graecum* (Fenugreek) seeds on glycaemic control and insulin resistance in type 2 diabetes. *Journal of Association Physicians India.* 2001;49:1057–61.
113. Losso JN, Holliday DL, Finley JW, Martin RJ, Rood JC, Yu Y, et al. Fenugreek bread: A treatment for diabetes mellitus. *Journal of Medicinal Food.* 2009;12(5):1046–9.
114. Robert SD, Ismail AA-s, Rosli WIW. Reduction of postprandial blood glucose in healthy subjects by buns and flatbreads incorporated with fenugreek seed powder. *European Journal of Nutrition.* 2016;55(7):2275–80.
115. Florentin M, Liberopoulos E, Elisaf MS, Tsimihodimos V. No effect of fenugreek, bergamot and olive leaf extract on glucose homeostasis in patients with prediabetes: A randomized double-blind placebo-controlled study. *Archives of Medical Sciences Atherosclerotic Diseases.* 2019;4:e162.
116. Kandhare AD, Rais N, Moulick N, Deshpande A, Thakurdesai P, Bhaskaran S. Efficacy and safety of herbal formulation rich in standardized fenugreek seed extract as add-on supplementation in patients with type 2 diabetes mellitus on sulfonylurea therapy: A 12-week, randomized, double-blind, placebo-controlled, multi-center study. *Pharmacognosy Magazine.* 2018;14(57):393.
117. Ghorbani A, Zarvandi M, Rakhshandeh H. A randomized controlled trial of a herbal compound for improving metabolic parameters in diabetic patients with uncontrolled dyslipidemia. *Endocrine, Metabolic & Immune Disorders-Drug Targets (Formerly Current Drug Targets-Immune, Endocrine & Metabolic Disorders).* 2019;19(7):1075–82.
118. Kiss R, Szabó K, Gesztelyi R, Somodi S, Kovács P, Szabó Z, et al. Insulin-sensitizer effects of fenugreek seeds in parallel with changes in plasma MCH levels in healthy volunteers. *International Journal of Molecular Sciences.* 2018;19(3):771.
119. Ranade M, Mudgalkar N. A simple dietary addition of fenugreek seed leads to the reduction in blood glucose levels: A parallel group, randomized single-blind trial. *Ayu.* 2017;38(1).
120. Rafraf M, Malekiyan M, Asghari-Jafarabadi M, Aliasgarzadeh A. Effect of fenugreek seeds on serum metabolic factors and adiponectin levels in type 2 diabetic patients. *International Journal for Vitamin and Nutrition Research.* 2014;84(3–4):196–205.
121. Abdel Barry J. Hypoglycaemic effect of aqueous extract of the leaves of *Trigonella foenum-graecum* in healthy volunteers. *EMHJ-Eastern Mediterranean Health Journal.* 2000;6(1): 83–88.
122. Sharma RD, Sarkar A, Hazara DK, Mishra B, Singh JB, Sharma SK, et al. Use of Fenuqreek seed powder in the management of non-insulin dependent diabetes mellitus. *Nutrition Research.* 1996;16(8):1331–9.

123. Sharma R, Raghuram T, Rao NS. Effect of fenugreek seeds on blood glucose and serum lipids in type I diabetes. *European Journal of Clinical Nutrition*. 1990;44(4):301–6.
124. Kochhar A, Nagi M. Effect of supplementation of traditional medicinal plants on blood glucose in non-insulin-dependent diabetics: A pilot study. *Journal of Medicinal Food*. 2005;8(4):545–9.
125. Kassaian N, Azadbakht L, Forghani B, Amini M. Effect of fenugreek seeds on blood glucose and lipid profiles in type 2 diabetic patients. *International Journal for Vitamin and Nutrition Research*. 2009;79(1):34–9.
126. Slivka D, Cuddy J, Hailes W, Harger S, Ruby B. Glycogen resynthesis and exercise performance with the addition of fenugreek extract (4-hydroxyisoleucine) to post-exercise carbohydrate feeding. *Amino Acids*. 2008;35(2):439–44.
127. Bawadi HA, Maghaydah SN, Tayyem RF, Tayyem RF. The postprandial hypoglycemic activity of fenugreek seed and seeds' extract in type 2 diabetics: A pilot study. *Pharmacognosy Magazine*. 2009;5(18):134.

11 Fenugreek in Management of Primary Hyperlipidaemic Conditions

Subhash L. Bodhankar, Amit D. Kandhare and Amol P. Muthal

CONTENTS

11.1 Introduction 150
    11.1.1 Fenugreek against Hyperlipidaemia: Preclinical (*In Vivo*) Evidence 151
        11.1.1.1 Fenugreek Seed Powder and Lipids 151
        11.1.1.2 Fenugreek Seed Extracts and Lipids 152
        11.1.1.3 Fenugreek Seed Galactomannan (GAL) and Lipids 156
        11.1.1.4 Fenugreek Saponins and Lipids 157
        11.1.1.5 Fenugreek Amino Acid and Alkaloid Constituent and Lipids ... 159
    11.1.2 Fenugreek against Hyperlipidaemia: Preclinical (*In Vitro*) Evidence 161
    11.1.3 Fenugreek against Hyperlipidaemia: Clinical Evidence 162
11.2 Conclusion 165
References 166

ABBREVIATIONS

4-HI: 4-Hydroxyisoleucine; ABCA1: $AlCl_3$—Aluminium chloride, ATP-Binding Cassette Transporter A1; ABCG5/G8: ATP-Binding Cassette G5/G8; ACC: Acetyl-CoA Carboxylase; AMPK: Adenosine Monophosphate-Activated Protein Kinase; aP2: adipocyte Fatty Acid Binding Protein; C/EBP-α: CCAAT/Enhancer Binding Protein; CCl_4—Carbon tetrachloride, Cd137: Tumour Necrosis Factor Receptor Superfamily Member 9 (TNFRSF9); Cited1: Cbp/p300-Interacting Transactivator; COX-2: Cylcooxgenase-2; Cox4: Cytochrome c Oxidase Subunit 4; CYP27A1: Cholesterol-27α-hydroxylase; CYP7A1: Cholesterol-7α-hydroxylase; DB: Double-Blind; DLL1: Delta Like Protein-1; FABP-4: Fatty Acids Binding Protein-4; FAS: Fatty Acid Synthase; FFA: Free Fatty Acid; FGF21: Fibroblast Growth Factor 21; FRU: High Fructose Diet; FSE: Fenugreek Seed Extract; FSP: Fenugreek Seed Powder; G6PD: Glucose-6-Phosphate Dehydrogenase; GAL: Galactomannans; HCD: High-Cholesterol Diet; HDL-C: High-Density Lipoprotein-Cholesterol; Hes1: Hairy and Enhancer of Split-1; Hey: Hes related with YRPW Motif Protein 1; HFD: High-Fat Diet; HFS: High-Fat High-Sucrose Diet; HMG-CoA: 3-Hydroxy-3-Methylglutaryl Coenzyme A; ILs: Interleukins; iNOS: inducible Nitric Oxide Synthase; Insig: Insulin-Inducible Gene; iRhom2: inactive Rhomboid 2; LCAT: Lecithin Cholesterol Acyl Transferase; LDL-C: Low-Density Lipoprotein-Cholesterol; LDLR: Low-Density Lipoprotein Receptor; LOX: Lipooxygenase; LXR-α: Liver X Receptor alpha; MPMs: Mouse Peritoneal Macrophages; MSG: Monosodium Glutamate; NF-κB: Nuclear Factor-Kappa B; NICD: Notch Intracellular Domain; NPC1L1: Niemann-Pick

DOI: 10.1201/9781003082767-15

Cl-Like 1; Nrf1: Nuclear Respiratory Factor 1; OO: One-arm, Open-labelled; PC: Placebo-Controlled; PGE2: Prostaglandin E2; PPARα: Peroxisome Proliferator-Activated Receptor alpha; R: Randomized; SLC16A1: Monocarboxylate Transporter 1; SREBP-1c: Sterol Regulatory Element-Binding Proteins; TACE: TNF-α Converting Enzyme; Tbx1: T-box Protein 1; TC: Total Cholesterol; Tfam: Mitochondrial Transcription Factor; TG: Triglyceride; Tmem26: Transmembrane Protein 26; TNF-α: Tumour Necrosis Factor-alpha; TRIP-Br2: Transcriptional Regulator Interacting with the PHD-Bromodomain 2; VCAM-1: Vascular Cell Adhesion Molecule-1; VLDL-C: Very Low-Density Lipoprotein-Cholesterol.

11.1 INTRODUCTION

Hyperlipidaemia is a chronic, complex condition well characterized by an abnormal increase in lipids including total cholesterol (TC) and triglyceride (TG) as well as lipoproteins such as low-density lipoprotein-cholesterol (LDL-C) and very-low-density lipoprotein-cholesterol (VLDL-C) with a decrease in high-density lipoprotein-cholesterol (HDL-C) levels in serum (Parhofer 2016). These elevated lipid levels are closely associated with increased risk of various disorders such as cardiovascular diseases, diabetes, insulin resistance, obesity, and hepatotoxicity. Furthermore, epidemiological studies have reported several unhealthy lifestyle factors, including cigarette smoking, alcohol consumption, poor diet, physical inactivity, and consumption of a diet with high saturated or trans fats responsible for hyperlipidaemia (Karr 2017; Writing Group et al. 2016). In addition, a genetic factor also plays an essential role in the development of familial hypercholesterolaemia where genetic mutations in the LDL receptor (LDLR) and apolipoprotein (apo) B gene caused significant elevations in TC and LDL-C (Nordestgaard et al. 2013; Sjouke et al. 2015).

The accumulation of free fatty acids (FFAs) due to diet-induced obesity causes elevated TG levels in non-adipose tissues controlled by leptin has been well documented. However, alteration in leptin signalling results in lipotoxicity, increasing the risk of cardiovascular diseases (Poetsch, Strano, and Guan 2020). Elevated hepatic FFAs result in impaired glucose metabolism, contributing to non-alcoholic steatohepatitis (Ai et al. 2015). Additionally, in renal tissue, elevated levels of FFAs induce toxicity in proximal tubular epithelial cells, leading to renal damage (Liu et al. 2019). Thus, elevated levels of lipid can cause multiorgan failure. Lifestyle modification plays a vital role in the management of lipid metabolism disorders. Interestingly, intake of the Mediterranean diet and/or nuts can reduce LDL-C levels (Bao et al. 2013; Estruch et al. 2013). However, long-term diet compliance limits its clinical implication for the treatment of chronic hyperlipidaemia. Statins (including simvastatin, atorvastatin, and lovastatin) have a predominant effect lowering TC and LDL-C through down-regulation of 3-hydroxy-3-methylglutaryl-coenzyme A (HMG-CoA) reductase enzymes (Parhofer 2016). Furthermore, some non-statin therapies such as ezetimibe, niacin, lomitapide, PCSK9 inhibitor (Alirocumab and Evolocumab) showed some positive benefits in halting the progression of atherosclerosis (Karr 2017). However, some patients produce adverse events or are intolerant or resistant to these therapies, or remain non-adherent, resulting in poor outcomes.

Nowadays, a shift has happened towards using natural products as a complementary and alternative medicine to prevent various disorders. Dietary phytochemicals have been identified with a wide range of chemical diversity with an ability to interact with multiple biological targets, offering them potential therapeutic agents for hyperlipidaemia management (Ansari et al. 2020; Thompson Coon and Ernst 2003). *Trigonella foenum-graecum* L. (fenugreek) is one of such medicinal plants that has been reported for the prevention and treatment of various diseases by its phytoconstituents, including dietary fibre, galactomannan (GAL), steroidal saponins (diosgenin), alkaloids (trigonelline), and amino acids (4-hydroxyisoleucine) (Bahmani et al. 2016; Fuller and Stephens 2015; Zameer et al. 2018). Fenugreeks seed powder (FSP), fenugreek seed extract (FSE), and its various phytoconstituents have been well documented for their anti-hyperlipidaemic and cholesterol-lowering potential in an array of experimental and clinical studies (Hasani-Ranjbar et al. 2010; Heshmat-Ghahdarijani et al. 2020; Khodamoradi et al. 2020). Several published reviews have described the lipid-lowering potential of fenugreek seeds in secondary hyperlipaemia (i.e., associated with other disease states such as diabetes) (Avalos-Soriano et al. 2016; Bahmani et al. 2016; Basch et al. 2003; Basu and Srichamroen 2010; Fuller and Stephens 2015; Goyal, Gupta, and Chatterjee 2016; Hasani-Ranjbar et al. 2010; Heshmat-Ghahdarijani et al. 2020; Khodamoradi et al. 2020; Kumar and Bhandari 2015; Roberts 2011; Srinivasan, Sambaiah, and Chandrasekhara 2004; Srinivasan 2006, 2013; Thompson Coon and Ernst 2003; Wu and Jiang 2019; Zameer et al. 2018). However, to date, no review has documented the effect of fenugreek seed on primary hyperlipaemia. The present chapter reviews the available evidence in the last 30 years (after 1991) of the potential of FSP, FSE, and its various phytoconstituents against primary hyperlipaemia in experimental and clinical settings.

11.1.1 Fenugreek against Hyperlipidaemia: Preclinical (*In Vivo*) Evidence

11.1.1.1 Fenugreek Seed Powder and Lipids

The potential of fenugreek seed powder has been evaluated by an array of researchers in various experimental animals by administrating it in their diet. Orally administered FSP (5% w/w, for 8 weeks) in Zucker obese rats was reported to significantly reduce hepatic TG level demonstrating its potential to reduce the hepatic TG accumulation and thus prevention in development of hepatic steatosis (Raju and Bird 2006). Another study has evaluated the influence of dietary FSP (5%, 6%, 10%, 12%, and 15% diet) for ten weeks on regression of preestablished cholesterol gallstones (CGS) (Reddy and Srinivasan 2009a, 2009b). FSP showed a significant reduction in serum, hepatic, and biliary TC levels, cholesterol: phospholipid ratio, biliary cholesterol: bile acid ratio, and cholesterol saturation index (Reddy and Srinivasan 2009a, 2009b). Furthermore, HCD-induced altered activities of hepatic enzymes, such as HMG-CoA reductase, cholesterol-7α-hydroxylase (CYP7A1), and cholesterol-27α-hydroxylase (CYP27A1) were effectively restored by FSP. It also decreased hepatic fat accumulation (Reddy and Srinivasan 2011a). In addition, FSP significantly prolonged the cholesterol nucleation time, reduced the vesicular form of cholesterol, and increased biliary phospholipid (Reddy and Srinivasan 2011b). These studies provided evidence of the potency of hypolipidaemic fenugreek seeds, which was modulated through influence on cholesterol metabolism and crystallization (Reddy and Srinivasan 2009a, 2009b, 2011a, 2011b). Furthermore, administration of FSP (0.5 g/kg, for 12 weeks) in high-fat high-sucrose (HFS) diet-fed male Sprague-Dawley rats showed an effective reduction of plasma TG and TC levels (Muraki et al. 2011).

In a separate study, the treatment of Egyptian FSP (0.5 and 1.0 g/kg, 4 weeks) in HCD fed obese rat (Ramadan, El-Beih, and Abd El-Kareem 2011) or FSP (0.5 g/kg, 4 weeks) in albino rabbits of European strains fed with HCD for 4 weeks (Sharma and Choudhary 2014) or 8 weeks (Sharma and Choudhary 2017) significantly alleviated serum total lipids, TC, TG, LDL, and atherogenic index, and thus indicated the usefulness of fenugreek as a dietary adjunct to control metabolic syndrome in obese patients. A separate study reported the hypolipidaemic potential of FSP (5% diet, for 8 weeks) in rats fed with aluminium chloride (Belaid-Nouira et al. 2012) or FSP (10% and 20% diet, for 9 weeks) in WNIN (GR-Ob) mutant obese rats (Ramulu, Giridharan, and Udayasekhararao 2011). This study suggested a significant reduction in serum TG, TC (Belaid-Nouira et al. 2012) via elevating HMG-CoA reductase activity (Ramulu, Giridharan, and Udayasekhararao 2011).

In an experimental study, Wistar albino rats were fed with HCD and then treated with FSP (4% and 8% diet, for 4 weeks) (Nader, Rasheed, and ALMusawi 2013) or FSP (5% and 10%, for 2 weeks) (Elmahdi and M. El-Bahr 2014). Findings suggested that FSP (4% and 8%) effectively reduced TC and LDL-C without altering HDL-C levels (Nader, Rasheed, and ALMusawi 2013). A lower dose of FSP (5%) effectively reduced elevated TC and LDL-C levels as well as increased HDL-C levels as compared to the high dose FSP (10%) thus, it was recommended to include a lower dose of fenugreek (5%) for its anti-hypercholesterolaemic potential (Elmahdi and M. El-Bahr 2014). However, another researcher showed that chronic treatment with FSP (10%, for 8 weeks) in HCD-fed (Mukthamba and Srinivasan 2015) and HFD-fed Wistar rats (Mukthamba and Srinivasan 2016a) offered additional protection against cardiac damage along with hepatic damage in terms of elevated levels of TG, LDL-C, cholesterol: phospholipid ratio and reduced HDL-C level. The authors concluded that dietary fenugreek might have cardioprotective potential and lipid-lowering effect (Mukthamba and Srinivasan 2015; Mukthamba and Srinivasan 2016a). Further, these researchers also documented the protective efficacy of FSP (10% diet, 8 weeks) against iron-induced (ferrous sulfate, 30 mg Fe^2/kg, i.p. and copper sulfate) LDL-oxidation in rats fed with HCD *in vivo* as well as *in vitro* (Mukthamba and Srinivasan 2016b).

The potential of FSP (2%) supplementation was evaluated in HFD-fed C57BL/6J mice for 16 weeks (Knott et al. 2017). FSP significantly improved HDL: LDL ratio without affecting TC, TG, or glycerol levels. Elevated expression of fatty acids binding protein-4 (FABP-4) in hepatic tissue was significantly decreased by FSP but failed to prevent hepatic steatosis. The authors attributed the effects of fenugreek to promoting metabolic resilience via selected effects on hyperlipidaemia (Knott et al. 2017). Furthermore, the same researchers recently reported a novel approach to study the effect of FSP on gut microbiota in mice (Bruce-Keller et al. 2020). The study showed a significant effect on gut microbiota with alterations in alpha and beta diversity and taxonomic redistribution under HFD conditions and suggested the beneficial effect of fenugreek in preserving the healthy populations of gut microbiota (Bruce-Keller et al. 2020).

In another study, the potential of FSP (2% diet, for 7 weeks) was reported against overproduction of VLDL in genetically hyperlipidaemia mice generated by depleting cAMP response binding protein H (CREBH) (Khound et al. 2018). The FSP significantly inhibited hepatic SREBP-1c (sterol regulatory element-binding proteins) activation, hepatic fat deposition, and VLDL secretion and increased insulin-inducible gene-1 (Insig-1), Insig-2, and PPARα (peroxisome proliferator-activated receptor alpha) expressions. These findings provide a novel mechanistic and strong rationale for utilizing fenugreek seed as a nutraceutical to prevent hyperlipidaemia (Khound et al. 2018).

These findings suggested that the administration of dietary fenugreek (2%–20%) for 4–16 weeks could reduce the elevated serum lipid levels either via regulating the activity of HMG-CoA reductase enzymes or inhibition of hepatic SREBP-1c (Table 11.1).

11.1.1.2 Fenugreek Seed Extracts and Lipids

Many investigators reported the anti-hyperlipidaemic potential of various extracts (aqueous, hydroalcoholic, ethanolic, methanolic, polyphenolic, and ethyl acetate) of fenugreek seeds. The potential of FSE (aqueous) determined against monosodium glutamate (MSG)-induced dyslipidaemia in neonatal Wistar rats (Kumar and Bhandari 2013, 2016). Hyperlipidaemia was induced by MSG (4 g/kg, s.c., from day 2 to 14 after birth, on every alternate day) and treated with FSE (0.5 and 1 g/kg, p.o., for 4 weeks) after six weeks of age. Administration of FSE significantly reduced weight of mesenteric, epididymal, and retroperitoneal fat, serum TC, TG, LDL-C, VLDL, atherogenic index, coronary risk index, activities of liver and epididymal FAS, and G6PD (glucose-6-phosphate dehydrogenase), and increased HDL-C levels (Kumar and Bhandari 2013, 2016). Furthermore, the inhibitory potential of FSE against fat accumulation and dyslipidaemia in obese female Wistar rats fed with HFD was noted in a separate study (Kumar, Bhandari, and Jamadagni 2014). Treatment with FSE significantly reduced serum lipids (TC, TG, LDL-C, and VLDL-C), leptin, lipase, apolipoprotein-B, FAS in the liver, and uterine white adipose tissue and increased adiponectin levels. Findings of these investigations demonstrated the preventive potential of FSE (aqueous) on fat accumulation and dyslipidaemia via inhibition of impaired lipid digestion and absorption (Kumar and Bhandari 2013; Kumar, Bhandari, and Jamadagni 2014; Kumar and Bhandari 2016). Furthermore, the anti-hyperlipidaemic potential against CCl_4-induced hepatotoxicity in albino rats is evidenced

TABLE 11.1
Hypolipidaemic Potential of Fenugreek Seed Powder in Experimental Animals

Study Model	Dose and Duration	Proposed Mechanism of Action	Reference
Zucker obese rats	5% w/w, 8 weeks	↓ hepatic TG	Raju and Bird (2006)
Mice fed with HCD	12% diet, 10 weeks	↓ serum, hepatic and biliary TC, cholesterol: phospholipid ratio, biliary cholesterol: bile acid ratio, and cholesterol saturation index	Reddy and Srinivasan (2009a, 2009b)
		↑ HMG-CoA reductase, CYP7A1, and CYP27A1	Reddy and Srinivasan (2011a)
		↑ cholesterol nucleation time, biliary phospholipid ↓ vesicular form of cholesterol	Reddy and Srinivasan (2011b)
Rats fed with HFS diet	0.5 g/kg, 12 weeks	↓ plasma TG and TC	Muraki et al. (2011)
Rats fed with HCD	0.5 and 1.0 g/kg, 4 weeks	↓ serum total lipids, TC, TG, and atherogenic index	Ramadan, El-Beih, and Abd El-Kareem (2011)
WNIN (GR-Ob) mutant obese rats	10% and 20% diet, 9 weeks	↓ serum TG, TC ↑ HMG-CoA reductase	Ramulu, Giridharan, and Udayasekhararao (2011)
Rats fed with aluminium chloride	5% diet, 8 weeks	↓ serum TC, TG, and LDL-C	Belaid-Nouira et al. (2012)
Rats fed with HCD	4% and 8% diet, 4 weeks	↓ serum TC, and LDL-C	Nader, Rasheed, and ALMusawi (2013)
Rabbits fed with HCD (4 weeks)	0.5 g/kg, 4 weeks	↓ serum TC, TG, LDL, VLDL-C, LDL-C: HDL-C	Sharma and Choudhary (2014)
Rabbits fed with HCD (8 weeks)		↓ serum TC, LDL, LDL-C: HDL-C ↑ HDL-C	Sharma and Choudhary (2017)
Rats fed with HCD	5% and 10% diet, 2 weeks	FSP (5%)—↓ serum TG, LDL, and ↑ HDL-C as compared to FSP (10%)	Elmahdi and M. El-Bahr (2014)
Rats fed with HCD or HFD	10% diet, 8 weeks	↓ serum TG, LDL-C, cholesterol: phospholipid ratio in serum, liver, and heart ↑ serum HDL-C	Mukthamba and Srinivasan (2015), Mukthamba and Srinivasan (2016a)
Rats fed with HCD and intoxicated with ferrous sulfate and copper sulfate		↓ Iron- as well as copper-induced LDL-oxidation	Mukthamba and Srinivasan (2016b)
Mice fed with HFD	2% diet, 16 weeks	↑ serum HDL: LDL ratio ↑ FABP-4	Knott et al. (2017)
		Affect gut microbiota with alterations in both alpha and beta diversity as well as taxonomic redistribution	Bruce-Keller et al. (2020)
Genetically hyperlipidaemia mice	2% diet, 7 weeks	↓ hepatic SREBP-1c activation, hepatic fat deposition, and VLDL recreation ↑ Insig-1, Insig-2, and PPARα	Khound et al. (2018)

CYP27A1: Cholesterol-27α-hydroxylase; CYP7A1: Cholesterol-7α-hydroxylase; FABP-4: Fatty Acids Binding Protein-4; HCD: High-Cholesterol Diet; HDL-C: High-Density Lipoprotein-Cholesterol; HFD: High-Fat Diet; HFS: High-Fat High-Sucrose Diet; HMG-CoA: 3-Hydroxy-3-Methylglutaryl Coenzyme A; Insig: Insulin-Inducible Gene; LDL-C: Low-Density Lipoprotein-Cholesterol; PPARα: Peroxisome Proliferator-Activated Receptor alpha; SREBP-1c: Sterol Regulatory Element-Binding Proteins; TG: Triglyceride, VLDL-C: Very Low-Density Lipoprotein-Cholesterol.

by a significant reduction of TG and TC by FSE (aqueous (4%), in drinking water, for 5 weeks) (Al-Sultan and El-Bahr 2015).

The potential of hydroalcoholic extract of fenugreek seeds, FSE (10 and 100 mg/300 g body weight, for 2 weeks) was evident from the reported study, in which FSE significantly decreased the serum TC, LDL, and VLDL, suggesting its putative anti-hypercholesterolaemia potential against various metabolic disorders (Petit et al. 1993). Thereafter, the curative and preventive potential of FSE (hydroalcoholic) against lipid metabolism in HFD-fed C57BL/6J mice has also been reported (Hamza et al. 2012). In a preventive study, FSE (2 g/kg, along with HFD for 20 weeks) effectively inhibited the elevation of serum TG level, whereas, in the curative study, FSE (2 g/kg, for 17 weeks after HFD administration) significantly reduced serum TG and TC levels (Hamza et al. 2012). Furthermore, inhibitory potential of FSE (hydroalcoholic) against lipid accumulation was reported by various researchers where treatment with FSE (hydroalcoholic, 200, 300, and 400 mg/kg, for 12 weeks) in HFD-fed female Wistar rats (Nagamma et al. 2019) or FSE (hydroalcoholic, 0.1 and 0.2 mL/g/day, for 12 weeks) in HFD-fed mice (Zhou et al. 2020) showed a significant decrease in serum TC, TG, and LDL-C and effective increase in serum HDL-C levels. Furthermore, the inhibition of inactive rhomboid 2 (iRhom2) and TNF-α converting enzyme (TACE) expression in subcutaneous adipose tissue by FSE, depicting the inhibitory potential of FSE against lipid accumulation in adipocytes, was reported (Zhou et al. 2020). Recently, the dietary supplementation of FSE (hydroalcoholic, 0.01%, 0.02%, 0.04%, 0.08%, and 0.16%, for 8 weeks) to juvenile blunt snout was reported a significant decrease in plasma TG levels, hepatic FAS, and SREBP1, which indicated a potential role of dietary FSE in regulating lipid metabolism (Yu et al. 2019).

Administration of FSE (ethanolic, 30 or 50 g/kg, 4 weeks) in hypercholesterolaemic rats (Stark and Madar 1993) or FSE (ethanolic, 200 mg/kg, 4 weeks) in Triton- or HFD-induced hyperlipidaemia in Charles Foster strain adult rats (Chaturvedi et al. 2013) or FSE (ethanolic, 200 mg/kg, 2 days) in Triton-induced hyperlipidaemia Wistar rats (Kaur et al. 2015) showed a significant reduction in serum and hepatic TG, TC, and LDL levels. The reduction in lipid profile was suggested to be mediated via activation of lecithin-cholesterol acetyltransferase, post heparin lipolytic activity, triglyceride lipase, lipoprotein lipase, and increased excretion of faecal bile acids (Chaturvedi et al. 2013). Authors attributed these effects to saponins (Stark and Madar 1993) or β-sitosterol and 4-HI (Chaturvedi et al. 2013) to interact with bile salts and contribute to the hypercholesterolaemic potential of fenugreek seeds. In addition, the protective efficacy of FSE (ethanolic, 0.3% and 1%, 3 weeks) in obese female ddY mice fed with HFD was reported (Handa et al. 2005). FSE significantly decreased weight of adipose tissue and liver TG levels suggesting the preventive potential of FSE against obesity induced by HFD (Handa et al. 2005).

The potential of FSE (aqueous and methanolic) was evaluated against aluminium chloride ($AlCl_3$)-induced hypercholesterolaemia in female Wistar rats (Belaid-Nouira et al. 2012). Hypercholesterolaemia was induced in rats by chronic administration of $AlCl_3$ (500 mg/kg, i.g., for one month, then 1600 ppm via drinking water for up to 5 months). Administration of FSE (100 mg/kg, i.g., for 8 weeks after 3 months of $AlCl_3$ administration) effectively diminished elevated levels of TC, TG, and LDL-C, suggesting its modulatory effect on plasmatic lipid metabolism (Belaid-Nouira et al. 2012).

The ethanol (6 g/kg per day) intoxicated rats treated with FSE (polyphenolic, 200 mg/kg, for 4 weeks) (Kaviarasan, Viswanathan, and Anuradha 2007), as well as rats fed with a high fructose diet (FRU), treated with FSE (60 g/100 g, 8 weeks) (Kannappan and Anuradha 2009), were reported to have experienced a beneficial influence on plasma TG and FFA, perhaps due to the presence of polyphenols in fenugreek seeds (Kannappan and Anuradha 2009; Kaviarasan, Viswanathan, and Anuradha 2007). Administration of FSE (thermostable, 1.5 and 15 mg/kg, for 2 weeks) in C57BL6/J mice fed with fat supplement (400 g groundnut and 200 g dried coconut/kg/day, for 15 days) resulted in a marked decrease in serum TG, TC, and LDL-C levels (Vijayakumar et al. 2010). The author suggested that FSE exerts its inhibitory potential against fat accumulation via upregulation in expressions of LDLR, suggesting its essential role in the management of dyslipidaemia and its associated metabolic disorders (Vijayakumar et al. 2010). Interestingly, the comparative hypercholesterolaemia potential of various FSE (water, methanol, ethyl acetate, hexane, dichloromethane, 0.125% each) was evaluated in male Wistar rats fed with HCD for 16 weeks (Belguith-Hadriche et al. 2010; Belguith-Hadriche et al. 2013). Results suggested that only ethyl acetate extract reduced TC, TG, LDL-C, and increased HDL-C significantly, depicting the presence of flavonoids, especially naringenin responsible for hypercholesterolaemic effects of FSE (Belguith-Hadriche et al. 2010; Belguith-Hadriche et al. 2013).

These experimental studies suggested that the alcoholic extract of fenugreek seed exert has maximum anti-hyperlipidaemic potential. Furthermore, the presence of various phytoconstituents such as polyphenols, flavonoids, saponins, β-sitosterol, and 4-HI may contribute to its protective anti-hyperlipidaemic efficacy (Table 11.2).

TABLE 11.2
Hypolipidaemic Potential of Fenugreek Seed Extracts in Experimental Animals

Study Model	Dose and Duration of Fenugreek Extract	Proposed Mechanism of Action	Reference
MSG-induced dyslipidaemia neonatal Wistar rats	Aqueous (0.5 and 1 g/kg, 4 weeks)	↓ serum TC, TG, LDL-C, VLDL and liver and epididymal FAS and G6PD ↑ serum HDL-C	Kumar and Bhandari (2013, 2016)
Rats fed with HFD		↓ serum TC, TG, LDL-C, VLDL-C, leptin, lipase, apolipoprotein-B, liver, and uterine FAS ↑ adiponectin	Kumar, Bhandari, and Jamadagni (2014)
Rats intoxicated with CCl_4	Aqueous (4%, 5 weeks)	↓ serum TC, TG	Al-Sultan and El-Bahr (2015)
Healthy rats	Hydroalcoholic (10 and 100 mg/300 g BW, 2 weeks)	↓ plasma TC, LDL-C, and VLDL-C	Petit et al. (1993)
Mice fed with HCD	Hydroalcoholic (2 g/kg, 20 weeks) (Curative)	↓ serum TG	Hamza et al. (2012)
	Hydroalcoholic (2 g/kg, 17 weeks) (Preventive)	↓ serum TG and TC	
Rats fed with HFD	Hydroalcoholic (200, 300, and 400 mg/kg, 12 weeks)	↓ serum TC, TG, TC: HDL ratio and LDL-C ↑ serum HDL-C	Nagamma et al. (2019)
Juvenile blunt snout bream	Hydroalcoholic (0.01%, 0.02%, 0.04%, 0.08% and 0.16%, 8 weeks)	↓ plasma TG levels, hepatic FAS and SREBP1	Yu et al. (2019)
Mice fed with HFD	Hydroalcoholic (0.1 and 0.2 mL/g/day, 12 weeks)	↓ plasma TC, TG, and LDL ↑ plasma HDL iRhom2 and TACE	Zhou et al. (2020)
Hypercholesterolaemic rats	Ethanolic (30 or 50 g/kg, 4 weeks)	↓ plasma and hepatic TC	Stark and Madar (1993)
Obese ddY mice fed with HFD	Ethanolic (0.3% and 1% diet, 3 weeks)	↓ liver TG	Handa et al. (2005)
Rats fed either with HFD or Triton WR-1339	Ethanolic (200 mg/kg, 4 weeks)	↓ plasma and hepatic lipid levels via activation of lecithin-cholesterol acetyltransferase, post heparin lipolytic activity, triglyceride lipase, lipoprotein lipase, ↑ excretion of faecal bile acids	Chaturvedi et al. (2013)

(Continued)

TABLE 11.2 (Continued)

Study Model	Dose and Duration of Fenugreek Extract	Proposed Mechanism of Action	Reference
Rats fed with Triton WR-1339	Ethanolic (200 mg/kg, 2 days)	↓ serum TC, TG, and LDL ↑ HDL-C	Kaur et al. (2015)
Rats fed with aluminium chloride	Methanolic (100 mg/kg, 8 weeks)	↓ serum TC, TG and LDL-C	Belaid-Nouira et al. (2012)
Rats intoxicated with ethanol	Polyphenolic (200 mg/kg, 4 weeks)	↓ plasma and liver lipids (TC, TG, LDL-C, VLDL-C, phospholipids, and free fatty acids)	Kaviarasan, Viswanathan, and Anuradha (2007)
Rats fed with FRU diet	Polyphenolic (60 g/100 g, 8 weeks)	↓ plasma TG and FFA	Kannappan and Anuradha (2009)
Mice fed with fat supplement	Thermostable (1.5 and 15 mg/kg, 2 weeks)	↓ serum TC, TG and LDL-C	Vijayakumar et al. (2010)
Rats fed with HCD	Ethyl acetate (0.125% diet, 16 weeks)	↓ serum TC, TG, LDL-C ↑ serum HDL-C	Belguith-Hadriche et al. (2010), (Belguith-Hadriche et al. 2013)

FFA: Free Fatty Acid; FRU: High Fructose Diet; G6PD: Glucose-6-Phosphate; HCD: High-Cholesterol Diet; HDL-C: High-Density Lipoprotein-Cholesterol; HFD: High-Fat Diet; iRhom2: inactive Rhomboid 2; LDL-C: Low-Density Lipoprotein-Cholesterol; MSG: Monosodium Glutamate; SREBP-1c: Sterol Regulatory Element-Binding Proteins; TACE: TNF-α Converting Enzyme; TC: Total Cholesterol; TG: Triglyceride, TNF-α: Tumour Necrosis Factor-alpha; VLDL-C: Very Low-Density Lipoprotein-Cholesterol.

11.1.1.3 Fenugreek Seed Galactomannan (GAL) and Lipids

Plant secretory cells synthesize mucilaginous fibres such as pectin, guar, and psyllium hydrocolloid or psyllium seed gum to prevent dryness of the endoplasm of the seed. Such GAL mucilage, hetero-polysaccharide in nature and easily water-soluble, was isolated from fenugreek seeds by researchers. Several studies have well documented the anti-hyperlipidaemic potential of fenugreek GAL in various experimental animal models.

Initially, the investigator determined the effect of various compositions and structures of GAL on its hypolipidaemic potential (Evans et al. 1992). A cholesterol and sodium cholate fed male Wistar rats were treated with a diet containing GAL (80 g/kg, 2 weeks) with different composition of galactose (G): mannose (M), namely, fenugreek (1G: 1M), guar gum (1G: 2M), and locust bean gum (1G: 4M). The results suggested that the higher the density of galactose side chains in GAL, the more markedly it decreases the serum TC, liver lipids and TC, rate of hepatic synthesis of cholesterol, and more significantly increased caecal volatile fatty acids (Evans et al. 1992).

Furthermore, administration of GAL (40 mg/kg, for 8 weeks) to healthy albino rats (Boban, Nambisan, and Sudhakaran 2006) or administration of GAL (40 mg/kg, for 12 weeks) to HCD-fed rabbits (Boban, Nambisan, and Sudhakaran 2009) showed a significant decrease of TC, TG, HDL, LDL, and VLDL levels in serum as well as TC and TG levels in liver and aorta. Furthermore, the author reported a significant inhibition in the synthesis and secretion of apoB-containing lipoproteins (VLDL) from hepatocytes by GAL treatment (Boban, Nambisan, and Sudhakaran 2006). The author attributed the hypolipidaemic effect of fenugreek GAL to inhibition in the synthesis and secretion of VLDL by hepatocytes (Boban, Nambisan, and Sudhakaran 2006). An array of studies further supported these findings where feeding of various concentration of GAL (2.5% and 5% diet, for 4 weeks) in rats fed with high sucrose (52% w/w) diets (Srichamroen et al. 2008) or GAL (2.5 and 5% diet, for 9 weeks) in WNIN (GR-Ob) mutant obese rats (Ramulu, Giridharan, and Udayasekhararao 2011) or GAL (15% diet, for 3 weeks) in healthy C57BL/6J mice (Shtriker et al.

2018) effectively reduced plasma TG and TC levels. Furthermore, the author attributed the hypolipidaemic effect of fenugreek seed to the presence of GAL executed via activation of the HMG-CoA reductase pathway (Ramulu, Giridharan, and Udayasekhararao 2011).

The possible interaction between fenugreek GAL and proteins (from albumin or casein) to reduce the risk of atherogenicity was studied (Dakam et al. 2009). Administration of formulations (250 mg/kg, for 4 weeks) containing fenugreek GAL along with albumin (FGA, 4% w/v) or with casein (FGC, 4% w/v) to male Wistar rats showed a significant reduction in plasma TC, LDL-C, atherogenicity indices (TC: HDL and LDL: HDL) and increased HDL-C (Dakam et al. 2009).

A preventive efficacy of GAL (0.5 g/kg, for 12 weeks) in high-fat high-sucrose (HFS) diet-fed male Sprague-Dawley rats is documented where it showed marked elevation in faecal lipid level and reduction in plasma and hepatic lipid levels (Muraki et al. 2009). Furthermore, administration of low molecular weight GAL (0.12%, 0.24%, 1.2%, 2.4%, and 4.8% for 12 weeks) in HFS-fed Sprague-Dawley rats also reported similar findings suggesting the potential of GAL, especially low molecular weight GAL, in amelioration of dyslipidaemia (Muraki et al. 2012). The anti-hyperlipidaemic potential of low molecular weight GAL was further supported by the report of GAL (200 mg/kg, i.p., twice per week for 15 weeks) in HFD-fed C57BL/6 mice showing an effective reduction of white and brown adipose tissue weights, serum TC and LDL-C (Cheng et al. 2018). These results were confirmed by the separate study of GAL (10, 30, and 100 mg/kg, for 12 weeks) in HFD-fed C57BL/6 mice where findings showed significantly inhibited HFD-induced elevated adipose tissue (brown, mesenteric, epididymal, and retroperitoneal) and liver weight, serum lipid (TC and TG), and mRNA expressions of fatty acid synthase (FAS), IL-6, leptin and transcriptional regulator interacting with the PHD-bromodomain 2 (TRIP-Br2) in brown adipose tissue, liver, and epididymal fat (Kandhare, Bandyopadhyay, and Thakurdesai 2018). The authors attributed the results to the ability of GAL to increase lipid mobilization and suggested GAL as a dietary supplement to ameliorate diet-induced obesity (Kandhare, Bandyopadhyay, and Thakurdesai 2018).

Recently, the researcher established the linkage between fenugreek GAL and lipid metabolism via a positive effect on gut microbiota reflected by a significant increase in counts of *Bacteroidetes* microbiome, monocarboxylate transporter 1 (SLC16A1, a promoter of short-chain fatty acid transport), and adenosine monophosphate-activated protein kinase (AMPK) post-GAL administration (Shtriker et al. 2018).

Taken together, GAL exerts its positive effect on cholesterol absorption and lipid metabolism via improving the gut microbiota, regulating the activity of HMG-CoA reductase, and FAS (Table 11.3).

11.1.1.4 Fenugreek Saponins and Lipids

Fenugreek seeds contain steroidal saponins, including diosgenin, yamogenin, gitogenin, tigogenin, and neotigogens, which are significant components behind their health benefits. The earliest studies in this regard reported the cholesterol-lowering potential of crude steroid saponins extract (12.5 mg/day per 300 g, for 2 weeks) in healthy Wistar rats (Petit et al. 1995) and hypolipidaemic potential of saponins fraction (30% diet, for 9 weeks) in New Zealand White rabbits fed with HFD (Al-Habori, Al-Aghbari, and Al-Mamary 1998). In healthy rats, saponins significantly reduced plasma TC whereas, in rabbits fed with HFD, it effectively reduced plasma TC, TG, LDL-C, TC: HDL-C, and improved HDL-C: LDL-C, suggesting the role of saponin in fenugreek seed towards its hypolipidaemic potential (Al-Habori, Al-Aghbari, and Al-Mamary 1998; Petit et al. 1995).

Later, researchers have determined that the potential standardized extract of fenugreek seed contains furostanolic saponins (Fenfuro™) (Hua et al. 2015). Treatment with Fenfuro™ (50 mg/kg, 24 weeks) in HFD-fed male C57BL/6J mice significantly reduced TG levels and hepatic fat accumulation evaluated by Oil Red O staining suggesting its anti-hyperlipidaemic potential (Hua et al. 2015). Further, it was reported the administration of saponins isolated from fenugreek seeds (6, 12, 24 mg/kg, for 8 weeks) to HFD-fed rats caused a significant improvement in serum lipid (TC, TG, LDL, and HDL), CYP7A1, HMG-CoA reductase levels, and hepatic lipid depositions in Sprague-Dawley rats fed with HFD (Chen et al. 2017).

TABLE 11.3
Hypolipidaemic Potential of Fenugreek Seed Galactomannan in Experimental Animals

Study Model	Dose and Duration of GAL	Proposed Mechanism of Action	Reference
Rats fed with cholesterol (10 g/kg) and sodium cholate (2 g/kg)	80 g/kg, 2 weeks	↓ serum TC. liver lipids. and TC ↓ hepatic synthesis of cholesterol	Evans et al. (1992)
Rats fed with standard normal diet	40 mg/kg, 8 weeks	↓ serum TC, TG, HDL, LDL, and VLDL, ↓ Liver and aorta TC and TG ↓ synthesis and secretion of apoB-containing lipoproteins	Boban, Nambisan, and Sudhakaran (2006)
Rats fed with high-sucrose (52% w/w) diets	2.5% and 5% diet, 4 weeks	↓ plasma TG and TC	Srichamroen et al. (2008)
Rabbits fed with HCD	40 mg/kg, 12 weeks	↓ serum TC, TG, HDL, LDL, and VLDL ↓ Liver and aorta TC and TG	Boban, Nambisan, and Sudhakaran (2009)
Rats fed with standard normal diet	4% w/v diet (with albumin or casein), 4 weeks	↓ plasma TC, LDL-C, and atherogenicity indices (TC: HDL and LDL: HDL)	Dakam et al. (2009)
Rats fed with HFS diet	0.5 g/kg, 12 weeks	↓ plasma and hepatic TC	Muraki et al. (2009)
WNIN (GR-Ob) mutant obese rats	2.5% and 5% diet, 9 weeks	↓ serum TG, TC ↑ HMG-CoA reductase	Ramulu, Giridharan, and Udayasekhararao (2011)
Rats fed with HFS diet	Low molecular weight GAL (0.12%, 0.24%, 1.2%, 2.4%, and 4.8%, 12 weeks)	↓ plasma and hepatic TC	Muraki et al. (2012)
Mice fed with HFD	Low molecular weight GAL (200 mg/kg, i.p. twice per week, 15 weeks)	↓ serum TG and LDL-C	Cheng et al. (2018)
	Low-molecular weight GAL (10, 30, and 100 mg/kg, 12 weeks)	↓ serum TC and TG ↓ FAS, IL-6, leptin, and TRIP-Br2	Kandhare, Bandyopadhyay, and Thakurdesai (2018)
Healthy C57BL/6J mice	15% diet, 3 weeks	↓ serum TC ↑ counts of *Bacteroidetes* microbiome, SLC16A1, and AMPK	Shtriker et al. (2018)

AMPK: Adenosine Monophosphate-Activated Protein Kinase; FAS: Fatty Acid Synthase; GAL: Galactomannans; HCD: High-Cholesterol Diet; HDL-C: High-Density Lipoprotein-Cholesterol; HFD: High-Fat Diet; HFS: High-Fat High-Sucrose Diet; HMG-CoA: 3-Hydroxy-3-Methylglutaryl Coenzyme A; ILs: Interleukins; LDL-C: Low-Density Lipoprotein-Cholesterol; SLC16A1: Monocarboxylate Transporter 1; TC: Total Cholesterol; TG: Triglyceride, TRIP-Br2: Transcriptional Regulator Interacting with the PHD-Bromodomain 2; VLDL-C: Very Low-Density Lipoprotein-Cholesterol

With increasing evidence for the hypolipidaemic potential for saponins from fenugreek seeds, a set of researchers studied the potential of diosgenin (a phytosteroidal sapogenin) in various experimental animals models. Administration of diosgenin (22.1, 44.2, and 88.4 mg/kg, for 6 weeks) (Gong et al. 2010) or diosgenin (0.15 and 0.3 g/kg, for 8 weeks) (Li et al. 2019) in HFD-fed rats effectively reduced serum TC, TG, and LDL-C. Another study reported a significant reduction in hepatic TC and free cholesterol levels, expression of intestinal Niemann-Pick C1-Like 1 (NPC1L1), liver X receptor-α (LXR-α), and a marked increase in the expression of ATP-binding cassette G5/

G8 (ABCG5/G8) in the liver and intestine by diosgenin (Li et al. 2019). The author concluded that diosgenin reduces cholesterol absorption by regulating intestinal NPC1L1 expression and promoting cholesterol efflux by regulating the hepatic ABCG5/G8 expression (Li et al. 2019).

The administration of diet containing diosgenin (0.5%, for 2 weeks) in Japanese quails fed with HCD (Al-Matubsi et al. 2011) or diosgenin (1.0%, for 8 weeks) in apoE KO mice fed with HFD (Lv et al. 2015) is reported to effectively decrease the TC and LDL-C levels and significantly increase the HDL levels. Additionally, diosgenin notably enhanced ABCA1 (ATP-binding cassette transporter A1)-dependent cholesterol efflux, faecal³ H-sterol originating from cholesterol-laden MPMs, and reduce aortic lipid deposition, plaque area, and collagen content (Lv et al. 2015). These results were attributed to the ability of diosgenin to inhibit aortic atherosclerosis via enhancing the ABCA1-dependent cholesterol efflux (Lv et al. 2015).

Researchers determined the potential of diosgenin (40 mg/kg, for 5 weeks) in nephrotoxic rats fed with adenine diet (Manivannan et al. 2013) or diosgenin (80 mg/kg, for 15 weeks) in rat fed with atherogenic diet (Binesh, Devaraj, and Devaraj 2018). Treatment with diosgenin effectively reduced the elevated levels of TC, TG, LDL-C, VLDL-C, and increased the HDL-C levels, and significantly restored the activity of HMG-CoA reductase and lecithin cholesterol acyl transferase (LCAT, an enzyme responsible for the conversion of cholesterol-to-cholesterol esters on the surface of HDL) (Manivannan et al. 2013). In a separate study, diosgenin inhibited the elevated expression of notch pathway proteins Notch1, DLL1 (Delta Like Protein-1), notch intracellular domain (NICD), Hes1 (hairy and enhancer of split-1), Hey (Hes related with YRPW motif protein 1), and Jagged1 in the aorta, differentiated and differentiation macrophage cells (Binesh, Devaraj, and Devaraj 2018). Furthermore, atherogenic diet-induced endothelial dysfunction, hepatic steatosis, and cellular stress were inhibited by diosgenin via inhibition of TNF-α and COX-2 expression in cardiac, hepatic, and brain tissue (Binesh, Devaraj, and Halagowder 2018).

Collectively, fenugreek saponins ameliorated hyperlipidaemia by their inhibitory potential against cholesterol synthesis and absorption. Fenugreek saponins, especially diosgenin, regulate the activity of HMG-CoA reductase, LCAT, NPC1L1, and ABCA1, which further inhibit cholesterol synthesis and absorption, thus exerting its anti-hyperlipidaemic potential (Table 11.4).

11.1.1.5 Fenugreek Amino Acid and Alkaloid Constituent and Lipids

A lipid-lowering potential of 4-HI, an amino acid from fenugreek seed, has been evaluated against the HFD-fed dyslipidaemia hamster model (Narender et al. 2006). Golden Syrian hamsters were fed with HFD and treated with 4-HI (50 mg/kg) for a week. HFD induced elevated levels of serum TG, TC, and free fatty acids (Narender et al. 2006). However, another researcher reported that administration of 4-HI (50 mg/kg, for 8 weeks) influenced elevated levels of HDL-C in the FRU-fed rats (Haeri et al. 2009).

The preventive efficacy of trigonelline, a major alkaloid of fenugreek seeds, was reported against the high-fat, high-sucrose (HFS) diet-induced metabolic diseases in male Sprague-Dawley rats (Muraki et al. 2009). In this study, the rats were treated either with HFS diet or HFS diet containing trigonelline (0.5 g/kg) for 12 weeks. Diets containing trigonelline showed marked elevation in faecal lipid level and reduced plasma and hepatic lipid levels, suggesting improved lipid metabolic disorders in plasma and the liver (Muraki et al. 2009).

A potential synergistic composition of FSE, that is, Sugaheal (containing 4-HI, trigonelline, and soluble low-molecular weight oligosaccharides), reported the anti-hyperlipidaemic effects in C57/BL6 mice fed with HFD (Thakurdesai, Mohan, and Kandhare 2015). HFD-fed mice treated with Sugaheal (30, 60, and 100 mg/kg) for 12 weeks showed a significant decrease in serum TC levels and SREBP-1c expressions in adipose tissue, liver, and pancreas without affecting TG, HDL, and LDL levels, and down inhibition of SREBPc. Researchers attributed the cholesterol-lowering potential of 4-HI, and trigonelline observed in the study to the down-regulation of SREBP-1c (Thakurdesai et al. 2015). The details of these studies with the prosed mechanisms are tabulated in Table 11.5.

TABLE 11.4
Hypolipidaemic Potential of Fenugreek Seed Saponins in Experimental Animals

Study Model	Dose and Duration of Saponins	Proposed Mechanism of Action	Reference
Healthy rats	Steroid saponins (12.5 mg/day per 300 g, 2 weeks)	↓ plasma TC	Petit et al. (1995)
Rabbits fed with HFD	Saponins fraction (30% diet, 9 weeks)	↓ plasma TC, TG, LDL-C, and TC: HDL-C ↑ HDL-C: LDL-C	Al-Habori, Al-Aghbari, and Al-Mamary (1998)
Mice fed with HFD	Fenugreek furostanolic saponins (Fenfuro™, 50 mg/kg, 24 weeks)	↓ serum TG and hepatic fat accumulation	Hua et al. (2015)
Rats fed with HFD	Fenugreek saponins (6, 12, 24 mg/kg, 8 weeks)	↓ serum lipid ↑ CYP7A1 ↓ HMG-CoA reductase	Chen et al. (2017)
Rats fed with HFD	Diosgenin (22.1, 44.2, and 88.4 mg/kg, 6 weeks)	↓ serum TC, TG, and LDL-C	Gong et al. (2010)
Japanese quails fed with HCD	Diosgenin (0.5% w/w, 2 weeks)	↓ serum TC, TG ↑ serum HDL-C	Al-Matubsi et al. (2011)
Rats fed with adenine diet	Diosgenin (40 mg/kg, 5 weeks)	↓ serum TC, TG, LDL-C, VLDL-C ↑ serum HDL-C ↑ HMG-CoA reductase and LCAT	Manivannan et al. (2013)
Mice fed with HFD	Diosgenin (1% diet, 8 weeks)	↑ HDL-C, ABCA1, faecal 3H-sterol ↓ serum TC, LDL-C, and cholesterol ester ↓ aortic lipid deposition, plaque area, and collagen	Lv et al. (2015)
Rats fed with atherogenic diet	Diosgenin (80 mg/kg, 15 weeks)	↓ serum TC, TG, LDL, and VLDL ↑ serum HDL-C ↓ Notch1, DLL1, NICD, Hes1, Hey, and Jagged1 in the aorta, macrophage cells	Binesh, Devaraj, and Devaraj (2018)
		↓ TNF-α and COX-2	Binesh, Devaraj, and Halagowder (2018)
Rats fed with HFD	Diosgenin (0.15 and 0.3 g/kg, 8 weeks)	↓ serum TC, TG, LDL-C ↓ hepatic TC and free cholesterol ↓ intestinal NPC1L1, hepatic LXR-α ↑ hepatic and intestinal ABCG5/G8	Li et al. (2019)

ABCA1: ATP-Binding Cassette Transporter A1; ABCG5/G8: ATP-Binding Cassette G5/G8; COX-2: Cylcooxgenase-2; CYP7A1: Cholesterol-7α-hydroxylase; DLL1: Delta Like Protein-1; HCD: High-Cholesterol Diet; HDL-C: High-Density Lipoprotein-Cholesterol; Hes1: Hairy and Enhancer of Split-1; Hey: Hes related with YRPW Motif Protein 1; HFD: High-Fat Diet; HMG-CoA: 3-Hydroxy-3-Methylglutaryl Coenzyme A; LCAT: Lecithin Cholesterol Acyl Transferase; LDL-C: Low-Density Lipoprotein-Cholesterol; LXR-α: Liver X Receptor alpha; NICD: Notch Intracellular Domain; NPC1L1: Niemann-Pick C1-Like 1; TC: Total Cholesterol; TG: Triglyceride, TNF-α: Tumour Necrosis Factor-alpha; VLDL-C: Very Low-Density Lipoprotein-Cholesterol

TABLE 11.5
Hypolipidaemic Potential of Fenugreek Seed Amino Acid and Alkaloid in Experimental Animals

Study Model	Dose and Duration	Proposed Mechanism of Action	Reference
Golden Syrian hamsters fed with HFD	4-HI (50 mg/kg, 1 week)	↓ serum TG, TC, and free fatty acids ↑ HDL-C: TC ratio	Narender et al. (2006)
Rats fed with FRU diet	4-HI (50 mg/kg, 8 weeks)	No effect on HDL-C	Haeri et al. (2009)
Rats fed with HFS diet	Trigonelline (0.5 g/kg, 12 weeks)	↓ plasma and hepatic lipid levels	Muraki et al. (2009)
Mice fed with HFD	Sugaheal (30, 60 and 100 mg/kg, 12 weeks)	↓ serum TC ↓ SREBP-1c in adipose tissue, liver, and pancreas	Thakurdesai, Mohan, and Kandhare (2015)

4-HI: 4-Hydroxyisoleucine; FRU: High Fructose Diet; HDL-C: High-Density Lipoprotein-Cholesterol; HFD: High-Fat Diet; HFS: High-Fat High-Sucrose Diet; SREBP-1c: Sterol Regulatory Element-Binding Proteins; TC: Total Cholesterol; TG: Triglyceride.

11.1.2 Fenugreek against Hyperlipidaemia: Preclinical (In Vitro) Evidence

In the past decade, various *in-vitro* cell-line models were used to evaluate fenugreek phytoconstituents' possible mechanisms of action for their anti-hyperlipidaemic potential. Incubation of FSE (thermostable, 0.5, 5 and 50 μg/ml) in sterol-enriched conditions showed an effective inhibition in expressions of PPAR-γ, SREBP-1, and c/EBP-α in differentiating and differentiated 3T3-L1 cells with decreased TG, TC levels and SREBP-1 expressions in HepG2 cells, suggesting the inhibitory potential against fat accumulation (Vijayakumar et al. 2010).

The efficacy of trigonelline (75 and 100 μM) and isoorientin (10 μM) from fenugreek seeds have been reported against the regulation of glycolipids metabolism in 3T3-L1 preadipocytes (Ilavenil et al. 2014; Luan et al. 2018). Trigonelline and isoorientin significantly inhibited PPARγ, C/EBP-α mRNA expressions. Additionally, trigonelline down-regulated adiponectin, adipogenin, leptin, resistin, and adipocyte fatty acid-binding protein (aP2) mRNA expressions (Ilavenil et al. 2014), suggesting the ability of trigonelline and isoorientin to down-regulate the PPARγ pathway to decrease adipogenesis. These effects were supported by a recent report on the anti-obesity potential of trigonelline on 3T3-L1 adipocytes (Choi, Mukherjee, and Yun 2021). Trigonelline (75 μM) induced the browning of 3T3-L1 white adipocytes by enhancing the expressions of brown-fat signature proteins and genes as well as beige-specific genes, including Cd137 (tumour necrosis factor receptor superfamily member 9, TNFRSF9), Cited1 (Cbp/p300-interacting transactivator 1), Tbx1 (T-box protein 1), and Tmem26 (transmembrane protein 26). Trigonelline decreases adipogenesis and lipogenesis while promoting lipolysis and fatty acid oxidation, thus improving lipid metabolism in white adipocytes (Choi, Mukherjee, and Yun 2021). Moreover, trigonelline increased the expression of Cox4 (cytochrome c oxidase subunit 4), Nrf1 (nuclear respiratory factor 1), and Tfam (mitochondrial transcription factor) genes that are responsible for mitochondrial biogenesis. Mechanistic studies revealed that the browning effect of trigonelline in 3T3-L1 white adipocytes is mediated by activating β3-AR and inhibiting PDE4 (phosphodiesterase 4), thereby stimulating the p38 MAPK (mitogen-activated protein kinase 14)/ATF-2 (Activating transcription factor 2) signalling pathway (Choi, Mukherjee, and Yun 2021).

The researcher evaluated the underlying mechanisms of diosgenin on macrophage cholesterol metabolism (Lv et al. 2015). The human THP-1 macrophages and mouse peritoneal macrophages (MPMs) were treated with diosgenin (100 mM). Findings suggested that diosgenin significantly up-regulated the expression of ATP-binding cassette transporter A1 (ABCA1) protein, whereas it down-regulated miR-19b levels. The author concluded that diosgenin inhibited macrophage cholesterol metabolism via activation of ABCA1 expression (Lv et al. 2015).

GAL (20, 40, and 80 µg/ml) were incubated with 3T3-L1 cells to study the mechanism of adipocyte differentiation and lipid metabolism (Cheng et al. 2018). Results in mature adipocytes suggested marked inhibition of lipid accumulation and increased lipolysis activity via activation of protein levels of peroxisome-proliferator-activated receptor-γ co-activator 1β (PGC1β), sirtuin 1 (SIRT1), SIRT3, and adenosine monophosphate-activated protein kinase (AMPK) by GAL (Cheng et al. 2018). The author concluded that activation of AMPK reduces lipid contents in adipocytes by GAL, which may be helpful for anti-obesity potential (Cheng et al. 2018).

The inhibitory potential of GAL against oxidation of LDL was confirmed by a study using human peripheral blood mononuclear cells (hPBMCs) (Saji et al. 2018). In this study, copper-mediated oxidation of LDL was performed in human peripheral blood mononuclear cells (hPBMCs), which were further incubated with different concentrations of curcumin-galactomannoside complex (6.25, 12.5, 25, and 50 µg/mL). The complex significantly inhibited elevated levels of an inflammatory response (COX (cyclooxygenase), LOX (lipooxygenase), PGE2 (prostaglandin E2), VCAM-1 (vascular cell adhesion molecule-1), IL-6 (interleukin-6), TNF-α (tumour necrosis factor-alpha), iNOS (inducible nitric oxide synthase), and NF-κB (nuclear factor kappa B)), depicting GAL's anti-atherogenic potential (Saji et al. 2018).

The anti-dyslipidaemic potential of FSE (hydroalcoholic), diosgenin, and 4-HI was reported on the HepG2 cell line (Hajizadeh et al. 2019). In this study, HepG2 cell lines were incubated with FSE (300 µg/mL), diosgenin (6.21 µg/mL), and 4-HI (1 µg/mL). The results showed that liver X receptor alpha (LXR-α), SREBP-1C, acetyl-CoA carboxylase (ACC), FAS were down-regulated significantly, and fibroblast growth factor 21 (FGF21), PPARγ, and low-density lipoprotein receptor (LDLR) were up-regulated significantly in HepG2 cells. The author suggested that FSE, diosgenin, and 4-HI possess the lipid metabolism modulatory potential (Hajizadeh et al. 2019).

These findings of *in-vitro* studies supported the conclusion of *in-vivo* experimental studies on diosgenin (regulates the cholesterol metabolism) and 4-HI with trigonelline (inhibited SREBP-1c to decrease cholesterol synthesis). Taken together, inhibition of SREBP-1 and c/EBP-α seems to be an emerging and cardinal mechanism of action by the fenugreek seed towards its anti-hypercholesterolaemic potential (Table 11.6).

11.1.3 Fenugreek against Hyperlipidaemia: Clinical Evidence

Based on the background of reports of efficacy from scientific studies, numerous researchers investigated the potential of FSP or FSE in healthy subjects and patients with hyperlipidaemia during randomized controlled clinical studies. The earliest study evaluated the hypolipidaemic potential of FSP in patients with hyperlipidaemia (n = 10) (Sharma, Raghuram, and Rao 1991). Patients were administered isocaloric diets with and without FSP (100 g debittered) for 3 weeks. Ingestion of FSP caused a significant reduction in the serum TC, LDL, and VLDL cholesterol and TG levels, whereas HDL: TC, HDL: LDL, and HDL: VLDL-C were increased significantly. There were no adverse effects associated with the experimental diet, and suggested the FSP as a possible dietary supplement in hyperlipidaemic patients (Sharma, Raghuram, and Rao 1991).

TABLE 11.6
Potential of Fenugreek against Hyperlipidaemia: Preclinical (*In Vitro*) Evidence

Study Model	Dose	Proposed Mechanism of Action	Reference
3T3-L1 preadipocytes	FSE (thermostable, 0.5, 5, and 50 µg/ml)	↓ PPAR-γ, SREBP-1, and c/EBP-α	Vijayakumar et al. (2010)
HepG2 cells		↓ TG and TC levels, SREBP-1	
3T3-L1 preadipocytes	Trigonelline (75 and 100 µM)	↓ PPARγ, C/EBP-α, adiponectin, adipogenin, leptin, resistin, and aP2	Ilavenil et al. (2014)
	Trigonelline (75 µM)	↑ Cd137, Cited1, Tbx1 and Tmem26, Cox4, Nrf1, and Tfam	Choi, Mukherjee, and Yun (2021)
THP-1 macrophages and MPMs	Diosgenin (100 mM)	↑ ABCA1 ↓ miR-19b	Lv et al. (2015)
3T3-L1 preadipocytes	GAL (20, 40, and 80 µg/ml)	↑ PGC1β, SIRT1, SIRT3, AMPK	Cheng et al. (2018)
3T3-L1 preadipocytes	Isoorientin (10 µM)	↓ PPARγ, C/EBP-α, and FAS	Luan et al. (2018)
hPBMCs	GAL with curcumin (6.25, 12.5, 25, and 50 µg/mL)	↓ COX, LOX, PGE2, VCAM-1, IL-6, TNF-α, iNOS, and NF-κB	Saji et al. (2018)
HepG2 cell line	FSE (hydroalcoholic, 300 µg/mL), diosgenin (6.21 µg/mL) and 4-HI (1 µg/mL)	↓ LXR-α, SREBP-1C, ACC, FAS ↑ FGF21, PPAR-γ, and LDLR	Hajizadeh et al. (2019)

4-HI: 4-Hydroxyisoleucine; ABCA1: ATP-Binding Cassette Transporter A1; ACC: Acetyl-CoA Carboxylase; aP2: adipocyte Fatty Acid Binding Protein; C/EBP-α: CCAAT/Enhancer Binding Protein; Cd137: Tumour Necrosis Factor Receptor Superfamily Member 9 (TNFRSF9); Cited1: Cbp/p300-Interacting Transactivator 1; COX: Cylcooxgenase; Cox4: Cytochrome c Oxidase Subunit 4; FGF21: Fibroblast Growth Factor 21; FSE: Fenugreek Seed Extract; GAL: Galactomannans; ILs: Interleukins; iNOS: inducible Nitric Oxide Synthase; LDLR: Low-Density Lipoprotein Receptor; LOX: Lipooxygenase; LXR-α: Liver X Receptor alpha; MPMs: Mouse Peritoneal Macrophages; NF-κB: Nuclear Factor-Kappa B; Nrf1: Nuclear Respiratory Factor 1; PGE2: Prostaglandin E2; SREBP-1: Sterol Regulatory Element-Binding Proteins; Tbx1: T-box Protein 1; TC: Total Cholesterol; Tfam: Mitochondrial Transcription Factor; TG: Triglyceride, Tmem26: Transmembrane Protein 26; TNF-α: Tumour Necrosis Factor-alpha; VCAM-1: Vascular Cell Adhesion Molecule-1.

The potential of FSP of germinated seeds has been evaluated in patients with hypercholesterolaemia (Sowmya and Rajyalakshmi 1999). Hypercholesterolaemic volunteers (n = 20) received supplementation of FSP (12.5 g and 18.0 g per day) for 4 weeks. FSP (18 g/day) consumption showed a hypercholesterolaemic effect reflected by a significant reduction in TC and LDL levels, whereas no significant changes in HDL and VLDL levels were found. These results were further supported by another study with the treatment of defatted FSP (25 or 50 gm, for 3 weeks) in a patient with hypercholesterolaemia (n = 18), which significantly resulted in decreased serum TC, TG, and VLDL levels (Prasanna 2000; Sowmya and Rajyalakshmi 1999). Therefore, the inclusion of fenugreek seed into the diet of hypercholesterolaemic patients was suggested for beneficial effects to reduce the risk for coronary artery disease (Prasanna 2000; Sowmya and Rajyalakshmi 1999).

During two different randomized, double-blind, placebo-controlled trials, overweight hyperlipidaemic subjects (TC > 200 mg/dl) (Pashine et al. 2012) or patients with borderline hyperlipidaemia (dyslipidemia: LDL > 135 mg and < 190 mg/dl; HDL < 40 mg/dl; TC > 200 mg/dl; TG >150 mg/dl) (Yousefi et al. 2017) were treated with FSP (1.6 g, n = 54, for 12 weeks) or FSP (8 g, n = 24, for 8 weeks), respectively. FSP significantly decreased serum TC, TG, LDL-C, and VLDL-C suggesting its lipid-lowering potential in hyperlipidaemic subjects (Pashine et al. 2012; Yousefi et al. 2017)

However, in healthy individuals, FSP (5 gm/day, for 12 weeks) (Bordia, Verma, and Srivastava 1997) or FSE (hydroalcoholic, 588 and 1176 mg, 7 weeks) (Chevassus et al. 2009) or FSE (methanolic, Furosap™, 500 mg/day, for 12 weeks) (Maheshwari et al. 2017) did not alter the blood lipids profiles. Similarly, in overweight subjects, a dose of FSE (hydroalcoholic, 1176 mg, 7 weeks) was found insufficient to ameliorate the elevated blood lipids profile (Chevassus et al. 2010).

In another randomized, double-blind, placebo-controlled clinical study, the efficacy of FSP was reported in subjects with hyperlipidaemia who were previously assigned to the American Heart Association (AHA) step 1 diet (Fedacko et al. 2016; Sajty et al. 2014). Patients with hyperlipidaemia (TC < 7.76 nmol/l) were supplemented with either FSP (60 g/day) or placebo (cellulose, 3.0 g/day) for 12 weeks. FSP showed a marked reduction in serum LDL, TC, TG, LDL: HDL-C, and an increase in serum HDL-C. FSP supplementation is associated with only mild gastrointestinal side effects. FSP supplementation showed potent anti-hypercholesterolaemic effects on the AHA step 1 diet (Fedacko et al. 2016; Sajty et al. 2014). A recently published meta-analysis supported the suggested efficacy of dietary fenugreek seed supplementation to improve the lipid parameters (Askarpour et al. 2020).

The anti-hyperlipidaemic efficacy of GAL-based FSE supplementation to resistance-trained healthy males was evaluated to obtain optimum body composition using a randomized, double-blind placebo-controlled design (Poole et al. 2010). In this study, GAL-based FSE (500 mg, once a day, 8 weeks) showed a significant reduction in body fat percentage and a significant increase in lean body mass compared to the placebo group (Poole et al. 2010). Furthermore, GAL-based standardized FSE (500 mg, once a day, 8 weeks) to female subjects with high-fat mass (overweight and obese) reported significant reduction in body fat related measures such as suprailiac skinfold thickness (v/s baseline) and abdominal skinfold thickness (v/s baseline and v/s placebo), and percent fat mass, (v/s baseline) during the randomized, double-blind, placebo-controlled clinical study (Deshpande et al. 2019). Furthermore, glycoside-based FSE reported a significant decrease in fat mass compared with matched placebo during the double-blind placebo-controlled study in college-age men (Wilborn et al. 2010).

Taken together, the FSP can be a useful dietary supplement in patients with hyperlipidaemia. However, the varying dosage of FSP (from 1.6 g to 100 g) with a wide range of duration of the study (from 3 weeks to 12 weeks) and inconsistent study protocols (Table 11.7) limits the recommendation

TABLE 11.7
Potential of Fenugreek against Hyperlipidaemia: Clinical Evidence

Constituent	Patient Population	Study Design	Dose per Day and Duration	Results	Reference
Fenugreek seed powder	Patient with hypercholesterolaemia	OO	100 g, 3 weeks	↓ serum TC, LDL and VLDL cholesterol and TG ↑ HDL: TC, HDL: LDL, and HDL: VLDL-C	Sharma, Raghuram, and Rao (1991)
Fenugreek seed powder	Healthy individuals	PC	5 g, 12 weeks	No alteration in blood lipids profile	Bordia, Verma, and Srivastava (1997)
Fenugreek seed powder	Patients with hypocholesterolaemia	OO	18 g, 4 weeks	↓ serum TC and LDL	Sowmya and Rajyalakshmi (1999)
Fenugreek seed powder	Patient with hypercholesterolaemia	R, PC	25 or 50 g, 3 weeks	↓ serum TC, TG, and VLDL	Prasanna (2000)
Fenugreek seed extract	Healthy individuals	DB	GAL-based (500 mg/day, 8 weeks)	↓ body fat mass	Poole et al. (2010)

Constituent	Patient Population	Study Design	Dose per Day and Duration	Results	Reference
Fenugreek seed extract	Healthy individuals	DB	Glycoside based (500 mg/day, 8 weeks)	↓ body fat	Wilborn et al. (2010)
Fenugreek seed powder	Overweight hyperlipidaemic subjects	R, DB, PC	1.6 g, 12 weeks	↓ serum TC, TG, LDL-C, and VLDL-C	Pashine et al. (2012)
Fenugreek seed powder	Patient with hyperlipidaemia previously assigned to American Heart Association step 1 diet	R, DB, PC	60 g, 12 weeks	↓ serum LDL, TC, TG, LDL: HDL-C ↑ serum HDL-C	Fedacko et al. (2016), Sajty et al. (2014)
Fenugreek seed powder	Patients with borderline hyperlipidaemia	R, DB, PC	8 g, 8 weeks	↓ TG, TC, and LDL	Yousefi et al. (2017)
Fenugreek seed extract	Healthy individuals	R, DB, PC	Hydroalcoholic (1176 mg, 7 weeks)	No alteration in blood lipids profile	Chevassus et al. (2009)
Fenugreek seed extract	Overweight subjects	R, DB, PC			Chevassus et al. (2010)
Fenugreek seed extract	Healthy individuals	OO	Methanolic (Furosap™, 500 mg/day, 12 weeks)		Maheshwari et al. (2017)
Fenugreek seed extract	Subjects with high-fat mass	DB	GAL-based (LMWGAL-TF, 500 mg/day, 8 weeks)	↓ body fat mass ↓ Skinfold thickness	Deshpande et al. (2019)

DB: Double-Blind; FSE: Fenugreek Seed Extract; FSP: Fenugreek Seed Powder; HDL-C: High-Density Lipoprotein-Cholesterol; LDL-C: Low-Density Lipoprotein-Cholesterol; OO: one-arm, open-labelled; PC: Placebo-Controlled; R: Randomized; TC: Total Cholesterol; TG: Triglyceride, VLDL-C: Very Low-Density Lipoprotein-Cholesterol

of the use of crude FSP as a potential agent for the management of hyperlipidaemia. However, repeated-dose supplementation of FSE, especially rich in crucial phytoconstituents such as soluble fibre or GA, demonstrated potential body fat and lipid-lowering effects in healthy individuals and overweight subjects (Table 11.7).

11.2 CONCLUSION

The available evidence suggests the potential of dietary fenugreek seeds, powder, extracts, and phytoconstituents (galactomannan, steroidal saponins (diosgenin), 4-HI, and trigonelline) to manage the elevated lipid levels via diverse mechanisms. The fenugreek fibre constituent, galactomannan, showed beneficial effects on cholesterol absorption and metabolism by improving the gut microbiota and regulating FAS and HMG-CoA reductase activity. The fenugreek saponins (diosgenin) were reported to regulate HMG-CoA reductase and lecithin cholesterol acyltransferase activity, inhibiting cholesterol synthesis and absorption. The fenugreek extract with marker amino acid constituent, 4-HI, and an alkaloid constituent, trigonelline, has cholesterol-lowering potential via downregulation of SREBP-1c. These results are supported by clinical evidence of the efficacy and safety of fenugreek supplementation, as demonstrated during the randomized placebo-controlled clinical studies. In conclusion, fenugreek and phytoconstituents can become a potential option to maintain healthy lipid levels and manage lipid-related disorders such as obesity, prediabetes, insulin resistance, and risk of cardiovascular diseases.

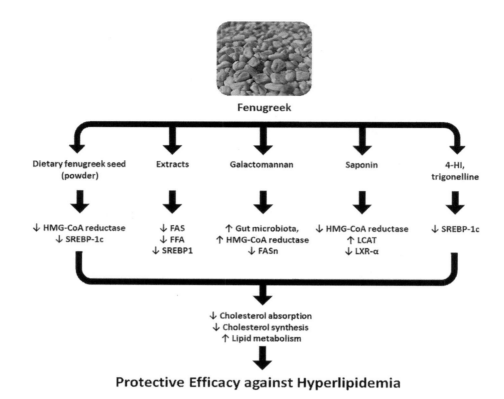

FIGURE 11.1 The anti-hyperlipidaemic potential of fenugreeks seed and its various phytoconstituents.

REFERENCES

Ai, L., Q. Xu, C. Wu, X. Wang, Z. Chen, D. Su, X. Jiang, A. Xu, Q. Lin, and Z. Fan. 2015. "A20 attenuates FFAs-induced lipid accumulation in nonalcoholic steatohepatitis." *International Journal of Biological Sciences* 11 (12):1436–46. doi: 10.7150/ijbs.13371.

Al-Habori, M., A. M. Al-Aghbari, and M. Al-Mamary. 1998. "Effects of fenugreek seeds and its extracts on plasma lipid profile: A study on rabbits." *Phytotherapy Research* 12 (8):572–5. doi: 10.1002/(sici)1099-1573(199812)12:8<572::Aid-ptr353>3.0.Co;2-e.

Al-Matubsi, H. Y., N. A. Nasrat, G. A. Oriquat, M. Abu-Samak, K. A. Al-Mzain, and M. Salim. 2011. "The hypocholesterolemic and antioxidative effect of dietary diosgenin and chromium chloride supplementation on high-cholesterol fed Japanese quails." *Pakistan Journal of Biological Sciences* 14 (7):425–32. doi: 10.3923/pjbs.2011.425.432.

Al-Sultan, S., and S. El-Bahr. 2015. "Effect of aqueous extract of fenugreek (*Trigonella foenum-graecum* L.) on selected biochemical and oxidative stress biomarkers in rats intoxicated with carbon tetrachloride." *International Journal of Pharmacology* 11 (1):43–9. doi: 10.3923/ijp.2015.43.49.

Ansari, B., M. Singh, S. Sharma, and B. Choudhary. 2020. "Preclinical anti hyperlipidemic effect of herbalism against lipid elevating agents: A review." *Biomedical and Pharmacology Journal* 13 (4):1695–707. doi: 10.13005/bpj/2044.

Askarpour, M., F. Alami, M. S. Campbell, K. Venkatakrishnan, A. Hadi, and E. Ghaedi. 2020. "Effect of fenugreek supplementation on blood lipids and body weight: A systematic review and meta-analysis of randomized controlled trials." *Journal of Ethnopharmacology* 253:112538. doi: 10.1016/j.jep.2019.112538.

Avalos-Soriano, A., R. De la Cruz-Cordero, J. L. Rosado, and T. Garcia-Gasca. 2016. "4-Hydroxyisoleucine from Fenugreek (*Trigonella foenum-graecum*): Effects on insulin resistance associated with obesity." *Molecules* 21 (11). doi: 10.3390/molecules21111596.

Bahmani, M., H. Shirzad, M. Mirhosseini, A. Mesripour, and M. Rafieian-Kopaei. 2016. "A Review on Ethnobotanical and Therapeutic Uses of Fenugreek (*Trigonella foenum-graceum* L)." *Evidence-Based Complementary and Alternative Medicine* 21 (1):53–62. doi: 10.1177/2156587215583405.

Bao, Y., J. Han, F. B. Hu, E. L. Giovannucci, M. J. Stampfer, W. C. Willett, and C. S. Fuchs. 2013. "Association of nut consumption with total and cause-specific mortality." *The New England Journal of Medicine* 369 (21):2001–11. doi: 10.1056/NEJMoa1307352.

Basch, E., C. Ulbricht, G. Kuo, P. Szapary, and M. Smith. 2003. "Therapeutic applications of fenugreek." *Alternative Medicine Review* 8 (1):20–7.

Basu, T. K., and A. Srichamroen. 2010. "Chapter 28—health benefits of fenugreek (*Trigonella foenum-graecum* leguminosse)." In *Bioactive Foods in Promoting Health*, edited by Watson, R.R. and Preedy, V.R., 425–35. San Diego: Academic Press.

Belaid-Nouira, Y., H. Bakhta, M. Bouaziz, I. Flehi-Slim, Z. Haouas, and H. Ben Cheikh. 2012. "Study of lipid profile and parieto-temporal lipid peroxidation in AlCl(3) mediated neurotoxicity. Modulatory effect of fenugreek seeds." *Lipids in Health and Disease* 11 (1):16. doi: 10.1186/1476-511X-11-16.

Belguith-Hadriche, O., M. Bouaziz, K. Jamoussi, A. El Feki, S. Sayadi, and F. Makni-Ayedi. 2010. "Lipid-lowering and antioxidant effects of an ethyl acetate extract of fenugreek seeds in high-cholesterol-fed rats." *Journal of Agricultural and Food Chemistry* 58 (4):2116–22. doi: 10.1021/jf903186w.

Belguith-Hadriche, O., M. Bouaziz, K. Jamoussi, M. S. Simmonds, A. El Feki, and F. Makni-Ayedi. 2013. "Comparative study on hypocholesterolemic and antioxidant activities of various extracts of fenugreek seeds." *Food Chemistry* 138 (2–3):1448–53. doi: 10.1016/j.foodchem.2012.11.003.

Binesh, A., S. N. Devaraj, and H. Devaraj. 2018. "Inhibition of nuclear translocation of notch intracellular domain (NICD) by diosgenin prevented atherosclerotic disease progression." *Biochimie* 148:63–71. doi: 10.1016/j.biochi.2018.02.011.

Binesh, A., S. N. Devaraj, and D. Halagowder. 2018. "Atherogenic diet induced lipid accumulation induced NFkappaB level in heart, liver and brain of Wistar rat and diosgenin as an anti-inflammatory agent." *Life Sciences* 196:28–37. doi: 10.1016/j.lfs.2018.01.012.

Boban, P. T., B. Nambisan, and P. R. Sudhakaran. 2006. "Hypolipidaemic effect of chemically different mucilages in rats: A comparative study." *British Journal of Nutrition* 96 (6):1021–9. doi: 10.1017/bjn20061944.

Boban, P. T., B. Nambisan, and P. R. Sudhakaran. 2009. "Dietary mucilage promotes regression of atheromatous lesions in hypercholesterolemic rabbits." *Phytotherapy Research* 23 (5):725–30. doi: 10.1002/ptr.2728.

Bordia, A., S. K. Verma, and K. C. Srivastava. 1997. "Effect of ginger (*Zingiber officinale* Rosc.) and fenugreek (*Trigonella foenum-graecum* L.) on blood lipids, blood sugar and platelet aggregation in patients with coronary artery disease." *Prostaglandins, Leukotrienes and Essential Fatty Acids* 56 (5):379–84. doi: 10.1007/s10565-007-9000-7.

Bruce-Keller, A. J., A. J. Richard, S. O. Fernandez-Kim, D. M. Ribnicky, J. M. Salbaum, S. Newman, R. Carmouche, and J. M. Stephens. 2020. "Fenugreek counters the effects of high fat diet on gut microbiota in mice: Links to metabolic benefit." *Scientific Reports* 10 (1):1245. doi: 10.1038/s41598-020-58005-7.

Chaturvedi, U., A. Shrivastava, S. Bhadauria, J. K. Saxena, and G. Bhatia. 2013. "A mechanism-based pharmacological evaluation of efficacy of *Trigonella foenum-graecum* (fenugreek) seeds in regulation of dyslipidemia and oxidative stress in hyperlipidemic rats." *Journal of Cardiovascular Pharmacology* 61 (6):505–12. doi: 10.1097/FJC.0b013e31828b7822.

Chen, Z., Y. L. Lei, W. P. Wang, Y. Y. Lei, Y. H. Liu, J. Hei, J. Hu, and H. Sui. 2017. "Effects of saponin from *Trigonella foenum-graecum* seeds on dyslipidemia." *Iranian Journal of Medical Sciences* 42 (6):577–85.

Cheng, C. Y., A. J. Yang, P. Ekambaranellore, K. C. Huang, and W. W. Lin. 2018. "Anti-obesity action of INDUS810, a natural compound from *Trigonella foenum-graecum*: AMPK-dependent lipolysis effect in adipocytes." *Obesity Research & Clinical Practice* 12 (6):562–9. doi: 10.1016/j.orcp.2018.08.005.

Chevassus, H., J. B. Gaillard, A. Farret, F. Costa, I. Gabillaud, E. Mas, A. M. Dupuy, F. Michel, C. Cantie, E. Renard, F. Galtier, and P. Petit. 2010. "A fenugreek seed extract selectively reduces spontaneous fat intake in overweight subjects." *European Journal of Clinical Pharmacology* 66 (5):449–55. doi: 10.1007/s00228-009-0770-0.

Chevassus, H., N. Molinier, F. Costa, F. Galtier, E. Renard, and P. Petit. 2009. "A fenugreek seed extract selectively reduces spontaneous fat consumption in healthy volunteers." *European Journal of Clinical Pharmacology* 65 (12):1175–8. doi: 10.1007/s00228-009-0733-5.

Choi, M., S. Mukherjee, and J. W. Yun. 2021. "Trigonelline induces browning in 3T3-L1 white adipocytes." *Phytotherapy Research* 35 (2):1113–24. doi: 10.1002/ptr.6892.

Dakam, W., R. Matsinkou, J. Ngondi, and J. Oben. 2009. "Abstract: P1116 effect of a protein supplementation on the weight control and anti-atherogenic properties of fenugreek galactomannan." *Atherosclerosis Supplements* 10 (2):e1142. doi: 10.1016/s1567-5688(09)71109-2.

Deshpande, P. O., V. G. Bele, K. Joshi, and P. A. Thakurdesai. 2019. "Effects of low molecular weight galactomannans based standardized fenugreek seed extract in subjects with high fat mass: A randomized, double-blind, placebo-controlled clinical study." *Journal of Applied Pharmaceutical Science* 10 (1):62–9. doi: 10.7324/JAPS.2020.101008.

Elmahdi, B., and S. M. El-Bahr. 2014. "Influence of dietary supplementation of fenugreek (*Trigonella foenum-graecum* L.) on serum biochemical parameters of rats fed high cholesterol diet." *International Journal of Biological Chemistry* 9 (1):1–10. doi: 10.3923/ijbc.2015.1.10.

Estruch, R., E. Ros, J. Salas-Salvado, M. I. Covas, D. Corella, F. Aros, E. Gomez-Gracia, V. Ruiz-Gutierrez, M. Fiol, J. Lapetra, R. M. Lamuela-Raventos, L. Serra-Majem, X. Pinto, J. Basora, M. A. Munoz, J. V. Sorli, J. A. Martinez, M. A. Martinez-Gonzalez, and P. S. Investigators. 2013. "Primary prevention of cardiovascular disease with a Mediterranean diet." *The New England Journal of Medicine* 368 (14):1279–90. doi: 10.1056/NEJMoa1200303.

Evans, A. J., R. L. Hood, D. G. Oakenfull, and G. S. Sidhu. 1992. "Relationship between structure and function of dietary fibre: A comparative study of the effects of three galactomannans on cholesterol metabolism in the rat." *British Journal of Nutrition* 68 (1):217–29. doi: 10.1079/bjn19920079.

Fedacko, J., R. Singh, M. Niaz, S. Ghosh, P. Fedackova, A. Tripathi, A. Etharat, E. Onsaard, V. Singh, and S. Shastun. 2016. "Fenugreek seeds decrease blood cholesterol and blood glucose as adjunct to diet therapy in patients with hypercholesterolemia." *World Heart Journal* 8 (3):239.

Fuller, S., and J. M. Stephens. 2015. "Diosgenin, 4-hydroxyisoleucine, and fiber from fenugreek: Mechanisms of actions and potential effects on metabolic syndrome." *Advances in Nutrition* 6 (2):189–97. doi: 10.3945/an.114.007807.

Gong, G., Y. Qin, W. Huang, S. Zhou, X. Wu, X. Yang, Y. Zhao, and D. Li. 2010. "Protective effects of diosgenin in the hyperlipidemic rat model and in human vascular endothelial cells against hydrogen peroxide-induced apoptosis." *Chemico-Biological Interactions* 184 (3):366–75. doi: 10.1016/j.cbi.2010.02.005.

Goyal, S., N. Gupta, and S. Chatterjee. 2016. "Investigating Therapeutic Potential of *Trigonella foenum-graecum* L. as Our Defense Mechanism against Several Human Diseases." *International Journal of Toxicology* 2016:1250387. doi: 10.1155/2016/1250387.

Haeri, M. R., M. Izaddoost, M. R. Ardekani, M. R. Nobar, and K. N. White. 2009. "The effect of fenugreek 4-hydroxyisoleucine on liver function biomarkers and glucose in diabetic and fructose-fed rats." *Phytotherapy Research* 23 (1):61–4. doi: 10.1002/ptr.2557.

Hajizadeh, M., M. Mohammad-Sadeghipour, M. Mahmoodi, S. Falahati-pour, A. Khoshdel, M. Fahmidehkar, M. Mirzaei, and M. Karimabad. 2019. "*Trigonella foenum-graecum* seed extract modulates expression of lipid metabolism- related genes in HepG2 cells." *Asian Pacific Journal of Tropical Biomedicine* 9 (6):240. doi: 10.4103/2221-1691.260396.

Hamza, N., B. Berke, C. Cheze, R. Le Garrec, A. Umar, A. N. Agli, R. Lassalle, J. Jove, H. Gin, and N. Moore. 2012. "Preventive and curative effect of *Trigonella foenum-graecum* L. seeds in C57BL/6J models of type 2 diabetes induced by high-fat diet." *Journal of Ethnopharmacology* 142 (2):516–22. doi: 10.1016/j.jep.2012.05.028.

Handa, T., K. Yamaguchi, Y. Sono, and K. Yazawa. 2005. "Effects of fenugreek seed extract in obese mice fed a high-fat diet." *Bioscience, Biotechnology, and Biochemistry* 69 (6):1186–8. doi: 10.1271/bbb.69.1186.

Hasani-Ranjbar, S., N. Nayebi, L. Moradi, A. Mehri, B. Larijani, and M. Abdollahi. 2010. "The efficacy and safety of herbal medicines used in the treatment of hyperlipidemia; a systematic review." *Current Pharmaceutical Design* 16 (26):2935–47. doi: 10.2174/138161210793176464.

Heshmat-Ghahdarijani, K., N. Mashayekhiasl, A. Amerizadeh, Z. Teimouri Jervekani, and M. Sadeghi. 2020. "Effect of fenugreek consumption on serum lipid profile: A systematic review and meta-analysis." *Phytotherapy Research* 34 (9):2230–45. doi: 10.1002/ptr.6690.

Hua, Y., S. Y. Ren, R. Guo, O. Rogers, R. P. Nair, D. Bagchi, A. Swaroop, and S. Nair. 2015. "Furostanolic saponins from *Trigonella foenum-graecum* alleviate diet-induced glucose intolerance and hepatic fat accumulation." *Molecular Nutrition & Food Research* 59 (10):2094–100. doi: 10.1002/mnfr.201500197.

Ilavenil, S., M. V. Arasu, J. C. Lee, D. H. Kim, S. G. Roh, H. S. Park, G. J. Choi, V. Mayakrishnan, and K. C. Choi. 2014. "Trigonelline attenuates the adipocyte differentiation and lipid accumulation in 3T3-L1 cells." *Phytomedicine* 21 (5):758–65. doi: 10.1016/j.phymed.2013.11.007.

Kandhare, A. D., D. Bandyopadhyay, and P. A. Thakurdesai. 2018. "Low molecular weight galactomannans-based standardized fenugreek seed extract ameliorates high-fat diet-induced obesity in mice via modulation of FASn, IL-6, leptin, and TRIP-Br2." *RSC Advances* 8 (57):32401–16. doi: 10.1039/c8ra05204b.

Kannappan, S., and C. V. Anuradha. 2009. "Insulin sensitizing actions of fenugreek seed polyphenols, quercetin & metformin in a rat model." *Indian Journal of Medical Research* 129 (4):401–8.

Karr, S. 2017. "Epidemiology and management of hyperlipidemia." *American Journal of Managed Care* 23 (9 Suppl):S139-S148.

Kaur, G., V. Wani, A. Dave, and P. Jadhav. 2015. "Effect of *Lagenaria siceraria* and *Trigonella foenum-graecum* on lipid absorption and excretion for modulation of lipid profile." *Science International* 3:18–24. doi: 10.17311/sciintl.2015.18.24.

Kaviarasan, S., P. Viswanathan, and C. V. Anuradha. 2007. "Fenugreek seed (*Trigonella foenum graecum*) polyphenols inhibit ethanol-induced collagen and lipid accumulation in rat liver." *Cell Biology and Toxicology* 23 (6):373–83. doi: 10.1007/s10565-007-9000-7.

Khodamoradi, K., M. H. Khosropanah, Z. Ayati, D. Chang, E. Nasli-Esfahani, M. H. Ayati, and N. Namazi. 2020. "The Effects of Fenugreek on Cardiometabolic Risk Factors in Adults: A Systematic Review and Meta-analysis." *Complementary Therapies in Medicine* 52:102416. doi: 10.1016/j.ctim.2020.102416.

Khound, R., J. Shen, Y. Song, D. Santra, and Q. Su. 2018. "Phytoceuticals in Fenugreek Ameliorate VLDL Overproduction and Insulin Resistance via the Insig Signaling Pathway." *Molecular Nutrition & Food Research* 62 (5). doi: 10.1002/mnfr.201700541.

Knott, E. J., A. J. Richard, R. L. Mynatt, D. Ribnicky, J. M. Stephens, and A. Bruce-Keller. 2017. "Fenugreek supplementation during high-fat feeding improves specific markers of metabolic health." *Scientific Reports* 7 (1):12770. doi: 10.1038/s41598-017-12846-x.

Kumar, P., and U. Bhandari. 2013. "Protective effect of *Trigonella foenum-graecum* Linn. on monosodium glutamate-induced dyslipidemia and oxidative stress in rats." *Indian Journal of Pharmacology* 45 (2):136–40. doi: 10.4103/0253-7613.108288.

Kumar, P., and U. Bhandari. 2015. "Common medicinal plants with antiobesity potential: A special emphasis on fenugreek." *Ancient Science of life* 35 (1):58–63. doi: 10.4103/0257-7941.165629.

Kumar, P., and U. Bhandari. 2016. "Fenugreek Seed Extract Prevents Fat Deposition in Monosodium Glutamate (MSG)-Obese Rats." *Drug Research (Stuttg)* 66 (4):174–80. doi: 10.1055/s-0035-1555812.

Kumar, P., U. Bhandari, and S. Jamadagni. 2014. "Fenugreek seed extract inhibit fat accumulation and ameliorates dyslipidemia in high fat diet-induced obese rats." *BioMed Research International* 2014:606021. doi: 10.1155/2014/606021.

Li, R., Y. Liu, J. Shi, Y. Yu, H. Lu, L. Yu, Y. Liu, and F. Zhang. 2019. "Diosgenin regulates cholesterol metabolism in hypercholesterolemic rats by inhibiting NPC1L1 and enhancing ABCG5 and ABCG8." *Biochimica et Biophysica Acta - Molecular and Cell Biology of Lipids* 1864 (8):1124–33. doi: 10.1016/j.bbalip.2019.04.010.

Liu, Z. X., Q. Hong, D. H. Peng, Y. Yang, W. L. Yu, H. Shui, X. Zhou, and S. M. Liu. 2019. "Evaluation of serum free fatty acids in chronic renal failure: Evidence from a rare case with undetectable serum free fatty acids and population data." *Lipids in Health and Disease* 18 (1):151. doi: 10.1186/s12944-019-1093-5.

Luan, G., Y. Wang, Z. Wang, W. Zhou, N. Hu, G. Li, and H. Wang. 2018. "Flavonoid Glycosides from Fenugreek Seeds Regulate Glycolipid Metabolism by Improving Mitochondrial Function in 3T3-L1 Adipocytes in Vitro." *Journal of Agricultural and Food Chemistry* 66 (12):3169–78. doi: 10.1021/acs.jafc.8b00179.

Lv, Y. C., J. Yang, F. Yao, W. Xie, Y. Y. Tang, X. P. Ouyang, P. P. He, Y. L. Tan, L. Li, M. Zhang, D. Liu, F. S. Cayabyab, X. L. Zheng, and C. K. Tang. 2015. "Diosgenin inhibits atherosclerosis via suppressing the MiR-19b-induced downregulation of ATP-binding cassette transporter A1." *Atherosclerosis* 240 (1):80–9. doi: 10.1016/j.atherosclerosis.2015.02.044.

Maheshwari, A., N. Verma, A. Swaroop, M. Bagchi, H. G. Preuss, K. Tiwari, and D. Bagchi. 2017. "Efficacy of Furosap(TM), a novel *Trigonella foenum-graecum* seed extract, in Enhancing Testosterone Level and Improving Sperm Profile in Male Volunteers." *International Journal of Medical Sciences* 14 (1):58–66. doi: 10.7150/ijms.17256.

Manivannan, J., E. Balamurugan, T. Silambarasan, and B. Raja. 2013. "Diosgenin improves vascular function by increasing aortic eNOS expression, normalize dyslipidemia and ACE activity in chronic renal failure rats." *Molecular and Cellular Biochemistry* 384 (1–2):113–20. doi: 10.1007/s11010-013-1788-2.

Mukthamba, P., and K. Srinivasan. 2015. "Hypolipidemic influence of dietary fenugreek (*Trigonella foenum-graecum*) seeds and garlic (*Allium sativum*) in experimental myocardial infarction." *Food and Function* 6 (9):3117–25. doi: 10.1039/c5fo00240k.

Mukthamba, P., and K. Srinivasan. 2016a. "Hypolipidemic and antioxidant effects of dietary fenugreek (*Trigonella foenum-graecum*) seeds and garlic (Allium sativum) in high-fat fed rats." *Food Bioscience* 14:1–9. doi: 10.1016/j.fbio.2016.01.002.

Mukthamba, P., and K. Srinivasan. 2016b. "Protective effect of dietary fenugreek (*Trigonella foenum-graecum*) seeds and garlic (*Allium sativum*) on induced oxidation of low-density lipoprotein in rats." *Journal of Basic and Clinical Physiology and Pharmacology* 27 (1):39–47. doi: 10.1515/jbcpp-2015-0037.

Muraki, E., E. Yamaguchi, N. Tsunoda, H. Chiba, and K. Kasono. 2009. "Therapeutic effects of fenugreek seeds and their active components on diet-induced metabolic disorders in rats—Part 1." *Chemistry and Physics of Lipids* 160S (Supplement):S45.

Muraki, E., H. Chiba, K. Taketani, S. Hoshino, N. Tsuge, N. Tsunoda, and K. Kasono. 2012. "Fenugreek with reduced bitterness prevents diet-induced metabolic disorders in rats." *Lipids in Health and Disease* 11:58. doi: 10.1186/1476-511X-11-58.

Muraki, E., H. Chiba, N. Tsunoda, and K. Kasono. 2011. "Fenugreek improves diet-induced metabolic disorders in rats." *Hormone and Metabolic Research* 43 (13):950–5. doi: 10.1055/s-0031-1291345.

Nader, M., M. N. Rasheed, and H. T. ALMusawi. 2013. "Effect of *Trigonella foenum* seeds powder on lipids profile in fed rats." *International Journal of Current Research* 5 (9):2443–5.

Nagamma, T., A. Konuri, C. D. Nayak, S. U. Kamath, P. E. G. Udupa, and Y. Nayak. 2019. "Dose-dependent effects of fenugreek seed extract on the biochemical and haematological parameters in high-fat diet-fed rats." *Journal of Taibah University Medical Sciences* 14 (4):383–9. doi: 10.1016/j.jtumed.2019.05.003.

Narender, T., A. Puri, Shweta, T. Khaliq, R. Saxena, G. Bhatia, and R. Chandra. 2006. "4-hydroxyisoleucine an unusual amino acid as antidyslipidemic and antihyperglycemic agent." *Bioorganic & Medicinal Chemistry Letters* 16 (2):293–6. doi: 10.1016/j.bmcl.2005.10.003.

Nordestgaard, B. G., M. J. Chapman, S. E. Humphries, H. N. Ginsberg, L. Masana, O. S. Descamps, O. Wiklund, R. A. Hegele, F. J. Raal, J. C. Defesche, A. Wiegman, R. D. Santos, G. F. Watts, K. G. Parhofer, G. K. Hovingh, P. T. Kovanen, C. Boileau, M. Averna, J. Boren, E. Bruckert, A. L. Catapano, J. A. Kuivenhoven, P. Pajukanta, K. Ray, A. F. Stalenhoef, E. Stroes, M. R. Taskinen, A. Tybjaerg-Hansen, and P. European Atherosclerosis Society Consensus. 2013. "Familial hypercholesterolaemia is underdiagnosed and undertreated in the general population: Guidance for clinicians to prevent coronary heart disease: Consensus statement of the European Atherosclerosis Society." *European Heart Journal* 34 (45):3478–90a. doi: 10.1093/eurheartj/eht273.

Parhofer, K. G. 2016. "The Treatment of Disorders of Lipid Metabolism." *Deutsches Ärzteblatt International* 113 (15):261–8. doi: 10.3238/arztebl.2016.0261.

Pashine, L., J. V. Singh, A. K. Vaish, S. K. Ojha, and A. Mehdi. 2012. "Hypolipidemic effects of *Trigonella foenum-graecum* on overweight Indian subjects: Double blind study." *Indian Journal of Preventive and Social Medicine* 43 (2):127–33.

Petit, P. R., Y. D. Sauvaire, D. M. Hillaire-Buys, O. M. Leconte, Y. G. Baissac, G. R. Ponsin, and G. R. Ribes. 1995. "Steroid saponins from fenugreek seeds: Extraction, purification, and pharmacological investigation on feeding behavior and plasma cholesterol." *Steroids* 60 (10):674–80. doi: 10.1016/0039-128x(95)00090-d.

Petit, P. R., Y. D. Sauvaire, G. Ponsin, M. Manteghetti, A. Fave, and G. Ribes. 1993. "Effects of a fenugreek seed extract on feeding behaviour in the rat: Metabolic-endocrine correlates." *Pharmacology Biochemistry and Behavior* 45 (2):369–74. doi: 10.1016/0091-3057(93)90253-p.

Poetsch, M. S., A. Strano, and K. Guan. 2020. "Role of Leptin in Cardiovascular Diseases." *Front Endocrinol (Lausanne)* 11:354. doi: 10.3389/fendo.2020.00354.

Poole, C., B. Bushey, C. Foster, B. Campbell, D. Willoughby, R. Kreider, L. Taylor, and C. Wilborn. 2010. "The effects of a commercially available botanical supplement on strength, body composition, power output, and hormonal profiles in resistance-trained males." *Journal of the International Society of Sports Nutrition* 7:34. doi: 10.1186/1550-2783-7-34.

Prasanna, M. 2000. "Hypolipidemic effect of fenugreek: A clinical study." *Indian journal of Pharmacology* 32 (1):34–6.

Raju, J., and R. P. Bird. 2006. "Alleviation of hepatic steatosis accompanied by modulation of plasma and liver TNF-alpha levels by *Trigonella foenum-graecum* (fenugreek) seeds in Zucker obese (fa/fa) rats." *International Journal of Obesity (Lond)* 30 (8):1298–307. doi: 10.1038/sj.ijo.0803254.

Ramadan, G., N. M. El-Beih, and H. F. Abd El-Kareem. 2011. "Anti-metabolic syndrome and immunostimulant activities of Egyptian fenugreek seeds in diabetic/obese and immunosuppressive rat models." *British Journal of Nutrition* 105 (7):995–1004. doi: 10.1017/S0007114510004708.

Ramulu, P., N. V. Giridharan, and P. Udayasekhararao. 2011. "Hypolipidemic effect of soluble dietary fiber (galactomannan) isolated from fenugreek seeds in WNIN (GR-Ob) obese rats." *Journal of Medicinal Plants Research* 5 (19):4804–13.

Reddy, R. L., and K. Srinivasan. 2009a. "Fenugreek seeds reduce atherogenic diet-induced cholesterol gallstone formation in experimental mice." *Canadian Journal of Physiology and Pharmacology* 87 (11):933–43. doi: 10.1139/y09-084.

Reddy, R. L., and K. Srinivasan. 2009b. "Dietary fenugreek seed regresses preestablished cholesterol gallstones in mice." *Canadian Journal of Physiology and Pharmacology* 87 (9):684–93. doi: 10.1139/y09-062.

Reddy, R. R., and K. Srinivasan. 2011a. "Dietary fenugreek and onion attenuate cholesterol gallstone formation in lithogenic diet-fed mice." *International Journal of Experimental Pathology* 92 (5):308–19. doi: 10.1111/j.1365-2613.2011.00782.x.

Reddy, R. R., and K. Srinivasan. 2011b. "Effect of dietary fenugreek seeds on biliary proteins that influence nucleation of cholesterol crystals in bile." *Steroids* 76 (5):455–63. doi: 10.1016/j.steroids.2010.12.015.

Roberts, K. T. 2011. "The potential of fenugreek (*Trigonella foenum-graecum*) as a functional food and nutraceutical and its effects on glycemia and lipidemia." *Journal of Medicinal Food* 14 (12):1485–9. doi: 10.1089/jmf.2011.0002.

Saji, S., S. Asha, P. J. Svenia, M. Ratheesh, S. Sheethal, S. Sandya, and I. M. Krishnakumar. 2018. "Curcumin-galactomannoside complex inhibits pathogenesis in Ox-LDL-challenged human peripheral blood mononuclear cells." *Inflammopharmacology* 26 (5):1273–82. doi: 10.1007/s10787-018-0474-0.

Sajty, M., L. Jackova, L. Merkovska, L. Jedlickova, J. Fedacko, M. Janicko, M. Vachalcova, and R. B. Singh. 2014. "Fenugreek seeds decrease oxidative stress and blood lipids and increase nitric oxide in patients with hyperlipidemia." *Atherosclerosis* 235 (2):e113. doi: 10.1016/j.atherosclerosis.2014.05.306.

Sharma, M. S., and P. R. Choudhary. 2014. "Hypolipidemic effect of fenugreek seeds and its comparison with atorvastatin on experimentally induced hyperlipidemia." *Journal of College of Physicians and Surgeons Pakistan* 24 (8):539–42. doi: 08.2014/JCPSP.539542.

Sharma, M. S., and P. R. Choudhary. 2017. "Effect of Fenugreek Seeds Powder (*Trigonella foenum-graecum* L.) on Experimental Induced Hyperlipidemia in Rabbits." *Journal of Dietary Supplements* 14 (1):1–8. doi: 10.3109/19390211.2016.1168905.

Sharma, R. D., T. C. Raghuram, and V. D. Rao. 1991. "Hypolipidaemic effect of fenugreek seeds. A clinical study." *Phytotherapy Research* 5 (3):145–7. doi: 10.1002/ptr.2650050313.

Shtriker, M. G., M. Hahn, E. Taieb, A. Nyska, U. Moallem, O. Tirosh, and Z. Madar. 2018. "Fenugreek galactomannan and citrus pectin improve several parameters associated with glucose metabolism and modulate gut microbiota in mice." *Nutrition* 46:134–42 e3. doi: 10.1016/j.nut.2017.07.012.

Sjouke, B., D. M. Kusters, I. Kindt, J. Besseling, J. C. Defesche, E. J. Sijbrands, J. E. Roeters van Lennep, A. F. Stalenhoef, A. Wiegman, J. de Graaf, S. W. Fouchier, J. J. Kastelein, and G. K. Hovingh. 2015. "Homozygous autosomal dominant hypercholesterolaemia in the Netherlands: Prevalence, genotype-phenotype relationship, and clinical outcome." *European Heart Journal* 36 (9):560–5. doi: 10.1093/eurheartj/ehu058.

Sowmya, P., and P. Rajyalakshmi. 1999. "Hypocholesterolemic effect of germinated fenugreek seeds in human subjects." *Plant Foods for Human Nutrition* 53 (4):359–65. doi: 10.1023/a:1008021618733.

Srichamroen, A., C. J. Field, A. B. Thomson, and T. K. Basu. 2008. "The Modifying Effects of Galactomannan from Canadian-Grown Fenugreek (*Trigonella foenum-graecum* L.) on the Glycemic and Lipidemic Status in Rats." *Journal of Clinical Biochemistry and Nutrition* 43 (3):167–74. doi: 10.3164/jcbn.2008060.

Srinivasan, K. 2006. "Fenugreek (*Trigonella foenum-graecum*): A Review of Health Beneficial Physiological Effects." *Food Reviews International* 22 (2):203–24. doi: 10.1080/87559120600586315.

Srinivasan, K. 2013. "Dietary spices as beneficial modulators of lipid profile in conditions of metabolic disorders and diseases." *Food and Function* 4 (4):503–21. doi: 10.1039/c2fo30249g.

Srinivasan, K., K. Sambaiah, and N. Chandrasekhara. 2004. "Spices as Beneficial Hypolipidemic Food Adjuncts: A Review." *Food Reviews International* 20 (2):187–220. doi: 10.1081/fri-120037160.

Stark, A., and Z. Madar. 1993. "The effect of an ethanol extract derived from fenugreek (*Trigonella foenum-graecum*) on bile acid absorption and cholesterol levels in rats." *British Journal of Nutrition* 69 (1):277–87. doi: 10.1079/bjn19930029.

Thakurdesai, P., V. Mohan, and A. Kandhare. 2015. "Synergistic composition of Fenugreek seed extract (Sugaheal®) ameliorates high-fat diet-induced insulin resistance in C57BL/6 mice via modulating the expression of leptin, Glut-3, Glut-4, IRS-2 and SREBP-1c." In 2nd International Conference on Biotechnology and Bioinformatics (ICBB-2015), Pune, India, February 6–8, 2015, p. 25.

Thakurdesai, P., V. Mohan, M. Adil, A. Kandhare, and S. L. Bodhankar. 2015. "Mechanism based efficacy study of 4-Hydroxyisolucine from fenugreeks seeds on glucose and lipid metabolism in streptozotocin-induced diabetic rats." In Fifth Euro-India International Conference on Holistic Medicine (ICHM-2015) Kottayam, India. p. IL3.

Thompson Coon, J. S., and E. Ernst. 2003. "Herbs for serum cholesterol reduction: A systematic view." *Journal of Family Practice* 52 (6):468–78.

Vijayakumar, M. V., V. Pandey, G. C. Mishra, and M. K. Bhat. 2010. "Hypolipidemic effect of fenugreek seeds is mediated through inhibition of fat accumulation and upregulation of LDL receptor." *Obesity (Silver Spring)* 18 (4):667–74. doi: 10.1038/oby.2009.337.

Wilborn, C., L. Taylor, C. Poole, C. Foster, D. Willoughby, and R. Kreider. 2010. "Effects of a Purported Aromatase and 5 α-Reductase Inhibitor on Hormone Profiles in College-Age Men." *International Journal of Sport Nutrition* 20 (6):457. doi: 10.1123/ijsnem.20.6.457.

Writing Group, M., D. Mozaffarian, E. J. Benjamin, A. S. Go, D. K. Arnett, M. J. Blaha, M. Cushman, S. R. Das, S. de Ferranti, J. P. Despres, H. J. Fullerton, V. J. Howard, M. D. Huffman, C. R. Isasi, M. C. Jimenez, S. E. Judd, B. M. Kissela, J. H. Lichtman, L. D. Lisabeth, S. Liu, R. H. Mackey, D. J. Magid, D. K. McGuire, E. R. Mohler, 3rd, C. S. Moy, P. Muntner, M. E. Mussolino, K. Nasir, R. W. Neumar, G. Nichol, L. Palaniappan, D. K. Pandey, M. J. Reeves, C. J. Rodriguez, W. Rosamond, P. D. Sorlie, J. Stein, A. Towfighi, T. N. Turan, S. S. Virani, D. Woo, R. W. Yeh, M. B. Turner, C. American Heart Association Statistics, and S. Stroke Statistics. 2016. "Heart Disease and Stroke Statistics-2016 Update: A Report From the American Heart Association." *Circulation* 133 (4):e338–60. doi: 10.1161/CIR.0000000000000350.

Wu, F. C., and J. G. Jiang. 2019. "Effects of diosgenin and its derivatives on atherosclerosis." *Food and Function* 10 (11):7022–36. doi: 10.1039/c9fo00749k.

Yousefi, E., S. Zareiy, R. Zavoshy, M. Noroozi, H. Jahanihashemi, and H. Ardalani. 2017. "Fenugreek: A therapeutic complement for patients with borderline hyperlipidemia: A randomised, double-blind, placebo-controlled, clinical trial." *Advances in Integrative Medicine* 4 (1):31–5. doi: 10.1016/j.aimed.2016.12.002.

Yu, H., H. Liang, M. Ren, K. Ji, Q. Yang, X. Ge, B. Xi, and L. Pan. 2019. "Effects of dietary fenugreek seed extracts on growth performance, plasma biochemical parameters, lipid metabolism, Nrf2 antioxidant capacity and immune response of juvenile blunt snout bream (Megalobrama amblycephala)." *Fish Shellfish Immunology* 94:211–9. doi: 10.1016/j.fsi.2019.09.018.

Zameer, S., A. K. Najmi, D. Vohora, and M. Akhtar. 2018. "A review on therapeutic potentials of *Trigonella foenum-graecum* (fenugreek) and its chemical constituents in neurological disorders: Complementary roles to its hypolipidemic, hypoglycemic, and antioxidant potential." *Nutritional Neuroscience* 21 (8):539–45. doi: 10.1080/1028415X.2017.1327200.

Zhou, C., Y. Qin, R. Chen, F. Gao, J. Zhang, and F. Lu. 2020. "Fenugreek attenuates obesity-induced inflammation and improves insulin resistance through downregulation of iRhom2/TACE." *Life Sciences* 258:118222. doi: 10.1016/j.lfs.2020.118222.

12 The Effects of Fenugreek on Cardiovascular Risk Factors
What Are Potential Mechanisms?

Nazli Namazi, Neda Roshanravan, Sayyedeh Fatemeh Askari, Nasrin Sayfouri and Mohammad Hossein Ayati

CONTENTS

12.1 Introduction	173
12.2 Fenugreek and Lipid Profile	173
12.2.1 How Does Fenugreek Act?	174
12.2.1.1 4-Hydroxyisoleucine (4-OH-Ile)	174
12.2.1.2 Soluble Fiber	175
12.2.1.3 Diosgenin	176
12.3 Fenugreek and Glycemic Status	176
12.4 Fenugreek and Hypertension	178
12.5 Fenugreek and Oxidative Stress	178
12.6 Fenugreek and Inflammation	180
12.7 Fenugreek, Obesity, and Metabolic Syndrome	180
12.8 Fenugreek and Cardiometabolic Risk Factors	180
12.9 Conclusion	181
References	182

12.1 INTRODUCTION

Cardiovascular diseases (CVDs) are critical public health concerns that are considerably growing across the world (1). The World Health Organization reported that annually about 30% of total mortality occurred due to such diseases (2). Disorders in lipid parameters, glycemic status, blood pressure, and anthropometric indices can enhance the risk of CVDs and the related complications (3).

Activation of inflammatory pathways (4) and disturbing the balance between oxidative and antioxidative agents (5) can be the key mechanisms to drive the occurrence of CVDs. Accordingly, therapeutic approaches focusing on such pathways can be effective for the prevention and treatment of CVDs (6). Potential therapeutic effects of medicinal herbs for the management of CVD risk factors have attracted the attention of researchers over the last several decades (7). One such herb is fenugreek (*Trigonella foenum-graecum L.*) (8). Given accumulating evidence on the modulation of cardiometabolic parameters using fenugreek, in the present chapter, recent evidence focusing on potential mechanisms of fenugreek in controlling CVD risk factors is discussed.

12.2 FENUGREEK AND LIPID PROFILE

Despite significant advances in the field of CVDs, hyperlipidemia is still the key contributor to the development of CVD in the setting of atherosclerotic CVDs (9); therefore, its treatment is crucial for reducing cardiovascular events and even premature death (10). There are accumulating scientific reports for preventative effects of herbs in terms of hyperlipidemia. In the last years, several studies have shown that heart disorders can be prevented or even reversed with folk medicine via ethnobotanical remedies. Recently, it has been proposed that fenugreek could be used as an adjuvant therapy for hyperlipidemia (11).

Fenugreek contains 58% carbohydrates (25% dietary fiber), 23–26% protein, and 6–7% fat (12). Beyond that, polyphenolic flavonoids, 4-hydroxyisoleucine, steroid saponins, polysaccharides, and mainly galactomannans are other active components of fenugreek seeds (13). Its seeds have been reported to have multiple pharmacological properties including anti-diabetic, anti-neoplastic, anti-inflammatory, antioxidant, and hypocholesterolemic effects (14, 15).

The lipid-lowering potency of fenugreek seed extract has been investigated in different experimental models. Available data suggested that administration of polyphenol extract of fenugreek seeds can alleviate hyperlipidemia and accumulation of lipids in the liver of ethanol-treated rats (16). Beyond that, other researchers revealed that the ethyl acetate extract of fenugreek seed, when administered for 16 weeks in Wistar rats, had shown significant hypocholesterolemic and significant antioxidant properties. Antioxidant effects are likely to be involved in controlling cholesterol levels (12). In the study by Belaïd-Nouira et al., supplementation with fenugreek seeds (fenugreek seed powder and fenugreek seed extract) decreased the levels of total cholesterol (TC), low-density lipoprotein (LDL-c), and triglycerides (TG) in aluminum chloride-treated rats (17). Furthermore, Khound et al. demonstrated that fenugreek seed and its bioactive compound (trigonelline) can ameliorate hyperlipidemia via the inhibition of lipid synthesis and enhancement of fatty acid β-oxidation as well as the inhibition of very low-density lipoprotein (VLDL) assembly and secretion in genetic hyperlipidemic mice (18). Similarly, Reddy et al. observed that application of the fenugreek seeds powder to the albino mice resulted in significant hypocholesterolemic effect (19). Generally, the effectiveness of fenugreek as an anti-hyperlipidemic agent was reported in other animal-based studies (20–22). In line with animal experiments, most human studies concentrated on the lipid-lowering effect of fenugreek.

In a meta-analysis by Gong and colleagues, supplementation with fenugreek significantly reduced the serum levels of TC and TG, while no significant changes were found on high-density lipoprotein (HDL-c) and LDL-c in patients with type 2 diabetes (T2DM) (23). Another systematic review and meta-analysis including 15 randomized clinical trials (RCT) with a total of 536 participants revealed that supplementation with fenugreek significantly improved lipid profile (reduced the TG and LDL-c and increased high-density lipoprotein cholesterol (HDL-c levels)) in patients with diabetes (24).

Generally, the findings of the investigations on the effectiveness of fenugreek on blood lipid profile mostly confirm its beneficial properties; nevertheless further studies are required to achieve more definitive results.

12.2.1 How Does Fenugreek Act?

While beneficial effects of fenugreek on hyperlipidemia have been widely reported in *in vivo* and clinical trials, the precise mechanisms of fenugreek-mediated beneficial actions on lipid profile remained unknown. Here, the proposed mechanisms of the effects of different components of fenugreek are discussed.

Fenugreek seed is a rich source of some bioactive chemical components including soluble dietary fibers, diosgenin, gum, alkaloids, some amino acids (4-hydroxyisoleucine amino acid (4-OH-Ile), lysine amino acid, and L-tryptophan amino acid), vitamin C, niacin, potassium, saponins, and flavonoids.

12.2.1.1 4-Hydroxyisoleucine (4-OH-Ile)

4-OH-Ile, a major branched-chain amino acid of fenugreek, has been proved to have an essential role in lipid level reduction (25). Some investigations have focused on insulinotropic effect of

4-OH-Ile. Enhancing insulin sensitivity and improved insulin secretion in hepatic and peripheral tissues are valuable treatment modalities of 4-OH-Ile. Based on *in vitro* and *in vivo* studies, this amino acid can be considered as an insulin-sensitizing agent via the insulin receptor substrate 1 (IRS-1)—related activation of phosphatidylinositide 3-kinase (PI3K) in insulin-sensitive tissues (26). Furthermore, in the cell culture models, it was shown that 4-OH-Ile treatment can ameliorate fatty acid-induced insulin resistance (27). As far as our knowledge is concerned, insulin resistance is most closely related to lipid abnormality. Epidemiological evidence has proved a strong correlation between abnormal lipid metabolism (high TG and low HDL-C levels) and insulin resistance (28, 29). Some experimental studies indicated the lipid-lowering effect of fenugreek's 4-OH-Ile (Table 12.1). Therefore, thanks to its potency in stimulating insulin secretion and lipid-lowering properties, fenugreek is suggested in health maintenance.

12.2.1.2 Soluble Fiber

Soluble fiber is a major compound found in fenugreek (45% and 50% of the seeds) (33). Studies with dietary fibers have suggested that this compound is associated with lipid metabolism. Most epidemiological observations showed a considerable link between the high consumption of compounds containing dietary fiber and lipid profile modulation. The role of dietary fibers and their hypolipidemic effect has remained under debate. One potential mechanism could be due to the inhibition of fat absorption because of the fiber contained in fenugreek (34). Another reason is that dietary fibers are fermented by gut microbiota to form short chain fatty acids (SCFA) such as propionate, butyrate, and acetate. Butyrate, as a main SCFA, has multiple roles in lipid-lowering effects (35). Other proposed mechanisms may be the body-weight-lowering effects of dietary fiber (36). Table 12.2 indicates the effects of soluble fiber extracted from fenugreek on lipid profile.

TABLE 12.1
Effects of 4-Hydroxyisoleucine (4-OH-Ile) Derived from Fenugreek on Lipid Profile

Fenugreek Derivative	Model	Outcome	Reference
4-OH-Ile	Hamsters	↓TG, TC, FFA ↑HDL:TC ratio	(22)
	Type 2 diabetic rats	↑HDL-c	(30)
	C57BL/db/db mice	↓TG, TC, LDL-c ↑HDL-c	(31)
	Type 1 diabetic rats	↓TC, LDL-c, TG ↑HDL-c	(32)

TC: serum total cholesterol; TG: triglycerides; HDL: high-density lipoprotein; LDL: low-density lipoprotein; ↑ increase in amount; ↓ decrease in amount.

TABLE 12.2
Effects of Soluble Fiber from Fenugreek on Lipid Profile

Fenugreek Derivative	Model	Outcome	Reference
Soluble fiber	Obese rats	↓TG, TC	(37)
		↓TC	(38)

TC: serum total cholesterol; TG: triglycerides; ↑ increase in amount; ↓ decrease in amount.

TABLE 12.3
Effects of Diosgenin Extracted from Fenugreek on Lipid Profile

Fenugreek Derivative	Model	Outcome	Reference
Diosgenin	Streptozotocin-induced diabetic rats	↓HDL-C, TC	(42)
	Streptozotocin-induced diabetic rats	↓TC, LDL-c ↑HDL-c	(40)
	NPC1L1-knockout C57BL6 mice	↓TC	(43)
	Diabetic obese KK-Ay mice	↓TG	(44)
	Chronic renal failure rats	↓TG	(45)
	Hypercholesterolemic rats	↓TC, TG, LDL-C	(46)

TC: serum total cholesterol; TG: triglycerides; HDL: high-density lipoprotein; LDL: low-density lipoprotein; ↑increase in amount; ↓ decrease in amount.

12.2.1.3 Diosgenin

Diosgenin is widely implicated as the chief biologically active steroid sapogenin component in fenugreek. Diosgenin treatment was selectively investigated due to its beneficial effects against a variety of pathologic conditions, including diabetes, cardiovascular disease, hyperlipidemia, and cancer (39). Its effectiveness in reducing blood glucose in diabetic animal models (both type 1 and type 2 diabetes) has been reported previously (40). Up regulation of hepatic glucokinase, down regulation of enzymes which are involved in hepatic gluconeogenesis and increased activity of the hepatic antioxidant enzymes are some proposed mechanisms of glucose-lowering actions attributed to diosgenin (40). Evidently, glucose-lowering agents can control lipid levels. Based on gene expression analysis, diosgenin suppresses hepatic mRNA expression of sterol regulatory element—binding protein-1c (SREBP-1c) and other lipogenic genes. As has been proved, SREBP-1c—due to its regulatory ability in the expression of several genes involved in fatty acid metabolism including acetyl-CoA carboxylase (ACC), fatty acid synthase, and stearoyl-CoA desaturase (SCD-1)—is the key element for TG synthesis (25). Clinical animal studies have also indicated that diosgenin is highly effective in the treatment of hyperlipidemia by regulating cholesterol homeostasis through enhancing the biliary cholesterol excretion (41).

Several studies reported that diosgenin can exert beneficial effects on lipid profile; however, more studies are needed to reach a more definitive conclusion. Some experimental studies that have investigated the lipid-lowering effect of diosgenin are shown in Table 12.3.

12.3 FENUGREEK AND GLYCEMIC STATUS

Diabetes mellitus characterized by high blood sugar can result in poor blood flow, cardiovascular and renal disorders, neuropathy, and retinopathy. Several studies have evaluated the antidiabetic effects of fenugreek (47, 48). In a review conducted by El-Abhar et al., fenugreek seeds in diabetic rats inhibited α-glucosidase and affected gluconeogenic, glucolytic, and glucose homeostasis (49). Based on evidence (50), medicinal properties of some major fenugreek compounds including 4-OH-Ile (affecting pancreatic β-cells and increasing glucose-induced insulin secretion) (51) and trigonelline (an alkaloid component affecting β-cell regeneration and insulin secretion) (52) can affect glucose level and other glycemic parameters. Furthermore, 4-OH-Ile exhibited a great role in insulin resistance associated with obesity because of increasing Akt[1] phosphorylation and reducing activation of JNK[2] 1/2, ERK[3] 1/2, MAPK,[4] and NF-κB[5](53).

In one single-arm 12-week clinical trial, two chapatis (Indian supplement containing *Nigella sativa* and fenugreek seed) were consumed twice a day and 6 days/week for a daily dose of 5.45 g by overweight patients with T2DM caused a reduction in HbA1c,[6] FBG,[7] and 2-h PPBG.[8] The researcher suggested that this formulation could affect the NAFLD[9] pathway related to liver disorders to pre-diabetes and T2DM (54). In a systematic review and meta-analysis on ten randomized controlled trials, the effects of supplementation with fenugreek were evaluated on glycemic status. Findings showed a significant reduction in FBG, 2-h PPBG, and HbA1c by administration of more than 5 g/day doses via reducing the rate of enzymatic digestion and glucose absorption from the GI tract (55). Another review article on *in vivo* and *in vitro* studies regarding 4-OH-Ile, revealed antidiabetic effects including decreasing serum levels of FBG, 2-h PPBG, and HbA1c. Potential mechanisms mentioned in the study were increasing the levels of antioxidant agents and changing hormones levels in patients with T2DM (56).

Several animal studies have evaluated the antidiabetic effects of fenugreek (57–59). A combination of fenugreek seeds and *Coriandrum sativum* leaves, for instance, showed a significant improvement in body weight (a marker of recovery from muscle depletion induced by glycemia) and attenuation in blood glucose in alloxan-induced diabetes mellitus in rats. The proposed mechanisms are as follows: increasing plasma insulin concentrations and transporting blood glucose in the peripheral cells, reducing levels of lipid peroxides and free radical scavenging features. Besides, histopathology of the pancreas showed enhancements in the number of islets and β cells of Langerhans and a reduction in fibrosis and inflammation of islets (57).

Regarding the effects of diosgenin, Kiss et al. demonstrated that diosgenin (1 mg/kg) and fenugreek seeds (0.2 g/kg) ameliorated elevated metabolic clearance rate of insulin and insulin sensitivity index in a rat insulin resistance model for 6 weeks. This study confirmed the correlation between the maintenance of normal glucose levels and IGF-1[10] because fenugreek and its isolated component could interact with a substrate of hypothalamic hypopituitary axis regulated hormones (58). In another study, genetic hyperlipidemic mice were fed by a chow containing 2% fenugreek seed for 7 weeks. The increased expression of Insig-1[11] and Insig-2[12] could reduce insulin resistance (59). Oral administration of fenugreek sprouts juice (1 ml, for 45 days) also decreased FBG and improved lipid profile in STZ-induced diabetes in rats. This extract also prevented some damages induced by diabetes such as reduction of Hb[13] levels and plasma α-amylase activities, elevation of MDA, reduction of CAT activities, and histopathological damages of the pancreas, liver, and kidney (60). Further studies on fenugreek and diabetes are provided in Table 12.4.

TABLE 12.4
Effects of Fenugreek on Diabetes Factors

Activity	Model	Outcome	Reference
Antidiabetic effect	STZ-induced diabetes rats, for 30 days, P.O.	↑insulin, ↑liver and muscle glycogen ↓glucose	(61)
Insulin resistance	3T3-L1 adipocytes	↑Phosphorylation of AKT and AMPK and amelioration insulin resistance	(62)
Insulin resistance	obese mice, twice per week for 15.5 weeks, i.p.	Improvement of insulin sensitivity	(63)
Insulin resistance	Human liver cells (HepG2), normoglycemic and hyperglycemic condition, for 72 hr	↑phosphorylation of IR-b, Akt and GSK-3a/b and GLUT2 and glucose uptake (the extract exerts more efficient activity)	(64)
Insulin resistance	Male mice were subjected to a high-fat diet, for 24 weeks	↓glucose intolerance and serum triglyceride and hepatic fat accumulation	(65)
Antidiabetic activity	A parallel, randomized, single-blind trial, 60 patients, for 6 months	↓FBG and HbA1C (in the 5th and 6th month, respectively)	(66)

AKT: protein kinase B; AMPK: AMP-activated protein kinase; i.p.: intraperitoneally; IR-b: insulin receptor-b; GSK-3a/b: glycogen synthase kinase-3a/b; GLUT2: glucose transporter 2; ↑ increase in amount; ↓ decrease in amount; P.O.: oral administration.

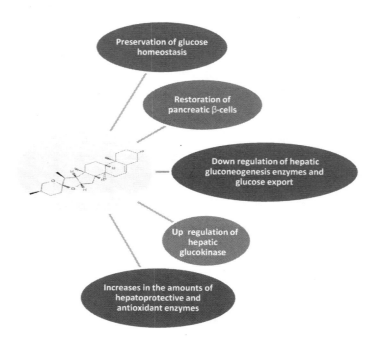

FIGURE 12.1 Potential mechanisms of diosgenin in controlling glucose level (67, 68).

Generally, clinical trials, animal, and *in vitro* studies showed positive effects of fenugreek and its components on glycemic status. The main mechanisms for this purpose by diosgenin are provided in Figure 12.1.

12.4 FENUGREEK AND HYPERTENSION

Hypertension, a fundamental determining factor in CVD, is defined as "office systolic blood pressure" (SBP) values ≥ 140 mmHg and/or diastolic blood pressure (DBP) values ≥ 90 mmHg (69). Some animal experiments have reported the beneficial effect of fenugreek on lowering blood pressure) BP(. Hamden et al. showed that administration of fenugreek essential oil from the seeds can modulate the key enzymes activities related to hypertension such as angiotensin-converting enzyme (ACE) by 37% (69). Interestingly, diosgenin in fenugreek has vasodilator properties via an increment in the mRNA expression of endothelial nitric oxide (NO) and even attenuating the aortic calcification (70). Another animal study also suggested that methanol extract of fenugreek seeds can modulate blood BP through 5-hydroxytryptamine receptor antagonism (71).

Taking fenugreek among patients under hypertension pharmacotherapy is widespread recently. It has been supposed that fenugreek administration can reduce high BP via preserving renal function parameters including decreasing albuminuria, blood nitrogen, and serum creatinine (72). Totally, due to limited human studies, further clinical trials should be conducted in order to elucidate its efficacy as an anti-hypertensive agent.

12.5 FENUGREEK AND OXIDATIVE STRESS

Oxidative stress is a disequilibrium of reactive oxygen species (as metabolism or environmental pollutants) and antioxidants, which can lead to biomolecules (protein, lipids, and nucleic acids) oxidation, enzyme inhibition, cell death, and disorders like cancer, inflammation, and diabetes (73).

Antioxidants play a key role in protecting the body against tissue damage (74). Different experimental models have shown the antioxidant effects of fenugreek seeds. Oral doses of fenugreek seeds (10%) and onion (*Allium cepa*, 3% powder) resulted in the relief of oxidative stress in cardiac tissue in STZ[14]-induced diabetic rats. This outcome might be due to reducing ACE[15] and AT1[16] (75). In another study on STZ-induced diabetes rats, the combination of oral fenugreek seed aqueous extract (0.87 g/kg) and swimming training increased the GPX,[17] CAT,[18] and SOD[19] activities compared to either of those alone. Therefore, a combination of fenugreek extract and swimming can be a potential candidate for controlling cardiac oxidative stress induced by type 1 diabetes mellitus (76). In another animal study, the effects of hydroalcoholic extract of fenugreek seeds on type 2 diabetic rats were investigated. The results suggested that the extract of fenugreek had hypoglycemic activities through modifying oxidative stress (GSX, SOD) and pro-inflammatory cytokines (TNF-α,[20] IL-6[21]) (77). Findings indicated that 200 mg/kg alcoholic extract of fenugreek seed exerted a potent antioxidant effect in both *in vitro* and *in vivo* studies. Antioxidant activities were recognized by inhibition of superoxide anion and hydroxyl free radicals generation *in vitro* and reduced levels of TBA,[22] TBARS[23] (indices for plasma lipid peroxidation), HNE,[24] and isoprostanes (the secondary products and key markers of lipid peroxidation) in triton-treated hyperlipidemic rats (78). Aqueous suspension of fenugreek (1 g/kg, for 30 days) reverted the decreased levels of GSX, GST,[25] and CAT in STZ-induced diabetic rats (61).

Findings of a review article revealed that the soluble components of fenugreek seeds are likely to be responsible for antioxidant properties by decreasing lipid peroxidation and peroxidative damages in alloxan-induced diabetic rats (79). Mucilage extracted from fenugreek could also increase the activities of vitamin C, antioxidant enzymes, and reduced glutathione levels in adjuvant-induced arthritic rats (80). Vitexin (a fenugreek flavonoid) is an antioxidant, cardioprotective, and renoprotective agent. Recently, one animal study showed significant inhibition of ethylene glycol-induced elevated oxido-nitrosative stress (SOD, GSX, MDA,[26] and NO[27]) in cardiac and renal tissue by vitexin (30 and 60 mg/kg) in uninephrectomized rats treatment for 28 days (81). In another study on adult male Sprague-Dawley rats, prepared glycosides from the hydroalcoholic extract of fenugreek seeds (20 and 40 mg/kg, P.O.[28]) significantly restored intratracheal bleomycin-induced elevated liver oxidative stress and decreased total antioxidant capacity via inhibition of oxido-nitrosative stress. These glycosides also reduced levels of AST,[29] ALT,[30] and GGT[31] and significantly increased the level of serum albumin (82). Akbari et al. revealed that phenolic and flavonoid contents of fenugreek seed oil are responsible for radical scavenging effects against DPPH[32] and ABTS+[33] radicals (73). Generally, the antioxidant activity of fenugreek has been confirmed in different model studies. Table 12.5 shows some other studies having investigated the effects of fenugreek on oxidative stress.

TABLE 12.5
Effects of Fenugreek on Oxidative Stress

Part of Plant (Extract)	Model	Outcome	Reference
Extract, 500 and 1000 mg/kg, 6 weeks	Alloxan-induced diabetic rats	↓Lipid peroxidation ↑SOD and GSH ↓Blood glucose	(83)
Aqueous extract (seeds)	Adriamycin-induced hepatotoxicity rats	↑SOD and CAT ↓MDA	(84)
Aqueous extract	Selenite-induced cataract in rats	↓MDA ↑GSH, SOD, and CAT	(85)

SOD: Superoxide dismutase; GSH: reduced glutathione; CAT: catalase; MDA: malondialdehyde; ↑ increase in amount; ↓ decrease in amount.

12.6 FENUGREEK AND INFLAMMATION

Inflammation is a complex physiopathological response to various stimuli participating in the activation of inflammatory mediators including cytokines prostaglandins. Overproduction of such mediators can result in tissue injury via damaging macromolecules and the lipid peroxidation of membranes (86). Tissue damage plays a pivotal role in the pathogenesis of a wide range of inflammatory diseases (87). Anti-inflammatory effects of fenugreek and its components have been showed in numerous experimental studies (88–93).

To identify the molecular mechanisms of fenugreek seed, Hassan et al. examined a crude methanolic extract of fenugreek on macrophage polarization in an *in vitro* model. They showed that fenugreek can regulate the expression of M1 (pro-inflammatory) and M2 (immunoregulatory) factors in THP-1 macrophages cells, potentially through the modulation of NF-kB activity (94).

According to Lee et al., a high-fat diet along with 2% fenugreek miniaturized the adipocytes and enhanced the levels of mRNA expression of differentiation-related genes in adipose tissues. They reported that this medicinal herb also decreased the mRNA expression levels of inflammatory genes and suppressed macrophage infiltration into adipose tissues (90). They found that the mentioned effects were induced by diosgenin. Fuller et al. in a review study focused on mechanisms of actions related to the main components of fenugreek (67). They concluded that diosgenin can exert anti-inflammatory characteristics through several pathways. It can reduce the levels of inflammatory factors including TNF-alpha, NF-KB, IL-6, and IL-1 beta. Furthermore, it showed positive effects on neutralizing lipid accumulation, decreasing macrophage infiltration and adipocyte inflammation. Beside this, diosgenin showed beneficial effects on enhancing anti-inflammatory parameters such as PPAR-gamma and reducing lipid peroxidation (67). Sharififar et al., for instance, concluded that fenugreek is one of the Iranian medicinal herbs used as an anti-inflammatory agent. Flavonoid components of fenugreek can inhibit both cyclooxygenase and lipoxygenase pathways at relatively high concentrations. They are also able to inhibit the nitric oxide synthase (95). Ethanol extract of fenugreek can also increase the peritoneal exudates and macrophage cell counts. Accordingly, one anti-inflammatory pathway for fenugreek is likely the activation of the macrophages (96). Another component of anti-inflammatory properties is 4-OH-Ile that affects parameters contributing to inflammatory responses such as NF-Kb and TNF-alpha and c-Jun N-terminal kinases (JNKs) (67).

12.7 FENUGREEK, OBESITY, AND METABOLIC SYNDROME

Diosgenin, 4-OH-Ile, and soluble fibers are the main bioactive constituents in fenugreek. They can exert beneficial effects for metabolic syndrome, insulin action, and cardiovascular health (67). They can also modulate the functions of adipose tissue. Dsouza et al. in an *in vitro* study showed that diosgenin, one steroidal saponin extracted from fenugreek, can help to delay glucose diffusion and provide sufficient time to eliminate glucose and prevent its uptake into the cells. It also showed positive effects to delay the absorption of fat and glucose. Therefore, it can be a good candidate for controlling metabolic syndrome (93). Fuller et al. also concluded that the soluble fiber of fenugreek induces beneficial effects on body weight via increasing satiety, delaying gastric emptying of carbohydrates, reducing lipase activity and lipid absorption, and decreasing intestinal glucose uptake (67). The soluble fiber content of fenugreek might also alter the gut microbiome. However, further investigation must be performed to clarify the efficacy and precise mechanisms of fenugreek fibers on gut flora (97). Another possible mechanism of fenugreek on body weight is reducing insulin resistance. A review study on the effects of 4-OH-Ile from fenugreek revealed that this component can reduce insulin resistance and increase insulin sensitivity via reduction in inflammatory and oxidative parameters (53).

12.8 FENUGREEK AND CARDIOMETABOLIC RISK FACTORS

The recent systematic review and meta-analysis concluded that fenugreek seed as an adjutant therapy can reduce serum levels of FBS, HbA1c, TC, and LDL-C compared to a placebo group.

However, no significant changes were observed in BMI, insulin resistance, 2-hour postprandial glucose, and serum levels of TG and HDL-C. However, the researchers mentioned that further placebo-controlled clinical trials are warranted to draw a fixed conclusion regarding the efficacy of fenugreek as a complementary therapy to control cardiometabolic risk factors (8).

12.9 CONCLUSION

Evidence shows that fenugreek can be an effective complementary therapy along with common treatments to control CVD risk factors. However, further high-quality studies in the form of experimental studies and clinical trials are needed for each parameter to shed more light on the most effective dosage and form of fenugreek as well as the duration of the intervention. Notably, precise mechanisms on controlling CVD risk factors have not been fully understood so far. Bioactive components of fenugreek including flavonoids, soluble dietary fibers, and other components with antioxidative and anti-inflammatory properties are likely to be responsible for the positive effects of fenugreek. The main suggested mechanisms of fenugreek extracted from *in vivo* and *in vitro* studies can be summarized as follows: (1) activating beta-cells functions, (2) reducing insulin resistance, (3) affecting the secretion of c-peptide, (4) reducing appetite and inducing delay in gastric emptying, (5) reducing the absorption of glucose and lipid, (6) affecting inflammatory and oxidative pathways, (7) modulating of glucagon-like peptide-1 (GLP-1), and (8) altering gut flora.

NOTES

1. protein kinase B
2. Jun N-terminal kinase 1/2
3. extracellular signal-regulated kinase 1/2
4. p38 mitogen-activated protein kinase
5. nuclear factor-κB
6. glycated hemoglobin
7. fasting blood glucose
8. postprandial blood glucose
9. nonalcoholic fatty liver disease
10. Insulin-like growth factor 1
11. insulin-inducible gene-1
12. insulin-inducible gene-2
13. hemoglobin
14. streptozotocin
15. angiotensin-converting enzyme
16. angiotensin type 1 receptor
17. glutathione peroxidase
18. catalase
19. superoxide dismutase
20. tumor necrosis factor α
21. interleukin 6
22. thiobarbituric acid
23. thiobarbituric acid reactive species
24. hepatic 4-hydroxynonenal
25. glutathione-S-transferase
26. malondialdehyde
27. nitric oxide
28. oral administration
29. serum aspartate transaminase
30. alanine transaminase
31. gamma-glutamyl transferase
32. 2,2-diphenyl-picrylhydrazyl
33. 2,20-Azino-bis-3-ethylbenzothiazoline-6-sulfonic acid

REFERENCES

1. Roth GA, Johnson C, Abajobir A, Abd-Allah F, Abera SF, Abyu G, et al. Global, regional, and national burden of cardiovascular diseases for 10 causes, 1990 to 2015. *Journal of the American College of Cardiology.* 2017;70(1):1–25.
2. Abedi Gaballu F, Abedi Gaballu Y, Moazenzade Khyavy O, Mardomi A, Ghahremanzadeh K, Shokouhi B, et al. Effects of a triplex mixture of Peganum harmala, Rhus coriaria, and Urtica dioica aqueous extracts on metabolic and histological parameters in diabetic rats. *Pharmaceutical Biology.* 2015;53(8):1104–9.
3. Upadhyay RK. Emerging risk biomarkers in cardiovascular diseases and disorders. *Journal of Lipids.* 2015;2015.
4. Montecucco F, Liberale L, Bonaventura A, Vecchiè A, Dallegri F, Carbone F. The role of inflammation in cardiovascular outcome. *Current Atherosclerosis Reports.* 2017;19(3):11.
5. Selvaraju V, Joshi M, Suresh S, Sanchez JA, Maulik N, Maulik G. Diabetes, oxidative stress, molecular mechanism, and cardiovascular disease—an overview. *Toxicology Mechanisms and Methods.* 2012;22(5):330–5.
6. Inagi R. Oxidative stress in cardiovascular disease: A new avenue toward future therapeutic approaches. *Recent Patents on Cardiovascular Drug Discovery.* 2006;1(2):151–9.
7. Al-Snafi AE. Medicinal plants for prevention and treatment of cardiovascular diseases-A review. *Respiration.* 2017;23:25.
8. Khodamoradi K, Khosropanah MH, Ayati Z, Chang D, Nasli-Esfahani E, Ayati MH, et al. The effects of fenugreek on cardiometabolic risk factors in adults: A systematic review and meta-analysis. *Complementary Therapies in Medicine.* 2020:102416.
9. Ritchie RH, Zerenturk EJ, Prakoso D, Calkin AC. Lipid metabolism and its implications for type 1 diabetes-associated cardiomyopathy. *Journal of Molecular Endocrinology.* 2017;58(4):R225–R40.
10. Karr S. Epidemiology and management of hyperlipidemia. *The American Journal of Managed Care.* 2017;23(9 Suppl):S139–S48.
11. Sandesara PB, Virani SS, Fazio S, Shapiro MD. The forgotten lipids: Triglycerides, remnant cholesterol, and atherosclerotic cardiovascular disease risk. *Endocrine Reviews.* 2019;40(2):537–57.
12. Belguith-Hadriche O, Bouaziz M, Jamoussi K, El Feki A, Sayadi S, Makni-Ayedi F. Lipid-lowering and antioxidant effects of an ethyl acetate extract of fenugreek seeds in high-cholesterol-fed rats. *Journal of Agricultural and Food Chemistry.* 2010;58(4):2116–22.
13. Genet S, Kale RK, Baquer NZ. Alterations in antioxidant enzymes and oxidative damage in experimental diabetic rat tissues: Effect of vanadate and fenugreek (Trigonella foenum graecum). *Molecular and Cellular Biochemistry.* 2002;236(1–2):7–12.
14. Dixit PP, Misar A, Mujumdar AM, Ghaskadbi S. Pre-treatment of Syndrex protects mice from becoming diabetic after streptozotocin injection. *Fitoterapia.* 2010;81(5):403–12.
15. Satheeshkumar N, Mukherjee PK, Bhadra S, Saha BP. Acetylcholinesterase enzyme inhibitory potential of standardized extract of Trigonella foenum-graecum L and its constituents. *Phytomedicine.* 2010;17(3–4):292–5.
16. Kaviarasan S, Viswanathan P, Anuradha CV. Fenugreek seed (Trigonella foenum graecum) polyphenols inhibit ethanol-induced collagen and lipid accumulation in rat liver. *Cell Biology and Toxicology.* 2007;23(6):373–83.
17. Belaïd-Nouira Y, Bakhta H, Bouaziz M, Flehi-Slim I, Haouas Z, Ben Cheikh H. Study of lipid profile and parieto-temporal lipid peroxidation in $AlCl_3$ mediated neurotoxicity. Modulatory effect of fenugreek seeds. *Lipids in Health and Disease.* 2012;11:16.
18. Khound R, Shen J, Song Y, Santra D, Su Q. Phytoceuticals in fenugreek ameliorate VLDL overproduction and insulin resistance via the Insig signaling pathway. *Molecular Nutrition & Food Research.* 2018;62(5).
19. Reddy RR, Srinivasan K. Dietary fenugreek and onion attenuate cholesterol gallstone formation in lithogenic diet-fed mice. *International Journal of Experimental Pathology.* 2011;92(5):308–19.
20. Annida B, Stanely Mainzen Prince P. Supplementation of fenugreek leaves lower lipid profile in streptozotocin-induced diabetic rats. *Journal of Medicinal Food.* 2004;7(2):153–6.
21. Xue WL, Li XS, Zhang J, Liu YH, Wang ZL, Zhang RJ. Effect of Trigonella foenum-graecum (fenugreek) extract on blood glucose, blood lipid and hemorheological properties in streptozotocin-induced diabetic rats. *Asia Pacific Journal of Clinical Nutrition.* 2007;16(Suppl 1):422–6.
22. Narender T, Puri A, Shweta, Khaliq T, Saxena R, Bhatia G, et al. 4-hydroxyisoleucine an unusual amino acid as antidyslipidemic and antihyperglycemic agent. *Bioorganic & Medicinal Chemistry Letters.* 2006;16(2):293–6.

23. Gong J, Fang K, Dong H, Wang D, Hu M, Lu F. Effect of fenugreek on hyperglycaemia and hyperlipidemia in diabetes and prediabetes: A meta-analysis. *Journal of Ethnopharmacology.* 2016;194:260–8.
24. Heshmat-Ghahdarijani K, Mashayekhiasl N, Amerizadeh A, Teimouri Jervekani Z, Sadeghi M. Effect of fenugreek consumption on serum lipid profile: A systematic review and meta-analysis. *Phytotherapy Research: PTR.* 2020.
25. Fuller S, Stephens JM. Diosgenin, 4-hydroxyisoleucine, and fiber from fenugreek: Mechanisms of actions and potential effects on metabolic syndrome. *Advances in nutrition (Bethesda, Md).* 2015;6(2):189–97.
26. Broca C, Breil V, Cruciani-Guglielmacci C, Manteghetti M, Rouault C, Derouet M, et al. Insulinotropic agent ID-1101 (4-hydroxyisoleucine) activates insulin signaling in rat. *American Journal of Physiology Endocrinology and Metabolism.* 2004;287(3):E463–E71.
27. Maurya CK, Singh R, Jaiswal N, Venkateswarlu K, Narender T, Tamrakar AK. 4-Hydroxyisoleucine ameliorates fatty acid-induced insulin resistance and inflammatory response in skeletal muscle cells. *Molecular and Cellular Endocrinology.* 2014;395(1–2):51–60.
28. Hua X, Hong Z. Effects of individualized prescriptive diet and single prescriptive diet on blood lipids and insulin resistance in type 2 diabetes mellitus patients. *Chinese General Practice.* 2011;14(1):133–5.
29. Yan YZ, Ma RL, Zhang JY, He J, Ma JL, Pang HR, et al. Association of insulin resistance with glucose and lipid metabolism: Ethnic heterogeneity in far western China. *Mediators of Inflammation.* 2016;2016:3825037.
30. Haeri MR, Izaddoost M, Ardekani MR, Nobar MR, White KN. The effect of fenugreek 4-hydroxyisoleucine on liver function biomarkers and glucose in diabetic and fructose-fed rats. *Phytotherapy Research: PTR.* 2009;23(1):61–4.
31. Singh AB, Tamarkar AK, Narender T, Srivastava AK. Antihyperglycaemic effect of an unusual amino acid (4-hydroxyisoleucine) in C57BL/KsJ-db/db mice. *Natural Product Research.* 2010;24(3):258–65.
32. Haeri MR, Limaki HK, White CJ, White KN. Non-insulin dependent anti-diabetic activity of (2S, 3R, 4S) 4-hydroxyisoleucine of fenugreek (Trigonella foenum graecum) in streptozotocin-induced type I diabetic rats. *Phytomedicine.* 2012;19(7):571–4.
33. Roberts KT. The potential of fenugreek (Trigonella foenum-graecum) as a functional food and nutraceutical and its effects on glycemia and lipidemia. *Journal of Medicinal Food.* 2011;14(12):1485–9.
34. Eidi A, Eidi M, Sokhteh M. Effect of fenugreek (Trigonella foenum-graecum L) seeds on serum parameters in normal and streptozotocin-induced diabetic rats. *Nutrition Research.* 2007;27(11):728–33.
35. Roshanravan N, Mahdavi R, Alizadeh E, Jafarabadi MA, Hedayati M, Ghavami A, et al. Effect of butyrate and inulin supplementation on glycemic status, lipid profile and glucagon-like peptide 1 level in patients with type 2 diabetes: A randomized double-blind, placebo-controlled trial. *Hormone and Metabolic Research = Hormon- und Stoffwechselforschung = Hormones et metabolisme.* 2017;49(11):886–91.
36. Handa T, Yamaguchi K, Sono Y, Yazawa K. Effects of fenugreek seed extract in obese mice fed a high-fat diet. *Bioscience, Biotechnology, and Biochemistry.* 2005;69(6):1186–8.
37. Ramulu P, Giridharan NV, Udayasekhararao P. Hypolipidemic effect of soluble dietary fiber (galactomannan) isolated from fenugreek seeds in WNIN (GR-Ob) obese rats. *Journal of Medicinal Plants Research.* 2011;5(19):4804–13.
38. Kumar P, Bhandari U, Jamadagni S. Fenugreek seed extract inhibit fat accumulation and ameliorates dyslipidemia in high fat diet-induced obese rats. *BioMed Research International.* 2014;2014:606021.
39. Das S, Dey KK, Dey G, Pal I, Majumder A, MaitiChoudhury S, et al. Antineoplastic and apoptotic potential of traditional medicines thymoquinone and diosgenin in squamous cell carcinoma. *PloS One.* 2012;7(10):e46641.
40. Kalailingam P, Kannaian B, Tamilmani E, Kaliaperumal R. Efficacy of natural diosgenin on cardiovascular risk, insulin secretion, and beta cells in streptozotocin (STZ)-induced diabetic rats. *Phytomedicine.* 2014;21(10):1154–61.
41. Temel RE, Brown JM, Ma Y, Tang W, Rudel LL, Ioannou YA, et al. Diosgenin stimulation of fecal cholesterol excretion in mice is not NPC1L1 dependent. *Journal of Lipid Research.* 2009;50(5):915–23.
42. Son IS, Kim JH, Sohn HY, Son KH, Kim JS, Kwon CS. Antioxidative and hypolipidemic effects of diosgenin, a steroidal saponin of yam (Dioscorea spp.), on high-cholesterol fed rats. *Bioscience, Biotechnology, and Biochemistry.* 2007;71(12):3063–71.
43. Tang W, Jia L, Ma Y, Xie P, Haywood J, Dawson PA, et al. Ezetimibe restores biliary cholesterol excretion in mice expressing Niemann-Pick C1-Like 1 only in liver. *Biochimica et biophysica acta.* 2011;1811(9):549–55.
44. Turer AT, Hill JA, Elmquist JK, Scherer PE. Adipose tissue biology and cardiomyopathy: Translational implications. *Circulation Research.* 2012;111(12):1565–77.

45. Manivannan J, Shanthakumar J, Arunagiri P, Raja B, Balamurugan E. Diosgenin interferes coronary vasoconstriction and inhibits osteochondrogenic transdifferentiation of aortic VSMC in CRF rats. *Biochimie*. 2014;102:183–7.
46. Li R, Liu Y, Shi J, Yu Y, Lu H, Yu L, et al. Diosgenin regulates cholesterol metabolism in hypercholesterolemic rats by inhibiting NPC1L1 and enhancing ABCG5 and ABCG8. *Biochimica et biophysica acta Molecular and cell biology of lipids*. 2019;1864(8):1124–33.
47. Suksomboon N, Poolsup N, Boonkaew S, Suthisisang CC. Meta-analysis of the effect of herbal supplement on glycemic control in type 2 diabetes. *Journal of Ethnopharmacology*. 2011;137(3):1328–33.
48. Baquer NZ, Kumar P, Taha A, Kale R, Cowsik S, McLean P. Metabolic and molecular action of Trigonella foenum-graecum (fenugreek) and trace metals in experimental diabetic tissues. *Journal of Biosciences*. 2011;36(2):383–96.
49. El-Abhar HS, Schaalan MF. Phytotherapy in diabetes: Review on potential mechanistic perspectives. *World Journal of Diabetes*. 2014;5(2):176.
50. Ríos JL, Francini F, Schinella GR. Natural products for the treatment of type 2 diabetes mellitus. *Planta Medica*. 2015;81.
51. Smith M. Therapeutic applications of fenugreek. *Alternative Medicine Review*. 2003;8(1):20–7.
52. Zhou J, Chan L, Zhou S. Trigonelline: A plant alkaloid with therapeutic potential for diabetes and central nervous system disease. *Current Medicinal Chemistry*. 2012;19(21):3523–31.
53. Avalos-Soriano A, la Cruz-Cordero D, Rosado JL, Garcia-Gasca T. 4-Hydroxyisoleucine from fenugreek (Trigonella foenum-graecum): Effects on insulin resistance associated with obesity. *Molecules*. 2016;21(11):1596.
54. Rao AS, Hegde S, Pacioretty LM, DeBenedetto J, Babish JG. Nigella sativa and Trigonella foenum-graecum supplemented chapatis safely improve HbA1c, body weight, waist circumference, blood lipids, and fatty liver in overweight and diabetic subjects: A twelve-week safety and efficacy study. *Journal of Medicinal Food*. 2020.
55. Neelakantan N, Narayanan M, de Souza RJ, van Dam RM. Effect of fenugreek (Trigonella foenum-graecum L.) intake on glycemia: A meta-analysis of clinical trials. *Nutrition Journal*. 2014;13(1):7.
56. Zafar MI, Gao F. 4-Hydroxyisoleucine: A potential new treatment for type 2 diabetes mellitus. *BioDrugs*. 2016;30(4):255–62.
57. Yella SST, Kumar RN, Ayyanna C, Varghese AM, Amaravathi P, Vangoori Y. The combined effect of Trigonella foenum seeds and Coriandrum sativum leaf extracts in alloxan-induced diabetes mellitus Wistar albino rats. *Bioinformation*. 2019;15(10):716.
58. Kiss R, Pesti-Asbóth G, Szarvas MM, Stündl L, Cziáky Z, Hegedűs C, et al. Diosgenin and its fenugreek based biological matrix affect insulin resistance and anabolic hormones in a rat based insulin resistance model. *BioMed Research International*. 2019;2019.
59. Khound R, Shen J, Song Y, Santra D, Su Q. Phytoceuticals in fenugreek ameliorate VLDL overproduction and insulin resistance via the insig signaling pathway. *Molecular Nutrition & Food Research*. 2018;62(5):1700541.
60. Mohamed RS, Marrez DA, Salem SH, Zaghloul AH, Ashoush IS, Farrag ARH, et al. Hypoglycemic, hypolipidemic and antioxidant effects of green sprouts juice and functional dairy micronutrients against streptozotocin-induced oxidative stress and diabetes in rats. *Heliyon*. 2019;5(2):e01197.
61. Marzouk M, Soliman A, Omar T. Hypoglycemic and antioxidative effects of fenugreek and termis seeds powder in streptozotocin-diabetic rats. *European Review for Medical and Pharmacological Sciences*. 2013;17(4):559–65.
62. Li G, Luan G, He Y, Tie F, Wang Z, Suo Y, et al. Polyphenol stilbenes from fenugreek (Trigonella foenum-graecum L.) seeds improve insulin sensitivity and mitochondrial function in 3T3-L1 adipocytes. *Oxidative Medicine and Cellular Longevity*. 2018;2018.
63. Cheng C-Y, Yang A-J, Ekambaranellore P, Huang K-C, Lin W-W. Anti-obesity action of INDUS810, a natural compound from Trigonella foenum-graecum: AMPK-dependent lipolysis effect in adipocytes. *Obesity Research & Clinical Practice*. 2018;12(6):562–9.
64. Naicker N, Nagiah S, Phulukdaree A, Chuturgoon A. Trigonella foenum-graecum seed extract, 4-hydroxyisoleucine, and metformin stimulate proximal insulin signaling and increase expression of glycogenic enzymes and GLUT2 in HepG2 cells. *Metabolic Syndrome and Related Disorders*. 2016;14(2):114–20.
65. Hua Y, Ren SY, Guo R, Rogers O, Nair RP, Bagchi D, et al. Furostanolic saponins from Trigonella foenum-graecum alleviate diet-induced glucose intolerance and hepatic fat accumulation. *Molecular Nutrition & Food Research*. 2015;59(10):2094–100.

66. Ranade M, Mudgalkar N. A simple dietary addition of fenugreek seed leads to the reduction in blood glucose levels: A parallel group, randomized single-blind trial. *Ayu.* 2017;38(1–2):24.
67. Fuller S, Stephens JM. Diosgenin, 4-hydroxyisoleucine, and fiber from fenugreek: Mechanisms of actions and potential effects on metabolic syndrome. *Advances in Nutrition.* 2015;6(2):189–97.
68. Ebrahimpour Koujan S, Gargari BP, Mobasseri M, Valizadeh H, Asghari-Jafarabadi M. Effects of Silybum marianum (L.) Gaertn. (silymarin) extract supplementation on antioxidant status and hs-CRP in patients with type 2 diabetes mellitus: A randomized, triple-blind, placebo-controlled clinical trial. *Phytomedicine.* 2015;22(2):290–6.
69. Williams B, Mancia G, Spiering W, Agabiti Rosei E, Azizi M, Burnier M, et al. 2018 ESC/ESH Guidelines for the management of arterial hypertension. *European Heart Journal.* 2018;39(33):3021–104.
70. Manivannan J, Balamurugan E, Silambarasan T, Raja B. Diosgenin improves vascular function by increasing aortic eNOS expression, normalize dyslipidemia and ACE activity in chronic renal failure rats. *Molecular and Cellular Biochemistry.* 2013;384(1–2):113–20.
71. Balaraman R, Dangwal S, Mohan M. Antihypertensive effect of Trigonella foenum-graecum. seeds in experimentally induced hypertension in rats. *Pharmaceutical Biology.* 2006;44(8):568–75.
72. Hadi A, Arab A, Hajianfar H, Talaei B, Miraghajani M, Babajafari S, et al. The effect of fenugreek seed supplementation on serum irisin levels, blood pressure, and liver and kidney function in patients with type 2 diabetes mellitus: A parallel randomized clinical trial. *Complementary Therapies in Medicine.* 2020;49:102315.
73. Akbari S, Abdurahman NH, Yunus RM, Alara OR, Abayomi OO. Extraction, characterization and antioxidant activity of fenugreek (Trigonella-foenum graecum) seed oil. *Materials Science for Energy Technologies.* 2019;2(2):349–55.
74. Ebadollahi SH, Pouramir M, Zabihi E, Golpour M, Aghajanpour-Mir M. The effect of arbutin on the expression of tumor suppressor P53, BAX/BCL-2 ratio and oxidative stress induced by tert-butyl hydroperoxide in fibroblast and LNcap cell lines. *Cell Journal.* 22(4):532.
75. Pradeep SR, Srinivasan K. Alleviation of cardiac damage by dietary fenugreek (Trigonella foenum-graecum) seeds is potentiated by onion (Allium cepa) in experimental diabetic rats via blocking renin—angiotensin system. *Cardiovascular Toxicology.* 2018;18(3):221–31.
76. Haghani K, Bakhtiyari S, Mohammadpour JD. Alterations in plasma glucose and cardiac antioxidant enzymes activity in streptozotocin-induced diabetic rats: Effects of Trigonella foenum-graecum extract and swimming training. *Canadian Journal of Diabetes.* 2016;40(2):135–42.
77. Joshi DV, Patil RR, Naik SR. Hydroalcohol extract of Trigonella foenum-graecum seed attenuates markers of inflammation and oxidative stress while improving exocrine function in diabetic rats. *Pharmaceutical Biology.* 2015;53(2):201–11.
78. Chaturvedi U, Shrivastava A, Bhadauria S, Saxena JK, Bhatia G. A mechanism-based pharmacological evaluation of efficacy of Trigonella foenum-graecum (fenugreek) seeds in regulation of dyslipidemia and oxidative stress in hyperlipidemic rats. *Journal of Cardiovascular Pharmacology.* 2013;61(6):505–12.
79. Srinivasan K. Antioxidant potential of spices and their active constituents. *Critical Reviews in Food Science and Nutrition.* 2014;54(3):352–72.
80. Sindhu G, Ratheesh M, Shyni G, Nambisan B, Helen A. Anti-inflammatory and antioxidative effects of mucilage of Trigonella foenum-graecum (Fenugreek) on adjuvant induced arthritic rats. *International Immunopharmacology.* 2012;12(1):205–11.
81. Zhang N, Chen Z, Shen J. Vitexin, a fenugreek flavonoid attenuated ethylene glycol induced urolithiasis in uninephrectomized rat via modulation of bikunin, iNOs, TNF-a, and osteopontin expressions. *Latin American Journal of Pharmacy.* 2020;39(1):90–101.
82. Kandhare AD, Bodhankar SL, Mohan V, Thakurdesai PA. Glycosides based standardized fenugreek seed extract ameliorates bleomycin-induced liver fibrosis in rats via modulation of endogenous enzymes. *Journal of Pharmacy & Bioallied Sciences.* 2017;9(3):185.
83. Middha S, Bhattacharjee B, Saini D, Baliga M, Nagaveni M, Usha T. Protective role of Trigonella foenum graecum extract against oxidative stress in hyperglycemic rats. *European Review for Medical and Pharmacological Sciences.* 2011;15(4):427.
84. Sakr SA, Abo-El-Yazid SM. Effect of fenugreek seed extract on Adriamycin-induced hepatotoxicity and oxidative stress in albino rats. *Toxicology and Industrial Health.* 2012;28(10):876–85.
85. Gupta SK, Kalaiselvan V, Srivastava S, Saxena R, Agrawal SS. Trigonella foenum-graecum (Fenugreek) protects against selenite-induced oxidative stress in experimental cataractogenesis. *Biological Trace Element Research.* 2010;136(3):258–68.
86. Dias IH, Milic I, Heiss C, Ademowo OS, Polidori MC, Devitt A, et al. Inflammation, lipid (per) oxidation, and redox regulation. *Antioxidants & Redox Signaling.* 2020.

87. Okin D, Medzhitov R. Evolution of inflammatory diseases. *Current Biology*. 2012;22(17):R733–R40.
88. Suresh P, Kavitha CN, Babu SM, Reddy VP, Latha AK. Effect of ethanol extract of Trigonella foenum-graecum (Fenugreek) seeds on Freund's adjuvant-induced arthritis in albino rats. *Inflammation*. 2012;35(4):1314–21.
89. Sayed AAR, Khalifa M, Abd el-Latif FF. Fenugreek attenuation of diabetic nephropathy in alloxan-diabetic rats. *Journal of Physiology and Biochemistry*. 2012;68(2):263–9.
90. Uemura T, Hirai S, Mizoguchi N, Goto T, Lee JY, Taketani K, et al. Diosgenin present in fenugreek improves glucose metabolism by promoting adipocyte differentiation and inhibiting inflammation in adipose tissues. *Molecular Nutrition & Food Research*. 2010;54(11):1596–608.
91. Sundaram G, Theagarajan R, Gopalakrishnan K, Babu GR, Murthy GD. Effect of fenugreek consumption with metformin treatment in improving plaque index in diabetic patients. *Journal of Natural Science, Biology and Medicine*. 2020;11(1):55.
92. Liu Y, Kakani R, Nair MG. Compounds in functional food fenugreek spice exhibit anti-inflammatory and antioxidant activities. *Food Chemistry*. 2012;131(4):1187–92.
93. Dsouza M, Rufina K, Hana D. Extraction of diosgenin from fenugreek and evaluation of its pharmacological role in alleviating metabolic syndrome in vitro. *Research Journal of Biotechnology*. 2018;13:12.
94. Hassan N, Withycombe C, Ahluwalia M, Thomas A, Morris K. A methanolic extract of Trigonella foenum-graecum (fenugreek) seeds regulates markers of macrophage polarization. *Functional Foods in Health and Disease*. 2015;5(12):417–26.
95. Sharififar F, Khazaeli P, Alli N, Talebian E, Zarehshahi R, Amiri S. Study of antinociceptive and anti-inflammatory activities of certain Iranian medicinal plants. *Journal of Complementary Medicine Research*. 2012;1(1):19–24.
96. Nathiya S, Durga M, Devasena T. Therapeutic role of Trigonella foenum-graecum [Fenugreek]—a review. *International Journal of Pharmaceutical Sciences Review and Research*. 2014;27(2):74–80.
97. Hasan M, Rahman M. Effect of fenugreek on type2 diabetic patients. *International Journal of Scientific and Research Publications*. 2016;6(1):251–5.

13 Trigonella foenum-graecum L. Therapeutic and Pharmacological Potentials—Special Emphasis on Efficacy against Inflammation and Pain

Sindhu G., Chithra K. Pushpan and Helen A.

CONTENTS

13.1	Introduction	187
13.2	Therapeutic Potential of Fenugreek	188
13.3	Pharmacological Potential of Fenugreek	189
13.4	Anti-Arthritic and Anti-Inflammatory Activity	189
13.5	Hypolipidemic Effects of Fenugreek	193
13.6	Anti-Diabetic Activity	195
13.7	Anti-Cancer Effect	197
13.8	Hepato-protective Effects of Fenugreek	198
13.9	Antimicrobial Effects of Fenugreek	198
13.10	Antioxidant and Immunomodulatory Properties of Fenugreek	199
13.11	Anti-Cataract Activity	200
13.12	Conclusion	200
Reference		200

13.1 INTRODUCTION

Plants, an essential source of medicine, play a crucial role in world health (Sandberg & Corrigan, 2001). For 1000 years, herbs have been used in all parts of the world as food and potent drugs, although they do not work like chemical drugs (Vuorelaa et al., 2004). Herbal medicines are demonstrated as effective remedies in the traditional medicine system and have been used extensively in medical practices. About 80% of the world's population uses medicinal plants, particularly in developing countries, for treatment and general health improvement, mainly due to the common belief that plant-derived drugs are devoid of any side effects along with being locally and economically accessible (Gupta & Raina, 1998). Traditional medicine garners comprehensive knowledge, skills, practices based on beliefs, experiences, and theories indigenous to different cultures and uses this to manage health, as well as to prevent, diagnose, improve, or treat physical and mental illness.

Fenugreek (*Trigonella foenum-graecum* L.) is an annual herb that belongs to the family Leguminosae grown in various countries around the world. It is also widely cultivated in India, China, northern and eastern Africa, and parts of Europe and Argentina (Acharya et al., 2007; Acharya et al., 2006). In the ancient world, fenugreek has a long history as both a medicinal and culinary herb. Fenugreek's functions were reported in ancient Egypt, where it was used to embalm mummies (Morcos et al., 1981). Also, Greeks and Romans utilized fenugreek for cattle fodder and it was used to aid labor and delivery. In Chinese traditional medicine, fenugreek seeds are used as a treatment for weakness and edema of the legs as well as a tonic (Yoshikawa et al., 1997). In India, fenugreek is used medicinally as a lactation stimulant and also commonly consumed as a condiment (Patil et al., 1997). The health-promoting properties of fenugreek have been long reported when it is eaten as a vegetable, taken as a food supplement or medicinal remedy. It is a famous spice in human food. Therefore, fenugreek is highly sought after as a chemurgic crop in the local, regional, and international nutraceutical, pharmaceutical, and functional food industries and markets as a medicinal herb (Basu et al., 2007; Shashikumar et al., 2018).

Scientists have reported several medicinal applications of fenugreek seeds, such as remedies for hypercholesterolemia and diabetes, protection against breast and colon cancer, and hepato-protection against free radicals (Al-Oqail et al., 2013). These protective roles are due to the non-nutritive secondary metabolites, also known as phytochemicals. Fenugreek contains a large number of chemical constituents like alkaloids, saponins, flavonoids, fibers, lipids, proteins, mucilage, vitamins, and coumarins. The stem of fenugreek contains alkaloids such as trigocoumarin, nicotinic acid, trimethyl coumarin, and trigonelline. Fenugreek contains protein (23–26%), fat (6–7%), and carbohydrates (58%). Furthermore, it is also rich in iron (33 mg/100 g dry weight) (Rao, 2004). Its leaves contain seven saponins, known as graecunins, and also contain moisture (86.1%), protein (4.4%), fat (0.9%), minerals (1.5%), fiber (1.1%), and carbohydrates (6%). The vitamins and minerals present in leaves include calcium, zinc, iron, phosphorus, riboflavin, carotene, thiamine, niacin, and vitamin C (Yadav & Sehgal, 1997). The aerial part of *Trigonella foenum-graecum* has a high polyphenol content which accounts for two flavonols namely quercetin and kaempferol, and both of them occur either in their free form or as glucosides (Gikas et al., 2011; Han et al., 2001). Benayad et al. (2014) separated and identified 24 flavonoid glycosides in the extract of fenugreek crude seeds. The seeds of fenugreek contain several coumarin compounds, as well as some alkaloids (e.g., trigonelline, gentianine, carpaine), diosgenin, and saponin (fenugrin B). Major bioactive compounds found in the seeds are polyphenols, such as rhaponticin and isovitexin, and also contain a small amount of volatile oils and fixed oil. In addition, seeds are excellent sources of mucilages (8.14%) which is a glycoprotein. It has been observed that the dry powder of fenugreek mucilage was galactomannan containing galactose and mannose in the ratio 1:1 (Boban et al., 2009). The fenugreek seed also contains 20–60% proteins, especially high in the amino acids lysine and tryptophan and 4-hydroxy-isoleucine, a free amino acid (Barnes et al., 2002; Van Wijk & Wink, 2005).

13.2 THERAPEUTIC POTENTIAL OF FENUGREEK

The medicinal importance of fenugreek seeds is mentioned in Ayurvedic texts as well as in Greek and Latin pharmacopoeia. In ancient Egypt, "methi" was taken to ease childbirth and to increase milk flow, and modern Egyptian women are still using it to relieve menstrual cramps, as well as abdominal pain. The Chinese mentioned fenugreek as "hu lu ba" and use it for treating abdominal pain. In India, "methi ka saag" (the stems and leaves) is commonly cooked as a winter vegetable. The seeds are eaten raw as sprouts and used medicinally. It is used as a flavoring agent for various dishes, and also in Ethiopia and Egypt, methi is used in bakery products. In the USA, it is mainly used to make spice blends for stews and soups (Passano, 1995). For centuries, European countries have used the herb of fenugreek as a cooking spice and it remains a popular ingredient in pickles, spice mixtures, and curry powders in India, Bangladesh, Pakistan, and other Asian countries. Moreover, fenugreek has been used in folk medicines to treat cellulitis, boils, and tuberculosis. It also has been recommended for the promotion of lactation. The fenugreek seeds have been used orally as an insulin substitute for a reduction in blood glucose, and the extracts from seed have been reported to lower blood glucose levels (Madar & Stark, 2002).

Fenugreek's maple aroma and flavor have led to its use in imitation maple syrup. It also has a beneficial effect on cleansing the blood, and as a diaphoretic, it can bring on a sweat and help to detox the body. A block in the lymphatic system can mean poor circulation of fluid, fluid retention, pain, energy loss, and disease anywhere in a person's body. Fenugreek's lymphatic cleansing activity is also known for its vital roles in irrigating the cells with nutrients and removing toxic wastes, dead cells, and trapped proteins from the body. It also maintains mucus conditions of the body, mostly the lungs, by helping to clear congestion and also acts as a throat cleanser and mucus solvent that eases the urge to cough. Drinking water in which fenugreek seeds have soaked, helps to soften and dissolve the accumulated and hardened cellular debris masses. In addition, fenugreek has been used to relieve colds, bronchial complaints, asthma, influenza, constipation, sinusitis, pleurisy, pneumonia, sore throat, laryngitis, hay fever, tuberculosis, smallpox, and emphysema (Anon, 2013). Being a natural health product, it can treat and cure diseases, thus providing medical and health benefits.

13.3 PHARMACOLOGICAL POTENTIAL OF FENUGREEK

Apart from the traditional medicinal uses, fenugreek is found to have many pharmacological properties (Umesh et al., 2014; Yadav & Baquer, 2014; Qamar et al., 2020; Saima et al., 2018; Srinivasan, 2006; Catherine et al., 2008) such as anti-arthritic, anti-inflammatory, hypolipidemic, anti-diabetic, anti-carcinogenic, anti-microbial effects. It also has anti-cataract, hepato-protective, antioxidant and immunomodulatory effects which are discussed later in detail. Although most of these studies have used whole seed powder or different forms of extracts, some have identified active constituents from seeds and attributed them medicinal values for different indications.

13.4 ANTI-ARTHRITIC AND ANTI-INFLAMMATORY ACTIVITY

Inflammation is a defense mechanism that is vital to health (Nathan & Ding, 2010) which acts by removing injurious stimuli and initiates the healing process (Ferrero-Miliani et al., 2007; Medzhitov, 2010). However, at the tissue level, inflammation is characterized by redness, swelling, heat, pain, and loss of tissue function, which result from local immune, vascular, and inflammatory cell responses to infection or injury (Takeuchi & Akira, 2010). Important microcirculatory events that occur during the inflammatory process include vascular permeability changes, leukocyte recruitment and accumulation, and inflammatory mediator release (Chertov et al., 2000). The inflammatory response is the coordinate activation of signaling pathways that regulate inflammatory mediator levels in resident tissue cells and inflammatory cells recruited from the blood (Lawrence, 2009). Although inflammatory response processes depend on the precise nature of the initial stimulus and its location in the body, they all share a common mechanism, which can be summarized as: (1) cell surface pattern receptors recognize detrimental stimuli; (2) inflammatory pathways are activated; (3) inflammatory markers are released; and (4) inflammatory cells are recruited. Normally, during acute inflammatory responses, cellular and molecular events and interactions efficiently minimize impending injury or infection. This mitigation process contributes to restoration of tissue homeostasis and resolution of the acute inflammation. However, uncontrolled acute inflammation may become chronic and contributes to a variety of chronic inflammatory diseases including cardiovascular and bowel diseases, diabetes, arthritis, and cancer (Zhou et al., 2016; Libby, 2007).

Rheumatoid arthritis (RA) is a chronic inflammatory, systemic autoimmune disease. RA primarily affects the lining of the synovial joints and is associated with progressive disability, premature death, and socioeconomic burdens. Of the numerous cell types that are involved, rheumatoid arthritis synovial fibroblasts (RASF), which in healthy joints are involved in homeostasis in the synovium, are thought to be central to the pathogenesis of joint destruction in cooperation with other mesenchymal and immunocompetent cells. The deleterious joint destruction is mediated by intracellular signaling pathways involving transcription factors, such as nuclear factor κB, cytokines, chemokines, growth factors, cellular ligands, and adhesion molecules in the rheumatoid synovium, which are initiated at the preclinical stage, leading to a cascade of pathophysiologic events that result in progressive and irreversible joint destruction.

The clinical symptoms of symmetrical joint involvement include swelling, redness, arthralgia, and even limiting the range of motion. Early diagnosis is designed as the principal improvement index for the most desirable outcomes (i.e., reduced joint destruction, no functional disability, less radiologic progression, and disease-modifying anti-rheumatic drugs (DMARD)-free remission) as well as cost effectiveness as the first 12 weeks after early symptoms occur is regarded as the optimal therapeutic window (van der Linden et al., 2010; Moura et al., 2015; Cho et al., 2019). With uncontrolled or severe disease, there is a risk that extra-articular indications such as keratitis, pericarditis/pleuritis, pulmonary granulomas (rheumatoid nodules), small vessel vasculitis, and other non-specific extra-articular symptoms will develop. Presently, there is no treatment for RA.

The treatment strategy aims to expedite diagnosis and rapidly achieve a low disease activity state (LDAS). Many composite scales analyze the disease activity, such as the Simplified Disease Activity Assessment Index (SDAI), Clinical Disease Assessment Index (CDAI), and Disease Activity Score using 28 joints (DAS-28) (Ometto et al., 2010). Universally applied pharmacologic therapy with corticosteroids, non-steroidal anti-inflammatory drugs (NSAIDs) and immunosuppressants have proven effective in relieving pain and stiffness but do not moderate disease progression. Over the last 20 years, the effectiveness of DMARDs has gained much attention as these can efficiently attenuate disease activity and substantially decrease and/or delay joint deformity (Grennan et al., 2001). The therapy classification includes traditional synthetic drugs, biological DMARDs, and novel potential small molecules. However, these drugs produce various serious and undesired side effects, including immunodeficiency, gastrointestinal disturbances, and humoral disturbances. In order to overcome these side effects, identification of new targets, safe and effective medicine, and naturally occurring antagonists of pro-inflammatory molecules are essential factors in the development of new drug discovery for RA (Olsen & Stein, 2004). Chopra et al. (2010) has explained a comprehensive therapeutic approach used in Ayurveda and modern medicine to treat arthritis. Rasayana and Kashayams (Shyni et al., 2015; Aswathy et al., 2019) are extensively used in Ayurvedic medicine for their immunomodulatory, anti-inflammatory, and hence anti-arthritic, and other biologic effects, which have been documented. However, the effect of fenugreek is commendable in terms of its being anti-inflammatory, arthralgic, and anti-arthritis.

Fenugreek at doses of 100 and 200 mg/kg reduced carrageenan-induced paw edema in rats (Rao et al., 2005). The alkaloid present in fenugreek extract has been reported to produce anti-inflammatory properties by reducing formalin-induced edema in rats and demonstrating an antipyretic effect by significantly reducing hyperthermia induced by brewer's yeast (Pandian et al., 2002). Another reason for the anti-inflammatory activity of fenugreek is probably due to the presence of saponins and flavonoids. Several reports also showed that flavonoids also act as potential inhibitors of cyclooxygenase, lipoxygenase, and nitric oxide synthase (Handa et al., 2005; Sharififara et al., 2009; Sharma et al., 1990; Weiser et al., 1997).

The main chemical constituents of fenugreek responsible for the anti-inflammatory activity are alkaloids, saponins, and flavonoids. Sharififara et al. (2009) studied the *in vivo* effect of methanolic extract using a cream-based system. Inflammation in terms of edema was induced in Wistar rats using carrageenan and an anti-inflammatory effect was observed both by intraperitoneal administration and by the topical application in the form of cream (Sharififara et al., 2009). Also, Kawabata et al. in 2011 studied the anti-inflammatory and anti-melanogenic effect in an *in vitro* system using a human monocytic cell line (THP-1). Production of inflammatory cytokines such as IL-1, IL-6, and TNF-α was initiated using phorbol myristate acetate. Inhibitory action of fenugreek extract with methanol as a solvent system was observed with suppression in TNF-α production. The extract was further subjected to the isolation of bioactive compounds such as saponin and two other compounds which were also found to inhibit other cytokines like IL-1 and IL-6 along with TNF-α. The inhibitory effects were concentration dependent. Some contrasting results as compared to the *in vitro* study were also evident in an *in vivo* system (Raju & Bird, 2006). Another study carried out in 2011 by Sumanth et al. (2011) reported an anti-inflammatory effect against the ulcer production. Immersion stress and indomethacin were used to induce ulcers in rats. The aqueous extract of fenugreek seeds showed the antiulcer effect as calculated by the ulcer index (Raju & Bird, 2006). Not only seeds but also antipyretic and anti-inflammatory activities of the leaves of *T. foenum-graecum* have been reported (Sumanth et al., 2011). Furthermore, Ravichandiran & Jayakumari (2013) compared the anti-inflammatory activity of a bioactive compound isolated from fenugreek seed and leaves extracts and its aqueous extracts both in *in vivo* and in *in vitro* systems. It was also observed that chloroform fraction of seeds and aqueous extract of leaves of fenugreek were effective against anti-inflammatory activity (Ahmadiani et al., 2001).

Recently, Pundarikakshudu (2019) revealed in his book chapter that fatty acids, flavonoids, saponins, alkaloids (trigonelline), and galactomannans present in different extracts of fenugreek seeds exhibited significant analgesic, anti-inflammatory, and anti-arthritic activities. However, Shishodia and Aggarwal (2006) tried to explore the effect of the diosgenin, steroid of fenugreek on the transcription factor NF κB-signaling pathway and on NF κB-regulated gene products on TNF-induced RAW 264.7 cells. Their data exhibited that diosgenin may mediate its anti-inflammatory, anticancer, and pro-apoptotic effects which were achieved through interruption of the NF κB pathway. On the other hand, the partially purified fraction (MTH) of the *Trigonella foenum-graecum* seed extract possesses potential analgesic and anti-inflammatory activities (Vyas et al., 2008). In a study, fenugreek seeds were extracted consecutively with petroleum ether, acidified chloroform (ACC), alkaline chloroform (AKC), methanol, and water. Among these fractions ACC fraction had exhibited the highest anti-inflammatory effect. By the spectrophotometric method the trigonelline content of the plant was identified. ACC fraction was further fractionated using column chromatography and four subfractions were selected for pharmacological study. Among them F3 subfraction exhibited the greatest inhibition at 15 mg/kg ($p < 0.001$) (Pournamdari et al., 2018). Kaempferol, a polyphenolic flavonoid extracted from fenugreek seeds, has been shown to have strong antioxidant and anti-inflammatory properties (Havsteen, 2002; Middleton et al., 2000). Hamalainen et al. (2007) reported in their *in vitro* studies that kaempferol inhibits both iNOS and cyclooxygenase enzymes. In addition, kaempferol produces its anti-inflammatory activity in rats by dual inhibition of prostaglandin generation and nitric oxide by modulation of the cyclooxygenase pathway via inhibition of nitric oxide production (Mahat et al., 2010).

Moreover, the apoptosis associated speck-like protein containing a caspase recruitment domain (CARD) (ASC) is central to inflammatory and cell death pathways in innate and adaptive immunity. Fenugreek seed extract provides cytoprotection to bacterial lipopolysaccharide (LPS) inflamed and nanosilica-treated fibroblasts via a ROS independent pathway (Sharma et al., 2017). All atom molecular dynamics simulations of ASC-ligand complex reveal that individual phytochemicals in fenugreek can bind to ASC via specific non-covalent interactions. Thus, the synergistic effect of fenugreek phytochemicals with the ASC protein alters its molecular properties resulting in altered cellular function and also such information is crucial to the development of targeted therapeutic interventions for inflammatory diseases.

Pain, one of the cardinal signs of inflammation, is an uncomfortable experience and remains a challenge for both physician and patient in spite of great advances in drug research. Apart from the pathophysiology of pain and individual experience, current drug regiments exhibit less or more side effects which might be more severe than the pain (Jorge et al., 2010). There are some reports concerning the antinociceptive effects of the plant fenugreek. Javan et al., 1997 have shown that the extract of TFG leaves produce antinociception in a dose dependent manner in both phases of formalin and tail-flick tests. Similar studies by Parvizpur et al. (2004, 2006) also demonstrated the analgesic effect of Trigonella *foenum-graecum* (TFG) leaves extract in both formalin and tail flick tests.

Another study involved the application of a standardized extract of fenugreek seeds which was used for fenugreek transdermal patch (10%, FDP) formulation and was treated in patients after inguinal hernia (IH) surgery. The pain intensity score was evaluated using a visual analogue score (VAS) and a significant reduction in FDP group was found (Ansari et al., 2019). This treatment was safe and free from allergic and/or severe early side effects. Fenugreek patches are a new option for topical purposes and can be a helpful way for overwhelming the pain and minimizing the side effects of antinociceptive drugs. Vyas et al. (2008) studied the analgesic effect of a partially purified fraction of the *Trigonella foenum-graecum* seed extract (MTH) at the dose of 40 mg/kg. This has shown significant analgesic activity as compared to diclofenac sodium and pentazocine. Further it is concluded that fenugreek has significant analgesic and antipyretic effects, and can be used as substituent for diclofenac sodium and indomethacin for minor pains (Abbas et al., 2016).

Freund's complete adjuvant-induced arthritis (AIA), a sub-chronic model of arthritis in an albino rats study by Suresh et al. (2012) on ethanolic extract of fenugreek exhibited a dose-dependent inhibition of cytokines level which was elevated by AIA induction. This inhibition of cytokine liberation may be the reason behind the anti-arthritic potential of fenugreek seed. Recently, Rishi et al. (2020) evaluated the protective role of ethanolic fenugreek seed extract in CFA-induced experimental arthritis. The activity of fenugreek seed extract was compared with methotrexate. The authors suggest that nitric oxide synthase inhibition leads to potentiated anti-arthritic effects of ethanolic fenugreek seed extract. That may be mediated through pro-inflammatory/anti-inflammatory cytokines imbalance. Moreover, Pundarikakshudu et al. (2016) reported that petroleum ether extract of fenugreek seeds has significant anti-inflammatory and anti-arthritic activities. They concluded their work by analyzing the petroleum ether extract of fenugreek seed on various *in vivo* models like carrageenan-induced paw edema, formaldehyde-induced paw edema, Freund's complete adjuvant-induced arthritis, and cotton pellet-induced granuloma and confirmed that anti-arthritic potential of fenugreek seed is due to the presence of linolenic and linoleic acids.

Furthermore, Baliga et al. (2015) reported in their book chapter that, in addition to reducing the inflammation and arthritic index and restoring body weight, fenugreek (200 and 400 mg/kg) was also effective in reducing the differential white blood cell (WBC) count, erythrocyte sedimentation rate (ESR), and WBC content, and increasing the red blood cell (RBC) count and hemoglobin (Hb). In the cytological study of synovial fluid's function and structure, which exhibits severe inflammation with usual synoviocytes (mesothelial cells), supplements of fenugreek mucilage were given to normalize the characters of these cells. Furthermore, in Ayurveda, methi (fenugreek) seed powder was used to cure rheumatic fever, rheumatoid arthritis due to their Aam-pachaka properties (detoxifying of the toxic particle). Vyas et al. (2010) reported that a combination of fenugreek, Boswellia, and Acacia shows the restoration of vascular and endothelial dysfunction caused during chronic arthritis in rats. It was also hypothesized that the anti-inflammatory and antioxidant activities of *T. foenum-graecum* L. might be a possible reason behind the observed anti-arthritic potential (Suresh et al., 2012). Treatment with the extract enhances both the peritoneal exudate cell and macrophage cell counts. The extract also exhibited a significant anti-inflammatory effect (Sur et al., 2001). Recently Pal et al. suggested that nitric oxide synthase inhibition leads to potentiated anti-arthritic, anti-inflammatory effects of ethanolic fenugreek seed extract (FSE) and that may be mediated through pro-inflammatory/anti-inflammatory cytokines imbalance (Pal et al., 2020).

In our study, fenugreek mucilage composed of galactomannan was evaluated against Freund's complete adjuvant-induced arthritis in albino rats for its anti-arthritic and antioxidant potential. It was observed that the activities of cyclooxygenase, lipoxygenase, and myeloperoxidase, and levels of nitrite and C-reactive protein were decreased by the oral administration of fenugreek mucilage at a dose of 75 mg/kg, which was elevated by the adjuvant induction. Moreover the antioxidant status during arthritic condition was also altered by increasing the concentration of thiobarbituric acid reactive substance and the reduced glutathione level and activities of antioxidant enzymes. Meanwhile the mucilage supplementation retains the antioxidant status. The increment in total white blood cells (WBC) and erythrocyte sedimentation rate (ESR), hemoglobin, and reduction in red blood cells (RBC) count and aberrant changes to the C-reactive protein (CRP) levels exhibited in the arthritic animals were significantly restored in fenugreek mucilage treated rats. Histopathological examination also revealed decreased edema formation and cellular infiltration on supplementation with fenugreek mucilage (Sindhu et al., 2012).

Sindhu et al. (2018) further reported that the administration of 75 mg/kg of fenugreek mucilage showed marked reduction in paw volume and arthritic index which is the average of the score given to the severity of the lesions compared to the arthritic rats. Thus, the selective reduction in the arthritis score distinguishes the immunosuppressive effects of fenugreek mucilage from its anti-inflammatory effects. In addition, fenugreek mucilage also reduced rheumatoid factor which shows the activation of B cells via TLRs and several genetic predispositions to arthritic diseases. This exhibits the mucilage inhibiting autoimmune stimulations in arthritic-induced rats. Also, elevated energy expenditures and pro-inflammatory cytokines such as tumor necrosis factor-α (TNF-α) and interleukin (IL-6) are considered to be one of the most important mediators that accelerate joint soreness and painfulness (Qamar Abbas Syed et al., 2020; Mclnnes & Schett, 2007). Finally, Sindhu et al. (2020) concluded the study that fenugreek mucilage significantly reduces cell influx, release of mediators, and B cell activation associated with arthritic conditions. Therefore fenugreek mucilage has the potential to be used as an anti-arthritic agent.

Another point regarding inflammatory response is the estrogen-like compounds that counteract autoimmune disorders through the binding of estrogen metabolites to DNA, thereby inhibiting tissue inflammation (Khan et al., 2011). Fenugreek has been shown to stall autoimmune diseases by acting as a mimic of estrogen. Sreeja et al. (2010) showed that fenugreek bound to estrogen receptors and functions as estrogen by influencing genetic activities by inducing the estrogen-responsive proteins.

13.5 HYPOLIPIDEMIC EFFECTS OF FENUGREEK

Abnormal lipid levels are a cause of concern as they are an evident potential risk to various complications like heart diseases. Global Health Observatory data indicates that increased cholesterol level is attributed to one-third of ischemic heart disease. It is also mentioned that a reduction of serum cholesterol levels by 10% reduced the heart disease in men of 40 years of age and 70 years of age by 50% and 20%. Reducing the cholesterol levels has for a long time been one of the targets for prevention and cure of heart diseases. Strategies in reducing cholesterol levels in blood includes reduction in consumption of lipid-containing foods, hypolipidemic medications that include HMG CoA reductase inhibitors, bile acid binding resins, CETP inhibitors, cholesterol absorption inhibitors, fibrates, nicotinic acid, etc. (Kramer, 2015). However, there are adverse effects from the treatment with these drugs making it challenging for clinical management of hypercholesterolemia. An important strategy in the management of hypercholesterolemia is through modification of diet and also inclusion of functional foods and nutraceuticals.

A wide variety of medicinal plants are studied for their therapeutic potential including lipid-lowering effects. One of the most important medicinal plants studied for its lipid-lowering effect is fenugreek. Several studies have shown its hypolipidemic properties. Here, a review of some of the important recent studies that support the hypolipidemic effect of fenugreek shall be made. There have been several studies on animal and human trials showing the hypolipidemic mechanism of its therapeutic action (Belguith-Hadriche et al., 2010; Belguith-Hadriche et al., 2013; Verma et al., 2016; Vijayakumar et al., 2010; Valette et al., 1984; Sui et al., 2012; Boban et al., 2009). Before analyzing the mechanism through which the lipid-lowering effect is induced by the plant and its products, some of the studies that have proven these are to be looked at. The effect of fenugreek seeds in reducing cholesterol levels in dogs was studied very early by Ribes et al. (1987). In this study fenugreek seed lipid extract and a defatted fraction was studied for its lipid-lowering effect. Lipid extract and defatted extract were fed to normal and diabetes-induced animals for 8 days. The study revealed the reduction in blood glucose and cholesterol levels in groups treated with defatted content of fenugreek. However, there were no changes in lipid extract treated groups. The study supports the role of fibers and saponins in defatted component of fenugreek as the major contributor to the sugar-lowering and lipid-lowering properties. They explain it might be through the inhibition of absorption of cholesterol from the intestine, thereby affecting cholesterol metabolism (Valette et al., 1984).

Bioactive compounds isolated from the leaves and seeds have been studied for their activity. Diosgenin has been widely studied for its vasoprotective effects. Fenugreek seed flour and diosgenin were studied for their ability to protect endothelial dysfunction in rat models. The animals were supplemented with fenugreek seed flour and diosgenin along with their diet in varying concentrations for a period of 6 weeks. The study showed protection against endothelial damage and antioxidant properties in both the fenugreek flour and diosgenin with fenugreek flour showing comparatively increased protection. This is explained as a cumulative effect of other bioactive components in fenugreek other than diosgenin. Their study shows an improvement in glutathione reductase activity, oxidized and reduced glutathione which is indicative of its ability to regulate the redox balance in blood, which can contribute to vasoprotective effects (Szabó et al., 2018). The vasoprotective effect by regulating lipid metabolism and molecules involved in cell signaling involved in vascular homeostasis has been studied with diosgenin (Manivannan et al., 2013).

An anticoagulant effect of fenugreek also has been reported in human subjects (Eldin et al., 2013). A reduction in blood glucose, cholesterol, and improved hematological parameters has been reported in various studies showing a cardioprotective effect of fenugreek (Xue et al., 2007). Some studies have focused on elucidating the molecular mechanism by which the vasoprotection and hypolipidemic properties are exhibited by fenugreek and the bioactive components of fenugreek. The elucidation of the molecular mechanism of the regulation of lipid metabolism by fenugreek extracts was studied on 3T3-L1 cells and HepG2 (human hepatoma cell line), and C57BL/6J mice (Vijayakumar et al., 2010). The lipid accumulation in cells, the cholesterol levels in animals treated with fenugreek, and various transcription factors and receptor molecules responsible for the molecular uptake of lipids were studied. The accumulation of lipid (triglycerides) in cells was found to be decreased in cells treated with extract of fenugreek seeds in a dose-dependent manner. The work also elucidates the mechanism through which the hypolipidemic effect is shown by the extract. The expression of genes responsible for uptake of lipids and lipogenesis was studied. A significant decrease in SREBP-1 in liver cells and fat cells was observed. The fatty acid synthase which is responsible for fatty acid synthesis was found to be inhibited upon treatment with fenugreek. The treatment with fenugreek extract was also shown to increase the LDL-R in hepatic cells, explaining one of the prime mechanisms of its ability to induce a hypolipidemic effect. The thermostable extract of fenugreek studied hence explores various strategies that lead to an atheroprotective effect (Vijayakumar et al., 2010).

Boban et al. (2006) demonstrated the hypolipidemic activity of mucilage isolated from the fenugreek seed. Rats fed with mucilages at a dose of 4 mg/100 g body weight per day for 8 weeks had changes in the levels of total cholesterol and triacylglycerols in serum, the liver, and aorta. Further, hepatocytes isolated from the liver of mucilage-fed rats showed a decrease in the synthesis and secretion of apoB containing lipoproteins, mainly VLDL. As elevated blood lipids are one of the major risk factors in the development of chronic conditions like atherosclerosis, diet-based lipid lowering is a useful approach, since fenugreek is used as a spice. Later in 2009, Boban et al. showed the anti-hypercholesterolemic and anti-atherogenic effect of the mucilage galactomannan from fenugreek seeds in experimental rabbits maintained on a high-cholesterol diet for 3 months. The regression studies showed that administration of mucilage significantly decreased the serum total and LDL cholesterol and aortic cholesterol. Also, significantly low sudanophilic staining confirmed that mucilage accelerated the regression of atheromatous lesions in the aorta.

Atherosclerosis is a multifactorial disease that involves chronic inflammation at every stage from initiation to progression and eventually, plaque rupture (Badimon & Vilahur, 2014; Chistiakov et al., 2016). Recently Sindhu et al. (2020) have demonstrated involvement of inflammatory response and fenugreek mucilage's efficacy for inhibiting the same during the development of atherosclerosis. In this research work the authors explained the therapeutic efficacy of fenugreek mucilage on chronic inflammation as a preventive regimen in rabbits fed a high-cholesterol diet. The uptake of fenugreek mucilage within a preventive diet regimen improved the functioning of the heart and the concentration of lipids. However the supplementation of mucilage along with an atherosclerotic diet reduced the activities of inflammatory enzymes like cyclooxygenase and lipoxygenase which were subclinical markers of atherosclerosis. This inhibited the macrophage adhesion and transmigration into the subintimal space which causes a reduction in atherosclerotic lesion size. Another useful risk marker and diagnostic tool in acute coronary syndromes was MPO whose elevation enhances the pro-atherogenic properties of LDL, but the supplementation of mucilage reduces the oxidative modification of LDL and plaque rupture. In addition, mucilage administration also suppresses the production of acute-phase proteins like CRP, and fibrinogen, released by the hepatocytes after cytokine stimulation.

Thereby, the administration of fenugreek can diminish the initiation and progression of atherosclerosis. At the same time, the size and shape of HDL particles in HCD-fed rabbits change, and this strongly influences the binding affinity and stability of PON1, responsible for the antioxidant activity of HDL. But the fenugreek mucilage regains serum HDL-C level along with upregulated PON1 gene expression, suggesting the retention of antioxidant and anti-atherogenic properties. This may be one of the mechanisms through which mucilage prevents atherosclerosis. Thus, from this evidence it was found that lipid-lowering and anti-atherogenic effects of mucilage from fenugreek used as a food flavoring spice highlight the importance of dietary intervention in the regression of atherosclerosis.

13.6 ANTI-DIABETIC ACTIVITY

Diabetes mellitus is a chronic metabolic disorder specified by derangements in the metabolism of carbohydrates, lipids, and protein caused by complete or relative insufficiency of insulin secretion and/or action (American Diabetes Association, 2009). Diabetes is a significant health concern, and its burden is increasing globally. Generally, after glucose intake, normal glucose homeostasis mainly occurs through three major pathways. First, insulin is secreted by the pancreas; second, glucose production by liver cells declines; third, the liver and peripheral muscles stimulate glucose uptake (Lillioja et al., 1988; DeFronzo, 1992). Shulman et al. (1990) have shown that muscle glycogen synthesis is the principal pathway of glucose disposal in both normal and diabetic subjects. This defect in muscle glycogen synthesis plays an important role in insulin resistance in T2DM. According to Warram et al. (1990) the symptoms of DM can be observed one to two decades before the actual occurrence of the disease in the offspring of diabetic patients. They have also reported that these patients have a hyperglycemic state, which is recompensed by hyperinsulinemia. The hyperglycemic state results in inflammation and stimulates β cells to produce Interleukin-1 (IL-1) at a higher rate. The lower concentration of IL-1 antagonists fails to compete and bind the receptors. This results in β cell destruction and loss of function. Saini (2010) has reviewed several such mechanisms involved in T2DM. Insulin resistance and pancreatic β cell dysfunction are significant factors for the cause of T2DM. It is also shown that there is reduction in β cell mass in severe DM. It has been well established that failure to maintain the targeted level of glycated hemoglobin (HbA1c) also resulted in progressive T2DM. Diabetes mellitus can be regulated by food habits that are rich in chemical constituents that will help maintain the blood glucose level, which offers an economical approach. Inadequate metabolic control leads to the development of complications associated with many functional limitations and health-related quality of life. Metformin and sulfonylureas and dipeptidyl-peptidase-4 (DPP-4) inhibitors are the first-line and second-line treatment for the T2DM prescribed worldwide. Though several forms of treatments are available as medications and injectable insulin, they are accompanied by side effects.

Herbal medicine plays an essential role in the treatment of diabetes (Bailey & Day, 1989; Jarald et al., 2008). Fenugreek is one of the well-studied herbal plants, which has been thoroughly researched with respect to its effect on diabetes. Fenugreek was used as an anti-diabetic in many model systems (Bawadi et al., 2009; Srinivasan, 2006; Abdel-Barry., 2000) and revealed that seeds, leaves, and extracts are good agents to fight against diabetes (Raju et al., 2004). Dietary fiber from fenugreek blunts glucose after a meal. Fenugreek seeds contain 45.4% dietary fiber (32% insoluble and 13.3% soluble), and the gum is composed of galactose and mannose. The gum is associated with a reduced glycemic effect. An active compound can also be isolated from the fenugreek's crude extract, which performs a beneficial role against a high glucose level (Moorthy et al., 1989). A study done by Moorthy et al. (2010) isolated GII from the aqueous extract of fenugreek seeds. The GII compound reduced blood glucose in the glucose tolerance test in sub-diabetic and moderately diabetic rabbits. GII compound even exhibited better results than the standard tolbutamide, sulfonylurea oral hypoglycemic medication (Moorthy et al., 2010). The work demonstrated that the hypoglycemic effect might be mediated through stimulating insulin synthesis and/or secretion from the pancreatic β cells. The prolonged administration of the same dose of the active principle for 30 days to the severely diabetic rabbits lowered fasting blood glucose significantly. Still, it could elevate the fasting serum insulin level to a much lower extent, suggesting an extra-pancreatic mode of action for the active principle. The effect may also be by increasing the sensitivity of tissues to the available insulin. The hypoglycemic effect was slow but sustained, without any risk of developing severe hypoglycemia (Puri et al., 2002). Even in Egyptian folk medicine, fenugreek held an essential place as a hypoglycemic agent. An *in vitro* study demonstrated by Gad et al. (2006) showed that the fenugreek extract in a dose-dependent manner was able to inhibit α-amylase activity. It showed the suppression of starch digestion and absorption in normal rats, suggesting that the plant extract's hypoglycemic effect was mediated through insulin mimetic effect (Ritika, 2016). Animal studies suggest the hypoglycemic effects of fenugreek (Broca et al., 1999; Al-Khateeb et al., 2012). Kiss et al. (2019) reported the increased insulin sensitivity index and high metabolic clearance rate of insulin found in the 1mg/kg diosgenin and the fenugreek seed treated group suggested an improved peripheral insulin sensitivity. Also their results from the 10mg/kg diosgenin group, however, suggest a marked insulin resistance.

Human studies also showed possible hypoglycemic properties of fenugreek. It was found that in Arab countries Iraq (Salih, & Al-Asadi, 2012; Al-Asadi, 2014) and Saudi Arabia (Al-Rowais, 2002), fenugreek was found to be the most common herbs used among people with diabetes. Gupta et al. (2001) reported that adjunct use of fenugreek seeds improves glycemic control and decreases insulin resistance in mild type 2 diabetic patients. In a study about healthy subjects, Shakib et al. showed a comparison between bread with and without fenugreek. They showed that those who consumed bread with fenugreek improved the postprandial glucose response (Shakib & Gabrial, 2010). In addition to the biochemical improvements, fenugreek seeds have been reported to significantly suppress the clinical symptoms of diabetes, such as polyuria, polydipsia, weakness, and weight loss (Sharma, 1986). A novel amino acid, 4-hydroxyleucine from fenugreek seeds, increased glucose-stimulated insulin release by islet cells in both rats and humans (Sauvaire et al., 1998). Owing to its particular insulinotropic action, 4-hydroxy isoleucine might be considered a novel secretagogue with a potential role in treating diabetes (Baquer et al., 2011). An *in vitro* study exposed that saponin compounds diosgenin, alkaloids, and trigonelline were associated with intestinal glucose uptake inhibition (Al-Habori et al., 2001). Moreover, fenugreek has been shown to increase erythrocyte insulin receptors and improve peripheral glucose utilization, thus showing potential pancreatic as well as extra-pancreatic effects (Raghuram et al., 1994).

However, fenugreek seed extracts have exhibited anti-diabetic potentials by producing delays in gastric emptying in the small intestine by its high-fiber content, slowing carbohydrate metabolism, and lowering blood glucose. Thus, restoring the function of pancreatic tissues (Patel et al., 2012; Gauttam & Kalia, 2013); protecting β cells; elevating serum insulin level, possibly through β cell regeneration or stimulation of insulin release from existing β cells of islets. It also stimulates the activity of glycogen synthetase and promotes the formation of liver and muscle glycogen (Bera et al., 2013), reducing the pro-inflammatory cytokines and pancreatic enzymes, and promoting the restoration of depleted glycogen (muscle and liver) (Joshi et al., 2014). Seed extract corrects the insulin-sensitive carbohydrate metabolic enzyme activities and serum lipid profiles (Bera et al., 2013). Furthermore, seed extract plays an important antioxidant status by preventing lipid peroxidation and restoring glutathione and superoxide dismutase (liver and pancreas) (Joshi et al.,

2014). Fenugreek seed promotes insulin sensitivity, improving insulin action at the cellular level (Patel et al., 2012; Gauttam & Kalia, 2013), and recovering the level of HbA1c by utilization of glucose in peripheral tissues whereby the blood glucose level is maintained (Bera et al., 2013). The diet containing fenugreek exhibited a marked reduction in signs and symptoms of diabetes like polyuria, urine sugar, renal hypertrophy, excessive thirst, and glomerular filtration rate (Shetty & Salimath, 2009).

13.7 ANTI-CANCER EFFECT

Cancer is one of the leading causes of mortality all around the world. Many reported studies have shown the protective effect of fenugreek seeds in experimental models of cancer using cell lines or experimental animals. In several studies, extracts of fenugreek seeds and some of their constituents have shown anti-carcinogenic potency. Consumption of fenugreek was followed by decreased polyamines (spermine, spermidine, putrescine) in tumor tissue (Zhilenko et al., 2012). Srinivasan et al. (2009) investigated the effect of diosgenin, a biologically active constituent of fenugreek seeds, on breast cancer cell lines. Diosgenin caused G1 cell cycle arrest by downregulating cyclin D1, cdk-2, and cdk-4 expression in both estrogen receptor-positive ER (+) and estrogen receptor-negative ER (−) breast cancer cells resulting in the inhibition of cell proliferation and induction of apoptosis. Furthermore, diosgenin was found to inhibit migration and invasion of prostate cancer PC-3 cells by reducing matrix metalloproteinases expression. It also inhibited extracellular signal-regulating kinase (ERK), c-Jun N-terminal kinase (JNK), and phosphatidylinositide-3 kinase (PI3K/Akt) signaling pathways as well as nuclear factor kappa B (NF-κB) activity (Chen et al., 2011). Das et al. (2012) also showed that diosgenin alone or in combination with Thymoquinone obtained from black cumin (*Nigella sativa* L.) inhibits cell proliferation and induces apoptosis in squamous cell carcinoma. Amin et al. (2005) showed that fenugreek seed extract significantly inhibited 7, 12-dimethylbenz (a) anthracene-induced mammary hyperplasia and reduces its incidence in rats and advised that the anti-breast cancer protective effects of fenugreek could be due to increased apoptosis. Protodioscin, a furostanol saponin isolated from fenugreek, also induces apoptotic changes leading to death in a leukemic cell line (HL–60). Several studies on anti-cancer properties of chemical constituents of fenugreek have been done and have shown positive results. Some alkaloids, called "trigonelline," have revealed the potential for use in cancer therapy (Koli & Ramkrishna, 2002).

To investigate the antineoplastic potential, *Trigonella foenum-graecum* seed extract has also been evaluated in the Ehrlich ascites carcinoma (EAC) model in Balb-C mice. Intraperitoneal administration of the alcohol extract of the seed both before and after the inoculation of the EAC cell in mice produced more than 70% inhibition of tumor cell growth compared to the control. Treatment with the extract enhances both the peritoneal exudate cell and macrophage cell counts. The extract also exhibited a significant anti-inflammatory effect (Sur et al., 2001). The effects of protodioscin (PD), which was purified from fenugreek, on cell viability in human leukemia HL-60 and human stomach cancer KATO III cells were investigated. A diet containing fenugreek seed powder decreased colon tumor incidence and hepatic lipid peroxidation in 1, 2-dimethylhydrazine treated rats. It also increased activities of catalase, superoxide dismutase, glutathione S-transferase, and glutathione peroxidase in the liver (Devasena & Menon, 2007). Li et al. (2010) showed that diosgenin could modulate the STAT3 signaling pathway in hepatocellular carcinoma by suppressing the activation of c-Src, JAK1, and JAK2. Diosgenin also downregulated the expression of various STAT3-regulated genes, inhibited proliferation, and potentiated the apoptotic effects of paclitaxel and doxorubicin, suggesting that diosgenin could be a novel and potential treatment option for hepatocellular carcinoma and other cancers. Fenugreek contains a crystalline steroid sapogenin, diosgenin, as a starting material for synthesizing steroid hormones such as cortisone and progesterone. It has the potential to prevent invasion, suppresses proliferation and osteoclastogenesis through inhibition of necrosis factor. It enhances apoptosis induced by cytokines and chemotherapeutic agents (Montgomery, 2009). On the other hand, chronic inflammatory processes elevate the risk of malignant transformation through increased cell necrosis and consequent stimulation of proliferation of cytokine activity supports regenerative processes. In an *in vivo* study, Varjas et al. (2011) induce cancer in mice of AKR/J H-2^k strain dimethylbenz[α]anthracene (DMBA) and treatment was carried out with fenugreek seeds. From the results and data obtained, authors found that fenugreek seed and its biologically active compounds could have a chemopreventive effect mediated through the suppression of lipoxygenase and cyclooxygenase genes.

13.8 HEPATO-PROTECTIVE EFFECTS OF FENUGREEK

Chronic alcoholism is one of the various associations with liver diseases and fibrosis (Bellentani et al., 1997; Patsenker et al., 2011). Hepatotoxicity/chronic liver injury due to various reasons is a major metabolic disorder affecting individuals of all ages (Dhiman & Chawla, 2005). Likewise, many allopathy drugs are also known to cause liver damage (Dhiman et al., 2012; Martinez et al., 2012). Traditionally, many herbal extracts were used as hepato-protective agents in the ancient Indian and Chinese medicine systems (Dhiman et al., 2012; Dhiman & Chawla, 2005). The digestion-stimulating effect of fenugreek may emerge from its hepato-protective role. In 2006, Kaviarasan et al. reported that alcohol treatment in human Chang liver cells suppressed the Chang liver cells' growth, induced cytotoxicity, oxygen-radical formation, and mitochondrial dysfunction compared with normal cells. Simultaneously, incubation of cells with a polyphenolic extract of fenugreek seeds and alcohol significantly elevated cell viability in a concentration-dependent manner, reduced lactate dehydrogenase leakage, and these cytoprotective effects of fenugreek were compared with that of a hepato-protective agent, silymarin. In addition, Kaviarasan & Anuradha in 2007 further demonstrated the hepato-protective effect of fenugreek seed using chronic ethanol-induced hepatic injury in rats. Its administration restored the altered levels of liver function enzymes, bilirubin, and gamma-glutamyl transferase and decreased liver glycogen. Also, it alleviates the alterations in detoxification and alcohol-metabolizing enzymes such as aldehyde dehydrogenase (ALDH) and alcohol dehydrogenase (ADH), and the electron transport component cytochrome c reductase. Furthermore, *ex vivo* hepatocyte treatment increased viability and reduced apoptotic nuclei.

In obese rats fed with fenugreek supplemented diets, hepatic triglyceride level, and the soluble and bound forms of TNF-α protein significantly decreased compared to control rats (Raju & Bird, 2006). This indicates that fenugreek dietary administration could reduce the triglyceride accumulation in the liver, a hallmark feature of hepatic steatosis. The effect of a well-known insulin-mimetic agent, fenugreek-derived 4-hydroxy isoleucine, was observed on blood glucose and liver function in fructose-fed, streptozotocin-induced diabetic rats, and insulin-resistant rats (Haeri et al., 2009). Here the authors showed that in fructose-fed rats, liver damage marker enzymes AST and ALT and levels of glucose were significantly elevated compared to controls, and that was restored to near control values by 4-hydroxy isoleucine. Also, 4-hydroxy isoleucine improves HDL-cholesterol levels in diabetic rats. The control animals tolerated the prolonged 4-hydroxyisoleucine treatments without any alteration in the levels of glucose or liver damage markers. These results suggest liver-protecting properties of fenugreek-derived 4-hydroxyisoleucine besides its usefulness in insulin resistance (Haeri et al., 2009).

13.9 ANTIMICROBIAL EFFECTS OF FENUGREEK

Herbal extracts are becoming popular as natural antimicrobial preservatives or additives (Pazos et al., 2008; Cox et al., 2010). The antimicrobial activities of plant extracts may reside in various components, including aldehyde and phenolic compounds. Many studies have reported the antibacterial activity of fenugreek. Omolosa and Vagi (2001) reported strong activity of *T. foenum-graecum* against 26 bacterial pathogens. Randhir et al. (2004) observed a high antimicrobial activity against peptic ulcer-linked *Helicobacter pylori* in the fenugreek sprout extract. They hypothesized that in fenugreek sprouts, simple free phenolics that are less polymerized have more antimicrobial function. Phenolic-type antimicrobial agents have long been used for their antiseptic, disinfectant, or preservative properties (Hugo & Bloomfield, 1971). The methanol extract of fenugreek (*Trigonella foenum-graecum* L.) and coriander (*Coriandrum sativum* L.) revealed an elevated antimicrobial activity against *Pseudomonas* spp., *Escherichia coli*, *Shigella dysentiriae*, and *Salmonella typhi* (Dash et al., 2011). Also, it was reported that the fatty oil of fenugreek seeds showed very significant antimycotic activity against *Aspergillus niger* and *A. fumigates* (Wagh et al., 2007).

13.10 ANTIOXIDANT AND IMMUNOMODULATORY PROPERTIES OF FENUGREEK

Reactive oxygen and nitrogen species like hydroxyl radical (.OH), superoxide ($O_2^{.-}$) and hydrogen peroxide (H_2O_2), nitric oxide (NO.), peroxynitrite (ONOO-) are some of the powerful free radicals involved in causing damage to biological molecules (Sergio Di Meo et al., 2016) causing oxidative modification and even resulting in complete damage of cells. Antioxidants scavenge the radicals, protecting the cells from oxidation and therefore an increased intake of antioxidants would reduce the oxidative damage caused by the free radicals (Xu et al., 2017). Antioxidants are present in a wide variety of fruits, vegetables, and medicinal plants. The main bioactive compounds in plants studied for their antioxidant properties in plants are polyphenols that include flavonoids, anthocyanins, etc., carotenoids like xanthophyll and carotenes, and some vitamins like vitamin C. These compounds have also been shown to have anti-inflammatory, anti-aging, anti-microbial, and anti-cancer activities (Deng et al., 2012; Li et al., 2014; Xu et al., 2017).

The anti-diabetic (Al-Habori & Raman, 1998; Ribes et al., 1986) lipid-lowering properties (Sharma et al., 1990; Kumar et al., 2014) in fenugreek have also been correlated to its antioxidant levels in several studies (Belguith-Hadriche et al., 2010; Murthy et al., 2010; Weerasingha & Atapattu, 2013; Kumar et al., 2014) as reactive oxygen species are involved in the progression of pathological conditions in diabetes and atherosclerosis. Hence the antioxidant properties have also been studied for their beneficial effects (Dixit et al., 2005; Kaviarasan et al., 2007). The antioxidant property of fenugreek seed powder in a rat model of high fat and low-dose streptozotocin-induced diabetes showed that the treatment with fenugreek regulated the blood glucose levels and improved the antioxidant activities (SOD, catalase, GPx, GRd, and GSH) compared to untreated groups. The study also showed the ability of fenugreek to prevent oxidative damage of tissues caused by diabetes (Sankar et al., 2012). The agent that increases or reduces the immune responses is known as an immunomodulator. This effect is referred to as an immunomodulatory effect. To access the immunomodulatory activity, male albino mice were treated with three doses of aqueous extract of fenugreek 50, 100, 250 mg/kg of body weight, respectively, for 10 days. The increase in thymus weight was due to an increase in cell counts. This may be due to the plant extract's stimulatory effect on the lymphocytes and bone marrow hematopoietic cells (VanFurth, 1982).

Aqueous fenugreek extract was tried to ameliorate the additive urotoxicity of buthionine, sulfoximine, and cyclophosphamide by restoring the antioxidant status and reversing the cyclophosphamide-induced cell death in free radical-mediated lipid peroxidation in the urinary bladder (Ahmadiani et al., 2001). Activated neutrophils infiltrating sites of inflammation are an important source of oxygen-derived free radicals and many pro-inflammatory mediators, which in turn cause inflammatory reactions. Meanwhile the fenugreek mucilage can retain the antioxidant status, thereby reducing release of inflammatory mediators and recovering from both acute and chronic inflammation associated disorders (Sindhu et al., 2012, 2020). The production of oxygen free radicals during arthritis leads to decreased GPx, SOD, and GSH levels as a consequence of their consumption during oxidative stress and cellular lysis (Sindhu et al., 2012). The mucilage isolated from fenugreek blunted the depletion of glutathione content and SOD, probably by competing in scavenging for free radicals, and as a result helped to preserve the integrity of cellular membranes. Another possibility for oxidant/antioxidant imbalance may be the decreased concentration of vitamin C due to its direct free radical quenching activity (Mottonen et al., 1999). Moreover the mucilage supplementation increased the concentration of vitamin C due to the sparing effect of the antioxidant defense system. Another reason for anti-inflammatory mechanisms of vitamin C is inactivation of NF-κB during the inflammation process (Bowie & O'Neill, 2000).

13.11 ANTI-CATARACT ACTIVITY

A cataract is characterized as the opacification in the eye lens, and it remains the leading cause of the visual abnormality, also contributing to 50% of blindness worldwide (WHO, 2005). Several risk factors have been known in the pathogenesis of senile cataract. Despite aging, diabetes, smoking, gender, steroids, and nitric oxide are liable for the growth of cataracts (Kim, 2006). Diabetic patient lenses are also prone to damage by enzymes that would usually protect against destructive free radicals. The anti-cataract potential of *Trigonella foenum-graecum* was explored in selenite induced in vitro medium. During the study, both selenite and aqueous extract of T. *foenum-graecum* to the test group were administered with the medium. The fenugreek supplement group significantly restored the glutathione and decreased malondialdehyde levels in a dose-dependent manner (Madar & Arad, 1989). It was reported that, in the presence of selenite stress, antioxidant enzymes were reduced as compared with the average group (Kirtikar & Basu, 1991). Fenugreek also plays an important part in the restoration of antioxidant enzymes such as superoxide dismutase, glutathione, peroxidase, catalase, and glutathione-S-transferase. Moreover, a reduction in cataract incidence has been demonstrated in diabetic rats receiving an extract of fenugreek seeds and leaves (Vats et al., 2004). Another *in vivo* study Gupta et al. (2009) reported a marked delay in the onset and progression of galactose-induced cataracts in rats fed on a 2.5% fenugreek diet. Yet another *in vitro* and *in vivo* study evaluated the anti-cataract potential of *Trigonella foenum-graecum* (TF) aqueous extract on galactose induced cataracts in a rat model (Gupta et al., 2009).

13.12 CONCLUSION

Herbs have gained more attention in medicine due to the rising popular concern about their safety. Fenugreek is an herb that may hold promise in this regard. It has an extensive variety of actions which are likely to protect the human body against a number of abnormalities. High fiber, protein content, and other bioactive compounds make it a naturally health-promoting herb. Depending on human and animal studies, fenugreek had been considered by many researchers as a potential medicinal herb particularly as an anti-diabetic, hypolipidemic, and as an antioxidant agent. It has an influence on immune functions, being anticancer and antimicrobial, and having hepato-, gastro, and cardioprotective effects as well as anti-arthritic and vascular protective effects. However, further research is needed to explore the modern isolation techniques for bioactive components for the development of novel functional foods and drugs.

ACKNOWLEDGEMENT

Sindhu G, acknowledges DHR-Young Scientist Fellowship under scheme of Human Resource Development for Health Research, Govt. of India for her financial support.

REFERENCES

Abbas, N., Naz, M., and AlSulaim, M., 2016. A comparative study of analgesic, antipyretic and anti-inflammatory effect of ethanolic extract of Trigonella foenum-graecum with indomethacin and diclofenac sodium. *Journal of Pharmaceutical Research International*, 10(5), pp. 1–8.

Abdel-Barry, J.A., 2000. Hypoglycaemic effect of aqueous extract of the leaves of Trigonella foenum-graecum in healthy volunteers. *Eastern Mediterranean Health Journal*, 6(1), pp. 83–88.

Acharya, S.N., Blade, S., Mir, Z., and Moyer, J.R., 2007. Tristar fenugreek. *Canadian Journal of Plant Science*, 87(4), pp. 901–903.

Acharya, S.N., Thomas, J.E., and Basu, S.K., 2006. Fenugreek: An "old world" crop for the "new world". *Biodiversity*, 7(3–4), pp. 27–30.

Ahmadiani, A., Javan, M., Semnanian, S., Barat, E., and Kamalinejad, M., 2001. Anti-inflammatory and antipyretic effects of *Trigonella foenum-graecum* leaves extract in the rat. *Journal of Ethnopharmacology*, 75(2–3), pp. 283–286.

Al-Asadi, J.N., 2014. Therapeutic uses of fenugreek (Trigonella foenum-graecum L.). *American Journal of Social Issues and Humanities*. pp. 21–36 : Fenugreek Special Issue, Editors: S. K. Basu & G. Agoramoorthy. AJSIH I ISSN: 2276–6928..

Al-Habori, M., and Raman, A., 1998. Antidiabetic and hypocholesterolaemic effects of fenugreek. *Phytotherapy Research: An International Journal Devoted to Pharmacological and Toxicological Evaluation of Natural Product Derivatives*, 12(4), pp. 233–242.

Al-Habori, M., Raman, A., Lawrence, M.J., and Skett, P., 2001. In vitro effect of fenugreek extracts on intestinal sodium-dependent glucose uptake and hepatic glycogen phosphorylase A. *International Journal of Experimental Diabetes Research*, 2(2), pp. 91–9.

Al-Khateeb, E., Hamadi, S.A., Al-Hakeemi, A.A., Abu-Taha, M., and Al-Rawi, N., 2012. Hypoglycemic effect of trigonelline isolated from Iraqi fenugreek seeds in normal and alloxan-diabetic rabbits. *European Scientific Journal*, 8(30), pp. 16–24.

Al-Oqail, M.M., Farshori, N.N., Al-Sheddi, E.S., Musarrat, J., Al-Khedhairy, A.A., and Siddiqui, M.A., 2013. In vitro cytotoxic activity of seed oil of fenugreek against various cancer cell lines. *Asian Pacific Journal of Cancer Prevention*, 14(3), pp. 1829–1832.

Al-Rowais, N.A., 2002. Herbal medicine in the treatment of diabetes mellitus. *Saudi Medical Journal*, 23(11), pp. 1327–1331.

American Diabetes Association, 2009. Diagnosis and classification of diabetes mellitus. *Diabetes Care* 2009; 32: pp. 62–267.

Amin, A., Alkaabi, A., Al-Falasi, S., and Daoud, S.A., 2005. Chemopreventive activities of Trigonella foenum-graecum (Fenugreek) against breast cancer. *Cell Biology International*, 29(8), pp. 687–694.

Anonymous, 2013. *Herbs are special*. Fenugreek. Available from: <www.herbsarespecial.com.au/free-sprout-information/fenugreek.html>.

Ansari, M., Sadeghi, P., Mahdavi, H., Fattahi-Dolatabadi, M., Mohamadi, N., Asadi, A., and Sharififar, F., 2019. Fenugreek dermal patch, a new natural topical antinociceptive medication for relieving the postherniotomy pain, a double-blind placebo controlled trial. *Journal of Complementary and Integrative Medicine*, 16(3).

Aswathy, I. S., Santhi, K., Jasmine, P., Vidya, S., and Helen, A., 2019. Scientific validation of Anti-arthritic effect of Kashayams—A polyherbal formulation in collagen induced arthritic rats. *Journal of Ayurveda and Integrative Medicine*, S0975–9476(17), pp. 30742–30748.

Badimon, L., and Vilahur, G., 2014. Thrombosis formation on atherosclerotic lesions and plaque rupture. *Journal of Internal Medicine*, 276(6), pp. 618–632.

Bailey, C.J., and Day, C., 1989. Traditional plant medicines as treatments for diabetes. *Diabetes Care*, 12(8), pp. 553–564.

Baliga, M.S., Mane, P.P., Timothy, Nallemgera, J., Thilakchand, K.R., and Kalekhan, F., 2015. Chapter 5 — dietary spices in the prevention of rheumatoid arthritis: Past, Present, and Future. In *Foods and dietary supplements in the prevention and treatment of disease in older adults*. San Diego: Academic Press.

Baquer, N.Z., Kumar, P., Taha, A., Kale, R.K., Cowsik, S.M., and McLean, P., 2011. Metabolic and molecular action of Trigonella foenum-graecum (fenugreek) and trace metals in experimental diabetic tissues. *Journal of Biosciences*, 36(2), pp. 383–396.

Barnes, J., Anderson, L., and Phillipson, J. D., 2002. *Herbal medicine—a guide for health care professionals*. London: Pharmaceutical Press.

Basu, S.K., Thomas, J.E., and Acharya, S.N., 2007. Prospects for growth in global nutraceutical and functional food markets: A Canadian perspective. *Australian Journal of Basic and Applied Sciences*, 1, pp. 637–649.

Bawadi, H. A., Maghaydah, S. N., Tayyem, R. F., and Tayyem, R. F., 2009. The postprandial hypoglycemic activity of fenugreek seed and seeds' extract in type 2 diabetics. *A Pilot Study*, 5(18), pp. 134–138.

Belguith-Hadriche, O., Bouaziz, M., Jamoussi, K., El Feki, A., Sayadi, S., and Makni-Ayedi, F., 2010. Lipid-lowering and antioxidant effects of an ethyl acetate extract of fenugreek seeds in high-cholesterol-fed rats. *Journal of Agricultural and Food Chemistry*, 58(4), pp. 2116–2122.

Belguith-Hadriche, O., Bouaziz, M., Jamoussi, K., Simmonds, Monique S.J., El Feki, Abdelfttah, Makni-Ayedi, Fatma, 2013. Comparative study on hypocholesterolemic and antioxidant activities of various extracts of fenugreek seeds. *Food Chemistry*, 138(2–3), pp.1448–1453.

Bellentani, S., Saccoccio, G., Costa, G., Tiribelli, C., Manenti, F., Sodde, M., Saveria Crocè, L., Sasso, F., Pozzato, G., Cristianini, G., and Brandi G., 1997. Drinking habits as cofactors of risk for alcohol induced liver damage. *The Dionysos Study Group. Gut*, 41, pp. 845–850.

Benayad, Z., Gómez-Cordovés, C., and Es-Safi, N.E., 2014. Characterization of flavonoid glycosides from fenugreek (*Trigonella foenum-graecum*) crude seeds by HPLC-DAD-ESI/MS analysis. International Journal of Molecular Sciences, 15(11), pp. 20668–20685.

Bera, T.K., Ali, K.M., Jana, K., Ghosh, A., and Ghosh, D., 2013. Protective effect of aqueous extract of seed of Psoralea corylifolia (Somraji) and seed of Trigonella foenum-graecum L.(Methi) in streptozotocin-induced diabetic rat: A comparative evaluation. *Pharmacognosy Research*, 5(4), pp. 277–285.

Boban, P.T., Nambisan, B., and Sudhakaran, P.R., 2006. Hypolipidaemic effect of chemically different mucilages in rats: A comparative study. *British Journal of Nutrition*, 96(6), pp. 1021–1029.

Boban, P.T., Nambisan, B., and Sudhakaran, P.R., 2009. Dietary mucilage promotes regression of atheromatous lesions in hypercholesterolemic rabbits. *Phytotherapy Research: An International Journal Devoted to Pharmacological and Toxicological Evaluation of Natural Product Derivatives*, 23(5), pp. 725–730.

Bowie, A.G., and O'Neill, L.A., 2000. Vitamin C inhibits NF-κB activation by TNF via the activation of p38 mitogen-activated protein kinase. *The Journal of Immunology*, 165(12), pp. 7180–7188.

Broca, C., Gross, R., Petit, P., Sauvaire, Y., Manteghetti, M., Tournier, M., Masiello, P., Gomis, R., and Ribes, G., 1999. 4-Hydroxyisoleucine: Experimental evidence of its insulinotropic and antidiabetic properties. *American Journal of Physiology-Endocrinology and Metabolism*, 277(4), pp. E617–E623.

Catherine, U., Ethan, B., Dilys, B., Lisa, C., Edzard, E., Nicole, G., Ivo, F., Paul, H., Sadaf, H., Grace, K., Michelle, M., Siddhartha, M., Michael, S., David, S., Shaina, T.C., Nazhiyath, V., and Wendy, W., 2008. Fenugreek (*Trigonella foenum-graecum* L. Leguminosae): An evidence-based systematic review by the natural standard research collaboration. *Journal of Herbal Pharmacotherapy*, 7(3–4), pp. 143–177.

Chen, P.S., Shih, Y.W., Huang, H.C., and Cheng, H.W., 2011. Diosgenin, a steroidal saponin, inhibits migration and invasion of human prostate cancer PC-3 cells by reducing matrix metalloproteinases expression. *PLoS One*, 6(5), p. e20164.

Chertov, O., Yang, D., Howard, O., and Oppenheim, J.J., 2000. Leukocyte granule proteins mobilize innate host defenses and adaptive immune responses. *Immunological Reviews*, 177, pp. 68–78.

Chistiakov, D.A., Orekhov, A.N., and Bobryshev, Y.V., 2016. Links between atherosclerotic and periodontal disease. *Experimental and Molecular Pathology*, 100(1), pp. 220–235.

Cho, S.K., Kim, D., Won, S., Lee, J., Choi, C.B., Choe, J.Y., Hong, S.J., Jun, J.B., Kim, T.H., Koh, E., and Lee, H.S., 2019. Factors associated with time to diagnosis from symptom onset in patients with early rheumatoid arthritis. *The Korean Journal of Internal Medicine*, 34(4), pp. 910–916.

Chopra, ASaluja, M., and Tillu, G., 2010. Ayurveda-modern medicine interface: A critical appraisal of studies of Ayurvedic medicines to treat osteoarthritis and rheumatoid arthritis. *Journal of Ayurveda and Integrative Medicine*, 1(3), pp. 190–198.

Cox, S., Abu-Ghannam, N., and Gupta, S., 2010. An assessment of the antioxidant and antimicrobial activity of six species of edible Irish seaweeds. *International Food Research Journal*, 17(1), pp. 205–220.

Das, S., Dey, K.K., Dey, G., Pal, I., Majumder, A., Maiti Choudhury, S., and Mandal, M., 2012. Antineoplastic and apoptotic potential of traditional medicines thymoquinone and diosgenin in squamous cell carcinoma. *PLoS One*, 7(10), p. e46641.

Dash, B., Sultana, S., and Sultana, N., 2011. Antibacterial activities of methanol and acetone extracts of fenugreek (Trigonella foenum) and coriander (Coriandrum sativum). *Life Sciences and Medicine Research*, pp. 1–8.

DeFronzo, R.A., Bonadonna, R.C., and Ferrannini, E., 1992. Pathogenesis of NIDDM: A balanced overview. *Diabetes Care*, 15(3), pp. 318–368.

Deng, G.F., Shen, C., Xu, X.R., Kuang, R.D., Guo, Y.J., Zeng, L.S., Gao, L.L., Lin, X., Xie, J.F., Xia, E.Q., and Li, S., 2012. Potential of fruit wastes as natural resources of bioactive compounds. *International Journal of Molecular Sciences*, 13(7), pp. 8308–8323.

Devasena, T., and Menon, P.V., 2007. Fenugreek seeds modulate 1, 2-dimethylhydrazine-induced hepatic oxidative stress during colon carcinogenesis. *Italian Journal of Biochemistry*, 56(1), pp. 28–34.

Dhiman, A., Nanda, A., and Ahmad, S., 2012. A recent update in research on the antihepatotoxic potential of medicinal plants. *Zhong Xi Yi Jie He Xue Bao*, 10, pp. 117–127.

Dhiman, R.K., and Chawla, Y.K., 2005. Herbal medicines for liver diseases. *Digestive Diseases and Sciences*, 50, pp. 1807–1812.

Dixit, P., Ghaskadbi, S., Mohan, H., and Devasagayam, T.P., 2005. Antioxidant properties of germinated fenugreek seeds. *Phytotherapy Research: An International Journal Devoted to Pharmacological and Toxicological Evaluation of Natural Product Derivatives*, 19(11), pp. 977–983.

Eldin, I.M.T., Abdalmutalab, M.M., and Bikir, H.E., 2013. An in vitro anticoagulant effect of Fenugreek (Trigonella foenum-graecum) in blood samples of normal Sudanese individuals. *Sudanese Journal of Paediatrics*, 13(2), pp. 52–56.

Ferrero-Miliani, L., Nielsen, O., Andersen, P., and Girardin, S., 2007. Chronic inflammation: Importance of NOD2 and NALP3 in interleukin-1β generation. *Clinical and Experimental Immunology*, 147, pp. 227–235.

Gad, M.Z., El-Sawalhi, M. M., Ismail, M. F., and El-Tanbouly, N. D., 2006. Biochemical study of the antidiabetic action of the Egyptian plants fenugreek and balanites. *Molecular and Cellular Biochemistry*, 281(1–2), pp. 173–183.

Gauttam, V.K., and Kalia, A.N., 2013. Development of polyherbal antidiabetic formulation encapsulated in the phospholipids vesicle system. *Journal of Advanced Pharmaceutical Technology & Research*, 4(2), pp. 108–117.

Gikas, E., Bazoti, F.N., Papadopoulos, N., Alesta, A., Economou, G., and Tsarbopoulos, A., 2011. Quantitation of the flavonols quercetin and kaempferol in the leaves of Trigonella foenum-graecum by high-performance liquid chromatography—diode array detection. *Analytical Letters*, 44(8), pp. 1463–1472.

Grennan, D.M., Gray, J., Loudon, J., and Fear, S., 2001. Methotrexate and early postoperative complications in patients with rheumatoid arthritis undergoing elective orthopaedic surgery. *Annals of the Rheumatic Diseases*, 60(3), pp. 214–217.

Gupta, A., Gupta, R., and Lal, B., 2001. Effect of Trigonella foenum-graecum (Fenugreek) seeds on glycaemic control and insulin resistance in type 2 diabetes. *Journal of Associations of Physicians India*, 49, pp. 1057–1061.

Gupta, L.M., and Raina, R., 1998. Side effects of some medicinal plants. *Current Science*, 75(9), pp. 897–900.

Gupta, S.K., Kalaiselvan, V., Srivastava, S., Saxena, R., and Agrawal, S.S., 2009. Inhibitory effect of Trigonella foenum-graecum on galactose induced cataracts in a rat model; in vitro and in vivo studies. *Journal of Ophthalmic & Vision Research*, 4(4), pp. 213–219.

Haeri, M.R., Izaddoost, M., Ardekani, M.R., Nobar, M.R., and White, K.N., 2009. The effect of fenugreek 4-hydroxyisoleucine on liver function biomarkers and glucose in diabetic and fructose-fed rats. *Phytotherapy Research*, 23, pp. 61–64.

Hämäläinen, M., Nieminen, R., Vuorela, P., Heinonen, M., and Moilanen, E., 2007. Anti-inflammatory effects of flavonoids: Genistein, kaempferol, quercetin, and daidzein inhibit STAT-1 and NF-κB activations, whereas flavone, isorhamnetin, naringenin, and pelargonidin inhibit only NF-κB activation along with their inhibitory effect on iNOS expression and NO production in activated macrophages. *Mediators of Inflammation*, p. 45673.

Han, Y., Nishibe, S., Noguchi, Y., and Jin, Z., 2001. Flavonol glycosides from the stems of Trigonella foenum-graecum. *Phytochemistry*, 58, pp. 577–580.

Handa, T., Yamaguchi, K., Sono, Y., and Yazawa, K., 2005. Effects of fenugreek seed extract in obese mice fed a high-fat diet. *Bioscience, Biotechnology, and Biochemistry*, 69, pp. 1186–1188.

Havsteen, B.H., 2002. The biochemistry and medical significance of the flavonoids. *Pharmacology & Therapeutics*, 96(2–3), pp. 67–202.

Hugo, W. B., and Bloomfield, S.F., 1971. Studies on the mode of action of the phenolic antibacterial agent fentichlor against *Staphylococcus aureus* and *Escherichia coli* II. The effects of fentichlor on the bacterial membrane and the cytoplasmic constituents of the cell. *Journal of Applied Bacteriology*, 34(3), pp. 579–591.

Jarald, E., Joshi, S.B., and Jain, D., 2008. Diabetes and herbal medicines. *Iranian Journal of Pharmacology and Therapeutics*, 7(1), pp. 97–100.

Javan, M., Ahmadiani, A., Semnanian, S., and Kamalinejad, M., 1997. Antinociceptive effects of Trigonella foenum-graecum leaves extract. *Journal of Ethnopharmacology*, 58(2), pp. 125–129.

Jorge, L.L., Feres, C.C., and Teles, V.E., 2010. Topical preparations for pain relief: Efficacy and patient adherence. *Journal of Pain Research*, 4, pp. 11–24.

Joshi, D.V., Patil, R.R., and Naik, S.R., 2014. Hydroalcohol extract of *Trigonella foenum-graecum* seed attenuates markers of inflammation and oxidative stress while improving exocrine function in diabetic rats. *Pharmaceutical Biology*, 53(2), pp. 201–211.

Kaviarasan, S., and Anuradha, C.V., 2007. Fenugreek (Trigonella foenumgraecum) seed polyphenols protect liver from alcohol toxicity: A role on hepatic detoxification system and apoptosis. *Pharmazie*, 62, pp. 299–304.

Kaviarasan, S., Naik, G.H., Gangabhagirathi, R., Anuradha, C.V., and Priyadarsini, K.I., 2007. In vitro studies on antiradical and antioxidant activities of fenugreek (Trigonella foenum graecum) seeds. *Food Chemistry*, 103, pp. 31–37.

Kaviarasan, S., Ramamurty, N., Gunasekaran, P., Elango, Varalakshmi, and Anuradha, C.V., 2006. Fenugreek (Trigonella foenum-graecum) seed extract prevents ethanol-induced toxicity and apoptosis in Chang liver cells. *Alcohol Alcoholism*, 41, pp. 267–273.

Kawabata, T., Cui, M.-Y., Hasegawa, T., Takano, F., and Ohta, T., 2011. Anti-inflammatory and anti-melanogenic steroidal saponin glycosides from fenugreek (*Trigonella foenum-graecum* L.) seeds. *Planta Medica*, 77(7), pp. 705–710.

Khan, W.A., and Assiri, A.S., 2011. Immunochemical studies on catechol-estrogen modified plasmid: Possible role in rheumatoid arthritis. *Journal of Clinical Immunology*, 31(1), pp. 22–29.

Kim, J.S., Ju, J.B., and Choi, C.W., 2006. Hypoglycemic and antihyperlipidemic effect of four Korean medicinal plants in alloxan induced diabetic rats. *American Journal of Biochemistry and Biotechnology*, 2(4), pp. 54–160.

Kirtikar, K.R., and Basu, B.D., 1991. *Indian medicinal plants*. Allahabad: Lalit Mohan Basu, pp. 700–701.

Kiss, R., Pesti-Asboth, G., Szarvas, M.M., Stundl, L., Cziaky, Z., Hegedus, C., Kovacs, D., Badale, A., Mathe, E., Szilvassy, Z., and Remenyik, J., 2019. Diosgenin and its fenugreek-based biological matrix affect insulin resistance and anabolic hormones in a rat-based insulin resistance model. *Biomed Research International*, p. 7213913.

Koli, N.R., and Ramkrishna, K., 2002. Frequency and spectrum of induced mutations and mutagenic effectiveness and efficiency in fenugreek (Trigonella foenum-graecum L.). *Indian Journal of Genetics*, 62(4), pp. 365–366.

Kramer, W., 2015. Antilipidemic drug therapy today and in the future. In *Metabolic control* (pp. 373–435). Cham: Springer.

Kumar, P., Bhandari, U., and Jamadagni, S., 2014. Fenugreek seed extract inhibit fat accumulation and ameliorates dyslipidemia in high fat diet-induced obese rats. *BioMed Research International*, 2014, 606021.

Lawrence, T., 2009. The nuclear factor NF-κB pathway in inflammation. *Cold Spring Harbor Perspectives in Biology*, 1(6), p. 1651.

Li, A.N., Li, S., Li, H.B., Xu, D.P., Xu, X.R., and Chen, F., 2014. Total phenolic contents and antioxidant capacities of 51 edible and wild flowers. *Journal of Functional Foods*, 6, pp. 319–330.

Li, F., Fernandez, P.P., Rajendran, P., Hui, K.M., and Sethi, G., 2010. Diosgenin, a steroidal saponin, inhibits STAT3 signaling pathway leading to suppression of proliferation and chemosensitization of human hepatocellular carcinoma cells. *Cancer Letters*, 292(2), pp. 197–207.

Libby, P., 2007. Inflammatory mechanisms: The molecular basis of inflammation and disease. *Nutrition Reviews*, 65, pp. S140–S146.

Lillioja, S., Mott, D.M., Howard, B.V., Bennett, P.H., Yki-Järvinen, H., Freymond, D., Nyomba, B.L., Zurlo, F., Swinburn, B., and Bogardus, C., 1988. Impaired glucose tolerance as a disorder of insulin action. *New England Journal of Medicine*, 318(19), pp. 1217–1225.

Madar, Z., and Arad, J., 1989. Effect of extract fenugreek on postprandial glucose levels in human diabetic subjects. *Nutrition Research*, 9(6), pp. 691–692.

Madar, Z., and Stark, A.H., 2002. New legume sources as therapeutic agents. *British Journal of Nutrition*, 88, pp. S287–S29.

Mahat, M.Y.A., Kulkarni, N.M., Vishwakarma, S.L., Khan, F.R., Thippeswamy, B.S., Hebballi, V., Adhyapak, A.A., Benade, V.S., Ashfaque, S.M., Tubachi, S., and Patil, B.M., 2010. Modulation of the cyclooxygenase pathway via inhibition of nitric oxide production contributes to the anti-inflammatory activity of kaempferol. *European Journal of Pharmacology*, 642(1–3), pp. 169–176.

Manivannan, J., Balamurugan, E., Silambarasan, T., and Raja, B., 2013. Diosgenin improves vascular function by increasing aortic eNOS expression, normalize dyslipidemia and ACE activity in chronic renal failure rats. *Molecular and Cellular Biochemistry*, 384(1–2), pp. 113–120.

Martinez, R.M., Nordt, S.P., and Cantrell, F.L., 2012. Prescription acetaminophen ingestions associated with hepatic injury and death. *Journal of Community Health*, 37, pp. 1249–1252.

McInnes, I.B., and Schett, G., 2007. Cytokines in the pathogenesis of rheumatoid arthritis. *Nature Reviews Immunology*, 7, pp. 429–442.

Medzhitov, R., Inflammation, 2010. New adventures of an old flame. *Cell*, 140, pp. 771–776.

Middleton, E., Kandaswami, C., and Theoharides, T.C., 2000. The effects of plant flavonoids on mammalian cells: Implications for inflammation, heart disease, and cancer. *Pharmacological Reviews*, 52(4), pp. 673–751.

Montgomery, J., 2009. *The potential of fenugreek (Trigonella foenum-graecum) as a forage for dairy herds in central Alberta*. MSc, University of Alberta, Edmonton, Alberta.

Moorthy, R., Prabhu, K.M., and Murthy, P.S., 1989. Studies on the isolation and effect of orally active hypoglycemic principle from the seeds of fenugreek (Trigonella foenum graecum). *Diabetes Bull*, 9, pp. 69–72.

Moorthy, R., Prabhu, K.M., and Murthy, P.S., 2010. Anti-hyperglycemic compound (GII) from fenugreek (Trigonella foenum-graecum Linn.) seeds, its purification and effect in diabetes mellitus. *IJEB*, 48(11), pp. 1111–1118.

Morcos, S.R., Elhawary, Z., and Gabrial, G.N., 1981. Protein-rich food mixtures for feeding the young in Egypt 1. Formulation. *Zeitschrift für Ernährungswissenschaft*, 20(4), pp. 275–282.

Möttönen, T., Hannonen, P., Leirisalo-Repo, M., Nissilä, M., Kautiainen, H., Korpela, M., Laasonen, L., Julkunen, H., Luukkainen, R., Vuori, K., and Paimela, L., 1999. Comparison of combination therapy with single-drug therapy in early rheumatoid arthritis: A randomised trial. *The Lancet*, 353(9164), pp. 1568–1573.

Moura, C.S., Abrahamowicz, M., Beauchamp, M.E., Lacaille, D., Wang, Y., Boire, G., Fortin, P.R., Bessette, L., Bombardier, C., Widdifield, J., and Hanly, J.G., 2015. Early medication use in new-onset rheumatoid arthritis may delay joint replacement: Results of a large population-based study. *Arthritis Research & Therapy*, 17(1), p. 197.

Murthy, P.S., Moorthy, R., Prabhu, K.M., and Puri, D., 2010. Anti-diabetic and cholesterol lowering preparation from fenugreek seeds. Diakron Pharmaceuticals Inc, U.S. Patent 7,815,946.

Nathan, C., and Ding, A., 2010. Non-resolving inflammation. *Cell*, 140, pp. 871–882.

Olsen, N.J., and Stein, C.M., 2004. New drugs for rheumatoid arthritis. *New England Journal of Medicine*, 351, pp. 937–938.

Ometto, F., Botsios, C., Raffeiner, B., Sfriso, P., Bernardi, L., Todesco, S., Doria, A., and Punzi, L., 2010. Methods used to assess remission and low disease activity in rheumatoid arthritis. *Autoimmunity Reviews*, 9(3), pp. 161–164.

Omoloso, A.D., and Vagi, J.K., 2001. Broad spectrum antibacterial activity of Allium cepa, Allium roseum, Trigonella foenum-graecum and Curcuma domestica. *Natural Product Sciences*, 7(1), pp. 13–16.

Pal, R., Kamal, P., and Nath, R., 2020. Role of nitric oxide (NO) in the modulation of anti-arthritic and anti-inflammatory effects of *Trigonella foenum-graecum* (fenugreek) ethanolic seed extract in rats. *The FASEB Journal*, 34(S1), pp. 1–1.

Pandian, R.S., Anuradha, C.V., and Viswanathan, P., 2002. Gastroprotective effect of fenugreek seeds (*Trigonella foenum graecum*) on experimental gastric ulcer in rats. *Journal of Ethnopharmacology*, 81, pp. 393–397.

Parvizpur, A., Ahmadiani, A., and Kamalinejad, M., 2004. Spinal serotonergic system is partially involved in antinociception induced by *Trigonella foenum-graecum* (TFG) leaf extract. *Journal of Ethnopharmacology*, 95(1), pp. 13–17.

Parvizpur, A., Ahmadiani, A., and Kamalinejad, M., 2006. Probable role of spinal purinoceptors in the analgesic effect of Trigonella foenum-graecum (TFG) leaves extract. *Journal of Ethnopharmacology*, 104(1–2), pp. 108–112.

Passano, P., 1995. The many uses of *Methi*. *Manushi*, 2, pp. 31–34.

Patel, D.K., Prasad, S.K., Kumar, R., and Hemalatha, S., 2012. An overview on antidiabetic medicinal plants having insulin mimetic property. *Asian Pacific Journal of Tropical Biomedicine*, 2(4), pp. 320–330.

Patil, S.P., Niphadkar, P.V., and Bapat, M.M., 1997. Allergy to fenugreek (*Trigonella foenum graecum*). *Annals of Allergy, Asthma & Immunology*, 78(3), pp. 297–300.

Patsenker, E., Stoll, M., Millonig, G., Agaimy, A., Wissniowski, T., Schneider, V., Mueller, S., Brenneisen, R., Seitz, H. K., Ocker, M., and Stickel, F., 2011. Cannabinoid receptor type I modulates alcohol-induced liver fibrosis. *Molecular Medicine*, 17, pp. 1285–1294.

Pazos, M., Alonso, A., Sánchez, I., and Medina, I., 2008. Hydroxytyrosol prevents oxidative deterioration in foodstuffs rich in fish lipids. *Journal of Agricultural and Food Chemistry*, 56(9), pp. 3334–3340.

Pournamdari, M., Mandegary, A., Sharififar, F., Zarei, G., Zareshahi, R., Asadi, A., and Mehdipour, M., 2018. Anti-inflammatory subfractions separated from acidified chloroform fraction of fenugreek seeds (Trigonella foenum-graecum L.). *Journal of Dietary Supplements*, 15(1), pp. 98–107.

Pundarikakshudu, K., 2019. *Chapter 4 — Antiinflammatory and antiarthritic activities of some foods and spices. in bioactive food as dietary interventions for arthritis and related inflammatory disease*. 2nd edition. Academic Press. DOI:10.1016/B978-0-12-813820-5.00004-0.

Pundarikakshudu, K., Deepak, H.S., Panchal, A.H., and Bhavsar, G.C., 2016. Anti-inflammatory activity of fenugreek (*Trigonella foenum-graecum* Linn) seed petroleum ether extract. *Indian Journal of Pharmacology*, 48(4), pp. 441–444.

Puri, D., Prabhu, K.M., and Murthy, P.S., 2002. Mechanism of action of ahypoglycemic principle isolated from fenugreek seeds. *Indian Journal of Physiology and Pharmacology*, 46(4), pp. 457–462.

Qamar, A.S., Zainab, R., Muhammad, H.A., Rizwan, S., Anum, I., Niaz, M., and Hafiz, U.U.R., 2020. Nutritional and therapeutic properties of fenugreek (*Trigonella foenum-graecum*): A review. *International Journal of Food Properties*, 23(1), pp. 1777–1791.

Raghuram, T.C., Sharma, R.D., Sivakumar, B., and Sahay, B.K., 1994. Effect of fenugreek seeds on intravenous glucose disposition in non-insulin dependent diabetic patients. *Phytotherapy Research*, 8(2), pp. 83–86.

Raju, J., and Bird, R.P., 2006. Alleviation of hepatic steatosis accompanied by modulation of plasma and liver TNF-?? levels by *Trigonella foenum-graecum* (fenugreek) seeds in Zucker obese (fa/fa) rats. *International Journal of Obesity*, 30(8), pp. 1298–1307.

Raju, J., Patlolla, J.M., Swamy, M.V., and Rao, C.V., 2004. Diosgenin, a steroid saponin of Trigonella foenum-graecum (Fenugreek), inhibits azoxymethane-induced aberrant crypt foci formation in F344 rats and induces apoptosis in HT-29 human colon cancer cells. *Cancer Epidemiology and Prevention Biomarkers*, 13(8), pp. 1392–1398.

Randhir, R., Lin, Y.T., Shetty, K., and Lin, Y.T., 2004. Phenolics, their antioxidant and antimicrobial activity in dark germinated fenugreek sprouts in response to peptide and phytochemical elicitors. *Asia Pacific Journal of Clinical Nutrition*, 13(3), pp. 295–307.

Rao, A.V., 2004. *Herbal cure for common diseases*. Diamond Pocket Books (P) Ltd. Adarsh Printers, Navin Shahdara Delhi.

Rao, Y.K., Fang, S.H., and Tzeng, Y.M., 2005. Antiinflammatory activities of flavonoids isolated from Caesalpinia pulcherrima. *Journal of Ethnopharmacology*, 10, pp. 249–253.

Ravichandiran, V., and Jayakumari, S., 2013. Comparative study of bioactive fraction of *Trigonella foenum-graecum* L. leaf and seed extracts for inflammation. *International Journal of Frontiers in Science and Technology*, 1(2), pp. 128–148.

Ribes, G., Da Costa, C., Loubatières-Mariani, M.M., Sauvaire, Y., and Baccou, J.C., 1987. Hypocholesterolaemic and hypotriglyceridaemic effects of subfractions from fenugreek seeds in alloxan diabetic dogs. *Phytotherapy Research*, 1(1), pp. 38–43.

Ribes, G., Sauvaire, Y., Costa, C.D., Baccou, J.C., and Loubatieres-Mariani, M.M., 1986. Antidiabetic effects of subtractions from fenugreek seeds in diabetic dogs. *Proceedings of the Society for Experimental Biology and Medicine*, 182(2), pp. 159–166.

Rishi, P., Parul, K., and Rajendra, N., 2020. Role of nitric oxide (NO) in the modulation of anti-arthritic and anti-inflammatory effects of *Trigonella foenum-graecum* (fenugreek) ethanolic seed extract in rats. *Pharmacology*, 34(Si), pp. 1–1.

Ritika., 2016. Trigonella foenum-graecum L. A review of its ethnobotany, pharmacology and phytochemistry. *International Journal of Advance Research in Science and Engineering*, 5, pp. 192–204.

Saima, Z., Abul Kalam, N., Divya, V., and Mohd, A., 2018. A review on therapeutic potentials of *Trigonella foenum-graecum* (fenugreek) and its chemical constituents in neurological disorders: Complementary roles to its hypolipidemic, hypoglycemic, and antioxidant potential. *Nutritional Neuroscience*, 21(8), pp. 539–545.

Saini, V., 2010. Molecular mechanisms of insulin resistance in type 2 diabetes mellitus. *World Journal of Diabetes*, 1(3), pp. 68–75.

Salih, N., and Al-Asadi, J.N., 2012. Herbal remedies use among diabetic patients in Nassyria, Iraq. *World Family Medicine Journal: Incorporating the Middle East. Journal of Family Medicine*, 99(316), pp. 1–7.

Sandberg, F., and Corrigan, D., 2001. *Natural remedies: Their origins and uses*. CRC Press. Taylor and Francis, London and New York.

Sankar, P., Subhashree, S., and Sudharani, S., 2012. Effect of Trigonella foenum-graecum seed powder on the antioxidant levels of high fat diet and low dose streptozotocin induced type II diabetic rats. *European Review for Medical and Pharmacological Sciences*, 16, pp. 10–17.

Sauvaire, Y., Petit, P., Broca, C., Manteghetti, M., Baissac, Y., Fernandez-Alvarez, J., Gross, R., Roye, M., Leconte, A., Gomis, R., and Ribes, G., 1998. 4-Hydroxyisoleucine: A novel amino acid potentiator of insulin secretion. *Diabetes*, 47(2), pp. 206–210.

Sergio Di, M., Tanea, T.R., Paola, V., and Victor, M.V., 2016. Role of ROS and RNS sources in physiological and pathological conditions. *Oxidative Medicine and Cellular Longevity*, p. 1245049.

Shakib, M.C., and Gabrial, S.G., 2010. Post-prandial responses to different bread products based on wheat, barley and fenugreek or ginger or both in healthy volunteers and their effect on the glycemic index of such products. *Journal of American Science*, 6(10), pp. 89–96.

Sharififara, F., Khazaelia, P., and Allib, N., 2009. In vivo evaluation of anti-inflammatory activity of topical preparations from fenugreek (*Trigonella foenum-graec*um L.) seeds in a cream base. *Iranian Journal of Pharmaceutical Research*, 5, pp. 157–162.

Sharma, N., Suresh, S., Debnath, A., and Jha, S., 2017. Trigonella seed extract ameliorates inflammation via regulation of the inflammasome adaptor protein, ASC. *Frontiers in Bioscience-Elite*, 9(2), pp. 246–257.

Sharma, R.D., 1986. Effect of fenugreek seeds and leaves on blood glucose and serum insulin responses in human subjects. *Nutrition Research*, 6(12), pp. 1353–1364.

Sharma, R.D., Raghuram, T.C., and Rao, N.S., 1990. Effect of fenugreek seeds on blood glucose and serum lipids in type I diabetes. *European Journal of Clinical Nutrition*, 44(4), pp. 301–306.

Shashikumar, J.N., Champawat, P.S., Mudgal, V.D., Jain, S.K., Deepak, S., and Mahesh, K., 2018. A review: Food, medicinal and nutraceutical properties of fenugreek (Trigonella Foenum-Graecum L.). *International Journal of Chemical Studies*, 6(2), pp. 1239–1245.

Shetty, A.K., and Salimath, P.V., 2009. Reno-protective effects of fenugreek (Trigonella foenum graecum) during experimental diabetes. *e-SPEN, the European e-Journal of Clinical Nutrition and Metabolism*, 4(3), pp. e137–e142.

Shishodia, S., and Aggarwal, B.B., 2006. Diosgenin inhibits osteoclastogenesis, invasion, and proliferation through the downregulation of Akt, I κ B kinase activation and NF-κ B-regulated gene expression. *Oncogene*, 25(10), pp. 1463–1473.

Shulman, G.I., Rothman, D.L., Jue, T., Stein, P., DeFronzo, R.A., and Shulman, R.G., 1990. Quantitation of muscle glycogen synthesis in normal subjects and subjects with non-insulin-dependent diabetes by 13C nuclear magnetic resonance spectroscopy. *New England Journal of Medicine*, 322(4), pp. 223–228.

Shyni, G.L., Sindhu, G., and Helen, A., 2015. Downregulation of inflammatory mediators and pro-inflammatory cytokines by alkaloids of Jeevaneeya Rasayana in adjuvant-induced arthritis. *Immunological Investigations*, 44(1), pp. 70–87.

Sindhu, G., Pushpan, Chitra K., Parvathy, S., Nambesan, Bala, and Helen, A., 2020. *Trigonella foenum-graecum* derived mucilage supplementation in diet alleviates the progression of atherosclerosis in high cholesterol diet-fed rabbit model by regulating inflammation. *Bioactive Carbohydrates and Dietary Fibre*, 24, p. 100246.

Sindhu, G., Ratheesh, M., Shyni, G.L., Nambisan, B., and Helen, A., 2012. Anti-inflammatory and antioxidative effects of mucilage of *Trigonella foenum-graecum* (Fenugreek) on adjuvant induced arthritic rats. *International Immunopharmacology*, 12(1), pp. 205–211.

Sindhu, G., Shyni, G.L., Pushpan, C.K., Nambisan, B., and Helen, A., 2018. Evaluation of anti-arthritic potential of *Trigonella foenum-graecum* L. (Fenugreek) mucilage against rheumatoid arthritis. *Prostaglandins & Other Lipid Mediators*, 138, pp. 48–53.

Sreeja, S., Anju, V.S., and Sreeja, S., 2010. In vitro estrogenic activities of fenugreek *Trigonella foenum-graecum* seeds. *Indian Journal of Medical Research*, 131(6), p. 814.

Srinivasan, K., 2006. Fenugreek (*Trigonella foenum-graecum*): A review of health beneficial physiological effects. *Food Reviews International*, 22(2), pp. 203–224.

Srinivasan, S., Koduru, S., Kumar, R., Venguswamy, G., Kyprianou, N., and Damodaran, C., 2009. Diosgenin targets Akt-mediated prosurvival signaling in human breast cancer cells. *International Journal of Cancer*, 125(4), pp. 961–967.

Sui, H., Nishimura, N., Yamamoto, A., Uemura, T., Okunishi, H., and Naora, K., 2012. Fenugreek seeds affect intestinal cholesterol transporters in caco-2 cells. *Shimane Journal of Medical Sciences*, 29, pp. 13–21.

Sumanth, M., Kapil, P., and Mihir, P., 2011. Screening of aqueous extract of *Trigonella foenum-graecum* seeds for its antiulcer activity. *International Journal of Research in Pharmaceutical and Biomedical Sciences*, 2(3), pp. 1085–1089.

Sur, P., Das, M., Gomes, A., Vedasiromoni, J.R., Sahu, N.P., Banerjee, S., Sharma, R.M., and Ganguly, D.K., 2001. *Trigonella foenum-graecum* (fenugreek) seed extract as an antineoplastic agent. *Phytotherapy Research: An International Journal Devoted to Pharmacological and Toxicological Evaluation of Natural Product Derivatives*, 15(3), pp. 257–259.

Suresh, P., Kavitha, C.N., Babu, S.M., Reddy, V.P., and Latha, A.K., 2012. Effect of ethanol extract of *Trigonella foenum-graecum* (Fenugreek) seeds on Freund's adjuvant-induced arthritis in albino rats. *Inflammation*, 35(4), pp. 1314–1321.

Szabó, K., Gesztelyi, R., Lampé, N., Kiss, R., Remenyik, J., Pesti-Asbóth, G., Priksz, D., Szilvássy, Z., and Juhász, B., 2018. Fenugreek (Trigonella foenum-graecum) seed flour and diosgenin preserve endothelium-dependent arterial relaxation in a rat model of early-stage metabolic syndrome. *International Journal of Molecular Sciences*, 19(3), p. 798.

Takeuchi, O., and Akira, S., 2010. Pattern recognition receptors and inflammation. *Cell*, 2010(140), pp. 805–820.

Umesh, C.S. Yadav, and Najma, Z.B., 2014. Pharmacological effects of *Trigonella foenum-graecum* L. in health and disease. *Pharmaceutical Biology*, 52(2), pp. 243–254.

Valette, G., Sauvaire, Y., Baccou, J.C., and Ribes, G., 1984. Hypocholesterolaemic effect of fenugreek seeds in dogs. *Atherosclerosis*, 50(1), pp. 105–111.

van der Linden, M.P., Le Cessie, S., Raza, K., van der Woude, D., Knevel, R., Huizinga, T.W., and van der Helm-van Mil, A.H., 2010. Long-term impact of delay in assessment of patients with early arthritis. *Arthritis & Rheumatism*, 62(12), pp. 3537–3546.

Van, Wijk, B.E., and Wink, M., 2005. *Medicinal plants of the world*. 1st edition. Pretoria: Briza.

VanFurth, R., 1982. Current view on the mononuclear phagocyte system. *Immunobiology*, 161(3–4), pp. 178–185.

Varjas, T., Nowrasteh, G., Budán, F., Horváth, G., Cseh, J., Gyöngyi, Z., Makai, S., and Ember, I., 2011. The effect of fenugreek on the gene expression of arachidonic acid metabolizing enzymes. *Phytotherapy Research*, 25(2), pp. 221–227.

Vats, V., Yadav, S.P., Biswas, N.R., and Grover, J.K., 2004. Anti-cataract activity of Pterocarpus marsupium bark and Trigonella foenum-graecum seeds extract in alloxan diabetic rats. *Journal of Ethnopharmacology*, 93(2–3), pp. 289–294.

Verma, N., Usman, K., Patel, N., Jain, A., Dhakre, S., Swaroop, A., Bagchi, M., Kumar, P., Preuss, H.G., and Bagchi, D., 2016. A multicenter clinical study to determine the efficacy of a novel fenugreek seed (Trigonella foenum-graecum) extract (Fenfuro™) in patients with type 2 diabetes. *Food & Nutrition Research*, 60(1), p. 32382.

Vijayakumar, M.V., Pandey, V., Mishra, G.C., and Bhat, M.K., 2010. Hypolipidemic effect of fenugreek seeds is mediated through inhibition of fat accumulation and upregulation of LDL receptor. *Obesity*, 18(4), pp. 667–674.

Vuorela, P., Leinonen, M., Saikku, P., Tammela, P., Rauha, J.P., Wennberg, T., and Vuorela, H., 2004. Natural products in the process of finding new drug candidates. *Current Medicinal Chemistry*, 11(11), pp. 1375–1389.

Vyas, A.S., Agrawal, R.P., Solanki, P., and Trivedi, P., 2008. Analgesic and anti-inflammatory activities of *Trigonella foenum-graecum* (seed) extract. *Acta Poloniae Pharmaceutica*, 65(4), pp. 473–476.

Vyas, A.S., Patel, N.G., Panchal, A.H., Patel, R.K., and Patel, M.M., 2010. Anti-arthritic and vascular protective effects of fenugreek, Boswellia serrata and acacia catechu alone and in combinations. *Pharma Science Monitor*, 1(2), pp. 95–111.

Wagh, P., Rai, M., Deshmukh, S.K., and Durate, M.C.T., 2007. Bio-activity of oils of Trigonella foenum-graecum and Pongamia pinnata. *African Journal of Biotechnology*, 6(13), pp. 1592–1596.

Warram, J.H., Martin, B.C., Krolewski, A.S., Soeldner, J.S., and Kahn, C.R., 1990. Slow glucose removal rate and hyperinsulinemia precede the development of type II diabetes in the offspring of diabetic parents. *Annals of Internal Medicine*, 113(12), pp. 909–915.

Weerasingha, A.S., and Atapattu, N.S.B.M., 2013. Effects of Fenugreek (Trigonella foenum-graecum L.) seed powder on growth performance, visceral organ weight, serum cholesterol levels and the nitrogen retention of broiler chicken. *Tropical Agricultural Research*, 24(3), pp. 289–295.

Weiser, M., Frishman, W.H., Michaelson, M.D., and Abdeen, M.A., 1997. The pharmacologic approach to the treatment of obesity. *Journal of Clinical Pharmacology*, 37, pp. 453–473.

WHO, 2005. Prevention of avoidable blindness and visual impairment, Provisional agenda item 4.9. EB117/35, 117th session 22nd December.

Xu, D.P., Li, Y., Meng, X., Zhou, T., Zhou, Y., Zheng, J., Zhang, J.J., and Li, H.B., 2017. Natural antioxidants in foods and medicinal plants: Extraction, assessment and resources. *International Journal of Molecular Sciences*, 18(1), p. 96.

Xue, W.L., Li, X.S., Zhang, J., Liu, Y.H., Wang, Z.L., and Zhang, R.J., 2007. Effect of Trigonella foenum-graecum (fenugreek) extract on blood glucose, blood lipid and hemorheological properties in streptozotocin-induced diabetic rats. *Asia Pacific Journal of Clinical Nutrition*, 16(Suppl 1), pp. 422–426.

Yadav, S.K., and Sehgal, S., 1997. Effect of home processing and storage on ascorbic acid and β-carotene content of bathua (Chenopodium album) and fenugreek (*Trigonella foenum graecum*) leaves. *Plant Foods for Human Nutrition*, 50(3), pp. 239–247.

Yadav, U.C., and Baquer, N.Z., 2014. Pharmacological effects of Trigonella foenum-graecum L. in health and disease. *Pharmaceutical Biology*, 52(2), pp. 243–254.

Yoshikawa, M., Murakami, T., Komatsu, H., Murakami, N., Yamahara, J., and Matsuda, H., 1997. Medicinal foodstuffs. IV. Fenugreek seed. (1): Structures of trigoneosides Ia, Ib, IIa, IIb, IIIa, and IIIb, new furostanol saponins from the seeds of Indian *Trigonella foenum-graecum* L. *Chemical and Pharmaceutical Bulletin*, 45(1), pp. 81–87.

Zhilenko, V.V., Zalietok, S.P., and Klenov, O.O., 2012. Effect of fenugreek on the growth of different genesis tumors. *Likars' ka Sprava*, 5, pp. 133–139.

Zhou, Y., Hong, Y., and Huang, H., 2016. Triptolide attenuates inflammatory response in membranous glomerulo-nephritis rat via downregulation of NF-κB signaling pathway. *Kidney and Blood Pressure Research*, 41, pp. 901–910.

14 Medicinal Potential of Fenugreek in Neuropathy and Neuroinflammation Associated Disorders

Aman Upaganlawar, Chandrashekhar Upasani and Mayur B. Kale

CONTENTS

14.1 Introduction 209
14.2 Fenugreek against Peripheral Neuropathy (PN) 211
    14.2.1 Anti-Inflammatory 211
    14.2.2 Antinociceptive 212
    14.2.3 Immunomodulator 213
    14.2.4 Antioxidant and Neuroprotective 214
14.3 Fenugreek against Diabetic Neuropathy (DN) 215
    14.3.1 Oxidative Stress and Neuroprotection in DN 215
    14.3.2 Inflammation and Pain in DN 216
14.4 Fenugreek against Chemotherapy-Induced Neuropathy 217
14.5 Fenugreek against Neuroinflammation of Neurodegenerative Disorders 217
    14.5.1 Fenugreek and Alzheimer's Disease (AD) 218
    14.5.2 Fenugreek and Parkinson Disease (PD) 218
    14.5.3 Fenugreek and Multiple Sclerosis 219
14.6 Conclusion 220
References 220

14.1 INTRODUCTION

Infection, tissue injury, or stress can trigger inflammatory responses, activating immune cells, enhancing the secretion of pro-inflammatory cytokines and reactive oxygen (Welcome 2020). Inflammatory responses, centralized within the brain and spinal cord, are generally referred to as "neuroinflammatory." Numerous pieces of evidence have shown that a range of diseases including diabetes; cardiovascular diseases; arthritis; CNS diseases such as neurodegenerative disorders, Alzheimer's disease (AD), and Parkinson's disease (PD); stroke; and depression (Aghadavod et al. 2016; Wang et al. 2016) can affect neurons to develop neuroinflammatory conditions. For example, inflammation plays a vital role in type 1 and type 2 diabetes (T1D and T2D) pathophysiology and associated metabolic disorders. Microglia, immunological cells of the CNS, play critical roles in mediating these neuroinflammatory responses (DiSabato, Quan, and Godbout 2016). The neuroinflammation is mediated by producing cytokines, chemokines, reactive oxygen species, and secondary messengers produced by resident CNS glia (microglia and astrocytes), endothelial cells, and peripherally derived immune cells (DiSabato, Quan, and Godbout 2016). Therefore, targeting inflammation and immunity to prevent and control the neuroinflammatory conditions is an effective strategy (Nasri and Shirzad 2013; Tsalamandris et al. 2019).

Although conventional synthetic drug treatments are available, the side effects limit their prolonged use. Therefore, natural treatment with potent efficacy with fewer or no side effects is gaining interest. The agents that can inhibit or prevent the secretion of pro-inflammatory immune and non-immune medicators and have antioxidant efficacy can be effective strategies for managing neuroinflammatory disorders. One such medicinal plant is fenugreek (*Trigonella foenum graceum* L., family: Fabaceae), which has been documented as having significant anti-inflammatory and antioxidant potential reported in traditional medicines (Srinivasan 2006). For centuries, fenugreek has been used as a spice and herb, a valuable source of protein, a leafy vegetable, and in traditional medicine to prevent and treat many disorders such as diabetes, pain, cancer, atherosclerosis, cardiovascular complications, and hypertension (Max 1992).

Fenugreek is documented as an important raw material for ayurvedic medicines for managing inflammation and pain-related disorders. Many polyherbal ayurvedic formulations, which the ayurvedic practitioner regularly prescribes for inflammatory disorders and joint pain in India, contain fenugreek (Balkrishna et al. 2019). Furthermore, IRA-01, an ayurvedic formulation rich in fenugreek seed extract, was reported with a significant reduction in joint pain, and swelling during three months, in a multicenter, double-blind, placebo-controlled study in 130 patients followed by a nine-months single-center, open-label study in 78 patients with rheumatoid arthritis (Chopra et al. 2004; Chopra, Saluja, and Tillu 2010). After three months, IRA-01 treated patients showed a significant fall in rheumatoid factor (RF) titer with 60% of patients showing ACR20 (20% improvement) as defined by the American College of Rheumatology (ACR). The open-label study reported significant improvement in ACR scores such as joint pain, swelling, and Health Assessment Questionnaire (HAQ) scores with 80% patients fulfilling ACR20 criteria and 40% patients showing ACR50 (50% improvement) (Pincus 2005).

A plethora of studies on animals and human patients indicated the potential efficacy of fenugreek seed extract as antidiabetic, hypocholesterolemic, analgesic, neuroprotective, and antioxidative, which can be helpful for the management of PN (Bahmani et al. 2016; Pal and Mukherjee 2020). The health benefits and disease prevention effects of fenugreek are attributed to the presence of diverse phytochemicals, including dietary fibers, amino acids, alkaloids, flavonoids, glycosides, and saponins (Rampogu et al. 2018; Salarbashi, Bazeli, and Fahmideh-Rad 2019; Sarwar et al. 2020; Shinde, Sancheti, and Prajapati 2015; Vígh et al. 2017; Wani and Kumar 2018; Zameer et al. 2018). Some of the significant fenugreek phytoconstituents, as reported recently (Wani and Kumar 2018), are tabulated in Table 14.1).

TABLE 14.1

Summary of Major Phytoconstituents in Fenugreek

Class	Phytoconstituents
Saponins	Yamogenin, diosgenin, smilagenin, sarsasapogenin, tigogenin, neotigogenin, gitogenin, neogitogenin, yuccagenin, saponaretin, graecunins, fenugrin B, fenugreekine, trigofoenosides A–G.
Alkaloids	Trimethylamine, neurin, trigonelline, choline, gentianine, carpaine, and betain.
Flavonoids	Quercetin, rutin, vitexin, isovitexin.
Fibers	Gum, neutral detergent fiber.
Amino acids	Isoleucine, 4-hydroxyisoleucine, histidine, leucine, lysine, L-tryptophan, arginine.
Other	Coumarin, lipids, vitamins, minerals. 28% mucilage; 22% proteins; 5% of stronger-smelling, bitter fixed oil.

Neuropathy and Neuroinflammation

At the same time, some of these fenugreek phytoconstituents were reported having both anti-inflammatory and antioxidant potential in various modern scientific studies as follows:

- **The steroidal saponin glycosides:**
  - Anti-inflammatory: (Kawabata et al. 2011; Madhava Naidu et al. 2011; Uemura et al. 2010; Zhu, Tang, and Su 2018)
  - Antioxidant: (Ahmad et al. 2016; Jagadeesan et al. 2012)
- **Fenugreek seed mucilage or gum:**
  - Anti-inflammatory: (Mishra et al. 2021; Sindhu et al. 2012, 2018)
  - Antioxidant: (Mishra et al. 2021; Sindhu et al. 2012)
- **4-hydroxyisoleucine (an amino acid):**
  - Anti-inflammatory: (Gautam et al. 2016; Maurya et al. 2014; Narayanaswamy, Wai, and Esa 2017; Yang et al. 2021; Zhou et al. 2020)
  - Antioxidant: (Dutta, Ghosh, Mohan, Mishra et al. 2014; Ibarra et al. 2008)
- **Trigonelline (an alkaloid):**
  - Anti-inflammatory: (Chowdhury et al. 2018; Khalili et al. 2018)
  - Antioxidant: (Chowdhury et al. 2018; Dutta, Ghosh, Mohan, Mishra et al. 2014; Dutta, Ghosh, Mohan, Thakurdesai et al. 2014; Hamadi 2012; Ibarra et al. 2008; Khalili et al. 2018; Yoshinari, Takenake, and Igarashi 2013)
- **Fenugreek seed oil**:
  - Anti-inflammatory: (Khaled et al. 2010; Pundarikakshudu et al. 2016)
  - Antioxidant: (Akbari et al. 2019; Ishtiaque et al. 2013)

Fenugreek is reported to reduce inflammation and related conditions by inhibiting pro-inflammatory mediators and antioxidant effects (Bae et al. 2012; Piao et al. 2017; Tavakoly et al. 2018; Uemura et al. 2010). Furthermore, many fenugreek extracts with different levels of these phytoconstituents have anti-inflammatory and antioxidant effects, making them suitable for managing neuroinflammatory conditions (Chauhan et al. 2010; Liu, Kakani, and Nair 2012). The present chapter reviews such scientific evidence of the efficacy of fenugreek and its constituents specifically against neuroinflammatory and neuropathy conditions with underlying cellular and molecular mechanisms.

14.2 FENUGREEK AGAINST PERIPHERAL NEUROPATHY (PN)

Peripheral neuropathy (PN) is the disease or dysfunction of peripheral nerves that often occurs outside the brain and spinal cord. PN is often classified according to the types or locations of nerves affected or causative disease. For example, neuropathy as a complication from diabetes is called diabetic neuropathy (DN). The neuropathy symptoms depend on the underlying cause and the individual but typically cause temporary or permanent numbness, tingling, prickling, or burning sensation, increased sensitivity to touch, pain, muscle weakness or wasting, paralysis, dysfunction in organs or glands, and impairment to urination and sexual functions.

Many natural medicines, including fenugreek, are explored as an option to manage PN by offering benefits against inflammation, pain and to provide neuroprotection or antioxidant properties. Several reports also suggested that fenugreek has anti-inflammatory, antipyretic, and antinociceptive properties in pain animal models of neuronal pain (Bhalke et al. 2009). This section of the chapter reviews present scientific evidence for the efficacy of fenugreek or its constituents against peripheral neuropathy through various mechanisms.

14.2.1 ANTI-INFLAMMATORY

Inflammation is considered a cardinal pathogenic mechanism for neuropathy. Thus, the anti-inflammatory approaches provide therapeutic targets for neuropathy development (Zhou and Zhou 2014).

The critical inflammatory molecules implicated in the development and progression of neuropathy include inflammatory cytokines, adhesion molecules, and chemokines.

The methanol extract of fenugreek seeds is reported to inhibit the production of inflammatory cytokines, tumor necrosis factor (TNF)-α, induced by phorbol 12-myristate 13-acetate in cultured THP-1 cells *in vitro* (Kawabata et al. 2011). The hexane, ethyl acetate, methanolic, and water extracts of fenugreek seeds and their isolates are reported to inhibit lipid peroxidation (LPO) and enzyme cyclooxygenase (COX) during *in vitro* assays (Liu, Kakani, and Nair 2012).

Anti-inflammatory activity of fenugreek leaves extract (1000 and 2000 mg/kg, intraperitoneal) was reported against the formalin-induced edema model in rats (Ahmadiani et al. 2001). Furthermore, the acute administration of partially purified fraction (10 and 20 mg/kg, intraperitoneal) demonstrated marked anti-inflammatory activity against the carrageenan-induced rat paw edema model (Vyas et al. 2008).

Amino acids, alkaloids, apigenin, and saponins isolated from fenugreek have displayed remarkable anti-inflammatory activity. Different extracts of fenugreek were reported to produce anti-inflammatory effects through different mechanisms (Ahmadiani et al. 2001; Morani et al. 2012; Pundarikakshudu et al. 2016; Vyas et al. 2008). The primary phytoconstituents contributing to the fenugreek's anti-inflammatory effect, 4-hydroxyisoleucine, was found to reverse the effects of TNF-α and normalize the glucose uptake (Gautam et al. 2016). The alkaloid and flavonoid fractions of fenugreek seeds inhibit pro-inflammatory enzymes, COXs, and lipoxygenases (Mandegary et al. 2012). The methanolic extract of fenugreek rich in saponins is reported to induce anti-inflammatory activity by inhibiting the production of inflammatory cytokines, such as IL-1β and TNF-α, and in cultured THP-1 cells *in vitro* (Kawabata et al. 2011). Fenugreek mucilage at a dose of 75 mg/kg reduces inflammatory enzymes COX, lipoxygenase, and myeloperoxidase (MPO) (Sindhu et al. 2012). The presence of linolenic and linoleic acids in fenugreek seeds is suggested to be responsible for significant anti-inflammatory activity (Pundarikakshudu et al. 2016). The linolenic and linoleic acids containing petroleum ether extract showed efficacy against acute (carrageenan-induced), subacute (formaldehyde-induced paw edema, and cotton pellets-induced granuloma), and sub-chronic (Freund's complete adjuvant-induced arthritis) model of inflammation (Pundarikakshudu et al. 2016).

In addition to these mechanisms, fenugreek also inhibits inflammation stimulated by insulin receptors. 4-Hydroxyisoleucine showed the inhibition on palmitate-induced production of reactive oxygen species (ROS) and associated inflammation, as the activation of NF-kappaB, JNK1/2, ERK1/2, and p38 MAPK was significantly reduced (Maurya et al. 2014). 4-Hydroxyisoleucine from fenugreek seeds is activated by AMP-activated protein kinase (AMPK) to enhance insulin-stimulated glucose transport rate in a dose-dependent manner (Gautam et al. 2016).

The steroidal saponin, diosgenin, present in the fenugreek also exhibits anti-inflammatory activity via altering inflammatory responses and reducing expression of Toll-like receptor 4 (TLR4), myeloid differentiation factor 88 (MyD88), nuclear factor-κB (NF-κB), transforming growth factor-β1 (TGF-β1), high-mobility group protein 1 (HMGB-1), interleukin-1 receptor-associated kinase 1 (IRAK1), and TNF-receptor-associated factor 6 (TRAF6) in the rat ischemic stroke model (Zhu, Tang, and Su 2018).

One of the mechanisms for peripheral neuropathy may include sciatic nerve demyelination and axonal loss due to inflammatory cytokines (Myers, Campana, and Shubayev 2006). Fenugreek was found to restore sciatic nerve properties with oligosaccharide-based (Preet et al. 2005; Thakurdesai et al. 2015) and trigonelline-based fenugreek seed extract in rats (Morani et al. 2012).

14.2.2 Antinociceptive

Neuropathic pain develops due to peripheral neuropathy and is usually challenging to treat due to complex and heterogeneous pathophysiological factors (Treede et al. 2008). The clinical features include the paradoxical combination of sensory loss in the affected area and hypersensitivity, such as allodynia and hyperalgesia (Ochoa 2009).

Several studies have reported the antinociceptive effects of fenugreek, suggesting a promising candidate for neuropathy treatment. In the earliest study, the antinociceptive potential of fenugreek leaves extract was reported during the tail-flick latency assay (at 1000 and 2000 mg/kg, intraperitoneal route) and formalin tests (at 300 mg, intraperitoneal route) (Javan et al. 1997). The anti-inflammatory and antinociceptive effects of alkaloid-rich fraction of methanol extract of fenugreek seeds were tested against formalin and carrageenan-induced paw edema in a dose-dependent manner (Mandegary et al. 2012). The researchers attributed the anti-inflammatory and analgesic effects of fenugreek seeds' alkaloid and flavonoid content (Mandegary et al. 2012). The antinociceptive effects of fenugreek against formalin-induced inflammation and pain were suggested to be through spinal serotonergic (5-HT) (Parvizpur, Ahmadiani, and Kamalinejad 2004) and purinoceptor systems (Parvizpur, Ahmadiani, and Kamalinejad 2006).

Some of the studies utilized the non-inflammatory animal models of pain such as acetic acid-induced writhing (chemically induced pain), hot-plate method (central medicated pain), and tail-flick model (spinal cord mediated pain). The partially purified fraction (10–40 mg/kg, oral) of the fenugreek seed extract was reported to have significant and dose-dependent analgesic effects against acetic acid-induced writhing (chemically induced pain) and hot-plate method (thermally induced pain) (Vyas et al. 2008). The petroleum ether, chloroform, ethyl acetate, and methanolic extracts of fenugreek leaves and seeds showed central and peripheral analgesic activity in a single dose of intraperitoneal administration (at 50 mg/kg) as evaluated in hot-plate-induced pain in mice (Bhalke et al. 2009).

Hydroalcoholic extract of fenugreek seeds (0.2, 2, and 20 mg/kg, oral for ten days) was reported to improve nerve healing, nerve conduction velocities, and voltage of action potential of the peripheral nerve against pyridoxine-induced peripheral neuropathy as shown during the tail-flick assay, electrophysiological and histological assays (Moghadam et al. 2013).

The subacute administration of standardized extract of fenugreek seeds (IDM-01) demonstrated excellent dose-dependent efficacy against established partial sciatic nerve ligation (PSNL) and sciatic nerve crush injury (SNCI) in rats (Morani et al. 2012). The fenugreek extract showed a significant reduction of pain threshold in thermal hyperalgesia (TH), improved the motor function test (MFT) score, and restored motor nerve conduction velocity (MNCV) within seven days of treatment in the SNCI model of peripheral neuropathic pain (Morani et al. 2012).

Diosgenin, the saponin glycoside constituent of many plants, including fenugreek seed, reported neuroprotective effects against allodynia and inflammatory mediator levels in rats against chronic contractile injury (CCI)-induced neuropathic pain (Zhao et al. 2017). In this study, diosgenin reversed CCI-induced changes, decreased mechanical and thermal withdrawal latency of rat paws, inhibited upregulated levels of the pro-inflammatory cytokines, TNF-α, interleukin (IL)-1β and IL-2, suppressed oxidative stress, and inhibited the expression of phosphorylated-p38 mitogen-activated protein kinase (MAPK) and NF-κB in the spinal cord (Zhao et al. 2017). Subsequently, seven days of oral supplementation of diosgenin following peripheral nerve injury induced by sciatic crushed nerve injury in rats showed pain control and functional recovery (Lee et al. 2018). Diosgenin treatment increased the sciatic function index value in walking track analysis, suppressed nerve injury-induced c-Fos expression in the relevant regions of the brain, and inhibited nerve injury-induced increase of brain-derived neurotrophic factor (BDNF), Tropomyosin receptor kinase B (TrkB), COX-2, and inducible nitric oxide synthase (iNOS) expressions (Lee et al. 2018).

Treatment with a major alkaloid of fenugreek seed, trigonelline (30 microM), showed an increased percentage of cells with neurites, enhancement of axonal extension, and suggested to promote the functional neurite outgrowth in human neuroblastoma SK-N-SH cells (Tohda et al. 1999).

14.2.3 Immunomodulator

Immune-mediated neuropathies are a diverse group of disorders caused by immune-mediated damage to peripheral nerves. Immune-mediated neuropathies most commonly occur when immunologic tolerance to key antigenic sites on myelin, axon, nodes of Ranvier, or ganglionic neurons is lost (Franssen and Straver 2013). Numerous pieces of evidence highlighted the role of the immune system in the initiation and maintenance of neuropathic pain (Watkins and Maier 2002).

Natural compounds from medicinal plants such as fenugreek have potential as therapeutic agents against neuropathies because of their antioxidant and immunomodulatory activities (Devasagayam and Sainis 2002). The oral treatment of aqueous extract of fenugreek (50, 100, and 250 mg/kg) for ten days was reported on the stimulatory effect on immune functions in male Swiss albino mice (Bin-Hafeez et al. 2003). Out of many fenugreek constituents, saponins are described as immunostimulant agents (Liu et al. 1995). Fenugreek seeds are rich sources of dietary fiber, which is suggested to have stimulatory effects on macrophages (Dakroury et al. 1986). Adding fenugreek leaves (100 grams per meal) to meals is reported to increase the total iron content in meals (3.24 mg to 9.12 mg) and facilitate hematopoietic stimulation in the bone marrow (Jonnalagadda and Seshdri 1994).

14.2.4 Antioxidant and Neuroprotective

Apart from inflammation, the oxidative damage-induced chronic stress in the brain includes increased production of 4-hydroxynonenal, results from LPO, which disturbs the integrity of the membrane and results in inhibition of Ca2+ATPase and MAO, which acts as a contributing factor in the progression of PN (Baquer et al. 2009).

Oxidative stress has been a major pathway for cellular injury in neuropathy, including damage to mitochondria and other cellular components (Tecilazich et al. 2013). Several factors contribute to ROS production in the brain that causes neuronal degeneration by augmenting the intracellular calcium ion level. Oxidative stress is implicated in many diseases, including diabetes, cancer, and neurological condition (Halliwell, Gutteridge, and Cross 1992). For example, enhanced glucose levels in diabetes provokes an abrupt increase in the free radical release, which can negatively impact synaptic plasticity and results in brain cell death and neurodegeneration (de la Monte 2017). Oxygen is a highly reactive atom that undergoes metabolism to generate reactive oxygen species (ROS) (Nishida et al. 2000). Mainly ROS include the hydroxyl radicals (OH), hydrogen peroxide (H_2O_2), and superoxide anions (O_2^-), which interact with various macromolecules, such as lipids, proteins, and nucleic acids, and thus participate in the pathophysiology of various diseases (Gutteridge and Professor Halliwell 1993). The oxidation and antioxidation are generally regulated in cells, but imbalance may lead to oxidative stress (Sies 1991). Therefore, the antioxidant compounds are believed to have an immunomodulating action and may help to bring back immune function to more optimum values (De La Fuente and Victor 2000).

When fenugreek seed powder was mixed into a standard diet of alloxan-treated diabetic rats, the animals showed a reversal of the disturbed antioxidant levels and peroxidative damage with moralization of SOD, GPx, and CAT levels (Genet, Kale, and Baquer 2002). The addition of ethanol extract fenugreek seed extract (rich in flavonoids and polyphenols) to glipalamide (antidiabetic drug) showed potent inhibition of ferrous sulfate (FeSo4), hydrogen peroxide (H_2O_2), and carbon tetrachloride (CCl_4)-induced LPO in lever tissues *in vitro* (Neha, Anand, and Sunanda 2019).

The phytochemical analysis of fenugreek seed extract revealed several constituents that possess antioxidant properties (Toda 2011). The antioxidant activity of fenugreek seeds husk and endosperm extract, rich in saponin and polyphenols, was reported (Naidu et al. 2011). A polyphenol-rich extract of fenugreek seeds protected human erythrocytes (RBCs) *ex vivo* of normal and diabetic subjects, against hydrogen peroxide (H_2O_2)-induced oxidative hemolysis and LPO to indicate potent antioxidant properties (Kaviarasan, Vijayalakshmi, and Anuradha 2004). Besides, polyphenol-rich fenugreek seeds extract showed the antioxidant (hydroxyl radicals (OH) scavenging and inhibition of hydrogen peroxide-induced lipid peroxidation in rat liver mitochondria *in vitro* (Kaviarasan et al. 2007). The fenugreek seed extract at high concentrations acted as a scavenger of 2,2'-diphenyl-1-picryl hydrazyl hydrate (DPPH), and 2,2'-azinobis 3-ethylbenzothiazoline-6-sulfonate (ABTS) radicals, suggesting antiradical activity (Kaviarasan et al. 2007).

Fenugreek seeds oil, containing many polyunsaturated fatty acids, acts as an antioxidant compound (Rekik et al. 2016). Fenugreek-incorporated silk fibroin nanofiber, prepared in four different ratios by the co-electrospinning method, reported having antioxidant properties during DPPH with good biocompatibility for topical wound-healing applications (Selvaraj and Fathima 2017). Edible

Neuropathy and Neuroinflammation

and biodegradable films from fenugreek-derived polysaccharide (FWEP) reinforced by poly vinyl alcohol (PVA) showed the antioxidant property *in vitro* during DPPH, hydroxyl, and ABTS radical-scavenging activity; reducing power assay; β-carotene bleaching method; and iron (Fe2+)-chelating activity (Feki et al. 2019).

The mechanisms responsible for antioxidant activity of fenugreek include decreased LPO, countering alteration in antioxidant molecules (Ravikumar and Anuradha 1999) (Kaviarasan et al. 2007), restoration of antioxidant molecules (Anuradha and Ravikumar 2001), a reversal of alterations in tissue antioxidant enzymes, peroxidative damage (Genet, Kale, and Baquer 2002), and antiradical effects (Kaviarasan et al. 2007).

14.3 FENUGREEK AGAINST DIABETIC NEUROPATHY (DN)

Diabetes mellitus is one of the progressive disorders that become life-threatening because of the risk of complications, including DN (Dixit et al. 2005; Gautam et al. 2016; Kashihara et al. 2010; Roberts 2011). Among the PN, diabetic neuropathy (DN) is a challenging and progressive condition (Callaghan et al. 2012) with a 30% pooled prevalence as reported from a recent systematic review and meta-analysis of 29 studies (Sun et al. 2020).

Fenugreek is a supplement to wheat and maize flour for breadmaking and a constituent of the general population's daily diet. Fenugreek seeds or extracts have shown to have anti-hyperglycemic effects in diabetes mellitus by various mechanisms like peripheral glucose utilization, insulin secretagogue actions, and the effect on the gum fiber in the intestines (Haeri et al. 2012; Kumar, Kale, and Baquer 2012b; Sankar, Subhashree, and Sudharani 2012; Tripathi, Uma, and Chandra 2010; Tripathi and Chandra 2009).

14.3.1 Oxidative Stress and Neuroprotection in DN

Several studies confirm that fenugreek's protective effects against DN's development and progression are mediated by alleviating renal oxidative stress and suppressing the TGF-??1/CTGF signaling pathway and many other mechanisms (Jin et al. 2014; Sayed, Khalifa, and Abd El-Latif 2012) (Figure 14.1).

Increased LPO and alterations in circulating antioxidants were observed in alloxan-induced diabetic rats (Anuradha and Ravikumar 2001; Ravikumar and Anuradha 1999). Supplementation of fenugreek seed powder in the diet to alloxan-induced diabetic rats is reported to reverse oxidative stress-induced changes to show effects such as lowered lipid peroxidation, increased glutathione,

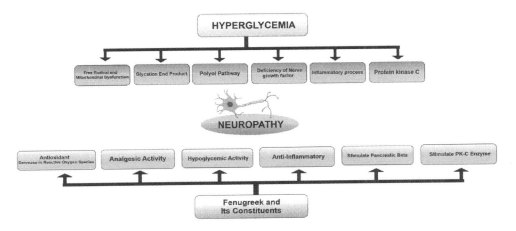

FIGURE 14.1 Major targets of fenugreek and its constituents in the treatment of diabetic neuropathy.

β-carotene, α-tocopherol content (Ravikumar and Anuradha 1999). In another similar study, fenugreek seed powder supplementation in the diet reported normalizing the alloxan-induced oxidative stress with reduction in peroxidation in the liver, kidney, and pancreas (Anuradha and Ravikumar 2001). In the same study, the aqueous extract of the seeds showed protective effects on the activity of calcium-dependent adenosinetriphosphatase (Ca2+ATPase) in liver homogenate in the presence of Fe2+/ascorbate *in vitro* (Anuradha and Ravikumar 2001).

Trigonelline, an alkaloid from fenugreek seeds, has neuroprotective effects against diabetic peripheral neuropathy induced by streptozotocin- and high-carbohydrate/high-fat diet-induced diabetic rats (Zhou and Zhou 2012). In this study, oral treatment of trigonelline (40 mg/kg) was reported to improve nerve conduction velocity, antioxidant enzyme activity, neuropathological changes of the sciatic nerve, decreasing LPO probably through the glucagon-like peptide-1 receptor/p38 mitogen-activated protein kinases signaling pathway (Zhou and Zhou 2012). Treatment with fenugreek and sodium orthovanadate and insulin was found to lower blood glucose levels, RESTORE LPO levels of membrane enzymes and neurolipofuscin accumulation, alter the activities of Ca2+ATPase, MAO, and normalize membrane fluidity in the brain tissues (Kumar et al. 2015). Accumulated findings indicated the role of fenugreek seed powder in reducing hyperglycemia in diabetes (Kumar, Kale, and Baquer 2012a), especially in the brain (Kumar, Kale, and Baquer 2012b), through reduction of LPO, neurolipofuscin in the deposition, intra-synaptosome Ca^{++} level, and restoration of synaptosome membrane fluidity (Zameer et al. 2018).

14.3.2 Inflammation and Pain in DN

The evidence also shows that diabetes-induced hyperalgesia is an attenuated leaf extract of fenugreek, and thus, it has potential benefit in painful diabetic neuropathy (Roghani et al. 2009). In this study, intraperitoneal injection of aqueous leaf extract of fenugreek (200 mg/kg every other day for one month) reported a lower nociceptive score during the formalin-induced nociceptive test in streptozotocin (STZ)-induced diabetic rats (Roghani et al. 2009).

Diabetic neuropathy is marked by various conditions such as demyelination, axonal degeneration, and atrophy with failed axonal regeneration. Multiple recent studies suggest that nerve growth factor (NGF) level reduction may have fundamental role pathogenesis of diabetic polyneuropathy (Pittenger and Vinik 2003). Diosgenin was reported to reverse functional and ultrastructural neuropathic changes (reduced disarrangement of the myelin sheath, increased area of myelinated axons), to induce neural regeneration to enhance nerve conduction velocity in a mouse model of diabetic neuropathy (Kang et al. 2011). In the same study, diosgenin treatment showed elevated NGF levels in sciatic nerve and neurite outgrowth in PC12 cells *in vitro* (Kang et al. 2011).

The subacute administration of low molecular weight galactomannans (LMWGAL) from fenugreek seed is reported to significantly reduce serum glucose level in alloxan-induced diabetic animals in a time-dependent manner, probably inhibiting pro-inflammatory cytokines and low-grade inflammation in diabetic conditions (Kamble et al. 2013).

The involvement of antioxidant and anti-inflammatory mechanisms in ameliorating diosgenin's potential, a saponin from fenugreek seeds, against streptozotocin (STZ)-induced diabetic neuropathic pain has been reported (Kiasalari et al. 2017). The subacute oral treatment of diosgenin (40 mg/kg daily for five weeks) increased mechanical and thermal injury thresholds with decreased pain scores against formalin tests in STZ-induced diabetic neuropathy (Kiasalari et al. 2017). Diosgenin was found to increase the mechanical and thermal nociceptive thresholds, and lowered pain score at a late phase of the formalin test, reduced the elevated levels of NF-κB, TNF α, interleukin 1β (IL-1β), and reduced the oxidative stress (Kiasalari et al. 2017).

Recently, the eight weeks of diosgenin treatment from fenugreek seeds to diabetic neuropathy induced by streptozotocin (STZ) in mice is reported to have dose-dependent efficacy in significantly alleviating the thermal hyperalgesia and paw tactile responses (Leng et al. 2020). In this study, diosgenin (50 and 100 mg/kg) showed a significant increase in the tail withdrawal latency during hot water tail-immersion test and increased tactile response in Von fray test, with reversal of STZ-induced morphological damage in sciatic nerves, improvement in antioxidant status (attenuated MDA levels, increased the activities of SOD and GP), and increased gene expression of the nuclear factor erythroid 2-related factor 2 (Nrf2), heme oxygenase-1 (HO-1), and (NAD(P)H Quinone Dehydrogenase 1) in the sciatic nerve of diabetic mice suggesting a role of antioxidant, cellular protection and prevention of inflammation in the mechanism of diosgenin against DN (Leng et al. 2020).

14.4 FENUGREEK AGAINST CHEMOTHERAPY-INDUCED NEUROPATHY

The use of neurotoxic chemotherapeutic agents, such as antimicrotubular agents, platinum compounds (cisplatin, carboplatin), vinca alkaloids, taxanes, can induce neuropathy by affecting motor or autonomic neurons, limiting these agents' use for treatment (Quasthoff and Hartung 2002). Repeated chemotherapy cycles damage ganglionic neurons and produce neuropathy symptoms, such as pain and paresthesia (Staff et al. 2017). Though the neuropathy mechanism is not precise, the chemotherapy shows selective vulnerability to cross the blood-nerve barrier near the dorsal root ganglion to produce neuropathic pain (Boyette-Davis, Walters, and Dougherty 2015). The complexity has hindered the understanding of the mechanisms underlying neuropathy. Despite intense research, the findings have not brought an effective mechanism-based treatment to prevent neuropathy induction and be beneficial for chemotherapy (Staff et al. 2017).

Current research implicates the different processes that affect cellular responses during chemotherapy that induce neuropathy, including elevated inflammatory cytokines that affect the neurons and glia, alteration in ion channels and neurotransmission, and oxidative stress-induced mitochondrial damage (Boyette-Davis, Walters, and Dougherty 2015).

The hydroalcoholic extract of fenugreek seed extract (0.2, 2 and 20 mg/kg, oral treatment for ten days) was reported to restore the function of nerve fibers and anti-neuropathic effects against pyridoxine-induced neuropathy in mice (Moghadam et al. 2013). Electrophysiological recordings of fenugreek extract-treated animals showed improved healing in terms of nerve conduction velocity, action potential, and pain in the tail-flick test (Moghadam et al. 2013).

14.5 FENUGREEK AGAINST NEUROINFLAMMATION OF NEURODEGENERATIVE DISORDERS

Neuroinflammation is a complex response to various mediators, pro-inflammatory molecules, and reactive oxygen species leading to tissue damage. Neuroinflammation is a prominent pathological feature of neurological disorders, characterized by activated microglia and infiltrating T-lymphocytes at neuronal injury (Skaper et al. 2018). In response to alterations induced in the innate immune system, T-lymphocytes, members of the adaptive immune system, infiltrate the CNS at sites of neuronal injury. Thus, active participation of inflammation in neurological disease pathogenesis and its contribution to the neurodegenerative pathology and tissue destruction is evident (Simon, Obst, and Gomez-Nicola 2019).

The uncontrolled neuroinflammatory response is noxious to neurons and may lead to disorders like AD (Bagyinszky et al. 2017; Fan and Pang 2017), Parkinson's disease (Rocha, De Miranda, and Sanders 2018), and multiple sclerosis (Koudriavtseva and Mainero 2016) via mobilization and activation of different types of resident cells within the central nervous system including microglia and astrocytes (McManus and Heneka 2017).

The pro-inflammatory response mediated by microglia secretes potent ROS such as superoxide radicals and nitric oxide, pro-inflammatory cytokines such as tumor necrosis factor-α (TNF-α), IL-6, and IL-1β, and neurotrophic factors (Wang et al. 2015). Other inflammatory mediators include the chemokine macrophage inflammatory protein-1α (MIP-1α), interferon-γ (IFN-γ), and compounds such as lipopolysaccharide (LPS) (Appel, Beers, and Zhao 2015).

Diosgenin, a major saponin phytoconstituent of oily embryos of fenugreek (Nishina, Verstuyft, and Paigen 1990), was reported to inhibit the inflammatory mediators such as TNF-α, NFκB,

P65, COX in the brain, during atherogenic-diet-induced neuroinflammation (Binesh, Devaraj, and Halagowder 2018).

The alkaloid constituent of fenugreek, trigonelline, has neuroprotective, anti-inflammatory, and antioxidant properties (Zhou, Chan, and Zhou 2012). The beneficial effect of trigonelline against lipopolysaccharide (LPS)-induced neuroinflammation was reported (Khalili et al. 2018). In this study, trigonelline (20, 40, or 80 mg/kg/day for one week) demonstrated antioxidant (lowered hippocampal MDA and acetylcholinesterase (AChE) activity, improved SOD, catalase, and GSH) and anti-inflammatory (depressed hippocampal NF-κB, toll-like receptor 4 (TLR4), and TNFα) effects in LPS-challenged rats (Khalili et al. 2018).

Myeloperoxidase (MPO) has been reported to play an important role in neuroinflammation following neurodegenerative diseases such as ischemic stroke (Pravalika et al. 2018). Trigonelline (100 mg/kg), post-ischemia administration, demonstrated excellent neuroprotection to the middle cerebral artery and related neuroinflammation (inhibition of glutathione-mediated MPO, cerebral infarction, motor, and neurological development deficits) (Pravalika et al. 2019).

14.5.1 Fenugreek and Alzheimer's Disease (AD)

The neuroprotective effect of fenugreek seed powder against AlCl3 induced memory deficits, amyloid and tau pathology, oxidative stress, and inflammation in AlCl3-induced rat models of AD. In this study, the 5% fenugreek seed powder in rat food to AlCl3-treated AD rats showed improved learning and memory impairments, reduced oxidative stress, and reversed alterations in the protein immunocontent patterns of IDE and CDK5 (enzymes involved in the metabolism of tau and amyloid proteins), pTau, GFAP, and Iba-1, IL-1β IL-6, TNFα, iNOS, NF-κB, COX-2, CDK5, BDNF, and STAT3 (Prema et al. 2017).

Furthermore, fenugreek saponin (with 0.05%, 0.1%, and 2%) supplemented AlCl3-induced AD rats to increase the AChEI and apoptosis activities and elevated the gene expression levels of Bax, Bcl2, and caspase-3 genes (Khalil, Roshdy, and Kassem 2016).

The protective effect of trigonelline (100 mg/kg, oral) against intracerebral Aβ(1–40) injected D rats was reported to improve cognition and is capable of alleviating neuronal loss through suppressing oxidative stress, astrocyte activity, and inflammation and also through preservation of mitochondrial integrity by ameliorating hippocampal levels of glial fibrillary acidic protein (GFAP), S100b, COX2, TNFα, and interleukin 6 (IL-6) (Fahanik-Babaei et al. 2019).

14.5.2 Fenugreek and Parkinson Disease (PD)

PD is a neurodegenerative disease with motor defects (Hirsch et al. 2016). PD pathophysiology includes the death of dopaminergic neurons in substantia nigra Pars compacta and the intracellular accumulation of Lewy bodies with alpha-synuclein protein (Appel, Beers, and Zhao 2015; Shulman, De Jager, and Feany 2011). The pathogenesis of PD also includes neuroinflammatory responses (Stojkovska, Wagner, and Morrison 2015; Wang, Tang, and Yenari 2007; Zhang et al. 2017). Alpha-synuclein protein contributes to PD progression by microglial activation via transcription factor NF-kB and pro-inflammatory cytokines like TNF-α, IL-1β (Kampinga et al. 2009). Genetic mutations lead to alpha-synuclein-induced microglial pathology and neuroinflammation due to microglial cells (Recchia et al. 2004; Schapansky, Nardozzi, and Lavoie 2015).

The effects of single-dose oral treatment of SFSE-T, trigonelline-based standardized fenugreek seed extract reported dose-dependent efficacy against 6-hydroxydopamine (6-OHDA)-induced unilateral lesions in rats, and 4-phenyl-1,2,3,6-tetrahydropyridine (MPTP)-induced neurodegeneration in C57BL/6 mice (Gaur et al. 2013). The neurotoxin MPTP increases the expression of pro-inflammatory cytokines and their receptors in the mouse brain (Lofrumento et al. 2011) and subsequent neuroinflammation (García-Domínguez et al. 2018). Acute and subacute administration of FSE-T (30 mg/kg, oral) showed significant reversal of MPTP-induced motor dysfunction, perhaps by

Neuropathy and Neuroinflammation

inhibiting neuroinflammation. Furthermore, the lesion induced by 6-OHDA to the nigrostriatal pathway is known to cause neuroinflammation and subsequent dopamine degeneration and is used as an animal model of PD (Cicchetti et al. 2002). The neuroprotective effect of trigonelline (50 and 100 mg/kg) was reported against unilateral nigrostriatal 6-OHDA-lesioned rats (Mirzaie et al. 2016). Trigonelline reversed the 6-OHDA-induced neuroinflammatory changes (reduced rotations, prevented the reduction of SNC neurons, prevented apoptosis, and restored the MDA level) in PD rats (Mirzaie et al. 2016).

14.5.3 Fenugreek and Multiple Sclerosis

Multiple sclerosis (MS) is the chronic, progressive, immune neuroinflammatory disease of the central nervous system (CNS), affecting over 2 million people worldwide (Reich, Lucchinetti, and Calabresi 2018). MS lesions appear in the entire CNS, easily recognized as demyelination, inflammation, and glial response lesions in the white matter, including spinal cord atrophy (Brownlee et al. 2017; Zhang et al. 2015).

Oligodendrocyte progenitor cells (OPC) differentiate into mature oligodendrocytes with remyelination. The impairment of this process is suggested to be a major reason for remyelination failure and disease such as multiple sclerosis (Kuhn et al. 2019). Diosgenin from fenugreek was reported to promote OPC differentiation without affecting migration, viability, and proliferation of rat primary oligodendrocyte progenitor cell culture *in vitro* (Xiao et al. 2012). In the same study, diosgenin was reported to significantly accelerate remyelination of cuprizone-induced demyelination in mice as shown by the increase in the number of mature oligodendrocytes in the corpus callosum without affecting the number of OPCs (Xiao et al. 2012). Subsequently, diosgenin was reported to alleviate experimental autoimmune encephalomyelitis (EAE) progression with reduced central nervous system inflammation and demyelination in a dose-dependent manner (Liu et al. 2017). In myelin oligodendrocyte glycoprotein EAE, diosgenin treatment significantly inhibited the activation of microglia and macrophages, suppressed CD4(+) T-cell proliferation, and hindered Th1/Th17 cell differentiation (Liu et al. 2017).

Some of the underlying mechanisms for fenugreek's efficacy against neuroinflammation and associated major neurological disorders are presented in Figure 14.2.

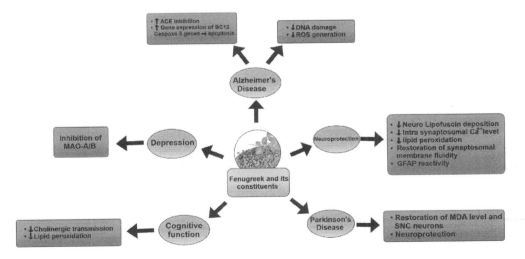

FIGURE 14.2 Possible mechanism of fenugreek and its constituents in the treatment of neuroinflammation and associated major neurological disorders.

14.6 CONCLUSION

The available experimental evidence suggested the therapeutic potential of fenugreek and constituents (e.g. diosgenin, trigonelline) to manage neuroinflammatory conditions such as peripheral neuropathy, diabetic neuropathy, and chemotherapeutic induced neuropathy, and neurovegetative diseases through antioxidant, neuroprotective, anti-inflammatory, antinociceptive, and immunomodulation mechanisms. However, further research on fenugreek towards clinical evaluation against neuroinflammatory disorders is needed.

REFERENCES

Aghadavod, E., S. Khodadadi, A. Baradaran, P. Nasri, M. Bahmani, and M. Rafieian-Kopaei. 2016. "Role of oxidative stress and inflammatory factors in diabetic kidney disease." *Iran J Kidney Dis* 10 (6):337–343.

Ahmad, R., T. Rahman, G. A. Froemming, and H. Nawawi. 2016. "Anti-oxidant properties of fenugreek and saponins, and their effects on eNOS expression on human coronary artery endothelial cells." *Atherosclerosis* 252:e158. doi: 10.1016/j.atherosclerosis.2016.07.763.

Ahmadiani, A., M. Javan, S. Semnanian, E. Barat, and M. Kamalinejad. 2001. "Anti-inflammatory and antipyretic effects of *Trigonella foenum-graecum* leaves extract in the rat." *J Ethnopharmacol* 75 (2–3):283–286. doi: 10.1016/S0378-8741(01)00187-8.

Akbari, S., N. H. Abdurahman, R. M. Yunus, O. R. Alara, and O. O. Abayomi. 2019. "Extraction, characterization and antioxidant activity of fenugreek (*Trigonella-Foenum graecum*) seed oil." *Materials Science for Energy Technologies* 2 (2):349–355. doi: 10.1016/j.mset.2018.12.001.

Anuradha, C. V., and P. Ravikumar. 2001. "Restoration on tissue antioxidants by fenugreek seeds (*Trigonella foenum graecum*) in alloxan-diabetic rats." *Indian J Physiol Pharmacol* 45 (4):408–420.

Appel, S. H., D. R. Beers, and W. Zhao. 2015. "Role of inflammation in neurodegenerative diseases." In *Neurobiology of Brain Disorders*, 380–395. Elsevier.

Bae, M. J., H. S. Shin, D. W. Choi, and D. H. Shon. 2012. "Antiallergic effect of *Trigonella foenum-graecum* L. extracts on allergic skin inflammation induced by trimellitic anhydride in BALB/c mice." *J Ethnopharmacol* 144 (3):514–22. doi: 10.1016/j.jep.2012.09.030.

Bagyinszky, E., V. V. Giau, K. Shim, K. Suk, S. S. A. An, and S. Kim. 2017. "Role of inflammatory molecules in the Alzheimer's disease progression and diagnosis." *J Neurol Sci* 376:242–254. doi: 10.1016/j.jns.2017.03.031.

Bahmani, M., H. Shirzad, M. Mirhosseini, A. Mesripour, and M. Rafieian-Kopaei. 2016. "A Review on Ethnobotanical and Therapeutic Uses of Fenugreek (*Trigonella foenum-graceum* L.)." *J Evid-Based Complementary Altern Med* 21 (1):53–62. doi: 10.1177/2156587215583405.

Balkrishna, A., R. Ranjan, S. S. Sakat, V. K. Sharma, R. Shukla, K. Joshi, R. Devkar, N. Sharma, S. Saklani, P. Pathak, P. Kumari, and V. R. Agarwal. 2019. "Evaluation of polyherbal ayurvedic formulation 'Peedantak Vati' for anti-inflammatory and analgesic properties." *J Ethnopharmacol* 235:361–374. doi: 10.1016/j.jep.2019.01.028.

Baquer, N. Z., A. Taha, P. Kumar, P. McLean, S. M. Cowsik, R. K. Kale, R. Singh, and D. Sharma. 2009. "A metabolic and functional overview of brain aging linked to neurological disorders." *Biogerontology* 10 (4):377–413. doi: 10.1007/s10522-009-9226-2.

Bhalke, R. D., S. Anarthe, K. Sasane, S. Satpute, S. Shinde, and V. Sangle. 2009. "Antinociceptive activity of *Trigonella foenum-graecum* leaves and seeds (Fabaceae)." *Iran J Pharmacol Ther* 8 (2):57–59.

Bin-Hafeez, B., R. Haque, S. Parvez, S. Pandey, I. Sayeed, and S. Raisuddin. 2003. "Immunomodulatory effects of fenugreek (*Trigonella foenum-graecum* L.) extract in mice." *Int Immunopharmacol* 3 (2):257–265. doi: 10.1016/S1567-5769(02)00292-8.

Binesh, A., S. N. Devaraj, and D. Halagowder. 2018. "Atherogenic diet induced lipid accumulation induced NFκB level in heart, liver and brain of Wistar rat and diosgenin as an anti-inflammatory agent." *Life Sci* 196:28–37. doi: 10.1016/j.lfs.2018.01.012.

Boyette-Davis, J. A., E. T. Walters, and P. M. Dougherty. 2015. "Mechanisms involved in the development of chemotherapy-induced neuropathy." *Pain Management* 5 (4):285–296. doi: 10.2217/pmt.15.19.

Brownlee, W. J., T. A. Hardy, F. Fazekas, and D. H. Miller. 2017. "Diagnosis of multiple sclerosis: Progress and challenges." *Lancet* 389 (10076):1336–1346. doi: 10.1016/s0140-6736(16)30959-x.

Callaghan, B. C., A. A. Little, E. L. Feldman, and R. A. C. Hughes. 2012. "Enhanced glucose control for preventing and treating diabetic neuropathy." *Cochrane Database Syst Rev* (6):CD007543. doi: 10.1002/14651858.cd007543.pub2.

Chauhan, G., M. Sharma, A. Varma, and H. Kharkwal. 2010. "Phytochemical analysis and anti-inflammatory potential of Fenugreek." *Med Plants - Int J Phytomed Relat Ind* 2 (1):39–44.

Chopra, A., M. Saluja, J. Patil, V. Anuradha, and L. Bichile. 2004. "IRA-01, an ayurvedic (Asian Indian) drug for rheumatoid arthritis (RA): Evaluation for efficacy and safety, and a probable lipid modifying effect." In *EULAR Annual European Congress of Rheumatology*, 9–12 June 2004.p. Abstract no: FRI0088.

Chopra, A., M. Saluja, and G. Tillu. 2010. "Ayurveda-modern medicine interface: A critical appraisal of studies of Ayurvedic medicines to treat osteoarthritis and rheumatoid arthritis." *J Ayurveda Integr Med* 1 (3):190–198. doi: 10.4103/0975-9476.72620.

Chowdhury, A. A., N. B. Gawali, R. Munshi, and A. R. Juvekar. 2018. "Trigonelline insulates against oxidative stress, proinflammatory cytokines and restores BDNF levels in lipopolysaccharide induced cognitive impairment in adult mice." *Metab Brain Dis* 33 (3):681–691. doi: 10.1007/s11011-017-0147-5.

Cicchetti, F., A. L. Brownell, K. Williams, Y. I. Chen, E. Livni, and O. Isacson. 2002. "Neuroinflammation of the nigrostriatal pathway during progressive 6-OHDA dopamine degeneration in rats monitored by immunohistochemistry and PET imaging." *Eur J Neurosci* 15 (6):991–998. doi: 10.1046/j.1460-9568.2002.01938.x.

Dakroury, A., A. A. El Galil, A. Darwish, and S. El Saadany. 1986. "Comparative analysis of total proteins, amino acids, carbohydrates, fibers, lipids, macro and micro minerals contents in different types of Egyptian bread." *Food/Nahrung* 30 (1):1–9.

De La Fuente, M., and V. M. Victor. 2000. "Anti-oxidants as modulators of immune function." *Immunology and Cell Biology* 78 (1):49–54. doi: 10.1046/j.1440-1711.2000.00884.x.

de la Monte, S. M. 2017. "Insulin resistance and neurodegeneration: Progress towards the development of new therapeutics for Alzheimer's disease." *Drugs* 77 (1):47–65. doi: 10.1007/s40265-016-0674-0.

Devasagayam, T. P. A., and K. B. Sainis. 2002. "Immune system and antioxidants, especially those derived from Indian medicinal plants." *Indian J Exp Biol* 40:639–655.

DiSabato, D. J., N. Quan, and J. P. Godbout. 2016. "Neuroinflammation: The devil is in the details." *J Neurochem* 139 (Suppl 2):136–153. doi: 10.1111/jnc.13607.

Dixit, P., S. Ghaskadbi, H. Mohan, and T. P. A. Devasagayam. 2005. "Antioxidant properties of germinated fenugreek seeds." *Phytother Res* 19 (11):977–983. doi: 10.1002/ptr.1769.

Dutta, M., A. K. Ghosh, V. Mohan, P. Mishra, V. Rangari, A. Chattopadhyay, T. Das, D. Bhowmick, and D. Bandyopadhyay. 2014. "Antioxidant mechanism (s) of protective effects of Fenugreek 4-hydroxyisoleucine and trigonelline enriched fraction (TF4H (28%)) Sugaheal against copper-ascorbate induced injury to goat cardiac mitochondria in vitro." *J Pharm Res* 8 (6):798–811.

Dutta, M., A. K. Ghosh, V. Mohan, P. Thakurdesai, A. Chattopadhyay, T. Das, D. Bhowmick, and D. Bandyopadhyay. 2014. "Trigonelline [99%] protects against copper-ascorbate induced oxidative damage to mitochondria: An in vitro study." *J Pharm Res* 8 (6):1694–1718.

Fahanik-Babaei, J., T. Baluchnejadmojarad, F. Nikbakht, and M. Roghani. 2019. "Trigonelline protects hippocampus against intracerebral Aβ(1–40) as a model of Alzheimer's disease in the rat: Insights into underlying mechanisms." *Metabolic Brain Disease* 34 (1):191–201. doi: 10.1007/s11011-018-0338-8.

Fan, L. W., and Y. Pang. 2017. "Dysregulation of neurogenesis by neuroinflammation: Key differences in neurodevelopmental and neurological disorders." *Neural Regen Res* 12 (3):366–371. doi: 10.4103/1673-5374.202926.

Feki, A., I. B. Amara, S. Bardaa, S. Hajji, N. Chabchoub, R. Kallel, T. Boudawara, S. Zghal, R. B. Salah, M. Nasri, and N. Ktari. 2019. "Preparation and characterization of polysaccharide based films and evaluation of their healing effects on dermal laser burns in rats." *Eur Polymer J* 115:147–156. doi: 10.1016/j.eurpolymj.2019.02.043.

Franssen, H., and D. C. Straver. 2013. "Pathophysiology of immune-mediated demyelinating neuropathies-part I: Neuroscience." *Muscle Nerve* 48 (6):851–64. doi: 10.1002/mus.24070.

García-Domínguez, I., K. Veselá, J. García-Revilla, A. Carrillo-Jiménez, M. A. Roca-Ceballos, M. Santiago, R. M. de Pablos, and J. L. Venero. 2018. "Peripheral inflammation enhances microglia response and nigral dopaminergic cell death in an in vivo MPTP model of Parkinson's disease." *Front Cell Neurosci* 12 (398). doi: 10.3389/fncel.2018.00398.

Gaur, V., S. L. Bodhankar, V. Mohan, and P. A. Thakurdesai. 2013. "Neurobehavioral assessment of hydroalcoholic extract of *Trigonella foenum-graecum* seeds in rodent models of Parkinson's disease." *Pharm Biol* 51 (5):550–557. doi: 10.3109/13880209.2012.747547.

Gautam, S., N. Ishrat, P. Yadav, R. Singh, T. Narender, and A. K. Srivastava. 2016. "4-Hydroxyisoleucine attenuates the inflammation-mediated insulin resistance by the activation of AMPK and suppression of SOCS-3 coimmunoprecipitation with both the IR-beta subunit as well as IRS-1." *Mol Cell Biochem* 414 (1–2):95–104. doi: 10.1007/s11010-016-2662-9.

Genet, S., R. K. Kale, and N. Z. Baquer. 2002. "Alterations in antioxidant enzymes and oxidative damage in experimental diabetic rat tissues: Effect of vanadate and fenugreek (*Trigonella foenum graecum*)." *Mol Cell Biochem* 236 (1–2):7–12. doi: 10.1023/a:1016103131408.

Gutteridge, J. M. C., and B. Professor Halliwell. 1993. "Invited review free radicals in disease processes: A compilation of cause and consequence." *Free Radical Res* 19 (3):141–158. doi: 10.3109/10715769309111598.

Haeri, M. R., H. K. Limaki, C. J. B. White, and K. N. White. 2012. "Non-insulin dependent anti-diabetic activity of (2S, 3R, 4S) 4-hydroxyisoleucine of fenugreek (*Trigonella foenum graecum*) in streptozotocin-induced type 1 diabetic rats." *Phytomedicine* 19 (7):571–574. doi: 10.1016/j.phymed.2012.01.004.

Halliwell, B., J. M. C. Gutteridge, and C. E. Cross. 1992. *Free Radicals, Antioxidants, and Human Disease: Where Are We Now?* Elsevier.

Hamadi, S. A. 2012. "Effect of trigonelline and ethanol extract of Iraqi Fenugreek seeds on oxidative stress in alloxan diabetic rabbits." *J. Assoc Arab Univ Basic Appl Sci.* 12 (1):23–26. doi: 10.1016/j.jaubas.2012.02.003.

Hirsch, L., N. Jette, A. Frolkis, T. Steeves, and T. Pringsheim. 2016. "The incidence of Parkinson's disease: A systematic review and meta-analysis." *Neuroepidemiology* 46 (4):292–300. doi: 10.1159/000445751.

Ibarra, A., K. He, N. Bai, A. Bily, M. Roller, A. Coussaert, N. Provost, and C. Ripoll. 2008. "Fenugreek extract rich in 4-hydroxyisoleucine and trigonelline activates PPARα and inhibits LDL oxidation: Key mechanisms in controlling the metabolic syndrome." *Natural Product Communications* 3 (9):1934578X0800300923. doi: 10.1177/1934578X0800300923.

Ishtiaque, S., N. Khan, M. A. Siddiqui, R. Siddiqi, and S. Naz. 2013. "Antioxidant potential of the extracts, fractions and oils derived from oilseeds." *Antioxidants (Basel)* 2 (4):246–256. doi: 10.3390/antiox2040246.

Jagadeesan, J., N. Nandakumar, T. Rengarajan, and M. P. Balasubramanian. 2012. "Diosgenin, a steroidal saponin, exhibits anticancer activity by attenuating lipid peroxidation via enhancing antioxidant defense system during NMU-induced breast carcinoma." *J Environ Pathol Toxicol Oncol* 31 (2):121–129. doi: 10.1615/jenvironpatholtoxicoloncol.v31.i2.40.

Javan, M., A. Ahmadiani, S. Semnanian, and M. Kamalinejad. 1997. "Antinociceptive effects of *Trigonella foenum-graecum* leaves extract." *J Ethnopharmacol* 58 (2):125–129. doi: 10.1016/s0378-8741(97)00089-5.

Jin, Y., Y. Shi, Y. Zou, C. Miao, B. Sun, and C. Li. 2014. "Fenugreek prevents the development of STZ-induced diabetic nephropathy in a rat model of diabetes." *Evid Based Complement Alternat Med* 2014:259368. doi: 10.1155/2014/259368.

Jonnalagadda, S. S., and S. Seshadri. 1994. "In vitro availability of iron from cereal meal with the addition of protein isolates and fenugreek leaves (Trigonella foenum-graecum)." *Plant Foods for Human Nutrition* 45 (2):119–125. doi: 10.1007/BF01088469.

Kamble, H., A. D. Kandhare, S. Bodhankar, V. Mohan, and P. Thakurdesai. 2013. "Effect of low molecular weight galactomannans from fenugreek seeds on animal models of diabetes mellitus." *Biomedicine and Aging Pathology* 3 (3):145–151. doi: 10.1016/j.biomag.2013.06.002.

Kampinga, H. H., J. Hageman, M. J. Vos, H. Kubota, R. M. Tanguay, E. A. Bruford, M. E. Cheetham, B. Chen, and L. E. Hightower. 2009. "Guidelines for the nomenclature of the human heat shock proteins." *Cell Stress and Chaperones* 14 (1):105–111. doi: 10.1007/s12192-008-0068-7.

Kang, T. H., E. Moon, B. N. Hong, S. Z. Choi, M. Son, J. H. Park, and S. Y. Kim. 2011. "Diosgenin from Dioscorea nipponica ameliorates diabetic neuropathy by inducing nerve growth factor." *Biol Pharm Bull* 34 (9):1493–1498. doi: 10.1248/bpb.34.1493.

Kashihara, N., Y. Haruna, V. K. Kondeti, and Y. S. Kanwar. 2010. "Oxidative stress in diabetic nephropathy." *Curr Med Chem* 17 (34):4256–4269. doi: 10.2174/092986710793348581.

Kaviarasan, S., G. H. Naik, R. Gangabhagirathi, C. V. Anuradha, and K. I. Priyadarsini. 2007. "In vitro studies on antiradical and antioxidant activities of fenugreek (*Trigonella foenum graecum*) seeds." *Food Chem* 103 (1):31–37. doi: 10.1016/j.foodchem.2006.05.064.

Kaviarasan, S., K. Vijayalakshmi, and C. V. Anuradha. 2004. "Polyphenol-rich extract of fenugreek seeds protect erythrocytes from oxidative damage." *Plant Foods for Human Nutr* 59 (4):143–147. doi: 10.1007/s11130-004-0025-2.

Kawabata, T., M. Y. Cui, T. Hasegawa, F. Takano, and T. Ohta. 2011. "Anti-inflammatory and anti-melanogenic steroidal saponin glycosides from Fenugreek (*Trigonella foenum-graecum* L.) seeds." *Planta Med* 77 (7):705–710. doi: 10.1055/s-0030-1250477.

Khaled, H., M. Hatem, C. Serge, and E. Abdalfatta. 2010. "Immunomodulatory, β-cell, and neuroprotective actions of fenugreek oil from alloxan-induced diabetes." *Immunopharmacol Immunotoxicol* 32 (November 2009):1–9. doi: 10.3109/08923970903490486.

Khalil, W. K. B., H. M. Roshdy, and S. M. Kassem. 2016. "The potential therapeutic role of Fenugreek saponin against Alzheimer's disease: Evaluation of apoptotic and acetylcholinesterase inhibitory activities." *J Applied Pharmaceutical Sci* 6 (9):166–173. doi: 10.7324/JAPS.2016.60925.

Khalili, M., M. Alavi, E. Esmaeil-Jamaat, T. Baluchnejadmojarad, and M. Roghani. 2018. "Trigonelline mitigates lipopolysaccharide-induced learning and memory impairment in the rat due to its antioxidative and anti-inflammatory effect." *Int Immunopharmacol* 61:355–362. doi: 10.1016/j.intimp.2018.06.019.

Kiasalari, Z., T. Rahmani, N. Mahmoudi, T. Baluchnejadmojarad, and M. Roghani. 2017. "Diosgenin ameliorates development of neuropathic pain in diabetic rats: Involvement of oxidative stress and inflammation." *Biomed Pharmacother* 86:654–661. doi: 10.1016/j.biopha.2016.12.068.

Koudriavtseva, T., and C. Mainero. 2016. "Neuroinflammation, neurodegeneration and regeneration in multiple sclerosis: Intercorrelated manifestations of the immune response." *Neural Regen Res* 11 (11):1727–1730. doi: 10.4103/1673-5374.194804.

Kuhn, S., L. Gritti, D. Crooks, and Y. Dombrowski. 2019. "Oligodendrocytes in development, myelin generation and beyond." *Cells* 8 (11). doi: 10.3390/cells8111424.

Kumar, P., R. K. Kale, and N. Z. Baquer. 2012a. "Antihyperglycemic and protective effects of *Trigonella foenum-graecum* seed powder on biochemical alterations in alloxan diabetic rats." *Eur Rev Med Pharmacol Sci* 16 (Suppl 3):18–27.

Kumar, P., R. Z. Kale, and N. Baquer. 2012b. "Antidiabetic and neuroprotective effects of *Trigonella foenum-graecum* seed powder in diabetic rat brain." *Prague Med Report* 113 (1):33–43. doi: 10.14712/23362936.2015.35.

Lee, B. K., C. J. Kim, M. S. Shin, and Y. S. Cho. 2018. "Diosgenin improves functional recovery from sciatic crushed nerve injury in rats." *J Exerc Rehabil* 14 (4):566–572. doi: 10.12965/jer.1836340.170.

Leng, J., X. Li, H. Tian, C. Liu, Y. Guo, S. Zhang, Y. Chu, J. Li, Y. Wang, and L. Zhang. 2020. "Neuroprotective effect of diosgenin in a mouse model of diabetic peripheral neuropathy involves the Nrf2/HO-1 pathway." *BMC Complement Med Ther* 20 (1):126. doi: 10.1186/s12906-020-02930-7.

Liu, J., S. Wang, H. Liu, L. Yang, and G. Nan. 1995. "Stimulatory effect of saponin from Panax ginseng on immune function of lymphocytes in the elderly." *Mech Ageing and Dev* 83 (1):43–53. doi: 10.1016/0047-6374(95)01618-A.

Liu, W., M. Zhu, Z. Yu, D. Yin, F. Lu, Y. Pu, C. Zhao, C. He, and L. Cao. 2017. "Therapeutic effects of diosgenin in experimental autoimmune encephalomyelitis." *J Neuroimmunol* 313:152–160. doi: 10.1016/j.jneuroim.2017.10.018.

Liu, Y., R. Kakani, and M. G. Nair. 2012. "Compounds in functional food fenugreek spice exhibit anti-inflammatory and antioxidant activities." *Food Chem* 131 (4):1187–1192. doi: 10.1016/j.foodchem.2011.09.102.

Lofrumento, D. D., C. Saponaro, A. Cianciulli, V. De Nuccio, V. Mitolo, G. Nicolardi, and M. A. Panaro. 2011. "MPTP-induced neuroinflammation increases the expression of pro-inflammatory cytokines and their receptors in mouse brain." *Neuroimmunomodulation* 18 (2):79–88. doi: 10.1159/000320027.

Madhava Naidu, M., B. N. Shyamala, J. Pura Naik, G. Sulochanamma, and P. Srinivas. 2011. "Chemical composition and antioxidant activity of the husk and endosperm of fenugreek seeds." *LWT—Food Sci Technol* 44 (2):451–456. doi: 10.1016/j.lwt.2010.08.013.

Mandegary, A., M. Pournamdari, F. Sharififar, S. Pournourmohammadi, R. Fardiar, and S. Shooli. 2012. "Alkaloid and flavonoid rich fractions of fenugreek seeds *(Trigonella foenum-graecum* L.) with antinociceptive and anti-inflammatory effects." *Food Chem Toxicol* 50 (7):2503–7. doi: 10.1016/j.fct.2012.04.020.

Maurya, C. K., R. Singh, N. Jaiswal, K. Venkateswarlu, T. Narender, and A. K. Tamrakar. 2014. "4-Hydroxyisoleucine ameliorates fatty acid-induced insulin resistance and inflammatory response in skeletal muscle cells." *Mol Cell Endocrinol* 395 (1–2):51–60. doi: 10.1016/j.mce.2014.07.018.

Max, B. 1992. "This and that: The essential pharmacology of herbs and spices." *Trends Pharmacol Sci* 13 (1):15–20. doi: 10.1016/0165-6147(92)90010-4.

McManus, R. M., and M. T. Heneka. 2017. "Role of neuroinflammation in neurodegeneration: New insights." *Alzheimers Res Ther* 9 (1):14. doi: 10.1186/s13195-017-0241-2.

Mirzaie, M., M. Khalili, Z. Kiasalari, and M. Roghani. 2016. "Neuroprotective and antiapoptotic potential of trigonelline in a striatal 6-hydroxydopamine rat model of Parkinson's disease." *Neurophysiology* 48 (3):176–183. doi: 10.1007/s11062-016-9586-6.

Mishra, P., A. K. Srivastava, T. C. Yadav, V. Pruthi, and R. Prasad. 2021. "Pharmaceutical and therapeutic applications of fenugreek gum." *Bioactive Natural Products for Pharmaceutical Applications*:379–408. doi: 10.1007/978-3-030-54027-2_11.

Moghadam, F. H., B. Vakili-Zarch, M. Shafiee, and A. Mirjalili. 2013. "Fenugreek seed extract treats peripheral neuropathy in pyridoxine induced neuropathic mice." *EXCLI J* 12:282.

Morani, A. S., S. L. Bodhankar, V. Mohan, and P. A. Thakurdesai. 2012. "Ameliorative effects of standardized extract from *Trigonella foenum-graecum* L. seeds on painful peripheral neuropathy in rats." *Asian Pacific J Tropical Med* 5 (5):385–390. doi: 10.1016/S1995-7645(12)60064-9.

Myers, R. R., W. M. Campana, and V. I. Shubayev. 2006. "The role of neuroinflammation in neuropathic pain: Mechanisms and therapeutic targets." *Drug Discovery Today* 11 (1–2):8–20. doi: 10.1016/S1359-6446(05)03637-8.

Naidu, M. M., B. Shyamala, J. P. Naik, G. Sulochanamma, and P. Srinivas. 2011. "Chemical composition and antioxidant activity of the husk and endosperm of fenugreek seeds." *LWT-Food Sci technol* 44 (2):451–456. doi: 10.1016/j.lwt.2010.08.013.

Narayanaswamy, R., L. K. Wai, and N. M. Esa. 2017. "Molecular docking analysis of phytic acid and 4-hydroxyisoleucine as cyclooxygenase-2, microsomal prostaglandin E synthase-2, tyrosinase, human neutrophil elastase, matrix metalloproteinase-2 and-9, xanthine oxidase, squalene synthase, nitric oxide synthase, human aldose reductase, and lipoxygenase inhibitors." *Pharmacognosy Magazine* 13 (Suppl 3):S512. doi: 10.4103/pm.pm_195_16.

Nasri, H., and H. Shirzad. 2013. "Toxicity and safety of medicinal plants." *J HerbMed Pharmacol* 2:21–22.

Neha, S., K. Anand, and P. Sunanda. 2019. "Administration of Fenugreek seed extract produces better effects in glibenclamide-induced inhibition in hepatic lipid peroxidation: An in vitro study." *Chin J Integr Med* 25 (4):278–284. doi: 10.1007/s11655-015-1793-z.

Nishina, P. M., J. Verstuyft, and B. Paigen. 1990. "Synthetic low and high fat diets for the study of atherosclerosis in the mouse." *J Lipid Res* 31 (5):859–869.

Nishida, P. M., Y. Maruyama, R. Tanaka, K. Kontani, T. Nagao, and H. Kurose. 2000. "Gα(i) and Gα(o) are target proteins of reactive oxygen species." *Nature* 408 (6811):492–495. doi: 10.1038/35044120.

Ochoa, J. L. 2009. "Neuropathic pain: Redefinition and a grading system for clinical and research purposes." *Neurology* 72 (14):1282–1283. doi: 10.1212/01.wnl.0000346325.50431.5f.

Pal, D., and S. Mukherjee. 2020. "Chapter 13 — Fenugreek (Trigonella foenum) seeds in health and nutrition." In *Nuts and Seeds in Health and Disease Prevention* (Second Edition), edited by Preedy, V.R. and Watson, R.R., 161–170. Academic Press.

Parvizpur, A., A. Ahmadiani, and M. Kamalinejad. 2004. "Spinal serotonergic system is partially involved in antinociception induced by *Trigonella foenum-graecum* (TFG) leaf extract." *J Ethnopharmacol* 95 (1):13–17. doi: 10.1016/j.jep.2004.05.020.

Parvizpur, A., A. Ahmadiani, and M. Kamalinejad. 2006. "Probable role of spinal purinoceptors in the analgesic effect of Trigonella foenum (TFG) leaves extract." *J Ethnopharmacol* 104 (1–2):108–112. doi: 10.1016/j.jep.2005.08.057.

Piao, C. H., T. T. Bui, C. H. Song, H. S. Shin, D. H. Shon, and O. H. Chai. 2017. "*Trigonella foenum-graecum* alleviates airway inflammation of allergic asthma in ovalbumin-induced mouse model." *Biochem Biophys Res Commun* 482 (4):1284–1288. doi: 10.1016/j.bbrc.2016.12.029.

Pincus, T. 2005. "The American College of Rheumatology (ACR) core data set and derivative" patient only" indices to assess rheumatoid arthritis." *Clin Exper Rheumatol* 23 (5):S109.

Pittenger, G., and A. Vinik. 2003. "Nerve growth factor and diabetic neuropathy." *Exper Diabesity Res* 4:271–285. doi: 10.1155/EDR.2003.271.

Pravalika, K., D. Sarmah, H. Kaur, M. Wanve, J. Saraf, K. Kalia, A. Borah, D. Yavagal, K. Dave, and P. Bhattacharya. 2018. "Myeloperoxidase and neurological disorder: A crosstalk." *ACS Chem Neurosci* 9. doi: 10.1021/acschemneuro.7b00462.

Pravalika, K., D. Sarmah, H. Kaur, K. Vats, J. Saraf, M. Wanve, K. Kalia, A. Borah, D. R. Yavagal, K. R. Dave, and P. Bhattacharya. 2019. "Trigonelline therapy confers neuroprotection by reduced glutathione mediated myeloperoxidase expression in animal model of ischemic stroke." *Life Sci* 216:49–58. doi: 10.1016/j.lfs.2018.11.014.

Preet, A., B. L. Gupta, M. R. Siddiqui, P. K. Yadava, and N. Z. Baquer. 2005. "Restoration of ultrastructural and biochemical changes in alloxan-induced diabetic rat sciatic nerve on treatment with Na3VO4 and Trigonella—a promising antidiabetic agent." *Molecular and Cellular Biochemistry* 278 (1–2):21–31. doi: 10.1007/s11010-005-7815-1.

Prema, A., A. Justin Thenmozhi, T. Manivasagam, M. Mohamed Essa, and G. J. Guillemin. 2017. "Fenugreek seed powder attenuated aluminum chloride-induced tau pathology, oxidative stress, and inflammation in a rat model of Alzheimer's disease." *J Alzheimer's Disease* 60 (s1):S209–S220. doi: 10.3233/JAD-161103.

Pundarikakshudu, K., D. H. Shah, A. H. Panchal, and G. C. Bhavsar. 2016. "Anti-inflammatory activity of fenugreek *(Trigonella foenum-graecum* Linn) seed petroleum ether extract." *Indian J Pharmacol* 48 (4):441. doi: 10.4103/0253-7613.186195.

Quasthoff, S., and H. P. Hartung. 2002. "Chemotherapy-induced peripheral neuropathy." *J Neurol* 249 (1):9–17. doi: 10.1007/PL00007853.

Rampogu, S., S. Parameswaran, M. R. Lemuel, and K. W. Lee. 2018. "Exploring the therapeutic ability of fenugreek against type 2 diabetes and breast cancer employing molecular docking and molecular dynamics simulations." *Evid-based Complement Alternat Med* 2018:1943203. doi: 10.1155/2018/1943203.

Ravikumar, P., and C. V. Anuradha. 1999. "Effect of fenugreek seeds on blood lipid peroxidation and antioxidants in diabetic rats." *Phytotherapy Res* 13 (3):197–201. doi: 10.1002/(SICI)1099-1573(199905)13:3<197::AID-PTR413>3.0.CO;2-L.

Recchia, A., P. Debetto, A. Negro, D. Guidolin, S. D. Skaper, and P. Giusti. 2004. "α-Synuclein and Parkinson's disease." *The FASEB Journal* 18 (6):617–626. doi: 10.1096/fj.03-0338rev.

Reich, D. S., C. F. Lucchinetti, and P. A. Calabresi. 2018. "Multiple Sclerosis." *N Engl J Med* 378 (2):169–180. doi: 10.1056/NEJMra1401483.

Rekik, D. M., S. B. Khedir, K. K. Moalla, N. G. Kammoun, T. Rebai, and Z. Sahnoun. 2016. "Evaluation of wound healing properties of grape seed, sesame, and fenugreek oils." *Evid-based Complement Alternat Med* 2016:7965689. doi: 10.1155/2016/7965689.

Roberts, K. T. 2011. "The potential of fenugreek (*Trigonella foenum-graecum*) as a functional food and nutraceutical and its effects on glycemia and lipidemia." *J Med Food* 14 (12):1485–1489. doi: 10.1089/jmf.2011.0002.

Rocha, E. M., B. De Miranda, and L. H. Sanders. 2018. "Alpha-synuclein: Pathology, mitochondrial dysfunction and neuroinflammation in Parkinson's disease." *Neurobiology of Disease* 109 (Pt B):249–257. doi: 10.1016/j.nbd.2017.04.004.

Roghani, M., M. Reza, V. Mahdavi, M. Khalili, and S. R. Miri. 2009. "The effect of fenugreek on nociceptive response in diabetic rats." *Basic and Clin Neurosci* 1 (1):22–25.

Salarbashi, D., J. Bazeli, and E. Fahmideh-Rad. 2019. "Fenugreek seed gum: Biological properties, chemical modifications, and structural analysis—A review." *Int J Biological Macromolecules* 138:386–393. doi: 10.1016/j.ijbiomac.2019.07.006.

Sankar, P., S. Subhashree, and S. Sudharani. 2012. "Effect of *Trigonella foenum-graecum* seed powder on the antioxidant levels of high fat diet and low dose streptozotocin induced type II diabetic rats." *Eur Rev Med Pharmacol Sci* 16 (Suppl. 3):10–17.

Sarwar, S., M. A. Hanif, M. A. Ayub, Y. D. Boakye, and C. Agyare. 2020. "Chapter 20 — Fenugreek." In *Medicinal Plants of South Asia*, edited by Hanif, M.A., Nawaz, H., Khan, M.M. and Byrne, H.J., 257–271. Elsevier.

Sayed, A. A. R., M. Khalifa, and F. F. Abd El-Latif. 2012. "Fenugreek attenuation of diabetic nephropathy in alloxan-diabetic rats: Attenuation of diabetic nephropathy in rats." *J Physiol Biochem* 68 (2):263–269. doi: 10.1007/s13105-011-0139-6.

Schapansky, J., J. D. Nardozzi, and M. J. Lavoie. 2015. "The complex relationships between microglia, alpha-synuclein, and LRRK2 in Parkinson's disease." *Neurosci* 302:74–88. doi: 10.1016/j.neuroscience.2014.09.049.

Selvaraj, S., and N. N. Fathima. 2017. "Fenugreek incorporated silk fibroin nanofibers a potential antioxidant scaffold for enhanced wound healing." *ACS Applied Materials & Interfaces* 9 (7):5916–5926. doi: 10.1021/acsami.6b16306.

Shinde, P. M., V. P. Sancheti, and A. D. Prajapati. 2015. "Phytochemical screening and study of antioxidant activity of fenugreek seed." *Int J Phytotherapy Res* 5 (1):1–13. doi: 10.15406/ppij.2019.07.00249.

Shulman, J. M., P. L. De Jager, and M. B. Feany. 2011. "Parkinson's disease: Genetics and pathogenesis." *Annu Rev Pathol: Mech of Disease* 6 (1):193–222. doi: 10.1146/annurev-pathol-011110-130242.

Sies, H. 1991. "Oxidative stress: From basic research to clinical application." *The American Journal of Medicine* 91 (3 SUPPL. 3):S31–S38. doi: 10.1016/0002-9343(91)90281-2.

Simon, E., J. Obst, and D. Gomez-Nicola. 2019. "The evolving dialogue of microglia and neurons in Alzheimer's disease: Microglia as necessary transducers of pathology." *Neurosci* 405:24–34. doi: 10.1016/j.neuroscience.2018.01.059.

Sindhu, G., M. Ratheesh, G. Shyni, B. Nambisan, and A. Helen. 2012. "Anti-inflammatory and antioxidative effects of mucilage of *Trigonella foenum-graecum* (Fenugreek) on adjuvant induced arthritic rats." *Int Immunopharmacol* 12 (1):205–211. doi: 10.1016/j.intimp.2011.11.012.

Sindhu, G., G. Shyni, C. K. Pushpan, B. Nambisan, and A. Helen. 2018. "Evaluation of anti-arthritic potential of *Trigonella foenum-graecum* L.(Fenugreek) mucilage against rheumatoid arthritis." *Prostaglandins & Other Lipid Mediators* 138:48–53. doi: 10.1016/j.prostaglandins.2018.08.002.

Skaper, S. D., L. Facci, M. Zusso, and P. Giusti. 2018. "An inflammation-centric view of neurological disease: Beyond the neuron." *Frontiers in Cellular Neuroscience* 12:12–72. doi: 10.3389/fncel.2018.00072.

Srinivasan, K. 2006. "Fenugreek (*Trigonella foenum-graecum*): A review of health beneficial physiological effects." *Food Rev Int* 22 (2):203–224. doi: 10.1080/87559120600586315.

Staff, N. P., A. Grisold, W. Grisold, and A. J. Windebank. 2017. "Chemotherapy-induced peripheral neuropathy: A current review." *Annals of Neurology* 81 (6):772–781. doi: 10.1002/ana.24951.

Stojkovska, I., B. M. Wagner, and B. E. Morrison. 2015. "Parkinson's disease and enhanced inflammatory response." *Exper Biology and Med* 240 (11):1387–1395. doi: 10.1177/1535370215576313.

Sun, J., Y. Wang, X. Zhang, S. Zhu, and H. He. 2020. "Prevalence of peripheral neuropathy in patients with diabetes: A systematic review and meta-analysis." *Primary Care Diabetes* 14 (5):435–444. doi: 10.1016/j.pcd.2019.12.005.

Tavakoly, R., M. R. Maracy, M. Karimifar, and M. H. Entezari. 2018. "Does fenugreek (*Trigonella foenum-graecum*) seed improve inflammation, and oxidative stress in patients with type 2 diabetes mellitus? A parallel group randomized clinical trial." *Eur J Integrative Med* 18:13–17. doi: 10.1016/j.eujim.2018.01.005.

Tecilazich, F., T. Dinh, T. E. Lyons, J. Guest, R. A. Villafuerte, C. Sampanis, C. Gnardellis, C. S. Zuo, and A. Veves. 2013. "Postexercise phosphocreatine recovery, an index of mitochondrial oxidative phosphorylation, is reduced in diabetic patients with lower extremity complications." *J Vascular Surgery* 57 (4):997–1005. doi: 10.1016/j.jvs.2012.10.011.

Thakurdesai, P., A. Kandhare, G. N. Zambare, S. Bodhankar, and V. Mohan. 2015. "Efficacy of oligosaccharides based standardized fenugreek seed extract in partial sciatic nerve ligation induced peripheral mononeuropathy in rats [NEU-46]." In 48th Annual Conference of Indian Pharmacological Society, IPSCON2015, Saurashtra University, Rajkot, India, Dec 28–20, 2005.p. S92.

Toda, S. 2011. "Polyphenol content and antioxidant effects in herb teas." *Chinese Medicine* 2 (01):29–31. doi: 10.4236/cm.2011.21005.

Tohda, C., N. Nakamura, K. Komatsu, and M. Hattori. 1999. "Trigonelline-induced neurite outgrowth in human neuroblastoma SK-N-SH cells." *Biol Pharm Bull* 22 (7):679–682. doi: 10.1248/bpb.22.679.

Treede, R. D., T. S. Jensen, J. N. Campbell, G. Cruccu, J. O. Dostrovsky, J. W. Griffin, P. Hansson, R. Hughes, T. Nurmikko, and J. Serra. 2008. "Neuropathic pain: Redefinition and a grading system for clinical and research purposes." *Neurology* 70 (18):1630–1635. doi: 10.1212/01.wnl.0000282763.29778.59.

Tripathi, U. N., and D. Chandra. 2009. "The plant extracts of *Momordica charantia* and *Trigonella foenum-graecum* have antioxidant and anti-hyperglycemic properties for cardiac tissue during diabetes mellitus." *Oxidative Medicine and Cellular Longevity* 2 (5):290–296. doi: 10.4161/oxim.2.5.9529.

Tripathi, U. N., and D. Chandra. 2010. "Anti-hyperglycemic and anti-oxidative effect of aqueous extract of *Momordica charantia* pulp and *Trigonella foenum-graecum* seed in alloxan-induced diabetic rats." *Indian J Biochem Biophys*:227–233.

Tsalamandris, S., A. S. Antonopoulos, E. Oikonomou, G. A. Papamikroulis, G. Vogiatzi, S. Papaioannou, S. Deftereos, and D. Tousoulis. 2019. "The role of inflammation in diabetes: Current concepts and future perspectives." *Eur Cardiol* 14 (1):50–59. doi: 10.15420/ecr.2018.33.1.

Uemura, T., S. Hirai, N. Mizoguchi, T. Goto, J. Y. Lee, K. Taketani, Y. Nakano, J. Shono, S. Hoshino, N. Tsuge, T. Narukami, N. Takahashi, and T. Kawada. 2010. "Diosgenin present in fenugreek improves glucose metabolism by promoting adipocyte differentiation and inhibiting inflammation in adipose tissues." *Mol Nutr Food Res* 54 (11):1596–1608. doi: 10.1002/mnfr.200900609.

Vígh, S., Z. Cziáky, L. T. Sinka, C. Pribac, L. Moş, V. Turcuş, J. Remenyik, and E. Máthé. 2017. "Analysis of phytoconstituent profile of fenugreek—*Trigonella foenuem-graecum* L.—seed extracts." *Studia Universitatis Babes-Bolyai Chemia* 62 (2):145–166. doi: 10.24193/subbchem.2017.2.11.

Vyas, S., R. P. Agrawal, P. Solanki, and P. Trivedi. 2008. "Analgesic and anti-inflammatory activities of *Trigonella foenum-graecum* (seed) extract." *Acta Pol Pharm* 65 (4):473–476.

Wang, Q., X. N. Tang, and M. A. Yenari. 2007. "The inflammatory response in stroke." *J Neuroimmunol* 184 (1–2):53–68. doi: 10.1016/j.jneuroim.2006.11.014.

Wang, S., L. Ding, H. Ji, Z. Xu, Q. Liu, and Y. Zheng. 2016. "The role of p38 MAPK in the development of diabetic cardiomyopathy." *Int J Mol Sci* 17 (7). doi: 10.3390/ijms17071037.

Wang, W. Y., M. S. Tan, J. T. Yu, and L. Tan. 2015. "Role of pro-inflammatory cytokines released from microglia in Alzheimer's disease." *Ann Transl Med* 3 (10):136. doi: 10.3978/j.issn.2305-5839.2015.03.49.

Wani, S. A., and P. Kumar. 2018. "Fenugreek: A review on its nutraceutical properties and utilization in various food products." *J Saudi Soc Agric Sci* 17 (2):97–106. doi: 10.1016/j.jssas.2016.01.007.

Watkins, L. R., and S. F. Maier. 2002. "Beyond neurons: evidence that immune and glial cells contribute to pathological pain states." *Physiol Rev* 82 (4):981–1011. doi: 10.1152/physrev.00011.2002.

Welcome, M. O. 2020. "Neuroinflammation in CNS diseases: Molecular mechanisms and the therapeutic potential of plant derived bioactive molecules." *PharmaNutrition* 11:100176. doi: 10.1016/j.phanu.2020.100176.

Xiao, L., D. Guo, C. Hu, W. Shen, L. Shan, C. Li, X. Liu, W. Yang, W. Zhang, and C. He. 2012. "Diosgenin promotes oligodendrocyte progenitor cell differentiation through estrogen receptor-mediated ERK1/2 activation to accelerate remyelination." *Glia* 60 (7):1037–52. doi: 10.1002/glia.22333.

Yang, J., Y. Ran, Y. Yang, S. Song, Y. Wu, Y. Qi, Y. Gao, and G. Li. 2021. "4-hydroxyisoleucine alleviates macrophage-related chronic inflammation and metabolic syndrome in mice fed a high-fat diet." *Frontiers in Pharmacol* 11:2186. doi: 10.3389/fphar.2020.606514.

Yoshinari, O., A. Takenake, and K. Igarashi. 2013. "Trigonelline ameliorates oxidative stress in type 2 diabetic Goto-Kakizaki rats." *J Med Food* 16 (1):34–41. doi: 10.1089/jmf.2012.2311.

Zameer, S., A. K. Najmi, D. Vohora, and M. Akhtar. 2018. "A review on therapeutic potentials of *Trigonella foenum-graecum* (fenugreek) and its chemical constituents in neurological disorders: Complementary roles to its hypolipidemic, hypoglycemic, and antioxidant potential." *Nutr Neurosci* 21 (8):539–545. doi: 10.1080/1028415X.2017.1327200.

Zhang, Q. S., Y. Heng, Y. H. Yuan, and N. H. Chen. 2017. "Pathological α-synuclein exacerbates the progression of Parkinson's disease through microglial activation." *Toxicol Lett* 265:30–37. doi: 10.1016/j.toxlet.2016.11.002.

Zhang, T., A. Shirani, Y. Zhao, M. E. Karim, P. Gustafson, J. Petkau, C. Evans, E. Kingwell, M. van der Kop, F. Zhu, J. Oger, and H. Tremlett. 2015. "Beta-interferon exposure and onset of secondary progressive multiple sclerosis." *Eur J Neurol* 22 (6):990–1000. doi: 10.1111/ene.12698.

Zhao, W. X., P. F. Wang, H. G. Song, and N. Sun. 2017. "Diosgenin attenuates neuropathic pain in a rat model of chronic constriction injury." *Mol Med Rep* 16 (2):1559–1564. doi: 10.3892/mmr.2017.6723.

Zhou, C., R. Chen, F. Gao, J. Zhang, and F. Lu. 2020. "4-Hydroxyisoleucine relieves inflammation through iRhom2-dependent pathway in co-cultured macrophages and adipocytes with LPS stimulation." *BMC Complement Med and Therapies* 20 (1):1–11.

Zhou, J. Y., L. Chan, and S. Zhou. 2012. "Trigonelline: A plant alkaloid with therapeutic potential for diabetes and central nervous system disease." *Current Medicinal Chem* 19:3523–31. doi: 10.2174/092986712801323171.

Zhou, J.-Y., and S.-W. Zhou. 2012. "Protection of trigonelline on experimental diabetic peripheral neuropathy." *Evid-Based Complement Alternat Med* 2012:164219. doi: 10.1155/2012/164219.

Zhou, J. Y., and S. W. Zhou. 2014. "Inflammation: Therapeutic targets for diabetic neuropathy." *Mol Neurobiol* 49 (1):536–546. doi: 10.1007/s12035-013-8537-0.

Zhu, S., S. Tang, and F. Su. 2018. "Dioscin inhibits ischemic stroke induced inflammation through inhibition of the TLR4/MyD88/NFkappaB signaling pathway in a rat model." *Mol Med Rep* 17 (1):660–666. doi: 10.3892/mmr.2017.7900.

15 Fenugreek in Management of Neurological and Psychological Disorders

Rohini Pujari and Prasad Thakurdesai

CONTENTS

15.1 Introduction 229
15.2 Management of Neurological Disorders 230
    15.2.1 Neuroprotective against Chemical-Induced Neurotoxicity 230
    15.2.2 Protective Effect against Neurodegenerative Disorders 231
    15.2.3 Management of Convulsive Disorders Including Epilepsy 234
    15.2.4 Management of Centrally Mediated Pain..234
    15.2.5 Management of Multiple Sclerosis 237
    15.2.6 Management of Cerebral Ischemia..237
    15.2.7 Management of Neuropathic Pain 237
15.3 Management of Psychological Disorders 238
    15.3.1 Management of Occupational Stress 239
    15.3.2 Management of Menopausal Stress 239
    15.3.3 Management of Anxiety Disorders 240
    15.3.4 Management of Depression and Mood Disorders 241
    15.3.5 Management of Cognitive (Learning and Memory) Disorders.........242
15.4 Management of Diabetes-Induced CNS Complications 246
15.5 Conclusions..248
References..248

Abbreviations
PDyn: prodynorphin; CO-1: Cytochrome c oxidase; PINK1: PTEN-induced putative kinase; CAT: catalase; SOD: superoxide dismutase; GPx: glutathione peroxidise; GSH: reduced glutathione; MAO-A and -B: Monoamino oxidase A and B; AChE: acetylcholineesterase; Aβ: Amyloid-beta; TNF-□: tumor necrosis factor-alpha; iNOS: inducible nitric oxide synthase; BDNF: brain derived neurotrophic factor, NF-kappaB; TLR4: toll-like receptor 4; GFAP: glial fibirally acidic protein; MDA: malondialdehyde; GFAP: glial fibrillary acidic protein; COX: cyclooxygenase; LOX: lipooxygenase; NO: nitric oxide; IL: interleukins; SNC: substantia nigra pars compacta; DAT: dopamine transporter

15.1 INTRODUCTION

More than 80% of the world's population suffers from several types of neurological and psychological disorders, emphasizing the requirement for therapeutic interventions for central nervous system (CNS) disorders (Zameer et al. 2018). The growing rate of neurological disorders is a significant concern today. The recent body of published research is focusing on therapeutic interventions providing symptomatic benefits in these disorders. However, these drugs are associated with a variety of endocrine, autonomic, hematopoietic, allergic, and neurological adverse effects. Therefore, the search continues for novel agents that can provide long-term relief with fewer or no adverse effects (Singh et al. 2020). The natural product can offer such options for managing diseases and disorders of the CNS.

Utilization of drugs based on natural origin as an adjuvant therapy with conventional drug therapy for CNS disorders has been increasing currently (Di Paolo et al. 2019). These adjuvant therapies help in reducing the dose of conventional synthetic drugs with dose-limiting adverse effects. Several natural products have been successful as adjuvants to synthetic agents. Most prominent amongst them are *Mucuna pruriens* for Parkinson's disease (PD) (Neta et al. 2018), Bramhi (*Bacopa monnieri*) as memory enhancer (Ingkaninan et al. 2013), and St John's Wort for depression (Solomon et al. 2011). Recently, one such promising natural product that has been demonstrated as an adjuvant in the management of CNS disorders is fenugreek (*Trigonella foenum-graecum* (L.), family: Fabaceae).

The whole plant and different parts of fenugreek, especially its leaves and seeds, have been used as food and in the traditional medicine system worldwide. The bread (Paratha) prepared from fenugreek seeds and maize and wheat is used by the Indian population as a protective food against various diseases, including Alzheimer's disease (Sarwar et al. 2020). Fenugreek has also been documented as a traditional Chinese herb to prevent and cure several CNS diseases (Zhou, Chan, and Zhou 2012). Fenugreek has been documented to possess several bioactive phytoconstituents, namely saponins and sapogenins (e.g., diosgenin); glycosides (e.g., graecunin, C-glycosylflavones, apigenin, and luteolin); amino acids (e.g., 4-hydroxyisoleucine); polysaccharides (e.g., galactomannan); alkaloids (e.g., trigonelline, carpaine, gentianine); flavonoids (e.g., quercetin and kaempferol), along with trigocoumarin, nicotinic acid, trimethylcoumarin, and polyphenols (Dar 2018). The diverse range of pharmacological actions of fenugreek has been attributed to its phytoconstituents (Dar 2018).

In the past, the positive impact of fenugreek and its phytoconstituents for general health as food supplements or medicine against many diseases has been covered by research reviews (Jabeen, Rani, and Mohammed 2018; Ulbricht et al. 2007; Wani and Kumar 2018; Zameer et al. 2018) as well as book chapters (Pal and Mukherjee 2020; Sarwar et al. 2020; Srinivasan 2019). However, fenugreek's potential in managing neuropsychiatric or CNS disorders has not been extensively reported yet. The present chapter reviews the available published literature on fenugreek's efficacy and safety (whole plant, seed, or other parts) against various psychological and neurological disorders during animal experimentation or clinical studies. Furthermore, this chapter attempts to cover the underlying mechanism of action of responsible phytochemicals, wherever such information is available.

15.2 MANAGEMENT OF NEUROLOGICAL DISORDERS

15.2.1 Neuroprotective against Chemical-Induced Neurotoxicity

The pathophysiology of chronic neurotoxicity induced by aluminum has been correlated with increased lipid peroxidation in the brain and hypertriglyceridemia and hypercholesterolemia (Kao et al. 2020). These factors have contributed to various pathological conditions such as atherosclerosis and neurodegenerative diseases, especially Alzheimer's disease (AD) (Ferretti et al. 2004).

The efficacy of fenugreek in ameliorating these aluminum-induced ailments has been studied extensively. The role of fenugreek seeds in powder and extract form against aluminum-induced neurodegeneration was studied (Belaid-Nouira et al. 2012) wherein 80% methanol extract of fenugreek seeds (100 mg/kg, orally for two months) exhibited significant neuroprotective activity by reducing lipid peroxidation in liver, plasma, and the brain, along with reduction in total cholesterol (TC), low-density lipoprotein-cholesterol (LDL-C), triglycerides (TG), and lactate dehydrogenase (LDH) activities in rats exposed to aluminum chloride (500 mg/kg, orally for one month followed by 1600 ppm via drinking water). Reduction in these fenugreek parameters confirmed its neuroprotective activity through its anti-oxidant, antihyperlipidemic, and antihyperglycemic actions. The antiperoxidative effect of fenugreek in the brain was attributed to plasma lipid metabolism modulation (Vijayakumar et al. 2010). Furthermore, anti-oxidant, antihyperlipidemic, and antihyperglycemic properties of the fenugreek phytoconstituents (diosgenin, trigonelline, and kaempferol) have been documented (Al-Matubsi et al. 2011; Helmy et al. 2007; Tripathi and Chandra 2009).

Various investigations have exhibited that both oxidative and inflammatory events play a crucial role in aluminum-induced neurotoxicity, proving aluminum to be a risk factor for AD's pathogenesis. The aluminum-induced neurotoxicity in animals mimics various pathophysiological characteristics of AD and is used as a model for investigating the underlying neurodegeneration mechanisms (Kakkar and Kaur 2011).

This study was further extended (Belaid-Nouira et al. 2013) towards reviewing the putative modulatory effect of fenugreek seeds on interleukin-6 (IL-6) hypoproduction, the decrease in glial fibrillary acidic protein (GFAP) reactivity (due to astrocytic dysfunction), and altered anti-oxidant status in the posterior brain directly induced by chronic aluminum on glial cells. Interleukin-6 (IL-6), a proinflammatory cytokine, has been documented to control Aβ deposition by increasing clearance of glia-mediated amyloid plaque from the brain, thus showing a beneficial role in the brain (Chakrabarty et al. 2010). Activated macrophages and microglia synthesize IL-6 in the brain in response to trauma and pathogenic invasion (Campbell et al. 2002). Any factor affecting astrocytes can cause disrupted IL-6 secretion. Glial fibrillary acidic protein (GFAP) immunostaining is usually used as an astroglia marker for understanding the cause of decreased production of IL-6 (O'Callaghan 1991). In this study, exposure of cortical and hippocampal regions to aluminum caused decreased immunoreactivity to GFAP, indicating dysfunction of astrocytes considered a major target for aluminum toxicity (Campbell et al. 2002; Guo-Ross et al. 1999).

Fenugreek seed powder (5% w/w in standard ground rat feed for two months) showed a neuroprotective effect against aluminum (500 mg/kg, orally for one month, followed by 1600 ppm in drinking water)-induced neurotoxicity by producing enhanced IL-6 expression, GFAP reactivity, and anti-oxidant status. The effect was attributed to astroglia cell protection, anti-oxidant, and immunomodulatory actions (Belaid-Nouira et al. 2013). The effects were also attributed to diosgenin, an important steroidal sapogenin present in fenugreek seeds which acts by modulating the production of inflammatory mediators (IL-6 and IL-1) *in vitro* through the NF-κB pathway (a potent regulator of IL-6) and other mechanisms such as decreased NO and reactive oxygen species production and decreased inducible nitric oxide synthase protein in inflammatory events (Jung et al. 2010). The flavonoid kaempferol present in fenugreek seeds has also been reported for modulating the cyclooxygenase pathway through nitric oxide synthesis inhibition in lipopolysaccharide-induced inflammation, which may also contribute to the neuroprotective mechanism of fenugreek (Mahat et al. 2010).

15.2.2 Protective Effect against Neurodegenerative Disorders

Parkinson's disease (PD) is a progressive, neurodegenerative disorder caused by the depletion of dopaminergic neurons and subsequent disruption of brain basal ganglia of substantia nigra pars compacta (SNC) (Massano and Bhatia 2012). It is characterized by various clinical manifestations such as tremors, muscle rigidity, autonomic dysfunction, bradykinesia, neuropsychiatric problems, sleep disorders, motor function impairment, gait, postural balance, and behavioral, cognitive functions, resulting in morbidity and low life quality (Foltynie and Kahan 2013). Despite the availability of conventional therapies for the treatment of PD through the improvement of the brain dopaminergic system, the remedies that impart neuroprotection remain an unmet need as neuroprotection of dopaminergic neurons has been suggested as a chief trait in the treatment of PD (Tohda, Kuboyama, and Komatsu 2005).

Oxidative stress is regarded as one of the chief etiological factors responsible for neuronal death in PD. Anti-oxidant and neuroprotective agents have shown fruitful results in preventing PD progression (Foley and Riederer 2000). Several studies have shown that various dietary ingredients have been supportive in reducing the progression of several neurological disorders, including PD, due to their chronic experience of dietary consumption and safety profiles, and are considered a new hope in PD management (Ciulla et al. 2019). Fenugreek and its major alkaloid, trigonelline, have shown a protective effect against oxidative stress (Chowdhury et al. 2018; Sreerama and Rachana 2019). Recently, aqueous extracts of fenugreek seeds provided marked neuroprotection to neuroblastoma cells in hydrogen peroxide (H_2O_2)-treated neuroblastoma (NB-41) cell line *in vitro* (Singh et al. 2020). The neuroprotective mechanisms were found to be due to the upregulation of neuronal nitric oxide synthase (nNOS) against hydrogen peroxide-generated stress in neuroblastoma cells. The results were confirmed by protein profiles, mRNA, nitric oxide (NO) levels, and cell viability analysis (Singh et al. 2020).

The protective effect of diosgenin (DG) against Parkinson's disease has been reported in several studies. DG is found to attenuate LPS-induced motor deficits in rats by suppression of the TLR4/NF-κB signaling pathway by reducing protein levels of toll-like receptors (TLR2, TLR4) nuclear factor kappa B (NF-κB) (Li et al. 2018). Anti-oxidant and neuroprotective effects of DG against 6-hydroxydopamine (6-OHDA)-induced PD model in Wistar rats is also reported. DG pretreatment improved motor behavior and asymmetry in 6-OHDA-lesioned rats and protected substantia nigra pars compacta neurons. The mechanism was attributed to attenuation of oxidative stress (reduced MDA and increased GSH) and astrogliosis (reduction of GFAP) (Ghasemi et al. 2017).

Trigonelline (N-methylnicotinic acid), a major alkaloid of fenugreek, is reported for its capability to provoke neurite outgrowth and reconstruct neuronal networks *in vitro* during the neuronal damage in the brain (Tohda, Kuboyama, and Komatsu 2005; Tohda et al. 1999). Subsequently, pre-lesion trigonelline treatment showed dose-dependent protective action in a 6-OHDA model of PD in rats (Mirzaie et al. 2016). Trigonelline (100 mg/kg, i.p., single dose) treatment reduced the number of rotations, prevented the reduction in neurons on the left side of the substantia nigra pars compacta (SNC), prevented apoptosis, and restored levels of MDA in 6-OHDA-induced brain lesions, and suggested anti-oxidant, antiapoptotic, and neuroprotective mechanism against PD (Mirzaie et al. 2016).

Trigonelline-based standardized hydroalcoholic extract of fenugreek seeds (SFSE-T) was reported as an antiparkinsonian compound using two animal models (Gaur et al. 2013). SFSE-T (30 mg/kg, orally, single dose) showed a significant increase in the number of ipsilateral rotations in 6-OHDA-induced unilateral lesioned rats indicated dopaminergic mediation. The 6-hydroxydopamine (6-OHDA) produces neurodegeneration in substantia nigra by accumulating and generating free radicals in dopaminergic neurons mimicking slow progressive neurodegenerative changes shown by clinical PD (Deumens, Blokland, and Prickaerts 2002). The 6-OHDA model induces slow and progressive loss of dopaminergic neurons; thus, it mimics the pathogenesis of Parkinson's disease (Deumens, Blokland, and Prickaerts 2002). Hence, the efficacy shown by SFSE-T against the 6-OHDA model indicated the antiparkinsonian efficacy by protecting dopaminergic neurons and possibly delaying the progression of PD.

Furthermore, SFSE-T also exhibited significant protection from 4-phenyl-1,2,3,6-tetrahydro pyridine (MPTP)-induced lesions of brain and motor dysfunction (indicated by speed, distance traveled, number of squares crossed, and spontaneous motor activity scores) in C57BL/6 mice (Gaur et al. 2013). The neurotoxin, MPTP, induces toxic effects on dopaminergic cells, especially in the brain's substantia nigra pars compacta region (Javitch and Snyder 1984), neuronal impairment and motor dysfunction similar to clinical PD. Neuronal impairment induced by MPTP occurs by converting MPTP to 1-ethyl-4-phenyl-2, 3-dihydropyridium ion (MPP+) by MAO-B in astrocytes (Nicklas, Vyas, and Heikkila 1985). MPP+ produces toxicity specifically in dopaminergic neurons due to its high affinity towards dopamine transporter (DAT) (Javitch and Snyder 1984). The neuroprotection against MPTP-induced neuronal damage by SFSE-T was attributed to the activation of DAT-activating electron transport system of mitochondrial complex-I (NADPH-ubiquinone oxidoreductase I). SFSE-T did not affect the oxotremorine-induced salivation and lacrimation, which showed a lack of anticholinergic effects (Gaur et al. 2013). Therefore, SFSE-T is suggested to provide significant safety benefits compared with many synthetic antiparkinsonian drugs concerning anticholinergic side effects, a major limitation in chronic management of PD (Tuite and Riss 2003).

However, the most significant development is reached when the clinical potential of fenugreek extract against PD is confirmed during the clinical study as a nutritional adjuvant to levodopa (L-DOPA), a drug of choice in the management of patients with PD (Nathan et al. 2014; Oliff 2014). In this double-blind placebo-controlled clinical study, IBHB, the capsules supplementation of SFSE-T, demonstrated excellent efficacy, safety, and tolerability in preventing PD progression (Nathan et al. 2014). Fifty PD patients with a stable dose of L-DOPA were supplemented with IBHB capsules (300 mg, twice daily, for six months' duration) (Nathan et al. 2014). A total score of the Unified Parkinson's Disease Rating Scale (UPDRS), an indicator of PD progression and severity, slowed down (0.098%) when IBHB was supplemented with L-DOPA as compared to the steep rise (13.36%) shown in the absence of IBHB supplementation. Clinicians and researchers use the UPDRS and the motor section to follow Parkinson's disease progression. The Movement Disorder Society Task Force highlighted the importance of identifying thresholds on the UPDRS as represented by clinically important differences (CID) between treatment and comparator (Fahn and Elton 1987; Shulman et al. 2010). The CID for total UPDRS scores and scores of the motor subsection of UPDRS was found to be 5.3 and 4.8, respectively, in favor of IBHB supplementation (Nathan et al. 2014). Therefore, IBHB supplementation can be a useful adjuvant treatment with L-DOPA in the management of PD patients.

The usefulness of IBHB in the management of PD as an adjuvant to L-DOPA is enhanced by the potential to prevent (Aswar et al. 2014) and ameliorate (Thakurdesai, Mohan, Kandhare et al. 2015) the experimental AIMs in the animal models of LID. L-DOPA induces abnormal involuntary movements (AIMs) in several patients following chronic exposure, commonly known as levodopa-induced dyskinesia (LIDs) (Goberman and Blomgren 2003; Papapetropoulos, Basel, and Mash 2007). Prophylactic efficacy of subacute treatment of IBHB (10, 30, or 60 mg/kg, p.o.) in 6-hydroxydopamine (6-OHDA, 4 μl, intracerebroventricular injection) lesioned rat model of PD with the induction of LID (14-day once-daily treatment of the combination of levodopa (20 mg/kg, i.p.) with benserazide (5 mg/kg, i.p.) is reported (Aswar et al. 2014). In this study, IBHB protected 6-OHDA-induced rats from AIMs and forelimb adjusting steps (FAS), a sign of dyskinesia, biochemical parameters (LID-induced increase in striatal FosB immunoreactivity) and molecular parameters (upregulation of mRNA expressions of FosB, prodynorphin (PDyn), Cytochrome c oxidase (CO-1), PTEN-induced putative kinase (PINK1), Parkin, DJ-1, and downregulation of JunD in rat striatum) were significantly prevented by IBHB treatment. These effects can be attributed to enhanced mitochondrial energy demand induced by L-DOPA (Aswar et al. 2014).

Subsequently, IBHB demonstrated therapeutic anti-dyskinetic efficacy against established LID in a separate set of experiments (Thakurdesai, Mohan, Kandhare et al. 2015). In this study, dyskinesia was induced similar to the earlier report (Aswar et al. 2014). However, subacute treatment of IBHB (15, 30, or 60 mg/kg, oral) for the next 45 days was administrated before performing the evaluations for behavioral, biochemical, and *ex vivo* molecular parameters. Subacute treatment of IBHB demonstrated a dose-dependent reduction in behavioral parameter scores (abnormal involuntary movements (AIMS), grid test, catalepsy bar test, FAS) as compared to the vehicle-treated LID group (Thakurdesai, Mohan, Kandhare et al. 2015). These effects were probably mediated through mechanisms such as restoration of upregulated neurotransmission (serotonin and dopamine turnover and 3-O-Methyldopa levels), immunoreactivity (FosB and 5HT2c), and mRNA expressions (c-FOS, CO-1, Homer-1, Parkin, PDyn, Penk, and PINK1). They downregulated mitochondrial respiratory chain complexes activity and mRNA expression of JunD in the striatum, as shown by *ex vivo* evaluations on the brain's striatum area (Thakurdesai, Mohan, Kandhare et al. 2015).

The usefulness of chronic L-DOPA treatment is compromised by the development of dyskinesias during the periods of the day when medication is at its most effective (Cotzias, Papavasiliou, and Gellene 1969; Farajdokht et al. 2020). Therefore, the prevention of LID is crucial for the success of long-term L-DOPA therapy in PD (Bishop et al. 2006). The preclinical evidence of efficacy against LID (Aswar et al. 2014; Thakurdesai, Mohan, Kandhare et al. 2015) supports the utility of IBHB as a more effective and safer option for the management of both LID and PD.

15.2.3 Management of Convulsive Disorders Including Epilepsy

Epilepsy is the second most significant, prevalent, frequently occurring, and progressive brain disorder featured by recurring seizures associated with or without loss of consciousness (Goldenberg 2010). It is caused due to abnormal electric discharge from cerebral neurons due to various etiological factors, including stroke, brain malignancy, trauma, and infection. Several mechanisms, such as low brain level catecholamines, the imbalance between glutamate and GABA-mediated neurotransmission, are known to be involved in epilepsy pathogenesis (Stafstrom and Carmant 2015). Currently available conventional antiepileptic therapies are associated with several adverse drug reactions exhibiting unpredictable pharmacological actions, thus enhancing the necessity to investigate newer antiepileptic medications with less adverse effects and increased pharmacological effects considering the chronic epilepsy therapy (Goldenberg 2010). Natural products are rich sources of bioactive substances authenticated to be safe and efficacious remedies for the treatment of epilepsy. Plants are rich sources of various bioactive compounds (Nsour, Lau, and Wong 2000). One such natural remedy, fenugreek (methanolic extract of seeds), was evaluated for its anticonvulsant action in rats against strychnine-induced convulsions at doses 50 mg, 100 mg, and 200 mg /kg (orally once daily for 14 days) (Khan, Assad, and Ali 2017). Strychnine-induced convulsions occur due to interference in postsynaptic inhibition, which is enhanced by glycine, an inhibitory neurotransmitter in the spinal cord's neurons. Strychnine thus acts as a competitive antagonist of glycine and abolishes glycine's inhibitory effects. It antagonizes the inhibitory reflexes from the brain and spinal cord and augments reflexes from the spinal cord, therefore induces convulsions (Larson and Beitz 1988). The levels of glutamic acid transmit excitatory nerve impulses and increase myocontraction and seizures (Amabeoku and Chandomba 1994).

In this study, fenugreek extract exhibited a significant and dose-dependent delay in the onset of seizures and reduction in mortality rate, frequency, and duration of convulsions with 90% percent protection against strychnine-induced convulsions, showing its efficacy as a potent anticonvulsant agent. The anticonvulsant activity of fenugreek extract was attributed to the presence of various phytoconstituents such as luteolin, gentianine, and diosgenin (Khan, Assad, and Ali 2017).

15.2.4 Management of Centrally Mediated Pain

Pain is a defensive reaction produced by the body in response to any invasive factor. Currently, available analgesic drugs produce a wide range of adverse drug reactions. The search for novel antinociceptive drugs with less adverse effects has increased in the past few years (Dubin and Patapoutian 2010). Traditional medicines and particularly medicinal plants are excellent sources that can fulfill these objectives. Herbal drugs have been utilized as major sources of novel chemical moieties with higher therapeutic efficacy (Biswal, Das, and Nayak 2003; Javan et al. 1997); fenugreek has been documented for its potency and efficacy as an antinociceptive drug in several studies (Table 15.1).

The evidence of studies of various extracts of different parts of fenugreek showed dose dependent of analgesic activities on central pain conditions including neuropathic pain and dysmenorrhea-induced pain. The mechanisms include involvement of the brain, spinal cord, and central and spinal and peripheral effects involved in analgesic effect-involved central and peripheral nerves through inhibition of endogenous opioids, 5-HT, COX and LOX enzymes, spinal purinoceptors by fenugreek phytochemicals such as rutin, apigenin, and saponins (Table 15.1).

TABLE 15.1
Antinociceptive Potential of Fenugreek Extract

Material (Extract and Dose)	Methods	Results and Proposed Mechanisms	References
Aqueous leaf extract (500, 1000, 2000 mg/kg, i.p.)	Formalin and tail-flick tests in rats	• Significant, dose-dependent analgesic effect • ↑ Tail-flick latency • Analgesia in formalin test phases • Central and peripheral mechanisms	Javan et al. (1997)
Ethanolic seed extract (5, 10, 20 mg/kg, i.p.)	Tail-flick latency (TFL) test in rats	• Significant, dose-dependent analgesic effect • ↑ TFL at all doses at all times • Endogenous opioid mechanism	Biswal, Das, and Nayak (2003)
Aqueous leaf extract Different doses by i.p., i.t., i.c.v. routes	Formalin test in male NMRI rats	• Analgesia by i.t. (0.5, 1, 2 mg/rat), i.p. (1 g/kg) route in first formalin test phase • ↓ Effect at low (0.5 mg/rat) and high (3 mg/rat) doses • Analgesia not affected by naloxone • Central action with partial involvement of spinal 5-HT	Parvizpur, Ahmadiani, and Kamalinejad (2004)
Aqueous leaf extract	(1) ADP-induced aggregation in rabbits (2) Alpha-beta-Me-ATP-induced mouse vas deferens contraction (3) Tail-flick test in male rats	• Significant dose-dependent effect at 0.5, 1, 1.5, 3 mg/ml • Inhibited ADP-induced platelet aggregation • Inhibited alpha-beta-Me-ATP-induced vas deferens contractions • Inhibited hyperalgesia in the tail-flick test • COX enzyme inhibition by blocking spinal purinoceptors	Parvizpur, Ahmadiani, and Kamalinejad (2006)
Fraction of methanolic seed extract (10–40 mg/kg, orally)	(1) Acetic acid-induced writhing in mice (2) Hot-plate method in mice	• Significant and dose-dependent analgesic effect • Suppression of synthesis and/or antagonistic effect of the fraction on physiological mediators	Vyas et al. (2008)
Methanolic seed extract (100 mg/kg, i.p.) Aqueous, n-hexane, dichloromethane, carbon tetrachloride, alkaline, and acidified chloroform fractions (5 mg/kg, i.p.)	Formalin test in mice	• Significant pain inhibition by aqueous, dichloromethane, chloroform fractions and extract in all formalin test phases • Opioid-like mechanism • Cyclooxygenases, lipoxygenases inhibitory mechanism	Mandegary et al. (2012)

(Continued)

TABLE 15.1 (Continued)

Material (Extract and Dose)	Methods	Results and Proposed Mechanisms	References
Standardized extract based on trigonelline (50, 100, 200 mg/kg orally, 30 days)	(1) Partial sciatic nerve ligation (PSNL) in rats (2) Sciatic nerve crush injury (SNCI) in rats (3) Hot plate test (4) Motor function test (MFT) in rats (5) Motor nerve conduction velocity (MNCV)	• Protected against thermal hyperalgesia • Increased MFT scores in both models • Restored MNCV reduction in SNCI model • Neuroprotection in painful peripheral neuropathy • Serotonergic and purinergic receptor pathways mechanism	Morani et al. (2012)
Hydroalcoholic seed extract (0.2, 2, and 20 mg/kg, i.p. once daily, 10 days)	(1) Pyridoxine-induced neuropathic mice (2) Tail-flick test in mice (3) Electrophysiological, histological assays	• Significant, dose-dependent anti-neuropathic effect • Restored sciatic nerve function • Most effective dose 20 mg/kg • Attributed to trigonelline, 4-hydroxyisoleucine	Moghadam et al. (2013)
Fenugreek transdermal patch 10%	(1) Double-blind placebo controlled clinical study (2) Post-operative pain of inguinal hernia (IH) (3) Visual analog score (VAS) for pain intensity	• ↓ Pain score significantly • ↓ Diclofene suppository demand • ↓ Morphine consumption • Attributed to the presence of rutin, apigenin, saponins	Ansari et al. (2019)
Alcoholic and hexane extracts (100, 200 mg/kg i.p., 28 days)	(1) STZ-induced diabetic neuropathy in rats (2) Acetic acid-induced writhing and hot-plate test in rats	• Significant analgesic effect by both extracts • Hexane extract (200 mg/kg) was the most effective	Nasri et al. (2013)
Fenugreek seed extract (2–3 tablets of 900 mg, orally, thrice a day)	(1) Dysmenorrhea among students—clinical study (2) Visual analog scale for severity of pain (3) Multidimensional verbal scale for systemic symptoms	• Significantly reduced pain severity, duration between two menstrual cycles • Reduced systemic symptoms of dysmenorrhea • Attributed to the presence of alkaloidal constituents	Younesy et al. (2014)
Fenugreek seed powder (3 gm, orally twice daily from day 1 to 3 of the menstrual cycle)	(1) Open-labeled, randomized, standard-controlled clinical study (2) Validated visual analog scale	• ↓ Pain intensity by 67% dysmenorrheal-induced pain • Attributed to chemical constituents, inhibitory effect on inflammatory mediators	Inanmdar et al. (2016)

15.2.5 MANAGEMENT OF MULTIPLE SCLEROSIS

Multiple sclerosis (MS) is a chronic inflammatory demyelinating disorder of the central nervous system (Mendell et al. 1987; Zhang 2020). The differentiation process of oligodendrocyte progenitor cells (OPCs) into mature oligodendrocytes is an essential step for remyelination (Bir, Chernyshev, and Minagar 2018; Goldman and Kuypers 2015). The series of experiments showed DG, a constituent of fenugreek seeds, increases OPC differentiation into mature oligodendrocyte through an ER-mediated ERK1/2 activation pathway to accelerate remyelination, which can be explored in the management of MS (Xiao et al. 2012). In another study, DG showed dose-dependent alleviation of the progression of experimental autoimmune encephalomyelitis with reduced demyelination and inflammation in CNS, inhibition of microglia and macrophages activation, Th1/Th17 cell differentiation suppression and CD4(+) T-cell proliferation suggesting protective efficacy against MS (Liu et al. 2017).

15.2.6 MANAGEMENT OF CEREBRAL ISCHEMIA

Cerebral ischemia, a severe condition that is commonly known as stroke, induces brain damage through a series of consequences followed by cerebral ischemia-reperfusion. The pathological changes during cerebral ischemia include mitochondrial dysfunction, inflammatory reactions, excitatory amino acids (EAA)-induced neurotoxicity, increased reactive oxygen species (ROS) generation, and neuronal cell death. Immediate restoration of the adequate blood supply to ischemic regions, reperfusion, is the best treatment strategy in the clinic (Cui et al. 2010).

The steroidal glycoside from fenugreek seeds, DG, decreases cerebral infarct size significantly, inhibits mortality, with motor impairment restoration and neurological deficit reduction (Zhang et al. 2016). The mechanism was attributed to reducing cellular apoptosis in the cortex and hippocampus CA1 through suppression of Bax/Bcl-2 ratio and caspase-3 activity. Besides, a mechanism such as inhibition of overproduction of proinflammatory cytokines (TNF-α, IL-1β, and IL-6), upregulation of protein expression of IκBα in the injured brain, and downregulation of p65 subunit of NF-κB in the nucleus are suggested as the mechanism of DG against cerebral ischemia (Zhang et al. 2016). The importance of inflammatory markers, especially interleukins in the pathophysiology of cerebral ischemia and reperfusion, is well known (Jayaraj et al. 2019; Rodriguez-Yañez and Castillo 2008). Therefore, the efficacy of DG against ischemia and reperfusion demonstrated by in the study (Zhang et al. 2016) can be useful in managing cerebral ischemia.

15.2.7 MANAGEMENT OF NEUROPATHIC PAIN

The chemical mediators of inflammation such as cytokines, chemokines, and lipid mediators stimulate or sensitize the central synaptic targets to promote long-term maladaptive plasticity resulting in persistent neuropathic pain (Ellis and Bennett 2013; Myers, Campana, and Shubayev 2006). The efficacy of DG against inflammatory mediators was reported on animal models of neuropathic pain (Kang et al. 2011; Zhao et al. 2017). DG showed ameliorative effects to increase the nerve conduction velocities and nerve growth factor (NGF) levels in a mouse model of diabetic neuropathy (Kang et al. 2011). In this study, DG reduced the myelin sheath's disorganization, increased the area of myelinated axons, improved damaged axons, and increased the neurite outgrowth in PC12 cells probably through increased expression of NGF (Kang et al. 2011). During another study, DG showed the protection to the sciatic nerve against chronic constriction injury (CCI)-induced thermal hyperalgesia and mechanical allodynia by inhibiting pro-inflammatory mediators such as cytokines (TNF-α, IL-1β, IL-2), and NF-κB and p38MAPK pathways with a reduction in oxidative stress (Zhao et al. 2017).

Taken together, evidence suggested the efficacy of fenugreek and its phytoconstituents against neurological disorders including PD, AD, epilepsy, central pain, multiple sclerosis, cerebral ischemia/reperfusion, neuropathic pain through multiple mechanism of action as presented in Table 15.2.

TABLE 15.2

Mechanisms of Fenugreek and Its Phytoconstituents against Neurological Disorders

Disorder (Constituent)	Action	Mechanism	Reference
Neurodegeneration (Diosgenin, trigonelline, and kaempferol)	Anti-oxidant	MDA, ↓ reactive oxygen species production)	Belaid-Nouira et al. (2012)
	Protecting astroglial cells	↑ clearance of glia-mediated amyloid plaque from the brain, ↑ IL-6 expression, ↑ GFAP reactivity	Belaid-Nouira et al. (2013)
	Reduced inflammation	↓ IL-6 and IL-1, COX pathway ↓ iNOS, ↓ NO production	Jung et al. (2010) Mahat et al. (2010)
Parkinson's disease (trigonelline)	Prevention of neuronal loss	In substantia nigra pars compacta (SNC) ↓ apoptosis	Mirzaie et al. (2016)
	Anti-oxidant activity	↓ MDA levels	Mirzaie et al. (2016)
	Reduced dopaminergic neurodegeneration	↑ DAT activation by activation electron transport system of mitochondrial complex-I, that is, NADPH- ubiquinone oxidoreductase I Prevention of downregulation of mRNA (JunD) in the striatum	Gaur et al. (2013) Aswar et al. (2014)
	↓ neurotransmitters	↓ Striatal FosB and 5HT2c immunoreactivity ↓ mRNA expressions of FosB, PDyn, CO-1, PINK1, Parkin, DJ-1, Homer-1, Penk, c- FOS ↓ Serotonin and dopamine turnover and 3-O-Methyldopa levels	Aswar et al. (2014) Thakurdesai, Mohan, Kandhare et al. (2015)
	Restoration of mitochondrial dysfunction	↑ Mitochondrial energy demand	
Epilepsy (luteolin, gentianine, and diosgenin)	Anti-oxidant effect	Anti-oxidant effect	Assad and Khan (2017)
Central pain (rutin, apigenin, saponins, trigonelline, 4-hydroxyisoleucine)	↑ pain latency	↑ Serotonergic and purinergic receptor pathways ↑ Endogenous opioid receptors ↓ COX and LOX enzymes	Biswal, Das, and Nayak (2003), Assad and Khan (2017), Parvizpur, Ahmadiani, and Kamalinejad (2004), Parvizpur, Ahmadiani, and Kamalinejad (2006), Mandegary et al. (2012), Morani et al. (2012)

15.3 MANAGEMENT OF PSYCHOLOGICAL DISORDERS

During the past few decades, pharmacotherapy with psychoactive drugs has been considered a primary treatment modality in managing various psychological disorders. Several natural products are reported for efficacy against many disorders of psychological disorders and the central nervous system (CNS) (Bharate, Mignani, and Vishwakarma 2018; Essa et al. 2012; Kumar 2006; Martucciello et al. 2020; McClatchey et al. 2009). They have shown promising results in overcoming anxiety, depression, mood disorders, and many more disorders.

The recent review reported the therapeutic potentials of fenugreek and its chemical constituents in neurological disorders (Zameer et al. 2018). Different extracts of fenugreek (alcoholic, petroleum ether, total glucosidal, total alkaloidal extracts, fenugreek oil) and phytoconstituents (diosgenin, trigonelline) at 50 and 100 mg/kg given intraperitoneally showed significant and dose-dependent general CNS stimulant activity during locomotor activity evaluation whereas aqueous extract alone produced a CNS depressant effect at the same doses (Natarajan et al. 2007). CNS stimulant activity of fenugreek extracts was attributed to phytoconstituents, quercetin, and kaempferol (Fujimori and Cobb 1955; Natarajan et al. 2007).

15.3.1 Management of Occupational Stress

Occupational stress is defined as harmful emotional and physical responses produced when the job's obligations, which in turn, produce a state of "burnout," resulting in injury, illness, and job incapability (Marotta et al. 2011). Chronic exposure to such stress can cause several health issues, including immune system suppression, cardiovascular disorders, and neuropsychological disorders due to impairment of neurotransmission (Caixeta et al. 2012). The conventional treatment strategies include treating with opioid drugs such as methadone and buprenorphine, but they are associated with addiction liabilities upon chronic use (Barry et al. 2011). Hence, additional options to manage occupational stress are needed.

A food-grade formulation containing the dietary fibers of fenugreek with curcumin (CurQfen, curcumagalactomannoside, CGM) reported efficacy against stress in randomized, double-blind, placebo-controlled clinical trials (Pandaran et al. 2016). Inventory scores show its effectiveness in reducing occupational stress and the associated consequences due to potent anti-oxidant activity and enhanced bioavailability compared to natural curcumin (Pandaran et al. 2016).

15.3.2 Management of Menopausal Stress

Menopause is the natural stoppage of the menstrual cycle in women, occurring in their 50s, leading to the end of fertility due to a decline in ovarian follicles' functioning. Menopause shows a rapid decrease in the blood estrogen levels (Barth, Villringer, and Sacher 2015), resulting in several neurological alterations in CNS such as depression, hot flushes, sleep disturbances, mood, and cognitive disorders (Lalo 2017). Reduction in estrogen levels during menopause is usually associated with increased oxidative stress, leading to learning and memory impairment (Borras et al. 2003; Delibasi et al. 2006). Hormone Replacement Therapy (HRT) with estrogen is considered the most effective therapy for treating postmenopausal manifestations (Doyle et al. 2009). However, severe adverse effects limit its acceptance and chronic use among postmenopausal women. Hence, there is a need for alternative therapies to treat postmenopausal cognitive deficits (Rozenberg, Vandromme, and Antoine 2013).

Fenugreek extract, known to contain phytoestrogens, was hypothesized to mimic estrogen (Abedinzade, Nasri, Jamal, Porramezan et al. 2015). The estrogenic activities of fenugreek have also been reported (Sreeja, Anju, and Sreeja 2010). The therapeutic efficacy of fenugreek seed extract in attenuating memory impairment, hippocampal neural cell deficits, and brain anti-oxidant status in ovariectomy-induced rats has been reported (Anjaneyulu et al. 2018). The ovariectomy procedure in animals mimics the clinical postmenopausal conditions concerning impairment in cognitive functions (Baeza et al. 2010; Dong et al. 2013). Hydroalcoholic extract of fenugreek seeds (200 mg/kg, orally for 30 days) alone, as well as in combination with choline-DHA and estradiol, exhibited significant efficacy to ameliorate the ovariectomy-induced changes such as decline of learning and memory, anti-oxidant efficacy (i.e. reversal of elevated malondialdehyde (MDA) and reduced glutathione (GSH) levels in the brain), neuronal degeneration and loss in the dentate gyrus and CA1, CA3 areas of the hippocampus in ovariectomized rats (Anjaneyulu et al. 2018). This positive effect of fenugreek on learning and memory was attributed to its increased cholinergic activity by an inhibitory effect on acetylcholinesterase, which hydrolyzes the neurotransmitter acetylcholine important for cognitive events (Prema et al. 2016). These findings suggested that fenugreek could be a potential therapeutic agent to ameliorate cognitive dysfunctions associated with menopause (Anjaneyulu et al. 2018).

Different extracts of fenugreek seeds (50 and 150 mg/dL of hexane extract and 150 mg/kg of ethanolic extract) administered intraperitoneally, once daily for 42 days, significantly reduced postmenopausal inflammatory and metabolic changes such as elevated body weight and fasting blood glucose levels in ovariectomized rats (Abedinzade, Nasri, Jamal, Ghasemi et al. 2015). These effects were attributed to the improvement of insulin resistance by fenugreek (Handa et al. 2005). Several researchers have reported that the ovariectomy may lead to inflammatory changes in the body due to increased production and accumulation of cytokines such as IL-1, TNF-α, and IL-6 (7, 33). The significant reduction of the elevated serum levels of pro-inflammatory cytokines in ovariectomized rats by fenugreek extract suggested efficacy in reducing postmenopausal, metabolic, and inflammatory alterations and its usefulness in the management of postmenopausal stress (Abedinzade, Nasri, Jamal, Ghasem et al. 2015). The immunomodulatory, anti-inflammatory, and anti-oxidant mechanisms and presence of phytoestrogen compounds like diosgenin of fenugreek might have contributed to reducing the changes in postmenopausal stress (Pfeilschifter et al. 2002; Uemura et al. 2010).

Various fenugreek extracts were found beneficial in managing postmenopausal discomforts during past clinical studies (Hakimi et al. 2005, 2006; Shamshad Begum et al. 2016; Steels et al. 2017). The fenugreek seeds (6 g of granulated fenugreek seed powder, 8 weeks) showed a significant decrease reduction in Greene score during a quasi-experimental study on early menopausal manifestations in perimenopausal women (Hakimi et al. 2005). Furthermore, treatment with fenugreek significantly reduced the frequency of vasomotor symptoms and hot flashes after four and eight weeks of effects in postmenopausal women (Hakimi et al. 2006).

In a recent randomized, double-blind, placebo-controlled clinical study, a standardized hydroethanolic extract of fenugreek husk, rich in phytoestrogens, was investigated for its safety and efficacy in the management of postmenopausal manifestations including vasomotor, physical, sexual, and psychological symptoms evaluated using Short Form-36 (SF-36), the Greene Climacteric Scale, anthropometric measurements, and biochemical/hematological analysis (Shamshad Begum et al. 2016). The standardized hydroethanolic extract of fenugreek husk (1000 mg/day, orally, once daily, for 90 days) produced a significant elevation in plasma estradiol concentration, marked improvements in quality of life by reducing postmenopausal discomforts such as psychological problems including depression, anxiety, mood swings, physical problems like hot flushes, night sweats, and sexual problems in extract-treated postmenopausal women without changes in biochemical and hematological parameters (Shamshad Begum et al. 2016).

In another double-blind, randomized, placebo-controlled clinical study, standardized fenugreek dehusked seed extract (600 mg/day, orally, once a day for 12 weeks) showed significant efficacy to overcome vasomotor manifestations (reduced frequency and intensity of the menopausal manifestations) and improve quality of life (using Menopause-Specific Quality of Life (MENQOL) questionnaire) in healthy women undergoing postmenopausal symptoms (Steels et al. 2017). The fenugreek extract successfully alleviated the psychosocial, physical, sexual, and vasomotor type of menopausal symptoms manifested by hot flushes, night sweats, depression, and anxiety by increasing serum estradiol levels, thus exhibiting its capability to overcome menopausal manifestations in healthy older women (Steels et al. 2017).

15.3.3 Management of Anxiety Disorders

Anxiety disorders are the most prevalent among all psychological disorders, characterized by extreme and unnecessary worry; continuous feelings of terror, doubt, and fear; intense nervousness, uneasiness, and tension; impending disaster and apprehension; sympathetic motor hyperactivity that disturbs usual emotional and physical functions (Cha et al. 2005; Chatterjee et al. 2013). Different anxiety disorders include traumatic stress, obsessive-compulsive disorder (OCD), phobias, generalized anxiety disorders, and panic disorders. Current therapies with conventional anxiolytic medicines are associated with several adverse drug reactions with enhanced tolerance and dependency upon chronic treatment (Bourin 2018). The design and development of newer anti-anxiety agents, especially those belonging to the herbal origin with lesser side effects, have become essential in neuropharmacology research (Faustino, Almeida, and Andreatini 2010). Fenugreek is one source of such natural products that was reported to have anxiolytic activity in several preclinical and clinical studies (Assad and Khan 2017; Dhananjaya et al. 2011; Hausenblas et al. 2020; Iyer et al. 2004; Mohan et al. 2006).

The methanol extract (30, 100 mg/kg) and ethyl acetate fraction of methanol extract (30, 100 mg/kg) of fenugreek seeds were reported as anxiolytic activity against meta-chlorophenylpiperazine (m-CPP, 1 mg/kg), a 5-HT2 receptor agonist, that induced anxiety in mice. Both extract and fraction showed a significant and dose-dependent increase in the number of entries and the time spent in open arms (elevated plus maze paradigm), the time spent in light zone (light-dark box), and the number of squares crossed and assisted rearing (open field apparatus) (Iyer et al. 2004). Ethyl acetate fraction of methanol extract showed antagonistic effects towards m-CPP (ED50 = 48.97 mg/kg) (Iyer et al. 2004). Subsequently, acetone soluble fraction (100, 200, and 400 mg/kg) of methanol extract of fenugreek seeds demonstrated significant and dose-dependent nootropic and anxiolytic activities against scopolamine-induced cognitive dysfunction during passive shock avoidance paradigm in rats in elevated plus maze paradigm in mice (Mohan et al. 2006). These results were attributed to alkaloids and saponins present in fenugreek (Mohan et al. 2006). Thereafter, the methanol extract of fenugreek seeds (100, 200, and 400 mg/kg) is reported as anxiolytic and skeletal muscle relaxant activity in mice following 15 days of oral feeding in mice in a dose-dependent manner when assessed using hole board test, light and dark test, phenobarbitone-stimulated sleeping time, and rotarod test (Assad and Khan 2017). The effects may be attributed to the presence of flavonoids, polysaccharides, saponins, flavonoids, and alkaloids in fenugreek extract (Assad and Khan 2017).

Besides, an ethanolic extract of fenugreek seeds (100 and 200 mg/kg, orally) was reported as anxiolytic without motor dysfunction in mice during studies with light-dark box test and elevated plus-maze test (Dhananjaya et al. 2011). Recently, the fenugreek's anxiolytic efficacy is also supported by a randomized, double-blind placebo-controlled clinical study in the aging male (Hausenblas et al. 2020). Fenugreek supplementation (500 mg per day) was found to be a useful nutritional intervention for anxiety levels, along with improving aging male symptoms and aspects of health-related quality of life (HRQoL) in healthy, recreationally active men without side effects (Hausenblas et al. 2020). Taken together, fenugreek is reported as a potent anxiolytic agent and showed promise as an effective and safer option alone or supplementation to existing anti-anxiety drugs.

15.3.4 Management of Depression and Mood Disorders

Depression is one of the significant and most frequently occurring, devastating, and life-threatening neuropsychiatric disorders characterized by a wide range of physical, behavioral, and cognitive symptoms such as apathy; low mood; pessimism; feeling of guilt; loss of interest, appetite, and libido; sleep disturbances; and absence of zeal and indecisiveness. The existing options of synthetic antidepressant drugs act via different mechanisms, including serotonergic, noradrenergic, and dopaminergic systems, are associated with several adverse drug reactions and unwanted drug interactions (Davies and Read 2019; Locher et al. 2017; Villas Boas et al. 2019). Therefore, traditional antidepressant therapies from natural or plant origin are gaining interest as a safer option as a nutritional supplement or add-on to existing drugs in the management of mood disorders including depression (Dome et al. 2019; Khan et al. 2018; Nabavi et al. 2017)

Various fenugreek extracts have demonstrated antidepressant potential during preclinical and clinical studies (Khursheed et al. 2014; Narapogu et al. 2018; Nayak, Adake, and Hafis 2019; Pawar et al. 2008). The significant and dose-dependent antidepressant activity of methanolic extract of fenugreek seeds (250 and 500 mg/kg, intraperitoneal, once daily for 7 days) was reported using a nonspecific animal model of depression (tail suspension test and forced swim test) (Pawar et al. 2008). The presence of saponin glycosides and flavonoids in fenugreek seeds and interaction with dopaminergic, serotonergic, and GABAergic systems was thought to be mechanisms behind the beneficial effects (Pawar et al. 2008).

The efficacy of ethanolic extract of fenugreek seeds was reported in normal rats, such as during the forced swimming test, tail suspension test, monoamine assay, and locomotor activity test (Khursheed et al. 2014). The antidepressant action was attributed to monoamine oxidase-A and MAO-B inhibition in the rat brain (Khursheed et al. 2014). Furthermore, aqueous extract of fenugreek seeds (200 mg/kg and 400 mg/kg, oral) exhibited antidepressant activity during forced swim and tail suspension tests, perhaps through anti-oxidant mechanism (estimation of serum levels of MDA and SOD) in Wistar rats (Narapogu et al. 2018).

The antidepressant activity of ethanolic fenugreek extract on subacute oral administration (100 mg, 200 mg, and 400 mg/kg, once daily, 14 days) showed significant and dose-dependent antidepressant activity (immobility time reduction in forced swim and tail suspension tests) possibly due to the presence of phytosterols, phenolic compounds, flavonoids, glycosides, and amino acid 4-hydroxyisoleucine, one of the active components of fenugreek (Nayak, Adake, and Hafis 2019).

The major amino acid constituent 4-hydroxyisoleucine (4-HI) obtained from fenugreek seeds demonstrated antidepressant-like effects on acute (Gaur et al. 2012) and subacute administration (Kalshetti et al. 2015) in animal models of depression. Oral acute administration of 4-HI (3, 10, and 30 mg/kg) showed the dose-dependent reversal in forced-swim induced increased immobility in reserpinized and non-reserpinized mice without affecting the spontaneous motor activity (SMA), suggesting central medicated mechanisms. Further, acute administration of 4-HI (30 mg/kg) increased the number of 5-HTP (precursor of serotonin)-induced head twitches, which suggested the brain serotonin turnover enhancement as a mechanism of antidepressant-like action of 4-HI. In the next series of experiments, antidepressant effects of 4-HI were reported in an animal model of isolation stress-induced depression, namely socially isolated olfactory bulbectomized (OBX-Iso) rats (Kalshetti et al. 2015). Behavioral depression- and anxiety-related parameters were evaluated using sucrose intake test, open field test (OFT), forced swim test (FST), and novelty suppressed feeding (NSF). Neurochemical evaluation for serotonin, nor-adrenaline, serum cortisol, and serotonin turnover was carried out. 4-HI (10, 30, 100 mg/kg, orally once daily from day 14 after bilateral olfactory bulbectomy) showed significant and dose-dependent protection from stress-induced depressive behavioral and neurochemical changes in OBX-Iso rats with possible involvement of multiple stress-relieving mechanisms such as prevention of hyperactive HPA axis (Kalshetti et al. 2015).

15.3.5 Management of Cognitive (Learning and Memory) Disorders

Learning is the process of acquiring a new aspect or information. At the same time, memory is known as encoding and storing the newly acquired information and remembrance of the same as per the need. These processes involve the functioning of several neurotransmitters such as serotonin, dopamine, and acetylcholine, along with long-lasting activation of receptor-linked enzymes that activate the intercellular messengers involved in the learning and memory process (Jiang et al. 2007; Maurer and Williams 2017).

Dementia is defined as a gradual impairment of learning and memory processes, which is highly prevalent in elderly people. Several pathological conditions such as cerebral ischemia, neurodegenerative diseases, anxiety, depression, tumors, hypoxia, heart surgery, head injury, untoward effects of drugs, undernourishment, old age, and attention deficit disorders are involved in the impairment of learning and memory processes (Budson and Price 2005). The resultant cognitive decline may lead to neurodegenerative disorders such as Alzheimer's disease (AD) and PD. This cognitive decline pathogenesis comprises various abnormalities such as ischemic damage, β-amyloidosis, abnormalities in tau proteins and neurotransmitters, synucleinopathy, and serotonergic impairment glutamatergic-cholinergic systems, oxidative stress, inflammatory reactions, neuroplasticity, and cell death (Ballard et al. 2007).

Pathogenesis of AD especially involves damage of cholinergic neurons in the cortical and hippocampal regions of the brain associated with the reduced choline acetyltransferase enzyme activity. This fact formed a base for the development of acetylcholinesterase (AChE) inhibitors to treat AD-associated dementia (Munoz-Torrero 2008). The cholinergic drugs such as butyrylcholinesterase inhibitors, acetylcholinesterase (AChE) inhibitors, and glutamate (NMDA) receptor antagonists have been developed for the treatment of dementia as well as AD but are associated with various side effects with limited efficacy (Orhan et al. 2009).

One of AD therapy's novel strategies is to enhance cholinergic activity during the pathway of AChE inhibition by using several natural products. The primary function of AChE is to transfer the impulse of the nerve cells through the cholinergic connection by the acetylcholine decomposition. For that reason, a search for natural products or extracts having the AChE inhibition properties to improve cognitive deficiency is vital. Several studies on various plants or plant-based natural products have shown therapeutic effectiveness as a learning and memory enhancer (Rabiei, Gholami, and Hojjati 2014). Fenugreek powder seed extracts, saponin fraction, and alkaloid constituent, trigonelline, are reported efficacious during preclinical studies on animal models of learning and memory disorders. The details are presented in the following paragraphs:

Fenugreek seed powder was reported to enhance learning and memory in a series of experiments (Prema et al. 2016, 2017; Saini et al. 2011). Fenugreek seed powder at 5%, 10%, and 15% concentrations decreased the cholinergic transmission, and MDA (lipid peroxidation) glucose and cholesterol levels in young mice indicated its multifaceted beneficial effects, including cholesterol-lowering, memory improvement, anti-oxidant and anticholinesterase properties (Saini et al. 2011). The neuroprotective effects of germinated fenugreek seed powder against chronic aluminum chloride ($AlCl_3$)-induced experimental AD model is reported (Prema et al. 2016). Administration of fenugreek seed powder (2.5%, 5%, and 10% mixed with ground standard feed) protected $AlCl_3$-induced memory and learning impairments, aluminum overload, AChE hyperactivity, amyloid-beta (Aβ) burden, and apoptosis in rats via activating Akt/GSK3beta pathway (Prema et al. 2016). During a detailed follow-up study, the fenugreek seed powder showed the neuroprotective and nootropic efficacy against the chronic administration of $AlCl_3$-induced cognitive decline (Prema et al. 2017). $AlCl_3$ ammonization to animals induces significant learning and memory impairments, oxidative stress, and alterations in the protein immunocontent, patterns of IDE and CDK5 (enzymes involved in the metabolism of tau and amyloid proteins), pTau, GFAP and Iba-1, IL-1beta, IL-6, TNF-alpha, iNOS, NF-kappaB, COX-2, CDK5, BDNF, and STAT3 (Valero et al. 2017). The co-administration of fenugreek seed powder significantly attenuated the $AlCl_3$-induced memory deficits, amyloid and tau pathology, oxidative stress, and inflammation in AD rats (Prema et al. 2017).

Alzheimer's disease (AD) is known as lacking in the brain cells' neurotransmitters due to an increase in the acetylcholinesterase (AChE) activity. Hence the use of AChE inhibitors (AChEI) is believed to be the best way to manage AD. The standardized hydroalcoholic extract (marker trigonelline), ethyl acetate fraction, total alkaloid fraction, and trigonelline showed significant acetylcholinesterase inhibitory activity *in vitro* (Satheeshkumar et al. 2010). In a separate study, hydroalcoholic extract of fenugreek seeds showed noticeable AChE inhibitory activity (TLC bioautography method) among three studied extracts of individual plants (Sharififar et al. 2012).

The methanolic extract of fenugreek seeds showed significant anti-amnesic efficacy against cognitive decline during interoceptive (passive avoidance apparatus, elevated plus maze) as well as exteroceptive models (scopolamine- or diazepam-induced amnesia) of amnesia (Assad, Khan, and Rajput 2018). The methanolic extract of fenugreek seeds (200 mg/kg, oral, 8 days) exhibited a highly noteworthy decline in transfer latency on both acquisition and retention days compared to control animals, suggestive of the improved learning-memory process (Assad, Khan, and Rajput 2018).

The supplementation of fenugreek saponin (0.05%, 0.1% and 2%) exhibited increased AChEI levels and apoptosis activities (elevated the gene expression levels of Bax; Bcl2 and caspase-3 genes in aluminum chloride-induced AD rats) (Khalil, Roshdy, and Kassem 2016). Fenugreek saponin is reported to decrease the intracellular ROS generation and DNA damage (an anti-oxidant mechanism). It enhances the apoptosis through regulation of caspase-3, Bax, and Bcl-2 in the apoptotic pathway (an antiapoptotic mechanism) in aluminum chloride-induced AD in rats (Khalil, Roshdy, and Kassem 2016).

The DG, a steroidal phytoconstituent of fenugreek, has shown reduced amyloid plaques/neurofibrillary tangles in the cerebral cortex and hippocampus, the number of degenerated axons and presynaptic terminals, and improved performance of object recognition memory in 5XFAD transgenic mouse model of AD study (Tohda et al. 2012). The activation of 1,25D$_3$-membrane-associated, rapid response steroid-binding protein (1,25D$_3$-MARRS) pathway as confirmed using rat and mouse cortical neuron primary culture study was the suggested mechanism of action of DG against AD (Tohda et al. 2012).

The beneficial effects of DG are also supported by the report on the Aβ-42 peptides and neurotoxicants-induced brain damage in transgenic 2576 (TG) mice injected with trimethyltin (TMT) (Koh et al. 2016). Treatment with DG significantly decreased the number of dead cells and Aβ-stained plaques in the dentate gyrus's granule cells layer. Additionally, a significant suppression of apoptosis (suppressed apoptotic Bax/Bcl-2 expression) and acetylcholinesterase (AChE) activity indicated antiapoptotic and enhanced cholinergic activity of DG. The nerve growth factor (NGF) concentration was observed to be markedly enhanced with the recovery of decreased phosphorylation of downstream members in the TrkA high-affinity receptor signaling pathway. A similar pattern was exhibited in p75NTR expression and JNK phosphorylation in the NGF low-affinity receptor signaling pathway. These outcomes are suggested to be mediated by the enhancement of NGF biosynthesis and SOD activity of DG (Koh et al. 2016).

DG has been reported to improve the cognitive deficit in human immunodeficiency virus (HIV) infection-induced dementia mode (*in vitro*) in human neuronal cultures with E4 allele of Apolipoprotein E (ApoE) (Turchan-Cholewo et al. 2006). The HIV-infected individuals with the E4 allele of Apolipoprotein E (ApoE) or a history of intravenous drug abuse are known to be at risk of developing dementia and peripheral neuropathy (Corder et al. 1998). *In vitro* studies showed that HIV proteins such as Tat and gp120, Tat + morphine produced increased neurotoxicity in human neuronal cultures with ApoE4 allele with differential alteration of transcripts involved in energy metabolism in cultures of ApoE3 and 4 neurons. DG showed marked neuroprotection against the Tat + morphine-induced neurotoxicity and oxidative stress and impaired morphine metabolism (Turchan-Cholewo et al. 2006). Furthermore, the neuroprotective and memory-enhancing efficacy of DG-rich yam was reported in senescent mice induced by D-galactose (D-gal). DG-rich yam and increased learning and memory abilities of the mice in the Morris water maze test reduced the oxidative stress in the brains of D-gal treated mice (Chiu et al. 2009). These results are further confirmed during a double-blind, placebo-controlled, randomized, crossover clinical study in healthy adults (Tohda et al. 2017).

Trigonelline, a major alkaloid of fenugreek seeds, is reported for a broad range of beneficial neuropsychological effects on CNS disorders (Zhou, Chan, and Zhou 2012). Recently, neuroprotective action of trigonelline against lipopolysaccharide (LPS (250 μg/kg) intraperitoneal) mediated cognitive impairment is reported (Chowdhury et al. 2018). Trigonelline (50 and 100mg/kg, oral, 7 days) reversed LPS-induced behavioral and cognitive disturbances, decreased the LPS-induced oxidative stress and AChE levels in the hippocampus and cortex, decreased proinflammatory cytokines (TNF-alpha and IL-6 levels), and caused a significant upregulation of BDNF levels, indicating putative anti-oxidant, anti-inflammatory, and BDNF restoration mechanisms (Chowdhury et al. 2018). Confirmation of these mechanisms was observed from another similar study, in which trigonelline (20, 40, or 80 mg/kg/day) found to significantly improve the memory dysfunction caused by LPS (500 μg/kg, i.p.) (Khalili et al. 2018). Trigonelline showed anti-oxidant (lowered hippocampal malondialdehyde (MDA) and acetylcholinesterase (AChE) activity, improved superoxide dismutase (SOD), catalase (CAT) and glutathione (GSH)) and anti-inflammatory (depressed hippocampal nuclear factor-kappaB (NF-kappaB), toll-like receptor 4 (TLR4), and tumor necrosis factor-alpha (TNF-alpha)) effects in LPS-challenged rats (Khalili et al. 2018). Collectively, trigonelline diminishes LPS-induced cognitive decline via suppression of hippocampal oxidative stress and inflammation with anticholinesterases mechanism (Khalili et al. 2018).

More recent evidence of trigonelline's potential against AD is obtained from a report of the protective effect of trigonelline isolated from fenugreek against intracerebral Aβ (1–40) as an AD model in the rat (Fahanik-Babaei et al. 2019). The results showed that trigonelline pretreatment of Aβ-microinjected rats significantly improved learning and memory performances by mitigating the hippocampal oxidative stress and improving mitochondrial membrane potential (MMP), ameliorating hippocampal levels of glial fibrillary acidic protein (GFAP), S100b, cyclooxygenase 2 (Cox2), tumor necrosis factor-alpha (TNF-alpha), and interleukin 6 (IL-6), and prevented the loss of hippocampal CA1 neurons in Aβ-microinjected rats (Fahanik-Babaei et al. 2019). Therefore, results suggested that trigonelline pretreatment in the Aβ model of AD improves cognition and alleviates neuronal loss by suppressing oxidative stress, inflammation maintain astrocyte activity, and mitochondrial integrity preservation (Fahanik-Babaei et al. 2019).

The reconstruction of neurons in the damaged brain is crucial in treating neurodegenerative diseases such as AD (Tohda, Kuboyama, and Komatsu 2005). Trigonelline is reported to have the neurite-regenerative activity (regeneration of dendrites and axons) in addition to memory improvement (Tohda, Kuboyama, and Komatsu 2005; Tohda et al. 1999). Treatment with trigonelline (30 microM) to SK-N-SH cells increased the percentage of neurons and neurites reacting positively to phosphorylated neurofilament-H at 6 days of treatment, suggesting regeneration dendrites and axons in addition to memory improvement (Tohda, Kuboyama, and Komatsu 2005). These results led to the potential of fenugreek (rich in trigonelline) as a possible option for managing neurodegenerative diseases, including AD, in the future.

Taken together, available published literature provided the evidence of efficacy of fenugreek and its phytoconstituents against psychological disorders including occupational and menopausal stress, anxiety, depression, and cognitive decline through multiple mechanisms of action as presented in Table 15.3.

TABLE 15.3
Mechanism of Fenugreek and Its Phytoconstituents against Psychological Disorders

Disorder (Constituent)	Action	Mechanism	References
Occupational stress (Galactomannan)	Anti-oxidant activity	↑ CAT, SOD, GPx, GSH ↓ lipid peroxidation	Pandaran et al. (2016)
Menopausal stress (Phytoestrogens)	Anti-oxidant activity in the brain	↓ MDA, ↑ GSH in brain	Anjaneyulu et al. (2018)
	↑ Cholinergic activity	Inhibition of acetylcholinesterase enzyme	Anjaneyulu et al. (2018)
	Neuroprotection	↓ Neuronal degeneration and loss in dentate gyrus and CA1, CA3 hippocampal areas	Anjaneyulu et al. (2018)
	Estrogen mimicking	↓ Ovariectomy-induced alterations	Anjaneyulu et al. (2018)
	↓ Inflammation	↓ Accumulation of cytokines such as IL-1, TNF-α, and IL-6	Abedinzade, Nasri, Jamal, Ghasemi et al. (2015)
	↓ Metabolic changes	↓ Blood glucose levels, lipids, and insulin	Handa et al. (2005) Abedinzade et al. (2012)

(Continued)

TABLE 15.3 (Continued)

Disorder (Constituent)	Action	Mechanism	References
Anxiety (Flavonoids, polysaccharides, saponins, alkaloids)	↑ Skeletal muscle relaxation	Antagonism of serotonin receptor	Assad and Khan (2017), Hausenblas et al. (2020), Iyer et al. (2004), Mohan et al. (2006), Dhananjaya et al. (2011)
Depression (Saponin glycosides, phytosterols, flavonoids, glycosides, amino acid (4-hydroxyisoleucine)	Restored neuro-transmission	↑ Serotonin, dopamine, and ↓ GABA	Pawar et al. (2008)
	Anti-oxidant activity	↓ MDA, ↑ SOD	Khursheed et al. (2014)
	↓ Hyperactive brain activity	↓ Hyperactive HPA axis, serum cortisol level, monoamines metabolism, and serotonin turnover	(Kalshetti et al. 2015)
	Reduced metabolism of dopamine	Inhibition of MAO-A and -B	Khursheed et al. (2014)
Learning and memory impairments (Trigonelline, saponins)	↑ Neurotraumatic ACh	↓ AChE hyperactivity	Prema et al. (2016)
	↓ Amyloid-beta (Aβ) burden	↓ Proteins patterns of pTau, enzymes, IDE, and CDK5 involved in the metabolism	Prema et al. (2017)
	↓ Apoptosis	↑ Akt/GSK3beta pathway ↑ Bax, Bcl2, and caspase-3 expression ↓ DNA damage ↑ Nerve regeneration	Prema et al. (2017), Khalil, Roshdy, and Kassem (2016), Khalili et al. (2018)
	Anti-oxidant activity	↓ MDA	Saini et al. (2011)
	↓ Neuro-inflammation	↓ Cox2, TNF-□, IL-6, IL-1beta, Iba-1, iNOS, NF-kappaB, TLR4, GFAP, STAT3, ↑ BDNF	Khalili et al. (2018), Chowdhury et al. (2018), Prema et al. (2017)
Cognitive disorders, Dementia, AD (Trigonelline)	↑ Cholinergic activity	↓ AChE	Khalili et al. (2018), Fahanik-Babaei et al. (2019)
	Anti-oxidant activity in the brain	↑ CAT, SOD, GSH, ↓ MDA	Fahanik-Babaei et al. (2019)
	↓ apoptosis	↑ Mitochondrial membrane potential	Fahanik-Babaei et al. (2019)
	↓ Neuro-inflammation	↓ Hippocampal levels of GFAP, S100b, Cox2, TNF-α, IL-6, protein carbonyl, LDH	Fahanik-Babaei et al. (2019)

15.4 MANAGEMENT OF DIABETES-INDUCED CNS COMPLICATIONS

Diabetes is a dreadful disease with a high prevalence rate characterized by metabolic abnormalities and hyperglycemia due to decreased insulin levels leading to metabolic, structural, and physiological complications in various organ systems, including CNS (Acherjya et al. 2017). Hyperglycemia and oxidative stress are the major factors responsible for electrophysiological, structural abnormalities, and neurodegeneration in the brain occur in DM (Thakur, Tyagi, and Shekhar 2019). The consequences of diabetes mellitus on the nervous system and research regarding medicinal remedies to overcome them are receiving remarkable attention (Bafadam et al. 2019).

Many natural products, including fenugreek, were explored to develop the CNS specifically to prevent diabetes-linked neurological and cognitive disorders (Patel and Udayabanu 2017). Fenugreek seed powder reported antihyperglycemic and neuroprotective activities in alloxan monohydrate (15 mg/100 g body weight)-induced diabetic rats (Kumar et al. 2012). In this study, 21 day supplementation of 5% finely powdered fenugreek seeds in rat feed (5 g of powder + 95 g of powdered rat feed) showed reversal of the hyperglycemia-induced changes (membrane linked enzymes (Na+K+ATPase, Ca2+ATPase), anti-oxidant enzymes (SOD, glutathione S-transferase), calcium (Ca2+) levels, and neurolipofuscin accumulation levels) to normal levels in the diabetic rat brain (Kumar et al. 2012). These parameters, especially the plasma membrane Ca2+-ATPase pumps, play an essential role in maintaining precise levels of intracellular Ca2+ essential to the functioning of neurons (Zaidi 2010). A decrease in synaptosome membrane fluidity is reported to influence the activity of membrane-linked enzymes in diabetes (Kamboj, Chopra, and Sandhir 2009). Additional experiments on the fenugreek powder (5% in feed) in combination with sodium orthovanadate (SOV), a competitive inhibitor of ATPases, reported similar reversal of the alloxan-induced changes in synaptosome membrane fluidity and histochemical localization and distribution of neurolipofuscin along with levels of membrane-linked enzymes (Na+/K+-ATPase and Ca2+ATPase activities) (Kumar et al. 2015).

The hydroalcoholic extract of fenugreek seed (50, 100, and 200 mg/kg, orally, daily for 6 weeks) was reported to prevent streptozotocin-induced (55 mg/kg, single intrapersonal injection) hyperglycemia, oxidative stress, and memory impairment in rats in a dose-dependent manner (Bafadam et al. 2019). The fenugreek seed extract prevented biochemical latencies (to enter the darkroom in passive avoidance test) and biochemical elevated levels (of NO metabolites and MDA and depleted levels of CAT, thiols, SOD in the cortical and hippocampal tissues) to suggest protective effects against streptozotocin-induced cognitive decline and oxidative stress (Bafadam et al. 2019). Similar ameliorative effects were reported by fenugreek seed extract against cognitive dysfunction in streptozotocin (60 mg/kg, single intrapersonal injection)-induced diabetic rats (Kodumuri et al. 2019). Fenugreek showed significantly improved performance in neurobehavioral tasks of cognitive functions (viz. T-maze and Morris water maze), reduced the oxidative stress in the brain hippocampal region (lipid peroxidation levels), and decreased neuronal loss from the CA1 and CA3 regions of the hippocampus, and suggested neuroprotective efficacy against streptozotocin-induced cognitive dysfunction in rats (Kodumuri et al. 2019).

A more recent study suggested the role of an anti-oxidant mechanism in neuroprotective activity of ethanolic extract of fenugreek seeds (250 mg/kg, oral, once daily for 4 weeks) against alloxan (120 mg/kg, single intraperitoneal injection)-induced diabetic rats (Pradeepkiran, Nandyala, and Bhaskar 2020). In this study, ethanolic extract of fenugreek seeds protected the rats from alloxan-induced alterations in the brain cytosolic enzymes involved in glucose homeostasis and neuronal integrity (succinate dehydrogenase, glutamate dehydrogenase, and glucose-6-phosphate dehydrogenase) (Pradeepkiran, Nandyala, and Bhaskar 2020).

In the reports of neuroprotective efficacy of fenugreek seed extracts (Kodumuri et al. 2019; Kumar et al. 2012, 2015; Pradeepkiran, Nandyala, and Bhaskar 2020), the responsible phytochemical has not been known. The preclinical evidence of efficacy for 4-hydroxyisoleucine (4HI), a major amino acid component of fenugreek seeds, against stress, anxiety, and depression (Gaur et al. 2012; Kalshetti et al. 2015) in non-diabetic animals, are important indicators in this regard. 4-HI has been reported to possess a glucose-dependent insulin-releasing activity in isolated islets (Broca et al. 2000), the pancreas (Sauvaire et al. 1998), and *in vivo* in rats and dogs (Broca et al. 1999).

The critical link between glucose metabolism and CNS complications in DM is the neurotransmitter, 5-hydroxytryptamine, also called serotonin. Serotonin plays a key role in controlling insulin secretion, and that its absence leads to diabetes (Robinson 2009). An increase in brain serotonin turnover reduces food intake (Garfield and Heisler 2009) and might reduce food craving in DM. Plasma serotonin levels are known to increase in DM (Martin et al. 1995), where glucose utilization needs to be increased in skeletal muscle. Furthermore, serotonin's ability to increase glucose uptake in skeletal muscle and lower blood glucose is a well-validated phenomenon (Guillet-Deniau, Burnol, and Girard 1997; Hajduch et al. 1999).

Therefore, the serotonergic medicated mechanism was expected to play a vital role in the mechanism of 4-HI for the management of CNS complications of DM. This notion was confirmed with a published study of the efficacy of 4HI against CNS complications in an animal model of established DM (Thakurdesai, Mohan, Adil et al. 2015). Subacute treatment of 4HI (10, 20, 40, and 80 mg/kg, once daily, 4 weeks) showed dose-dependent reversal of STZ 45 mg/kg, intraperitoneal)-induced depression and anxiety parameters such as increased anxiety (elevated plus maze, EPM), increased depressive behavior (tail suspension test, TST), and reduced locomotion (open field test, OFT) and reduced brain neurotransmitters (GABA, dopamine, serotonin, noradrenaline) levels in rats reported (Thakurdesai, Mohan, Adil et al. 2015). In an additional *in vitro* study, the goat brain mitochondria, when co-incubated with 4HI, showed significant protection against CuAs-induced oxidative stress parameters alterations such as levels of biomarkers, enzymes of Kreb's cycle and mitochondrial respiratory chain, and maintained mitochondrial morphology (Thakurdesai, Mohan, Adil et al. 2015). These reports suggested that the neuroprotective and restorative action of fenugreek seed extracts against the CNS complications of DM is, at least partly, attributed to 4HI, probably through serotoninergic and anti-oxidant defense system mechanisms.

15.5 CONCLUSIONS

The documented literature of fenugreek seed powder, extract, and phytoconstituents revealed beneficial effects in managing neurological and psychological disorders through a diverse set of mechanisms. The phytoconstituents, such as 4HI (major amino acid) and trigonelline (major alkaloid) from fenugreek seed, demonstrated beneficial effects on psychological (stress, mood, anxiety, depression) and neurological (PD, AD, and cognitive dysfunctions) conditions, respectively. This information will help design and develop fenugreek-based complementary medicines and food supplements to manage neuropsychological disorders.

REFERENCES

Abedinzade, M., S. Nasri, Omodi M. Jamal, Bizhan Porramezan, and Korosh Khanaki. 2012. "The effect of Fenugreek (*Trigonella foenum-graecum*) Seed and 17-β estradiol on serum apelin, glucose, lipids and insulin in ovariectomized rats." *Biotechnology and Health Sciences* 2. doi: 10.17795/bhs-30402.

Abedinzade, Mahmood, Sima Nasri, Omidi Masome Jamal, Bizhan Porramezan, and Korosh Khanaki. 2015. "The effect of fenugreek (*Trigonella foenum-graecum*) seed and 17-β estradiol on serum apelin, glucose, lipids, and insulin in ovariectomized rats." *Biotechnology and Health Sciences* 2 (3). doi: 10.17795/bhs-30402.

Abedinzade, M., S. Nasri, Omodi M. Jamal, E. Ghasemi, and A. Ghorbani. 2015. "Efficacy of *Trigonella foenum-graecum* seed extract in reducing metabolic and inflammatory alterations associated with menopause." *Iran Red Crescent Med J* 17 (11):e26685. doi: 10.5812/ircmj.26685.

Acherjya, Goutam, Md Uddin, M. A. Chowdhury, and A. V. Srinivasan. 2017. "Central nervous system manifestations in diabetes mellitus—A review." *J Med* 18:109. doi: 10.3329/jom.v18i2.33689.

Al-Matubsi, H. Y., N. A. Nasrat, G. A. Oriquat, M. Abu-Samak, K. A. Al-Mzain, and M. Salim. 2011. "The hypocholesterolemic and antioxidative effect of dietary diosgenin and chromium chloride supplementation on high-cholesterol fed Japanese quails." *Pak J Biol Sci* 14 (7):425–32. doi: 10.3923/pjbs.2011.425.432.

Amabeoku, G., and R. Chandomba. 1994. "Strychnine-induced seizures in mice: The role of noradrenaline." *Prog Neuropsychopharmacol Biol Psychiatry* 18 (4):753–63. doi: 10.1016/0278-5846(94)90082-5.

Anjaneyulu, K., S. Rai Kiranmai, T. Rajesh, T. Nagamma, and Kumar M. R. Bhat. 2018. "Therapeutic efficacy of fenugreek extract or/and choline with docosahexaenoic acid in attenuating learning and memory deficits in ovariectomized rats." *Journal of Krishna Institute of Medical Sciences University* 7 (2):10–20.

Ansari, M., P. Sadeghi, H. Mahdavi, M. Fattahi-Dolatabadi, N. Mohamadi, A. Asadi, and F. Sharififar. 2019. "Fenugreek dermal patch, a new natural topical antinociceptive medication for relieving the postherniotomy pain, a double-blind placebo controlled trial." *J Complement Integr Med* 16 (3):1–8. doi: 10.1515/jcim-2018-0082.

Assad, T., and R. A. Khan. 2017. "Effect of methanol extract of *Trigonella foenum-graecum* L. seeds on anxiety, sedation and motor coordination." *Metab Brain Dis* 32 (2):343–9. doi: 10.1007/s11011-016-9914-y.

Assad, T., R. A. Khan, and M. A. Rajput. 2018. "Effect of *Trigonella foenum-graecum* Linn. seeds methanol extract on learning and memory." *Metab Brain Dis* 33 (4):1275–80. doi: 10.1007/s11011-018-0235-1.

Aswar, Manoj, Piyush Upadhyay, Amit Kandhare, Subhash L. Bodhankar, Vishwaraman Mohan, and Prasad Thakurdesai. 2014. "Mechanism based efficacy study of trigonelline based standardized fenugreek seed extract (IBHB) on levodopa induced dyskinesia in 6-hydroxydopamine lesioned rat model of Parkinson's disease [P-81]." 2nd National Conference on "Herbal and Synthetic Drug Studies (HSDS-2014), Azam Campus, Camp, Pune, 10–12 Feb 2014.

Baeza, I., N. M. De Castro, L. Gimenez-Llort, and M. De la Fuente. 2010. "Ovariectomy, a model of menopause in rodents, causes a premature aging of the nervous and immune systems." *J Neuroimmunol* 219 (1–2):90–9. doi: 10.1016/j.jneuroim.2009.12.008.

Bafadam, S., F. Beheshti, T. Khodabakhshi, A. Asghari, B. Ebrahimi, H. R. Sadeghnia, M. Mahmoudabady, S. Niazmand, and M. Hosseini. 2019. "*Trigonella foenum-graecum* seed (Fenugreek) hydroalcoholic extract improved the oxidative stress status in a rat model of diabetes-induced memory impairment." *Horm Mol Biol Clin Investig* 39 (2):1–13. doi: 10.1515/hmbci-2018-0074.

Ballard, C. G., K. A. Chalmers, C. Todd, I. G. McKeith, J. T. O'Brien, G. Wilcock, S. Love, and E. K. Perry. 2007. "Cholinesterase inhibitors reduce cortical Abeta in dementia with Lewy bodies." *Neurology* 68 (20):1726–9. doi: 10.1212/01.wnl.0000261920.03297.64.

Barry, D. T., M. Beitel, T. Breuer, C. J. Cutter, J. Savant, R. S. Schottenfeld, and B. J. Rounsaville. 2011. "Conventional and unconventional treatments for stress among methadone-maintained patients: Treatment willingness and perceived efficacy." *Am J Addict* 20 (2):137–42. doi: 10.1111/j.1521-0391.2010.00109.x.

Barth, C., A. Villringer, and J. Sacher. 2015. "Sex hormones affect neurotransmitters and shape the adult female brain during hormonal transition periods." *Front Neurosci* 9:37. doi: 10.3389/fnins.2015.00037.

Belaid-Nouira, Y., H. Bakhta, M. Bouaziz, I. Flehi-Slim, Z. Haouas, and H. Ben Cheikh. 2012. "Study of lipid profile and parieto-temporal lipid peroxidation in AlCl(3) mediated neurotoxicity. Modulatory effect of fenugreek seeds." *Lipids Health Dis* 11 (16):16. doi: 10.1186/1476-511X-11-16.

Belaid-Nouira, Y., H. Bakhta, S. Samoud, M. Trimech, Z. Haouas, and H. Ben Cheikh. 2013. "A novel insight on chronic AlCl3 neurotoxicity through IL-6 and GFAP expressions: Modulating effect of functional food fenugreek seeds." *Nutr Neurosci* 16 (5):218–24. doi: 10.1179/1476830512Y.0000000048.

Bharate, Sonali S, Serge Mignani, and Ram A Vishwakarma. 2018. "Why are the majority of active compounds in the CNS domain natural products? A critical analysis." *J Medicinal Chem* 61 (23):10345–74.

Bir, Shyamal C., Oleg Y. Chernyshev, and Alireza Minagar. 2018. "Chapter 7 — remyelination in multiple sclerosis: A mechanistic look." In *Neuroinflammation*, edited by Alireza Minagar, 163–74. Academic Press.

Bishop, C., J. L. Taylor, D. M. Kuhn, K. L. Eskow, J. Y. Park, and P. D. Walker. 2006. "MDMA and fenfluramine reduce L-DOPA-induced dyskinesia via indirect 5-HT1A receptor stimulation." *Eur J Neurosci* 23 (10):2669–76. doi: 10.1111/j.1460-9568.2006.04790.x.

Biswal, S., M. C. Das, and P. Nayak. 2003. "Antinociceptive activity of seeds of *Trigonella foenum gracecum* in rats." *Indian J Physiol Pharmacol* 47 (4):479–80.

Borras, C., J. Sastre, D. Garcia-Sala, A. Lloret, F. V. Pallardo, and J. Vina. 2003. "Mitochondria from females exhibit higher antioxidant gene expression and lower oxidative damage than males." *Free Radic Biol Med* 34 (5):546–52. doi: 10.1016/s0891-5849(02)01356-4.

Bourin, M. 2018. "Clinical pharmacology of anxiolytics." *Arch Depress Anxiety* 4 (1):21–5.

Broca, C., R. Gross, P. Petit, Y. Sauvaire, M. Manteghetti, M. Tournier, P. Masiello, R. Gomis, and G. Ribes. 1999. "4-Hydroxyisoleucine: Experimental evidence of its insulinotropic and antidiabetic properties." *Am J Physiol* 277 (4):E617–E23. doi: 10.1152/ajpendo.1999.277.4.E617.

Broca, C., M. Manteghetti, R. Gross, Y. Baissac, M. Jacob, P. Petit, Y. Sauvaire, and G. Ribes. 2000. "4-Hydroxyisoleucine: Effects of synthetic and natural analogues on insulin secretion." *Eur J Pharmacol* 390 (3):339–45. doi: 10.1016/s0014-2999(00)00030-3.

Budson, A. E., and B. H. Price. 2005. "Memory dysfunction." *N Engl J Med* 352 (7):692–9. doi: 10.1056/NEJMra041071.

Caixeta, L., G. M. N. da Silva Junior, V. M. Caixeta, C. H. R. Reimer, and Pvbe Azevedo. 2012. "Occupational health, cognitive disorders and occupational neuropsychology." *Dement Neuropsychol* 6 (4):198–202. doi: 10.1590/S1980-57642012DN06040002.

Campbell, A., E. Y. Yang, M. Tsai-Turton, and S. C. Bondy. 2002. "Pro-inflammatory effects of aluminum in human glioblastoma cells." *Brain Res* 933 (1):60–5. doi: 10.1016/s0006-8993(02)02305-3.

Cha, H. Y., J. H. Park, J. T. Hong, H. S. Yoo, S. Song, B. Y. Hwang, J. S. Eun, and K. W. Oh. 2005. "Anxiolytic-like effects of ginsenosides on the elevated plus-maze model in mice." *Biol Pharm Bull* 28 (9):1621–5. doi: 10.1248/bpb.28.1621.

Chakrabarty, P., K. Jansen-West, A. Beccard, C. Ceballos-Diaz, Y. Levites, C. Verbeeck, A. C. Zubair, D. Dickson, T. E. Golde, and P. Das. 2010. "Massive gliosis induced by interleukin-6 suppresses Abeta deposition in vivo: Evidence against inflammation as a driving force for amyloid deposition." *FASEB J* 24 (2):548–59. doi: 10.1096/fj.09-141754.

Chatterjee, M., R. Verma, V. Lakshmi, S. Sengupta, A. K. Verma, A. A. Mahdi, and G. Palit. 2013. "Anxiolytic effects of *Plumeria rubra* var. acutifolia (Poiret) L. flower extracts in the elevated plus-maze model of anxiety in mice." *Asian J Psychiatr* 6 (2):113–8. doi: 10.1016/j.ajp.2012.09.005.

Chiu, C. S., J. S. Deng, M. T. Hsieh, M. J. Fan, M. M. Lee, F. S. Chueh, C. K. Han, Y. C. Lin, and W. H. Peng. 2009. "Yam (Dioscorea pseudojaponica Yamamoto) ameliorates cognition deficit and attenuates oxidative damage in senescent mice induced by D-galactose." *Am J Chin Med* 37 (5):889–902. doi: 10.1142/s0192415x09007296.

Chowdhury, A. A., N. B. Gawali, R. Munshi, and A. R. Juvekar. 2018. "Trigonelline insulates against oxidative stress, proinflammatory cytokines and restores BDNF levels in lipopolysaccharide induced cognitive impairment in adult mice." *Metab Brain Dis* 33 (3):681–91. doi: 10.1007/s11011-017-0147-5.

Ciulla, M., L. Marinelli, I. Cacciatore, and A. D. Stefano. 2019. "Role of dietary supplements in the management of Parkinson's disease." *Biomolecules* 9 (7):1–23. doi: 10.3390/biom9070271.

Corder, E. H., K. Robertson, L. Lannfelt, N. Bogdanovic, G. Eggertsen, J. Wilkins, and C. Hall. 1998. "HIV-infected subjects with the E4 allele for APOE have excess dementia and peripheral neuropathy." *Nat Med* 4 (10):1182–4. doi: 10.1038/2677.

Cotzias, G. C., P. S. Papavasiliou, and R. Gellene. 1969. "Modification of Parkinsonism—chronic treatment with L-dopa." *N Engl J Med* 280 (7):337–45. doi: 10.1056/nejm196902132800701.

Cui, Y., J. Wu, S. C. Jung, D. B. Park, Y. H. Maeng, J. Y. Hong, S. J. Kim, S. R. Lee, S. J. Kim, S. J. Kim, and S. Y. Eun. 2010. "Anti-neuroinflammatory activity of nobiletin on suppression of microglial activation." *Biol Pharm Bull* 33 (11):1814–21. doi: 10.1248/bpb.33.1814.

Dar, Tariq Ahmad. 2018. "Fenugreek: A miraculous medicinal herb." *J Complement Med Alternat Healthcare* 7 (2):1–3. doi: 10.19080/jcmah.2018.07.555710.

Davies, James, and John Read. 2019. "A systematic review into the incidence, severity and duration of antidepressant withdrawal effects: Are guidelines evidence-based?" *Addictive Behaviors* 97:111–21. https://doi.org/10.1016/j.addbeh.2018.08.027.

Delibasi, T., C. Kockar, A. Celik, and O. Kockar. 2006. "Antioxidant effects of hormone replacement therapy in postmenopausal women." *Swiss Med Wkly* 136 (31–32):510–4. doi: 2006/31/smw-11390.

Deumens, R., A. Blokland, and J. Prickaerts. 2002. "Modeling Parkinson's disease in rats: An evaluation of 6-OHDA lesions of the nigrostriatal pathway." *Exp Neurol* 175 (2):303–17. doi: 10.1006/exnr.2002.7891.

Dhananjaya, D. R., K. S. Vijay, G. P. Chandrashekar, I. K. Makhija, and Shivakumara S. 2011. "Anxiolytic activity of ethanolic extract of *Trigonella foenum-graecum* seeds." *Arch Appl Sci Res* 3 (1):91–5.

Di Paolo, M., L. Papi, F. Gori, and E. Turillazzi. 2019. "Natural products in neurodegenerative diseases: A great promise but an ethical challenge." *Int J Mol Sci* 20 (20). doi: 10.3390/ijms20205170.

Dome, Peter, Laszlo Tombor, Judit Lazary, Xenia Gonda, and Zoltan Rihmer. 2019. "Natural health products, dietary minerals and over-the-counter medications as add-on therapies to antidepressants in the treatment of major depressive disorder: A review." *Brain Research Bulletin* 146:51–78. https://doi.org/10.1016/j.brainresbull.2018.12.015.

Dong, Y., Y. Wang, Y. Liu, N. Yang, and P. Zuo. 2013. "Phytoestrogen alpha-zearalanol ameliorates memory impairment and neuronal DNA oxidation in ovariectomized mice." *Clinics (Sao Paulo)* 68 (9):1255–62. doi: 10.6061/clinics/2013(09)13.

Doyle, B. J., J. Frasor, L. E. Bellows, T. D. Locklear, A. Perez, J. Gomez-Laurito, and G. B. Mahady. 2009. "Estrogenic effects of herbal medicines from Costa Rica used for the management of menopausal symptoms." *Menopause* 16 (4):748–55. doi: 10.1097/gme.0b013e3181a4c76a.

Dubin, A. E., and A. Patapoutian. 2010. "Nociceptors: The sensors of the pain pathway." *J Clin Invest* 120 (11):3760–72. doi: 10.1172/JCI42843.

Ellis, A., and D. L. Bennett. 2013. "Neuroinflammation and the generation of neuropathic pain." *Br J Anaesth* 111 (1):26–37. doi: 10.1093/bja/aet128.

Essa, Musthafa M, Reshmi K Vijayan, Gloria Castellano-Gonzalez, Mustaq A Memon, Nady Braidy, and Gilles J Guillemin. 2012. "Neuroprotective effect of natural products against Alzheimer's disease." *Neurochem Res* 37 (9):1829–42.

Fahanik-Babaei, J., T. Baluchnejadmojarad, F. Nikbakht, and M. Roghani. 2019. "Trigonelline protects hippocampus against intracerebral Abeta(1–40) as a model of Alzheimer's disease in the rat: Insights into underlying mechanisms." *Metab Brain Dis* 34 (1):191–201. doi: 10.1007/s11011-018-0338-8.

Fahn, S., and R. L. Elton. 1987. "Members of the UPDRS development committee. Unified Parkinson's disease rating scale." In *Recent Developments in Parkinson's Disease*, edited by S. Fahn, C. D. Marsden, D. B. Calne and M. Goldstein, 153–64. Florham Park, NJ: Macmillan Health Care Information.

Farajdokht, F., S. Sadigh-Eteghad, A. Majdi, F. Pashazadeh, S. M. Vatandoust, M. Ziaee, F. Safari, P. Karimi, and J. Mahmoudi. 2020. "Serotonergic system modulation holds promise for L-DOPA-induced dyskinesias in hemiparkinsonian rats: A systematic review." *EXCLI J* 19:268–95. doi: 10.17179/excli2020-1024.

Faustino, T. T., R. B. Almeida, and R. Andreatini. 2010. "Medicinal plants for the treatment of generalized anxiety disorder: A review of controlled clinical studies." *Braz J Psychiatry* 32 (4):429–36. doi: 10.1590/s1516-44462010005000026.

Ferretti, G., T. Bacchetti, C. Marchionni, and N. Dousset. 2004. "Effect of non-enzymatic glycation on aluminium-induced lipid peroxidation of human high density lipoproteins (HDL)." *Nutr Metab Cardiovasc Dis* 14 (6):358–65. doi: 10.1016/s0939-4753(04)80026-7.

Foley, P., and P. Riederer. 2000. "Influence of neurotoxins and oxidative stress on the onset and progression of Parkinson's disease." *J Neurol* 247 Suppl 2:II82–94. doi: 10.1007/pl00007766.

Foltynie, T., and J. Kahan. 2013. "Parkinson's disease: An update on pathogenesis and treatment." *J Neurol* 260 (5):1433–40. doi: 10.1007/s00415-013-6915-1.

Fujimori, H., and D. Cobb. 1955. "Potentiation of barbital hypnosis as an evaluation method for central nervous system." *Psychopharmacology* 7:374–7.

Garfield, A. S., and L. K. Heisler. 2009. "Pharmacological targeting of the serotonergic system for the treatment of obesity." *J Physiol* 587 (Pt 1):49–60. doi: 10.1113/Jphysiol.2008.164152 [pii].

Gaur, Vaibhav, Subhash L. Bodhankar, Vishwaraman Mohan, and P. A. Thakurdesai. 2012. "Antidepressant-like effect of 4-hydroxyisoleucine from *Trigonella foenum-graecum* L. seeds in mice." *Biomedicine & Aging Pathology* 2 (3):121–5. doi: 10.1016/j.biomag.2012.07.002.

Gaur, V., S. L. Bodhankar, V. Mohan, and P. A. Thakurdesai. 2013. "Neurobehavioral assessment of hydroalcoholic extract of *Trigonella foenum-graecum* seeds in rodent models of Parkinson's disease." *Pharm Biol* 51 (5):550–7. doi: 10.3109/13880209.2012.747547.

Ghasemi, Zahra, Zahra Kiasalari, Fatemeh Ebrahimi, Fariba Ansari, Maryam Sharayeli, and Mehrdad Roghani. 2017. "Neuroprotective effect of diosgenin in 6- hydroxydopamine-induced model of Parkinson's disease in the rat." *Daneshvar Medicine* 24 (129).

Goberman, A. M., and M. Blomgren. 2003. "Parkinsonian speech disfluencies: Effects of L-dopa-related fluctuations." *J Fluency Disord* 28 (1):55–70. doi: 10.1016/s0094-730x(03)00005-6.

Goldenberg, M. M. 2010. "Overview of drugs used for epilepsy and seizures: Etiology, diagnosis, and treatment." *P T* 35 (7):392–415.

Goldman, S. A., and N. J. Kuypers. 2015. "How to make an oligodendrocyte." *Development* 142 (23):3983–95. doi: 10.1242/dev.126409.

Guillet-Deniau, I., A. F. Burnol, and J. Girard. 1997. "Identification and localization of a skeletal muscle secretonin 5-HT2A receptor coupled to the Jak/STAT pathway." *J Biolog Chem* 272 (23):14825–9.

Guo-Ross, S. X., E. Y. Yang, T. J. Walsh, and S. C. Bondy. 1999. "Decrease of glial fibrillary acidic protein in rat frontal cortex following aluminum treatment." *J Neurochem* 73 (4):1609–14. doi: 10.1046/j.1471-4159.1999.0731609.x.

Hajduch, E., F. Rencurel, A. Balendran, I. H. Batty, C. P. Downes, and H. S. Hundal. 1999. "Serotonin (5-Hydroxytryptamine), a novel regulator of glucose transport in rat skeletal muscle." *J Biolog Chem* 274 (19):13563–8.

Hakimi, S., M. R. Siahi Shadbad, R. Bamdad Moghadam, F. Abbasalizadeh, P. Mustafa Ghrebaghi, Hossein Babaei, S. Bamdad Moghadam, and A. Delazar. 2005. "Effect of fenugreek seed on early menopausal symptoms." *Pharmaceut Sci* (2):83–90.

Hakimi, S., S. Mohammad Alizadeh, A. Delazar, F. Abbasalizadeh, R. Bamdad Mogaddam, M. R. Siiahi, and P. Mostafa Garabagi. 2006. "Probable effects of fenugreek seed on hot flash in menopausal women." *JMPIR* 5 (19):9–14.

Handa, T., K. Yamaguchi, Y. Sono, and K. Yazawa. 2005. "Effects of fenugreek seed extract in obese mice fed a high-fat diet." *Biosci Biotechnol Biochem* 69 (6):1186–8. doi: 10.1271/bbb.69.1186.

Hausenblas, H. A., K. L. Conway, K. R. M. Coyle, E. Barton, L. D. Smith, M. Esposito, C. Harvey, D. Oakes, and D. R. Hooper. 2020. "Efficacy of fenugreek seed extract on men's psychological and physical health: A randomized placebo-controlled double-blind clinical trial." *J Complement Integr Med*. doi: 10.1515/jcim-2019-0101.

Helmy, Neveen, Neveen Abou El-Soud, Mona Khalil, Jihan Hussein, F. S. H. Oraby, and Abdel Razik Farrag. 2007. "Antidiabetic effects of fenugreek alkaloid extract in streptozotocin induced hyperglycemic rats." *J Appl Sci Res* 3:1073–83.

Inanmdar, W., A. Sultana, U. Mubeen, and K. Rahman. 2016. "Clinical efficacy of *Trigonella foenum-graecum* (Fenugreek) and dry cupping therapy on intensity of pain in patients with primary dysmenorrhea." *Chin J Integr Med*:1–8. doi: 10.1007/s11655-016-2259-x.

Ingkaninan, Kornkanok, Nanteetip Limpeanchop, Sakchai Wittaya-areekul, Krongkarn Chootip, Pornnarin Taepavarapruk, Nuwat Taepavarapruk, Jintanaporn Wattanathorn, Seewaboon Sireeratawong, Waraporn Putalun, and Watoo Phrompittayarat. 2013. "Brahmi, a medicinal plant for memory improvement." *The Open Conference Proceedings Journal* 4 (1):85.

Iyer, M., H. Belapurkar, Dr Omkar Sherikar, and Sanjay Kasture. 2004. "Anxiolytic activity of *Trigonella foenum-greacum* seeds." *J Natur Remedies* 4:61–65. doi: 10.18311/jnr/2004/384.

Jabeen, Asra, S. Rani, and Ibrahim Mohammed. 2018. "Pharmacognostic and therapeutic importance of fenugreek (*Trigonella foenum graecium* L.)." *Indo Am J Pharmaceutical Sci* 5 (6):5253–62. doi: 10.5281/zenodo.1291099.

Javan, M., A. Ahmadiani, S. Semnanian, and M. Kamalinejad. 1997. "Antinociceptive effects of *Trigonella foenum-graecum* leaves extract." *J Ethnopharmacol* 58 (2):125–9. doi: 10.1016/s0378-8741(97)00089-5.

Javitch, J. A., and S. H. Snyder. 1984. "Uptake of MPP(+) by dopamine neurons explains selectivity of parkinsonism-inducing neurotoxin, MPTP." *Eur J Pharmacol* 106 (2):455–6. doi: 10.1016/0014-2999(84)90740-4.

Jayaraj, Richard L., Sheikh Azimullah, Rami Beiram, Fakhreya Y. Jalal, and Gary A. Rosenberg. 2019. "Neuroinflammation: Friend and foe for ischemic stroke." *J Neuroinflammation* 16 (1):142. doi: 10.1186/s12974-019-1516-2.

Jiang, T., B. K. Huang, Q. Y. Zhang, T. Han, H. C. Zheng, and L. P. Qin. 2007. "Effect of Liriope platyphylla total saponin on learning, memory and metabolites in aging mice induced by D-galactose." *Zhong Xi Yi Jie He Xue Bao* 5 (6):670–4. doi: 10.3736/jcim20070614.

Jung, D. H., H. J. Park, H. E. Byun, Y. M. Park, T. W. Kim, B. O. Kim, S. H. Um, and S. Pyo. 2010. "Diosgenin inhibits macrophage-derived inflammatory mediators through downregulation of CK2, JNK, NF-kappaB and AP-1 activation." *Int Immunopharmacol* 10 (9):1047–54. doi: 10.1016/j.intimp.2010.06.004.

Kakkar, V., and I. P. Kaur. 2011. "Evaluating potential of curcumin loaded solid lipid nanoparticles in aluminium induced behavioural, biochemical and histopathological alterations in mice brain." *Food Chem Toxicol* 49 (11):2906–13. doi: 10.1016/j.fct.2011.08.006.

Kalshetti, P. B., R. Alluri, V. Mohan, and P. A. Thakurdesai. 2015. "Effects of 4-hydroxyisoleucine from fenugreek seeds on depression-like behavior in socially isolated olfactory bulbectomized rats." *Pharmacogn Mag* 11 (Suppl 3):S388–96. doi: 10.4103/0973-1296.168980.

Kamboj, S. S., K. Chopra, and R. Sandhir. 2009. "Hyperglycemia-induced alterations in synaptosomal membrane fluidity and activity of membrane bound enzymes: Beneficial effect of N-acetylcysteine supplementation." *Neuroscience* 162 (2):349–58. doi: 10.1016/j.neuroscience.2009.05.002.

Kang, Tong Ho, Eunjung Moon, Bin Na Hong, Sang Zin Choi, Miwon Son, Ji-Ho Park, and Sun Yeou Kim. 2011. "Diosgenin from Dioscorea nipponica ameliorates diabetic neuropathy by inducing nerve growth factor." *Biological & Pharmaceutical Bulletin* 34 (9):1493–8. doi: 10.1248/bpb.34.1493.

Kao, Y. C., P. C. Ho, Y. K. Tu, I. M. Jou, and K. J. Tsai. 2020. "Lipids and Alzheimer's disease." *Int J Mol Sci* 21 (4). doi: 10.3390/ijms21041505.

Khalil, Wagdy, Hanaa Roshdy, and Salwa Kassem. 2016. "The potential therapeutic role of Fenugreek saponin against Alzheimers disease: Evaluation of apoptotic and acetylcholinesterase inhibitory activities." *J Appl Pharmaceutical Sci* 6 (9):166–73. doi: 10.7324/japs.2016.60925.

Khalili, M., M. Alavi, E. Esmaeil-Jamaat, T. Baluchnejadmojarad, and M. Roghani. 2018. "Trigonelline mitigates lipopolysaccharide-induced learning and memory impairment in the rat due to its anti-oxidative and anti-inflammatory effect." *Int Immunopharmacol* 61:355–62. doi: 10.1016/j.intimp.2018.06.019.

Khan, Alam Rafeeq, Tahira Assad, and Rajput Muhammad Ali. 2017. "Anticonvulsant effects of *Trigonella foenum-graecum* L. in strychnine induced epilepsy model." *J Nutr Health & Food Sci* 5 (7):1–6. doi: 10.15226/jnhfs.2017.001115.

Khan, Haroon, Sadia Perviz, Antoni Sureda, Seyed M. Nabavi, and Silvia Tejada. 2018. "Current standing of plant derived flavonoids as an antidepressant." *Food and Chem Toxicol* 119:176–88. https://doi.org/10.1016/j.fct.2018.04.052.

Khursheed, R., G. H. Rizwani, V. Sultana, M. Ahmed, and A. Kamil. 2014. "Antidepressant effect and categorization of inhibitory activity of monoamine oxidase type A and B of ethanolic extract of seeds of *Trigonella foenum-graecum* Linn." *Pak J Pharm Sci* 27 (5 Spec no):1419–25.

Kodumuri, P. K., C. Thomas, R. Jetti, and A. K. Pandey. 2019. "Fenugreek seed extract ameliorates cognitive deficits in streptozotocin-induced diabetic rats." *J Basic Clin Physiol Pharmacol* 30 (4). doi: 10.1515/jbcpp-2018-0140.

Koh, E. K., W. B. Yun, J. E. Kim, S. H. Song, J. E. Sung, H. A. Lee, E. J. Seo, S. W. Jee, C. J. Bae, and D. Y. Hwang. 2016. "Beneficial effect of diosgenin as a stimulator of NGF on the brain with neuronal damage induced by Aβ-42 accumulation and neurotoxicant injection." *Lab Anim Res* 32 (2):105–15. doi: 10.5625/lar.2016.32.2.105.

Kumar, P., A. Taha, N. Kumar, V. Kumar, and N. Z. Baquer. 2015. "Sodium orthovanadate and *Trigonella foenum-graecum* prevents neuronal parameters decline and impaired glucose homeostasis in alloxan diabetic rats." *Prague Med Rep* 116 (2):122–38. doi: 10.14712/23362936.2015.51.

Kumar, P., R. K. Kale, P. McLean, and N. Z. Baquer. 2012. "Antidiabetic and neuroprotective effects of *Trigonella foenum-graecum* seed powder in diabetic rat brain." *Prague Med Rep* 113 (1):33–43. doi: 10.14712/23362936.2015.35.

Kumar, Vikas. 2006. "Potential medicinal plants for CNS disorders: An overview." *Phytotherapy Research: An International Journal Devoted to Pharmacological and Toxicological Evaluation of Natural Product Derivatives* 20 (12):1023–35.

Lalo, Rezarta. 2017. "Menopausal symptoms and women's quality of life outcomes: Literature review." *EC Gynaecology* 6:167–72.

Larson, A. A., and A. J. Beitz. 1988. "Glycine potentiates strychnine-induced convulsions: Role of NMDA receptors." *J Neurosci* 8 (10):3822–6. doi: 10.1523/jneurosci.08-10-03822.1988.

Li, B., P. Xu, S. Wu, Z. Jiang, Z. Huang, Q. Li, and D. Chen. 2018. "Diosgenin attenuates lipopolysaccharide-induced Parkinson's disease by inhibiting the TLR/NF-κB pathway." *J Alzheimers Dis* 64 (3):943–55. doi: 10.3233/jad-180330.

Liu, W., M. Zhu, Z. Yu, D. Yin, F. Lu, Y. Pu, C. Zhao, C. He, and L. Cao. 2017. "Therapeutic effects of diosgenin in experimental autoimmune encephalomyelitis." *J Neuroimmunol* 313:152–60. doi: 10.1016/j.jneuroim.2017.10.018.

Locher, Cosima, Helen Koechlin, Sean R. Zion, Christoph Werner, Daniel S. Pine, Irving Kirsch, Ronald C. Kessler, and Joe Kossowsky. 2017. "Efficacy and safety of selective serotonin reuptake inhibitors, serotonin-norepinephrine reuptake inhibitors, and placebo for common psychiatric disorders among children and adolescents: A systematic review and meta-analysis." *JAMA Psychiatry* 74 (10):1011–20. doi: 10.1001/jamapsychiatry.2017.2432.

Mahat, M. Y., N. M. Kulkarni, S. L. Vishwakarma, F. R. Khan, B. S. Thippeswamy, V. Hebballi, A. A. Adhyapak, V. S. Benade, S. M. Ashfaque, S. Tubachi, and B. M. Patil. 2010. "Modulation of the cyclooxygenase pathway via inhibition of nitric oxide production contributes to the anti-inflammatory activity of kaempferol." *Eur J Pharmacol* 642 (1–3):169–76. doi: 10.1016/j.ejphar.2010.05.062.

Mandegary, A., M. Pournamdari, F. Sharififar, S. Pournourmohammadi, R. Fardiar, and S. Shooli. 2012. "Alkaloid and flavonoid rich fractions of fenugreek seeds (*Trigonella foenum-graecum* L.) with antinociceptive and anti-inflammatory effects." *Food Chem Toxicol* 50 (7):2503–7. doi: 10.1016/j.fct.2012.04.020.

Marotta, F., Y. Naito, F. Padrini, X. Xuewei, S. Jain, V. Soresi, L. Zhou, R. Catanzaro, K. Zhong, A. Polimeni, and D. H. Chui. 2011. "Redox balance signalling in occupational stress: Modification by nutraceutical intervention." *J Biol Regul Homeost Agents* 25 (2):221–9.

Martin, F. J., J. M. Miguez, M. Aldegunde, and G. Atienza. 1995. "Effect of streptozotocin-induced diabetes mellitus on serotonin measures of peripheral tissues in rats." *Life Sci* 56 (1):51–9. doi: 10.1016/0024-3205(94)00407-j.

Martucciello, Stefania, Milena Masullo, Antonietta Cerulli, and Sonia Piacente. 2020. "Natural products targeting ER stress, and the functional link to mitochondria." *Int J Mol Sci* 21 (6):1905.

Massano, J., and K. P. Bhatia. 2012. "Clinical approach to Parkinson's disease: Features, diagnosis, and principles of management." *Cold Spring Harb Perspect Med* 2 (6):a008870. doi: 10.1101/cshperspect.a008870.

Maurer, S. V., and C. L. Williams. 2017. "The cholinergic system modulates memory and hippocampal plasticity via its interactions with non-neuronal cells." *Front Immunol* 8 (1489):1489. doi: 10.3389/fimmu.2017.01489.

McClatchey, Will C, Gail B Mahady, Bradley C Bennett, Laura Shiels, and Valentina Savo. 2009. "Ethnobotany as a pharmacological research tool and recent developments in CNS-active natural products from ethnobotanical sources." *Pharmacology & Therapeutics* 123 (2):239–54.

Mendell, J. R., S. Kolkin, J. T. Kissel, K. L. Weiss, D. W. Chakeres, and K. W. Rammohan. 1987. "Evidence for central nervous system demyelination in chronic inflammatory demyelinating polyradiculoneuropathy." *Neurology* 37 (8):1291–391. doi: 10.1212/wnl.37.8.1291.

Mirzaie, M., M. Khalili, Z. Kiasalari, and M. Roghani. 2016. "Neuroprotective and antiapoptotic potential of trigonelline in a striatal 6-hydroxydopamine rat model of Parkinson's disease." *Neurophysiology* 48 (3):176–83. doi: 10.1007/s11062-016-9586-6.

Moghadam, F. H., B. Vakili-Zarch, M. Shafiee, and A. Mirjalili. 2013. "Fenugreek seed extract treats peripheral neuropathy in pyridoxine induced neuropathic mice." *EXCLI J* 12:282–90.

Mohan, Mahalaxmi, A. Banekar, T. Birdi, Pradnya Bharambe, S. Kaul, and A. Patel. 2006. "Nootropic and anxiolytic activity of fenugreek seeds." *Journal of Natural Remedies* 6:153–56. doi: 10.18311/jnr/2006/459.

Morani, A. S., S. L. Bodhankar, V. Mohan, and P. A. Thakurdesai. 2012. "Ameliorative effects of standardized extract from *Trigonella foenum-graecum* L. seeds on painful peripheral neuropathy in rats." *Asian Pac J Trop Med* 5 (5):385–90. doi: 10.1016/S1995-7645(12)60064-9.

Munoz-Torrero, D. 2008. "Acetylcholinesterase inhibitors as disease-modifying therapies for Alzheimer's disease." *Curr Med Chem* 15 (24):2433–55. doi: 10.2174/092986708785909067.

Myers, R. R., W. M. Campana, and V. I. Shubayev. 2006. "The role of neuroinflammation in neuropathic pain: Mechanisms and therapeutic targets." *Drug Discov Today* 11 (1–2):8–20. doi: 10.1016/s1359-6446(05)03637-8.

Nabavi, Seyed Mohammad, Maria Daglia, Nady Braidy, and Seyed Fazel Nabavi. 2017. "Natural products, micronutrients, and nutraceuticals for the treatment of depression: A short review." *Nutri Neurosci* 20 (3):180–94. doi: 10.1080/1028415X.2015.1103461.

Narapogu, Venkatanarayana, Naveen Aalsyam, Naveen Pokala, T. Jayasree, and John Premendran. 2018. "Evaluation of antidepressant and antioxidant activity of fenugreek (*Trigonella foenum-greacum*) seed extract in wistar rats." *J Chem Pharmaceutical Res* 10 (6):135–40.

Nasri, S., M. Abedinzade, Omidi M. Jamal, and F. Noursabaghi. 2013. "Comparison of analgesic and hypoglycemic effect of hexanic and aolhlic extract of fenugreek seed in male diabetic rats." *J Ardabil University of Med Sci* 13 (1):102–9.

Natarajan, B., A. Muralidharan, R. Satish, and R. Dhananjayan. 2007. "Neuropharmacological activity of *Trigonella foenum-graecum* Linn. seeds." *J Natur Remedies* 7 (1):160–5.

Nathan, J., Siddika Panjwani, V. Mohan, Veena Joshi, and P. A. Thakurdesai. 2014. "Efficacy and safety of standardized extract of *Trigonella foenum-graecum* L seeds as an adjuvant to L-dopa in the management of patients with Parkinson's disease." *Phytotherapy Res* 28:172–8.

Nayak, Roopa P., Prabhakar Adake, and K. T. Hafis. 2019. "Antidepressant activity of *Trigonella foenum* leaves in Wistar albino rats." *Int J Basic & Clin Pharmacol* 8 (5):963. doi: 10.18203/2319-2003.ijbcp20191584.

Neta, Francisca, Ianara Da Costa, Francisca Lima, Luciana Fernandes, José Cavalcanti, Marco Freire, Eudes De Souza Lucena, Amália Meneses Do Rêgo, Eduardo De Azevedo, and Fausto Guzen. 2018. "Effects of Mucuna pruriens (L.) supplementation on experimental models of Parkinson's disease: A systematic review." *Pharmacognosy Reviews* 12 (23).

Nicklas, W. J., I. Vyas, and R. E. Heikkila. 1985. "Inhibition of NADH-linked oxidation in brain mitochondria by 1-methyl-4-phenyl-pyridine, a metabolite of the neurotoxin, 1-methyl-4-phenyl-1,2,5,6-tetrahydropyridine." *Life Sci* 36 (26):2503–8. doi: 10.1016/0024-3205(85)90146-8.

Nsour, W. M., C. B. Lau, and I. C. Wong. 2000. "Review on phytotherapy in epilepsy." *Seizure* 9 (2):96–107. doi: 10.1053/seiz.1999.0378.

O'Callaghan, J. P. 1991. "Assessment of neurotoxicity: Use of glial fibrillary acidic protein as a biomarker." *Biomed Environ Sci* 4 (1–2):197–206.

Oliff, Heather S. 2014. "Parkinson's disease progression slowed by adjunctive fenugreek extract "*HerbalGram* Fall (103):33.

Orhan, G., I. Orhan, N. Subutay-Oztekin, F. Ak, and B. Sener. 2009. "Contemporary anticholinesterase pharmaceuticals of natural origin and their synthetic analogues for the treatment of Alzheimer's disease." *Recent Pat CNS Drug Discov* 4 (1):43–51. doi: 10.2174/157488909787002582.

Pal, Dilipkumar, and Souvik Mukherjee. 2020. "Chapter 13 — Fenugreek (*Trigonella foenum*) seeds in health and nutrition." In *Nuts and Seeds in Health and Disease Prevention* (Second Edition), edited by Victor R. Preedy and Ronald Ross Watson, 161–70. Cambridge, MA: Academic Press.

Pandaran, Sudheeran, S., D. Jacob, Mulakal Natinga, J., Gopinathan G. Nair, A. Maliakel, B. Maliakel, R. Kuttan, and K. Im. 2016. "Safety, tolerance, and enhanced efficacy of a bioavailable formulation of curcumin with fenugreek dietary fiber on occupational stress: A randomized, double-blind, placebo-controlled pilot study." *J Clin Psychopharmacol* 36 (3):236–43. doi: 10.1097/JCP.0000000000000508.

Papapetropoulos, S., M. Basel, and D. C. Mash. 2007. "Dopaminergic innervation of the human striatum in Parkinson's disease." *Movement Disorders* 22 (2):286–8. doi: 10.1002/mds.21196 [doi].

Parvizpur, A., A. Ahmadiani, and M. Kamalinejad. 2004. "Spinal serotonergic system is partially involved in antinociception induced by *Trigonella foenum-graecum* (TFG) leaf extract." *J Ethnopharmacol* 95 (1):13–7. doi: 10.1016/j.jep.2004.05.020.

Parvizpur, A., A. Ahmadiani, and M. Kamalinejad. 2006. "Probable role of spinal purinoceptors in the analgesic effect of *Trigonella foenum* (TFG) leaves extract." *J Ethnopharmacol* 104 (1–2):108–12. doi: 10.1016/j.jep.2005.08.057.

Patel, S. S., and M. Udayabanu. 2017. "Effect of natural products on diabetes associated neurological disorders." *Rev Neurosci* 28 (3):271–93. doi: 10.1515/revneuro-2016-0038.

Pawar, Vinod S, Shivakumar Hugar, Bhagyashri Gawade, and RN Patil. 2008. "Evaluation of antidepressant like activity of *Trigonella foenum-graecum* Liinn. seeds in mice." *Pharmacology Online* 1:455–65.

Pfeilschifter, J., R. Koditz, M. Pfohl, and H. Schatz. 2002. "Changes in proinflammatory cytokine activity after menopause." *Endocr Rev* 23 (1):90–119. doi: 10.1210/edrv.23.1.0456.

Pradeepkiran, Jangampalli Adi, Venkata Subbaiah Nandyala, and Matcha Bhaskar. 2020. "*Trigonella foenum-graecum* seeds extract plays a beneficial role on brain antioxidant and oxidative status in alloxan-induced Wistar rats." *Food Quality and Safety* 4 (2):83–9. doi: 10.1093/fqsafe/fyaa015.

Prema, A., A. J. Thenmozhi, T. Manivasagam, M. M. Essa, M. D. Akbar, and M. Akbar. 2016. "Fenugreek seed powder nullified aluminium chloride induced memory loss, biochemical changes, abeta burden and apoptosis via regulating Akt/GSK3beta signaling pathway." *PLoS One* 11 (11):e0165955. doi: 10.1371/journal.pone.0165955.

Prema, A., A. J. Thenmozhi, T. Manivasagam, M. Mohamed Essa, and G. J. Guillemin. 2017. "Fenugreek seed powder attenuated aluminum chloride-induced tau pathology, oxidative stress, and inflammation in a rat model of Alzheimer's disease." *J Alzheimers Dis* 60 (s1):S209–S20. doi: 10.3233/JAD-161103.

Rabiei, Z, M Gholami, and MR Hojjati. 2014. "The effect of *Cyperus rotundus* ethanolic extract on motor coordination in a rat model of Alzheimer." *J Adv Med Biomed Res* 22 (92):43–54.

Robinson, R. 2009. "Serotonin's role in the pancreas revealed at last." *PLoS Biol* 7 (10):e1000227. doi: 10.1371/journal.pbio.1000227.

Rodriguez-Yañez, Manuel, and Jose Castillo. 2008. "Role of inflammatory markers in brain ischemia." *Current Opinion in Neurology* 21:353–7. doi: 10.1097/WCO.0b013e3282ffafbf.

Rozenberg, S., J. Vandromme, and C. Antoine. 2013. "Postmenopausal hormone therapy: Risks and benefits." *Nat Rev Endocrinol* 9 (4):216–27. doi: 10.1038/nrendo.2013.17.

Saini, Dinesh, Ashwani Dhingra, Bhawna Chopra, and Milind Parle. 2011. "Psychopharmacological Investigation of the Nootropic potential of Trigonella foenum Linn in mice." *Asian J Pharmaceutical and Clin Res* 4:76–84.

Sarwar, Sidra, Muhammad Asif Hanif, Muhammad Adnan Ayub, Yaw Duah Boakye, and Christian Agyare. 2020. "Chapter 20 — Fenugreek." In *Medicinal Plants of South Asia*, edited by Muhammad Asif Hanif, Haq Nawaz, Muhammad Mumtaz Khan and Hugh J. Byrne, 257–71. London: Elsevier.

Satheeshkumar, N., P. K. Mukherjee, S. Bhadra, and B. P. Saha. 2010. "Acetylcholinesterase enzyme inhibitory potential of standardized extract of Trigonella foenum-graecum L and its constituents." *Phytomedicine* 17 (3–4):292–5. doi: 10.1016/j.phymed.2009.06.006.

Sauvaire, Y., P. Petit, C. Broca, M. Manteghetti, Y. Baissac, J. Fernandez-Alvarez, R. Gross, M. Roye, A. Leconte, R. Gomis, and G. Ribes. 1998. "4-Hydroxyisoleucine: A novel amino acid potentiator of insulin secretion." *Diabetes* 47 (2):206–10. doi: 10.2337/diab.47.2.206.

Shamshad Begum, S., H. K. Jayalakshmi, H. G. Vidyavathi, G. Gopakumar, I. Abin, M. Balu, K. Geetha, S. V. Suresha, M. Vasundhara, and I. M. Krishnakumar. 2016. "A novel extract of fenugreek husk (FenuSMART) alleviates postmenopausal symptoms and helps to establish the hormonal balance: A randomized, double-blind, placebo-controlled study." *Phytother Res* 30 (11):1775–84. doi: 10.1002/ptr.5680.

Sharififar, F., M. H. Moshafi, E. Shafazand, and A. Koohpayeh. 2012. "Acetyl cholinesterase inhibitory, antioxidant and cytotoxic activity of three dietary medicinal plants." *Food Chemistry* 130 (1):20–3. doi: 10.1016/j.foodchem.2011.06.034.

Shulman, Lisa M., Ann L. Gruber-Baldini, Karen E. Anderson, Paul S. Fishman, Stephen G. Reich, and William J. Weiner. 2010. "The Clinically Important Difference on the Unified Parkinson's Disease Rating Scale." *Archives of Neurology* 67 (1):64–70. doi: 10.1001/archneurol.2009.295.

Singh, S., V. Kumar, N. Kumar, P. Sharma, and S. M. Waheed. 2020. "Protective and modulatory effects of *Trapa bispinosa* and *Trigonella foenum-graecum* on neuroblastoma cells through neuronal nitric oxide synthase." *Assay Drug Dev Technol* 18 (1):64–74. doi: 10.1089/adt.2018.912.

Solomon, D., E. Ford, J. Adams, and N. Graves. 2011. "Potential of St John's Wort for the treatment of depression: The economic perspective." *Aust N Z J Psychiatry* 45 (2):123–30. doi: 10.3109/00048674.2010.526094.

Sreeja, S., V. S. Anju, and S. Sreeja. 2010. "In vitro estrogenic activities of fenugreek *Trigonella foenum-graecum* seeds." *Indian J Med Res* 131:814–9.

Sreerama, Yadahally N., and Ma Rachana. 2019. "Fenugreek (*Trigonella foenum-graecum* L) seed phytochemicals protect DNA from free radical induced oxidative damage and inhibit dipeptidyl peptidase-IV an enzyme associated with hyperglycemia."

Srinivasan, Krishnapura. 2019. "Chapter 3.15 — Fenugreek (Trigonella foenum-graecum L.) seeds used as functional food supplements to derive diverse health benefits." In *Nonvitamin and Nonmineral Nutritional Supplements*, edited by Seyed Mohammad Nabavi and Ana Sanches Silva, 217–21. Cambridge, MA: Academic Press.

Stafstrom, C. E., and L. Carmant. 2015. "Seizures and epilepsy: An overview for neuroscientists." *Cold Spring Harb Perspect Med* 5 (6):a022426. doi: 10.1101/cshperspect.a022426.

Steels, E., M. L. Steele, M. Harold, and S. Coulson. 2017. "Efficacy of a proprietary *Trigonella foenum-graecum* L. de-husked seed extract in reducing menopausal symptoms in otherwise healthy women: A double-blind, randomized, placebo-controlled study." *Phytother Res* 31 (9):1316–22. doi: 10.1002/ptr.5856.

Thakur, Ajit Kumar, Sakshi Tyagi, and Nikhila Shekhar. 2019. "Comorbid brain disorders associated with diabetes: Therapeutic potentials of prebiotics, probiotics and herbal drugs." *Translational Medicine Communications* 4 (1):12.

Thakurdesai, Prasad, V. Mohan, Mohammad Adil, Amit Kandhare, Subhash Laxman Bodhankar, Arnab Kumar Ghosh, and Debasish Bandyopadhyay. 2015. "Efficacy of 4-Hydroxyisoleucine from Fenugreek seeds against central nervous system complications of Diabetes: *In vivo* and *in vitro* evidence." First International Conference on Novel Frontiers in Pharmaceutical & Health Sciences (INNOPHARM1), Bhopal.

Thakurdesai, Prasad, Vishwaraman Mohan, Amit Kandhare, and Subhash L. Bodhankar. 2015. "Therapeutic efficacy of trigonelline-based standardized extract of fenugreek seeds on Levodopa induced dyskinesia in 6-OHDA lesioned rat model of Parkinson's disease." Journal of Natural Product and Chemistry Research, HICC Hyderabad, India, October 26–28, 2015.

Tohda, C., T. Kuboyama, and K. Komatsu. 2005. "Search for natural products related to regeneration of the neuronal network." *Neurosignals* 14 (1–2):34–45. doi: 10.1159/000085384.

Tohda, C., N. Nakamura, K. Komatsu, and M. Hattori. 1999. "Trigonelline-induced neurite outgrowth in human neuroblastoma SK-N-SH cells." *Biological & Pharmaceutical Bulletin* 22 (7):679–82.

Tohda, Chihiro, Takuya Urano, Masahito Umezaki, Ilka Nemere, and Tomoharu Kuboyama. 2012. "Diosgenin is an exogenous activator of 1,25D_3-MARRS/Pdia3/ERp57 and improves Alzheimer's disease pathologies in 5XFAD mice." *Scientific reports* 2:535–35. doi: 10.1038/srep00535.

Tohda, C., X. Yang, M. Matsui, Y. Inada, E. Kadomoto, S. Nakada, H. Watari, and N. Shibahara. 2017. "Diosgenin-rich yam extract enhances cognitive function: A placebo-controlled, randomized, double-blind, crossover study of healthy adults." *Nutrients* 9 (10). doi: 10.3390/nu9101160.

Tripathi, U. N., and D. Chandra. 2009. "The plant extracts of *Momordica charantia* and *Trigonella foenum-graecum* have anti-oxidant and anti-hyperglycemic properties for cardiac tissue during diabetes mellitus." *Oxid Med Cell Longev* 2 (5):290–6. doi: 10.4161/oxim.2.5.9529.

Tuite, P., and J. Riss. 2003. "Recent developments in the pharmacological treatment of Parkinson's disease." *Expert Opin Investig Drugs* 12 (8):1335–52. doi: 10.1517/13543784.12.8.1335.

Turchan-Cholewo, Jadwiga, Yiling Liu, Suzanne Gartner, Rollie Reid, Chunfa Jie, Xuejun Peng, Kuey Chen, Ashok Chauhan, Norman Haughey, Roy Cutler, Mark Mattson, Carlos Pardo, katherine Conant, Ned Sacktor, Justin McArthur, Kurt Hauser, Chandra Gairola, and Avindra Nath. 2006. "Increased vulnerability of ApoE4 neurons to HIV proteins and opiates: Protection by diosgenin and L-deprenyl." *Neurobiology of Disease* 23:109–19. doi: 10.1016/j.nbd.2006.02.005.

Uemura, T., S. Hirai, N. Mizoguchi, T. Goto, J. Y. Lee, K. Taketani, Y. Nakano, J. Shono, S. Hoshino, N. Tsuge, T. Narukami, N. Takahashi, and T. Kawada. 2010. "Diosgenin present in fenugreek improves glucose metabolism by promoting adipocyte differentiation and inhibiting inflammation in adipose tissues." *Mol Nutr Food Res* 54 (11):1596–608. doi: 10.1002/mnfr.200900609.

Ulbricht, C., E. Basch, D. Burke, L. Cheung, E. Ernst, N. Giese, I. Foppa, P. Hammerness, S. Hashmi, G. Kuo, M. Miranda, S. Mukherjee, M. Smith, D. Sollars, S. Tanguay-Colucci, N. Vijayan, and W. Weissner. 2007. "Fenugreek (*Trigonella foenum-graecum* L. Leguminosae): An evidence-based systematic review by the natural standard research collaboration." *J Herb Pharmacother* 7 (3–4):143–77. doi: 10.1080/15228940802142852.

Valero, J., L. Bernardino, F. L. Cardoso, A. P. Silva, C. Fontes-Ribeiro, A. F. Ambrosio, and J. O. Malva. 2017. "Impact of neuroinflammation on hippocampal neurogenesis: Relevance to aging and Alzheimer's disease." *J Alzheimers Dis* 60 (s1):S161–S8. doi: 10.3233/JAD-170239.

Vijayakumar, M. V., V. Pandey, G. C. Mishra, and M. K. Bhat. 2010. "Hypolipidemic effect of fenugreek seeds is mediated through inhibition of fat accumulation and upregulation of LDL receptor." *Obesity (Silver Spring)* 18 (4):667–74. doi: 10.1038/oby.2009.337.

Villas Boas, Gustavo Roberto, Roseli Boerngen de Lacerda, Marina Meirelles Paes, Priscila Gubert, Wagner Luis da Cruz Almeida, Vanessa Cristina Rescia, Pablinny Moreira Galdino de Carvalho, Adryano Augustto Valladao de Carvalho, and Silvia Aparecida Oesterreich. 2019. "Molecular aspects of depression: A review from neurobiology to treatment." *European Journal of Pharmacology* 851:99–121. https://doi.org/10.1016/j.ejphar.2019.02.024.

Vyas, S., R. P. Agrawal, P. Solanki, and P. Trivedi. 2008. "Analgesic and anti-inflammatory activities of *Trigonella foenum-graecum* (seed) extract." *Acta Pol Pharm* 65 (4):473–6.

Wani, Sajad Ahmad, and Pradyuman Kumar. 2018. "Fenugreek: A review on its nutraceutical properties and utilization in various food products." *Journal of the Saudi Society of Agricultural Sciences* 17 (2):97–106. doi: 10.1016/j.jssas.2016.01.007.

Xiao, L., D. Guo, C. Hu, W. Shen, L. Shan, C. Li, X. Liu, W. Yang, W. Zhang, and C. He. 2012. "Diosgenin promotes oligodendrocyte progenitor cell differentiation through estrogen receptor-mediated ERK1/2 activation to accelerate remyelination." *Glia* 60 (7):1037–52. doi: 10.1002/glia.22333.

Younesy, S., S. Amiraliakbari, S. Esmaeili, H. Alavimajd, and S. Nouraei. 2014. "Effects of fenugreek seed on the severity and systemic symptoms of dysmenorrhea." *J Reprod Infertil* 15 (1):41–8.

Zaidi, A. 2010. "Plasma membrane Ca-ATPases: Targets of oxidative stress in brain aging and neurodegeneration." *World J Biol Chem* 1 (9):271–80. doi: 10.4331/wjbc.v1.i9.271.

Zameer, S., A. K. Najmi, D. Vohora, and M. Akhtar. 2018. "A review on therapeutic potentials of *Trigonella foenum-graecum* (fenugreek) and its chemical constituents in neurological disorders: Complementary roles to its hypolipidemic, hypoglycemic, and antioxidant potential." *Nutr Neurosci* 21 (8):539–45. doi: 10.1080/1028415X.2017.1327200.

Zhang, X., X. Xue, J. Zhao, C. Qian, Z. Guo, Y. Ito, and W. Sun. 2016. "Diosgenin attenuates the brain injury induced by transient focal cerebral ischemia-reperfusion in rats." *Steroids* 113:103–12. doi: 10.1016/j.steroids.2016.07.006.

Zhang, Y. 2020. *Better Understanding of Progressive Multiple Sclerosis*. PhD Thesis, Hobart, Australia: University of Tasmania.

Zhao, Wei-Xin, Peng-Fei Wang, Hui-Gang Song, and Nai Sun. 2017. "Diosgenin attenuates neuropathic pain in a rat model of chronic constriction injury." *Mol Med Rep* 16 (2):1559–64. doi: 10.3892/mmr.2017.6723.

Zhou, J., L. Chan, and S. Zhou. 2012. "Trigonelline: A plant alkaloid with therapeutic potential for diabetes and central nervous system disease." *Curr Med Chem* 19 (21):3523–31. doi: 10.2174/092986712801323171.

16 Fenugreek in Management of Female-Specific Health Conditions

Urmila Aswar and Deepti Rai

CONTENTS

16.1 Introduction 259
16.2 Fenugreek as Galactagogue 260
    16.2.1 Fenugreek as a Galactagogue in Breastfeeding Mothers 260
    16.2.2 Fenugreek for Animal Milk Production 262
    16.2.3 Probable Mechanisms of Galactagogue Action of Fenugreek 262
16.3 Fenugreek in Management of Polycystic Ovarian Syndrome (PCOS) 262
16.4 Fenugreek as a Female Libido Enhancer 264
16.5 Fenugreek in the Management of Post-Menopausal Symptoms (PMS) 264
    16.5.1 Fenugreek in Dysmenorrhea 265
    16.5.2 Fenugreek on Post-Menopausal Cognitive Decline 266
    16.5.3 Fenugreek in Management of Menopause-Induced Osteoporosis 267
16.6 Fenugreek in Management of Female-Specific Cancers 267
16.7 Safety of Fenugreek in Females 269
16.8 Conclusion 271
References 272

16.1 INTRODUCTION

Worldwide, women are the center of the home and society and play a vital role in their overall health. Women undergo many physiological changes starting with the initiation of menses in the teens, pubertal changes, sexual life, motherhood, and the post-menopausal period. World Health Organization's report show women outlive men, although they show a need for better health care options (Thornton 2019). Though they have more life expectancy, females also suffer from poor health, prominently in underdeveloped and developing countries. The significant causes of death amongst women are breast cancer, maternal conditions, cervical cancer, and Alzheimer's disease (Thornton 2019). In developing and underdeveloped countries, women's health problems are often linked to nutritional deficiencies and low and inadequate prenatal and antenatal care due to poor socio-economic conditions. Women in developing countries are 25 times more prone to death due to improper pregnancy than women in developed countries (Picetti et al. 2020).

Traditional or herbal medicines for specific use in females are well documented and have been passed to the next generations, as evident by use in the current generations of women (Raja 2015). In both rural areas and urban setups, women trust herbal and kitchen remedies to overcome female-related disorders such as urinary tract infections, post-menopausal syndrome, hot flushes, menopause, polycystic ovarian syndrome (PCOS), bacterial vaginosis, yeast infections, infertility, delayed labor, low breast milk production, abortion, and other female disorders.

Medicinal use of many kitchen spices, including fenugreek (scientific name: *Trigonella foenum-graecum* L.), have been reported by women from rural areas (Bora et al. 2016) and the Western world (Hegde, Hegde, and Kholkute 2007). Worldwide, fenugreek is consumed as a food ingredient as a flavor in cooking and as a green leafy vegetable for centuries (Srinivasan 2005). Depending on the country, fenugreek seeds have been recommended in the traditional medicinal system as an antispasmodic; externally for abscesses, boils, galactagogues (El-Kamali and Khalid 1996); as an appetite stimulant; for blood cleansing; as a laxative and tonic (Ziyyat et al. 1997); as a demulcent, emollient, expectorant, aphrodisiac (Foster and Tyler 1999); and for restorative and nutritive properties (Petit et al. 1995). In the Indian traditional medicine system, ground fenugreek seeds mixed with the jaggary are given to females after childbirth to develop and strengthen the muscles (Choudhary et al. 2001; Mittal and Gopaldas 1986).

Fenugreek seeds and leaves contain many biologically and medicinally active phytoconstituents, such as mucilaginous fiber, proteins, saponins, lipids, and carbohydrates. The defatted seeds are a rich source of steroids, especially saponins-rich fractions (Petit et al. 1995), trigonelline, flavonoids, and carotenoids (Al-Habori and Raman 2002; Basch et al. 2003). The phytoestrogens present are claimed to be useful in disease prevention and health promotion in females of all ages (Nagulapalli Venkata et al. 2017). Fenugreek seed contains 50% of edible fiber, out of which 20% of fibers are insoluble, while 30% make up a unique soluble gum, mainly galactomannan (Thomas et al. 2011). Besides its glucose and lipid-lowering property, it is also reported to exhibit an anabolic effect as demonstrated by an increase in weight of *levator ani*, a skeletal muscle, of rats when treated with galactomannan (35 mg/kg, p.o.) (Aswar et al. 2008).

Apart from traditional medicinal literature, many scientific studies have reported on the beneficial health effects of various parts of fenugreek such as its leaf, seeds, the whole plant, and derived phytoconstituents—the body of evidence is available in modern scientific literature demonstrating a broad spectrum of activities. Substantial scientific evidence is available to support the benefits of fenugreek for female-specific disorders. The present chapter summarizes such evidence from experimental and clinical studies and putative mechanisms of fenugreek and its phytoconstituents for potential use in managing female-specific disorders. The scientific evidence on the beneficial effects of fenugreek as a galactagogue (Table 16.1) and in management of other female-specific conditions (Table 16.2) is also presented.

16.2 FENUGREEK AS GALACTAGOGUE

The breastfeeding mother's milk is easily digested by the newborn, directly promoting their growth and health (Damanik, Wahlqvist, and Wattanapenpaiboon 2006). However, a medical management option, the galactagogue, is needed in nursing mothers who have insufficient milk production (Khan, Wu, and Dolzhenko 2018). The synthetic galactagogues such as metoclopramide, domperidone, chlorpromazine, and sulpiride increase milk production through serum prolactin stimulation and dopamine D2 receptor antagonism but have unacceptable side effects (Gabay 2002). Therefore, natural galactagogues have become the preference of lactating mothers.

16.2.1 Fenugreek as a Galactagogue in Breastfeeding Mothers

Like many other plants, fenugreek contains many useful components to enhance breast milk supply (Zapantis, Steinberg, and Schilit 2012). Fenugreek seeds have been recommended as one of the herbs taken by breastfeeding mothers to support lactation and increase milk production in traditional medicinal system worldwide, especially in countries like India, China, Italy, Sudan, and Egypt (Al-Asadi 2014; Forinash et al. 2012; Gabay 2002; Morcos and Gabrial 1983). In the Indian culture, these supplementations are provided in the form of food preparations such as besan methi (vegetable), methi raita (a curd preparation), methi ladoo (ball-shaped preparation), and sprouted methi (Arora and Ramawat 2018).

A case report summarized the anecdotal use of fenugreek in 1200 women, who reported an increase in milk supply within 24 h to 72 h (Dog 2009). The usual dose of fenugreek dried powdered seeds for galactagogue effect is 1 to 2 g, to be taken 3 times per day (Dog 2009). The beneficial effects of fenugreek in lactating mothers were demonstrated during a randomized, double-blind, placebo-controlled study (Turkyılmaz et al. 2011). The herbal tea (3 cups, total 200 ml) containing granules of galactagogue fenugreek and another as compared with placebo (apple) containing tea to lactating mothers showed convincing beneficial effects on breast milk volume and time to regain weight without adverse effects (Turkyılmaz et al. 2011). A population-based retrospective survey performed in Australian breastfeeding mothers revealed fenugreek as the most used galactagogue amongst 59% of breastfeeding women who took herbs for lactation stimulation (Sim et al. 2013).

During a triple-blind, placebo-controlled study, the oral supplementation of fenugreek seeds (3 capsules each containing 610 mg, 3 times per day) for 5 days, to 60 postpartum mothers at a tertiary private hospital was reported to significantly increase mean breast milk volume production (ml/hr) without serious side effects (Brillante and Mantaring 2014). Another clinical research in Sudanese mothers revealed increased prolactin and milk secretion on supplementation of 500 mg of fenugreek seed powder 3 times a day for 3 months (Ahmed 2015).

The galactagogue effects of 2-week treatment of herbal tea (1 cup 3 times a day containing 2 g of fenugreek) in terms of a significant increase in breast milk volume from the third day onwards from postpartum, was reported during a randomized controlled study on 75 breastfeeding women (El Sakka, Salama, and Salama 2014). A significant increase in the overall development of 78 girl infants (0–4 months) and increased breastfeeding frequency by mothers was reported in mothers who consumed fenugreek-containing herbal tea for 4 months (Ghasemi, Kheirkhah, and Vahedi 2015). A study conducted in healthy female infants and their mothers showed consumption of herbal tea containing 7.5 g of fenugreek seed powder (3 times per day, 1 month) improved breast milk sufficiency Hudson (2017). During a recent study, overnight soaked fenugreek water (7.5 g) for seven days administration to 60 postnatal women could enrich the breast milk production in mothers and achieve significant weight gain in the infants in the initial week of their life (Ravi and Joseph 2020). However, the meta-analysis of many clinical studies on the galactagogue effects of fenugreek

TABLE 16.1
Applications of Fenugreek and Phytoconstituents as Galactagogue

Fenugreek Constituent (Dose)	Material and Methods	Results and Proposed Mechanisms	References
Dried fenugreek leaves (1–2 g, 3 times a day, 72 h)	Case report, 1200 women	↑ breast milk	(Dog 2009)
Fenugreek herbal tea (3 cups, 200 ml)	RCT v/s apple tea, lactating mothers	↑ on breast milk ↓ time to regain weight No adverse effects	(Turkyılmaz et al. 2011)
Fenugreek seed powder (500 mg, 3 times a day, 3 months)	Open-label study, Sudanese breastfeeding mothers	↑ prolactin and milk secretion	(Ahmed 2015)
Fenugreek seed powder (7.5 g as herbal tea, 3 times a day for 1 month)	RCT, 72 Iranian breastfeeding mothers	↑ Breast milk	(Ghasemi, Kheirkhah, and Vahedi 2015)
Fenugreek-water (7.5 g overnight soaked, once a day for 1 week)	RCT, 60 Indian breastfeeding mothers	↑ Milk production and quality ↑ Infants weight	(Ravi and Joseph 2020)
Fenugreek seed (2 g in herbal tea, 14 days)	RCT, 75 puerperal women	↑ Breast milk ↑ % weight	(El Sakka, Salama, and Salama 2014)
Ground fenugreek seeds (50 g/kg for 30 days)	Anatolian water buffaloes	↑ Feed intake and milk production	(Degirmencioglu et al. 2016)
Fenugreek in diet (1 g/kg per day, during location period)	72 rats with lactation challenge	↑ Milk production (16%) ↑ Pup growth (11%)	(Sevrin et al. 2019)

RCT—Randomized controlled trial

compared to placebo control provided the most convincing evidence (Khan, Wu, and Dolzhenko 2018).

16.2.2 FENUGREEK FOR ANIMAL MILK PRODUCTION

Fenugreek was traditionally used as a forage crop in the Mediterranean region (Acharya, Thomas, and Basu 2006). This fenugreek-containing forage (named as "greek hay") can yield high-quality forage suitable for milking animals (Petropoulos 2002). Various veterinary studies have shown improvement in milk production in ruminant animals in terms of quantity and milk quality (Alemu and Doepel 2011; Moyer et al. 2003; Sah et al. 2007; Tiran 2003). The Anatolian water buffaloes were fed with 50 g/kg of ground fenugreek seeds in their diet for 30 days, resulting in increased feed intake and milk production (increased from 7.34 kg/day to 8.01 kg/day), without improvement in fat % and protein content. The increased milk production was attributed to the diosgenin content of seeds (Degirmencioglu et al. 2016). During a similar study, a significant rise in milk production was found in dairy buffaloes fed with a fenugreek seed-containing diet (Sah et al. 2007).

These results are supported by the recent preclinical studies in goats (El-Tarabany et al. 2018) and rats (Sevrin et al. 2019). The supplementation of dietary fenugreek seeds (50 g and 100 g per animal) improved the milk yield, milk protein, serum globulin, and thyroxine levels, but fat, serum glucose, triglycerides, cholesterol, and triiodothyronine levels were reduced (El-Tarabany et al. 2018). Besides, hematological parameters and the antioxidant capacity of heat-stressed goats improved (El-Tarabany et al. 2018). The rat dams fed with 1g/kg/day dietary fenugreek resulted in increased milk production (16%), improved pup growth (11%), and increased lactose (27%) without exhibiting any deleterious effect on dam litter metabolism of lipids and protein content (Sevrin et al. 2019).

16.2.3 PROBABLE MECHANISMS OF GALACTAGOGUE ACTION OF FENUGREEK

There are few mechanisms proposed by researchers to explain fenugreek's galactagogue action. Fenugreek is considered to induce milk production in females with less or no lactation (nonpuerperal women) by stimulating perspiration that might be responsible for improving blood and fluid circulation to increase breast milk supply (Betzold 2004; Gabay 2002; Mortel and Mehta 2013). The increased sweat production leads to an increase in milk supply, as the breast is considered a modified sweat gland and fenugreek phytoestrogens and diosgenin are steroid sapogenin components (Forinash et al. 2012; Turkyılmaz et al. 2011).

The most critical factor affecting milk production and supply is the hormone prolactin (PRL) because of optimum secretion during the postnatal period (Riordan 2005; Walker 2006). Serum prolactin levels rise rapidly during pregnancy, increasing the size and number of lactotrophs (Saleem, Martin, and Coates 2018). Other hormones such as oxytocin, somatotropin, cortisol, insulin, leptin, estrogen, and progesterone also play an essential role as galactagogues. Oral administration of fenugreek seeds powder (500 mg, 3 times daily, for 3 months) to lactating Sudanese mothers resulted in a highly significant increase in prolactin, milk production, and body weight as compared with respective control (Ahmed 2015). Insulinotropic effects of fenugreek seed extracts are well reported (Bashtian et al. 2013; Sharma 1986). Recently, water extract of fenugreek seeds is suggested to extend the duration of peak milk synthesis through modulation of the insulin-growth hormone (GH)-insulin like growth factor-1 (IGF-1) axis and increase milk ejection by activation of oxytocin secretion (Sevrin et al. 2020).

16.3 FENUGREEK IN MANAGEMENT OF POLYCYSTIC OVARIAN SYNDROME (PCOS)

A hormonal imbalance in estrogen, prolactin, or progesterone during reproductive years can result in irregular menstrual cycles, excessive body or facial hair, miscarriage, and polycystic ovarian syndrome (PCOS) (Ndefo, Eaton, and Green 2013). Globally, 22%–26% of women are affected with PCOS (Singh, Vijaya, and Laxmi 2018) with clinical features such as pelvic pain, depression, anxiety, obesity, and disorders of lipid metabolism, menstrual irregularities, hirsutism, acne, male pattern baldness, infertility, and weight gain (Reddy et al. 2016; Rocca et al. 2015). The use of existing non-pharmacological therapies and pharmacological medications reduce body fat and restore hormonal balance. However, the use of these medications is accompanied by side effects, limiting their usage (Arentz et al. 2014). Therefore, natural products from medicinal herbs as potential PCOS management options are gaining interest (Ashkar et al. 2020).

Infertility is a severe, significant consequence of PCOS (Balen and Rutherford 2007; Brassard, AinMelk, and Baillargeon 2008; Gorry, White, and Franks 2006). Excessive luteinizing hormone (Kassem et al. 2006), higher estrogen (Jelodar and Askari 2012), and insulin resistance are known to accelerate the progression of PCOS (Bashtian et al. 2013). The activated mammalian target of rapamycin (mTOR) is reported as a common cause of diabetes and PCOS (Ko and Kim 2020).

The close relationship between insulin resistance and infertility has been confirmed in idiopathic (Saleh et al. 2020) PCOS-associated (Al-Jefout, Alnawaiseh, and Al-Qtaitat 2017) and obesity-associated (Gambineri et al. 2019) infertility and PCOS (Kamble et al. 2020). Furthermore, oxidative stress is reported as the prominent connecting link between insulin resistance, obesity, diabetes, infertility, and PCOS (Bhardwaj, Panchal, and Saraf 2020; Özer et al. 2016). Recently, a large body of evidence confirmed the beneficial effects of fenugreek seeds and their phytoconstituents against insulin resistance (Kandhare et al. 2015a; Khound et al. 2018; Mohan and Balasubramanyam 2001), hyperglycemia (Gupta, Gupta, and Lal 2001), increased body fat (Avalos-Soriano et al. 2016; Zhou et al. 2020), and associated oxidative stress (Chaturvedi et al. 2013; Mohamed et al. 2019; Pradeep and Srinivasan 2017). Therefore, fenugreek seeds are expected to provide benefits for management of infertility conditions.

Many scientific reports supported the estrogenic potential of fenugreek seed phytoestrogens to provide better hormonal balance and management of several health problems such as fertility and menopause-related disorders. A fenugreek seed extract has shown restoration of hormonal balance between LH and FSH and reduced the number and size of the ovarian cysts in PCOS patients during a 90-day consumption period (Sies, Berndt, and Jones 2017).

The fenugreek seed extract supplementation with metformin for eight weeks, is reported to normalize the reproductive cycle by reducing the number of cysts in 58 women with oligo-anovulatory PCOS and typical ovaries during the randomized, double-blind, placebo-control study (Bashtian et al. 2013). Clinical efficacy of fenugreek extract enriched with 40% furostanolic saponins (Furocyst) in 50 premenopausal women diagnosed with PCOS was reported for 3 months in an open-label, one-arm, non-randomized post-marketing surveillance study (Swaroop et al. 2015). Fenugreek extract has reported a reduction in size or abolitions, return of regular menses, increased LH and FSH levels, and 12% of subjects conceived (Swaroop et al. 2015). During a subsequent 12-week, open-label, non-comparative study in 107 females with PCOS with fenugreek extract (Furocyst) supplementation showed a reduction of ovary volume and the number of ovarian cysts, HOMA index, hirsutism score with regular menses (Sankhwar, Goel, and Tiwari 2018).

The beneficial effects of fenugreek are supported by ameliorating the animal model of PCOS in female albino rats (Magdy Mohamady et al. 2018). Twenty-one days of oral administration of fenugreek seed extract (100 mg/kg, once a day, for 21 days) to letrozole-induced in albino rats showed the disappearance of these cysts and increased number of corpora lutea indicating successful ovulation, with normal follicular maturation during microscopic evaluations (Magdy Mohamady et al. 2018). However, the gold standard for evaluating pathophysiology of the absence of gonadal hormone effects such as infertility or menopause in female animal models and possible treatment options is the ovariectomy (OVX) or surgical removal of the ovaries (Koebele and Bimonte-Nelson 2016). Aqueous extract of fenugreek seed in 5 ml of drinking water and fenugreek seed powder in 5 g of food at a dose equivalent to 450 mg/kg/day and 900 mg/kg/day, demonstrated estrogenic effects such as prominent vaginal opening, increased thickening of uterine tissue, uterine weight, epithelial cell proliferation and rapid extension of stromal and glandular cell proliferation (Brogi et al. 2019). These results supported the estrogenic potential of fenugreek phytoestrogens with interaction with estrogen receptors (Brogi et al. 2019).

16.4 FENUGREEK AS A FEMALE LIBIDO ENHANCER

Females sexual dysfunction (FSD) is defined as decreased sex drive, reduced engagement in sexual activity, difficulty in arousal, inability to reach orgasm, painful intercourse with significant negative consequences for the quality of life (Allahdadi, Tostes, and Webb 2009). Sexual arousal is characterized by engorgement of the clitoris and the labia minora due to increased blood supply with an increase in vaginal and clitoral size (LoPiccolo and Stock 1986). The female's sexual response was proposed in the 4-stage model (excitement, plateau, orgasmic, and resolution phases) (Bitzer 2016; LoPiccolo and Stock 1986) or three-phase model (desire, arousal, and orgasm), with desire being the factor inciting the overall response cycle (Berman, Berman, and Goldstein 1999).

Several premenopausal and post-menopausal females prefer natural products to improve sexual function (Dording and Sangermano 2018). The plants containing phytoestrogens are most common amongst them (Kellogg-Spadt and Albaugh 2003). It is speculated that phytoestrogens increase serum estradiol within the reference range contradictory to HRT treatment, associated with severe side effects on long-term use (Dording and Sangermano 2018). However, only a few well-designed, double-blind, placebo-controlled trials of natural remedies for sexual dysfunction among women are reported.

Fenugreek seeds and whole plants have demonstrated efficacy towards the management of FSD (Aswar, Mohan, and Bodhankar 2009; Bhaskaran, Venkatesh, and Veeravalli 2018). During the double-blind placebo-controlled clinical study, oral supplementation of specialized fenugreek extract (Libifem, two capsules each 300 mg/day, 8 weeks, with 50% saponin glycoside) demonstrated improved sexual arousal and derived in 80 healthy menstruating women (Rao et al. 2015). The supplementation showed a significant increase in the serum levels of female sex and libido-related hormones, including estrogen and free testosterone. Furthermore, the supplementation showed increase in the frequency of sexual intercourse, scores in sexual functions (Derogatis interview for sexual functioning—self-reported, DISF-SR, and female sexual function index, FSFI), reduced stress (Perceived Stress Scale), and fatigue (Multidimensional Fatigue Inventory, MFI-20), and quality of life measures (Dyadic Adjustment Scale, DAS) as compared with matching placebo (Rao et al. 2015). DISF-SR questionnaire is used to evaluate sexual function (Derogatis 1997) with scoring on 25 interview items into five domains of sexual functioning, namely sexual cognition/fantasy, sexual arousal, sexual behavior/experience, orgasm, and sexual drive/relationship. The FSFI is a 19-item questionnaire that measures six domains: desire, arousal, lubrication, orgasm, satisfaction, and pain (Meston 2003). The increased serum estrogen levels are suggested to be responsible for better vaginal lubrication and blood flow, leading to sexual arousal and orgasm (Simon 2011), further supporting the benefit of specialized fenugreek extract for sexual function support in premenopausal women.

16.5 FENUGREEK IN THE MANAGEMENT OF POST-MENOPAUSAL SYMPTOMS (PMS)

The reproductive cycle of women comprises two major turning points—menarche and menopause. Menarche is the start, and menopause is the termination of the menstrual cycle and reproductive capacity (Karapanou and Papadimitriou 2010). Menopause is the loss of function of the ovarian follicle at the end of the reproductive cycle (Greendale, Lee, and Arriola 1999). Menopause is associated with varied unfavorable consequences in women's lives such as anxiety, hot flushes, irritability, mood swings, dysmenorrhea, vaginal dryness, urinary incontinence, reduced sexual function, obesity, and depression (Burger et al. 2002; Marjoribanks et al. 2012). Menopausal symptoms have been associated with the decrease in the level of inhibin and anti-Mullerian hormone and subsequent hormonal imbalance or decreased estrogen and androgen levels and increased FSH level (Sowers et al. 2008).

Much evidence is available to demonstrate the usefulness of fenugreek or its phytoconstituents in improving menopause-related problems, perhaps through estrogenic properties (Sreeja, Anju, and Sreeja 2010). During a double-blind placebo-controlled, crossover study, four weeks of 6 g of fenugreek extract showed a significant decrease in Greene menopausal score compared with matching placebo, but less than HRT (0.625 mg conjugated estrogen and 10 mg medroxyprogesterone acetate) (Mostafa-Gharabaghi et al. 2012).

The Greene Climacteric Scale (GCS) provides a brief measure of menopause symptoms on 21 items, each with 0 to 3 score on main areas of post-menopausal symptoms (Greene 1998). The 36-Item Short-Form Health Survey (SF-36) is a set of generic, coherent, and easily administered quality-of-life measures. A randomized, double-blind, placebo-controlled study reported the 90-days of oral supplementation novel extract of fenugreek husk (FenuSMART™), at 1000 mg/day of showed increase in plasma estradiol (120%) and quality of life and improvements in various PMS discomforts (GCS and SF-36) (Shamshad Begum et al. 2016).

On the other hand, oral supplementation proprietary fenugreek de-husked seed extract, Libifem® (600 mg/day, 12 weeks), demonstrated efficacy against post-menopausal symptoms in healthy aging women in a double-blind, randomized, placebo-controlled trial (Steels et al. 2017). The results revealed supplementation benefits to reduce total and vasomotor, psychosocial, and sexual symptoms measured by Menopause-Specific Quality of Life (MENQOL) questionnaire, lesser daytime hot flushes, lesser night sweats with average estradiol levels as compared with placebo (Steels et al. 2017). The MENQOL is a self-administered, 29-item questionnaire that scores to assesses four domains of menopausal symptoms in a Likert-scale format (Hilditch et al. 1996).

Recently, a unique hydroalcoholic extract of fenugreek (FHE) demonstrated benefits to 48 post-menopausal women against post-menopausal discomfort (hot flushes, irritable mood, vaginal dryness, and pain in joints and legs) and hormone imbalance during a randomized, double-blind, placebo-controlled study (Thomas et al. 2020). In the study, 42 days of supplementation improved somatic, psychological, and urogenital scores as measured by a validated menopause rating scale (MRS) questionnaire, which measures the severity of menopause-related symptoms and health-related quality of life. Reestablishing the hormonal balance due to decreased FSH and SHBG with an increase in estradiol, free testosterone, and progesterone concentrations were observed (Thomas et al. 2020).

16.5.1 Fenugreek in Dysmenorrhea

Dysmenorrhea is defined as painful cramps of uterine origin that occur during menstruation and represents one of the most common causes of pelvic pain and menstrual disorder (Bernardi et al. 2017). Hypersecretion of prostaglandins (an inflammatory mediator) and increased uterine contractility are significant causes behind severe pain (Bernardi et al. 2017). Recently, the changes in concentration of many proinflammatory mediators such as prostaglandins, Tumor Necrosis Factor α (TNF α), Interleukin-6 (IL6); Vascular Endothelial Growth Factor (VEGF), C-Reactive Protein (CRP) in the pathogenesis of dysmenorrhea and menopause are confirmed (Barcikowska et al. 2020). Furthermore, increased incidences of arthritis, osteoporosis, and inflammatory conditions are associated with post-menopausal women (Ginaldi et al. 2019).

The hexane and ethanolic extract of fenugreek seed extract-treated ovariectomized rats showed reduced proinflammatory mediators such as Interleukin-1 (IL-1), IL-6, and TNF-α as compared with control rats (Abedinzade et al. 2015). These effects were suggested to be due to the presence of both polar (tannins and glycosides) (Seidel 2006) and nonpolar constituents (fatty acids, alkanes, sterols, coumarins, and some terpenoids) (Raju and Rao 2012) of the seed, including the phytoestrogens and diosgenin (Ginaldi et al. 2019).

A fenugreek seed extract has shown efficacy in reducing the pain in women with dysmenorrhea amongst many natural alternatives during the systematic review (Pattanittum et al. 2016). The evidence from the double-blind placebo-controlled study reported significant pain reduction by fenugreek seed powder supplementation (900 mg, 3 times a day for 2 consecutive menstrual cycles) as compared to matching placebo (Starch) in 101 unmarried girl students with severe dysmenorrhea (Younesy et al. 2014). In this study, the fenugreek seed powder supplementation reduced severity and duration of pain as measured by the Visual Analog Scale (VAS) and other symptoms such as fatigue, lack of energy, mood swings as measured by a multidimensional verbal scoring system (Younesy et al. 2014). Researchers attributed these results to diosgenin (which has structural similarity with cortisol), calcium, iron, β-carotene, and vitamins to alleviate dysmenorrhea and associated symptoms (Younesy et al. 2014).

Simultaneously, a prospective, open-labeled, randomized study on oral supplementation of fenugreek seed alone and with the dry cupping method (as prescribed by the Unani system of medicine) in women with dysmenorrhea reported significant benefits in attenuating lower abdominal pain during menses (Inanmdar et al. 2016). Fenugreek seed powder (3 capsules of 1 g each, twice daily, from day 1 to 3 of the menstrual cycle) showed a higher reduction of pain (VAS) as compared with cup therapy and pain killer regimen without adverse effects (Inanmdar et al. 2016).

A large body of scientific evidence demonstrated the anti-inflammatory potential of diverse phytoconstituents, namely alkaloids (Mandegary et al. 2012), saponins (Kawabata et al. 2011), mucilage (Sindhu et al. 2012), linolenic and linoleic acids (Pundarikakshudu 2019), through inhibition inflammatory mediators to relieve pain. Besides, the antinociceptive efficacy of fenugreek and its phytoconstituents has also been reported (Mandegary et al. 2012; Vyas et al. 2008). These phytoconstituents are suggested to contribute to the efficacy of fenugreek against dysmenorrhea-induced pain (Abedinzade et al. 2015; Younesy et al. 2014).

16.5.2 Fenugreek on Post-Menopausal Cognitive Decline

The brain is a vital target organ for estrogen. In addition to direct effects, estrogen influences brain function by affecting the neurotransmitters, vasculature, and the immune system (Henderson 2008). Natural menopause is not associated with substantial cognitive change (Henderson 2008). However, the severe estrogen deficiency in menopause or surgical menopause leads to brain hypometabolism proceeding into cognitive impairment and early AD (Qu et al. 2013; Zárate, Stevnsner, and Gredilla 2017). At the time of the menopausal transition, many memory problems, perhaps through hormonal changes associated with menopause, are reported by many women (Shaywitz et al. 1999). Especially for midlife menopausal women, neuropathological changes in the brain and hormonal changes are significant causes of cognitive disorders, mild cognitive impairment to dementia (Kim and Jung 2015)

Estrogen is a potent antioxidant, anti-inflammatory, and neuroprotective in nature. (Behl 2002; Petrovska, Dejanova, and Jurisic 2012; Villa et al. 2016). Therefore, estrogenic agents that offer neuroprotective effects during menopause are essential for managing post-menopausal cognitive decline. Hormone replacement therapy (HRT) is known to treat vasomotor symptoms successfully, is thought to have neuroprotection to restore cognition during menopause. However, evidence for this is inconsistent and contradictory (Clement et al. 2011).

There is a good body of evidence for herbal remedies and dietary supplements to manage menopausal symptoms, including cognitive problems (Clement et al. 2011). The phytoestrogens from natural sources are known to provide excellent estrogenic activity (Adlercreutz et al. 1992; Messina, Barnes, and Setchell 1997; Pitkin 2012). The presence of phytoestrogens in the hydroalcoholic extract of fenugreek seed demonstrated cognitive improvement, neuroprotective effect, and antioxidant effect in ovariectomized (OVX) rats (Anjaneyulu et al. 2017). In rats, ovariectomy mimics natural and surgical menopause, making it a suitable animal model for menopause and cognitive aging research (Zakaria et al. 2019). The oral supplementation of fenugreek seed hydroalcoholic extract (200 mg/kg/day for 30 days) could prevent OVX-induced learning and memory decline and shrinkage and death of hippocampal CA3 neurons (Anjaneyulu et al. 2017).

16.5.3 Fenugreek in Management of Menopause-Induced Osteoporosis

The reduced estrogen levels after menopause have been linked to women's bone-related issues (Ji and Yu 2015; Møller et al. 2020). Estrogen deficiency can lead to excessive bone resorption, leading to osteoporosis (Cortet et al. 2011; Pavone et al. 2017). Nutritional supplements are an effective management option for post-menopausal osteoporosis (Camacho et al. 2016; Pavone et al. 2017).

A recent study reported beneficial anti-osteoporotic activity of hydroalcoholic extract of fenugreek seeds on bone structure and strength in ovariectomy rats with menopause-induced osteopenia (Anjaneyulu et al. 2018). The addition of 1% fenugreek seed added to food increased the tibial metaphysis's strength (cancellous bone) in ovariectomized rats (Folwarczna et al. 2014). Recently, three months of oral supplementation of fenugreek alcoholic extract (200 mg/kg oral) showed synergistic effects combined with purified curcumin to improve bone mineral density, bone turnover, bone strength, and bone mass in ovariectomized rats (Esmat et al. 2018). Ovariectomized rats are a well-validated animal model of post-menopausal osteoarthritis (Høegh-Andersen et al. 2004; Sniekers et al. 2008; Yousefzadeh et al. 2020). The beneficial effects of phytoestrogens' efficacy in maintaining bone mass in post-menopausal women have been reported in many clinical studies (Abdi et al. 2016). Therefore, the efficacy of fenugreek seed to increase the bone mechanical properties against ovariectomized rats can be attributed to phytoestrogen presence such as diosgenin (Anjaneyulu et al. 2018). Diosgenin, steroidal sapogenin glycosides from fenugreek seeds, has already been reported to have favorable effects on bone tissue (Alcantara et al. 2011; Higdon et al. 2001; Shishodia and Aggarwal 2006). Some of the probable mechanisms of diosgenin against osteoporosis are increasing bone matrix protein synthesis and bone-specific transcription factor Runx2 (Alcantara et al. 2011), osteoclastogenesis, invasion, and proliferation through the downregulation of Akt, κB kinase activation, and NF-κB-regulated gene expression (Shishodia and Aggarwal 2006).

Another constituent of fenugreek seed, trigonelline, has been recognized as a novel phytoestrogen (Allred et al. 2009). Recently, trigonelline's phytoestrogen potential has been demonstrated with an anti-osteoporotic effect against dexamethasone-induced osteoporotic rats (Rathi et al. 2020). Oral supplementation of trigonelline (63.3 mg/kg) for 8 weeks showed positive effects on bone mineral density (BMD), bone microarchitecture, bone volume, trabecular thickness and number, serum calcium and serum phosphorus, and serum estradiol (Rathi et al. 2020).

16.6 FENUGREEK IN MANAGEMENT OF FEMALE-SPECIFIC CANCERS

Cancer is a disease in which cells in the body grow out of control. Although breast cancer does not only affect women, it is the most common in women with cervical cancer. The natural phytoestrogens are reported to inhibit breast cancer development and progression in multiple targets within cells with supraphysiological doses, perhaps due to structural similarity with estrogen (Bilal et al. 2014).

Diosgenin, a phytoestrogen from the fenugreek seeds, reported to reduce cell viability of the ER+ (estrogen receptor) MCF-7 cells and induce apoptosis on the breast cancer cell lines in a dose dependent manner in the presence or absence of estrogen receptors (ER) *in vitro* (Sowmyalakshmi et al. 2005). These results were confirmed during another *in vitro* study, where fenugreek seed extract (50 µ/ml) could kill the MCF-7 human immortalized breast cells line with 23.2% and 73.8% apoptosis in cells after 24 h and 48 h respectively (Khoja et al. 2011). The increased expression of a crucial component of the apoptotic pathway (caspase (3,8,9), p53, fas *FADD*, *Bax*, and *Bak*) was suggested as a probable mechanism (Khoja et al. 2011). These anti-cancer properties of fenugreek are attributed to phytoestrogens such as diosgenin, furanones, dioscin, protodioscin, and trigonelline (Khoja et al. 2011).

The apoptotic mechanism is supported during *in vivo* report of protective effects of aqueous fenugreek seed extract (200 mg/kg/day, in olive oil, 7 days of pretreatment) against 7,12-dimethylbenz(α) anthracene (DMBA)-induced breast cancer in rats (Amin et al. 2005). DMBA affects cellular signaling pathways to induce apoptosis and induce cytotoxicity in various cell types (Ko et al. 2004). The significant protection of the DMBA-induced mammary hyperplasia can be attributed to apoptotic induction by fenugreek extract pretreatment (Amin et al. 2005). Recently, protodioscin, a steroidal saponin compound of fenugreek seed, has been hypothesized to induce apoptosis through ROS-mediated endoplasmic reticulum stress via the JNK/p38 activation pathways in human cervical cancer cells (Lin et al. 2018). The recent report of the efficacy of protodioscin (2–16 μM) against the HeLa and C33A human cervical cancer cell line further supported the apoptosis mechanism (Hernández-Vázquez et al. 2020).

TABLE 16.2
Evidence of Oral Supplementation of Fenugreek or Its Phytoconstituents for Use in Female-Specific Conditions

Condition	Fenugreek Constituent (Dose)	Material and Methods	Results and Proposed Mechanisms	References
Post-menopausal symptoms	Novel fenugreek husk extract (1000 mg/day, 90 days).	RCT, 88 women post-menopausal women with discomfort	↓ Vasomotor symptoms—hot flushes ↑ Estradiol level ↑ Serum calcium	(Shamshad Begum et al. 2016)
Post-menopausal symptoms	Fenugreek hydroalcoholic extract (250 mg × 2/day, 42 days).	RCT, 40 post-menopausal women	↓ Hot flush, night sweats and pain on leg muscles and joints ↓ irritability and vaginal dryness, pain Attained hormonal balance	(Thomas et al. 2020)
Post-menopausal symptoms	Fenugreek seeds (granulated, 6 g per day, 8 weeks).	RCT, 50 peri-menopausal women	↓ Greene menopausal score	(Mostafa-Gharabaghi et al. 2012)
Post-menopausal symptoms	Fenugreek seed extract (50 and 150 mg/kg, once a day for 2 days).	Animal model, 41 ovariectomized female rats	↓ metabolic and inflammatory alternations ↓ Interleukin 1 & 6 & TNF-α	(Abedinzade et al. 2015)
Dysmenorrhea	Fenugreek seed powder (2 capsules of 900 mg, 3 times a day for 2 consecutive menstrual cycles).	RCT, 101 unmarried girl students	↓ Pain (Severity and duration of pain) ↓ Fatigue, headache, nausea, vomiting, lack of energy, syncope	(Younesy et al. 2014)
Dysmenorrhea	Fenugreek seed powder (3 g, twice a day, day 1 to 3 of menstrual cycle).	RCT, open label, 60 women with dysmenorrhea	↓ Lower abdominal pain during menses	(Inanmdar et al. 2016)
Osteoporosis	Pulverized fenugreek seed (1% in the diet per day, for 4 weeks).	Animal model, ovariectomized rats	↑ Strength of cancellous and compact bones No change in bone mineralization and serum turnover markers	(Folwarczna et al. 2014)
Osteoporosis	Fenugreek alkaloid trigonelline (63.3 mg/kg, 12 weeks).	Animal model, dexamethasone-induced rats	Prevent the progression. ↑ Bone mineral density Restored bone physiology	(Rathi et al. 2020)

Condition	Fenugreek Constituent (Dose)	Material and Methods	Results and Proposed Mechanisms	References
Menopause-induced Osteopenia	Hydroalcoholic fenugreek seed extract (200 mg/kg/day for 30 days).	Animal model, ovariectomized rats	↑ Bone structure and biomechanical properties	(Anjaneyulu et al. 2018)
Menopause-induced neurocognitive deficit	Hydroalcoholic fenugreek seed extract (200 mg/kg/day for 30 days).	Animal model, ovariectomized rats, passive avoidance test and histology of hippocampal neurons	↑ Improved the memory and learning ↑ Protected the hippocampal neuronal architecture and survival.	(Anjaneyulu et al. 2017)
Menopause-related disorders	Fenugreek seed extract (450 mg/kg/day) and fenugreek seed powder (900 mg/kg/day) for 15 days).	Animal model—ovariectomized rats	↑ Estrogenic activity	(Brogi et al. 2019)
Fertility and safety	Trigonelline from fenugreek seed (75 mg/kg, day 1 to 7 of pregnancy).	Animal model—female rats during pregnancy	↑ No alteration of estrous cycle, number of implants and number of pups (v/s control) Safe during pregnancy	(Aswar, Mohan, and Bodhankar 2009)
Polycystic Ovarian Syndrome	Lyophilized hydroalcoholic extract of fenugreek seeds 525 mg, 2 times a day) with metformin (500 mg, once or twice a day) for 8 weeks.	RCT, 58 women with PCOS	↓ in PAO (polycystic appearing ovaries)	(Bashtian et al. 2013)
Polycystic Ovarian Syndrome	Fenugreek seed extract (Furocyst, 2 capsules of 500 mg per day).	Open-label, non-randomized, post-marketing surveillance study in 50 premenopausal women	↓ Ovary volume ↓ Cyst size or dissolution Return of menstrual cycle (71%) Pregnancy (12%)	(Swaroop et al. 2015)
Polycystic Ovarian Syndrome	Fenugreek seed extract (Furocyst) (2 capsules of 500 mg per day for 12 weeks).	Open-label, non-randomized, single-arm, 107 women with Polycystic Ovarian Syndrome	↓ Cyst size in both the ovaries ↓ Hirsutism score Pregnancy (15%) ↓ Insulin resistance	(Sankhwar, Goel, and Tiwari 2018)
Polycystic Ovarian Syndrome	Fenugreek seed extract (100 mg/kg, once a day, 21 days).	Animal model—letrozole-induced PCOS in female albino rats	↓ Cysts ↑ Corpora lutea with ovulation	(Magdy Mohamady et al. 2018)
Female libido	Standardized extract of fenugreek, Libifem (600 mg per day, 8 weeks).	RCT, double-blind 80 healthy menstruating women	↑ Sexual arousal and drive ↑ Frequency of sexual intercourse ↑ Estrogen	(Rao et al. 2015)

RCT—Randomized controlled trial

16.7 SAFETY OF FENUGREEK IN FEMALES

The physiological differences between males and females can be critical in response to any desirable (Soldin and Mattison 2009) or undesirable (Parekh et al. 2011) treatment. These differences are appropriately considered for risk assessment by regulatory bodies worldwide to ensure safe and effective treatments. As dietary supplements are consumed without medical supervision, the broad margin of safety is paramount. Therefore, many new products containing fenugreek seed powder, extract, or phytoconstituents for managing female disorders were introduced into the global market (Yao et al. 2020).

Fenugreek and its phytoconstituents from seeds and leaves have been extensively studied and reported as medicine, spices, and vegetables in various female-specific disorders as pharmaceuticals, nutraceuticals, and functional food (Nagulapalli Venkata et al. 2017; Pal and Mukherjee 2020; Wani and Kumar 2018; Yao et al. 2019). The US Food and Drug Administration has rated fenugreek as Generally Regarded as Safe (GRAS), based on the documented broad margin of safety for fenugreek seeds.

Many products containing fenugreek as a bioactive ingredient have been present in the international markets for a long time without serious safety concerns (Yao et al. 2020). The available body of evidence of fenugreek toxicological data has been reviewed and highlighted the need for female-specific safety considerations for fenugreek-containing ingredients and products (Kandhare et al. 2019; Ulbricht et al. 2007). The published literature on safety evaluations in female animals of fenugreek demonstrated contrasting conclusions about safe dosages, especially during breeding times. A large number of studies concluded good safety for fenugreek in female animals (Aswar, Mohan, and Bodhankar 2009; Deshpande et al. 2016a, 2016b; Deshpande, Mohan, Pore et al. 2017; Deshpande, Mohan, Reddy et al. 2017; Folwarczna et al. 2014; Kandhare et al. 2015b, 2016) and female human subjects (Lu et al. 2008; Nathan et al. 2014; Raju et al. 2001; Rao et al. 2015; Steels et al. 2017; Swaroop et al. 2015; Verma et al. 2016). Few reports suggested adverse teratogenic or abortifacient effects (Ahirwar, Ahirwar, and Kharya 2010; Araee et al. 2009; Dande and Patil 2012; Khalki et al. 2012; Modaresi, Jalalizand, and Mahdian 2012).

These contrasting conclusions are due to large variations in content and composition of fenugreek, due to diverse growing and harvesting conditions of the plant, soil quality, plant parts, storage conditions, storage duration, cleaning and disinfection processes (if any), extraction methods, choice of solvents, formulation by the manufacturer, route of administrations (oral v/s injections), etc. (Altemimi et al. 2017; Kunle, Egharevba, and Ahmadu 2012; Zhang, Lin, and Ye 2018). More than 100 phytochemicals have been isolated and identified from fenugreek seeds, mainly polysaccharides, saponins, alkaloids, polyphenols, and flavonoids (Mandegary et al. 2012). Fenugreek seeds are composed of soluble galactomannan (about 30%), insoluble fiber (about 20%), protein (20–30%), fat (5–10%), alkaloids, which are mainly composed of trigonelline (0.2–0.38%), diosgenin (0.6–1.7%) and yamogenin-based saponins (about 4.8%), 4-hydroxyisoleucine (about 0.09%), volatile oil (about 1.25%), and C-glycosylflavones of apigenin and luteolin (about 0.10%) (Yao et al. 2020). Therefore, standardization of ingredients based on relevant bioactive phytoconstituent is necessary for consistent conclusions regarding the fenugreek's safety or toxicological properties.

A recent systematic review on fenugreek toxicity literature using ToxRTool found the study's quality and reliability as a significant limitation for the reliable toxicological conclusions and risk assessments (Kandhare et al. 2019). During this systematic review, fenugreek was found safe on female animals and human subjects as concluded in "Reliable Without Restriction" studies with all of them including authentication of the source material, standardization, and characterization of product, and following international standard procedures (OECD guidelines) for such toxicological studies (Kandhare et al. 2019).

Section 4 of a toxicological testing guideline issued by the Organisation for Economic Co-operation and Development (OECD) is an internationally accepted standard method for safety testing among industries, academia, and governments to assess chemicals on health (Organisation for Economic Co-operation and Development 1998). The prenatal developmental toxicity testing (OECD Test No. 414) specifically addresses the concerns of reproductive and teratogenic effects on prenatal exposure on the pregnant test animal and the developing organism (Barrow 2013; OECD

2018). This test includes comprehensive assessments on dams (weight, clinical observations, uterine examination, endocrine-related measurements), live fetuses (weights, soft tissue, and skeletal changes), and fetuses after cesarean sections (gravid uterine weight, implantation sites, early and late resorptions, live and dead fetuses) (OECD 2018).

Many prenatal developmental toxicity reports on embryo-fetal development during the organogenesis period (20 days) in pregnant female rats conducted as per guidelines (OECD Test No. 414) confirmed the absence of maternal or developmental (fetotoxicity or teratogenicity) toxicity to oral supplementation of fenugreek extracts with marker compounds standardized to furostenol glycosides (FENU-FG) (Deshpande et al. 2016b), glycosides (SFSE-G) (Deshpande, Mohan, Pore et al. 2017), low molecular weight galactomannan (LMWGAL-TF) (Deshpande et al. 2016a) at no-adverse-event level (NOAEL) more than 1000 mg/kg. Furthermore, IDM01, the botanical composition of 4-hydroxyisoleucine- and trigonelline-based standardized fenugreek seed extract, was found safe during pregnancy in rats with NOAEL more than 500 mg/kg (Deshpande, Mohan, Reddy et al. 2017).

The acute (OECD Test No. 423 or 425), 28-days repeated-dose, subacute (OECD Test No. 407), and 90-days repeated-dose sub-chronic (OECD Test No. 408) toxicological evaluations need to include an equal number of female animals as a requirement (OECD 1997a, 1997b, 1998, 2001, 2002, 2008). The comprehensive set of clinical and functional observations, body weights and food/water consumption measurements, hematology, clinical biochemistry, and gross necropsy and histopathology are evaluated during testing as per OECD guidelines.

The excellent safety profile of on acute, subacute, and sub-chronic exposure of the oral supplementation of many standardized fenugreek seed extracts such as FENU-FG (Deshpande, Mohan, Ingavale et al. 2017), SFSE-G (Deshpande, Mohan, and Thakurdesai 2016b; Kandhare et al. 2015b), LMWGAL-TF (Deshpande, Mohan, and Thakurdesai 2016a), IDM01 (Deshpande, Mohan, and Thakurdesai 2017), furostanolic saponins (> 60% w/w), enriched (Fenfuro) (Swaroop et al. 2014) and isolated phytoconstituents such as Vicenin-1 (Kandhare et al. 2016) were reported.

For regulatory purposes, a chemical's mutagenicity potential has mainly been evaluated using *in vitro* assays, such as the Bacterial Reverse Mutation Test, AMES test (Mortelmans and Zeiger 2000). Mutagens are agents that can cause heritable changes in DNA, and their capacity to cause mutations is defined as mutagenicity (Cvetković, Takić Miladinov, and Stojanović 2018). As all information for the proper development, functioning, and reproduction of organisms is coded in DNA, mutations can result in harmful effects and play a role in genetic disorders (Verheyen 2017), especially for food-related products (Mandal et al. 2018; Weisburger 1999). The structural chromosomal abnormalities such as chromosomal aberration (a missing, extra, or irregular portion of chromosomal DNA) result from breakage and incorrect rejoining of chromosomal segments and result in many genetic diseases disorders such as cancer (Nguyen 2020). The standardized fenugreek seed extracts with markers such as furostenol saponins, glycosides, or low molecular galactomannans such as FENU-FG, SFSE-G, and LMWGAL-TF, were found safe and devoid of mutagenicity (OECD Test No. 471) and genotoxicity (Mammalian Chromosomal Aberration, OECD Test No. 473) potential during these studies (Deshpande, Mohan, and Thakurdesai 2016a; Deshpande, Mohan, Ingavale et al. 2017).

16.8 CONCLUSION

The present evidence indicated excellent efficacy of fenugreek in the form of seed or leaves powder, extracts, or phytoconstituents for many female-specific applications (Figure 16.1) such as galactagogue/lactational aid, libido enhancer, and provide a option for managing PCOS, post-menopausal conditions, including dysmenorrhea, cognitive decline, osteoporosis, and female-specific cancers with broad margin of safety.

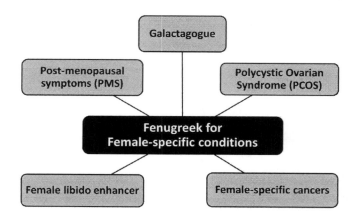

FIGURE 16.1 Fenugreek application in management of female-specific conditions.

REFERENCES

Abdi, F., Z. Alimoradi, P. Haqi, and F. Mahdizad. 2016. "Effects of phytoestrogens on bone mineral density during the menopause transition: A systematic review of randomized, controlled trials." *Climacteric* 19 (6):535–545. doi: 10.1080/13697137.2016.1238451.

Abedinzade, M., S. Nasri, M. Jamal Omodi, E. Ghasemi, and A. Ghorbani. 2015. "Efficacy of Trigonella foenum-graecum seed extract in reducing metabolic and inflammatory alterations associated with menopause." *Iran Red Crescent Medical Journal* 17 (11):e26685. doi: 10.5812/ircmj.26685.

Acharya, S., J. Thomas, and S. Basu. 2006. "Fenugreek: An "old world" crop for the "new world"." *Biodiversity* 7 (3–4):27–30.

Adlercreutz, H., E. Hämäläinen, S. Gorbach, and B. Goldin. 1992. "Dietary phyto-oestrogens and the menopause in Japan." *Lancet* 339 (8803):1233. doi: 10.1016/0140-6736(92)91174-7.

Ahirwar, D., B. Ahirwar, and M. Kharya. 2010. "Evaluation of antifertility activity of *Trigonella foenum-graecum* seeds." *Der Pharmacia Sinica* 1 (3):33–39.

Ahmed, M. E. M. 2015. "The effect of fenugreek seeds powder on prolactin level in lactating Sudanese mothers." Thesis submitted for the degree of Master of Science, College of Science, Sudan University of Science and Technology.

Al-Asadi, J. N. 2014. "Therapeutic uses of fenugreek (Trigonella foenum-graecum L.)." *American Journal of Humanities and Social Sciences Research* (March/April):21–36.

Al-Habori, M., and A. Raman. 2002. "Pharmacological Properties in Fenugreek-The genus Trigonella by GA Petropoulos." *Taylor and Francis, London and New York* 10:163–182.

Al-Jefout, M., N. Alnawaiseh, and A. Al-Qtaitat. 2017. "Insulin resistance and obesity among infertile women with different polycystic ovary syndrome phenotypes." *Scientific Reports* 7 (1):5339. doi: 10.1038/s41598-017-05717-y.

Alcantara, E. H., M. Y. Shin, H. Y. Sohn, Y. M. Park, T. Kim, J. H. Lim, H. J. Jeong, S. T. Kwon, and I. S. Kwun. 2011. "Diosgenin stimulates osteogenic activity by increasing bone matrix protein synthesis and bone-specific transcription factor Runx2 in osteoblastic MC3T3-E1 cells." *J Nutr Biochem* 22 (11):1055–1063. doi: 10.1016/j.jnutbio.2010.09.003.

Alemu, A., and L. Doepel. 2011. "Fenugreek (Trigonella foenum-graecum L.) as an alternative forage for dairy cows." *Animal: An international journal of animal bioscience* 5 (9):1370.

Allahdadi, K. J., R. C. Tostes, and R. C. Webb. 2009. "Female sexual dysfunction: Therapeutic options and experimental challenges." *Cardiovasc Hematol Agents Med Chem* 7 (4):260–269. doi: 10.2174/187152509789541882.

Allred, K. F., K. M. Yackley, J. Vanamala, and C. D. Allred. 2009. "Trigonelline is a novel phytoestrogen in coffee beans." *J Nutr* 139 (10):1833–1838. doi: 10.3945/jn.109.108001.

Altemimi, A., N. Lakhssassi, A. Baharlouei, D. G. Watson, and D. A. Lightfoot. 2017. "Phytochemicals: Extraction, Isolation, and Identification of Bioactive Compounds from Plant Extracts." *Plants (Basel)* 6 (4). doi: 10.3390/plants6040042.

Amin, A., A. Alkaabi, S. Al-Falasi, and S. A. Daoud. 2005. "Chemopreventive activities of Trigonella foenum-graecum (Fenugreek) against breast cancer." *Cell Biology International* 29 (8):687–694.

Anjaneyulu, K., K. M. Bhat, S. R. Srinivasa, R. A. Devkar, and T. Henry. 2018. "Beneficial Role of Hydro-alcoholic Seed Extract of Trigonella foenum-graecum on Bone Structure and Strength in Menopause Induced Osteopenia." *Ethiop J Health Sci* 28 (6):787–794. doi: 10.4314/ejhs.v28i6.14.

Anjaneyulu, K., K. S. Rai, R. Jetti, and K. M. Bhat. 2017. "Effect of fenugreek seed extract on menopause induced neurocognitive deficit." *Adv Sci Lett* 23 (3):1864–1868.

Araee, M., M. Norouzi, G. Habibi, and M. Sheikhvatan. 2009. "Toxicity of Trigonella foenum-graecum (Fenugreek) in bone marrow cell proliferation in rat." *Pak J Pharm Sci* 22 (2):126–130.

Arentz, S., J. A. Abbott, C. A. Smith, and A. Bensoussan. 2014. "Herbal medicine for the management of polycystic ovary syndrome (PCOS) and associated oligo/amenorrhoea and hyperandrogenism; a review of the laboratory evidence for effects with corroborative clinical findings." *BMC Complementary and Alternative Medicine* 14 (1):511.

Arora, J., and K. G. Ramawat. 2018. "Bioactive molecules, nutraceuticals, and functional foods in Indian vegetarian diet and during postpartum healthcare." In *Bioactive Molecules in Food*, edited by Mérillon, J.-M. and Ramawat, K.G., 1–30. Cham: Springer International Publishing.

Ashkar, F., S. Rezaei, S. Salahshoornezhad, F. Vahid, M. Gholamalizadeh, S. M. Dahka, and S. Doaei. 2020. "The Role of medicinal herbs in treatment of insulin resistance in patients with Polycystic Ovary Syndrome: A literature review." *Biomol Concepts* 11 (1):57–75. doi: 10.1515/bmc-2020-0005.

Aswar, U., V. Mohan, and S. Bodhankar. 2009. "Effect of trigonelline on fertility in female rats." *International Journal of Green Pharmacy* 3 (3).

Aswar, U., V. Mohan, S. Bhaskaran, and S. Bodhankar. 2008. "Study of galactomannan on androgenic and anabolic activity in male rats." *PharmacologyOnline* 2:56–65.

Avalos-Soriano, A., R. De la Cruz-Cordero, J. L. Rosado, and T. Garcia-Gasca. 2016. "4-Hydroxyisoleucine from Fenugreek (Trigonella foenum-graecum): Effects on insulin resistance associated with obesity." *Molecules* 21 (11). doi: 10.3390/molecules21111596.

Balen, A. H., and A. J. Rutherford. 2007. "Managing anovulatory infertility and polycystic ovary syndrome." *BMJ* 335 (7621):663–666. doi: 10.1136/bmj.39335.462303.80.

Barcikowska, Z., E. Rajkowska-Labon, M. E. Grzybowska, R. Hansdorfer-Korzon, and K. Zorena. 2020. "Inflammatory markers in dysmenorrhea and therapeutic options." *Int J Environ Res Public Health* 17 (4). doi: 10.3390/ijerph17041191.

Barrow, P. 2013. *Teratogenicity Testing: Methods and Protocols, Methods in Molecular Biology*. New York: Humana Press; Springer.

Basch, E., C. Ulbricht, G. Kuo, P. Szapary, and M. Smith. 2003. "Therapeutic applications of fenugreek." *Altern Med Rev* 8 (1):20–27.

Bashtian, M. H., S. A. Emami, N. Mousavifar, H. A. Esmaily, M. Mahmoudi, and A. H. M. Poor. 2013. "Evaluation of fenugreek (Trigonella Foenum-graecum L.), effects seeds extract on insulin resistance in women with polycystic ovarian syndrome." *Iran J Pharmaceut Res: IJPR* 12 (2):475.

Behl, C. 2002. "Estrogen as a neuroprotective hormone." *Nature Reviews. Neuroscience* 3:433–442. doi: 10.1038/nrn846.

Berman, J. R., L. Berman, and I. J. U. Goldstein. 1999. "Female sexual dysfunction: Incidence, pathophysiology, evaluation, and treatment options." *Urology* 54 (3):385–391.

Bernardi, M., L. Lazzeri, F. Perelli, F. M. Reis, and F. Petraglia. 2017. "Dysmenorrhea and related disorders." *F1000Res* 6:1645. doi: 10.12688/f1000research.11682.1.

Betzold, C. M. 2004. "Galactagogues." *Journal of Midwifery and Womens Health* 49 (2):151–154.

Bhardwaj, J. K., H. Panchal, and P. Saraf. 2020. "Ameliorating effects of natural antioxidant compounds on female infertility: A review." *Reproductive Sciences*. doi: 10.1007/s43032-020-00312-5.

Bhaskaran, S., R. V. Venkatesh, and J. Veeravalli. 2018. *Use of Fenugreek Extract to Enhance Female Libido*. PTO, U. USA: GE Nutrients Inc. US14/097,465.

Bilal, I., A. Chowdhury, J. Davidson, and S. Whitehead. 2014. "Phytoestrogens and prevention of breast cancer: The contentious debate." *World J Clin Oncol* 5 (4):705–712. doi: 10.5306/wjco.v5.i4.705.

Bitzer, J. 2016. "The female sexual response: Anatomy and physiology of sexual desire, arousal, and orgasm in women."199–212. doi: 10.1007/978-1-4939-3100-2_18.

Bora, D., S. Mehmud, K. K. Das, and H. Medhi. 2016. "Report on folklore medicinal plants used for female health care in Assam (India)." *Int J Herbal Med* 4 (6):4–13.

Brassard, M., Y. AinMelk, and J. P. Baillargeon. 2008. "Basic infertility including polycystic ovary syndrome." *Med Clin North Am* 92 (5):1163–1192, xi. doi: 10.1016/j.mcna.2008.04.008.

Brillante, C., and J. B. Mantaring. 2014. "O-203 Triple-blind, randomised controlled trial on the use of fenugreek (*Trigonella foenum-graecum* L.) for augmentation of breastmilk volume among postpartum mothers." *Archives of Disease in Childhood* 99 (Suppl 2):A101.

Brogi, H., H. Elbachir, N. El Amrani, S. Amsaguine, and D. Radallah. 2019. "Fenugreek seeds estrogenic activity in ovariectomized female rats." *Current Issues in Pharmacy and Medical Sciences* 32 (3):138–145. doi: 10.2478/cipms-2019-0026.

Burger, H. G., E. C. Dudley, D. M. Robertson, and L. Dennerstein. 2002. "Hormonal changes in the menopause transition." *Recent Progress in Hormone Research* 57:257–276.

Camacho, P. M., S. M. Petak, N. Binkley, B. L. Clarke, S. T. Harris, D. L. Hurley, M. Kleerekoper, E. M. Lewiecki, P. D. Miller, and H. S. Narula. 2016. "American association of clinical endocrinologists and American college of endocrinology clinical practice guidelines for the diagnosis and treatment of postmenopausal osteoporosis—2016." *Endocrine Practice* 22 (s4):1–42.

Chaturvedi, U., A. Shrivastava, S. Bhadauria, J. K. Saxena, and G. Bhatia. 2013. "A mechanism-based pharmacological evaluation of efficacy of Trigonella foenum-graecum (fenugreek) seeds in regulation of dyslipidemia and oxidative stress in hyperlipidemic rats." *J Cardiovasc Pharmacol* 61 (6):505–512. doi: 10.1097/FJC.0b013e31828b7822.

Choudhary, D., D. Chandra, S. Choudhary, and R. Kale. 2001. "Modulation of glyoxalase, glutathione S-transferase and antioxidant enzymes in the liver, spleen and erythrocytes of mice by dietary administration of fenugreek seeds." *Food and Chemical Toxicology* 39 (10):989–997.

Clement, Y. N., I. Onakpoya, S. K. Hung, and E. Ernst. 2011. "Effects of herbal and dietary supplements on cognition in menopause: A systematic review." *Maturitas* 68 (3):256–263. doi: 10.1016/j.maturitas.2010.12.005.

Cortet, B., F. Blotman, F. Debiais, D. Huas, F. Mercier, C. Rousseaux, V. Berger, A.-F. Gaudin, and F.-E. Cotté. 2011. "Management of osteoporosis and associated quality of life in post menopausal women." *BMC Musculoskeletal Disorders* 12 (1):7. doi: 10.1186/1471-2474-12-7.

Cvetković, V. J., D. Takić Miladinov, and S. Stojanović. 2018. "Genotoxicity and mutagenicity testing of biomaterials." In *Biomaterials in Clinical Practice: Advances in Clinical Research and Medical Devices*, edited by Zivic, F., Affatato, S., Trajanovic, M., Schnabelrauch, M., Grujovic, N. and Choy, K.L., 501–527. Cham: Springer International Publishing.

Damanik, R., M. L. Wahlqvist, and N. Wattanapenpaiboon. 2006. "Lactagogue effects of Torbangun, a Bataknese traditional cuisine." *Asia Pacific Journal of Clinical Nutrition* 15 (2):267.

Dande, P., and S. Patil. 2012. "Evaluation of Saponins from *Trigonella foenum-graecum* seeds for its antifertility activity." *Asian J Pharm Clin Res* 5 (3):154–157.

Degirmencioglu, T., H. Unal, S. Ozbilgin, and H. Kuraloglu. 2016. "Effect of ground fenugreek seeds (Trigonella foenum-graecum) on feed consumption and milk performance in Anatolian water buffaloes." *Archives Animal Breeding* 59:345–349. doi: 10.5194/aab-59-345-2016.

Derogatis, L. R. 1997. "The Derogatis interview for sexual functioning (DISF/DISF-SR): An introductory report." *J Sex Marital Ther* 23 (4):291–304. doi: 10.1080/00926239708403933.

Deshpande, P., V. Mohan, D. Ingavale, J. Mane, M. Pore, and P. Thakurdesai. 2017. "Preclinical safety assessment of furostanol glycoside based standardized fenugreek seed extract in laboratory rats." *Journal of Dietary Supplements* 14 (5):521–541. http://dx.doi.org/10.1080/19390211.2016.1272659.

Deshpande, P., V. Mohan, M. P. Pore, S. Gumaste, and P. A. Thakurdesai. 2016a. "Prenatal developmental toxicity evaluation of low molecular weight galactomannans based standardized fenugreek seed extract during organogenesis period of pregnancy in rats." *International Journal of Pharmacy and Pharmaceutical Sciences* 8 (5):248–253.

Deshpande, P., V. Mohan, M. P. Pore, S. Gumaste, and P. A. Thakurdesai. 2016b. "Prenatal developmental toxicity evaluation of furostanol saponin glycoside based standardized fenugreek seed extract during organogenesis period of pregnancy in rats." *International Journal of Pharmacy and Pharmaceutical Sciences* 8 (12):124–129.

Deshpande, P., V. Mohan, M. P. Pore, S. Gumaste, and P. A. Thakurdesai. 2017. "Prenatal developmental toxicity study of glycosides based standardized fenugreek seed extract in rats." *Pharmacognosy Magazine* 13 (49):135–141.

Deshpande, P., V. Mohan, K. R. R. Reddy, V. R. L. N. Manjunath, and P. Thakurdesai. 2017. "Prenatal developmental toxicity evaluation of IDM01, a botanical composition of 4-hydroxyisoleucine- and trigonelline-based standardized fenugreek seed extract, during organogenesis period of pregnancy in rats." *Journal of Applied Pharmaceutical Science* 7 (10):62–69.

Deshpande, P., V. Mohan, and P. Thakurdesai. 2016a. "Preclinical safety evaluation of low molecular weight galactomannans based standardized fenugreek seeds extract." *EXCLI J* 15:446–459. doi: 10.17179/excli2016-461.

Deshpande, P., V. Mohan, and P. Thakurdesai. 2016b. "Preclinical safety assessment of glycosides based standardized fenugreek seeds extract: Acute, subchronic toxicity and mutagenicity studies." *J Appl Pharm Sci* 6 (9):179–188.

Deshpande, P., V. Mohan, and P. Thakurdesai. 2017. "Preclinical toxicological evaluation of IDM01: The botanical composition of 4-Hydroxyisoleucine- and trigonelline-based standardized fenugreek seed extract." *Pharmacognosy Research* 9 (2):138–150. doi: 10.4103/0974-8490.204649.

Dog, T. L. 2009. "The use of botanicals during pregnancy and lactation." *Alternative therapies in Health and Medicine* 15 (1):54–58.

Dording, C. M., and L. Sangermano. 2018. "Female Sexual Dysfunction: Natural and Complementary Treatments." *Focus (Am Psychiatr Publ)* 16 (1):19–23. doi: 10.1176/appi.focus.20170049.

El-Kamali, H., and S. Khalid. 1996. "The most common herbal remedies in Central Sudan." *Fitoterapia(Milano)* 67 (4):301–306.

El Sakka, A., M. Salama, and K. Salama. 2014. "The effect of fenugreek herbal tea and palm dates on breast milk production and infant weight." *Journal of Pediatric Sciences* 6:e202. doi: 10.17334/jps.30658.

El-Tarabany, A., F. E. Teama, M. A. Atta, and M. El-Tarabany. 2018. "Impact of dietary fenugreek seeds on lactational performance and blood biochemical and hematological parameters of dairy goats under hot summer conditions." *Izvorni znanstveni rad* 68 (3):214–223.

Esmat, N. M., A. El-Nasr, E.-D. Ezz Eddin, S. M. Sawsan, and M. N. Salwa. 2018. "Osteo-protective effect of Curcumin; Fenugreek and their combination on ovariectomized female rats." *Egyptian Journal of Veterinary Sciences* 49 (1):1–12. doi: 10.21608/ejvs.2018.2312.1024.

Folwarczna, J., M. Zych, B. Nowińska, and M. Pytlik. 2014. "Effects of fenugreek (Trigonella foenum-graecum L.) seed on bone mechanical properties in rats." *Eur Rev Med Pharmacol Sci* 18 (13):1937–1947.

Forinash, A. B., A. M. Yancey, K. N. Barnes, and T. D. Myles. 2012. "The use of galactogogues in the breastfeeding mother." *Annals of Pharmacotherapy* 46 (10):1392–1404. doi: 10.1345/aph.1R167.

Foster, S., and V. Tyler. 1999. *Tyler's Honest Herbal: A Sensible Guide to the Use of Herbs and Related Remedies*. Oxfordshire, UK: Routledge.

Gabay, M. P. 2002. "Galactogogues: Medications that induce lactation." *Journal of Human Lactation* 18 (3):274–279. doi: 10.1177/089033440201800311.

Gambineri, A., D. Laudisio, C. Marocco, S. Radellini, A. Colao, S. Savastano, E. R. on behalf of the Obesity Programs of nutrition, and g. Assessment. 2019. "Female infertility: Which role for obesity?" *International Journal of Obesity Supplements* 9 (1):65–72. doi: 10.1038/s41367-019-0009-1.

Ghasemi, V., M. Kheirkhah, and M. Vahedi. 2015. "The effect of herbal tea containing fenugreek seed on the signs of breast milk sufficiency in Iranian girl infants." *Iran Red Crescent Med J.* 17 (8):e21848. doi: 10.5812/ircmj.21848.

Ginaldi, L., M. De Martinis, S. Saitta, M. M. Sirufo, C. Mannucci, M. Casciaro, F. Ciccarelli, and S. Gangemi. 2019. "Interleukin-33 serum levels in postmenopausal women with osteoporosis." *Sci Rep* 9 (1):3786. doi: 10.1038/s41598-019-40212-6.

Gorry, A., D. M. White, and S. Franks. 2006. "Infertility in polycystic ovary syndrome." *Endocrine* 30 (1):27–33. doi: 10.1385/ENDO:30:1:27.

Greendale, G. A., N. P. Lee, and E. R. Arriola. 1999. "The menopause." *The Lancet* 353 (9152):571–580.

Greene, J. G. 1998. "Constructing a standard climacteric scale." *Maturitas* 29 (1):25–31. doi: 10.1016/s0378-5122(98)00025-5.

Gupta, A., R. Gupta, and B. Lal. 2001. "Effect of Trigonella foenum-graecum (fenugreek) seeds on glycaemic control and insulin resistance in type 2 diabetes mellitus: A double blind placebo controlled study." *J Assoc Physicians India* 49:1057–1061.

Hegde, H., G. Hegde, and S. Kholkutec. 2007. "Herbal care for reproductive health: Ethno medicobotany from Uttara Kannada district in Karnataka, India." *Complement Ther Clin Pract* 13 (1):38–45. doi: 10.1016/j.ctcp.2006.09.002.

Henderson, V. W. 2008. "Cognitive changes after menopause: Influence of estrogen." *Clin Obstet Gynecol* 51 (3):618–626. doi: 10.1097/GRF.0b013e318180ba10.

Hernández-Vázquez, J. M. V., H. López-Muñoz, M. L. Escobar-Sánchez, F. Flores-Guzmán, B. Weiss-Steider, J. C. Hilario-Martínez, J. Sandoval-Ramírez, M. A. Fernández-Herrera, and L. S. Sánchez. 2020. "Apoptotic, necrotic, and antiproliferative activity of diosgenin and diosgenin glycosides on cervical cancer cells." *European Journal of Pharmacology* 871:172942.

Higdon, K., A. Scott, M. Tucci, H. Benghuzzi, A. Tsao, A. Puckett, Z. Cason, and J. Hughes. 2001. "The use of estrogen, DHEA, and diosgenin in a sustained delivery setting as a novel treatment approach for osteoporosis in the ovariectomized adult rat model." *Biomed Sci Instrum* 37:281–286.

Hilditch, J. R., J. Lewis, A. Peter, B. van Maris, A. Ross, E. Franssen, G. H. Guyatt, P. G. Norton, and E. Dunn. 1996. "A menopause-specific quality of life questionnaire: Development and psychometric properties." *Maturitas* 24 (3):161–175. doi: 10.1016/s0378-5122(96)82006-8.

Høegh-Andersen, P., L. B. Tankó, T. L. Andersen, C. V. Lundberg, J. A. Mo, A. M. Heegaard, J. M. Delaissé, and S. Christgau. 2004. "Ovariectomized rats as a model of postmenopausal osteoarthritis: Validation and application." *Arthritis Res Ther* 6 (2):R169–R80. doi: 10.1186/ar1152.

Hudson, T. 2017. "Fenugreek seed tea improves breast milk sufficiency in infants." *Townsend Letter* 109.

Inanmdar, W., A. Sultana, U. Mubeen, and K. Rahman. 2016. "Clinical efficacy of Trigonella foenum-graecum (Fenugreek) and dry cupping therapy on intensity of pain in patients with primary dysmenorrhea." *Chinese Journal of Integrative Medicine*:1–8.

Jelodar, G., and K. Askari. 2012. "Effect of Vitex agnus-castus fruits hydroalcoholic extract on sex hormones in rat with induced polycystic ovary syndrome (PCOS)." *Physiology and Pharmacology* 16 (1):62–69.

Ji, M. X., and Q. Yu. 2015. "Primary osteoporosis in postmenopausal women." *Chronic Dis Transl Med* 1 (1):9–13. doi: 10.1016/j.cdtm.2015.02.006.

Kamble, A., S. Dhamane, A. Kulkarni, and V. Potnis. 2020. "Review on effects of herbal extract for the treatment of polycystic ovarian syndrome (PCOS)." *Plant Archives* 20 (1):1189–1195.

Kandhare, A., S. L. Bodhankar, V. Mohan, and P. A. Thakurdesai. 2015a. "Prophylactic efficacy and possible mechanisms of oligosaccharides based standardized fenugreek seed extract on high-fat diet-induced insulin resistance in C57BL/6 mice." *Journal of Applied Pharmaceutical Science* 5 (3):35–45.

Kandhare, A. D., S. L. Bodhankar, V. Mohan, and P. A. Thakurdesai. 2015b. "Acute and repeated doses (28 days) oral toxicity study of glycosides based standardized fenugreek seed extract in laboratory mice." *Regul Toxicol Pharmacol* 72 (2):323–234. doi: 10.1016/j.yrtph.2015.05.003.

Kandhare, A. D., S. L. Bodhankar, V. Mohan, and P. A. Thakurdesai. 2016. "Acute and repeated doses (28 days) oral toxicity study of Vicenin-1, a flavonoid glycoside isolated from fenugreek seeds in laboratory mice." *Regul Toxicol Pharmacol* 81:522–531. doi: 10.1016/j.yrtph.2016.10.013.

Kandhare, A. D., P. A. Thakurdesai, P. Wangikar, and S. L. Bodhankar. 2019. "A systematic literature review of fenugreek seed toxicity by using ToxRTool: Evidence from preclinical and clinical studies." *HELIYON* 5:e01536. doi: 10.1016/j.heliyon.2019. e01536.

Karapanou, O., and A. Papadimitriou. 2010. "Determinants of menarche." *Reproductive Biology and Endocrinology* 8 (1):115.

Kassem, A., A. Al-Aghbari, A.-H. Molham, and M. Al-Mamary. 2006. "Evaluation of the potential antifertility effect of fenugreek seeds in male and female rabbits." *Contraception* 73 (3):301–306.

Kawabata, T., M. Y. Cui, T. Hasegawa, F. Takano, and T. Ohta. 2011. "Anti-inflammatory and anti-melanogenic steroidal saponin glycosides from Fenugreek (Trigonella foenum-graecum L.) seeds." *Planta Med* 77 (7):705–710. doi: 10.1055/s-0030-1250477.

Kellogg-Spadt, S., and J. Albaugh. 2003. "Herbs, amino acids, and female libido." *Urol Nurs* 23 (2):160–161.

Khalki, L., M. Bennis, Z. Sokar, and S. Ba-M'hamed. 2012. "The developmental neurobehavioral effects of fenugreek seeds on prenatally exposed mice." *Journal of Ethnopharmacology* 139 (2):672–677. doi: 10.1016/j.jep.2011.12.011.

Khan, T. M., D. B.-C. Wu, and A. V. Dolzhenko. 2018. "Effectiveness of fenugreek as a galactagogue: A network meta-analysis." *Phytotherapy Research* 32 (3):402–412. https://doi.org/10.1002/ptr.5972.

Khoja, K. K., G. Shaf, T. N. Hasan, N. A. Syed, A. S. Al-Khalifa, A. H. Al-Assaf, and A. A. Alshatwi. 2011. "Fenugreek, a naturally occurring edible spice, kills MCF-7 human breast cancer cells via an apoptotic pathway." *Asian Pacific Journal of Cancer Prevention* 12 (12):3299–3304.

Khound, R., J. Shen, Y. Song, D. Santra, and Q. Su. 2018. "Phytoceuticals in Fenugreek Ameliorate VLDL Overproduction and Insulin Resistance via the Insig Signaling Pathway." *Mol Nutr Food Res* 62 (5). doi: 10.1002/mnfr.201700541.

Kim, S. A., and H. Jung. 2015. "Prevention of cognitive impairment in the midlife women." *J Menopausal Med* 21 (1):19–23. doi: 10.6118/jmm.2015.21.1.19.

Ko, C. B., S. J. Kim, C. Park, B. R. Kim, C. H. Shin, S. Choi, S. Y. Chung, J. H. Noh, J. H. Jeun, N. S. Kim, and R. Park. 2004. "Benzo(a)pyrene-induced apoptotic death of mouse hepatoma Hepa1c1c7 cells via activation of intrinsic caspase cascade and mitochondrial dysfunction." *Toxicology* 199 (1):35–46. doi: 10.1016/j.tox.2004.01.039.

Ko, S. H., and H. S. Kim. 2020. "Menopause-associated lipid metabolic disorders and foods beneficial for postmenopausal women." *Nutrients* 12 (1):202. doi: 10.3390/nu12010202.

Koebele, S. V., and H. A. Bimonte-Nelson. 2016. "Modeling menopause: The utility of rodents in translational behavioral endocrinology research." *Maturitas* 87:5–17. doi: 10.1016/j.maturitas.2016.01.015.

Kunle, O. F., H. O. Egharevba, and P. O. Ahmadu. 2012. "Standardization of herbal medicines-A review." *International Journal of Biodiversity and Conservation* 4 (3):101–112.

Lin, C. L., C. H. Lee, C. M. Chen, C. W. Cheng, P. N. Chen, T. H. Ying, and Y. H. Hsieh. 2018. "Protodioscin induces apoptosis through ROS-mediated endoplasmic reticulum stress via the JNK/p38 activation pathways in human cervical cancer cells." *Cell Physiol Biochem* 46 (1):322–334. doi: 10.1159/000488433.

LoPiccolo, J., and W. E. Stock. 1986. "Treatment of sexual dysfunction." *J Consult Clin Psychol* 54 (2):158–167. doi: 10.1037//0022-006x.54.2.158.

Lu, F. R., L. Shen, Y. Qin, L. Gao, H. Li, and Y. Dai. 2008. "Clinical observation on trigonella foenum-graecum L. total saponins in combination with sulfonylureas in the treatment of type 2 diabetes mellitus." *Chin J Integr Med* 14 (1):56–60. doi: 10.1007/s11655-007-9005-3.

Magdy Mohamady, N., S. Hassan Refaat, E. K. Habib, and G. Taha El-Sayed. 2018. "Effect of fenugreek seed extract (*Trigonella foenum-graecum*) in letrozole induced polycystic ovary syndrome in female albino rat." *QJM: An International Journal of Medicine* 111 (suppl 1):hcy200.087.

Mandal, P., A. Rai, S. Mishra, A. Tripathi, and M. Das. 2018. "Mutagens in food."133–160. doi: 10.1016/b978-0-12-809252-1.00007-9.

Mandegary, A., M. Pournamdari, F. Sharififar, S. Pournourmohammadi, R. Fardiar, and S. Shooli. 2012. "Alkaloid and flavonoid rich fractions of fenugreek seeds (Trigonella foenum-graecum L.) with antinociceptive and anti-inflammatory effects." *Food Chem Toxicol* 50 (7):2503–2507. doi: 10.1016/j.fct.2012.04.020.

Marjoribanks, J., C. Farquhar, H. Roberts, and A. Lethaby. 2012. "Long term hormone therapy for perimenopausal and postmenopausal women." *Cochrane Database of Systematic Reviews* (7).

Messina, M., S. Barnes, and K. D. Setchell. 1997. "Phyto-oestrogens and breast cancer." *Lancet* 350 (9083):971–972. doi: 10.1016/s0140-6736(05)64062-7.

Meston, C. M. 2003. "Validation of the Female Sexual Function Index (FSFI) in women with female orgasmic disorder and in women with hypoactive sexual desire disorder." *J Sex Marital Ther* 29 (1):39–46. doi: 10.1080/713847100.

Mittal, N., and T. Gopaldas. 1986. "Effect of fenugreek (Trigonella foenum graceum) seeds based on the birth outcome in albino rats." *Nutrition and Reproduction International* 33:362–369.

Modaresi, M., A. Jalalizand, and B. Mahdian. 2012. "The effect of hydro-alcoholic extract of fenugreek seeds on female reproductive hormones in mice." In International Conference on Applied Life Sciences (ICALS2012), Turkey, September 10–12, 2012, pp. 437–442.

Mohamed, R. S., D. A. Marrez, S. H. Salem, A. H. Zaghloul, I. S. Ashoush, A. R. H. Farrag, and A. M. Abdel-Salam. 2019. "Hypoglycemic, hypolipidemic and antioxidant effects of green sprouts juice and functional dairy micronutrients against streptozotocin-induced oxidative stress and diabetes in rats." *Heliyon* 5 (2):e01197. doi: 10.1016/j.heliyon.2019.e01197.

Mohan, V., and M. Balasubramanyam. 2001. "Fenugreek and insulin resistance." *J Assoc Physicians India* 49:1055–1056.

Møller, A. M. J., J.-M. Delaissé, J. B. Olesen, J. S. Madsen, L. M. Canto, T. Bechmann, S. R. Rogatto, and K. Søe. 2020. "Aging and menopause reprogram osteoclast precursors for aggressive bone resorption." *Bone research* 8 (1):1–11.

Morcos, S. R., and G. N. Gabrial. 1983. "Protein-rich food mixtures for feeding the young in Egypt II-chemical and biological evaluation." *Plant Foods for Human Nutrition* 32 (1):75–81. doi: 10.1007/BF01093932.

Mortel, M., and S. D. Mehta. 2013. "Systematic review of the efficacy of herbal galactogogues." *J Hum Lact* 29 (2):154–162. doi: 10.1177/0890334413477243.

Mortelmans, K., and E. Zeiger. 2000. "The Ames Salmonella/microsome mutagenicity assay." *Mutat Res* 455 (1–2):29–60. doi: 10.1016/s0027-5107(00)00064-6.

Mostafa-Gharabaghi, P., S. Hakimi, A. Delazar, M. Sayyah-Melli, and M. Jafari-Shobeiri. 2012. "Is fenugreek seed effective on early menopausal symptoms?" *International Journal of Gynecology and Obstetrics* (119S3):S558. doi: 10.1016/S0020-7292(12)61276-4.

Moyer, J., S. Acharya, Z. Mir, and R. Doram. 2003. "Weed management in irrigated fenugreek grown for forage in rotation with other annual crops." *Canadian Journal of Plant Science* 83 (1):181–188.

Nagulapalli Venkata, K. C., A. Swaroop, D. Bagchi, and A. Bishayee. 2017. "A small plant with big benefits: Fenugreek (Trigonella foenum-graecum Linn.) for disease prevention and health promotion." *Mol Nutr Food Res* 61 (6):1600950. doi: 10.1002/mnfr.201600950.

Nathan, J., S. Panjwani, V. Mohan, V. Joshi, and P. A. Thakurdesai. 2014. "Efficacy and safety of standardized extract of *Trigonella foenum-graecum* L seeds as an adjuvant to L-dopa in the management of patients with Parkinson's disease." *Phytotherapy Research* 28:172–178.

Ndefo, U. A., A. Eaton, and M. R. Green. 2013. "Polycystic ovary syndrome: A review of treatment options with a focus on pharmacological approaches." *Pharmacy and Therapeutics* 38 (6):336.

Nguyen, K. V. 2020. "Impacts & effects of chromosome abnormalities." *Journal of Down Syndrome & Chromosome Abnormalities* 6 (2):135. doi: 10.4172/2472-1115.20.6.135.

OECD. 1997a. *Test No. 473: In vitro Mammalian Chromosome Aberration Test*. Paris, France: OECD Publishing.

OECD. 1997b. *Test No. 471: Bacterial Reverse Mutation Test*. Paris, France: OECD Publishing.

OECD. 1998a. *Test No. 408: Repeated Dose 90-Day Oral Toxicity Study in Rodents*. Paris, France: OECD Publishing.

Organisation for Economic Co-operation and Development. 1998b. *OECD Guidelines for the Testing of Chemicals, Section 4: Health Effects*. Paris: OECD Publishing.

OECD. 2001. *Test No. 414: Prenatal Development Toxicity Study in Section 4: Health Effects*. Paris: OECD Publishing.

OECD. 2002. "Test No. 423: Acute Oral toxicity—Acute Toxic Class Method." In *OECD Guidelines for the Testing of Chemicals, Section 4: Health Effects*. Paris: OECD Publishing.

OECD. 2008. *Test No. 425: Acute Oral Toxicity: Up-and-Down Procedure*. Paris, France: OECD Publishing.

OECD. 2018. "Test No. 414: Prenatal Development Toxicity Study." In *OECD Guidelines for the Testing of Chemicals, Section 4: Health Effects*. Paris: OECD Publishing.

Özer, A., M. Bakacak, H. Kıran, Ö. Ercan, B. Köstü, M. Kanat-Pektaş, M. Kılınç, and F. Aslan. 2016. "Increased oxidative stress is associated with insulin resistance and infertility in polycystic ovary syndrome." *Ginekologia Polska* 87 (11):733–738. doi: 10.5603/gp.2016.0079.

Pal, D., and S. Mukherjee. 2020. "Chapter 13 — Fenugreek (Trigonella foenum) Seeds in Health and Nutrition." In *Nuts and Seeds in Health and Disease Prevention* (Second Edition), edited by Preedy, V.R. and Watson, R.R., 161–170. Cambridge: Academic Press.

Parekh, A., E. O. Fadiran, K. Uhl, and D. C. Throckmorton. 2011. "Adverse effects in women: Implications for drug development and regulatory policies." *Expert Rev Clin Pharmacol* 4 (4):453–466. doi: 10.1586/ecp.11.29.

Pattanittum, P., N. Kunyanone, J. Brown, U. S. Sangkomkamhang, J. Barnes, V. Seyfoddin, and J. Marjoribanks. 2016. "Dietary supplements for dysmenorrhoea." *Cochrane Database of Systematic Reviews* (3). doi: 10.1002/14651858.CD002124.pub2.

Pavone, V., G. Testa, S. M. C. Giardina, A. Vescio, D. A. Restivo, and G. Sessa. 2017. "Pharmacological therapy of osteoporosis: A systematic current review of literature." *Front Pharmacol* 8:803. doi: 10.3389/fphar.2017.00803.

Petit, P., Y. Sauvaire, D. Hillaire-Buys, O. Leconte, Y. Baissac, G. Ponsin, and G. Ribes. 1995. "Steroid saponins from fenugreek seeds: Extraction, purification, and pharmacological investigation on feeding behavior and plasma cholesterol." *Steroids* 60 (10):674–680.

Petropoulos, G. A., ed. 2002. *Fenugreek: The Genus Trigonella*. 1st ed. London: Taylor & Francis.

Petrovska, S., B. Dejanova, and V. Jurisic. 2012. "Estrogens: Mechanisms of neuroprotective effects." *Journal of Physiology and Biochemistry* 68 (3):455–460. doi: 10.1007/s13105-012-0159-x.

Picetti, R., L. Miller, H. Shakur-Still, T. Pepple, D. Beaumont, E. Balogun, E. Asonganyi, R. Chaudhri, M. El-Sheikh, B. J. B. P. Vwalika, and Childbirth. 2020. "The WOMAN trial: Clinical and contextual factors surrounding the deaths of 483 women following post-partum haemorrhage in developing countries." *BMC Pregnancy Childbirth* 20 (1):1–9. doi: 10.1186/s12884-020-03091-8.

Pitkin, J. 2012. "Alternative and complementary therapies for the menopause." *Menopause Int* 18 (1):20–27. doi: 10.1258/mi.2012.012001.

Pradeep, S. R., and K. Srinivasan. 2017. "Amelioration of oxidative stress by dietary fenugreek (Trigonella foenum-graecum L.) seeds is potentiated by onion (Allium cepa L.) in streptozotocin-induced diabetic rats." *Appl Physiol Nutr Metab* 42 (8):816–828. doi: 10.1139/apnm-2016-0592.

Pundarikakshudu, K. 2019. "Chapter 4 — antiinflammatory and antiarthritic activities of some foods and spices." In *Bioactive Food as Dietary Interventions for Arthritis and Related Inflammatory Diseases (Second Edition)*, edited by Watson, R.R. and Preedy, V.R., 51–68. Cambridge: Academic Press.

Qu, N., L. Wang, Z.-C. Liu, Q. Tian, and Q. Zhang. 2013. "Oestrogen receptor α agonist improved long-term ovariectomy-induced spatial cognition deficit in young rats." *International Journal of Neuropsychopharmacology* 16 (5):1071–1082.

Raja, R. 2015. "Vital role of herbal medicines in womens health: A perspective review." *African Journal of Plant Science* 9 (8):320–326. doi: 10.5897/AJPS2015.1315.

Raju, J., D. Gupta, A. R. Rao, P. K. Yadava, and N. Z. Baquer. 2001. "*Trigonella foenum-graecum* (fenugreek) seed powder improves glucose homeostasis in alloxan diabetic rat tissues by reversing the altered glycolytic, gluconeogenic and lipogenic enzymes." *Molecular and Cellular Biochemistry* 224 (1–2):45–51. doi: 10.1023/a:1011974630828.

Raju, J., and C. V. Rao. 2012. "Diosgenin, a steroid saponin constituent of yams and fenugreek: Emerging evidence for applications in medicine." *Bioactive Compounds in Phytomedicine* 125:143.

Rao, A., E. Steels, G. Beccaria, W. J. Inder, and L. Vitetta. 2015. "Influence of a Specialized *Trigonella foenum-graecum* Seed Extract (Libifem), on Testosterone, Estradiol and Sexual Function in Healthy Menstruating Women, a Randomised Placebo Controlled Study." *Phytotherapy Research* 29 (8):1123–1130. doi: 10.1002/ptr.5355.

Rathi, A., M. Ishaq, A. K. Najmi, and M. Akhtar. 2020. "Trigonelline demonstrated ameliorative effects in dexamethasone induced osteoporotic rats." *Drug Research* 70 (06):257–264. doi: 10.1055/a-1147-5724.

Ravi, R., and J. Joseph. 2020. "Effect of fenugreek on breast milk production and weight gain among Infants in the first week of life." *Clinical Epidemiology and Global Health* 8 (3):656–660.

Reddy, P. S., N. Begum, S. Mutha, and V. Bakshi. 2016. "Beneficial effect of curcumin in letrozole induced polycystic ovary syndrome." *Asian Pacific Journal of Reproduction* 5 (2):116–122.

Riordan, J. 2005. *Breastfeeding and Human Lactation*. 3rd ed. Boston and London: Jones and Bartlett, pp. 75–77.

Rocca, M. L., R. Venturella, R. Mocciaro, A. Di Cello, A. Sacchinelli, V. Russo, S. Trapasso, F. Zullo, and M. Morelli. 2015. "Polycystic ovary syndrome: Chemical pharmacotherapy." *Expert Opinion on Pharmacotherapy* 16 (9):1369–1393.

Sah, A. E.-N., H. Khattab, H. Al-Alamy, F. Salem, and M. Abdou. 2007. "Effect of some medicinal plants seeds in the rations on the productive performance of lactating buffaloes." *International Journal of Dairy Science* 3:31–41.

Saleem, M., H. Martin, and P. Coates. 2018. "Prolactin Biology and Laboratory Measurement: An Update on Physiology and Current Analytical Issues." *Clin Biochem Rev* 39 (1):3–16.

Saleh, A., E. M. Amin, A. A. Elfallah, and A. M. Hamed. 2020. "Insulin resistance and idiopathic infertility: A potential possible link." *Andrologia* 52 (11):e13773. doi: 10.1111/and.13773.

Sankhwar, P., A. Goel, and K. Tiwari. 2018. "Clinical Evaluation of Furostanolic Saponins and Flavanoids in Polycystic Ovarian Syndrome (PCOS) Patients." *Gynecology & Reproductive Health* 2 (4):1–7.

Seidel, V. 2006. "Initial and bulk extraction." In *Natural Products Isolation*, edited by Sarker, S.D., Latif, Z. and Gray, A.I., 27–46. Totowa, NJ: Humana Press.

Sevrin, T., M.-C. Alexandre-Gouabau, B. Castellano, A. Aguesse, K. Ouguerram, P. Ngyuen, D. Darmaun, and C.-Y. Boquien. 2019. "Impact of Fenugreek on milk production in rodent models of lactation challenge." *Nutrients* 11 (11):2571.

Sevrin, T., C. Y. Boquien, A. Gandon, I. Grit, P. de Coppet, D. Darmaun, and M. C. Alexandre-Gouabau. 2020. "Fenugreek stimulates the expression of genes involved in milk synthesis and milk flow through modulation of insulin/GH/IGF-1 Axis and Oxytocin secretion." *Genes (Basel)* 11 (10). doi: 10.3390/genes11101208.

Shamshad Begum, S., H. K. Jayalakshmi, H. G. Vidyavathi, G. Gopakumar, I. Abin, M. Balu, K. Geetha, S. V. Suresha, M. Vasundhara, and I. M. Krishnakumar. 2016. "A novel extract of Fenugreek Husk (FenuSMART) alleviates postmenopausal symptoms and helps to establish the hormonal balance: A randomized, double-blind, placebo-controlled study." *Phytother Research* 30 (11):1775–1784. doi: 10.1002/ptr.5680.

Sharma, R. 1986. "Effect of fenugreek seeds and leaves on blood glucose and serum insulin responses in human subjects." *Nutr Res* 6 (12):1353.

Shaywitz, S. E., B. A. Shaywitz, K. R. Pugh, R. K. Fulbright, P. Skudlarski, W. E. Mencl, R. T. Constable, F. Naftolin, S. F. Palter, K. E. Marchione, L. Katz, D. P. Shankweiler, J. M. Fletcher, C. Lacadie, M. Keltz, and J. C. Gore. 1999. "Effect of estrogen on brain activation patterns in postmenopausal women during working memory tasks." *JAMA* 281 (13):1197–1202. doi: 10.1001/jama.281.13.1197.

Shishodia, S., and B. B. Aggarwal. 2006. "Diosgenin inhibits osteoclastogenesis, invasion, and proliferation through the downregulation of Akt, I kappa B kinase activation and NF-kappa B-regulated gene expression." *Oncogene* 25 (10):1463–1473. doi: 10.1038/sj.onc.1209194.

Sies, H., C. Berndt, and D. P. Jones. 2017. "Oxidative stress." *Annual Review of Biochemistry* 86:715–748.

Sim, T. F., J. Sherriff, H. L. Hattingh, R. Parsons, and L. B. Tee. 2013. "The use of herbal medicines during breastfeeding: A population-based survey in Western Australia." *BMC Complement Altern Med* 13:317. doi: 10.1186/1472-6882-13-317.

Simon, J. A. 2011. "Identifying and treating sexual dysfunction in postmenopausal women: The role of estrogen." *J Womens Health (Larchmt)* 20 (10):1453–1465. doi: 10.1089/jwh.2010.2151.

Sindhu, G., M. Ratheesh, G. L. Shyni, B. Nambisan, and A. Helen. 2012. "Anti-inflammatory and antioxidative effects of mucilage of Trigonella foenum-graecum (Fenugreek) on adjuvant induced arthritic rats." *Int Immunopharmacol* 12 (1):205–211. doi: 10.1016/j.intimp.2011.11.012.

Singh, A., K. Vijaya, and K. S. Laxmi. 2018. "Prevalence of polycystic ovarian syndrome among adolescent girls: A prospective study." *Int J Reprod Contracept Obstet Gynecol* 7 (11):4375–4378.

Sniekers, Y. H., H. Weinans, S. M. Bierma-Zeinstra, J. P. T. M. van Leeuwen, and G. J. V. M. van Osch. 2008. "Animal models for osteoarthritis: The effect of ovariectomy and estrogen treatment—a systematic approach." *Osteoarthritis and Cartilage* 16 (5):533–541. https://doi.org/10.1016/j.joca.2008.01.002.

Soldin, O. P., and D. R. Mattison. 2009. "Sex differences in pharmacokinetics and pharmacodynamics." *Clin Pharmacokinet* 48 (3):143–157. doi: 10.2165/00003088-200948030-00001.

Sowers, M. F. R., A. D. Eyvazzadeh, D. McConnell, M. Yosef, M. L. Jannausch, D. Zhang, S. Harlow, J. F. Randolph Jr. 2008. "Anti-mullerian hormone and inhibin B in the definition of ovarian aging and the menopause transition." *J Clin Endocrinol Metab* 93 (9):3478–3483. doi: 10.1210/jc.2008-0567.

Sowmyalakshmi, S., R. Ranga, C. G. Gairola, and D. Chendil. 2005. *Effect of diosgenin (Fenugreek) on breast cancer cells. Cancer Research* 65 (9 Supplement):1382–1382.

Sreeja, S., V. S. Anju, and S. Sreeja. 2010. "In vitro estrogenic activities of fenugreek Trigonella foenum-graecum seeds." *Indian Journal of Medical Research* 131:814–819.

Srinivasan, K. 2005. "Role of spices beyond food flavoring: Nutraceuticals with multiple health effects." *Food Reviews International* 21 (2):167–188.

Steels, E., M. L. Steele, M. Harold, and S. Coulson. 2017. "Efficacy of a proprietary Trigonella foenum-graecum L. De-husked seed extract in reducing menopausal symptoms in otherwise healthy women: A double-blind, randomized, placebo-controlled study." *Phytother Res* 31 (9):1316–1322. doi: 10.1002/ptr.5856.

Swaroop, A., M. Bagchi, P. Kumar, H. G. Preuss, K. Tiwari, P. A. Marone, and D. Bagchi. 2014. "Safety, efficacy and toxicological evaluation of a novel, patented anti-diabetic extract of Trigonella Foenum-Graecum seed extract (Fenfuro)." *Toxicol Mech Methods* 24 (7):495–503. doi: 10.3109/15376516.2014.943443.

Swaroop, A., A. S. Jaipuriar, S. K. Gupta, M. Bagchi, P. Kumar, H. G. Preuss, and D. Bagchi. 2015. "Efficacy of a novel fenugreek seed extract (Trigonella foenum-graecum, FurocystTM) in polycystic ovary syndrome (PCOS)." *International Journal of Medical Sciences* 12 (10):825.

Thomas, J. E., M. Bandara, D. Driedger, and E. L. Lee. 2011. "Fenugreek in western Canada." *Am. J. Plant Sci Biotechnol* 5:32–44.

Thomas, J. V., J. Rao, F. John, S. Begum, B. Maliakel, K. Im, and A. Khanna. 2020. "Phytoestrogenic effect of fenugreek seed extract helps in ameliorating the leg pain and vasomotor symptoms in postmenopausal women: A randomized, double-blinded, placebo-controlled study." *PharmaNutrition* 14:100209. https://doi.org/10.1016/j.phanu.2020.100209.

Thornton, J. 2019. "WHO report shows that women outlive men worldwide." *BMJ* 365:l1631. doi: 10.1136/bmj.l1631.

Tiran, D. 2003. "The use of fenugreek for breast feeding women." *Complement Ther Nurs Midwifery* 9 (3):155–156. doi: 10.1016/s1353-6117(03)00044-1.

Turkyılmaz, C., E. Onal, I. M. Hirfanoglu, O. Turan, E. Koç, E. Ergenekon, and Y. Atalay. 2011. "The effect of galactagogue herbal tea on breast milk production and short-term catch-up of birth weight in the first week of life." *The Journal of Alternative and Complementary Medicine* 17 (2):139–142.

Ulbricht, C., E. Basch, D. Burke, L. Cheung, E. Ernst, N. Giese, I. Foppa, P. Hammerness, S. Hashmi, G. Kuo, M. Miranda, S. Mukherjee, M. Smith, D. Sollars, S. Tanguay-Colucci, N. Vijayan, and W. Weissner. 2007. "Fenugreek (Trigonella foenum-graecum L. Leguminosae): An evidence-based systematic review by the natural standard research collaboration." *J Herb Pharmacother* 7 (3–4):143–177. doi: 10.1080/15228940802142852.

Verheyen, G. R. 2017. "Testing the mutagenicity potential of chemicals." *Journal of Genetics and Genome Research* 4 (1). doi: 10.23937/2378-3648/1410029.

Verma, N., K. Usman, N. Patel, A. Jain, S. Dhakre, A. Swaroop, M. Bagchi, P. Kumar, H. G. Preuss, and D. Bagchi. 2016. "A multicenter clinical study to determine the efficacy of a novel fenugreek seed (Trigonella foenum-graecum) extract (Fenfuro™) in patients with type 2 diabetes." *Food Nutr Res* 60:32382. doi: 10.3402/fnr.v60.32382.

Villa, A., E. Vegeto, A. Poletti, and A. Maggi. 2016. "Estrogens, neuroinflammation, and neurodegeneration." *Endocrine Reviews* 37 (4):372–402. doi: 10.1210/er.2016-1007.

Vyas, S., R. P. Agrawal, P. Solanki, and P. Trivedi. 2008. "Analgesic and anti-inflammatory activities of Trigonella foenum-graecum (seed) extract." *Acta Pol Pharm* 65 (4):473–476.

Walker, M. 2006. *Breastfeeding Management for the Clinician: Using the Evidence*. Sudbury, MA: Jones and Bartlett, pp. 63–66.

Wani, S. A., and P. Kumar. 2018. "Fenugreek: A review on its nutraceutical properties and utilization in various food products." *Journal of the Saudi Society of Agricultural Sciences* 17 (2):97–106. doi: 10.1016/j.jssas.2016.01.007.

Weisburger, J. H. 1999. "Carcinogenicity and mutagenicity testing, then and now1I dedicate this paper to the memory of Ernst L. Wynder, MD, Founder and President of the American Health Foundation, a specialized research center on the causes, modulators and prevention of the main chronic diseases, with emphasis on cancer prevention. Ernst Wynder died on July 14, 1999.1." *Mutation Research/Reviews in Mutation Research* 437 (2):105–112. https://doi.org/10.1016/S1383-5742(99)00077-0.

Yao, D., B. Zhang, J. Zhu, Q. Zhang, Y. Hu, S. Wang, Y. Wang, H. Cao, and J. Xiao. 2019. "Advances on application of fenugreek seeds as functional foods: Pharmacology, clinical application, products, patents and market." *Crit Rev Food Sci Nutr*:1–11. doi: 10.1080/10408398.2019.1635567.

Yao, D., B. Zhang, J. Zhu, Q. Zhang, Y. Hu, S. Wang, Y. Wang, H. Cao, and J. Xiao. 2020. "Advances on application of fenugreek seeds as functional foods: Pharmacology, clinical application, products, patents and market." *Critical Reviews in Food Science and Nutrition* 60 (14):2342–2352. doi: 10.1080/10408398.2019.1635567.

Younesy, S., S. Amiraliakbari, S. Esmaeili, H. Alavimajd, and S. Nouraei. 2014. "Effects of fenugreek seed on the severity and systemic symptoms of dysmenorrhea." *Journal of Reproduction and Fertility* 15 (1):41–48.

Yousefzadeh, N., K. Kashfi, S. Jeddi, and A. Ghasemi. 2020. "Ovariectomized rat model of osteoporosis: A practical guide." *EXCLI Journal* 19:89. doi: 10.17179/excli2019-1990.

Zakaria, R., B. Al-Rahbi, A. Ahmad, R. Mohd Said, Z. Othman, F. Khairunnuur, and C. Badariah. 2019. "Menopause rodent models: Suitability for cognitive aging research." *International Medical Journal* 26:450–452.

Zapantis, A., J. G. Steinberg, and L. Schilit. 2012. "Use of herbals as galactagogues." *Journal of Pharmacy Practice* 25 (2):222–231.

Zárate, S., T. Stevnsner, and R. Gredilla. 2017. "Role of estrogen and other sex hormones in brain aging. Neuroprotection and DNA repair." *Frontiers in Aging Neuroscience* 9:430.

Zhang, Q. W., L. G. Lin, and W. C. Ye. 2018. "Techniques for extraction and isolation of natural products: A comprehensive review." *Chin Med* 13:20. doi: 10.1186/s13020-018-0177-x.

Zhou, C., Y. Qin, R. Chen, F. Gao, J. Zhang, and F. Lu. 2020. "Fenugreek attenuates obesity-induced inflammation and improves insulin resistance through downregulation of iRhom2/TACE." *Life Sci* 258:118222. doi: 10.1016/j.lfs.2020.118222.

Ziyyat, A., A. Legssyer, H. Mekhfi, A. Dassouli, M. Serhrouchni, and W. Benjelloun. 1997. "Phytotherapy of hypertension and diabetes in oriental Morocco." *J Ethnopharmacol* 58 (1):45–54.

17 Potential of Fenugreek in Management of Kidney and Lung Disorders

Amit D. Kandhare, Anwesha A. Mukherjee-Kandhare and Subhash L. Bodhankar

CONTENTS

17.1 Introduction 283
17.2 Fenugreek in Management of Renal Diseases 284
    17.2.1 Efficacy against Diabetic Nephropathy 284
    17.2.2 Efficacy against Urolithiasis 288
    17.2.3 Efficacy against Drug or Chemical-Induced Nephrotoxicity 291
    17.2.4 Efficacy against Immunity-Induced Nephrotoxicity 293
17.3 Fenugreek in Management of Pulmonary Diseases 295
    17.3.1 Efficacy against Acute Lung Disorder 295
    17.3.2 Efficacy against Asthma 296
17.4 Conclusion 298
References 298

ABBREVIATIONS

AHR: Airway Hyperresponsiveness; ALI: Acute Lung Injury; ARDS: Acute Respiratory Distress Syndrome; BUN: Blood Urea Nitrogen; CKD: Chronic Kidney Diseases; COX-2: Cyclooxygenase-2; GBM: Glomerular Basement Membrane; GLUT: Glucose Transporters; GRAS: Generally Recognized as Safe; GPx: Glutathione peroxidase; GSH: Reduced Glutathione; GSTs: Glutathione S-transferases; IFN: Interferon; ILs: Interleukins; IgE: Immunoglobulin E; LDH: Lactate dehydrogenase; LPO: Lipid peroxidation; LPS: Lipopolysaccharide; MDA: Malondialdehyde; MPO: Myeloperoxidase; NF-kB: Nuclear Factor kappa-light-chain-enhancer of activated B cells; OVA: Ovalbumin; ROS: Reactive Oxygen Species; SFSE-G: Glycosides-based standardized fenugreek seed extract; SFSE-T: Trigonelline rich standardized fenugreek seed extract; SOD: Superoxide Dismutase; STZ: Streptozotocin; TBARS: Thiobarbituric acid reactant substances; TLR4: Toll-like receptor-4; TNF-α: Tumor necrosis factor-α

17.1 INTRODUCTION

Occurrences of renal and pulmonary diseases constitute a global health problem. Health system planning needs a practical assessment of diseases keeping in view their epidemiology, morbidity, and mortality data. Chronic kidney disease (CKD) is considered one of the essential contributing factors for metabolic syndrome. In contrast, pulmonary diseases are among the risk factors for cardiovascular disorders (GBD Chronic Kidney Disease Collaboration 2020). The prevalence of CKD in the general population is about 14%, with hypertension and diabetes as important underlying causes (GBD Chronic Kidney Disease Collaboration 2020). CKD is often referred to as a silent disease because of undetected symptoms until it progresses into an advanced stage requiring significant treatment attention. A report estimated the average annual medical expenditure for CKD patients on dialysis to be US$15,066, and for kidney transplant patients to be US$21,027, both of which are remarkably high (Zhang et al. 2020). Thus, millions of patients preferred herbal medicine as an alternative treatment option due to its economic and cultural advantages. As an added advantage, these medicinal plants have a long history of safe usage with minimal side effects.

Chronic respiratory diseases such as interstitial lung diseases asthma, chronic obstructive pulmonary disease (COPD), pneumoconiosis, and pulmonary sarcoidosis are among the leading causes of morbidity worldwide. Cases of chronic respiratory diseases have increased by 40% over the last couple of decades (Xie et al. 2020). A study suggested that several risk factors contribute to the disability-adjusted life-years in COPD, which include air pollution (53.7%), use of tobacco (25.4%), and occupational risks (16.5%) (Salvi et al. 2018). COPD and asthma are the leading causes of death among non-communicable diseases globally. Several National Health Policies, including India, have suggested reducing mortality by 25% by 2025 (Salvi et al. 2018). A systematic review of clinical studies conducted on herbal medicine as an adjunct therapy during COPD found significant improvement in clinical symptoms and quality of life (Chen et al. 2014), which led to a growing interest in adjunct herbal remedies.

Fenugreek (*Trigonella foenum graecum*) has a long history of use since ancient times for its culinary and medicinal potential. It is native to Western Asia, the Mediterranean region, and southern Europe; its seed is widely used for cooking and traditionally used to treat various maladies, including diabetes, inflammation, and hyperlipidemia (Murakami et al. 2000; Patil, Niphadkar, and Bapat 1997). An array of investigators documented its antioxidant and antiinflammatory benefits during various experimental and clinical studies (Chevassus et al. 2009; Chevassus et al. 2010; Emtiazy et al. 2018; Khan, Negi, and Kumar 2018). Fenugreek seed contains many phytoconstituents, including alkaloid (trigonelline), furostanol and flavonol glycosides, and steroidal saponins. Researchers have established a key role of fenugreek phytoconstituents in managing various maladies (Wani and Kumar 2018; Zameer et al. 2018).

Long-term consumption of fenugreek seed and its isolated phytoconstituents such as trigonelline and glycosides have been deemed safe. Oral and subcutaneous administration of trigonelline reported LD_{50} of 5000 mg/kg (Zeiger and Tice 1997). Additionally, it does not produce any carcinogenicity and mutagenicity, nor does the administration of trigonelline (3,500 mg) for 70 days produce any visible effects in experimental animals (Zeiger and Tice 1997). Furthermore, fenugreek seed is recognized as a Generally Recognized as Safe (GRAS) ingredient as per 21 CFR21 § 182.20 (US Food and Drug Administration 2010). Thus, its broad margin of safety has inspired researchers to utilize the antioxidant and antiinflammatory potential of fenugreek against the management of various renal and respiratory disorders (Baset et al. 2020; Darwish et al. 2020; Emtiazy et al. 2018; Thakurdesai, Mohan et al. 2015). This chapter reviews the experimental and clinical studies on the beneficial effects of fenugreek seed and its phytoconstituents in various renal and pulmonary diseases.

17.2 FENUGREEK IN MANAGEMENT OF RENAL DISEASES

17.2.1 Efficacy against Diabetic Nephropathy

Diabetic nephropathy (DN) is life-threatening and progressive, apart from being one of the leading causes of renal failure during chronic diabetes mellitus (DM) condition (Gheith et al. 2016; Zhou et al. 2017). The current estimate of about 350 million (8%) people affected by DM worldwide is expected to rise to ~550 million by 2025 as per World Health Organization (Gheith et al.

2016). Additionally, more than 40% of DM patients eventually develop DN and end-stage kidney disease, which require dialysis or renal transplantation (Gheith et al. 2016). It has been suggested that DN is associated with a significant economic burden, and the annual cost of treatment of DN ranges from US$362 to US$3,716 (Zhou et al. 2017). Current therapeutic regimens are aimed to reduce blood sugar levels, blood pressure, cholesterol levels, and proteinuria. The standard treatment includes antihypertensive agents (such as angiotensin-converting enzyme (ACE), Angiotensin II receptor blockers (ARBs), α-blockers, β-blockers, diuretics) and hyperglycemic agents (including sodium-glucose cotransporter 2 (SGLT2) inhibitors and dipeptidyl peptidase 4 (DPP-4) inhibitors), along with recent advanced therapies including mineralocorticoid receptor antagonists (MRAs) and Endothelin 1 (ET-1) (Foggensteiner, Mulroy, and Firth 2001). However, the effectiveness of these therapies are restricted to making the patients symptom-free. There is no contribution towards inhibition or reversal of DN progression.

Traditionally, herbal moieties with medicinal values have been widely used to prevent and treat various disease states. Fenugreek is one such plant with a 3000-year-old history of medicinal use possessing a wide range of pharmacological potential. Its efficacy as an antidiabetic in both type 1 and type 2 DM (Geberemeskel, Debebe, and Nguse 2019; Gupta, Gupta, and Lal 2001; Najdi et al. 2019) patients along with experimental animal models (Al-Habori and Raman 1998; Baset et al. 2020; Li et al. 2018) has been well established and reiterated over decades. Additionally, elevated glucose levels have been associated with an increased risk of renal failure (Alsahli and Gerich 2015; Pecoits-Filho et al. 2016). Thus, the antihyperglycemic potential of fenugreek has led to a host of studies evaluating its nephroprotective effects during diabetic nephropathy (Arora et al. 2012; Baset et al. 2020; Ghule, Jadhav, and Bodhankar 2012; Hamden et al. 2010, 2013, 2017; Jin et al. 2014; Konopelniuk et al. 2017; Pradeep, Barman, and Srinivasan 2019; Sayed, Khalifa, and Abd el-Latif 2012; Shetty and Salimath 2009; Thakurdesai, Mohan et al. 2015; Xue et al. 2011). It has been well documented that the administration of compounds such as streptozotocin (STZ) or alloxan induces selective toxicity to pancreatic β cells and induces oxidative damage to islets of Langerhans, which further results in elevated blood sugar levels (Sasase et al. 2019). This hyperglycemic condition induces structural/functional perturbation in glomerular basement membrane (GBM)-associated glycosaminoglycan heparan sulfate (HS) and type IV collagen, which in turn contribute to renal toxicity. These insights helped elucidate the molecular mechanism underlying the nephroprotective potential of fenugreek during diabetic condition.

The exploration of renoprotective efficacy and possible mechanism of action of fenugreek against STZ-induced diabetic rats is reported (Shetty and Salimath 2009). Diabetes was induced in male Wistar rats by intraperitoneal injection of STZ (55 mg/kg body weight), and animals were fed with fenugreek powder (5% diet). Results suggested that fenugreek administration marked attenuated diabetes-induced polydipsia, polyuria, renal hypertrophy, glomerular filtration rate, and renal enzymes (N-acetyl glucosaminidase, L-glutamine fructose-6-phosphate aminotransferase, and β-glucuronidase). Fenugreek treatment also significantly increased glycosaminoglycan heparan sulfate content in renal tissue. Furthermore, immunohistological analysis of kidney tissue revealed that fenugreek administration effectively reduced type IV collagen deposition in the glomerulus region. The authors mentioned that the beneficial effects of fenugreek in ameliorating diabetic nephropathy could be attributed to the presence of 4-hydroxy isoleucine (4-HI), which is reported to decrease hyperglycemia via stimulation of insulin release (Shetty and Salimath 2009).

Potential of fenugreek oil was evaluated in alloxan-induced diabetic rats (Hamden et al. 2010). The diabetic rats were treated with fenugreek oil (10% diet), and various biochemical and molecular parameters evaluated in renal tissue. Administration of fenugreek oil significantly reduced elevated blood glucose levels, subsequently increasing insulin sensitivity and glucose intolerance when compared with the diabetic control group. Moreover, diabetes-induced decreased superoxide dismutase (SOD), glutathione peroxidase (GPx), and catalase levels in renal tissue were effectively restored by fenugreek oil treatment. Additionally, it significantly attenuated elevated levels of renal lipid peroxidation (LPO). Histopathological studies revealed that fenugreek oil restored renal architecture. The author concluded that fenugreek oil ameliorated diabetes-induced renal toxicity via its antioxidant potential (Hamden et al. 2010).

Xue et al. (2011) assessed the potential of fenugreek seed aqueous extract in an experimental model of high-lipid-feeding-induced insulin resistance followed by STZ-induced diabetic rats. Insulin-resistant diabetic rats were fed with fenugreek seed aqueous extract (440, 870, and 1740 mg/kg) for 6 weeks. Diabetic rats exhibited significant increases in blood glucose, relative kidney weight, serum creatinine, BUN (Blood Urea Nitrogen), urinary protein, and creatinine clearance, which was significantly restored by fenugreek treatment. Decreased renal SOD and catalase levels, and increased renal malondialdehyde (MDA), urinary 8-hydroxy-2'-deoxyguanosine and renal cortex DNA content were markedly attenuated by fenugreek. Additionally, transmission electron microscopy analysis suggested that the glomerulus basement's diabetes-induced thickening was markedly reduced by fenugreek treatment. Thus, these findings concluded that fenugreek protects diabetes-induced renal dysfunction, possibly via protecting the DNA damage as well as its antioxidant activity (Xue et al. 2011).

Researchers have determined the nephroprotective efficacy of reconstructed composition of fenugreek seeds in alloxan-induced diabetic rats (Arora et al. 2012). The reconstructed composition of fenugreek seeds extract (IND01) contains 4-hydroxyisoleucine, trigonelline, and galactomannan (40%:30%:30%). The renoprotective effect of oral administration of IND01 (50, 100, and 200 mg/kg) was evaluated in alloxan-induced (160 mg/kg, i.p.) diabetic rats. Administration of IND01 improved creatinine clearance and decreased BUN levels. Furthermore, histopathological analyses of kidney tissue showed protection from interstitial inflammation, tubular necrosis, glomerular matrix formation, and fibrosis by IND01 treatment. The study concluded that IND01 exerts its renoprotective efficacy against diabetic nephropathy owing to its antidiabetic potential (Arora et al. 2012).

The effect of fenugreek alkaloid trigonelline has been determined against nephropathy in neonatal non-insulin-dependent DM rats (Ghule, Jadhav, and Bodhankar 2012). The non-insulin-dependent DM was induced in neonatal Wistar rat pups via intraperitoneal administration of STZ (50 mg/kg). After 16 weeks of development, period animals were treated with trigonelline (50 and 100 mg/kg). Various parameters were evaluated in renal tissue, including renal function, renal apoptosis, and histomorphology changes. The administration of trigonelline significantly lowered serum creatinine levels, BUN, LPO, tumor necrotic factor (TNF-α), and hydroxyproline content, whereas glomerular filtration and activities of renal antioxidant enzymes (SOD and GSH), as well as a membrane-bound enzyme, were increased after trigonelline treatment. Flow cytometric analysis revealed that trigonelline significantly ameliorated diabetes-induced apoptosis in renal tissue. Furthermore, histological analysis (inflammatory and fibrotic changes) of kidney tissue from trigonelline-treated rats were also devoid of biochemical aberrations. The author concluded that trigonelline ameliorated diabetes-induced kidney damage in neonatal rats due to its antioxidant, antiinflammatory, and antiapoptotic potential (Ghule, Jadhav, and Bodhankar 2012).

A study investigated the protective efficacy of fenugreek seed against diabetes-induced renal dysfunction in experimental rats (Sayed, Khalifa, and Abd el-Latif 2012). Fenugreek seed powder (5% diet, for 12 weeks) was administered in alloxan-induced diabetic rats and evaluated for antioxidant and antiinflammatory potential against DN. Fenugreek treatment effectively attenuated elevated levels of serum glucose, creatinine, urea, sodium, potassium, and IL-6, as well as renal MDA and IL-6 levels. Whereas SOD, GSH, and catalase activities were markedly increased by fenugreek treatment. Furthermore, histological analysis of renal tissue suggested reduction in glomerular mesangial expansion after fenugreek administration. Results indicate that the antioxidant and antiinflammatory properties of fenugreek play a crucial role during amelioration of diabetic nephropathy (Sayed, Khalifa, and Abd el-Latif 2012).

A study suggested that diabetes and its complications, including DN, can be ameliorated by inhibiting intestinal carbohydrate-hydrolyzing enzymes α-amylase, maltase, and lipase (Hamden et al. 2010). An older study reported that trigonelline from fenugreek inhibited glycolysis enzymes, glycogen metabolism, gluconeogenesis, and the polyol pathway (Moorthy, Prabhu, and Murthy 2010). Subsequently, the investigation was proposed to evaluate trigonelline's role in amelioration of aberrations in renal function of alloxan-induced diabetic rats by key metabolic enzymes (Hamden et al. 2013). Diabetic rats treated with trigonelline (50 mg/kg) showed a significant decrease in intestinal α-amylase and maltase activity, shielding pancreas β-cells from damage. Trigonelline further improved the results of renal function tests reflected by decreased LDH (Lactate dehydrogenase), creatinine, urea, and albumin levels. The histological analysis of kidney tissues treated with trigonelline showed protection against diabetes-induced damage. The findings thus established that trigonelline protects renal function in diabetic rats (Hamden et al. 2013).

The underlying mechanism of fenugreek seeds' powder is suggested in the study against STZ-induced diabetic nephropathy in experimental rats (Jin et al. 2014). Fenugreek powder was standardized to polysaccharides (0.74%) and administered in diabetic rats at a dose of 9 gm/kg orally. Various renal functions were evaluated including albuminuria, BUN, serum creatinine, kidney index, renal oxidative stress, and accumulation of extracellular matrix. Fenugreek treatment significantly inhibited STZ-induced alterations in renal functions and elevated oxidative stress (catalase, reduced glutathione (GHS) and MDA). Immunohistochemistry and quantitative (q) polymerase chain reaction analysis showed that upregulated fibronectin, type IV collagen, connective growth factor, and transforming growth factor-beta 1 (TGF-β1) were decreased by fenugreek treatment. These findings were further confirmed by western blot analysis of renal TGF-β1 protein expressions. Transmission electron microscopy suggested that STZ-induced glomerular morphological changes in kidneys were ameliorated by fenugreek treatment. Based on these findings, the authors concluded that fenugreek prevented development of renal toxicity via restrained TGF-β1/connective tissue growth factor (CTGF) signaling pathway STZ-induced diabetic nephropathy (Jin et al. 2014).

4-Hydroxyisoleucine (4-HI), a major amino acid from fenugreek seeds, has been well documented for its antihyperglycemic potential. The renoprotective efficacy and mechanism of action of 4-HI have been studied *in vivo* (in alloxan-induced DN in rats) and *in vitro* (in copper ascorbate induced renal mitochondrial damage). 4-HI (80 mg/kg) was administered in alloxan-induced diabetic rats, and various biochemical and histological parameters were evaluated. Whereas, for the *in vitro* experiment, 4-HI (0.01, 0.02, 0.04, and 0.08 μg/ml) was co-incubated with goat kidneys to evaluate its protective efficacy against copper ascorbate-induced oxidative stress. 4-HI significantly protected against diabetes-induced increased 24-h urine volume, creatinine clearance, serum creatinine, and BUN, as well as histological aberration induced in renal tissue. The *in vitro* analysis showed significant protection against copper ascorbate-induced elevated oxidative damage (LPO, nitric oxide, protein carbonyl content), improved antioxidant status (GSH, SOD, and GPx), Krebs cycle enzymes (α-ketoglutarate dehydrogenase, succinate dehydrogenase), and mitochondrial electron transport chain enzymes (nicotinamide adenine dinucleotide (NADH)-Cytochrome C reductase and Cytochrome C oxidase). This investigation concluded that 4-HI exerts its antihyperglycemic effect and provides protection to the mitochondrion electron transport chain maintaining energy balance in renal tissue, thus contributing towards amelioration of DN (Thakurdesai, Mohan et al. 2015).

The renoprotective potential of fenugreek seed oil and its pure triglyceride were evaluated in alloxan-induced diabetic male Wistar rats (Hamden et al. 2017). Fenugreek seed oil contains linoleic acid (C18:2 41.13%), oleic acid (C18:1 17.07%), and linolenic acid (C18:3 26.14%) as major fatty acids, whereas β-sitosterol (85.3%) as a major sterol. Diabetic rats were treated with fenugreek triglycerides (50 mg/kg), which showed inhibition of intestinal α-amylase and lipase activity, and improved insulin sensitivity. Administration of fenugreek triglycerides decreased serum and renal angiotensin-converting enzyme levels. Interestingly, the fenugreek triglycerides ameliorated renal indices dysfunction reflected by a correction in diabetes-induced alterations of creatinine, urea, and albumin levels. Moreover, histological changes induced by diabetes was ameliorated by fenugreek triglycerides administration. This study demonstrated the beneficial effects of fenugreek triglycerides in attenuation of diabetes and its related disorders (Hamden et al. 2017).

High calorie diet-induced obesity is known to contribute towards metabolic syndrome and CKD owing to its association with type 2 diabetes (Camara et al. 2017). Consequently, the potential of fenugreek-based nanocomposite was evaluated against high caloric diet-induced renal dysfunction in experimental rats. Rats were administered with a high caloric diet for 98 days and concomitantly treated with fenugreek-based nanocomposite (150 mg/kg) for 21 days. Findings revealed that a high caloric diet is associated with increased levels of endogenous intoxication syndrome markers such as oligopeptides, low and middle molecules, which activate proteolytic enzymes and increase oxidative stress. However, the activity of these intoxication syndrome markers were attenuated by fenugreek administration. Furthermore, fenugreek significantly ameliorated high caloric diet-induced alteration in the levels of renal function markers (urea, creatinine, and uric acid). Ultrasound data depicted the attenuation of high caloric diet-induced renal injury after fenugreek administration. The author concluded that fenugreek-based nanocomposite inhibited endogenous intoxication syndrome markers and kidney dysfunction evoked by high calorie diet-induced obesity (Konopelniuk et al. 2017).

A study demonstrated the potential of dietary fenugreek seed (3%) administration against renin-angiotensin system-mediated renal damage in diabetic rats (Pradeep, Barman, and Srinivasan 2019). Fenugreek seed administration showed profound down-regulation in renal glucose transporters (GLUT-1 and GLUT-2), renal angiotensin-converting enzyme (ACE activity) and AT1 receptor expression, metabolites of the polyol pathway, and N-acetyl-β-d-glycosaminidase activity. The upregulated expression of kidney injury molecule-1, inducible nitric oxide, and type I collagen were markedly ameliorated by dietary fenugreek. Furthermore, podocyte damage was partially restored by fenugreek administration reflected by a correction in urinary nephrin, podocin, and podocalyxin markers. Diabetes-induced renal aberrations were also reduced by dietary fenugreek intake. The researcher concluded that dietary intake of fiber-rich fenugreek seeds was associated with inhibition of glucose translocation and renin-angiotensin system, which halted the development of diabetic nephropathy (Pradeep, Barman, and Srinivasan 2019).

Recently, the renoprotective efficacy of fenugreek seed extract in diabetic rats after its oral and intraperitoneal administration has been reported (Baset et al. 2020). Fenugreek seed extract (100 mg/kg) was administered in STZ-induced diabetic rats either by intraperitoneal means (daily or every other day) or orally (every day) for four weeks. Results showed that fenugreek administration was associated with a significant reduction of urea levels (intraperitoneal administration) and creatinine (oral administration). Catalase and glutathione S-transferase (GST) activity elevated significantly after fenugreek treatment, whereas it markedly attenuated levels of lipid peroxidase. Diabetes-induced histological alteration in kidney tissue was protected by fenugreek administration. The author concluded that fenugreek exerts a beneficial effect against renal damage. Additionally, long-term administration of fenugreek was recommended for optimum renal tissue protection (Baset et al. 2020).

Overall, the findings suggest that 4-hydroxyisoleucine, trigonelline, linoleic acid, oleic acid, linolenic acid, and β-sitosterol are the major phytoconstituents of fenugreek seed that contribute to its nephroprotective potential. Fenugreek exerts its beneficial effect on amelioration of diabetic nephropathy via inhibition of elevated blood sugar levels, oxidative stress, α-amylase and maltase activity, angiotensin-converting enzyme activity, and collagen deposition (Table 17.1).

17.2.2 Efficacy against Urolithiasis

Urolithiasis or kidney stone is a common prevalent disorder of the urinary system caused by many etiological factors. Calcium oxalate-associated urolithiasis is the predominant (approximately 80%) amongst various pathologies of kidney stone (Butterweck and Khan 2009; Coe, Evan, and Worcester 2005). Evidence suggests that renal calculi formation is a complex process involving a series of physicochemical reactions within renal tubules, which include supersaturation, nucleation, development, aggregation, and retention (Alelign and Petros 2018). The current treatment options

TABLE 17.1
Potential of Fenugreek Extract against Diabetic Nephropathy

Active Phytoconstituent(s) or Fenugreek Extract	Animal Model	Dose	Proposed Mechanism of Action	Reference
Fenugreek seed	STZ-induced DN	Fenugreek (5% diet)	Inhibited type IV collagen deposition	Shetty and Salimath (2009)
Fenugreek oil	Alloxan-induced DN	Fenugreek oil (10% diet)	Increased SOD, GPx, and catalase levels in kidney Decreased renal LPO	Hamden et al. (2010)
Fenugreek seed aqueous extract	STZ-induced DN	Fenugreek seed aqueous extract (440, 870, and 1740 mg/kg)	Inhibited oxidative stress (SOD, GSH, and MDA), and DNA damage (8-hydroxy-2'-deoxyguanosine)	Xue et al. (2011)
Fenugreek extract (IND01) (4-hydroxyisoleucine, trigonelline, and galactomannan (40%:30%:30%))	Alloxan-induced DN	IND01 (50, 100, and 200 mg/kg)	Inhibited elevated blood sugar levels	Arora et al. (2012)
Trigonelline	STZ-induced DN	Trigonelline (50 and 100 mg/kg)	Inhibited oxidative stress (SOD, GSH, and MDA), inflammatory markers (TNF-α), collagen synthesis (hydroxyproline), and apoptosis	Ghule, Jadhav, and Bodhankar (2012).
Fenugreek seed	Alloxan-induced DN	Fenugreek (5% diet)	Inhibited oxidative stress (SOD, GSH, catalase, and MDA) and inflammatory markers (IL-6)	Sayed, Khalifa, and Abd el-Latif (2012)
Trigonelline	Alloxan-induced DN	Trigonelline (50 mg/kg)	Inhibited intestinal α-amylase and maltase activity	Hamden et al. (2013)
Fenugreek seed	STZ-induced DN	Fenugreek seed (9 g/kg)	Inhibited TGF-β1/CTGF signaling pathway	Jin et al. (2014)
4-hydroxyisoleucine	Alloxan-induced DN	4-HI (80 mg/kg)	Inhibited mitochondrial oxidative stress Protected mitochondrial electron transport chain enzymes	Thakurdesai, Mohan et al. (2015)
Fenugreek seed oil and its triglyceride	Alloxan-induced DN	Fenugreek triglycerides (50 mg/kg contains linoleic acid, oleic acid, linolenic acid, and β-sitosterol)	Inhibited intestinal α-amylase and lipase activity Inhibited ACE	Hamden et al. (2017)
Fenugreek seed	High caloric diet-induced DN	Fenugreek-based nanocomposite (150 mg/kg)	Inhibited oligopeptides, low and middle molecules	Konopelniuk et al. (2017)
Fenugreek seed	STZ-induced DN	Fenugreek seed (3%)	Inhibited renal GLUT-1 and GLUT-2, renal ACE activity), AT1 receptor expression, metabolites of the polyol pathway, N-acetyl-β-d-glucosaminidase, kidney injury molecule-1, inducible nitric oxide, and type I collagen activity	Pradeep, Barman, and Srinivasan (2019)
Fenugreek seed extract	STZ-induced DN	Fenugreek seed extract (100 mg/kg)	Inhibited oxidative stress (Catalase, GST, and lipid peroxidase)	Baset et al. (2020)

4-HI: 4-hydroxyisoleucine; ACE: Angiotensin-Converting Enzyme; AT-1: Angiotensin 1; CTGF (connective tissue growth factor); DN: Diabetic Nephropathy; GSH: Glutathione; GLUT: Glucose transporter; ILs: Interleukins; LPO: Lipid peroxidation; MDA: Malonaldehyde; SOD: Superoxide dismutase; STZ: Streptozotocin; TGF-β1: Transforming growth factor-beta 1; TNF-α: Tumor necrosis factor-alpha.

for urolithiasis management include non-pharmacological interventions, extracorporeal shockwave lithotripsy (ESWL), and laser of high magnitude. However, these surgical procedures may cause acute renal injury, and infections resulting in renal dysfunction and sometimes calculi recurrence (Shafi et al. 2016). The long-term treatment strategies include thiazide diuretics and alkali-citrate (K-citrate), which have been implicated for the prevalence of nephrolithiasis recurrence; however, these therapies showed poor efficacy (Shafi et al. 2016). A recent development in therapeutic research suggested that herbs played a significant role in treating kidney stones. The potential of Cystone, a polyherbal formulation, against urolithiasis has been well established, and thus approved by India's drug regulatory authority for the treatment of urolithiasis (Erickson, Vrtiska, and Lieske 2011). Traditionally, fenugreek has also been suggested useful for treatment of kidney stone (Ahsan et al. 1989).

Fenugreek seeds have been used by traditional Chinese herbalists for kidney problems and conditions affecting the male reproductive tract. Over the last few decades, various studies have investigated the potential of fenugreek against management of urolithiasis (Ahsan et al. 1989; Kapase et al. 2013; Laroubi et al. 2007; Shekha et al. 2014).

Studies have widely utilized ethylene glycol (EG)-induced urolithiasis animal model to evaluate the antiurolithiatic activity of fenugreek. Chronic administration of EG in rats caused hyperoxaluria that elevated renal retention, and excessive accumulation and secretion of urinary oxalate, contributing to the formation of urinary calculi (Liu et al. 2020; Susilo et al. 2021).

The potential of fenugreek seed against glycolic acid-induced experimental renal calculi in animals is reported (Ahsan et al. 1989). Calcium oxalate urolithiasis was induced in male Sprague-Dawley rats by glycolic acid (3% diet) and administered with a daily dose of fenugreek seed powder (500 mg/kg, 4 weeks). This study's findings indicate that the daily administration of fenugreek provided significant protection against deposited renal calcium oxalate. This study provided further support for the traditional use of fenugreek to prevent and treat kidney stones in folk medicine.

Aqueous extract of fenugreek seed was examined for its ameliorative potential against EG-induced calcium oxalate renal calculi (Laroubi et al. 2007). Chronic administration of EG (with 2% ammonium chloride) significantly induced hyperoxaluria and calcium oxalate deposits in renal tissue, which was inhibited by the administration of fenugreek extract (100 and 200 mg/kg). This was reflected by a decrease in the concentration of urea, creatinine, and calcium in serum. Additionally, fenugreek treatment significantly decreased calcium content in renal tissue, which was further supported by histopathological evaluation. It showed a reduction of crystalline deposits in the renal tissues after fenugreek treatment. The author mentioned that flavonoids in fenugreek extract exert diuretic and antioxidant activity, which contribute to its antiurolithiatic property (Laroubi et al. 2007).

Prophylactic potential of fenugreek seed against EG-induced urolithiasis was confirmed by Shekha et al. (2014). Fenugreek (10 gm in 100 ml of water and 10% standard diet) was concomitantly administered in Wistar rats for 28 days, along with EG (1%). Treatment with fenugreek significantly attenuated EG-induced increased kidney weight and MDA levels. Urinary microscopic examination also supported the notion that fenugreek administration reduced calcium oxalate crystal concentration in urine. Based on these findings, the authors concluded that the antiurolithiasis effects of fenugreek that prevented kidney stone formation was a result of its antioxidative potential (Shekha et al. 2014).

Antiurolithiasis potential of hydroalcoholic extract of fenugreek seed was investigated using EG-induced urolithiasis rats (Kapase et al. 2013). This standardized fenugreek seed extract is enriched in trigonelline (82%) which possess established nephroprotective potential (Arora et al. 2012; Ghule, Jadhav, and Bodhankar 2012). Administration of trigonelline-rich standardized fenugreek seed extract (SFSE-T, 30 and 60 mg/kg) significantly inhibited EG-induced alterations in urinary (urine output, pH, citrate, creatinine, chloride, oxalate, and uric acid) and serum (creatine, BUN, and uric acid) parameters. Histopathological aberrations induced in renal tissue (cellular and tubulointerstitial damage) were reduced by SFSE-T treatment. The trigonelline's zwitterionic

Management of Kidney and Lung Disorders

nature helped maintain urine pH, inhibited mineral supersaturation and crystallization, which further contributed to the prevention of stone formation. The authors concluded that trigonelline from fenugreek seed extracts was mainly responsible for its antiurolithiatic potential, which was executed through the maintenance of solute balance, urinary pH, and calcium oxalate crystal deposition inhibition in an animal model of EG-induced urolithiasis (Kapase et al. 2013).

The market compound in SFSE-T, trigonelline, is a major alkaloid of fenugreek seeds. Trigonelline is a zwitterion and formed by the methylation of the nitrogen atom of niacin (vitamin B_3) (Yuyama and Suzuki 1991). Furthermore, trigonelline is known to remain unabsorbed from GIT (Yuyama 1999), excreted unchanged in urine, and bioavailable for action in kidneys (Yuyama and Suzuki 1985). The zwitterionic nature of trigonelline helps to maintain pH of urine favorable for dissolution of solutes and can prevent stone formation.

Oxalate along with calcium oxalate (CaOx) monohydrate induces the generation of free radicals, the major mediators of pathologic consequences behind kidney stone formation (Dal Moro et al. 2005). Increased free radical generation and associated injury to the renal tubules are often considered as a prelude to the complex sequel of stone formation (Scheid et al. 1996). Studies have pointed out that mitochondria are the major contributor of free radicals in oxalate toxicity, and the contribution of other non-mitochondrial sources is minimal (Khand et al. 2002). Mitochondrial dysfunction is the primary event in oxalate toxicity (Cao et al. 2004; McMartin and Wallace 2005). Further, trigonelline is reported to have ameliorating effects on oxidative stress via down-regulation of mitochondrial electron transfer system-related gene expressions (Yoshinari, Takenake, and Igarashi 2013). Therefore, mechanisms such as Osmolytic (restoration of electrolyte balance and buffering mechanism), and antioxidant mechanisms seems to contribute to efficacy of trigonelline in ameliorating effects against urolithiasis.

17.2.3 Efficacy against Drug or Chemical-Induced Nephrotoxicity

Drug- or chemical-induced nephrotoxicity is a commonly associated disorder during certain clinical conditions, including cardiovascular disease and diabetes. The incidence of drug-related acute kidney injury in older adults is almost 66%, amongst which approximately 20% are hospital-acquired episodes (Ghane Shahrbaf and Assadi 2015). Although these drug-induced renal toxicities are often reversible with discontinuation of drug treatment, many times, worsening of these conditions may require multiple interventions and sometimes hospitalization, placing a significant economic burden (Naughton 2008).

Gentamicin is a widely used aminoglycoside antibiotic for managing gram-negative microorganisms, whereas cisplatin is an antineoplastic agent against solid tumors. However, both gentamicin and cisplatin are the most associated with nephrotoxicity, limiting their clinical efficacy (Sales and Foresto 2020). These agents induce kidney damage through multiple pathogenic mechanisms, including inflammation, oxidative stress, apoptosis, tubular cell toxicity, thrombotic microangiopathy, crystal nephropathy, and intraglomerular hemodynamics (Naughton 2008).

Renal toxicity can be associated with exposure to many chemicals or pesticides (Scammell et al. 2019; Valcke et al. 2017). Aluminum is one of the metals widely distributed in the environment and routinely used in daily life. It is also utilized for the purification of drinking water. However, an overdose of aluminum may lead to the generation of reactive oxygen species (ROS) and elevated oxidative stress, damaging renal tubular cells (Al Dera 2016). Similarly, sodium nitrite, widely used as a food preservative, is reported to induce oxidative stress, which causes DNA damage in kidney tissue (Uslu, Uslu, and Adalı 2019). Additionally, chronic exposure to various pesticides such as cypermethrin, paraquat dichloride, and captan are also documented to induce nephrotoxicity via elevated ROS in renal tissue (Sushma and Devasena 2010).

Therapeutic measures for acute renal toxicity arising from these intoxicants include administering appropriate antidotes, extracorporeal blood purification therapy, intermittent or continuous hemodialysis (Petejova et al. 2019). Additionally, in chronic renal toxicity, renal replacement therapy or transplantations using specific immunosuppressants have been suggested. However, these pharmacological and non-pharmacological options are associated with cost and surgical complications. Thus, many researchers have explored therapeutic regimens of herbal origin, and fenugreek is one of the promising herbal treatment options for nephrotoxicity (Belaid-Nouira et al. 2013; Darwish et al. 2020; Hilmi, Dewan, and Kabir 2018; Kaur et al. 2016; Pribac et al. 2015; Sushma and Devasena 2010; Uslu, Uslu, and Adalı 2019).

The potential of aqueous extract of germinated fenugreek seeds against cypermethrin-induced renal toxicity is reported (Sushma and Devasena 2010). Fenugreek has been established to possess antioxidant potential; thus, in the present investigation, cypermethrin-intoxicated rats (25 mg/kg in corn oil) were treated with aqueous fenugreek extract (10%) for 60 days. Chronic cypermethrin administration caused renal damage reflected by elevated oxidative stress in renal tissue. The concomitant administration of fenugreek extract caused a marked increase in SOD, catalase, GPx, and GSH activity, whereas it decreased thiobarbituric acid reactive substances (TBARS) in renal tissue. Reduced level of total phospholipids and increased phospholipases A and C in renal tissue were ameliorated by fenugreek administration. The study concluded that fenugreek plays a vital role in attenuating pesticide-induced toxicity in experimental rats (Sushma and Devasena 2010).

Researchers have assessed the efficacy of fenugreek in aluminum chloride-induced nephrotoxicity in experimental animals (Belaid-Nouira et al. 2013). Nephrotoxicity was induced in female Wistar rats by oral administration of aluminum chloride (500 mg/kg for 1 month, then 1600 ppm through drinking water for 4 months) and treated with fenugreek seed powder (5% diet) for the last 2 months. Chronic administration of aluminum chloride induced a significant decrease in total antioxidant status, increasing serum markers of renal toxicity (urea and creatinine) and LPO. However, treatment with fenugreek markedly decreased LPO and renal toxicity markers in serum, subsequently improving total antioxidant status. Additionally, fenugreek attenuated aluminum chloride-induced renal atrophy and histological alterations, including Bowman's capsule, the glomerulus, and renal tubules. Taken together, this study highlighted the beneficial effects of dietary fenugreek seed supplementation in ameliorating nephrotoxicity induced by chronic ingestion of aluminum salts (Belaid-Nouira et al. 2013).

The nephroprotective potential of fenugreek was evaluated in alcohol-intoxicated rats (Pribac et al. 2015). Renal toxicity was induced in Wistar rats by chronic ethanol administration (10% v/v, for 31 days), and rats were concomitantly treated with fenugreek seed (10% diet). Transmission electron microscopy (TEM) analysis revealed that ethanol intoxication caused renal damage reflected by cytoplasm vacuolation, diffused mitochondrial matrix, nephrotic edemas, diffused cytoplasm, increased lysozymes, and peroxisomes with capillary congestion. However, fenugreek administration showed amelioration of ultrastructural damage induced by ethanol in renal morphology, suggesting its nephroprotective efficacy. These findings suggested that dietary intake of fenugreek prevents cellular damage induced in renal morphology (Pribac et al. 2015).

The nephroprotective effect of fenugreek seed aqueous extract against gentamicin-induced renal nephropathy is reported (Kaur et al. 2016). Gentamicin (100 mg/kg, i.p., for 8 days) was administered in Wistar rats to induce renal injury, and animals were concomitantly treated with fenugreek seed aqueous extract (200, 400, and 800 mg/kg, p.o.). Fenugreek treatment markedly inhibited gentamicin-induced serum biochemical alterations (BUN, serum creatinine and protein). Gentamicin-induced elevated LPO and decreased GSH and catalase levels were significantly attenuated by fenugreek treatment. Furthermore, histological aberrations induced by gentamicin in renal tissue were profoundly ameliorated by fenugreek treatment. The authors concluded that the antioxidant potency of fenugreek seeds contributed towards its protective efficacy against gentamicin-induced nephrotoxicity in the experimental animal model (Kaur et al. 2016).

A study subjected ethanolic extract of fenugreek seed to experimental gentamicin-induced nephrotoxicity (Hilmi, Dewan, and Kabir 2018). Gentamicin-intoxicated rats were concomitantly treated with ethanol extract (70%) of fenugreek seed (500 mg/kg/day). Biochemical analysis of gentamicin-intoxicated rats suggested elevated serum creatinine and serum urea levels were ameliorated by fenugreek treatment. Additionally, gentamicin-induced elevated oxidative stress (GSH and LPO) was also decreased by fenugreek treatment. Histological changes in renal architecture induced by chronic administration of gentamicin were significantly ameliorated by fenugreek treatment. The study concluded that fenugreek ameliorated gentamicin-induced tubular damage by its antioxidant potential (Hilmi, Dewan, and Kabir 2018).

The correlation between chronic exposure to many synthetic food additives, including sodium nitrite, and the incidence of nephrotoxicity have been well established (Alyoussef and Al-Gayyar 2016). The chronic administration of sodium nitrite (80 mg/kg, for 3 months) is reported to cause a significant increase in serum urea and creatinine levels along with upregulated pro-inflammatory cytokine (TNF-α, IL-6, IL-1α) levels in renal tissue (Uslu, Uslu, and Adalı 2019). The protective efficacy of hydroalcoholic extract of fenugreek seed (150 mg/kg, oral) against sodium nitrite-induced nephrotoxicity in experimental rats was reported due to antioxidant and antiinflammatory properties (Uslu, Uslu, and Adalı 2019). Fenugreek extract significantly attenuated sodium nitrite-induced elevated serum urea, creatinine, and pro-inflammatory cytokine levels, along with protection from renal damage during histopathological examinations (Uslu, Uslu, and Adalı 2019).

A recent study evaluated the nephroprotective potential of dried and germinated fenugreek seeds against gentamicin-induced renal toxicity (Darwish et al. 2020). Gentamicin intoxicated mice were treated with either dried or germinated fenugreek (500 mg/kg, 5 days), and various renal function tests, oxidative stress, and histopathological tests were assessed. Results suggested a gentamicin-induced significant increase in blood urea, serum creatinine, uric acid, lipid peroxidase levels, and decreased GST, GSH, glutathione reductase, SOD, and catalase activity. These alterations in biochemical parameters were significantly ameliorated by fenugreek co-administration. Gentamicin-induced renal architectural damage was also prevented by fenugreek. The study concluded that concomitant administration of fenugreek prevented gentamicin-induced nephrotoxicity by its antioxidant effect (Darwish et al. 2020).

In summary, attenuation of drug or chemical-induced nephrotoxicity by fenugreek can be attributed to its antioxidant and antiinflammatory potential. However, further studies are needed to elucidate fenugreek's chemical constituents responsible for this pharmacological potential. The probable mechanisms of action of fenugreek extracts against urolithiasis and nephropathy are presented as Table 17.2.

17.2.4 Efficacy against Immunity-Induced Nephrotoxicity

Glomerulonephritis or nephritis is a severe and life-threatening illness that occurs due to inflammation of the glomeruli. Although prevalence is low, nephritis can be rapidly progressive and the patient may need immediate treatment. Additionally, damage to the glomerulus results in arterial hypertension and renal failure. Glomerulonephritis includes many diseases, namely anti- GBM antibody disease, IgA nephropathy, lupus nephritis, and ANCA-associated vasculitis (McAdoo and Pusey 2017). The pathophysiology remains unknown for glomerulonephritis; however, bacterial and viral infections have been encountered frequently during a clinical investigation (Couser and Johnson 2014). The recommended treatment regimen for glomerulonephritis includes daily administration of oral steroids such as cyclophosphamide and plasma exchange to decrease the serum levels of anti-GBM antibodies. Although these therapies are more efficient in removing antibodies from serum, their cost and availability have acted as limitations in widespread clinical practice.

The potential of glycoside-based standardized extract of fenugreek seeds (SFSE-G) against GBM-induced glomerulonephritis in experimental rats has been reported (Thakurdesai, Vichare et al. 2015). Glomerulonephritis was induced in Wistar rats by sensitizing and boosting rabbit IgG antibody (5 mg/kg, s.c.) followed by a challenge with anti-rat-GBM antibody obtained from rabbits (2.5 ml/kg, i.v.). Rats were orally treated with SFSE-G (10, 30, and 100 mg/kg, b.i.d.) for the next 28 days. The results suggested that administration of SFSE-G significantly improved creatinine clearance and decreased BUN, creatinine, and urinary protein extraction. Furthermore, elevated levels of

TABLE 17.2
Potential Mechanisms of Fenugreek Extract against Urolithiasis and Nephropathy

Phytoconstituent(s)	Dose	Animal Model	Proposed Mechanism	Reference
Fenugreek seed	Fenugreek seed powder (500 mg/kg)	Glycolic acid-induced urolithiasis	Inhibited calcium oxalate deposition	Ahsan et al. (1989)
Fenugreek seed aqueous extract	Fenugreek extract (100 and 200 mg/kg)	EG-induced urolithiasis	Inhibited calcium concentration	Laroubi et al. (2007)
Fenugreek seed hydroalcoholic extract	Trigonelline rich standardized fenugreek seed extract (30 and 60 mg/kg)	EG-induced urolithiasis	Maintained solute balance and urinary pH Inhibited calcium oxalate crystal deposition by osmolytes action	Kapase et al. (2013)
Fenugreek seed	Fenugreek (10 gm in 100 ml of water and 10% standard diet)	EG-induced urolithiasis	Inhibited MDA levels	Shekha et al. (2014)
Germinated fenugreek seeds	Aqueous fenugreek seeds extract (10%)	Cypermethrin-induced nephropathy	Inhibited oxidative stress (SOD, catalase, GPx, GSH, and TBARS), phospholipases A and C	Sushma and Devasena (2010)
Fenugreek seed	Fenugreek seed powder (5% diet)	Aluminum chloride-induced nephropathy	Inhibited LPO Improved total antioxidant status	Belaid-Nouira et al. (2013)
Fenugreek seed	Fenugreek seed powder (10% diet)	Ethanol-induced nephropathy	Inhibited ultrastructural changes	Pribac et al. (2015)
Fenugreek seed aqueous extract	Aqueous extract of fenugreek seeds (200, 400, and 800 mg/kg)	Gentamicin-induced nephropathy	Inhibited oxidative stress (Catalase, GSH, and LPO)	Kaur et al. (2016)
Fenugreek seed ethanolic extract	Ethanolic extract of fenugreek seeds (500 mg/kg)	Gentamicin-induced nephropathy	Inhibited oxidative stress (GSH and LPO)	Hilmi, Dewan, and Kabir (2018)
Fenugreek seed hydroalcoholic extract	Fenugreek seed hydroalcoholic extract (150 mg/kg)	Sodium nitrite-induced nephropathy	Inhibited pro-inflammatory cytokine (TNF-α, IL-6, and IL-1α) expressions	Uslu, Uslu, and Adalı (2019)
Fenugreek seed	Dried or germinated fenugreek (500 mg/kg)	Gentamicin-induced nephropathy	Inhibited oxidative stress (SOD, GST GSH, glutathione reductase, catalase, and LPO)	Darwish et al. (2020)
Glycoside-based standardized fenugreek seeds extract	SFSE-G (10, 30, and 100 mg/kg, b.i.d.)	Anti-GBM antibody-induced nephropathy	Inhibited pro-inflammatory cytokine (TNF-α and IL-1β) levels	Thakurdesai, Vichare et al. (2015)

EG: Ethylene glycol; GSH: Glutathione; GPx: Glutathione peroxidase; ILs: Interleukins; LPO: Lipid peroxidation; SOD: Superoxide dismutase; TNF-α: Tumor necrosis factor-alpha; TBARs: Thiobarbituric acid reactant substances.

serum TNF-α and IL-1β were markedly decreased by SFSE-G treatment. Additionally, histopathological alteration induced in renal tissue was also ameliorated by SFSE-G treatment. The authors concluded that administration of glycoside-based standardized fenugreek seeds extract exerted its antiinflammatory potential to protect against anti-GBM antibody-induced glomerulonephritis in experimental rats (Thakurdesai, Vichare et al. 2015).

17.3 FENUGREEK IN MANAGEMENT OF PULMONARY DISEASES

17.3.1 Efficacy against Acute Lung Disorder

Pulmonary edema is an initial stage of acute lung injury (ALI) resulting from vasoconstriction of blood vessels in the pulmonary area (Piantadosi and Schwartz 2004). It advances towards excessive production of pro-inflammatory cytokines (including TNF-α), interleukin-1β (IL-1β), IL-6, and chemokines accompanied by accumulation of neutrophils in the respiratory system (Dushianthan et al. 2011). The interaction between neutrophils and cytokines results in elevated production of ROS. This leads to increased neutrophil production and a rapid deterioration of the pulmonary environment. Thus, elevated circulatory and pulmonary pro-inflammatory cytokine production is associated with a higher incidence of acute respiratory distress syndrome (ARDS), one of the critical risk factors for increased mortality rates (Moldoveanu et al. 2009).

Numerous animal models have been developed to understand the pathophysiology of ALI, which were instrumental in developing various novel therapeutic moieties against ALI. Amongst the various animal models of ALI, acid aspiration is a widely used experimental model where acid induces disruption of the alveolar/capillary barrier followed by neutrophilic infiltration in the pulmonary area, whereas administration of intravenous or intrapulmonary bacteria causes induction of interstitial edema, congestion, polymorphonuclear leukocytes infiltration, and alterations in permeability (Matute-Bello, Frevert, and Martin 2008). Furthermore, chemical agents such as lipopolysaccharide (LPS) induce a neutrophilic inflammatory response in lung tissue, further aggravating the production of pro-inflammatory cytokines. In contrast, antineoplastic antibiotics such as bleomycin induce an acute inflammatory response in lung tissue, contributing to pulmonary fibrosis (Matthay and Howard 2012).

Evidence suggests that bacterial infection is an essential pathophysiological factor responsible for activating LPS and releasing many inflammatory factors (Krupa et al. 2014). LPS is the primary component of G-bacillus bacteria, which activates several signaling molecules, including mitogen-activated protein kinase (MAPK) and factor kappa-light-chain-enhancer of activated B cells (NF-kB), resulting in various inflammatory response factor expressions. This vicious cycle is responsible for systemic inflammatory response syndrome (SIRS), and immoderate SIRS causes multiple organ syndromes dysfunction leading to ALI/ARDS in pulmonary tissue (Rafat et al. 2014; Xu et al. 2016). During the initial stage of ALI, neutrophil manifestation plays a significant role in lung epithelial injury via inflammatory mediators (McKallip, Hagele, and Uchakina 2013). Inflammatory cell-induced activation and formation of LPS/CD14 complex, in turn, stimulate endogenous antigens like TLR4 (toll-like receptor 4) present on the cell surface (Krupa et al. 2013). TLR4 initiates MyD88-dependent signal transduction pathways, which activate NF-κB aiding the release of several inflammatory mediators, including cyclooxygenase (COX)-2, pro-inflammatory cytokines (TNF-α, and ILs), resulting in neutrophil infiltration that contributes to endothelial cell injury (Han et al. 2016). Steroidal saponins of herbal origin have demonstrated their efficacy against LPS-induced acute lung injury in experimental animals (Chen, Rong, and Qiao 2014; Park et al. 2018). Thus, Zeng et al. (2018) investigated the efficacy and possible mechanism of dioscin, a steroidal saponin of fenugreek seed, against LPS-induced lung injury (Zeng et al. 2018). Dioscin (20, 40, and 60 mg/kg) was administered in LPS-injured mice, and various molecular and histological analyses were performed on lung tissue. Findings suggested that dioscin significantly attenuated alveolar macrophages infiltration, which is the preliminary defensive substance for the pulmonary system.

Furthermore, dioscin markedly down-regulated the protein levels of TNF-α, ILs, and NF-kB, whereas it reserved IFN-γ (Interferon-gamma), MPO (myeloperoxidase), and ICAM-1 (Intercellular Adhesion Molecule 1) activity. Additionally, western blot analysis suggested that diosgenin inhibits COX-2, TLR-4, and MyD88 protein expression in lung tissue. An older study had also documented similar findings where dioscin treatment inhibited TNF-α-induced activation of the NF-κB and ICAM-1 pathway (Wu et al. 2015). The author concluded that Dioscin notably inhibited lung inflammation and pulmonary edema via inhibition of the TLR4/MyD88 signaling pathway (Zeng et al. 2018).

Recently, fenugreek seed powder nanoparticle has been evaluated for its potential against hydrochloric acid (HCl)-induced structural aberration in pulmonary tissue, goblet cell proliferation, and DNA damage during ALI in mice (Hamad and El-Naggar 2020). Acid aspiration, such as HCl, is known to activate pro-inflammatory cytokines as well as LPO (i.e., MDA), which further contribute to edema and thus acute lung inflammation (Zhang, Du, and Deng 2018). Studies suggested that HCl causes accumulation of DNA content into the pulmonary area and induces pathological alterations such as goblet cell proliferation in the bronchiole wall, reflected by thick interalveolar septa with congested alveolar sacs edematous alveoli (Hamad and El-Naggar 2020). However, pretreatment with fenugreek nanoparticles (2.5 mg/kg) in ALI mice showed significant improvement of the lung's histopathology, evident by the prevention of goblet cell proliferation and decreased DNA content. Several studies have documented the antioxidant, antiinflammatory, antidiabetic, and antihypercholestemic potential of fenugreek seed nanoparticle (Li 2009; Murugesan, Revathi, and Manju 2011; Naidu et al. 2011; Thirunavukkarasu, Anuradha, and Viswanathan 2003). Presence of flavonoid moieties in fenugreek is suggested to play a vital role in LPO inhibition and subsequently protection of airways from oxidative damage (Rice-Evans 2001; Tariq et al. 2003). Additionally, these flavonoids may also inhibit the activity of mast cells and basophils, which further protect the lung tissue (Naidu et al. 2011). In conclusion, fenugreek seeds possess the ability to balance oxidant/antioxidant ratio owing to these phytoconstituents.

17.3.2 Efficacy against Asthma

Allergic asthma is primarily characterized by airway inflammation (Honmore et al. 2016). This asthmatic condition is a chronic, complex, severe respiratory system disorder resulting from various

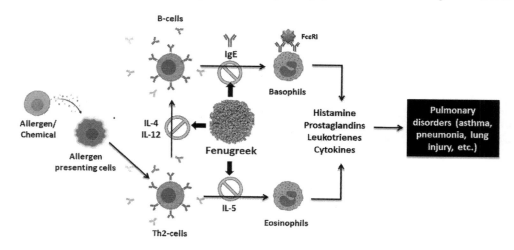

FIGURE 17.1 Potential mechanisms of fenugreek against pulmonary disorders.

pathological alterations, including mucus hypersecretion, obstruction of the pulmonary airway, and high infiltration of inflammatory cells, eosinophils, lymphocytes, and mast cells (Tagaya and Tamaoki 2007). A gradual increase in asthma prevalence has become a global burden, and it has been estimated that almost 300 million people are affected by this disease worldwide (Koul and Dhar 2018). Asthma, an immune-inflammatory disease, imposes significant economic and social impact on the healthcare system with an estimation of the treatment cost in the range of $US1,900–$US3,100 per patient per year (Nunes, Pereira, and Morais-Almeida 2017). Repeated exposure to various allergens such as tree pollen, mold spores, dust mite, tobacco smoke, and air pollution induce airway inflammation and remodeling during allergic asthma (Win and Hussain 2008). This chronic exposure to allergens disintegrates immune tolerance generating T helper type 2 (Th2)-based immune response (Bogaert et al. 2009). Th2 cytokines further produce allergen-specific immunoglobulin (Ig)-E antibodies from B cells. Moreover, these cytokines also induce activated mast cells to release pro-inflammatory mediators such as TNF-α and ILs. These mediators induce a structural modification in the pulmonary area, including subepithelial fibrosis, epithelial denudation, smooth muscle proliferation, and angiogenesis, thus contributing to allergic asthma (Holgate et al. 2009).

Existing treatment strategies for allergic asthma management include synthetic moieties such as inhaled corticosteroid, leukotriene antagonist, beta-blocker, etc., which improve lung functions (Rai et al. 2007). However, these treatment regimens provide relief in a fraction of patients and are also associated with various side effects. Resultantly, these disadvantages limit their clinical implication against asthma. Over the past couple of decades, research conducted in herbal medicines has proven their safety and effectiveness for allergic asthma treatment. Fenugreek is one such herbal remedy that has been traditionally used to treat various respiratory disorders, including asthma and bronchitis (Khan, Negi, and Kumar 2018). Glycoside-based standardized fenugreek seed extract is known to exert protective efficacy against pulmonary fibrosis via its antiinflammatory, antiapoptotic, and antifibrotic potential (Ketkar et al. 2015). Recently anti-asthmatic potential of fenugreek has also been empirically documented in an experimental animal model of OVA (Ovalbumin)-induced asthma (Piao et al. 2017) as well as in patients with mild asthma (Emtiazy et al. 2018).

OVA-induced allergic asthma is a widely used, reproducible, and well-established animal model of airway hyperresponsiveness (AHR), which closely replicates most of the pathophysiological features of clinical allergic asthma (Kandhare et al. 2019; National Asthma Education and Prevention Program 2007). A recent study reported the efficacy of Fenugreek against OVA-induced airway inflammation in BALB/c mice through antiinflammatory properties (Piao et al. 2017). In this study, the intratracheal challenge with OVA induces late phase allergic reaction reflected by pulmonary airways inflammation followed by mucus hypersecretion and goblet cell hyperplasia. Treatment of fenugreek (20 mg/mL) by oral gavage in OVA-challenged mice resulted in the reversal of elevated eosinophils, total cell counts, and infiltration of inflammatory cells Bronchoalveolar lavage fluid (BALF). In addition, administration of fenugreek significantly attenuated altered levels of TNF-α; ILs (IL-1β, IL-5, IL-6, IL-10); IFN-γ in lung tissue; TNF-α, IL-10, and IFN-γ in BALF; along with IgE and anti-OVA IgG1 levels in serum. Furthermore, OVA-induced histopathological alterations such as mucus hypersecretion, subepithelial bronchial fibrosis, hyperplasia of goblet cells, eosinophils infiltration, and collagen deposition in the pulmonary area were significantly reduced by fenugreek treatment (Piao et al. 2017).

The efficacy of aqueous extract of fenugreek seed was reported in mild asthmatic patients during a randomized, double-blind, placebo-controlled clinical trial (Emtiazy et al. 2018). Patients (49 female and 30 male; aged 20–70 years) with mild asthma were administered fenugreek syrup (10 ml, with 50% honey, twice daily) for 4 weeks. Changes in quality of life, lung function tests, and serum IL-4 levels were determined at baseline and after fenugreek treatment. Administration of fenugreek significantly improved their quality of life and lung function tests, whereas serum IL-4 was significantly reduced. Spirometry analysis indicated that fenugreek seed extract administration was associated with a 10% improvement in FEV1 (forced expiratory volume in 1 s) and FEV1/

TABLE 17.3
Potential of Fenugreek Extract and Its Phytoconstituents against Pulmonary Disorders

Phytoconstituent(s) or Extract	Dose	Animal Model	Proposed Mechanism	Reference
Dioscin	Dioscin (20, 40, and 60 mg/kg)	LPS-induced ALI in C57BL/6J mice (8 weeks old)	Down-regulate TNF-α, ILs, NF-kB, COX-2, TLR4, and MyD88 levels Upregulation of IFN-γ, MPO, and ICAM-1 levels	Zeng et al. (2018)
Nano-fenugreek seed powder	Fenugreek nanoparticle (2.5 mg/kg)	HCL-induced ALI in Swiss albino mice (9–12 weeks old)	Inhibited DNA damage	Hamad and El-Naggar (2020)
Fenugreek powder	Fenugreek (20 mg/mL)	OVA-induced AHR in BALB/c mice (5-week-old)	Inhibited TNF-α, IL-1β, IL-5, IL-6, IgE, and anti-OVA IgG1 levels Upregulated IL-10 and IFN-γ levels	
Fenugreek seed aqueous extract	Fenugreek syrup (10 ml, with 50% honey, twice daily)	Patients with mild asthma (49 female and 30 male; aged 20–70 years)	Inhibited serum IL-4 levels	Emtiazy et al. (2018)

COX-2: Cyclooxygenase-2; GSH: Glutathione; IFN-γ: Interferon-gamma; IgE: Immunoglobulin E; ILs: Interleukins; MDA: Malondialdehyde; MPO: Myeloperoxidase; NF-kB: Nuclear Factor kappa-light-chain-enhancer of activated B cells; SOD: Superoxide dismutase; STZ: Streptozotocin; TNF-α: Tumor necrosis factor-alpha; TLR4: Toll-like receptor 4.

FVC (forced expiratory volume in 1 s/forced vital capacity) levels. Fenugreek treatment did not induce any severe side effects, and it was well tolerated. These findings are in line with other study results where the administration of fenugreek seed showed anti-asthmatic potential via inhibition of elevated inflammatory influx (Piao et al. 2017). Based on these findings, fenugreek seeds can be considered a safe and effective treatment for mild asthma.

Based on existing preclinical and clinical evidence, it can be summarized that the antiinflammatory potential of fenugreek underlies its efficacy against various pulmonary disorders, including lung injury and asthma (Figure 17.1 and Table 17.3).

17.4 CONCLUSION

The available evidence demonstrated the potential of fenugreek and its phytoconstituents for the management of many renal and pulmonary disorders as complementary medicine or an adjunct to mainline therapy. The standardized extracts and phytoconstituents such as 4-hydroxyisoleucine, trigonelline, and glycosides showed promising efficacy through multiple mechanisms during animal studies. The clinical validation of these studies can provide better and safer options for managing renal and pulmonary disorders.

REFERENCES

Ahsan, S. K., M. Tariq, A. M. Ageel, M. A. al-Yahya, and A. H. Shah. 1989. "Effect of Trigonella foenum-graecum and Ammi majus on calcium oxalate urolithiasis in rats." *J Ethnopharmacol* 26 (3):249–54. doi: 10.1016/0378-8741(89)90097-4.

Al Dera, H. S. 2016. "Protective effect of resveratrol against aluminum chloride induced nephrotoxicity in rats." *Saudi Med J* 37 (4):369–78. doi: 10.15537/smj.2016.4.13611.

Alelign, T., and B. Petros. 2018. "Kidney stone disease: An update on current concepts." *Adv Urol* 2018:3068365. doi: 10.1155/2018/3068365.

Al-Habori, M., and A. Raman. 1998. "Antidiabetic and hypocholesterolaemic effects of fenugreek." *Phytotherapy Research* 12 (4):233–42. doi: 10.1002/(sici)1099-1573(199806)12:4<233::aid-ptr294>3.0.co;2-v.

Alsahli, M., and J. E. Gerich. 2015. "Hypoglycemia in patients with diabetes and renal disease." *J Clin Med* 4 (5):948–64. doi: 10.3390/jcm4050948.

Alyoussef, A., and M. M. H. Al-Gayyar. 2016. "Thymoquinone ameliorated elevated inflammatory cytokines in testicular tissue and sex hormones imbalance induced by oral chronic toxicity with sodium nitrite." *Cytokine* 83:64–74. doi: 10.1016/j.cyto.2016.03.018.

Arora, S., S. L. Bodhankar, V. Mohan, and P. A. Thakurdesai. 2012. "Renoprotective Effects of Reconstructed Composition of Trigonella foenum-graecum L. Seeds in Animal Model of Diabetic Nephropathy with and without Renal Ischemia Reperfusion in Rats." *Int J Pharmacol* 8 (5):321–32. doi: 10.3923/ijp.2012.321.332.

Baset, M., T. Ali, H. Elshamy, A. El Sadek, D. Sami, M. Badawy, S. Abou-Zekry, H. Heiba, M. Saadeldin, and A. Abdellatif. 2020. "Anti-diabetic effects of fenugreek (Trigonella foenum-graecum): A comparison between oral and intraperitoneal administration—an animal study." *Int J Functional Nutr* 1 (1):2. doi: 10.3892/ijfn.2020.2.

Belaid-Nouira, Y., H. Bakhta, Z. Haouas, I. Flehi-Slim, and H. Ben Cheikh. 2013. "Fenugreek seeds reduce aluminum toxicity associated with renal failure in rats." *Nutr Res Pract* 7 (6):466–74. doi: 10.4162/nrp.2013.7.6.466.

Bogaert, P., K. G. Tournoy, T. Naessens, and J. Grooten. 2009. "Where asthma and hypersensitivity pneumonitis meet and differ: Noneosinophilic severe asthma." *Am J Pathol* 174 (1):3–13. doi: 10.2353/ajpath.2009.071151.

Butterweck, V., and S. R. Khan. 2009. "Herbal medicines in the management of urolithiasis: Alternative or complementary?" *Planta Med* 75 (10):1095–103. doi: 10.1055/s-0029-1185719.

Camara, N. O., K. Iseki, H. Kramer, Z. H. Liu, and K. Sharma. 2017. "Kidney disease and obesity: Epidemiology, mechanisms and treatment." *Nat Rev Nephrol* 13 (3):181–90. doi: 10.1038/nrneph.2016.191.

Cao, L. C., T. W. Honeyman, R. Cooney, L. Kennington, C. R. Scheid, and J. A. Jonassen. 2004. "Mitochondrial dysfunction is a primary event in renal cell oxalate toxicity." *Kidney Int* 66 (5):1890–900. doi: 10.1111/j.1523-1755.2004.00963.x.

Chen, H. Y., C. H. Ma, K. J. Cao, J. Chung-Man Ho, E. Ziea, V. T. Wong, and Z. J. Zhang. 2014. "A systematic review and meta-analysis of herbal medicine on chronic obstructive pulmonary diseases." *Evid Based Complement Alternat Med* 2014:925069. doi: 10.1155/2014/925069.

Chen, Y. Q., L. Rong, and J. O. Qiao. 2014. "Antiinflammatory effects of Panax notoginseng saponins ameliorate acute lung injury induced by oleic acid and lipopolysaccharide in rats." *Mol Med Rep* 10 (3):1400–8. doi: 10.3892/mmr.2014.2328.

Chevassus, H., J. B. Gaillard, A. Farret, F. Costa, I. Gabillaud, E. Mas, A. M. Dupuy, F. Michel, C. Cantie, E. Renard, F. Galtier, and P. Petit. 2010. "A fenugreek seed extract selectively reduces spontaneous fat intake in overweight subjects." *Eur J Clin Pharmacol* 66 (5):449–55. doi: 10.1007/s00228-009-0770-0.

Chevassus, H., N. Molinier, F. Costa, F. Galtier, E. Renard, and P. Petit. 2009. "A fenugreek seed extract selectively reduces spontaneous fat consumption in healthy volunteers." *Eur J Clin Pharmacol* 65 (12):1175–8. doi: 10.1007/s00228-009-0733-5.

Coe, F. L., A. Evan, and E. Worcester. 2005. "Kidney stone disease." *J Clin Invest* 115 (10):2598–608. doi: 10.1172/JCI26662.

Couser, W. G., and R. J. Johnson. 2014. "The etiology of glomerulonephritis: Roles of infection and autoimmunity." *Kidney Int* 86 (5):905–14. doi: 10.1038/ki.2014.49.

Dal Moro, F., M. Mancini, I. M. Tavolini, V. De Marco, and P. Bassi. 2005. "Cellular and molecular gateways to urolithiasis: A new insight." *Urol Int* 74 (3):193–7. doi: 10.1159/000083547.

Darwish, M., S. Shaalan, M. Amer, and S. Hamad. 2020. "Ameliorative effects of dried and germinated fenugreek seeds on kidney failure induced by gentamicin in male mice." *Am J Biomed Sci Res* 9 (6):459–66. doi: 10.34297/AJBSR.2020.09.001452.

Dushianthan, A., M. P. Grocott, A. D. Postle, and R. Cusack. 2011. "Acute respiratory distress syndrome and acute lung injury." *Postgrad Med J* 87 (1031):612–22. doi: 10.1136/pgmj.2011.118398.

Emtiazy, M., L. Oveidzadeh, M. Habibi, L. Molaeipour, D. Talei, Z. Jafari, M. Parvin, and M. Kamalinejad. 2018. "Investigating the effectiveness of the Trigonella foenum-graecum L. (fenugreek) seeds in mild asthma: A randomized controlled trial." *Allergy Asthma Clin Immunol* 14 (1):19. doi: 10.1186/s13223-018-0238-9.

Erickson, S. B., T. J. Vrtiska, and J. C. Lieske. 2011. "Effect of Cystone(R) on urinary composition and stone formation over a one year period." *Phytomedicine* 18 (10):863–7. doi: 10.1016/j.phymed.2011.01.018.

Foggensteiner, L., S. Mulroy, and J. Firth. 2001. "Management of diabetic nephropathy." *J R Soc Med* 94 (5):210–7. doi: 10.1177/014107680109400504.

GBD Chronic Kidney Disease Collaboration. 2020. "Global, regional, and national burden of chronic kidney disease, 1990–2017: A systematic analysis for the global burden of disease study 2017." *The Lancet* 395 (10225):709–33. doi: 10.1016/s0140-6736(20)30045-3.

Geberemeskel, G. A., Y. G. Debebe, and N. A. Nguse. 2019. "Antidiabetic Effect of Fenugreek Seed Powder Solution (Trigonella foenum-graecum L.) on Hyperlipidemia in Diabetic Patients." *J Diabetes Res* 2019:8507453. doi: 10.1155/2019/8507453.

Ghane Shahrbaf, F., and F. Assadi. 2015. "Drug-induced renal disorders." *J Renal Inj Prev* 4 (3):57–60. doi: 10.12861/jrip.2015.12.

Gheith, O., N. Farouk, N. Nampoory, M. A. Halim, and T. Al-Otaibi. 2016. "Diabetic kidney disease: World wide difference of prevalence and risk factors." *J Nephropharmacol* 5 (1):49–56.

Ghule, A. E., S. S. Jadhav, and S. L. Bodhankar. 2012. "Trigonelline ameliorates diabetic hypertensive nephropathy by suppression of oxidative stress in kidney and reduction in renal cell apoptosis and fibrosis in streptozotocin induced neonatal diabetic (nSTZ) rats." *Int Immunopharmacol* 14 (4):740–8. doi: 10.1016/j.intimp.2012.10.004.

Gupta, A., R. Gupta, and B. Lal. 2001. "Effect of Trigonella foenum-graecum (fenugreek) seeds on glycaemic control and insulin resistance in type 2 diabetes mellitus: A double blind placebo controlled study." *J Assoc Physicians India* 49:1057–61.

Hamad, S. R., and M. E. El-Naggar. 2020. "Blocking of gastric acid induced histopathological alterations, enhancing of DNA content and proliferation of goblet cells in the acute lung injury mice models by nano-fenugreek oral administration." *Toxicol Mech Methods* 30 (2):153–8. doi: 10.1080/15376516.2019.1669249.

Hamden, K., B. Jaouadi, T. Salami, S. Carreau, S. Bejar, and A. Elfeki. 2010. "Modulatory effect of fenugreek saponins on the activities of intestinal and hepatic disaccharidase and glycogen and liver function of diabetic rats." *Biotechnol Bioprocess Eng* 15 (5):745–53. doi: 10.1007/s12257-009-3159-0.

Hamden, K., H. Keskes, O. Elgomdi, A. Feki, and N. Alouche. 2017. "Modulatory effect of an isolated triglyceride from fenugreek seed oil on of alpha-amylase, lipase and ACE activities, liver-kidney functions and metabolic disorders of diabetic rats." *J Oleo Sci* 66 (6):633–45. doi: 10.5650/jos.ess16254.

Hamden, K., K. Mnafgui, Z. Amri, A. Aloulou, and A. Elfeki. 2013. "Inhibition of key digestive enzymes related to diabetes and hyperlipidemia and protection of liver-kidney functions by trigonelline in diabetic rats." *Sci Pharm* 81 (1):233–46. doi: 10.3797/scipharm.1211-14.

Han, L.-p., C.-j. Li, B. Sun, Y. Xie, Y. Guan, Z.-j. Ma, and L.-m. Chen. 2016. "Protective effects of celastrol on diabetic liver injury via TLR4/MyD88/NF-κB signaling pathway in type 2 diabetic rats." *Journal of Diabetes Research* 2016. doi: 10.1155/2016/2641248.

Hilmi, S. R., Z. F. Dewan, and A. K. M. N. Kabir. 2018. "Effect of ethanol extract of Trigonella foenum-graecum on gentamicin-induced nephrotoxicity in rat." *Bangabandhu Sheikh Mujib Medical University Journal* 11 (2):107. doi: 10.3329/bsmmuj.v11i2.35778.

Holgate, S. T., H. S. Arshad, G. C. Roberts, P. H. Howarth, P. Thurner, and D. E. Davies. 2009. "A new look at the pathogenesis of asthma." *Clin Sci (Lond)* 118 (7):439–50. doi: 10.1042/CS20090474.

Honmore, V. S., A. D. Kandhare, P. P. Kadam, V. M. Khedkar, D. Sarkar, S. L. Bodhankar, A. A. Zanwar, S. R. Rojatkar, and A. D. Natu. 2016. "Isolates of *Alpinia officinarum* Hance as COX-2 inhibitors: Evidence from anti-inflammatory, antioxidant and molecular docking studies." *Int Immunopharmacol* 33:8–17. doi: 10.1016/j.intimp.2016.01.024.

Jin, Y., Y. Shi, Y. Zou, C. Miao, B. Sun, and C. Li. 2014. "Fenugreek Prevents the Development of STZ-Induced Diabetic Nephropathy in a Rat Model of Diabetes." *Evid-Based Complement Alternat Med* 2014:259368. doi: 10.1155/2014/259368.

Kandhare, A. D., Z. Liu, A. A. Mukherjee, and S. L. Bodhankar. 2019. "Therapeutic potential of morin in ovalbumin-induced allergic asthma via modulation of SUMF2/IL-13 and BLT2/NF-kB signaling pathway." *Curr Mol Pharmacol* 12 (2):122–38. doi: 10.2174/1874467212666190102105052.

Kapase, C., S. Bodhankar, V. Mohan, and P. Thakurdesai. 2013. "Therapeutic effects of standardized fenugreek seed extract on experimental urolithiasis in rats." *J Applied Pharmaceut Sci* 3:29–35. doi: 10.7324/japs.2013.3906.

Kaur, H., A. Singh, S. K. Singh, A. Bhatia, and B. Kumar. 2016. "Attenuation of gentamicin induced nephrotoxicity in rats by aqueous extract of Trigonella foenum graceum seeds." *Int J Res Ayurveda & Pharmacy* 7 (4):178–83. doi: 10.7897/2277-4343.074180.

Ketkar, S., A. Rathore, A. D. Kandhare, S. Lohidasan, S. L. Bodhankar, A. Paradkar, and K. Mahadik. 2015. "Alleviating exercise-induced muscular stress using neat and processed bee pollen: Oxidative markers, mitochondrial enzymes, and myostatin expression in rats." *Integr Med Res* 4 (3):147–60. doi: 10.1016/j.imr.2015.02.003.

Khan, F., K. Negi, and T. Kumar. 2018. "Effect of sprouted fenugreek seeds on various diseases: A review." *J Diabetes, Metabolic Disorders & Control* 5 (4):119–25. doi: 10.15406/jdmdc.2018.05.00149.

Khand, F. D., M. P. Gordge, W. G. Robertson, A. A. Noronha-Dutra, and J. S. Hothersall. 2002. "Mitochondrial superoxide production during oxalate-mediated oxidative stress in renal epithelial cells." *Free Radic Biol Med* 32 (12):1339–50. doi: 10.1016/s0891-5849(02)00846-8.

Konopelniuk, V. V., Goloborodko, II, T. V. Ishchuk, T. B. Synelnyk, L. I. Ostapchenko, M. Y. Spivak, and R. V. Bubnov. 2017. "Efficacy of Fenugreek-based bionanocomposite on renal dysfunction and endogenous intoxication in high-calorie diet-induced obesity rat model-comparative study." *EPMA J* 8 (4):377–90. doi: 10.1007/s13167-017-0098-2.

Koul, P. A., and R. Dhar. 2018. "Economic burden of asthma in India." *Lung India* 35 (4):281–3. doi: 10.4103/lungindia.lungindia_220_18.

Krupa, A., M. Fol, M. Rahman, K. Y. Stokes, J. M. Florence, I. L. Leskov, M. V. Khoretonenko, M. A. Matthay, K. D. Liu, C. S. Calfee, A. Tvinnereim, G. R. Rosenfield, and A. K. Kurdowska. 2014. "Silencing Bruton's tyrosine kinase in alveolar neutrophils protects mice from LPS/immune complex-induced acute lung injury." *Am J Physiol Lung Cell Mol Physiol* 307 (6):L435–L48. doi: 10.1152/ajplung.00234.2013.

Krupa, A., R. Fudala, J. M. Florence, T. Tucker, T. C. Allen, T. J. Standiford, R. Luchowski, M. Fol, M. Rahman, and Z. Gryczynski. 2013. "Bruton's tyrosine kinase mediates FcγRIIa/Toll-like receptor—4 receptor crosstalk in human neutrophils." *American Journal of Respiratory Cell and Molecular Biology* 48 (2):240–9. doi: 10.1165/rcmb.2012-0039oc.

Laroubi, A., M. Touhami, L. Farouk, I. Zrara, R. Aboufatima, A. Benharref, and A. Chait. 2007. "Prophylaxis effect of Trigonella foenum-graecum L. seeds on renal stone formation in rats." *Phytother Res* 21 (10):921–5. doi: 10.1002/ptr.2190.

Li, X. M. 2009. "Complementary and alternative medicine in pediatric allergic disorders." *Curr Opin Allergy Clin Immunol* 9 (2):161–7. doi: 10.1097/ACI.0b013e328329226f.

Li, X. Y., S. S. Lu, H. L. Wang, G. Li, Y. F. He, X. Y. Liu, R. Rong, J. Li, and X. C. Lu. 2018. "Effects of the fenugreek extracts on high-fat diet-fed and streptozotocin-induced type 2 diabetic mice." *Animal Model Exp Med* 1 (1):68–73. doi: 10.1002/ame2.12004.

Liu, J., X. C. Han, D. Wang, A. D. Kandhare, A. A. Mukherjee-Kandhare, S. L. Bodhankar, and K. M. Wang. 2020. "Elucidation of molecular mechanism involved in nephroprotective potential of naringin in ethylene glycol-induced urolithiasis in experimental uninephrectomized hypertensive rats." *Lat Am J Pharm* 39 (5):991–9.

Matthay, M. A., and J. P. Howard. 2012. "Progress in modelling acute lung injury in a pre-clinical mouse model." *Eur Respir J* 39 (5):1062–3. doi: 10.1183/09031936.00204211.

Matute-Bello, G., C. W. Frevert, and T. R. Martin. 2008. "Animal models of acute lung injury." *Am J Physiol Lung Cell Mol Physiol* 295 (3):L379–L99. doi: 10.1152/ajplung.00010.2008.

McAdoo, S. P., and C. D. Pusey. 2017. "Anti-glomerular basement membrane disease." *Clin J Am Soc Nephrol* 12 (7):1162–72. doi: 10.2215/cjn.01380217.

McKallip, R. J., H. F. Hagele, and O. N. Uchakina. 2013. "Treatment with the hyaluronic acid synthesis inhibitor 4-methylumbelliferone suppresses SEB-induced lung inflammation." *Toxins (Basel)* 5 (10):1814–26. doi: 10.3390/toxins5101814.

McMartin, K. E., and K. B. Wallace. 2005. "Calcium oxalate monohydrate, a metabolite of ethylene glycol, is toxic for rat renal mitochondrial function." *Toxicol Sci* 84 (1):195–200. doi: 10.1093/toxsci/kfi062.

Moldoveanu, B., P. Otmishi, P. Jani, J. Walker, X. Sarmiento, J. Guardiola, M. Saad, and J. Yu. 2009. "Inflammatory mechanisms in the lung." *J Inflamm Res* 2:1–11.

Moorthy, R., K. M. Prabhu, and P. S. Murthy. 2010. "Anti-hyperglycemic compound (GII) from fenugreek (Trigonella foenum-graecum Linn.) seeds, its purification and effect in diabetes mellitus." *Indian J Exp Biol* 48 (11):1111–8.

Murakami, T., A. Kishi, H. Matsuda, and M. Yoshikawa. 2000. "Medicinal foodstuffs. XVII. Fenugreek seed. (3): Structures of new furostanol-type steroid saponins, trigoneosides Xa, Xb, XIb, XIIa, XIIb, and XIIIa, from the seeds of Egyptian *Trigonella foenum-graecum* L." *Chem Pharm Bull* 48 (7):994–1000. doi: 10.1248/cpb.48.994.

Murugesan, M., R. Revathi, and V. Manju. 2011. "Cardioprotective effect of fenugreek on isoproterenol-induced myocardial infarction in rats." *Indian J Pharmacol* 43 (5):516–9. doi: 10.4103/0253-7613.84957.

Naidu, M. M., B. Shyamala, J. P. Naik, G. Sulochanamma, and P. Srinivas. 2011. "Chemical composition and antioxidant activity of the husk and endosperm of fenugreek seeds." *LWT-Food Sci Technol* 44 (2):451–6.

Najdi, R. A., M. M. Hagras, F. O. Kamel, and R. M. Magadmi. 2019. "A randomized controlled clinical trial evaluating the effect of Trigonella foenum-graecum (fenugreek) versus glibenclamide in patients with diabetes." *Afr Health Sci* 19 (1):1594–601. doi: 10.4314/ahs.v19i1.34.

National Asthma Education and Prevention Program. 2007. "Section 2, definition, pathophysiology and pathogenesis of asthma, and natural history of asthma." In *Expert Panel Report 3: Guidelines for the Diagnosis and Management of Asthma*. Bethesda, MD: National Heart, Lung, and Blood Institute.

Naughton, C. A. 2008. "Drug-induced nephrotoxicity." *Am Fam Physician* 78 (6):743–50.

Nunes, C., A. M. Pereira, and M. Morais-Almeida. 2017. "Asthma costs and social impact." *Asthma Res Pract* 3 (1):1. doi: 10.1186/s40733-016-0029-3.

Park, B. K., K. S. So, H. J. Ko, H. J. Kim, K. S. Kwon, Y. S. Kwon, K. H. Son, S. Y. Kwon, and H. P. Kim. 2018. "Therapeutic potential of the rhizomes of anemarrhena asphodeloides and timosaponin A-III in an animal model of lipopolysaccharide-induced lung inflammation." *Biomol Ther (Seoul)* 26 (6):553–9. doi: 10.4062/biomolther.2017.249.

Patil, S. P., P. V. Niphadkar, and M. M. Bapat. 1997. "Allergy to fenugreek (Trigonella foenum graecum)." *Ann Allergy Asthma Immunol* 78 (3):297–300. doi: 10.1016/S1081-1206(10)63185-7.

Pecoits-Filho, R., H. Abensur, C. C. Betonico, A. D. Machado, E. B. Parente, M. Queiroz, J. E. Salles, S. Titan, and S. Vencio. 2016. "Interactions between kidney disease and diabetes: Dangerous liaisons." *Diabetol Metab Syndr* 8 (1):50. doi: 10.1186/s13098-016-0159-z.

Petejova, N., A. Martinek, J. Zadrazil, and V. Teplan. 2019. "Acute toxic kidney injury." *Ren Fail* 41 (1):576–94. doi: 10.1080/0886022X.2019.1628780.

Piantadosi, C. A., and D. A. Schwartz. 2004. "The acute respiratory distress syndrome." *Ann Intern Med* 141 (6):460–70. doi: 10.7326/0003-4819-141-6-200409210-00012.

Piao, C. H., T. T. Bui, C. H. Song, H. S. Shin, D. H. Shon, and O. H. Chai. 2017. "Trigonella foenum-graecum alleviates airway inflammation of allergic asthma in ovalbumin-induced mouse model." *Biochem Biophys Res Commun* 482 (4):1284–8. doi: 10.1016/j.bbrc.2016.12.029.

Pradeep, S. R., S. Barman, and K. Srinivasan. 2019. "Attenuation of diabetic nephropathy by dietary fenugreek (Trigonella foenum-graecum) seeds and onion (Allium cepa) via suppression of glucose transporters and renin-angiotensin system." *Nutrition* 67–68:110543. doi: 10.1016/j.nut.2019.06.024.

Pribac, G. C., M. F. Sferdian, C. Neamtu, C. Craciun, C. L. Rosioru, A. Ardelean, and B. D. Totolici. 2015. "Fenugreek powder exerts protective effects on alcoholised rats' kidney, highlighted using ultrastructural studies." *Rom J Morphol Embryol* 56 (2):445–51. doi: 10.3390/molecules21010064.

Rafat, N., C. Dacho, G. Kowanetz, C. Betzen, B. Tonshoff, B. Yard, and G. Beck. 2014. "Bone marrow-derived progenitor cells attenuate inflammation in lipopolysaccharide-induced acute respiratory distress syndrome." *BMC Res Notes* 7 (1):613. doi: 10.1186/1756-0500-7-613.

Rai, S. P., A. P. Patil, V. Vardhan, V. Marwah, M. Pethe, and I. M. Pandey. 2007. "Best treatment guidelines for bronchial asthma." *Med J Armed Forces India* 63 (3):264–8. doi: 10.1016/S0377-1237(07)80151-1.

Rice-Evans, C. 2001. "Flavonoid antioxidants." *Curr Med Chem* 8 (7):797–807. doi: 10.2174/0929867013373011.

Sales, G. T. M., and R. D. Foresto. 2020. "Drug-induced nephrotoxicity." *Rev Assoc Med Bras (1992)* 66 (Suppl 1):s82–s90. doi: 10.1590/1806-9282.66.S1.82.

Salvi, S., G. A. Kumar, R. S. Dhaliwal, K. Paulson, A. Agrawal, P. A. Koul, P. A. Mahesh, S. Nair, V. Singh, A. N. Aggarwal, D. J. Christopher, R. Guleria, B. V. M. Mohan, S. K. Tripathi, A. G. Ghoshal, R. V. Kumar, R. Mehrotra, D. K. Shukla, E. Dutta, M. Furtado, D. Bhardwaj, M. Smith, R. S. Abdulkader, M. Arora, K. Balakrishnan, J. K. Chakma, P. Chaturvedi, S. Dey, D. Ghorpade, S. Glenn, P. C. Gupta, T. Gupta, S. C. Johnson, T. K. Joshi, M. Kutz, M. R. Mathur, P. Mathur, P. Muraleedharan, C. M. Odell, S. Pati, Y. Sabde, D. N. Sinha, K. R. Thankappan, C. M. Varghese, G. Yadav, S. S. Lim, M. Naghavi, R. Dandona, K. S. Reddy, T. Vos, C. J. L. Murray, S. Swaminathan, and L. Dandona. 2018. "The burden of chronic respiratory diseases and their heterogeneity across the states of India: The Global Burden of Disease Study 1990–2016." *The Lancet Global Health* 6 (12):e1363–e74. doi: 10.1016/s2214-109x(18)30409-1.

Sasase, T., F. Fatchiyah, K. Miyajima, and M. Koide. 2019. "Animal models of diabetes and related metabolic diseases." *Int J Endocrinol* 2019:6147321. doi: 10.1155/2019/6147321.

Sayed, A. A., M. Khalifa, and F. F. Abd el-Latif. 2012. "Fenugreek attenuation of diabetic nephropathy in alloxan-diabetic rats: Attenuation of diabetic nephropathy in rats." *J Physiol Biochem* 68 (2):263–9. doi: 10.1007/s13105-011-0139-6.

Scammell, M. K., C. M. Sennett, Z. E. Petropoulos, J. Kamal, and J. S. Kaufman. 2019. "Environmental and occupational exposures in kidney disease." *Seminars in Nephrology* 39(3):230–43 doi: 10.1016/j.semnephrol.2019.02.001.

Scheid, C., H. Koul, W. A. Hill, J. Luber-Narod, L. Kennington, T. Honeyman, J. Jonassen, and M. Menon. 1996. "Oxalate toxicity in LLC-PK1 cells: Role of free radicals." *Kidney Int* 49 (2):413–9. doi: 10.1038/ki.1996.60.

Shafi, H., B. Moazzami, M. Pourghasem, and A. Kasaeian. 2016. "An overview of treatment options for urinary stones." *Caspian J Intern Med* 7 (1):1–6.

Shekha, M. S., A. B. Qadir, H. H. Ali, and X. E. Selim. 2014. "Effect of Fenugreek (Trigonella Foenum-Graecum) on ethylene glycol induced kidney stone in rats." *Jordan Journal of Biological Sciences* 7 (4):257–60. doi: 10.12816/0008248.

Shetty, A. K., and P. V. Salimath. 2009. "Reno-protective effects of fenugreek (*Trigonella foenum greacum*) during experimental diabetes." *e-SPEN, the European e-Journal of Clinical Nutrition and Metabolism* 4 (3):e137–e42. doi: 10.1016/j.eclnm.2009.02.002.

Sushma, N., and T. Devasena. 2010. "Aqueous extract of Trigonella foenum-graecum (fenugreek) prevents cypermethrin-induced hepatotoxicity and nephrotoxicity." *Hum Exp Toxicol* 29 (4):311–9. doi: 10.1177/0960327110361502.

Susilo, J., B. Purwanto, M. Doewes, and D. Indarto. 2021. "Calcium oxalate crystals: Epidemiology, causes, modeling of crystal formation and treatment management." *J Pharmaceut Sci Res* 13 (2):118–23.

Tagaya, E., and J. Tamaoki. 2007. "Mechanisms of airway remodeling in asthma." *Allergol Int* 56 (4):331–40. doi: 10.2332/allergolint.R-07-152.

Tariq, M., A. A. Moutaery, M. Arshaduddin, H. A. Khan, D. P. Evans, and S. Jacobs. 2003. "Fluconazole attenuates lung injury and mortality in a rat peritonitis model." *Intensive Care Med* 29 (11):2043–9. doi: 10.1007/s00134-003-1960-3.

Thakurdesai, P. A., V. Mohan, S. Naqvi, S. L. Bodhankar, A. K. Ghosh, M. Dutta, V. Mohan, and D. Bandyopadhyay. 2015. "Renoprotective effects of 4-hydroxyisoleucine from fenugreek seeds on diabetic nephropathy in rats: Role of mitochondrial antioxidant defense system." In *International Symposium on New Perspectives in Modern Biotechnology (SAB-NPMB)*. Puducherry, India, March 23–25, 2015, Puducherry: Society of Applied Biotechnology. pp. OP-54.

Thakurdesai, P. A., R. Vichare, S. Bodhankar, U. Aswar, and V. Mohan. 2015. "Glycosides based standardized extract of fenugreek seeds ameliorates anti-glomerular basement membrane antibody-induced glomerulonephritis in rats." Paper presented at 2nd International Congress of Society for Ethnopharmacology (SFEC — 2015) Nagpur, India.

Thirunavukkarasu, V., C. V. Anuradha, and P. Viswanathan. 2003. "Protective effect of fenugreek (Trigonella foenum graecum) seeds in experimental ethanol toxicity." *Phytother Res* 17 (7):737–43. doi: 10.1002/ptr.1198.

US Food and Drug Administration. 2010. *21 CFR § 182.20, Part 182 — Substances Generally Recognized as Safe, Subpart A—General Provisions., Sec. 182.20 Essential Oils, Oleoresins (solvent-free), and Natural Extractives (Including Distillates)*. Washington, DC: Department of Health and Human Services, Food and Drug Administration.

Uslu, G. A., H. Uslu, and Y. Adalı. 2019. "Hepatoprotective and nephroprotective effects of Trigonella foenum-graecum L.(Fenugreek) seed extract against sodium nitrite toxicity in rats." *Biomed. Res. Ther* 6:3142–50. doi: 10.3390/molecules25163592.

Valcke, M., M. E. Levasseur, A. Soares da Silva, and C. Wesseling. 2017. "Pesticide exposures and chronic kidney disease of unknown etiology: An epidemiologic review." *Environ Health* 16 (1):49. doi: 10.1186/s12940-017-0254-0.

Wani, S. A., and P. Kumar. 2018. "Fenugreek: A review on its nutraceutical properties and utilization in various food products." *J the Saudi Soc Agr Sci* 17 (2):97–106. doi: 10.1016/j.jssas.2016.01.007.

Win, P. H., and I. Hussain. 2008. "Asthma triggers: What really matters?" *Clin Asthma*:149–56. doi: 10.1016/b978-032304289-5.10017-7.

Wu, S., H. Xu, J. Peng, C. Wang, Y. Jin, K. Liu, H. Sun, and J. Qin. 2015. "Potent anti-inflammatory effect of dioscin mediated by suppression of TNF-α-induced VCAM-1, ICAM-1and EL expression via the NF-κB pathway." *Biochimie* 110:62–72. doi: 10.1016/j.biochi.2014.12.022.

Xie, M., X. Liu, X. Cao, M. Guo, and X. Li. 2020. "Trends in prevalence and incidence of chronic respiratory diseases from 1990 to 2017." *Respir Res* 21 (1):49. doi: 10.1186/s12931-020-1291-8.

Xu, X., N. Liu, Y. X. Zhang, J. Cao, D. Wu, Q. Peng, H. B. Wang, and W. C. Sun. 2016. "The protective effects of HJB-1, a derivative of 17-Hydroxy-Jolkinolide B, on LPS-induced acute distress respiratory syndrome mice." *Molecules* 21 (1):77. doi: 10.3390/molecules21010077.

Xue, W., J. Lei, X. Li, and R. Zhang. 2011. "Trigonella foenum-graecum seed extract protects kidney function and morphology in diabetic rats via its antioxidant activity." *Nutr Res* 31 (7):555–62. doi: 10.1016/j.nutres.2011.05.010.

Yoshinari, O., A. Takenake, and K. Igarashi. 2013. "Trigonelline ameliorates oxidative stress in type 2 diabetic Goto-Kakizaki rats." *J Med Food* 16 (1):34–41. doi: 10.1089/jmf.2012.2311.

Yuyama, S. 1999. "Absorption of trigonelline from the small intestine of the specific pathogen-free (SPF) and germ-free (GF) rats in vivo." *Adv Exper Med Biol* 467:723–7.

Yuyama, S., and T. Suzuki. 1985. "Isolation and identification of N1-methylnicotinic acid (trigonelline) from rat urine." *J Nutr Sci Vitaminology* 31 (2):157–67. doi: 10.3177/JNSV.31.157.

Yuyama, S., and T. Suzuki. 1991. "The excretion of N1-methyl-2-pyridone-5-carboxylic acid and related compounds in human subjects after oral administration of nicotinic acid, trigonelline and N1-methyl-2-pyridone-5-carboxylic acid." *Adv Exper Med Biol* 294:475–9. doi: 10.1007/978-1-4684-5952-4_48.

Zameer, S., A. K. Najmi, D. Vohora, and M. Akhtar. 2018. "A review on therapeutic potentials of Trigonella foenum-graecum (fenugreek) and its chemical constituents in neurological disorders: Complementary roles to its hypolipidemic, hypoglycemic, and antioxidant potential." *Nutr Neurosci* 21 (8):539–45. doi: 10.1080/1028415X.2017.1327200.

Zeiger, E., and R. Tice. 1997. "Trigonelline; Review of toxicological literature." *Res Triangle Park, North Carolina: Nation Institute of Env Health Sci Integrated Laboratory Systems* 27. doi: 10.22037/ijpr.2019.1100790.

Zeng, H., L. Yang, X. Zhang, Y. Chen, and J. Cai. 2018. "Dioscin prevents LPS-induced acute lung injury through inhibiting the TLR4/MyD88 signaling pathway via upregulation of HSP70." *Molecular Medicine Reports* 17 (5):6752–8. doi: 10.3892/etm.2020.9097.

Zhang, H., C. Zhang, S. Zhu, H. Ye, and D. Zhang. 2020. "Direct medical costs of end-stage kidney disease and renal replacement therapy: A cohort study in Guangzhou City, southern China." *BMC Health Serv Res* 20 (1):122. doi: 10.1186/s12913-020-4960-x.

Zhang, Y., J.-M. Du, and X.-M. Deng. 2018. "Glycyrrhizin acid prevent hydrochloric acid-induced inhalational lung injury in mice through inhibition of MAPK pathway." *Int J Clin Exp Med* 11 (9):9264–71.

Zhou, Z., P. Chaudhari, H. Yang, A. P. Fang, J. Zhao, E. H. Law, E. Q. Wu, R. Jiang, and R. Seifeldin. 2017. "Healthcare resource use, costs, and disease progression associated with diabetic nephropathy in adults with type 2 diabetes: A retrospective observational study." *Diabetes Ther* 8 (3):555–71. doi: 10.1007/s13300-017-0256-5.

18 Potential of Fenugreek in Management of Fibrotic Disorders

Amit D. Kandhare, Sunil Bhaskaran and Subhash L. Bodhankar

CONTENTS

18.1 Introduction 305
18.2 Fenugreek against Pulmonary Fibrosis 307
18.3 Fenugreek against Hepatic Fibrosis 309
18.4 Fenugreek against renal fibrosis 312
18.5 Fenugreek against Cardiac Fibrosis 312
18.6 Fenugreek against Systemic Sclerosis 313
18.7 Safety of Anti-Fibrotic Fenugreek Phytoconstituents 313
18.8 Conclusions 314
References 314

ABBREVIATIONS

α-SMA: α-Smooth Muscle Actin; ALP: Alkaline phosphatase; ALT: Alanine transaminase; ASMC: Airway Smooth Muscle Cells; AST: Aspartate transaminase; BALF: Bronchoalveolar lavage fluid; BUN: Blood urea nitrogen; ECM: Extracellular Matrix; FPS: Fenugreek powder supplementation; FSP: Fenugreek seed powder; FSPE: Fenugreek seed polyphenol extract; GGT: Gamma-glutamyl transferase; GRAS: Generally Recognized as Safe; GSH: Glutathione; ILs: Interleukins; IPF: Idiopathic Pulmonary Fibrosis; MDA: Malondialdehyde; MPO: Myeloperoxidase; MMP: Matrix Metalloproteinase; Nrf2: Nuclear factor erythroid 2-related factor 2; ROS: Reactive Oxygen Species; SFSE-G: Glycosides-based standardized fenugreek seed extract; SOD: Superoxide Dismutase; STZ: Streptozotocin; TEFS: Thermostable extract of fenugreek seeds; TGF-β: Transforming growth factor-β; TNF-α: Tumor necrosis factor-α.

18.1 INTRODUCTION

Fenugreek (*Trigonella foenum-graecum* L.), an aromatic crop and traditional spice, is widely grown in many areas of the world (Zandi et al. 2017). In traditional medicine in many countries such as India, Egypt, and Iran, common people have used fenugreek to manage many disorders, including inflammation and pain (Murakami et al. 2000; Patil, Niphadkar, and Bapat 1997).

The inflammation plays a significant role in the body's defense mechanisms against several harmful stimuli. An array of stimuli such as autoimmune reactions, infections, toxins, and mechanical injuries induces tissue damage via activation of inflammatory pathways, and plays a vital role in the development and progression of numerous chronic inflammatory diseases (Lee and Kalluri 2010). During the pathogenesis of inflammatory reactions, several mediators are reported to release, including chemokines, cytokines, and adhesion molecules (Abdulkhaleq et al. 2018).

The anti-inflammatory activity of fenugreek seed and its isolated bioactive compounds from *in vivo* and *in vitro* systems have been reviewed (Ravichandiran and Jayakumari 2013). In one of the earliest reports, the fenugreek leaves also have been reported to exhibit anti-inflammatory effects against the experimental model of formalin-induced edema (Ahmadiani et al. 2001). The administration of fenugreek seed exerts its anti-inflammatory property in Zucker fatty rats via inhibition of tumor necrosis factor-α (TNF-α) expressions during hepatic steatosis (Raju and Bird 2006). Topical application of a cream containing methanolic extract of fenugreek seed exerts its anti-inflammatory effect in carrageenan-induced rat paw edema *in vivo* (Sharififar, Khazaeli, and Alli 2009). The potential of ethanol extract of fenugreek in an animal model of chronic inflammation, Freund's complete adjuvant-induced arthritis in rats through inhibition of release of inflammatory mediators such as TNF-α and interleukins (ILs), was reported (Suresh et al. 2012).

Numerous investigators have documented that alkaloids, flavonoids, glycosides, and saponins in fenugreek are mainly responsible for its anti-inflammatory activity (Liu, Kakani, and Nair 2012; Mandegary et al. 2012; Pundarikakshudu et al. 2016; Sindhu et al. 2012; Wilborn et al. 2017). The research has shown that flavonoid-C-glycosides isolates from fenugreek seed inhibited lipoxygenase (LPO) and Cyclooxygenase (COX)-1 and COX-2 enzymes to exert their anti-inflammatory benefits (Liu, Kakani, and Nair 2012). Flavonoid components from aqueous extract of fenugreek seeds showed significant anti-inflammatory activity against the experimental model of carrageenan-induced rat paw edema (Mandegary et al. 2012). Fenugreek mucilage showed anti-inflammatory potential against complete Freund's adjuvant-induced arthritis in experimental rats (Sindhu et al. 2012). The potential of linoleic and linolenic acid-rich extract of fenugreek seeds is reported with significant anti-inflammatory against acute, subacute, and subchronic models of inflammation and arthritis in rats (Pundarikakshudu et al. 2016). Furthermore, the glycoside-based standardized fenugreek seed extract exerts its inflammatory potential during acute eccentric resistance exercise in young exercising human subjects through cytokine inhibition mechanisms (Wilborn et al. 2017). These findings suggest the potential of fenugreek and its constituents in ameliorating various chronic inflammatory conditions.

One of the most striking features of chronic inflammatory conditions is fibrosis. The term fibrosis refers to the development and excessive deposition of extracellular matrices (ECM) such as fibrillar collagens, elastic fibers, proteoglycans, and fibronectin by the proliferation of fibroblasts in response to tissue damage or injury (Jefferies 2018; Kandhare, Mukherjee et al. 2016). The chronic inflammations caused activation of myofibroblast, which is a crucial mediator of fibrosis. An array of stimuli, such as allergic responses, autoimmune reactions, chemical insults, persistent infections, and radiation, cause a series of inflammatory reactions that resulted in fibrosis (Jefferies 2018). Several studies have identified that myofibroblast, an essential mediator of fibrosis, is responsible for activating various collagen-producing cells. An array of cells, including epithelial, mesenchymal, and endothelial cells, contain myofibroblasts, and its activation by a variety of stimuli results in release from these cells known as epithelial/endothelial-mesenchymal (EMT/EndMT) transition (Gibb, Lazaropoulos, and Elrod 2020; Kendall and Feghali-Bostwick 2014). The levels of reactive oxygen species (ROS) such as superoxide, peroxides, hydroxyl radicals inducing oxidative stress are elevated in the fibroblasts of damaged tissue (Gibb, Lazaropoulos, and Elrod 2020; Xalxo and Keshavkant 2019). In addition, various other mediators such as chemokines (monocyte chemoattractant protein-1 (MCP-1), macrophage inflammatory protein-1β (MIP-1 β)), cytokines (IL-13, IL-21, transforming growth factor (TGF)-β1), angiogenic factors (vascular endothelial growth factor (VEGF)), peroxisome proliferator-activated receptors (PPARs), growth factors (platelet-derived growth factor (PDGF)), pro-inflammatory cytokines (TNF-α and IL-1β) are responsible for excessive ECM production to regulate fibrosis (Xalxo and Keshavkant 2019).

The significant body of evidence suggested the efficacy of fenugreek and its constituents against various fibrotic disorders of various organs such as the lungs, liver, kidneys. This chapter reviews

18.2 FENUGREEK AGAINST PULMONARY FIBROSIS

Lung fibrosis or idiopathic pulmonary fibrosis (IPF) is a complex, chronic, progressive interstitial lung disease with a poor prognosis and mean survival of 3 to 5 years after its diagnosis (Wuyts et al. 2020). During pulmonary fibrosis, normal alveolar space is replaced by extracellular matrix and mesenchymal cells, resulting in loss of lung elasticity, leading to symptomatic conditions, including in breathing, chest pain, persistent cough, and fatigue (Jefferies 2018). The report suggested that the incidence of idiopathic pulmonary fibrosis is 7–16 per 100,000 persons, with almost 5 million people suffering from IPF worldwide (Nalysnyk et al. 2012). Excessive exposure to mineral dust, asbestos, tobacco smoking, and the adverse event of neoplastic therapy, including bleomycin, infections such as pneumonia and tuberculosis, contribute to pulmonary induction and the development of pulmonary fibrosis (Harkema and Wagner 2019). Additionally, patients with pulmonary fibrosis exhibit subsequent comorbidity, such as pulmonary hypertension, and significantly elevated mortality rates (Hutchinson et al. 2015).

Many investigators have documented the vital role of fenugreek and constituents during the management of lung fibrosis using experimental animal models (Boominathan 2015; Kandhare et al. 2015; Yacoubi et al. 2011, 2012). These researchers have implemented an animal model of bleomycin-induced pulmonary fibrosis for evaluation of the potential of fenugreek. Bleomycin is a member of the glycopeptides group of antibiotics, which has been widely implicated as an effective antineoplastic agent for the clinical management of various malignancies (Liu, De Los Santos, and Phan 2017). However, repeated bleomycin administration is associated with pulmonary fibrosis due to overproduction of ROS, oxidative stress, activation of TGF-β1, and DNA damage (Boominathan 2015; Kandhare et al. 2015; Yacoubi et al. 2011, 2012).

The potential of fenugreek seed polyphenol extract (FSPE 200 mg/kg/day, equivalents to 15 mg of gallic acid) as well as fenugreek powder supplementation (FPS, 20% diet) in bleomycin-administered male Wistar rats was reported (Yacoubi et al. 2011, 2012). FSPE and FPS significantly improved circulatory total antioxidant status and decreased malondialdehyde (MDA) levels and suggested fenugreek's efficacy in restoring oxidant and antioxidant balance. Furthermore, reduction in inflammation in lung tissue during histological analysis was observed in FSPE-treated rats. However, TGF-β in macrophages, peribronchial, and bronchial inflammatory infiltrate in lung tissues remained unaffected. Thus, based on these findings, the author concluded that the polyphenolic content of fenugreek seed is responsible for inhibiting pulmonary inflammation via its anti-inflammatory and antioxidant potential (Yacoubi et al. 2011, 2012).

Subsequently, the potential of glycosides-based standardized fenugreek seed extract (SFSE-G) against experimental pulmonary fibrosis induced by bleomycin was reported (Kandhare et al. 2015). In this study, SFSE-G (5, 10, 20, and 40 mg/kg) orally administered in fibrotic rats for 28 days ameliorated bleomycin-induced pulmonary fibrosis biphasic response. In the initial phase (16 days), the elevated inflammatory response was significantly reduced by SFSE-G via inhibition of the release of various mediators, including 5-hydroxytryptamine, TNF-α, ILs in the lungs, and bronchoalveolar lavage fluid (BALF). The attenuation of elevated oxido-nitrosative stress (superoxide dismutase (SOD), glutathione (GSH), MDA, myeloperoxidase (MPO), nitric oxide (NO), nuclear factor E2-related factor (Nrf2), and Heme oxygenase 1 (HO-1)) was observed. Furthermore, SFSE-G treatment inhibited the second phase of fibrogenic changes (day 16–28) via fibrotic and apoptotic mediators such as collagen-1, endothelin-1 (ET-1), mucin 5ac (Muc5ac), mothers against decapentaplegic homolog-3 (Smad3), nuclear factor-kappa B (NF-kB), TGF-β, VEGF, B-cell lymphoma 2 (Bcl-2), Bcl-2-associated x protein (Bax), and Caspase-3. The histological findings from lung tissue corroborated with the biochemical results and indicated the anti-fibrotic efficacy of SFSE-G against bleomycin-induced pulmonary fibrosis. The author concluded that glycosides-based standardized fenugreek seed extract ameliorated pulmonary fibrosis via activation of the Nrf2 pathway and inhibition of TGF-β (Kandhare et al. 2015).

The thermostable extract of fenugreek seeds (TEFS) was reported to halt the progression of pulmonary fibrosis through up-regulation in the expression of Toll-like receptor 4 (TLR4) and HA synthase 2 (HAS2) (Boominathan 2015). TEFS is a protein-based extract of fenugreek seeds with a protein concentration of 0.5–1.5 mg/ml (Vijayakumar et al. 2010). Increased expression of HAS2 plays a vital role in renewing alveolar progenitor/stem cell and lung injury repair through increased extracellular matrix glycosaminoglycan hyaluronan level on the cell surface (Boominathan 2015).

Apigenin (4',5,7-trihydroxyflavone) is another flavonoid from fenugreek seed with various pharmacological activity (Rayyan, Fossen, and Andersen 2010; Salehi et al. 2019). The anti-fibrotic potential of fenugreek apigenin (5, 10, and 20 μM) was demonstrated using primary human bronchial fibroblasts (HBFs) isolated from bronchial biopsies derived from asthmatic patients *ex vivo* (Wojcik et al. 2013). In the same study, the human embryonic lung IMR-90 fibroblasts cultured with apigenin (10 μM) *in vitro* inhibited TGF-β1-induced fibroblast-to-myofibroblast transition in HBFs cultures. The TGF-β1-induced elevated mRNA expressions of α-smooth muscle actin (α-SMA) and tenascin C were down-regulated by apigenin treatment. Apigenin was found to be non-cytotoxic and non-cytostatic. These findings suggested that apigenin interferes in remodeling a bronchial wall during the asthmatic process via inhibition of TGF-β1-induced activation of α-SMA and tenascin C *in vivo* (Wojcik et al. 2013). The subsequent supporting study reported the efficacy of apigenin to inhibit the proliferation and migration of TGF-β1-induced airway smooth muscle cells (ASMCs) during airway remodeling in asthma through TGF-β1-induced ASMCs migration through the Smad signaling pathway (Li et al. 2015). In this study, human ASMCs were incubated with apigenin (1, 10, and 50 nM) followed by TGF-β1 (10 ng/ml) stimulation. Results suggested that treatment with apigenin halted the progression of ASMC proliferation at G1/S-interphase via Smad2 and Smad3 phosphorylation inhibition (Li et al. 2015).

The evidence of the efficacy of fenugreek and constituent against experimental pulmonary fibrosis with their possible mechanisms of action are summarized in Table 18.1 and Figure 18.1.

TABLE 18.1
Fenugreek Extracts and Their Phytoconstituents against Pulmonary Fibrosis

Phytoconstituent(s)	Dose	Proposed Mechanism	Reference
Polyphenols (gallic acid)	FSPE (200 mg/kg/day) FPS (20% diet)	Amelioration of pulmonary inflammation via its anti-inflammatory and antioxidant potential	Yacoubi et al. (2011), Yacoubi et al. (2012)
Flavonoid (apigenin)	Apigenin (5, 10, and 20 μM)	Inhibited TGF-β1-induced activation of α-SMA and tenascin C	Wojcik et al. (2013)
Furostanol glycosides (trigoneoside Ib) and flavonoid glycosides (vicenin 1)	SFSE-G (5, 10, 20, 40 mg/kg)	Amelioration of pulmonary fibrosis via activation of the Nrf2 pathway and inhibition of TGF-β	Kandhare et al. (2015)
Protein	TEFS (1.5 mg/kg/day or 15 mg/kg/day)	Halt the progression of pulmonary fibrosis through up-regulation in the expression of TLR4 and HA synthase 2 HAS2	Boominathan (2015)
Flavonoid (apigenin)	Apigenin (1, 10, and 50 nM)	Inhibited TGF-β1-induced ASMCs migration through Smad signaling pathway	Li et al. (2015)

FSPE: Fenugreek Seed Polyphenol Extract; FPS: Fenugreek Powder Supplementation; SFSE-G: Glycosides-based Standardized Fenugreek Seed Extract; TEFS: Thermostable Extract of Fenugreek Seeds.

Management of Fibrotic Disorders

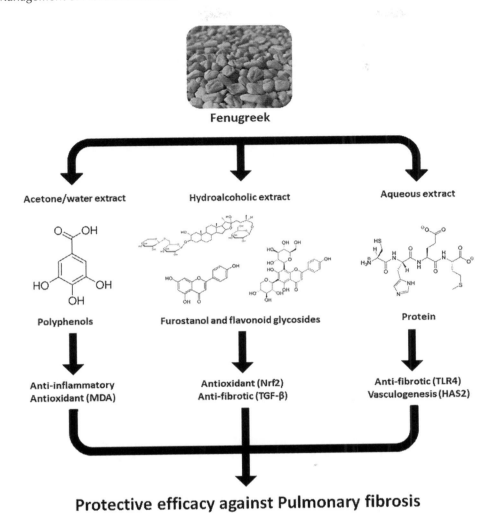

FIGURE 18.1 Molecular mechanism of fenugreek phytoconstituents against pulmonary fibrosis.

18.3 FENUGREEK AGAINST HEPATIC FIBROSIS

Hepatic fibrosis (liver fibrosis) is another critical progressive medical condition that arises from repetitive damage to liver tissue, followed by the accumulation of excessive ECM protein (Nathwani et al. 2019). Chronic hepatic fibrosis leads to liver cirrhosis and portal hypertension, which often needs hepatic transplantation. Numerous evidence suggested that chronic hepatic C virus infection, nonalcoholic steatohepatitis, and alcohol abuse are the primary cause of liver fibrosis (Caballeria et al. 2018; Nathwani et al. 2019). The presence of type 2 diabetes, obesity, arterial hypertension, hyperlipidemia, and alcohol consumption are common risk factors for liver fibrosis (Caballeria et al. 2018). The patients with at least one risk factor exhibit a 5% prevalence of hepatic fibrosis (Caballeria et al. 2018).

The anti-fibrotic potential of ethanolic extract of the whole plant of fenugreek (100 mg/kg) against CCl_4 (carbon tetrachloride)-induced hepatotoxicity in experimental animals was reported (Thaakur et al. 2014). The CCl_4 is a toxicant and documented to induce hepatotoxicity via the formation of trichloromethyl free radical after its cytochrome P450-induced biotransformation. The trichloromethyl free radical is responsible for adipose tissue peroxidative degradation in the presence of ROS, which culminates in the loss of cellular integrity of hepatocytes and hepatic fibrosis (Unsal, Cicek, and Sabancilar 2020). Biochemical estimation of a battery of liver function tests, such as aspartate transaminase (AST), alanine transaminase (ALT), alkaline phosphatase (ALP), and total bilirubin levels in serum, as well as liver hydroxyproline levels, showed reduction after administration of fenugreek extract depicted its inhibitory potential against CCl_4-induced liver fibrosis (Thaakur et al. 2014). The flavonoids and glycosides content of the ethanolic extract of fenugreek were suggested as the potential phytoconstituents responsible for anti-fibrotic potential (Dong et al. 2016).

The subsequent study evaluated the anti-fibrotic potential of fenugreek seed powder (FSP, 5% w/w diet) against cadmium-induced hepatotoxicity in experimental rats (Arafa, Mohammad, and Atteia 2014). The liver plays a cardinal role in the rapid clearance of cadmium; however, chronic administration of cadmium resulted in its hepatic accumulation, which induces various toxic responses such as elevated ROS, oxidative stress, and α1-collagen (I) (Arafa, Mohammad, and Atteia 2014). In this investigation, FSP significantly reduced hepatic serum markers (AST, ALT, and gamma-glutamyl transferase (GGT)), hepatic oxidative stress (catalase (CAT), SOD, GSH, and MDA), hepatic inflammatory markers (NO, IL-4, TNF-α), and hepatic fibrotic markers (TGF-β1 and hydroxyproline) in cadmium-intoxicated rats (Arafa, Mohammad, and Atteia 2014). The researchers attributed the anti-fibrotic effects of FSP to the presence of flavonoids (vitexin, naringenin, and quercetin) to provide hepatoprotection from cadmium toxicity through modulation of TGF-β1 (Arafa, Mohammad, and Atteia 2014).

Further evidence of the hepatoprotective and anti-fibrotic effect of fenugreek constituent was reported to SFSE-G against bleomycin-induced hepatic fibrosis in Sprague-Dawley rats (Kandhare et al. 2017). In this study, SFSE-G reduces bleomycin-induced elevated levels of serum AST, ALT, bilirubin, and GGT, and hepatic oxidative stress. Bleomycin has been documented to induce fibrosis via DNA/Fe^{2+}/bleomycin complex after interaction with ROS. Reverse transcription-polymerase chain reaction (RT-PCR) analysis revealed that SFSE-G attenuated bleomycin-induced up-regulated hepatic Farnesoid X receptor (FXR) mRNA expression. The flavonoid glycosides from fenugreek seed were suggested to be responsible for its hepatoprotective potential, perhaps through inhibition of FXR expression during liver fibrosis (Kandhare et al. 2017).

A more recent study confirmed the hepatoprotective effect of the fenugreek-containing herbal nutritional mixture (HNM) against CCl_4-induced liver cirrhosis in rats (Nithyananthan et al. 2020). The aqueous extract of HNM contains fenugreek (10%) along with germinated horse gram (*Macrotyloma uniflorum*), green gram (*Vigna radiata*), and *Curcuma longa*, attenuated elevated serum liver markers (AST, ALT, ALP, and GGT) and hepatic oxidative stress (CAT, SOD, GSH, and MDA), and down-regulated liver matrix metalloproteinase (MMP) and α-SMA activity, as determined by zymography and western blot analysis (Nithyananthan et al. 2020). The researchers attributed the exhibited hepatoprotective property of HNM to the presence of flavonoids through the modulation of endogenous biomarkers (MMP and α-SMA) (Nithyananthan et al. 2020).

Taken together, these available pieces of evidence suggested that fenugreek and its glycoside constituents demonstrated anti-fibrotic efficacy through modulation of TGF-β1 and hydroxyproline markers during hepatotoxicity as described in Table 18.2 and Figure 18.2.

Renal fibrosis is a life-threatening disease that mainly affects three functional sets of architecture of renal tissues (glomerulus, vasculature, and tubulointerstitial), affecting the glomerular filtration rate, gradually resulting in irreversible renal dysfunction leading to end-stage renal disease (ESRD) (Hewitson 2012). ESRD management requires life-long dialysis or renal transplantation, which has a significant economic burden on the patient (Lv et al. 2018).

Trigonelline, a major alkaloid constituent of fenugreek seed, has several pharmacological activities, including anti-fibrotic potential (Zhou, Chan, and Zhou 2012). Fenugreek seed endosperm contains a higher amount of alkaloids (35%), with a primary component of these alkaloids (Jani, Udipi, and Ghugre 2009).

TABLE 18.2
Fenugreek Extract and Its Phytoconstituents against Hepatic Fibrosis

Phytoconstituent (s)	Animal Model	Dose	Proposed Mechanism	Reference
Flavonoids and glycosides	CCl_4-induced	Ethanolic extract of fenugreek (100 mg/kg)	Attenuation of hepatic fibrosis via reduction of hydroxyproline levels	Thaakur et al. (2014)
Flavonoids	Cadmium-induced	FSP (5% w/w diet)	Anti-fibrotic through modulation of TGF-β1 and hydroxyproline levels	Arafa, Mohammad, and Atteia (2014)
Furostanol glycosides (trigoneoside Ib) and Flavonoid glycosides (vicenin 1)	Bleomycin-induced	SFSE-G (5, 10, 20, 40 mg/kg)	Hepatoprotective through inhibition of FXR expression	Kandhare et al. (2017)
Flavonoids	CCl_4-induced	Aqueous extract of HNM contain fenugreek (10%)	Hepatoprotective through the modulation of MMP and α-SMA	Nithyananthan et al. (2020)

FPS: Fenugreek Powder Supplementation; SFSE-G: Glycosides-based Standardized Fenugreek Seed Extract.

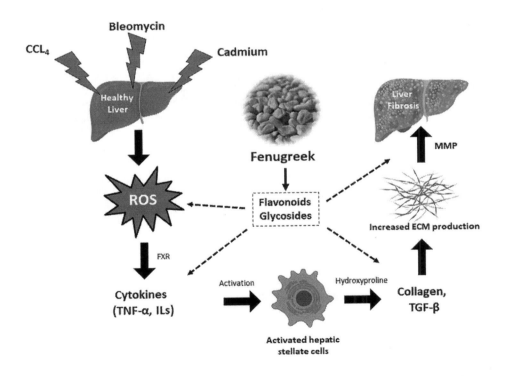

FIGURE 18.2 Molecular mechanisms of fenugreek phytoconstituents against hepatic fibrosis.

18.4 FENUGREEK AGAINST RENAL FIBROSIS

The anti-fibrotic potential of trigonelline against streptozotocin-induced diabetic nephropathy in neonatal diabetic rats is reported (Ghule, Jadhav, and Bodhankar 2012). In this study, oral treatment of trigonelline (50 and 100 mg/kg, orally) significantly ameliorated streptozotocin-induced fibrotic changes in renal tissues, such as reduction of peritubular fibrosis through attenuation of renal hydroxyproline content and blood TNF-α levels (Ghule, Jadhav, and Bodhankar 2012). The prevention of streptozotocin (STZ)-induced disturbance in kidney function biochemical parameters (serum creatinine and blood urea nitrogen (BUN)) and renal antioxidant enzymes (SOD, GSH, and MDA) were observed in trigonelline-treated rats (Ghule, Jadhav, and Bodhankar 2012).

Another constituent of fenugreek seed, SESF-G, demonstrated the nephroprotective and anti-fibrotic potential against unilateral ureteral obstruction (UUO)-induced renal interstitial fibrosis in mice through antioxidants ECM deposition modulation (Thakurdesai and Mohan 2014). The induction of UUO is known to activate endothelin-1 (ET-1) and renin-angiotensin-aldosterone system (RAS), which significantly contributes to renal vasoconstriction, elevated inflammatory influx, oxidative stress, and renal fibrosis (Martinez-Klimova et al. 2019). Subacute oral treatment of SESF-G (30, 60, and 100 mg/kg, twice daily) in mice showed significant prevention of UUO surgery-induced changes such as elevated collagen deposition with interstitial fibrosis (Sirius red staining of kidney tissues), up-regulated Nrf2 and HO-1 mRNA expressions, with down-regulated α-SMA mRNA expression (Thakurdesai and Mohan 2014).

Fibrosis plays a crucial role in the progression of nephropathy during microvascular diseases, including diabetes (Hewitson 2012; Pradeep, Barman, and Srinivasan 2019). Apigenin, a fenugreek glycoside constituent, has been reported for its anti-fibrotic and antidiabetic potential against the development and progression of diabetes-induced renal fibrosis in rats (Malik et al. 2017). The STZ-induced elevated glucose levels initiate a vicious cycle of oxidative stress, activate TGF-β1 and pro-inflammatory cytokines, and ECM accumulates, resulting in tubulointerstitial tissues to mimic renal fibrosis (Hewitson 2012; Pradeep, Barman, and Srinivasan 2019). Diabetes was induced in laboratory rats by administering STZ (55 mg/kg, i.p.) and then they were treated with apigenin (5, 10, and 20 mg/kg, p.o., 8 months). Apigenin (20 mg/kg) treatment attenuated STZ-induced elevated oxidative stress, renal fibrotic (TGF-β1, fibronectin, and type IV collagen), inflammatory (TNF-α, IL-6, and NF-kB), and apoptosis (Bax and caspase-3) markers and collagen deposition in renal tissues (Malik et al. 2017). The researchers suggested the nephroprotective potential of apigenin against diabetes-induced renal fibrosis to inhibit the TGF-β1-MAPK (mitogen-activated protein kinase)-fibronectin pathway (Malik et al. 2017).

Recently, the renoprotective potential of fenugreek against diabetes-induced renal fibrosis was reported (Pradeep, Barman, and Srinivasan 2019). In this study, the dietary fenugreek seed (10%) to STZ-induced diabetic rats reduced the degree of fibrosis (as measured by renal collagen I protein expression and collagen fibers, types I and IV) and deposition in the kidney (from Masson's trichrome (MT) staining) (Pradeep, Barman, and Srinivasan 2019). The inhibition of accumulation of ECM and collagen fibers in the renal basement membrane through modulation of endogenous type I collagen IV was suggested as a possible mechanism behind anti-fibrotic effects of fenugreek against the STZ-induced renal fibrosis (Pradeep, Barman, and Srinivasan 2019).

Taken together, fenugreek and its phytoconstituents showed good promising efficacy in alleviating renal fibrosis, as summarized in Table 18.3.

18.5 FENUGREEK AGAINST CARDIAC FIBROSIS

Cardiovascular diseases, including cardiac fibrosis, are recognized as vital predisposition factors for cardiac failure (Hinderer and Schenke-Layland 2019; Mozaffarian et al. 2015). The abnormal and excessive ECM deposition (such as type I collagen) results in impaired myocardial tissue function during cardiac remodeling and fibrosis (Hinderer and Schenke-Layland 2019). Furthermore, type

TABLE 18.3
Potential of Fenugreek Extract and Its Phytoconstituents against Renal Fibrosis

Phytoconstituent(s)	Animal Model	Dose	Proposed Mechanism	References
Trigonelline	Streptozotocin-induced	Trigonelline (50 and 100 mg/kg)	Anti-fibrotic through attenuation of renal hydroxyproline content	Ghule, Jadhav, and Bodhankar (2012)
Furostanol glycosides (trigoneoside Ib) and flavonoid glycosides (vicenin 1)	Unilateral ureteral obstruction-induced	SESF-G (30, 60, and 100 mg/kg)	Antioxidant (Nrf2 and HO-1) potential and modulation of ECM deposition (α-SMA)	Thakurdesai and Mohan (2014)
Flavonoid (Apigenin)	Streptozotocin-induced	Apigenin (5, 10, and 20 mg/kg)	Anti-fibrotic through inhibition of TGF-β1-MAPK-fibronectin pathway	Malik et al. (2017)
Fenugreek diet	Streptozotocin-induced	Fenugreek (10%) diet	Anti-fibrotic through endogenous inhibition type I collagen and ECM deposition	Pradeep, Barman, and Srinivasan (2019)

SFSE-G: Glycosides-based Standardized Fenugreek Seed Extract.

I and type III collagen and their ratio during the myocardial infarction and ischemic cardiomyopathy are essential factors for cardiac fibrosis (Hinderer and Schenke-Layland 2019).

Isoproterenol, a β-adrenergic agonist, induces irreversible damage to the myocardium, is used as an animal model of cardiac ischemia and fibrosis (Dai et al. 2019). The subacute intragastric treatment of fenugreek seed powder (250 mg/kg) to isoproterenol-induced cardiac ischemia and fibrosis in rats showed protective anti-fibrotic effects on cardiac tissues with antioxidant mechanisms (Murugesan, Revathi, and Manju 2011).

18.6 FENUGREEK AGAINST SYSTEMIC SCLEROSIS

Systemic sclerosis is a chronic autoimmune disease where fibrosis with disturbances of immune responses are prominent features (Sobolewski et al. 2019). Chronic progressive fibrosis in various organs such as lungs, heart, joints, skin, and kidneys in systemic sclerosis results in deterioration of the organs' function, leading to its failure (Sierra-Sepulveda et al. 2019). Several signaling pathways such as TGF-β, ROS, Wnts, and innate immune response were suggested to cause the fibroblast activation (Bhattacharyya et al. 2012; Sierra-Sepulveda et al. 2019), NAD^+-dependent deacetylases Sirtuin 1 (SIRT1) and Sirtuin 3 (SIRT3) (Peclat et al. 2020), and TGF-β-induced collagen synthesis, and α-SMA stimulated fiber formation (Peclat et al. 2020) that play a vital role in the progression of systemic sclerosis.

The inhibitory potential of apigenin against NAD^+-dependent deacetylases SIRT1 and CD38 is reported *in vitro* (Escande et al. 2013). The high-throughput analysis of the same study suggested inhibition of the CD38 overexpression by apigenin, a half-maximal inhibitory concentration (IC_{50}) = 10.3 ± 2.4 μmol/L for the NAD^+ase activity (Escande et al. 2013).

18.7 SAFETY OF ANTI-FIBROTIC FENUGREEK PHYTOCONSTITUENTS

The available scientific evidence strongly indicates the potential of fenugreek seed and phytoconstituents (e.g., trigonelline, glycosides, and apigenin) in the clinical management of fibrotic disorders.

Therefore, the safety aspects of fenugreek extracts, especially in the standardized form based on specific constituents, are essential before possible clinical evaluations in patients.

A recent systematic review suggested the broad margin of safety to many standardized extracts of fenugreek seeds (Kandhare et al. 2019). Furostenol glycoside-based standardized fenugreek seed extract (Fenu-FG) showed No Observed Adverse Effect Level (NOEAL) of 1,000 mg/kg/day without mutagenicity (up to a concentration of 5,000 μg/plate) or structural chromosomal aberrations possibility (up to 2,000 μg/ml) (Deshpande, Mohan, and Thakurdesai 2016; Wilborn et al. 2017).

Furthermore, experimental studies suggested that daily intake of anti-fibrotic fenugreek seed phytoconstituents (trigonelline, glycosides, apigenin) did not show any toxic effect during long-term administration (Kandhare, Bodhankar et al. 2016; Kandhare et al. 2019; Salehi et al. 2019; Zeiger and Tice 1997). Trigonelline demonstrated excellent safety (LD_{50} > 5000 mg/kg) with no adverse effects at doses 50 mg/kg for 21 days or 3500 mg for 70 days, no carcinogenicity or mutagenicity (Zeiger and Tice 1997). Acute oral toxicity study suggested flavonoid glycosides (Vicenin-1) exhibited median lethal dose (LD_{50}) of 4837.5 mg/kg and NOAEL of 75 mg/kg during a subacute toxicity study (Kandhare, Bodhankar et al. 2016). Apigenin is considered safe, and evidence from experimental studies has shown that apigenin is non-mutagenic and non-genotoxic (Salehi et al. 2019). Moreover, fenugreek seed is a certified GRAS (Generally Recognized as Safe) ingredient (21 CFR § 182.20 2010). Thus, fenugreek seed and its phytoconstituents have a broad margin of safety and can be evaluated for further clinical development to manage fibrotic disorders.

18.8 CONCLUSIONS

In conclusion, the available evidence suggests the potential anti-fibrotic efficacy of fenugreek phytoconstituents, like trigonelline and glycosides, to manage various fibrotic disorders as drugs or dietary supplements after suitable clinical studies.

REFERENCES

21 CFR § 182.20.2010. *Part 182 — Substances Generally Recognized as Safe, Subpart A—General Provisions. Sec. 182.20 Essential Oils, Oleoresins (Solvent-Free), and Natural Extractives (Including Distillates)*. Washington, DC: Department of Health and Human Services, US FDA.

Abdulkhaleq, L. A., M. A. Assi, R. Abdullah, M. Zamri-Saad, Y. H. Taufiq-Yap, and M. N. M. Hezmee. 2018. "The crucial roles of inflammatory mediators in inflammation: A review." *Vet World* 11 (5):627–35. doi: 10.14202/vetworld.2018.627-635.

Ahmadiani, A., M. Javan, S. Semnanian, E. Barat, and M. Kamalinejad. 2001. "Anti-inflammatory and antipyretic effects of *Trigonella foenum-graecum* leaves extract in the rat." *J Ethnopharmacol* 75 (2–3):283–6. doi: 10.1016/s0378-8741(01)00187-8.

Arafa, M. H., N. S. Mohammad, and H. H. Atteia. 2014. "Fenugreek seed powder mitigates cadmium-induced testicular damage and hepatotoxicity in male rats." *Exp Toxicol Pathol* 66 (7):293–300. doi: 10.1016/j.etp.2014.04.001.

Bhattacharyya, S., J. Wei, W. G. Tourtellotte, M. Hinchcliff, C. G. Gottardi, and J. Varga. 2012. "Fibrosis in systemic sclerosis: Common and unique pathobiology." *Fibrogenesis Tissue Repair* 5 (Suppl 1):S18. doi: 10.1186/1755-1536-5-S1-S18.

Boominathan, L. 2015. "Thermostable extract of fenugreek seeds inhibits the progression of pulmonary fibrosis via up-regulation of TLR4 and Has2." https://genomediscovery.org/2016/10/natural-product-therapy-for-pulmonary-fibrosis-thermostable-extract-of-fenugreek-seeds-tefs-increases-toll-like-receptor-4-expression-and-extracellular-matrix-glycosaminoglycan-hyaluronan-levels-p/ (Accessed: 1st May 2021).

Caballeria, L., G. Pera, I. Arteaga, L. Rodriguez, A. Aluma, R. M. Morillas, N. de la Ossa, A. Diaz, C. Exposito, D. Miranda, C. Sanchez, R. M. Prats, M. Urquizu, A. Salgado, M. Alemany, A. Martinez, I. Majeed, N. Fabrellas, I. Graupera, R. Planas, I. Ojanguren, M. Serra, P. Toran, J. Caballeria, and P. Gines. 2018. "High prevalence of liver fibrosis among European adults with unknown liver disease: A population-based study." *Clin Gastroenterol Hepatol* 16 (7):1138–45 e5. doi: 10.1016/j.cgh.2017.12.048.

Dai, H., L. Chen, D. Gao, and A. Fei. 2019. "Phosphocreatine attenuates isoproterenol-induced cardiac fibrosis and cardiomyocyte apoptosis." *Biomed Res Int* 2019:5408289. doi: 10.1155/2019/5408289.

Deshpande, P., V. Mohan, and P. Thakurdesai. 2016. "Preclinical safety assessment of glycosides based standardized fenugreek seeds extract: Acute, subchronic toxicity and mutagenicity studies." *J Appl Pharm Sci* 6 (9):179–88. doi: 10.7324/japs.2016.60927.

Dong, S., Q. L. Chen, Y. N. Song, Y. Sun, B. Wei, X. Y. Li, Y. Y. Hu, P. Liu, and S. B. Su. 2016. "Mechanisms of CCl_4-induced liver fibrosis with combined transcriptomic and proteomic analysis." *J Toxicol Sci* 41 (4):561–72. doi: 10.2131/jts.41.561.

Escande, C., V. Nin, N. L. Price, V. Capellini, A. P. Gomes, M. T. Barbosa, L. O'Neil, T. A. White, D. A. Sinclair, and E. N. Chini. 2013. "Flavonoid apigenin is an inhibitor of the NAD+ase CD38: Implications for cellular NAD+ metabolism, protein acetylation, and treatment of metabolic syndrome." *Diabetes* 62 (4):1084–93. doi: 10.2337/db12-1139.

Ghule, A. E., S. S. Jadhav, and S. L. Bodhankar. 2012. "Trigonelline ameliorates diabetic hypertensive nephropathy by suppression of oxidative stress in kidney and reduction in renal cell apoptosis and fibrosis in streptozotocin induced neonatal diabetic (nSTZ) rats." *Int Immunopharmacol* 14 (4):740–8. doi: 10.1016/j.intimp.2012.10.004.

Gibb, A. A., M. P. Lazaropoulos, and J. W. Elrod. 2020. "Myofibroblasts and fibrosis: Mitochondrial and metabolic control of cellular differentiation." *Circ Res* 127 (3):427–47. doi: 10.1161/CIRCRESAHA.120.316958.

Harkema, J. R., and J. G. Wagner. 2019. "Pathology of the respiratory system." In *Toxicologic Pathology for Non-Pathologists*, edited by Steinbach, T.J., Patrick, D.J. and Cosenza, M.E., 311–54. New York: Springer.

Hewitson, T. D. 2012. "Fibrosis in the kidney: Is a problem shared a problem halved?" *Fibrogenesis Tissue Repair* 5 (Suppl 1):S14. doi: 10.1186/1755-1536-5-S1-S14.

Hinderer, S., and K. Schenke-Layland. 2019. "Cardiac fibrosis—A short review of causes and therapeutic strategies." *Adv Drug Deliv Rev* 146:77–82. doi: 10.1016/j.addr.2019.05.011.

Hutchinson, J., A. Fogarty, R. Hubbard, and T. McKeever. 2015. "Global incidence and mortality of idiopathic pulmonary fibrosis: A systematic review." *Eur Respir J* 46 (3):795–806. doi: 10.1183/09031936.00185114.

Jani, R., S. A. Udipi, and P. S. Ghugre. 2009. "Mineral content of complementary foods." *Indian J Pediatr* 76 (1):37–44. doi: 10.1007/s12098-009-0027-z.

Jefferies, C. A. 2018. "19 — Pathogenesis of fibrosis—the lung as a model." In *Dubois' Lupus Erythematosus and Related Syndromes*, edited by Wallace, D.J. and Hahn, B.H., 261–8. London: Elsevier.

Kandhare, A. D., S. L. Bodhankar, V. Mohan, and P. A. Thakurdesai. 2015. "Effect of glycosides based standardized fenugreek seed extract in bleomycin-induced pulmonary fibrosis in rats: Decisive role of Bax, Nrf2, NF-kappaB, Muc5ac, TNF-alpha and IL-1beta." *Chem Biol Interact* 237:151–65. doi: 10.1016/j.cbi.2015.06.019.

Kandhare, A. D., S. L. Bodhankar, V. Mohan, and P. A. Thakurdesai. 2016. "Acute and repeated doses (28 days) oral toxicity study of Vicenin-1, a flavonoid glycoside isolated from fenugreek seeds in laboratory mice." *Regul Toxicol Pharmacol* 81:522–31. doi: 10.1016/j.yrtph.2016.10.013.

Kandhare, A. D., S. L. Bodhankar, V. Mohan, and P. A. Thakurdesai. 2017. "Glycosides based standardized fenugreek seed extract ameliorates bleomycin-induced liver fibrosis in rats via modulation of endogenous enzymes." *J Pharm Bioallied Sci* 9 (3):185–94. doi: 10.4103/0975-7406.214688.

Kandhare, A. D., A. Mukherjee, P. Ghosh, and S. L. Bodhankar. 2016. "Efficacy of antioxidant in idiopathic pulmonary fibrosis: A systematic review and meta-analysis." *EXCLI J* 15:636–51. doi: 10.17179/excli2016-619.

Kandhare, A. D., P. A. Thakurdesai, P. Wangikar, and S. L. Bodhankar. 2019. "A systematic literature review of fenugreek seed toxicity by using ToxRTool: Evidence from preclinical and clinical studies." *Heliyon* 5 (4):e01536. doi: 10.1016/j.heliyon.2019.e01536.

Kendall, R. T., and C. A. Feghali-Bostwick. 2014. "Fibroblasts in fibrosis: Novel roles and mediators." *Front Pharmacol* 5:123. doi: 10.3389/fphar.2014.00123.

Lee, S. B., and R. Kalluri. 2010. "Mechanistic connection between inflammation and fibrosis." *Kidney Int Suppl* (119):S22–6. doi: 10.1038/ki.2010.418.

Li, L. H., B. Lu, H. K. Wu, H. Zhang, and F. F. Yao. 2015. "Apigenin inhibits TGF-beta1-induced proliferation and migration of airway smooth muscle cells." *Int J Clin Exp Pathol* 8 (10):12557–63. doi: 10.18632/oncotarget.15771.

Liu, T., F. G. De Los Santos, and S. H. Phan. 2017. "The bleomycin model of pulmonary fibrosis." *Methods Mol Biol* 1627:27–42. doi: 10.1007/978-1-4939-7113-8_2.

Liu, Y., R. Kakani, and M. G. Nair. 2012. "Compounds in functional food fenugreek spice exhibit anti-inflammatory and antioxidant activities." *Food Chemistry* 131 (4):1187–92. doi: 10.1016/j.foodchem.2011.09.102.

Lv, W., G. W. Booz, F. Fan, Y. Wang, and R. J. Roman. 2018. "Oxidative stress and renal fibrosis: Recent insights for the development of novel therapeutic strategies." *Front Physiol* 9 (105):105. doi: 10.3389/fphys.2018.00105.

Malik, S., K. Suchal, S. I. Khan, J. Bhatia, K. Kishore, A. K. Dinda, and D. S. Arya. 2017. "Apigenin ameliorates streptozotocin-induced diabetic nephropathy in rats via MAPK-NF-kappaB-TNF-alpha and TGF-beta1-MAPK-fibronectin pathways." *Am J Physiol Renal Physiol* 313 (2):F414-F422. doi: 10.1152/ajprenal.00393.2016.

Mandegary, A., M. Pournamdari, F. Sharififar, S. Pournourmohammadi, R. Fardiar, and S. Shooli. 2012. "Alkaloid and flavonoid rich fractions of fenugreek seeds *(Trigonella foenum-graecum* L.) with antinociceptive and anti-inflammatory effects." *Food Chem Toxicol* 50 (7):2503–7. doi: 10.1016/j.fct.2012.04.020.

Martinez-Klimova, E., O. E. Aparicio-Trejo, E. Tapia, and J. Pedraza-Chaverri. 2019. "Unilateral ureteral obstruction as a model to investigate fibrosis-attenuating treatments." *Biomolecules* 9 (4):141. doi: 10.3390/biom9040141.

Mozaffarian, D., E. J. Benjamin, A. S. Go, D. K. Arnett, M. J. Blaha, M. Cushman, S. de Ferranti, J. P. Despres, H. J. Fullerton, V. J. Howard, M. D. Huffman, S. E. Judd, B. M. Kissela, D. T. Lackland, J. H. Lichtman, L. D. Lisabeth, S. Liu, R. H. Mackey, D. B. Matchar, D. K. McGuire, E. R. Mohler, 3rd, C. S. Moy, P. Muntner, M. E. Mussolino, K. Nasir, R. W. Neumar, G. Nichol, L. Palaniappan, D. K. Pandey, M. J. Reeves, C. J. Rodriguez, P. D. Sorlie, J. Stein, A. Towfighi, T. N. Turan, S. S. Virani, J. Z. Willey, D. Woo, R. W. Yeh, M. B. Turner, C. American Heart Association Statistics, and S. Stroke Statistics. 2015. "Heart disease and stroke statistics—2015 update: A report from the American heart association." *Circulation* 131 (4):e29–322. doi: 10.1161/CIR.0000000000000152.

Murakami, T., A. Kishi, H. Matsuda, and M. Yoshikawa. 2000. "Medicinal foodstuffs. XVII. Fenugreek seed. (3): Structures of new furostanol-type steroid saponins, trigoneosides Xa, Xb, XIb, XIIa, XIIb, and XIIIa, from the seeds of Egyptian *Trigonella foenum-graecum* L." *Chem Pharm Bull* 48 (7):994–1000. doi: 10.1248/cpb.48.994.

Murugesan, M., R. Revathi, and V. Manju. 2011. "Cardioprotective effect of fenugreek on isoproterenol-induced myocardial infarction in rats." *Indian J Pharmacol* 43 (5):516–9. doi: 10.4103/0253-7613.84957.

Nalysnyk, L., J. Cid-Ruzafa, P. Rotella, and D. Esser. 2012. "Incidence and prevalence of idiopathic pulmonary fibrosis: Review of the literature." *Eur Respir Rev* 21 (126):355–61. doi: 10.1183/09059180.00002512.

Nathwani, R., B. H. Mullish, D. Kockerling, R. Forlano, P. Manousou, and A. Dhar. 2019. "A review of liver fibrosis and emerging therapies." *EMJ* 4 (4):105–16.

Nithyananthan, S., P. Keerthana, S. Umadevi, S. Guha, I. H. Mir, J. Behera, and C. Thirunavukkarasu. 2020. "Nutrient mixture from germinated legumes: Enhanced medicinal value with herbs-attenuated liver cirrhosis." *J Food Biochem* 44 (1):e13085. doi: 10.1111/jfbc.13085.

Patil, S. P., P. V. Niphadkar, and M. M. Bapat. 1997. "Allergy to fenugreek *(Trigonella foenum graecum)*." *Ann Allergy Asthma Immunol* 78 (3):297–300. doi: 10.1016/S1081-1206(10)63185-7.

Peclat, T. R., B. Shi, J. Varga, and E. N. Chini. 2020. "The NADase enzyme CD38: An emerging pharmacological target for systemic sclerosis, systemic lupus erythematosus and rheumatoid arthritis." *Curr Opin Rheumatol* 32 (6):488–96. doi: 10.1097/BOR.0000000000000737.

Pradeep, S. R., S. Barman, and K. Srinivasan. 2019. "Attenuation of diabetic nephropathy by dietary fenugreek *(Trigonella foenum-graecum)* seeds and onion (Allium cepa) via suppression of glucose transporters and renin-angiotensin system." *Nutrition* 67–68:110543. doi: 10.1016/j.nut.2019.06.024.

Pundarikakshudu, K., D. H. Shah, A. H. Panchal, and G. C. Bhavsar. 2016. "Anti-inflammatory activity of fenugreek *(Trigonella foenum-graecum* Linn) seed petroleum ether extract." *Indian J Pharmacol* 48 (4):441–4. doi: 10.4103/0253-7613.186195.

Raju, J., and R. P. Bird. 2006. "Alleviation of hepatic steatosis accompanied by modulation of plasma and liver TNF-alpha levels by *Trigonella foenum-graecum* (fenugreek) seeds in Zucker obese (fa/fa) rats." *Int J Obes (Lond)* 30 (8):1298–307. doi: 10.1038/sj.ijo.0803254.

Ravichandiran, V., and S. Jayakumari. 2013. "Comparative study of bioactive fraction of *Trigonella foenum-graecum* L. leaf and seed extracts for inflammation." *Int J Front Sci Tech* 1 (2):128–48.

Rayyan, S., T. Fossen, and M. Andersen. 2010. "Flavone C-glycosides from seeds of fenugreek, *Trigonella foenum-graecum* L." *J Agric Food Chem* 58 (12):7211–7. doi: 10.1021/jf100848c.

Salehi, B., A. Venditti, M. Sharifi-Rad, D. Kregiel, J. Sharifi-Rad, A. Durazzo, M. Lucarini, A. Santini, E. B. Souto, E. Novellino, H. Antolak, E. Azzini, W. N. Setzer, and N. Martins. 2019. "The therapeutic potential of apigenin." *Int J Mol Sci* 20 (6):1305. doi: 10.3390/ijms20061305.

Sharififar, F., P. Khazaeli, and N. Alli. 2009. "In vivo evaluation of anti-inflammatory activity of topical preparations from Fenugreek *(Trigonella foenum-graecum* L.) seeds in a cream base." *Iran J Pharm Sci* 5 (3):157–62.

Sierra-Sepulveda, A., A. Esquinca-Gonzalez, S. A. Benavides-Suarez, D. E. Sordo-Lima, A. E. Caballero-Islas, A. R. Cabral-Castaneda, and T. S. Rodriguez-Reyna. 2019. "Systemic sclerosis pathogenesis and emerging therapies, beyond the fibroblast." *Biomed Res Int* 2019:4569826. doi: 10.1155/2019/4569826.

Sindhu, G., M. Ratheesh, G. L. Shyni, B. Nambisan, and A. Helen. 2012. "Anti-inflammatory and antioxidative effects of mucilage of *Trigonella foenum-graecum* (Fenugreek) on adjuvant induced arthritic rats." *Int Immunopharmacol* 12 (1):205–11. doi: 10.1016/j.intimp.2011.11.012.

Sobolewski, P., M. Maślińska, M. Wieczorek, Z. Łagun, A. Malewska, M. Roszkiewicz, R. Nitskovich, E. Szymańska, and I. Walecka. 2019. "Systemic sclerosis—multidisciplinary disease: Clinical features and treatment." *Reumatologia* 57 (4):221–33. doi: 10.5114/reum.2019.87619.

Suresh, P., N. Kavitha Ch, S. M. Babu, V. P. Reddy, and A. K. Latha. 2012. "Effect of ethanol extract of *Trigonella foenum-graecum* (Fenugreek) seeds on Freund's adjuvant-induced arthritis in albino rats." *Inflammation* 35 (4):1314–21. doi: 10.1007/s10753-012-9444-7.

Thaakur, S. R., G. R. Saraswathy, E. Maheswari, N. Kumar, T. Hazarathiah, K. Sowmy, P. Reddy, and B. Kumar. 2014. "Inhibition of CCl_4-induced liver fibrosis by *Trigonella foenum-graecum* Linn." *Ind J Nat Prod Res* 6:11–17.

Thakurdesai, P., and V. Mohan. 2014. "Renoprotective and anti-fibrotic efficacy of glycosides based standardized fenugreek seed extract (INDUS830) on unilateral ureteral obstruction induced renal interstitial fibrosis in mice." In 2nd National Conference on "Herbal and Synthetic Drug Studies (HSDS-2014), Azam Campus, Camp, Pune, 10–12 Feb 2014, p. 42.

Unsal, V., M. Cicek, and İ. Sabancilar. 2020. "Toxicity of carbon tetrachloride, free radicals and role of antioxidants." *Rev Environ Health*. doi: 10.1515/reveh-2020-0048.

Vijayakumar, M. V., V. Pandey, G. C. Mishra, and M. K. Bhat. 2010. "Hypolipidemic effect of fenugreek seeds is mediated through inhibition of fat accumulation and upregulation of LDL receptor." *Obesity (Silver Spring)* 18 (4):667–74. doi: 10.1038/oby.2009.337.

Wilborn, C., S. Hayward, L. Taylor, S. Urbina, C. Foster, P. Deshpande, V. Mohan, and P. Thakurdesai. 2017. "Effects of a proprietary blend rich in glycoside based standardized fenugreek seed extract (IBPR) on inflammatory markers during acute eccentric resistance exercise in young subjects." *Asian J Pharm Clin Res* 10 (10):99–104. doi: 10.22159/ajpcr.2017.v10i10.18811.

Wojcik, K. A., M. Skoda, P. Koczurkiewicz, M. Sanak, J. Czyz, and M. Michalik. 2013. "Apigenin inhibits TGF-beta1 induced fibroblast-to-myofibroblast transition in human lung fibroblast populations." *Pharmacol Rep* 65 (1):164–72. doi: 10.1016/s1734-1140(13)70974-5.

Wuyts, W. A., M. Wijsenbeek, B. Bondue, D. Bouros, P. Bresser, C. Robalo Cordeiro, O. Hilberg, J. Magnusson, E. D. Manali, A. Morais, S. Papiris, S. Shaker, M. Veltkamp, and E. Bendstrup. 2020. "Idiopathic pulmonary fibrosis: Best practice in monitoring and managing a relentless fibrotic disease." *Respiration* 99 (1):73–82. doi: 10.1159/000504763.

Xalxo, R., and S. Keshavkant. 2019. "Melatonin, glutathione and thiourea attenuates lead and acid rain-induced deleterious responses by regulating gene expression of antioxidants in *Trigonella foenum-graecum* L." *Chemosphere* 221:1–10. doi: 10.1016/j.chemosphere.2019.01.029.

Yacoubi, L., A. Abidi, N. Kourda, H. Mohamed Hédi, S. Fattouch, and S. Khamsa. 2012. "Activities extract from *Trigonella foenum-graecum* L. (Fenugreek) seed on experimental pulmonary fibrosis." *Fund Clin Pharmacol* 26 (91):P335.

Yacoubi, L., L. Rabaoui, M. H. Hamdaoui, S. Fattouch, R. Serairi, N. Kourda, and S. B. Khamsa. 2011. "Antioxidative and anti-inflammatory effects of *Trigonella foenum-graecum* Linnaeus, 1753 (Fenugreek) seed extract in experimental pulmonary fibrosis." *J Med Plants Res* 5 (17):4315–25.

Zandi, P., S. K. Basu, W. Cetzal-Ix, M. Kordrostami, S. K. Chalaras, and L. B. Khatibai. 2017. "Fenugreek (*Trigonella foenum-graecum* L.): An important medicinal and aromatic crop." *Active Ingredients from Aromatic and Medicinal Plants*:207–24. doi: 10.5772/66506.

Zeiger, E., and R. Tice. 1997. "Trigonelline; Review of toxicological literature." *Research Triangle Park, North Carolina: National Institute of Environmental Health Sciences and Integrated Laboratory Systems* 27.

Zhou, J., L. Chan, and S. Zhou. 2012. "Trigonelline: A plant alkaloid with therapeutic potential for diabetes and central nervous system disease." *Curr Med Chem* 19 (21):3523–31. doi: 10.2174/092986712801323171.

19 Fenugreek in Management of Immunological, Infectious, and Malignant Disorders

Rohini Pujari and Prasad Thakurdesai

CONTENTS

19.1 Introduction 320
19.2 Application of Fenugreek in Growth and Development 321
19.3 Application of Fenugreek in Immunological Disorders 323
    19.3.1 Immunity Enhancement Potential of Fenugreek 323
    19.3.2 Immunostimulant in Normal Status 323
    19.3.3 Immunostimulant against the Drug-Induced Immunocompromised Status 324
    19.3.4 Fenugreek against Allergic Disorders 324
    19.3.5 Fenugreek against Autoimmune Disorder (Rheumatoid Arthritis) 325
19.4 Fenugreek in Management of Infectious Diseases 327
    19.4.1 Fenugreek against Bacterial Infections 327
    19.4.2 Fenugreek against Fungal Infections 330
    19.4.3 Fenugreek against Protozoal Infections 331
    19.4.4 Fenugreek against Helminth Infections 331
    19.4.5 Fenugreek against Vector-Borne Diseases 332
19.5 Fenugreek for Management of Malignancy Disorders (Cancers) 332
    19.5.1 Fenugreek against Multiple Cancerous Cell Lines *In Vitro* 333
    19.5.2 Fenugreek against Ehrlich Ascites Carcinoma (EAC) 334
    19.5.3 Fenugreek against Skin Carcinogenesis 334
    19.5.4 Fenugreek against Leukemia (Blood Cancer) 340
    19.5.5 Fenugreek against Hepatocellular Carcinoma (Liver Cancer) 340
    19.5.6 Fenugreek against T-Cell Lymphoma 341
    19.5.7 Fenugreek against Breast Cancer 341
    19.5.8 Fenugreek against Colon Cancer 342
19.6 Conclusion 343
References 344

ABBREVIATIONS

Akt: Ak strain thymoma (serine/threonine-specific protein kinase); ALT: alanine transaminase; ASK: apoptosis signal-regulating kinase; AST, aspartate transaminase; Bak, Bcl-2 homologous antagonist/killer; Bax: Bcl-2-associated X protein; Bcl-2: B-cell lymphoma 2; Bcl-xL: B-cell lymphoma-extra large; CAT: Catalase; Cdc42: cell division cycle 42; cdk: Cyclin-dependent kinases; COX: Cyclooxygenase; COX-2: cyclooxygenase-2; cPLA2: Phospholipase A2; CSC: Cancer stem cells; Src; DDX3: DEAD-box RNA helicase; DNA:

deoxyribonucleic acid; ED50: median effective dose; EMMPRIN: Extracellular matrix metalloproteinase inducer; EMT: epithelial-mesenchymal transition; ER: estrogen receptor; ERK: extracellular signal-regulated kinase; FADD: fas-associated protein with death domain; Fas: factor ligand superfamily (tumor necrosis factor ligand superfamily); GPX: Glutathione peroxidase; GSK: Glycogen synthase kinase; HER2: human epidermal receptor 2; HIF: Hypoxia-Inducible Factor; hsp70: heat shock protein70; hTERT: human telomerase reverse transcriptase gene; IkBα: nuclear factor of kappa light polypeptide gene enhancer in B-cells inhibitor, alpha; IKK-b: inhibitor of nuclear factor kappa-B kinase subunit beta; IL-6: interleukin-6; IL-1β: interleukin-1beta; IRE-1α: Inositol-requiring transmembrane kinase/endoribonuclease 1α; JAK: janus kinase; JNK: c-Jun N-terminal kinase; LDH: lactate dehydrogenase; LPO: Lipoxygenases; LPS: Lipopolysaccharides; MAPK: mitogen-activated protein kinase; MEK: meiotic chromosome-axis-associated kinase; mRNA: micro ribonucleic acid; MMP: matrix metalloproteinase; mTOR: mammalian target of rapamycin; MTT: 3-(4,5-dimethylthiazol-2-yl)-2,5-diphenyltetrazolium bromide; NEDD: neural precursor cell expressed developmentally down-regulated protein; NF-kB: nuclear factor kappa-light-chain-enhancer of activated B cells; NO: Nitric oxide; Nrf-2: nuclear factor (erythroid-derived 2)-like 2; OATP1B1: organic anion transporting polypeptide 1B1; p21: cyclin-dependent kinase inhibitor 1; p53: phosphoprotein 53; PARP: poly (ADP-ribose) polymerase; PCNA: proliferating cell nuclear antigen; PD-1: programmed cell death protein-1; PGE: prostaglandins E; PI3K: phosphatidylinositide-3 kinase; ROCK: Rho-associated protein kinase; Raf: v-Raf murine sarcoma viral oncogene homolog B; Ras: Kirsten rat sarcoma; RT-PCR: reverse transcription polymerase chain reaction; SH-PTP: Protein tyrosine phosphatase; SOD: Superoxide dismutase; STAT: Signal transducer and activator of transcription; STAT-3: signal transducer and activator of transcription-3; TAZ: transcriptional co-activator with PDZ binding motif; TGF-β: transforming growth factor-β; TIMP: tissue inhibitor of metalloproteinase; TIMP-2: tissue inhibitor of metalloproteinases 2; TNF: Tumor necrosis factor; TNF-α: tumor necrosis factor-α; TRAIL: TNF-related apoptosis-inducing ligand; Vav2: guanine nucleotide exchange factor VAV2; VEGF: vascular endothelial growth factor; Waf1: cyclin-dependent kinase inhibitor 1; XIAP: X-linked inhibitor of apoptosis protein.

19.1 INTRODUCTION

Food is an essential part of our existence to ensure growth in children and youth, maintain good health throughout life, meet unique needs of pregnancy and lactation, and provide immunity to protect and recover from illness. Nutrients are components of foods needed for the body in adequate amounts for average growth and reproduction for a healthy life (Misra et al. 2011). Micronutrient deficiencies specifically have effects such as poor growth, impaired intellect, and increased mortality and susceptibility to infection (Childs, Calder, and Miles 2019).

A well-functioning immune system is critical for survival. Malnutrition is the primary cause of immunodeficiency worldwide, with infants, children, adolescents, and the elderly most affected. There is a strong relationship between malnutrition and infection and infant mortality because poor nutrition leaves children underweight, weakened, and vulnerable (Katona and Katona-Apte 2008). Complete nutrition has the reversal potential to fight effectively against malnutrition-induced immune deficiencies (Bourke, Berkley, and Prendergast 2016; Maggini, Pierre, and Calder 2018) and cancer (Zitvogel, Pietrocola, and Kroemer 2017).

There is substantial indirect and supportive evidence that the immune response plays a vital role in cancer etiology (Gonzalez, Hagerling, and Werb 2018; Michaud et al. 2015; Valdes-Ramos and Benitez-Arciniega 2007). Carcinogenesis involves three stages: cancer initiation (by the generation of genotoxic stress), promotion (by induction of cellular proliferation), and cancer progression (by enhancement of angiogenesis and tissue invasion). The potential mechanisms of cancer causation include chronic inflammation (Anand et al. 2008) and immunosurveillance (Mortaz et al. 2016). Tumor immunity and inflammation are mutually exclusive processes, and cancer immunosurveillance and tumor-promoting inflammation co-exist in the same tumor microenvironment (Bui and Schreiber 2007; Chow, Möller, and Smyth 2012; Mlynska 2018). Chronic inflammation has been documented for its contribution in all the stages of carcinogenesis (Fernandes et al. 2015; Landskron et al. 2014; Multhoff, Molls, and Radons 2011).

The immune system's demands for energy and nutrients can be met from exogenous sources, such as the diet, or if dietary sources are inadequate, from endogenous sources such as body stores. Some micronutrients and dietary components have specific roles in developing and maintaining an effective immune system throughout the life course (França et al. 2009). Such nutritional interventions enhancing the immune function in normal individuals and preventing the onset of infections or chronic inflammatory diseases, ultimately preventing cancer are gaining interest (Patel et al. 2018). The utilization of immunostimulants as food supplements has been considered a promising practice that helps maintain a healthy environment in the body with increasing growth performance (Newaj-Fyzul and Austin 2015).

Any malfunction or defects in either the innate or acquired immune response can provoke illness or diseases that are closely related (Castro and Gourley 2010; Chinen and Shearer 2010). For example, the inadequate immune responses are called immunodeficiency. An overactive immune response is known as hypersensitivity reactions (e.g., allergic disorders). In addition, the disease that responds to an exaggerated immune reaction to the self is called autoimmunity (e.g., rheumatoid arthritis, psoriasis).

Fenugreek (*Trigonella foenum-graecum*, family: Fabaceae) is a common spice with nutritional and health benefits, documented in various traditional medicines worldwide. Due to several phytoconstituents (steroidal sapogenins, dietary fibers, galactomannans, alkaloids, and amino acids), it was found effective for the prevention or treatment of many diseases (Sarwar et al. 2020; Srinivasan 2019; Srivastava et al. 2020; Yao et al. 2019) including immunodeficiency, inflammation, and cancer (Goyal, Gupta, and Chatterjee 2016). The present chapter focuses on available modern scientific evidence of beneficial effects of fenugreek in average growth and development, immunity, and associated disorders such as malnutrition, immunosuppression, infections, and cancer (malignancy) with the probable mechanisms of actions.

19.2 APPLICATION OF FENUGREEK IN GROWTH AND DEVELOPMENT

Several phytoconstituents of fenugreek have been reported for growth-promoting activities. Bioassay-guided fractionation of methanolic extract from fenugreek seeds and saponin I (1) and dioscin (9), along with two new (i.e., 2 and 3) and five known analogs (i.e., 4–8) are reported as growth-hormone release stimulators *in vitro* (Shim et al. 2008). The structures of the new steroidal saponins, fenugreek saponins I, II, and III (1–3, resp.), were determined as gitogenin 3-O-beta-D-xylopyranosyl-(1→6)-beta-D-glucopyranoside, sarsasapogenin 3-O-beta-D-xylopyranosyl-(1→6)-beta-D-glucopyranoside, and gitogenin 3-O-alpha-L-rhamnopyranosyl-(1→2)-beta-D-glucopyranoside, respectively. Fenugreek saponin I and dioscin reported 12.5- and 17.7-fold stimulation, respectively, to enhance growth hormone release from rat pituitary cells where gitogenin (5) showed moderate activity (Shim et al. 2008).

Fenugreek seed powder (0, 5, 10, 20, and 40 g doses per kg added in commercial broiler diet for 42 days) reduced the appetite and growth performance in the broiler chicks, which was attributed to the enhanced blood glucose levels (Duru et al. 2013). The results showed that breast weight, leg weight, body weight, feed intake, feed efficiency were reduced by fenugreek powder. Treatment with fenugreek powder also enhanced the blood glucose and LDL level but reduced the triglyceride levels (Duru et al. 2013). A subsequent study reported dose-dependent immunostimulatory action of the dietary administration of fenugreek seed powder to enhance gilthead seabream's immune status and growth (*Sparus aurata* L), a species with one of the highest rates of production in marine aquaculture (Awad, Cerezuela, and Esteban 2015). Fenugreek seed powder (1%, 5%, and 10%, in the diet for four weeks) showed dose-dependent stimulation of immunity (leucocyte peroxidase and content respiratory burst activity), humoral immunity (IgM level, complement activity, total protein, antiprotease, and peroxidase), and hematological parameters (WBC and RBC counts) which indicate growth-promoting potential. Furthermore, significant up-regulation of the expression of various immunity-related genes in head-kidney (CSF-1R, IgM, IL-8, and MHC1) and antioxidant enzyme in the liver (SOD, CAT, and GR) of seabream were observed with fenugreek treatment (Awad, Cerezuela, and Esteban 2015).

Furthermore, dietary administration of fenugreek seeds, alone or combined with probiotic strains (i.e., *Bacillus subtilis* (B46) *Lactobacillus plantarum Bacillus licheniformis* (TSB27)), was reported to enhance the immune response of gilthead seabream (Bahi et al. 2017). The fenugreek seed (50 g per kg, in the pelleted diet for three weeks)-containing diet enhanced the growth performance, humoral immune parameters (peroxidase, natural hemolytic complement, total IgM levels, antiproteases, and proteases activities), and the expression of immune-relevant genes such as tcr-β, igm, bd, and csfrl (Bahi et al. 2017). In the subsequent study, powdered fenugreek seed (1%, 5%, and 10% in the diet for eight weeks)-containing diet produced marked improvement in its metabolic and immune status of gilthead seabream (Guardiola, Bahi, and Esteban 2018). In this study, fenugreek seed-containing diet produced significant improvement of metabolic (reduction in serum levels of albumin/globulin ratio, creatine kinase, aspartate, potassium, and aminotransferase) and the immune (hemolytic complement and peroxidase activity, antiprotease activity) status (Guardiola, Bahi, and Esteban 2018). In the concluding study, the fenugreek seed (1%, 5%, and 10% for eight weeks)-containing diet demonstrated the antioxidant status and immune system stimulation in skin mucosa of gilthead seabream (Guardiola et al. 2018). In this study, the dietary powdered fenugreek seed-fed fishes demonstrated improved hydroxyl radical scavenging capacity and augmented the peroxidase and protease activities with a general decrease in the antiprotease activity, indicating enhanced mucosal skin immunity and antioxidant status of gilthead seabream (Guardiola et al. 2018).

Many infections are known to produce oxidative stress and cause further damage (Butcher et al. 2017; Ivanov, Bartosch, and Isaguliants 2017). Conversely, enhanced oxidative status can be helpful to build immunity against infections (Grant and Hung 2013; Gruden et al. 2012; Wang, Zhang, and Bai 2020). The dietary fenugreek seed powder (3% for eight weeks) was reported to have antioxidant activity, enhance the oxidative status and immune response in a species of fish, Nile tilapia (*Oreochromis niloticus*), infected with *Aeromonas hydrophila*, a gram-positive bacteria (Moustafa et al. 2020). The fish fed with a fenugreek seed powder diet exhibited improved antioxidant parameters (malondialdehyde, glutathione peroxidase, and superoxide dismutase), immune parameters (immunoglobulin, lysozyme, and respiratory burst activity), up-regulated the immune-related gene expressions in the kidney and liver (IL-1β and TNF-α), and decreased the alanine aminotransferase (ALT) and blood aspartate aminotransferase (AST), indicating enhanced resistance and survival rate and protection against immune and bacterial infection challenge (Moustafa et al. 2020).

The study in another fish, juvenile blunt snout bream (*Megalobrama amblycephala*), supports the efficacy of fenugreek seed extracts to protect against oxidative stress and inflammation without affecting growth, development, and immune response (Yu et al. 2019). In this study, the fenugreek seed extract (0.04%, 0.08%, and 0.16% in diet for eight weeks) significantly lowered the lipid content, immunoglobulin M (IgM), plasma complement component 3 (C3), total protein (TP), and albumin (ALB) levels and reduced the relative expressions of fatty acid synthase (FAS), acetyl CoA carboxylase (ACC), and sterol regulatory element-binding protein-1 (SREBP1) mRNA in the fish liver. Furthermore, fenugreek seed extract diet showed reduced pro-inflammatory genes expressions by regulating interleukin 8 (IL-8) mRNA levels and tumor necrosis factor-α (TNF-α) and kelch-like ECH-associated protein 1 (Keap1) and enhanced the expressions of anti-inflammatory genes by regulating transforming growth factor (TGF-β) and interleukin 10 (IL-10), the target of rapamycin (TOR), S6 kinase-polypeptide 1 (S6K1) and growth factor-1 (IGF-1), glutathione peroxidase-1 (GPx1), and nuclear factor erythroid 2-related factor 2 (Nrf2). Dietary Fenugreek seed extract showed antioxidant properties without any effect on specific growth rate (SGR), feed conversion ratio (FCR), weight gain (WG), and final weight (FW) (Yu et al. 2019).

Dietary fenugreek seeds (or extract) demonstrated benefits towards growth and performance against higher species of animals such as pigs (Begum, Hossain, and Kim 2016) and rabbits (Mabrouki, Chalghoumi, and Abdouli 2017). In weanling pigs, fenugreek seed alcoholic extract (0.1% and 0.2%, oral or in diet for 42 days) has been reported to enhance growth performance and reduce fecal gas emission (Begum, Hossain, and Kim 2016). The fenugreek seed extract-containing diet enhanced total tract digestibility, red blood cells (RBC), immunoglobulin G (IgG), high-density lipoprotein cholesterol (HDL-C), reduced fecal ammonia (NH_3) and hydrogen sulfide (H_2S) gas emission, suggesting growth, immunity, and energy digestibility enhancement potential (Begum, Hossain, and Kim 2016). In newly weaned rabbits frequently suffering from digestive disorders that were fed with fenugreek seed-rich dietary fibers, the following was reported: decreased feed intake, improved feed conversion ratio (FCR), with no effect on carcass characteristics and meat composition, suggesting enhanced growth performance, health, carcass parameters, and digestibility (Mabrouki, Chalghoumi, and Abdouli 2017).

19.3 APPLICATION OF FENUGREEK IN IMMUNOLOGICAL DISORDERS

19.3.1 Immunity Enhancement Potential of Fenugreek

The immune system is a collective terminology comprising of chemicals, cells, and processes involved in the protection of the human body from foreign antigens, such as microorganisms like bacteria, viruses, parasites, fungi, and toxins as well as cancer cells. (Marshall et al. 2018). The immune system can be divided into two types, namely innate (antigen-independent non-specific) and acquired (antigen-specific and dependent) immunity (Bonilla and Oettgen 2010; Turvey and Broide 2010). The deficiencies in the immune system's functioning result in the dysfunction of one's immune system, which is called immunocompromised or immunosuppressed conditions. Primary immune delicacies such as genetic disorders or secondary immune deficiencies are the most common cause of immunosuppression. Most common causes of immunosuppression include medications (glucocorticoids, chemotherapy drugs), surgery (e.g., splenectomy, bone marrow transplantation), trauma/injury, radiation, low oxygen, metabolic disease (e.g., diabetes), infections, old age, and malnutrition.

Immunostimulants (immunostimulatory) are substances (drugs and nutrients) that stimulate the immune system by inducing activation or increasing the activity of any of its components. Several studies have documented the immunostimulatory activity of fenugreek (Bin-Hafeez et al. 2003; Hamden et al. 2010; Ramadan, El-Beih, and Abd El-Kareem 2011; Tripathi et al. 2012) (Anarthe et al. 2019; Kumar et al. 2017; Nimbalkar et al. 2018; Walaa, Emeish, and El-Deen 2016; Zentek et al. 2013). This section of the chapter reviews the evidence of immunostimulatory efficacy in normal and immunocompromised status.

19.3.2 Immunostimulant in Normal Status

Fenugreek galactomannan-rich fraction was reported to stimulate the immune system via phagocytosis in rat peritoneal macrophages and proliferation and IgM secretion in a human-human hybridoma, HB4C5 cells (Ferreira et al. 2015; Ramesh, Yamaki, and Tsushida 2002). Aqueous extract of fenugreek seeds showed a dose-dependent immunostimulatory effect on immune functions in mice (Bin-Hafeez et al. 2003). In this study, fenugreek seed extract (50 and 100 mg/kg, once a day, orally) demonstrated a significant increase in relative organ weights (thymus and liver), cellularity of thymus and bone marrow, delayed-type hypersensitivity (DTH) response, humoral immunity, phagocytic index and phagocytic capacity of macrophages and lymphoproliferation (Bin-Hafeez et al. 2003).

Similar immunostimulant activity of ethanolic extract of fenugreek seed (200 and 400 mg/kg)-treated mice were reported during carbon clearance test, delayed type of hypersensitivity (DTH), T-cell population test, and sheep erythrocyte agglutination test (Tripathi et al. 2012). The study results indicated stimulation of the reticuloendothelial system, anti-inflammatory and stimulation of cell-mediated and humoral immune system by ethanolic extract of fenugreek seeds (Tripathi et al. 2012).

Another study using fenugreek seed fed to piglets after weaning demonstrated enhancements in immunological responses without affecting their health status and performance (Zentek et al. 2013). The diet containing fenugreek seeds (1.5 g/kg for 28 days) showed a reduction in pH of the caecum and colon content and antigen-presenting cells (MHCII+CD5-) of the piglet (indicating immune stimulation). It also increased L-lactic acid and n-butyric acid concentration (which indicate enhanced metabolism in the colon) and relative concentrations of the gamma delta T-cell population (TCR1+CD8alpha-) (Zentek et al. 2013).

19.3.3 IMMUNOSTIMULANT AGAINST THE DRUG-INDUCED IMMUNOCOMPROMISED STATUS

Drug-induced immune suppression (or medication-induced immunosuppression) refers to impaired immune system function due to medications used to treat systemic diseases. It manifests as recurrent, severe, and opportunistic infections (Giancane et al. 2020; Wiseman 2016). Egyptian fenugreek seed powder was reported to show dose-dependent alleviation in cyclophosphamide (CP)-induced experimental immunosuppression and delayed burn-healing in rats (Ramadan, El-Beih, and Abd El-Kareem 2011). In this study, fenugreek seed powder (0.5 and 1 g/kg, oral suspension, four weeks) significantly ameliorated CP-induced weight loss, and cellularity of lymphoid organs leucopenia (resulting from lymphopenia and neutropenia), serum γ-globulin levels, delay in the skin-burning healing process, and delayed type of hypersensitivity response (Ramadan, El-Beih, and Abd El-Kareem 2011). Oral subacute (21 days) treatment of diosgenin, a steroid saponin constituent of fenugreek seeds, was documented to dose-dependent stimulatory effect on specific and non-specific immune response in cyclophosphamide (100 mg/kg, p.o.)-induced immunosuppression in laboratory rats (Nimbalkar et al. 2018). Diosgenin (50, 100, and 150 mg/kg, for 21 days) to CP-induced rats exhibited a significant increase in antibody titer production, primary HA titer and secondary HA titer, DTH response, and a phagocytic index indicating the stimulation of the reticuloendothelial system (Nimbalkar et al. 2018). In recent evidence of immunostimulatory activity of fenugreek, the whole plant, and methanolic extract of fenugreek showed dose-dependent prevention of the azathioprine-induced myelosuppression, T-cell population, carbon clearance, and delayed hypersensitivity at the doses of 100, 300, and 500 mg/kg in laboratory rats (Anarthe et al. 2019).

19.3.4 FENUGREEK AGAINST ALLERGIC DISORDERS

Allergy is an immunity-associated disease resulting from sensitization and hypersensitive immune response to harmless substances in the environment called allergens. Asthma is one of the allergic, severe, chronic, progressive, and inflammatory bronchial diseases. Allergic asthma involves the symptoms such as dyspnea (shortness of breath) and wheezing (high-pitched whistling sound), resulting from increased bronchial hyperreactivity to a variety of allergenic and non-allergenic stimuli (Bosnjak et al. 2011). Many patients with chronic allergic conditions, such as allergic rhinitis and asthma seek complementary alternative medicine to attain better control of symptoms due to limitations of existing options (Amaral-Machado et al. 2020; Hussain et al. 2017; Koshak 2019).

T-helper 2 (TH2) cells priming plays a cardinal role in sensitization by allergens, and its consequences such as the production of TH2 cytokines (such as IL-4 and IL-13), immunoglobulin E (IgE) by B cells, more mucus production, activation of endothelial cells, and eosinophil migration to tissues result in allergic symptoms (Maeda, Caldez, and Akira 2019; O'Connor and Nichol 2015). Fenugreek was reported to possess anti-allergic efficacy as revealed against experimental allergic conditions induced by Th2-related allergens, such as trimellitic anhydride (TMA) and ovalbumin

(OVA) (Bae et al. 2012; Piao et al. 2017). The TMA-induced hypersensitivity model of Balb/c mice mimics atopic dermatitis and eczema, the allergic disorders of skin that show inflamed skin, intense infiltration with T-cells, major histocompatibility complex II-positive cells, eosinophils, mast cells, a T-helper cell-type (Th) 2 cytokine profile, and a substantial increase of serum IgE levels (Schneider et al. 2009). The OVA-induced allergic asthma is well-established and validated animal model of lung allergy, which mimics the pathophysiology of clinical allergic asthma (Du et al. 1992; Elwood et al. 1991; Shilovskiy et al. 2019; Shum, Rolph, and Sewell 2008).

The initial evidence of the anti-allergic potential of ethanolic extract of fenugreek seed (250 mg/kg for 7 days) was reported against experimental allergy models *in vivo* (TMA-induced contact hypersensitivity in BALB/c mice) and *ex vivo* (OVA-induced lung hypersensitiveness) (Bae et al. 2012). In this study, oral treatment with ethanolic extract of fenugreek seeds showed a reduction of allergic parameters such as ear thickness, infiltration of eosinophils and mast cells, the production of interleukins (IL-4, IL-5, IL-13, and IL-1β). Researchers suggested the mechanism such as the suppression of the IL-4 secretion and mRNA expression of GATA-binding protein 3 (GATA-3, an IL-4 transcription factor), and promotion of Th1 differentiation (enhanced production of IFN-γ and by IFN-γ-producing CD4þ T-cells), which is dependent on the mRNA expression of T-box transcription factor 21 gene (T-bet, and IFN-γ transcription factor) (Bae et al. 2012).

During another study, fenugreek-fed mice were reported to be inhibiting the production of allergen-specific cytokine response, airway inflammation, and lung pathology in OVA-induced allergic hyperresponsiveness in BALB/c mice (Piao et al. 2017). In this study, allergic asthma was initiated in BALB/c mice by sensitizing them with OVA emulsified in aluminum on days 1 and 14, then aerosol challenged with OVA on days 27, 28, and 29. Fenugreek treatment before the OVA challenge significantly alleviated the number of inflammatory cells in bronchoalveolar lavage fluid (BALF), ameliorated lung inflammation, reduced the goblet cells and collagen deposition, down-regulated the expression of Th2 cytokines, and enhanced Th1 cytokines (inflammatory mediators in asthma) in lung homogenates and BALF and inhibited serum IgE and anti-OVA IgG (Piao et al. 2017). Cytokines and immunoglobulin E (IgE) play a crucial role in orchestrating the chronic inflammation, structural changes of the respiratory tract in both asthma and COPD, and type 1 hypersensitivity reactions (e.g., allergic asthma, allergic rhinitis) and become essential targets for the development of new therapeutic strategies in these diseases (Barnes 2008; Platts-Mills 2001). The reported efficacy of fenugreek in this study suggests the excellent potential for additional management options for chronic allergic inflammatory diseases (Piao et al. 2017).

However, the first and sole clinical evidence of the efficacy of fenugreek-based syrup preparation on 79 patients with mild asthma during the randomized, double-blind, placebo-controlled study is reported (Emtiazy et al. 2018). The four weeks of treatment of syrup (50 g aqueous extract of fenugreek seeds in 100 ml syrup) showed 10% increase in forced expiratory volume in 1 (FEV1) and forced expiratory volume/forced vital capacity (FEV1/FVC) levels, with a significant reduction in serum cytokine IL-4 levels, improvement in Quality of life scores (Activity, Impact, Symptom, and total) without serious side effects (Emtiazy et al. 2018).

19.3.5 Fenugreek against Autoimmune Disorder (Rheumatoid Arthritis)

Autoreactivity is a crucial mechanism that contributes to immunologic and non-immunologic disease pathogenesis (Rosenblum, Remedios, and Abbas 2015). Disordered innate immunity, including immune complex-mediated complement activation, adaptive immune responses against 'self'-antigens, post-translationally modified proteins, dysregulated cytokine networks, osteoclast or chondrocyte activation, and imprinting of resident stromal cells are some of the factors responsible for most progressive autoimmune disorders (Firestein and McInnes 2017), which can lead to organ system dysfunction and premature death (Meyer, Decker, and Baughman 2010).

One of the most common, chronic, progressive, and disabling autoimmune diseases is rheumatoid arthritis (RA), characterized by chronic synovial inflammation and proliferative synovitis with severe cartilage and bone destruction, causing significant joint damage and disability (Guo et al. 2018). Dysregulation of the immune system plays an essential role in the pathogenesis of rheumatoid arthritis (RA) (Firestein and McInnes 2017; Yap et al. 2018). Furthermore, several pro-inflammatory mediators, such as interleukin-6, interleukin-1β, tumor necrosis factor (TNF-α), prostaglandins, leukotriene, nitric oxide (NO), platelet-activating factor, enzymes (cyclooxygenases, phospholipases, lipoxygenases), play a crucial role in the inflammatory pathophysiology of the synovial membrane and bone destruction in RA (Kany, Vollrath, and Relja 2019).

The limitation of synthetic anti-arthritic drugs, such as side effects, led the researchers to investigate natural anti-inflammatory alternatives (Choudhary et al. 2015). Fenugreek is a dietary spice ingredient traditionally used to treat inflammation (Baliga et al. 2015). Furthermore, fenugreek containing Ayurvedic formulation was reported to have anti-arthritic efficacy during randomized placebo-controlled clinical studies (Almuhareb et al. 2019; Chopra, Saluja, and Tillu 2010).

In the first study, IRA-01, an Ayurvedic formulation rich in fenugreek seed extract, demonstrated significant ameliorative efficacy during the 3-month, multicenter, double-blind, placebo-controlled phase, followed by a 9-month, single-center, open-label phase (Chopra, Saluja, and Tillu 2010). During the double-blind phase, IRA-01 showed excellent efficacy in terms of physician and patient global assessment score, rheumatoid factor (RF), and improvement in 60% of patients with American College of Rheumatology (ACR20) criteria (v/s 53% patients with ACR20 in placebo). In an open-label phase, significant improvement in all efficacy variables, namely scores of joint pains, swelling in the Health Assessment Questionnaire (HAQ), was found in IRA-01 treated patients with 80% and 40% of patients achieving improvement as per ACR20 and ACR50 (Chopra, Saluja, and Tillu 2010). An ACR20 (at least 20% improvement and ACR50 (at least 50% improvement) in both the tender joint count and the swollen joint and other core sets of ACR measures are considered as standard efficacy criteria of effectiveness of agents against RA during clinical studies (Felson et al. 1995). The rheumatoid factors (RF), which is a characteristic serological feature of RA, is a titer of circulating antibodies directed against antigenic sites on the immunoglobulin G (IgG) molecule (anti-IgG antibodies) (Johnson and Faulk 1976). For many decades, RF has been considered a prognostic rather than diagnostic marker of numerous autoimmune diseases, including RA (Dörner et al. 2004). Furthermore, the Indian HAQ is a validated instrument for the measurement of functional disability (and quality of life) routinely done in all outcome studies on patients with rheumatoid arthritis (RA) (Kumar et al. 2002). In the second study, a complementary and alternatives medicine (CAM) formulation containing fenugreek was reported to have significant benefits such as enhanced erythrocyte sedimentation rate (ESR) levels in addition to the reduction in duration of RA symptoms during the cross-sectional study in 438 patients (Almuhareb et al. 2019).

Fenugreek or its constituents reported anti-inflammatory and disease-modifying potential in numerous studies involving the animal model of inflammation, rheumatoid arthritis, and osteoarthritis (Folwarczna et al. 2014; Pundarikakshudu 2019; Sindhu et al. 2012, 2018; Suresh et al. 2012). Fenugreek mucilage exhibited potent anti-inflammatory effects in Freund's complete adjuvant (FCA)-induced arthritis (AIA) in rats (Sindhu et al. 2012). The AIA model is a well-validated model that has been widely used to induce a rheumatoid arthritis-like inflammation in rats and has excellent clinical correlation, especially with chronic and progressive nature (Rainsford 1982; Wooley 1991). Subacute oral treatment of fenugreek mucilage (75 mg/kg) to AIA rats showed the prevention of arthritic-induced changes such as paw edema (swelling), and joint histology changes such as cellular infiltration and edema formation, with the restoration of the blood levels of C-reactive protein (CRP), cyclooxygenase-2 enzyme, ESR, and WBC (Sindhu et al. 2012). The subsequent advanced study reported a significant reduction in the concentration of pro-inflammatory enzymes (cyclooxygenase, lipoxygenase), and cytokines (IL-6, TNF-α), arthritic index, and rheumatoid factor by subacute oral treatment of fenugreek mucilage (75 mg/kg) to AIA rats (Sindhu et al. 2018).

In another similar study, oral treatment of ethanolic extract of fenugreek (200 and 400 mg/kg) to AIA rats reported significant dose-dependent prevention of the AIA-induced changes: paw edema, weight loss, elevated levels of blood WBC and ESR, and interleukins (IL-1β, IL-2, IL-1α,

IL-6), and TNF-α in cartilage tissues (Suresh et al. 2012). Furthermore, daily oral supplementation of fenugreek seed (1% pulverized seeds in the diet) and fenugreek seed extract standardized to 4-hydroxy-L-isoleucine (50 mg of 4-hydroxy-L-isoleucine/kg) for four weeks reportedly increased the mechanical strength of the tibial metaphysis (cancellous bone) in 3-month-old non-ovariectomized and femoral diaphysis (compact bone) in ovariectomized rats (decreased estrogen level and developing osteoporosis) (Folwarczna et al. 2014).

19.4 FENUGREEK IN MANAGEMENT OF INFECTIOUS DISEASES

Infectious diseases are one of the most important causes of morbidity and mortality worldwide. (Mathers 2020). Despite the significant progress made in microbiology and the control of microorganisms, sporadic epidemics and pandemics incidents emerge worldwide due to drug-resistant microorganisms and unknown disease-causing microbes (Mahady 2005; Weledji, Weledji, and Assob 2017). Fenugreek has traditionally been documented for efficacy in several infectious diseases (Malik et al. 2013), perhaps due to the presence of many antimicrobial phytochemicals such as phytoalexin, allicins, isothiocyanates, anthocyanins, tannins, essential oils, polyphenols, terpenoids (Salah, Bestoon, and Osman 2010), lysine and L-tryptophan, mucilaginous fiber, saponins, coumarin, trigonelline (Walli et al. 2015). This section of the chapter reviews the applications of fenugreek or constituents in the management of infectious diseases caused by worms and bacterial, fungal, protozoal, and vector-borne infections.

19.4.1 FENUGREEK AGAINST BACTERIAL INFECTIONS

Bacterial infections remain the leading cause of death in children, the elderly, and immunocompromised patients, especially in developing countries. Although many antibiotics are available, the chronic treatment with antibiotics is limited due to the resistance, allergic reactions, toxicity, and even dual infections (Blumenthal et al. 2019; Fair and Tor 2014; Ventola 2015). Hence, a search for less resistant or non-resistant, effective antibacterial drugs from natural sources has attracted scientists' attention (Fernández et al. 2020).

Aeromonas hydrophila is a heterotrophic, gram-negative, rod-shaped bacterium found in fresh or brackish water. Because of resistance to most common antibiotics, the *A. hydrophila* challenge test is used *in vivo* preclinical models to evaluate the various treatments on immunity and survival ability of fishes. The supplementation of 1% fenugreek for 30 successive days to *C. gariepinus* (sharptooth catfish) challenged with *A. hydrophila* reported to stimulate the immune system and decrease the mortality rate (Walaa, Emeish, and El-Deen 2016). Fenugreek supplementation markedly enhanced fish's immune parameters, including serum globulins, phagocytic index, catalase (CAT), glutathione peroxidase, monocyte counts, weight, and mortality (Walaa, Emeish, and El-Deen 2016). Another similar study, with fenugreek seed (5, 10, or 20 g/kg) in the diet for 90 days to common carp (*Cyprinus carpio* Linn.) fingerlings, reported the dose-dependent enhancement of the non-specific immunity and reduced mortality against infection-induced *A. hydrophila* (Kumar et al. 2017). Furthermore, methanolic extract of fenugreek showed antibacterial effects were observed against *A hydrophila* and *Pseudomonas liquefaciens* (Kumar et al. 2017).

The fenugreek seed essential oil alone and with gentamycin cream have reported excellent antibacterial effects (reduced skin lesion size, swelling, redness, amount of puss) against wounds inoculated with *Pseudomonas aeroginosa* in mice skin than gentamycin alone (Kahaleq, Abu-Raghif, and Kadhim 2016). The alcoholic extract of fenugreek seeds (100, 200, 300 mg/ml, orally, once daily for four weeks) was reported to have efficacy against oral infection of *Klebsiella* pneumonia (isolated from diarrheal cases in children)-induced diarrhea in mice (Rana 2016). The alcoholic extract of fenugreek seeds could prevent infection-induced mucosal ulceration and damage with inflammatory cells inside the villi and restore the near-normal structure of the intestine in infected mice (Rana 2016).

Apart from these reports, fenugreek extract or constituents were reported to have antibacterial properties in many *in vitro* evaluations as listed in Table 19.1.

TABLE 19.1
Antibacterial Efficacy of Fenugreek (Chronological Sorted)

Fenugreek Constituent (Dose)	Bacterial strains	Remarks	References
Germinating sprouts (50, 100, 150 ppm per L)	*Helicobacter pylori*	Antibacterial presence of phenolics	Randhir, Lin, and Shetty (2004)
Seeds extract (hexane, ethanol, chloroform) and fractions of ethanol extract (5, 50, 100 mg/ml)	*S. aureus* *B. subtilis* *M. luteus* *P. aeruginosa* *E. coli* *S. choleraesuis*	Antibacterial against *P. aeruginosa* and *E. coli*	Saleem et al. (2008)
Seed oil in hexane (50%, 90%, and 100%)	*E coli* *S. aureus* *S. typhimurium*	Antibacterial activity against all strains Highest inhibition of *E. coli*	Abdel, Ahmed, and Awad (2008)
Seeds extract (acetone, water, chloroform, diethyl ether)	*E. coli* *B. cereus* *L. acidophilus* *Pneumococcus*	Antibacterial—extracts of acetone > diethyl ether > chloroform > water	Upadhyay et al. (2008)
Seeds extract (60% ethanolic or aqueous) (5, 10, 20, and 40 mg/mL)	*S. typhimurium* *L. monocytogenes*, *E. coli*	Antibacterial activity—mild	Over et al. (2009)
Seeds extract (methanol, acetone, 2–512 µg/ml)	*Pseudomonas spp.* *Escherichia coli* *S. dysentiriae* *S. typhi*	Methanol extract: highest on *Pseudomonas spp.* Acetone extract: highest on *E. coli*	Dash, Sultana, and Sultana (2011)
Leaves extract (chloroform, hexane, methanol, ethanol, water)	*E, coli* *P. mirabilis* *Klebsiella spp,* *S. aureus* *P. aeruginosa* *E. erogens*	Antibacterial by all extracts on all strains Ethanol extract—highest Water extract—lowest	Premnath, Lakshmidevi, and Aradhya (2011)
Seeds extract (EthOH:water)	*S. aureus* *B. cereus* *E. coli* *P. aeruginosa* *K. pneumoni* *S. typhi*	Antibacterial on all strains Highest inhibition of *P. aeruginosa*	Abd-Alrahman and Salem-Bekhet (2013)
Seeds extract (methanol, acetone)	*E. coli* *P. aeruginosa* *L. lactis* *B. amylo-liquifaciens*	Antibacterial on all strains by both Highest: acetone extract on *P. aeruginosa*, *E. coli*	Malik et al. (2013)
Seeds extract (aqueous ethanolic, 50%)	*S. aureus* *B. cereus* *E. coli* *P. aeruginosa* *K. pneumonia* *P. vulgaris* *S. typhi*	Strong antibacterial against all	Sobhy et al. (2013)
Seeds extract (water, acetone, chloroform, ethanol, methanol)	*E. coli* *P. aeruginosa* *S. sonnei* *S. aureus*	Antibacterial on all strains by all extracts Highest: chloroform, and methanol extract	Alwahibi and Soliman (2014)

Fenugreek Constituent (Dose)	Bacterial strains	Remarks	References
Seeds extract (aqueous, hexane, chloroform, acetone, ethanol, methanol (2.5, 5, 10 mg/ml)	*S. typhi* *K. pneumonia* (Clinical isolates) *E. coli* (Morbid rabbit isolate)	Antibacterial activity by all extracts except aqueous extract	Chalghoumi and Hedi (2016)
Leaves extract (hexane, ethyl acetate, methanol, aqueous)	*E. coli* *P. aeruginosa* *S. marcescens* *B. cereus*	Antibacterial activity by all extracts *P. aeruginosa* was most resistant	Dharajiya et al. (2016)
Essential oil extract from seed	*P. aeruginosa*, 10 isolates	Strong antibacterial effect	Kahaleq, Abu-Raghif, and Kadhim (2016)
Seeds extract (methanol, hot) (75%, 50%, 25% v/v)	*S. aureus* *B. subtilus* *P. aeruginosa* *E. coli*	Strong antibacterial by both extracts, Highest: against *P. aeruginosa*	Alhan, Jassim, and Jasim (2017)
Leaves + seeds + stem extracts (methanol, acetone, aqueous)	*E. coli* *S. aureus*	Methanol extract: highest inhibition of both strains	Sharma, Singh, and Rani (2017)
Silver nanoparticles synthesized using fenugreek seeds	*K. pneumoniae,* *B. subtilis* *S. aureus* *E. coli*	Antibacterial effect against all strains Highest: *K. pneumoniae,* *B. subtilis, S. aureus*	Alwahibi et al. (2018)
Seeds extract (ethanol, water, chloroform, benzene, acetone)	*S. aureus* *E. coli* *S. typhi* *P. aeruginosa* *K. pneumonia*	Antibacterial effect against all strains by all extracts Ethanol extract: highest	Nandagopal et al. (2018)
Seeds extract (petroleum ether, methanol, and water)	*E. coli* *S. typhi* *S. aureus*	Antibacterial against *E. coli*	Abdelmageed et al. (2019)
Silver nanoparticles synthesized using fenugreek seeds	*E. coli* *S. aureus*	Antibacterial effect against both strains	Deshmukh, Gupta, and Kim (2019)
Seeds extract (70% methanol)	*E. coli* *Pseudomonas*	Antibacterial against both strains	Jatav et al. (2019)
Seeds extracts (ethanol and methanol)	*E. coli* *S. aureus*	Antibacterial against both strains by both extracts	Kaveri and Sincy (2019)
Seeds extract (ethanol)	*B. subtilis* *S. aureus* *E. coli* *P. aeruginosa*	Antibacterial against *B. subtilis* only	Khan et al. (2019)
Seeds extract (50% ethanol-water)	*S. aureus* *P. aeruginosa* *S. typhi* *P. mirabilis* *E. coli* *V. parahaemolyticus*	Highest against *S. aureus* and *P. aeruginosa*	Al-Timimi (2019)
Seeds and roots—callus roots, cotyledons (methanol)	*P. aeruginosa* *E. coli* *S. typhi* *S. aureus*	Seed extract: highest against *P. aeruginosa* hypocotyl callus extract: highest against *E. coli* and *S. aureus*	Magdoleen et al. (2020)

19.4.2 Fenugreek against Fungal Infections

Fungal infections represent an example of neglected emerging diseases, accounting for approximately 150 million severe cases of fungal infections that occur worldwide, resulting in approximately 1.7 million deaths per year (Kainz et al. 2020). Chronic fungal diseases result from evasion of immune system responses through different processes such as mitosis, recombination, and expression of genes (Sepahvand et al. 2018). Fenugreek and constituents have been reported for excellent antifungal properties from several studies (Alluri and Majumdar 2014; Alwahibi et al. 2018; Dharajiya et al. 2016; Elnour et al. 2015; Haouala et al. 2008; Kulkarni et al. 2019; Kumari, Rao, and Gajula 2016; Oddepally and Guruprasad 2015; Sudan, Goswami, and Singh 2020; Varadarajan et al. 2015; Walli et al. 2015), which can provide a required alternative for a better, cheaper, and safer antifungal agent (Sepahvand et al. 2018)

The initial evidence shows antifungal effects of aqueous extracts of various plant parts of fenugreek (3%) and methanolic fractions of the aerial parts against pathogenic mycelial fungi (Haouala et al. 2008). In this study, *Rhizoctinia solani*, *Fusarium graminearum*, and *Alternaria sp.* were the most sensitive species, and *Pythium aphanidermatum* was most resistant to fenugreek (Haouala et al. 2008). Subsequently, the antifungal activity of methanolic extracts of fenugreek seeds (against *Aspergillus flavus*, *Trichophyton rubrum*, and *Candida albicans*) (Alluri and Majumdar 2014) and leaves (against *Trichoderma viridae*) was reported (Dharajiya et al. 2016). Furthermore, ethyl acetate extract of fenugreek leaves (6.25 mg/ml) also showed potent antifungal properties against *T. viridae* (Dharajiya et al. 2016). Recently, the petroleum ether, aqueous, ethyl acetate, and ethanol extracts of fenugreek seed powder reported potent antifungal activity at 25, 50, and 100 µml against *Microsporum gypseum* (Sudan, Goswami, and Singh 2020). In addition, a defensin-like antifungal peptide (Tf-AFP) isolated from fenugreek seed, was reported to inhibit the growth of fungal species such as *Rhizoctonia solani*, *Fusarium solani*, and *Fusarium oxysporum* during *in vitro* testing (Oddepally and Guruprasad 2015).

The silver nanoparticles (AgNPs) synthesized from silver nitrate ($AgNO_3$) solution using fenugreek seed extract showed significant fungicidal effect against *F. equiseti* and *A. alternata* (Alwahibi et al. 2018). Synthesizing nanoparticles by biological (green synthesis) methods using plant extracts for many medical and other applications has attracted considerable attention recently because of their cost-effective, eco-friendly, and nontoxic nature (Singh et al. 2018).

Candidiasis is one of the most frequent opportunistic infections caused by *Candida albicans*. The high prevalence, substantial morbidity, and economic losses of recurrent vulvovaginal (Denning et al. 2018), invasive (Ben-Ami 2018), and cutaneous (Taudorf et al. 2019) candidiasis require better solutions and improved quality of care. The available evidence of the antifungal potential of fenugreek against *C. albicans* (Elnour et al. 2015; Kumari, Rao, and Gajula 2016; Varadarajan et al. 2015; Walli et al. 2015) indicated a potential solution for candidiasis.

In the earliest study, the boiling (but not cold or hot) water extract of fenugreek seeds showed antifungal activity against *C. albicans* during a disc diffusion method *in vitro* assay (Walli et al. 2015). At the same time, the methanol and petroleum ether extracts of fenugreek seeds reported potent inhibition of fungal species *Candida albicans* and *Aspergillus niger* during *in vitro* tests (Elnour et al. 2015). Similar antifungal activity was reported by the methanolic extract of fenugreek seeds against *Candida parapsilosis*, another major cause of invasive candida disease (Kumari, Rao, and Gajula 2016).

Antifungal resistance represents a major clinical challenge to clinicians responsible for treating invasive fungal infections due to the limited arsenal of systemically available antifungal agents (Arendrup 2014; Kontoyiannis 2017; Pai, Ganavalli, and Kikkeri 2018). Amongst them, fluconazole resistance *Candida* and *non-candida* species are of a particular concern due to the increased incidence of infections worldwide (Wiederhold 2017). The hydroalcoholic extracts of fenugreek seeds were found effective and showed a dose-dependent inhibition against fluconazole-resistant *Candida albicans* during *in vitro* assay with minimum inhibitory concentration (MIC) of 15.62 µg/ml (Varadarajan et al. 2015).

Immunological, Infectious, Malignant

Malassezia spp. (especially *Malassezia furfur*) are the yeasts capable of causing cutaneous ailments such as seborrheic dermatitis, resulting in increased cell turnover, scaling, and inflammation in the epidermis, called dandruff (Saunte, Gaitanis, and Hay 2020). Recently, ethanolic and aqueous extracts of fenugreek leaves exhibited potent antifungal activity against clinical isolate and commercial strain of *M. furfur*, along with other pathogenic fungi such as *C. albicans* and *Aspergillus niger* (Kulkarni et al. 2019). In the same study, the gel formulation prepared from 30% aqueous fenugreek leaf extract showed a protective effect against *M. furfur* infections in New Zealand rabbits (Kulkarni et al. 2019). Thus, the topical gel formulation containing fenugreek leaf aqueous extract can provide effective and safe herbal treatment for various cutaneous fungal infections, including dandruff (Kulkarni et al. 2019).

19.4.3 Fenugreek against Protozoal Infections

Protozoal infections are one of the leading causes of morbidity and mortality worldwide for many years (Fletcher et al. 2012). One of the significant causes of protozoal infection is malnutrition, making the individual vulnerable to diarrheal diseases and enteric infections through various mechanisms (Berhe et al. 2020; Siddiqui, Belayneh, and Bhutta 2021). Infectious diarrhea is responsible for more deaths than other gastrointestinal tract diseases such as gastrointestinal cancers, peptic ulcers, or inflammatory bowel disease (Siciliano et al. 2020). However, the actual burden remains unknown as many of the incidences of enteric protozoa are often ignored as a cause of diarrheal illness and never reported (Fletcher et al. 2012).

Amebiasis and giardiasis are the most severe gastrointestinal diseases caused by protozoa *Entamoeba histolytica* and *Giardia lamblia*, respectively, responsible for acute and chronic diarrhea in humans, especially in developing countries (Tenali et al. 2018). There are only two reports of the efficacy of fenugreek against various protozoal infections such as amebiasis (Kaya et al. 2019) and giardiasis (Kayondo, Jumaa, and Elnazeer 2019). In one report, methanol extract of fenugreek seeds (1, 2, 4, 8, 16, and 32 mg/ml) reported the dose-dependent prevention of the proliferation of the Acanthamoeba cysts (Kaya et al. 2019). In the other study, the ethanolic extract of fenugreek seeds has been reported for antigiardial activity *in vitro* against *Giardia lamblia* trophozoites isolated from the fecal sample (Kayondo, Jumaa, and Elnazeer 2019). The ethanolic extract of fenugreek seed (5 mg/ml)-treated sample resulted in 98% mortality in *Giardia lamblia* trophozoites within 96 h (Kayondo, Jumaa, and Elnazeer 2019).

19.4.4 Fenugreek against Helminth Infections

Helminthiasis or worm infections (e.g., roundworms, tapeworms, flukeworms) are among the most persistent health problems causing human debility, cognitive deficits, and sometimes death, with huge economic losses, especially in animal husbandry. Furthermore, helminthiasis is reported to exacerbate other severe infections such as tuberculosis and the human immunodeficiency virus (HIV) (Ishnava and Konar 2020). Fenugreek was being explored in the last few years as a potential option for conventional anthelmintic agents to overcome significant challenges such as adverse effects and resistance (Lalthanpuii and Lalchhandama 2020).

Adult Indian *Pheretima posthuma* has been utilized as an organism of choice for evaluating new anthelmintic agents because of anatomical and physiological resemblance to human intestinal pathogenic roundworms parasites and easy availability (Mali and Mehta 2008; Murugamani et al. 2012). In the earlier study on fenugreek, the significant anthelmintic activity of methanolic, ethyl acetate, chloroform, and petroleum ether extracts of fenugreek leaves and seeds was reported against Indian adult earthworms, a roundworm, *Pheretima posthuma in vitro* (Bhalke et al. 2008). In another similar study, the hydroalcoholic extracts of fenugreek seeds exhibited *in vitro* anthelmintic activity against Indian adult earthworms (Khadse and Kakde 2010). In this study, the hydroalcoholic extracts of fenugreek seeds showed a dose-dependent inhibition of spontaneous motility (paralysis) of earthworms, which is more than a conventional drug, albendazole (Khadse and Kakde 2010). The most recent study reported potent and concentration-dependent helmintholytic activity (inhibition of spontaneous motility and death) of aqueous and methanol extracts of fenugreek leaves against Indian adult earthworms as models of roundworms (Putta 2018). In another study, significant dose-dependent *in vitro* anthelmintic efficacy of methanolic extract of fenugreek seeds was reported on *Haemanthus sp.* (abomasal parasites, present in cattle stomach), which suggest its broad spectrum of activity.

19.4.5 Fenugreek against Vector-Borne Diseases

Vectors are living organisms such as flies, centipedes, bugs, spiders, fleas, crayfish, mosquitoes, shrimps, freshwater aquatic snails, and ticks that can transmit infectious and life-threatening diseases between humans or from animals to humans. The vector-borne diseases comprise 17% of infectious diseases and nearly 7,00,000 deaths annually and pose heavy losses to humans (Warpeha et al. 2020). The mosquito, the most common vector transmitting infectious diseases, is responsible for malaria, yellow fever, chikungunya, West Nile, dengue fever, filariasis, zika, and other arboviruses. Recently, the active ingredients in plant extracts have been explored to interfere in the mosquito's life cycle and physiological processes and prevent transmission of vector-borne diseases (Iwuagwu et al. 2020).

Mosquitos harbor the parasites such as *Plasmodium falciparum*, and *P. vivax*, and cause malaria, one of the most severe vector-borne disease of humankind (Shoemaker et al. 2002), especially in low-income and developing countries (Lewnard and Reingold 2019). At present, the treatment of malaria is facing a grave challenge because of the alarming spread of a drug-resistant strain of *Plasmodium* and a limited number of effective anti-malarial drugs (Shibeshi, Kifle, and Atnafie 2020; Tang et al. 2020), and high cost and adverse effects of some of the antimalarial agents (Uzor, Prasasty, and Agubata 2020).

The present research evidence suggests natural products such as fenugreek as a new ray of hope towards the development of safe, effective, and non-resistant antimalarial agents. In fact, the two most successful antimalarial drugs artemisinin and quinine were sourced from medicinal plants (Uzor, Prasasty, and Agubata 2020). Fenugreek has been documented as a traditional remedy against headache, fever associated with a cold, sore throat, hay fever, and sinusitis (Moradi-Kor, Didarshetaban, and Saeid Pour 2013). The fenugreek seeds contain oil and its leaves have been used for a long time in many forms such as herbal tea, dried powder, fresh leaves, or mixed with ghee in traditional remedies for numerous fever-related diseases, including malaria (Singh et al. 2008).

Culex quinquefasciatus, commonly known as the southern house mosquito, is a medium-sized mosquito found in tropical and subtropical regions. The 3rd–4th instar larvae of *Cx. quinquefasciatus* laboratory colony was reported with high susceptibility during *in vitro* assay with 100% and 98% mortality (to the water and ethanol extract of fenugreek at 30% and 20% concentration respectively) (Fallatah 2010). In this study, fenugreek extracts cellular vacuolization and disintegration and rupture of the epithelial layer of muscles, nerve ganglia, midgut, hindgut of larvae with significant protein loss to suggest high larvicidal properties against mosquitos (Fallatah 2010). In another study, ethanolic extracts of fenugreek leaves and two fractions (ethanol and butanol) demonstrated potent *in vitro* anti-plasmodial assay against laboratory-adapted chloroquine-sensitive and -resistant *P. falciparum* isolates during Schizont maturation inhibition assay (Palaniswamy et al. 2010).

19.5 FENUGREEK FOR MANAGEMENT OF MALIGNANCY DISORDERS (CANCERS)

Malignancy disorders or cancer are the second most leading cause of death worldwide behind cardiovascular disease (Li et al. 2019; Nagai and Kim 2017). There are an estimated about 19.3 million new cases of cancer and 10.0 million deaths from cancer worldwide in 2020 (Ferlay et al.). The present cancer management options such as radiotherapy, chemotherapy, and surgery have a cytotoxic nature and severe side effects (Ashraf 2020; Nessa, Rahman, and Kabir 2020). The natural products, including dietary components have been a mainstay of anti-cancer medication for many decades because of safety, non-toxic nature, and easy availability (Demain and Vaishnav 2011; Huang, Lu, and Ding 2021; Mann 2002; Slichenmyer and Von Hoff 1990). At the same time, the potential of fenugreek and constituents against multiple or specific cancer cell lines has been studied *in vitro* or in an animal model of cancer *in vivo*. Mechanisms such as oxidative stress, inflammation, inhibition of proliferation, apoptosis, and invasion are suggested as possible cellular and molecular mechanisms of anti-cancer fenugreek and constituents (El Bairi et al. 2017). This section of the chapter describes an update on the available evidence of fenugreek for applications of specific types of cancers or malignancy disorders.

19.5.1 Fenugreek against Multiple Cancerous Cell Lines *In Vitro*

Fenugreek seed extract (10–15 μg/ml) on 72-h exposure, reported concentration-dependent cytotoxic effect *in vitro* against a panel of cancer cell lines including prostate, pancreatic, and breast cancer cell lines but did not affect normal cells (Shabbeer et al. 2009). In the study, fenugreek seed extract caused the down-regulation of mutant p53 expression in prostate and breast cancer cell lines and up-regulation of p21 expression in PC-3 cells with inhibition of TGFβ-induced phosphorylation of Akt. Furthermore, the potential to differentiate between the malignant and primary cancer cells was shown by seed extract and not its components like diosgenin (Shabbeer et al. 2009). In another set of studies that utilized multiple cancer cell lines, exposure to fenugreek seed oil (1000 μg/ml for 24 h) showed a cytotoxic effect *in vitro* against cancer cell lines of epidermoid (HEp-2), breast (MCF-7), uterine (WISH) cancer and a non-cancerous, normal cell line (Vero cells, normal kidney epithelial cell) during MTT (assessing cell metabolic activity), NRU assays, and cellular morphological alterations (phase-contrast light microscopy). MTT assay assesses cell metabolic activity whereas NRU (neutral red uptake) assay indicates cell viability or cytotoxicity (Al-Oqail et al. 2013). Subsequently, fenugreek seed extract exposure (100 μg/ml, 200 μg/ml, and 300 μg/ml) for different time points (0, 24, 48, 72, and 96 h) exhibited dose-dependent cytotoxic effects (early and late apoptosis) *in vitro* to a panel of cancer cell lines, including T-cell lymphoma and T-cell lymphoma-TCP, and human thyroid papillary carcinoma-FRO, without affecting the normal cell line (Alsemari et al. 2014).

Ultraviolet exposure and sunburn have proven vital in developing melanoma, the most dangerous type of skin cancer (Lopes et al. 2021; Park et al. 2020). Methanolic extract of fenugreek seed is reported to control the intracellular synthesis of melanin in murine melanoma B16F1 cells (Kawabata et al. 2011). B16 melanoma is a murine tumor cell line and established as a valuable model for studying human skin cancers, metastasis, and solid tumor formation. In the same study, the methanolic extract of fenugreek seed inhibited the formation of phorbol-12-myristate-13-acetate-induced inflammatory cytokines such as tumor necrosis factor (TNF)-α in cultured THP-1 cells (Kawabata et al. 2011). The THP-1 cells, a human leukemia monocytic cell line, has been extensively used to study immunological pathways and mechanisms, especially monocyte/macrophage functions, and has become a good *in vitro* model to evaluate anti-leukemia agents (Chanput, Mes, and Wichers 2014). These effects were attributed to the three active steroidal saponin glycosides phytoconstituents from fenugreek seed extracts (namely pseudoprotodioscin, 26- O-β-D-glucopyranosyl-(25 R)-furost-5(6)-en-3 β,22 β,26-triol-3-O-α-L-rhamno-pyranosyl-(1"→2')-O-[β-D-glucopyranosyl-(1"→6')-O]-β-D-glucopyranoside and minutoside B) also demonstrated moderate anti-inflammatory effects (Kawabata et al. 2011),

Fenugreek powder was reported to be effective against experimental gliomas and carcinomas through a mechanism such as NF-κB-dependent signaling pathways and subsequent PA synthesis and DNA methylation (Bentrad, Zaletok, and Zelena 2015). Fenugreek powder demonstrated efficacy against the Wistar rats with intracranial grafted C6 glioma, subcutaneously grafted Guerin carcinoma, and Guerin carcinoma substrains (resistant to doxorubicin and cisplatin); C57Bl/6 mice with grafted Ca755 mammary carcinoma and grafted Lewis lung carcinoma were induced in 6 different groups of animals. Fenugreek powder (250 mg/kg of body weight) was reported to increase the lifetime of animals by 15–50%, inhibit tumor growth by 25–48%, and decrease the average volume of the metastases by 18–86%, with increased global DNA methylation in tumor cells, and reduce NF-κB (p50/p65) expression in nuclei, diminish NF-κB-dependent genes expression (c-myc, bcl-xl, and COX-2) and expression of odc (ornithine decarboxylase), decrease level of polyamine: putrescine ratio (30–77%) with spermidine (11–26%) and spermine (12–24%) (Bentrad, Zaletok, and Zelena 2015).

Some of the constituents that are isolated from fenugreek seeds demonstrated anti-cancer effects on multiple cell lines during *in vitro* studies. The protodioscin (a steroidal saponin bioactive compound of fenugreek) was reported to exhibit strong inhibitory activity against the leukemic cell line HL-60 (by activating apoptosis) and a weak inhibitory effect on the gastric cancer cell line, KAT, proliferation (Hibasami et al. 2003). Three-day treatment (2.5, 5, and 10 microM) with PD produced both time- and concentration-dependent and fragmentation of DNA into oligonucleosomal-sized fragments (characteristic of apoptosis) and gradual increase in hypodiploid nuclei in the HL-60 cells (Hibasami et al. 2003). Dioscin is a spirostanyl glycoside from fenugreek seed, is reported with multiple pharmacological effects, including anti-cancer potential (Yang et al. 2019). The anti-cancer activities of dioscin against a diverse variety of cancers such as lung cancer (Hsieh et al. 2013), colon cancer (Li et al. 2018; Raju et al. 2004), melanoma (Xiao et al. 2016), gall bladder cancer (Song et al. 2017), colon cancer (Wu et al. 2020), and other cancers have been reviewed in the past (Xu et al. 2016; Yang et al. 2019; Yum, You, and Ji 2010). The role of multiple pathways are suggested to be involved for apoptotic effects (through intrinsic mitochondrial apoptosis activation of caspase-9 and caspase-3 and antiapoptotic proteins, PI3K/Akt/mTOR and p38 MAPK and JNK signaling pathways); activation of ERK1/2 and AIF pathway; the increase in the levels of NO and inducible NO synthase; and inhibition of migration, invasion, and angiogenesis; autophagy; and a few others (Yang et al. 2019; Zhu et al.). Furthermore, dioscin could overcome multidrug resistance and enhance the antitumor activity of other drugs (Wang et al. 2016; Wang et al. 2013; Wang et al. 2014).

Another constituent of fenugreek seed, diosgenin, a phytosteroidal saponin, is reported to have anti-cancer efficacy with the large body of evidence during *in vitro* and animal models of cancer, as a potent inhibitor of cancer cell invasion, migration, and tumor-associated angiogenesis, perhaps through amelioration of chronic inflammatory pathology and apoptosis (Sethi et al. 2018). This evidence of anti-cancer activities of diosgenin with probable mechanisms of action is summarized in Table 19.2.

19.5.2 Fenugreek against Ehrlich Ascites Carcinoma (EAC)

Ehrlich carcinoma, the cultured and undifferentiated tumor, can be converted to ascites to form EAC (Ozaslan et al. 2011). EAC resembles human tumor, has a rapid growth rate, is most sensitive to chemotherapy and common chemotherapeutic agents (Ozaslan et al. 2011). Alcoholic extract of fenugreek seeds to EAC inoculated mice showed 70% inhibition of tumor cell growth, with cytostatic effects in terms of ribosome depletion and detachment, cytoplasmic and nuclear edema, inhibition of protein and enzymes synthesis, expansion of rough endoplasmic reticulum and mitochondria, hypertrophy of nucleoli and diminished smooth endoplasmic reticulum (Sur et al. 2001). In another study, ten days of oral treatment of ethanolic extract of fenugreek leaves (100, 200, and 400 mg/kg) reported a dose-dependent increase in survival rate and life span with restored hematological parameters in EAC-inoculated mice (Prabhu and Krishnamoorthy 2010).

19.5.3 Fenugreek against Skin Carcinogenesis

Skin carcinogenesis or cancer is the most common type of cancer, and its incidence has gradually increased in recent years (Park et al. 2020). The ultraviolet radiation-induced immunosuppression (Hart and Norval 2018) and oxidative stress (Xian et al. 2019) are major risk factors for skin cancer.

TABLE 19.2
Anti-cancer Efficacy of Diosgenin (a Fenugreek Constituent)

Cell Line	Results and Proposed Mechanisms	References
Human osteosarcoma 1547	↓ Cell growth ↑ Apoptosis G1 phase cell cycle arrest ↑ p53 and p21, mRNA expression ↓ mRNA expression of Bax and bcl-2 ↑ NF-κB, PGE2, hsp70	Moalic et al. (2001)
Human 1547 osteosarcoma	↓ Cell proliferation ↑ Apoptosis cell cycle distribution ↑ NF-κB binding to DNA ↑ p53 protein expression Cell cycle arrest	Corbiere et al. (2003)
Azoxymethane-induced colon carcinogenesis, HT-29 human colon cancer cell	↓ Number of multicrypt foci and aberrant crypt foci ↓ cell growth ↑ Apoptosis, caspase-3, NF-κB, p53, p21 ↓ bcl-2	Raju et al. (2004)
Human erythroleukemia (HEL)	↑ Apoptosis Cell cycle arrest ↑ Intracellular calcium ↑ cPLA2 activation ↑ COX-2	Leger et al. (2004)
Polyvalent human erythroleukemia HEp-2 human laryngocarcinoma M4Beu human melanoma	↑ Cell apoptosis and differentiation ↓ Proliferation of cell lines ↑ Apoptosis, Caspase-3 ↓ Mitochondrial membrane potential ↑ p53 activation Arrest: HEp-2 cells in S phase, M4Beu cells in G2/M phase Nuclear localization of apoptosis-inducing factor (AIF) and poly (ADP-ribose) polymerase cleavage	Leger et al. (2004) Corbiere et al. (2004)
Human chronic myelogenous leukemia HL-60, K562 cells	Multinucleation and apoptosis in K562 Cells Mitotic arrest	Liu et al. (2004)
HeLa cells	↓ HeLa cell growth ↑ Apoptosis ↓ Mitochondrial membrane potential ↓ Bcl-2 expression ↑ Caspase pathway	Hou et al. (2004)
Human cancer cells A431, K562, A2780, A549, HCT-15	↑ Apoptosis on HCT-15 cells ↑ Mitochondria-controlled apoptotic pathway ↓ Mitochondrial potential (deltapsim) ↓ Mitochondrial cytochrome-c release in cytosol ↓ Bcl-2/Bax expression level ratio	Wang et al. (2004)
Human chronic myelogenous leukemia K562 cell	↑ Apoptosis Arrest of cell cycle in G2/M phase ↑ Mitochondrial dysfunction ↓ Cyclin B1 and p21Cip1/Waf1 ↑ cdc2, Caspase-3, Bax ↓ Intracellular Ca2+ concentration Hyperpolarization and depolarization of MMP ↓ Generation of ROS ↓ Antiapoptotic proteins Bcl-2, Bcl-xL	Liu et al. (2005)

(Continued)

TABLE 19.2 (Continued)

Cell Line	Results and Proposed Mechanisms	References
Erythroleukemia K562 and HEL cells	↑ Apoptosis in both cell lines ↓ DNA binding of NF-κB ↑ COX-2 in HEL cells ↓ ERK ↑ p38, MAPK ↑ c-jun NH2-terminal kinases (JNKs) in HEL cells	Liagre et al. (2005)
Human myeloid KBM-5 cells Mouse macrophage raw 264.7 cells Human embryonic kidney A293 cells	↓ Tumor cell proliferation ↓ Invasion ↓ Steoclastogenesis ↑ Apoptosis ↓ Receptor-activated NF-κB-induced osteoclastogenesis ↓ TNFα ↑ IκBα kinase, IκBα phosphorylation IκBα degradation, p65 phosphorylation, ↓ Akt activation ↓ TNFα-induced expression of IκBα-regulated gene products (cyclin D1, COX-2, c-myc), antiapoptotic factors, MMP-9	Shishodia and Aggarwal (2006)
HEL cells	↑ Megakaryocytic HEL cell differentiation ↑ Apoptosis ↑ ERK ↓ p38 MAPK pathway ↑ ERK by MEK inhibitor ↓ NF-κB, Akt and Bcl-xL ↑ Caspase-3 with PARP cleavage	Leger, Liagre, and Beneytout (2006)
HER2-overexpressing breast cancer cells	↓ Cell proliferation ↑ Apoptosis ↑ Paclitaxel-induced cytotoxicity ↓ Fatty acid synthase (FAS) ↓ Akt and mTOR phosphorylation ↑ JNK phosphorylation	Chiang et al. (2007)
B16 melanoma cells	↓ Melanin content ↓ Melanogenesis ↑ PI3K pathway ↓ Akt and GSK 3 ↑ Tyrosinase, MITF (microphthalmia-associated transcription factor) ↑ PI3K pathway	Lee et al. (2007)
HCT-116 human colon carcinoma cells	↓ Cell growth, viability ↑ Apoptosis, cytotoxicity ↓ p21, ras, beta-catenin	Raju and Bird (2007)
Estrogen receptor positive (ER1) and estrogen receptor negative (ER2) breast cancer (BCa) cells MCF-7 and MDA-231 xenografts in nude mice (*in vivo*)	↓ Cell proliferation ↑ Apoptosis ↓ Tumor growth ↓ pAkt, Cyclin D1, cdk-2, cdk-4 ↓ Akt kinase activity ↓ NF-κB, Bcl-2, survivin, XIAP ↓ Raf/MEK/ERK pathway by ER+ cells G1 cell cycle arrest	Srinivasan et al. (2009)

Cell Line	Results and Proposed Mechanisms	References
Human hepatocellular carcinoma cell (HCC) C3A and HepG2 cells	↓ Cell proliferation ↑ Apoptotic effects of doxorubicin and paclitaxel in C3A cells ↓ Phosphorylation of STAT3 in C3A cells ↓ Nuclear pool of STAT3 in HCC cells ↓ IL-6-inducible Akt phosphorylation and STAT3 inducible phosphorylation in HCC cells ↓ IL-6-induced STAT3-dependent reporter gene expression ↓ Constitutive activation of c-Src, JAK1 and JAK2 ↓ STAT3 activation through Tyrosine phosphatases ↑ SH-PTP2 expression in HCC cells ↓ Cyclin D1, Bcl-2, Bcl-xL, survivin, VEGF expression ↑ Accumulation of cells in G1 phase of cell cycle ↑ Caspase-3 activation and PARP cleavage	Li et al. (2010)
Colorectal cancer cell lines HCT-116 and HT-29	↑ Apoptosis ↑ TRAIL-induced apoptosis ↑ Death receptor-5 ↑ p38 MAPK pathway ↑ COX-2 expression	Lepage et al. (2011)
Human prostate cancer PC-3 cells	↓ Cell migration and invasion ↓ MMP-2 and MMP-9 ↑ EMMPRIN ↑ TIMP-2 ↓ VEGF ↓ Tube formation of endothelial cells ↓ PI3K, Akt, ERK, and JNK ↓ NF-kB	Chen et al. (2011)
Squamous cell carcinoma—A431 and Hep2 cell lines Sarcoma 180-induced tumors (*in vivo*)	↑ Cytotoxicity and apoptosis ↓ Cell proliferation ↓ Tumor mass, volume ↑ Apoptosis ↑ Sub-G1 population ↑ Chromatin condensation ↑ DNA laddering and TUNEL-positive cells ↑ Bax/Bcl-2 ratio ↑ Caspases ↑ Poly ADP ribose polymerase cleavage ↓ Akt and JNK phosphorylation	Das et al. (2012)
N-Methyl-N-nitrosourea (NMU)-induced mammary carcinogenesis (*in vivo*)	↓ Peroxidation reaction ↓ Tumor marker enzymes ↑ Intrinsic antioxidant defense system ↓ LPO, LDH ↑ SOD, CAT, GPX	Jagadeesan et al. (2012)
Hypoxia-mimic (cobalt chloride) sensitive gastric cancer cell line BGC-823 with HIF-1α shRNAs	↓ Ability of invasion and survival of cells ↓ HIF-1α expression ↑ Integrinα5, E-cadherin, integrinβ6 expression	Mao et al. (2012)

(*Continued*)

TABLE 19.2 (Continued)

Cell Line	Results and Proposed Mechanisms	References
Sarcoma-180 tumor cells	↓ Tumor growth ↑ Thymus and spleen weights ↑ Cellular immune responses ↑ NO and TNFα secretion ↑ Lymphocyte transformation ↑ Phagocytic capability of macrophages ↑ Secretion of TNFα in macrophages	He et al. (2012)
HepG2 human hepatocellular carcinoma cells	↑ Dose-dependent apoptosis Cleavage of PARP) ↑ Caspase-3, -8, -9, Cytochrome c release, Bax, Bid, Bcl-2, Bax/Bcl-2 ratio, phosphorylation of JNK, p38 MAPK, ASK-1 ↓ ROS	Kim et al. (2012)
A549 lung cancer cell lines	↓ hTERT	Mohammad et al. (2013)
Human THP-1 monocytic cells exposed to tumor necrosis factor-α (TNF-α)	↓ TNFα-induced TF activity and expression in monocytes, procoagulant activity ↓ Protein expression and mRNA accumulation of tissue factor (TF) ↓ TNF-α induced phosphorylation of NF-κB/p65, IKK-β, Akt, ERK, and JNK	Yang et al. (2013)
Human breast cancer MDA-MB-231 cells	↓ Cell migration and metastasis ↓ Actin polymerization, Vav2 phosphorylation and Cdc42 activation	He et al. (2014)
A549 lung cancer cell line	↑ Cytotoxicity effects ↓ hTERT gene expression	Rahmati-Yamchi et al. (2014)
Bel-7402, SMMC-7721, and HepG2 hepatocellular carcinoma cell lines	↑ Apoptosis G2/M cell cycle arrest ↓ Akt phosphorylation ↑ p21 and p27 ↑ Caspase cascades -3, -8 and -9	Li et al. (2015)
RAW 264.7 macrophage culture	↓ Cell proliferation ↑ Apoptosis ↓ LPS-induced TNF-α ↑ Antioxidant activity	Selim and Al Jaouni (2016)
Chronic myeloid leukemia (CML) cells	↑ Autophagy and cytotoxicity ↓ Oxidative stress ↑ Accumulation of autophagic flux, autophagosomes autophagosome-lysosome fusion ↑ Degradation of autophagosomes ↓ mTOR signaling pathway	Jiang et al. (2016)
DU145 human prostate cancer cells	↓ Cell proliferation ↑ Apoptosis and autophagy ↓ PI3K/Akt/mTOR	Nie et al. (2016)
Oral squamous cell cancer (OSCC) in PE/CA-PJ15 cells	↓ Cell viability ↑ Apoptosis ↓ Cell migration	Pons-Fuster López et al. (2017)
Cancer stem cells (CSCs) from breast cancer cell lines—MCF7, T47D, and MDA-MB-231	↑ Apoptosis ↓ CSC-associated phenotypes ↓ Wnt β-Catenin Signaling via the Wnt Antagonist Secreted Frizzled Related Protein-4	Bhuvanalakshmi et al. (2017)

Cell Line	Results and Proposed Mechanisms	References
Human ovarian cancer SKOV3 cells	↓ Cell viability ↑ Apoptosis ↓ VEGF-2 ↓ Phosphoinositide 3-kinase (PI3K) ↓ Phosphorylated AKT expression ↓ Expression of phosphorylated p38 MAPK) ↑ Caspase-3 and caspase-9 activity ↑ Protein expression of Bax and cleaved poly (ADP-ribose) polymerase	Guo and Ding (2018)
Hepatocellular carcinoma cells (HCC) HepG2 and SMMC-7721 cells	↓ Cell proliferation ↑ Apoptotic cell death G2/M phase arrest ↓ Cell migration and invasion abilities ↑ DDX3 expression	Yu et al. (2018)
Melanoma-bearing C57BL/6 mice	Anti-melanoma effect ↑ Antitumor immunity ↑ Intestinal microbiota ↑ Immune checkpoint antibody PD-1 ↑ CD4+/CD8+, T-cell infiltration, IFN-γ expression in tumor tissues	Dong et al. (2018)
Liver cancer	↓ Cell growth ↑ Apoptosis, suppressed cell migration and invasion ↓ TAZ	Chen et al. (2018)
Human hepatocellular carcinoma cell lines—HepG2 and Bel-7402	↑ Autophagy and apoptosis ↑ Cell entry through OATP1B1 transporter ↑ ER swelling, Mitochondrial damage, autophagosome ↑ IRE-1α, Caspase-8	Meng et al. (2019)
Human prostate cancer PC-3 cells	↓ Cell growth ↑ Apoptosis Cell cycle arrest ↓ NEDD4	Zhang et al. (2019)
Human breast cancer cell lines MCF-7 and Hs578T	↓ Cell growth Cell cycle arrest ↑ Apoptosis in both cell lines ↑ Ratio of phosphorylated cyclin checkpoint1 (p-Chk1Ser345) to cyclin B expression G2/Mphase blockade ↑ Cdc25C-Cdc2 signaling ↓ Mitochondrial membrane potential ↓ Anti-apoptotic protein, Bcl-2, Cytochrome-c release ↑ Caspase signaling cascade	Liao et al. (2019)
AGS and SGC-7901 gastric cancer cells	↓ Cell viability ↑ Apoptosis ↓ Cell proliferation G0/G1 phase cell cycle arrest ↓ Rho/ROCK signaling ↓ EMT	Liu et al. (2020)

Akt: Protein kinase B; ASK: apoptosis signal-regulating kinase; Bax: BCL2 Associated X, Apoptosis Regulator; bcl-2: B-cell lymphoma 2; Bcl-xL: B-cell lymphoma-extra large; CAT: Catalase; cdk: Cyclin-dependent kinases; COX: Cyclo-

(Continued)

TABLE 19.2 (Continued)

oxygenase; cPLA2: Phospholipase A; CSC: Cancer stem cells; c-Src: Proto-oncogene tyrosine-protein kinase, Src; DDX3: DEAD-box RNA helicase; EMMPRIN: Extracellular matrix metalloproteinase inducer; EMT: epithelial-mesenchymal transition; ERK: extracellular signal-regulated kinase; GPX: Glutathione peroxidase; GSK: Glycogen synthase kinase; HIF: Hypoxia-Inducible Factor; hsp70: heat shock protein70; hTERT: human telomerase reverse transcriptase gene; IκBα: nuclear factor of kappa light polypeptide gene enhancer in B-cells inhibitor, alpha; IRE-1α: Inositol-requiring transmembrane kinase/endoribonuclease 1α; JAK: Janus Kinase; JNK: c-Jun N-terminal kinase; LDH: Lactate dehydrogenase; LPO: Lipoxygenases; LPS: Lipopolysaccharides; MAPK: A mitogen-activated protein kinase; MAPK: mitogen-activated protein kinase; MEK: meiotic chromosome-axis-associated kinase; MMP: Metalloproteinase; mTOR: mammalian target of rapamycin; NEDD: neural precursor cell expressed developmentally down-regulated protein; NF-κB:Nuclear Factor kappa-light-chain-enhancer of activated B cells; NO: Nitric oxide; OATP1B1:organic anion transporting polypeptide 1B1; PARP: poly-ADP-ribose polymerase; PD-1: programmed cell death protein-1; PGE: prostaglandins E; PI3K: phosphatidylinositide-3 kinase; ROCK: Rho-associated protein kinase; SH-PTP: Protein tyrosine phosphatase; SOD: Superoxide dismutase; STAT: Signal transducer and activator of transcription; TAZ: transcriptional co-activator with PDZ binding motif; TIMP: tissue inhibitor of metalloproteinase; TNF: Tumor necrosis factor; TRAIL: Tumor necrosis factor-related apoptosis-inducing ligand; VEGF: vascular endothelial growth factor; Waf1:cyclin-dependent kinase inhibitor 1; XIAP:X-linked inhibitor of apoptosis protein.

Tumor formation, induced by the topical application of a combination of two different chemicals, 7,12-dimethylbenz[a]anthracene (DMBA) and 12-O-tetradecanoyl phorbol-13-acetate (TPA), to cause papilloma formation in the skin is the most common model of two-stage skin carcinogenesis (Au—Vähätupa et al. 2019). The methanol extract of fenugreek seeds demonstrated anti-cancer efficacy against two-stage skin carcinogenesis induced by DMBA and TPA in mice (Chatterjee, Kumar, and Kumar 2012). Seven days of oral supplementation of methanol extract of fenugreek seeds (400 mg/kg) demonstrated significant reduction of papillomas, the rate of tumor incidence, tumor burden and tumor yield, a cumulative number of papillomas in all stages of tumorigenesis during DMBA+TPA induced two-stage tumor-induced mice (Chatterjee, Kumar, and Kumar 2012).

19.5.4 Fenugreek against Leukemia (Blood Cancer)

Chronic lymphocytic leukemia (CLL) is a disease with proliferative diseases with heterogenous outcome and multiple organ involvement (Jain et al. 2015; Moreno et al. 2010; Strati et al. 2016). Despite new treatment options, CLL is considered difficult to manage due to the lack of multifaceted therapeutic agents that restore apoptosis, inhibit proliferation, increase chemosensitivity, low toxicity, and lack of resistance (Mattsson et al. 2020; Sharma and Rai 2019).

The cytotoxic efficacy (apoptosis induction) of 48-h exposure of an aqueous extract of fenugreek seeds derived from five different genotypes of Canada origin against chronic lymphocytic leukemia cell lines (CLL) were reported during flow cytometry analysis (Acharya et al. 2011). This study was further supported by *in vivo* evidence of the efficacy of fenugreek seed extract against lymphocytic leukemia (L1210)-induced ascites in mice, an animal model for CLL (Zailei et al. 2016). In this study, 21 days of oral treatment of 20% (w/v) of boiling water infusion of the fenugreek seeds showed anti-leukemic effects, that is, prevention of CLL-induced changes, and resulted in a decrease in weight, decreased amount of ascetic fluid and less tendency to a solid tumor, reduced tumor invasion, improved the morphological changes, reduced tissue damage, and rebuilt the surrounding visceral organs (Zailei et al. 2016).

19.5.5 Fenugreek against Hepatocellular Carcinoma (Liver Cancer)

Worldwide, hepatocellular carcinoma (HCC) is the sixth most common cancer and the third most common cause of cancer deaths (Bertuccio et al. 2017; El-Serag 2020). Recent reports of the efficacy of several natural plant extracts or their bioactive constituents in inhibition of liver cancer development and progression, protection against liver carcinogens, enhancement of effects of chemotherapeutic drugs, and inhibition of tumor cell growth, metastasis, and suppression of oxidative stress and chronic inflammation (Abdel-Hamid et al. 2018) suggested use of natural products as a co-treatment option with synthetic chemotherapeutic agents.

The exposure of the methanolic extract of fenugreek seeds (100 to 500 ??g/mL) to HepG2 cell line is reported to result in dose-dependent caspase-mediated apoptosis as evidenced by MTT assay, cell morphology alterations, enzyme-linked immunosorbent assay (ELISA), flow cytometric analysis, caspase-3 activity, and expression of p53, a proapoptotic protein, Bax, and proliferating cell nuclear antigen (PCNA) (Khalil et al. 2015). Inactivation of the p53 tumor suppressor is a frequent event in tumorigenesis, and apoptosis, including hepatocellular carcinoma. The p53 gene is documented to play a crucial role as a transcriptional activator in inducing the transcription of several genes, including apoptosis-related genes (Hussain et al. 2007; Rivlin et al. 2011). The Bax gene is a proapoptotic gene that acts as a transcriptional target of p53 (Benchimol 2001; Moxley and Reisman 2021). PCNA is responsible for several cellular mechanisms, including apoptosis, cell cycle regulation, DNA synthesis and repair, and interaction with various proteins in a p53-dependent manner (Chen 2016). Another study reported the dose-dependent cytotoxicity efficacy of ethanolic extract of fenugreek in a reduction in cell viability, increasing the cell mortality of HepG2 cell lines during *in vitro* MTT assay, perhaps through antioxidant properties (Al-Dabbagh et al. 2018).

Trigonelline, the major alkaloid in fenugreek seeds, has been reported to significantly suppress the ROS-induced increase of invasive capacity of hepatoma AH109A cells without altering the cell proliferation (El Bairi et al. 2017). Furthermore, trigonelline is reported to inhibit the Hep3 B cell migration through down-regulation of nuclear factor E2-related factor 2—dependent antioxidant enzymes activity with Raf/ERK/Nrf2 signaling pathway role MMP-7 gene expression (Liao et al. 2015). At the same time, the primary amino acid of fenugreek seeds, 4-hydroxyisoleucine (0.5–64 mM), has been reported to demonstrate concentration- and time-dependent cytotoxic effects against HepG2 and Huh-7 hepatoma cell lines (Babaei et al. 2015).

19.5.6 Fenugreek against T-Cell Lymphoma

T-cell lymphoid malignancies are a heterogeneous cohort of highly aggressive cancers with variations in clinical characteristics (such as rash-like skin redness, scaly round patches on the skin), generally resistant to conventional treatment modalities, and sometimes lead to skin cancers (Tang et al. 2010). The ethanolic extract of fenugreek seeds (30 to 1500 µg/ml) is reported to have a time- and dose-dependent autophagy response of T-cell lymphoma Jurkat cells through enhanced expression of autophagic marker LC3 transcripts (confirmed from trypan blue assay), membrane disintegration, induction of vacuolization, and other morphological observation during microscopical examination (Al-Daghri et al. 2012).

19.5.7 Fenugreek against Breast Cancer

Breast cancer is documented as one of the most prevalent cancers with rising disease burden worldwide in females at premenopausal and postmenopausal ages (Heer et al. 2020). Breast cancer recently became the most prevalent cancer and replaced lung cancer worldwide (Cao et al. 2021). Lifestyle and nutrition are considered significant factors worldwide governing breast cancer-induced mortality (Clemons and Goss 2001).

Several reports on breast cancer have exhibited that the risk of occurrence of breast cancer increases with increased levels of estrogens in the human body (Albini et al. 2014; Howland et al. 2020; Samavat and Kurzer 2015; Tin, Reeves, and Key 2021). On the other hand, natural plant-based estrogen-like compounds, such as phytoestrogens, can restore endogenous balanced estrogens and have been explored as a possible management option to prevent and treat breast cancer (Basu and Maier 2018; Branca and Lorenzetti 2005; Chen and Chien 2019; Hsieh et al. 2018; Tanwar et al. 2020).

Fenugreek itself or any of its constituents are reported in scientific literature as phytoestrogens and have demonstrated their efficacy to restore female hormonal balance in many female-related conditions, including offering benefits as a galactagogue (Ahmed 2015; Brillante and Mantaring 2014; Dog 2009; Khan, Wu, and Dolzhenko 2018; Sim et al. 2013; Turkyılmaz et al. 2011) and lactation aid (Forinash et al. 2012; Turkyılmaz et al. 2011) and managing conditions such as polycystic ovarian syndrome (PCOS) (Bashtian et al. 2013; Brogi et al. 2019; Sies, Berndt, and Jones 2017) and post-menopausal symptoms (PMS) (Shamshad Begum et al. 2016; Sreeja, Anju, and Sreeja 2010; Thomas et al. 2020). Furthermore, the soluble fiber constituent of fenugreek seeds, galactomannan, was suggested as a potential drug candidate against breast cancer because of optimum mannose/galactose ratios, excellent higher molecular dock scores, stable molecular dynamics (MD) simulations with lower binding energy calculations during the computational study (Wu et al. 2012).

The evidence of efficacy on fenugreek seed extracts from many *in vitro* studies against breast cancer cell lines indicated apoptosis as an effective mechanism of anti-cancer action (Alrumaihi et al. 2021; Khoja et al. 2011; Sebastian and Thampan 2007). The chemical DMBA is a potent immunosuppressor that metabolites into the active form to damage the DNA and induces physiological changes that closely mimic human breast cancer (Neagu et al. 2016; Vinothini, Murugan, and Nagini 2009). The aqueous extract of fenugreek seed (200 mg/kg, once a day for 7 days) has been reported effective *in vivo* against DMBA induced breast cancer in rats with inhibition of mammary hyperplasia and incidence probably through involvement of apoptosis (Amin et al. 2005). Ethanolic extract of fenugreek seeds showed growth-preventive effects against MCF-7 cells, an estrogen receptor-positive breast cancer cell line, with a significant reduction in cell viability by inducing apoptotic alterations at initial stages such as reduction in mitochondrial membrane potential, disruption of cellular DNA into several fragments, and flipping of phosphatidylserine with the presence of a subG1 population and cell arrest at G2/M phase (Sebastian and Thampan 2007). These results are supported by another study of chloroform extract of fenugreek seeds against MCF-7 breast cancer cell lines (Khoja et al. 2011). In this study, chloroform extract of fenugreek seeds showed time- and dose-dependent enhancement of apoptotic gene expressions of p53, Bak, Fas, FADD, Caspases 3, 8, 9, and Bax as evidenced by RT-PCR and terminal deoxynucleotidyl transferase-mediated dUTP nick end labeling (TUNEL) assay (Khoja et al. 2011). The subsequent study reported the efficacy of methanolic extract of fenugreek plant (65 µg/mL for 24-h and 48-h exposure) against MCF-7 human breast cancer cells mediated through death receptor pathways of apoptosis, Fas receptors, without the involvement of p53, Caspase 3, 8, 9, or Bax gene expression (Alshatwi et al. 2013). Another study reported insignificant necrosis or apoptosis of fenugreek seed extract (400 µg/ml and 72-h exposure) against MCF-7 cell lines as evidenced through colorimetric MTT assay (Al-Timimi 2019). However, a very recent study on methanolic fenugreek seed extract reported the cytotoxic efficacy against MCF-7 and SK-BR3 breast cancer cells with IC50 of 150 and 40 µg/mL respectively (Alrumaihi et al. 2021). The dose-dependent inhibition of the migration and adhesion, a significant shift from G2/M, and polyploidy (>G) at higher concentrations, the activation of p53-mediated mitotic catastrophe, increased Bax/Bcl-2 ratio and apoptosis was suggested as the mechanism of action (Alrumaihi et al. 2021).

19.5.8 Fenugreek against Colon Cancer

Colorectal cancer is a common preventable cancer with increased incidence and mortality, especially over the last 25 years due to many factors, including changing lifestyles and unbalanced nutrition (Ahmed 2020). The role of oxidative stress and ROS in pathogenesis (Perše 2013), progression (Inokuma et al. 2009), and risk assessment (Mandal 2017) has been confirmed in the past. A hydrazine compound, 1,2-dimethylhydrazine (DMH), is known to produce reactive oxygen species (ROS) in circulation to initiate colon carcinogenesis in animals by alkylating the DNA (Venkatachalam et al. 2020). Fenugreek seed powder (2 g/kg, for 30 weeks) included in a regular diet to DMH-induced carcinogenic colon rats reported the increased circulatory levels of vitamins (ascorbic acid, vitamin E) and enzymes (reduced glutathione, glutathione peroxidase, glutathione S-transferase, superoxide dismutase, and catalase) (Devasena and Menon 2002). Furthermore, two fenugreek protein hydrolysates were reported to significantly reduce the progression of colorectal cancer in colonic adenocarcinoma Caco2/TC7 cells *in vitro* (Allaoui et al. 2019). The researcher attributed this efficacy to cytochrome C release to the cytoplasm, mitochondrial membrane permeabilization and caspase-3 activation, subsequent inhibition of undifferentiated cell proliferation by early apoptosis, and cell cycle arrest in phase G1 (Allaoui et al. 2019).

19.6 CONCLUSION

In the last few years, fenugreek and its constituents demonstrated beneficial effects on growth and immunity enhancement, as evident from the traditional and modern scientific literature. Furthermore, significant body evidence from several preclinical, clinical, and mechanism-based scientific studies indicated the potential of fenugreek seed extracts and constituents, such as diosgenin, trigonelline, dioscin, galactomannan, protodioscin, 4-hydroxyisoleucine, for prevention and treatment of malnutrition, immunodeficiency, infections, and cancer through the diverse mechanism (Figure 19.1). In conclusion, fenugreek and constituents, either alone or as an adjuvant, can be explored to manage immunological, infectious, or malignant disorders after suitable clinical development studies.

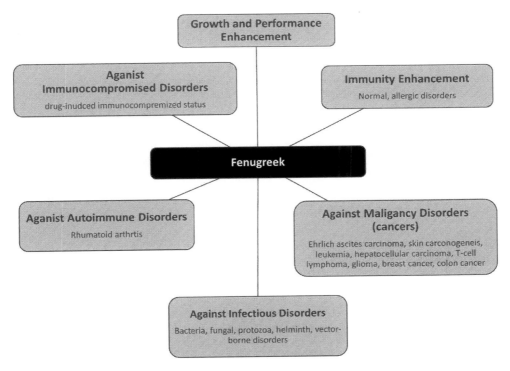

FIGURE 19.1 Fenugreek in management of immunological, infectious, and malignancy-related disorders.

REFERENCES

Abd-Alrahman, S., and M. Salem-Bekhet. 2013. "Phytochemical screening and antimicrobial activity of EthOH/water *Trigonella foenum-graecum* (fenugreek) extracts." *J Pure Appl Microbiol* 7:417–22. doi: 10.4103/0250-474X.108420.

Abdel, M. E. S., H. E. Ahmed, and A. Awad. 2008. "The chemical composition of fenugreek (*Trigonella foenum graceum* L) and the antimicrobial properties of its seed oil." *Gezira J of Eng and Applied Sci* 3:52–71.

Abdel-Hamid, N. M., S. A. Abass, A. A. Mohamed, and D. Muneam Hamid. 2018. "Herbal management of hepatocellular carcinoma through cutting the pathways of the common risk factors." *Biomed Pharmacother* 107:1246–58. doi: 10.1016/j.biopha.2018.08.104.

Abdelmageed, M. E., H. O. Abdalla, O. H. Arabi, M. Babiker, and E. S. Abdel Moneim. 2019. "Evaluation of in vitro utilization of some medicinal plants seeds as source antimicrobials." *EC Microbiol* 15 (1):2–8.

Acharya, S. N., K. Acharya, S. Paul, and S. K. Basu. 2011. "Antioxidant and antileukemic properties of selected fenugreek (*Trigonella foenum-graecum* L.) genotypes grown in western Canada." *Can J. Plant Sci* 91 (1):99–105. doi: 10.4141/cjps10025.

Ahmed, M. E. M. 2015. "The effect of fenugreek seeds powder on prolactin level in lactating Sudanese mothers." Thesis submitted for the degree of Master of Science, College of Science, Sudan University of Science and Technology.

Ahmed, M. E. M. 2020. "Colon cancer: A clinician's perspective in 2019." *Gastroenterology Res* 13 (1):1–10. doi: 10.14740/gr1239.

Al-Daghri, N. M., M. S. Alokail, K. M. Alkharfy, A. K. Mohammed, S. H. Abd-Alrahman, S. M. Yakout, O. E. Amer, and S. Krishnaswamy. 2012. "Fenugreek extract as an inducer of cellular death via autophagy in human T lymphoma Jurkat cells." *BMC Complement Altern Med* 12:202. doi: 10.1186/1472-6882-12-202.

Al-Dabbagh, B., I. A. Elhaty, A. Al Hrout, R. Al Sakkaf, R. El-Awady, S. S. Ashraf, and A. Amin. 2018. "Antioxidant and anticancer activities of *Trigonella foenum-graecum, Cassia acutifolia* and *Rhazya stricta*." *BMC Complement Altern Med* 18 (1):240. doi: 10.1186/s12906-018-2285-7.

Al-Oqail, M. M., N. N. Farshori, E. S. Al-Sheddi, J. Musarrat, A. A. Al-Khedhairy, and M. A. Siddiqui. 2013. "In vitro cytotoxic activity of seed oil of fenugreek against various cancer cell lines." *Asian Pac J Cancer Prev* 14 (3):1829–32. doi: 10.7314/apjcp.2013.14.3.1829.

Al-Timimi, L. A. N. 2019. "Antibacterial and anticancer activities of fenugreek seed extract." *Asian Pac J Cancer Prev* 20 (12):3771–6. doi: 10.31557/apjcp.2019.20.12.3771.

Albini, A., C. Rosano, G. Angelini, A. Amaro, A. I. Esposito, S. Maramotti, D. M. Noonan, and U. Pfeffer. 2014. "Exogenous hormonal regulation in breast cancer cells by phytoestrogens and endocrine disruptors." *Curr Med Chem* 21 (9):1129–45. doi: 10.2174/09298673113206660291.

Alhan, A., I. M. Jassim, and G. M. Jasim. 2017. "Study of antibacterial activities of seeds extract of fenugreek (*Trigonella foenum-graecum*)." *Diyala J Med* 13 (1):62–67.

Allaoui, A., S. Gascón, S. Benomar, J. Quero, J. Osada, M. Nasri, M. J. Rodríguez-Yoldi, and A. Boualga. 2019. "Protein hydrolysates from fenugreek (*Trigonella foenum graecum*) as nutraceutical molecules in colon cancer treatment." *Nutrients* 11 (4). doi: 10.3390/nu11040724.

Alluri, N., and M. Majumdar. 2014. "Phytochemical analysis and in vitro antimicrobial activity of calotropis gigantea, lawsonia inermis and trigonella foecum-graecum." *Int J Pharm. Pharm Sci* 6 (4):524–7.

Almuhareb, A. M., T. M. Alhawassi, A. A. Alghamdi, M. A. Omair, H. Alarfaj, A. Alarfaj, B. A. Alomari, M. S. Alblowi, and H. M. Almalag. 2019. "Prevalence of complementary and alternative medicine use among rheumatoid arthritis patients in Saudi Arabia." *Saudi Pharm J* 27 (7):939–44. doi: 10.1016/j.jsps.2019.07.002.

Alrumaihi, F. A., M. A. Khan, K. S. Allemailem, M. A. Alsahli, A. Almatroudi, H. Younus, S. A. Alsuhaibani, M. Algahtani, and A. Khan. 2021. "Methanolic Fenugreek seed extract induces p53-dependent mitotic catastrophe in breast cancer cells, leading to apoptosis." *J Inflamm Res* 14:1511–35. doi: 10.2147/JIR.S300025.

Alsemari, A., F. Alkhodairy, A. Aldakan, M. Al-Mohanna, E. Bahoush, Z. Shinwari, and A. Alaiya. 2014. "The selective cytotoxic anti-cancer properties and proteomic analysis of *Trigonella Foenum-Graecum*." *BMC Complement Altern Med* 14:114. doi: 10.1186/1472-6882-14-114.

Alshatwi, A. A., G. Shafi, T. N. Hasan, N. A. Syed, and K. K. Khoja. 2013. "Fenugreek induced apoptosis in breast cancer MCF-7 cells mediated independently by fas receptor change." *Asian Pac J Cancer Prev* 14 (10):5783–8. doi: 10.7314/apjcp.2013.14.10.5783.

Alwahibi, M., and D. Soliman. 2014. "Evaluating the antibacterial activity of fenugreek (*Trigonella foenum-graecum*) seed extract against a selection of different pathogenic bacteria." *J Pure Appl Microbiol* 8 (2):817–21.

Alwahibi, M., D. Soliman, M. Awad, H. Rizwana, and N. Marraiki. 2018. "Biosynthesis of silver nanoparticles using fenugreek seed extract and evaluation of their antifungal and antibacterial activities." *J Comput Theor Nanosci* 15:1255–60. doi: 10.1166/jctn.2018.7301.

Amaral-Machado, L., W. N. Oliveira, S. S. Moreira-Oliveira, D. T. Pereira, É. N. Alencar, N. Tsapis, and E. S. T. Egito. 2020. "Use of Natural Products in Asthma Treatment." *Evid-Based Complement Alternat Med* 2020:1021258. doi: 10.1155/2020/1021258.

Amin, A., A. Alkaabi, S. Al-Falasi, and S. A. Daoud. 2005. "Chemopreventive activities of *Trigonella foenum-graecum* (fenugreek) against breast cancer." *Cell Biol Int* 29 (8):687–94. doi: 10.1016/j.cellbi.2005.04.004.

Anand, P., A. B. Kunnumakkara, C. Sundaram, K. B. Harikumar, S. T. Tharakan, O. S. Lai, B. Sung, and B. B. Aggarwal. 2008. "Cancer is a preventable disease that requires major lifestyle changes." *Pharm Res* 25 (9):2097–116. doi: 10.1007/s11095-008-9661-9.

Anarthe, S., D. Sunitha, R. Sandhya, and M. Raju. 2019. "Immunomodulatory activity for methanolic extract of *Trigonella foenum-graecum* whole plant in Wistar albino rats." *Am J Phytomed Clin Ther* 2 (9):1081–92.

Arendrup, M. 2014. "Update on antifungal resistance in Aspergillus and Candida." *Clin Microbiol Infect* 20:42–8. doi: 10.1111/1469-0691.12513.

Ashraf, M. A. 2020. "Phytochemicals as potential anticancer drugs: Time to ponder nature's bounty." *BioMed Research International* 2020:8602879. doi: 10.1155/2020/8602879.

Au—Vähätupa, M., T. Au—Pemmari, I. Au—Junttila, M. Au—Pesu, and T. A. H. Au—Järvinen. 2019. "Chemical-induced skin carcinogenesis model using dimethylbenz[a]Anthracene and 12-O-tetradecanoyl phorbol-13-acetate (DMBA-TPA)." *JoVE* (154):e60445. doi: 10.3791/60445.

Awad, E., R. Cerezuela, and M. Á. Esteban. 2015. "Effects of fenugreek (Trigonella foenum graecum) on gilthead seabream (Sparus aurata L.) immune status and growth performance." *Fish & Shellfish Immunology* 45 (2):454–64. doi: 10.1016/j.fsi.2015.04.035.

Babaei, M., M. R. Haeri, S. Z. Mousavi, and M. Khazeni. 2015. "Cytotoxic effect of 4-Hydroxyisoleucine on liver cancer cell line." *Qom Univ Med Sci J* 9 (1):1–6.

Bae, M. J., H. S. Shin, D. W. Choi, and D. H. Shon. 2012. "Antiallergic effect of Trigonella foenum-graecum L. extracts on allergic skin inflammation induced by trimellitic anhydride in BALB/c mice." *J Ethnopharmacol* 144 (3):514–22. doi: 10.1016/j.jep.2012.09.030.

Bahi, A., F. A. Guardiola, C. Messina, A. Mahdhi, R. Cerezuela, A. Santulli, A. Bakhrouf, and M. A. Esteban. 2017. "Effects of dietary administration of fenugreek seeds, alone or in combination with probiotics, on growth performance parameters, humoral immune response and gene expression of gilthead seabream (Sparus aurata L.)." *Fish Shellfish Immunol* 60:50–8. doi: 10.1016/j.fsi.2016.11.039.

Baliga, M. S., P. P. Mane, J. Timothy Nallemgera, K. R. Thilakchand, and F. Kalekhan. 2015. "Chapter 5 — dietary spices in the prevention of rheumatoid arthritis: Past, present, and Future." In *Foods and Dietary Supplements in the Prevention and Treatment of Disease in Older Adults*, edited by Watson, R.R., 41–9. San Diego: Academic Press.

Barnes, P. J. 2008. "The cytokine network in asthma and chronic obstructive pulmonary disease." *J Clin Investig* 118 (11):3546–56. doi: 10.1172/JCI36130.

Bashtian, M. H., S. A. Emami, N. Mousavifar, H. A. Esmaily, M. Mahmoudi, and A. H. M. Poor. 2013. "Evaluation of fenugreek (Trigonella Foenum-graceum L.), effects seeds extract on insulin resistance in women with polycystic ovarian syndrome." *Iran J Pharm Sci: IJPR* 12 (2):475.

Basu, P., and C. Maier. 2018. "Phytoestrogens and breast cancer: In vitro anticancer activities of isoflavones, lignans, coumestans, stilbenes and their analogs and derivatives." *Biomed Pharmacother* 107:1648–66. doi: 10.1016/j.biopha.2018.08.100.

Begum, M., M. M. Hossain, and I.-S. Kim. 2016. "Effects of fenugreek seed extract supplementation on growth performance, nutrient digestibility, diarrhoea scores, blood profiles, faecal microflora and faecal noxious gas emission in weanling piglets." *J Anim Physiol Anim Nutr* 100. doi: 10.1111/jpn.12496.

Ben-Ami, R. 2018. "Treatment of invasive candidiasis: A narrative review." *J Fungi (Basel)* 4 (3). doi: 10.3390/jof4030097.

Benchimol, S. 2001. "p53-dependent pathways of apoptosis." *Cell Death Differ* 8 (11):1049–51. doi: 10.1038/sj.cdd.4400918.

Bentrad, V., S. Zaletok, and O. Zelena. 2015. "P1–9–61Antitumor activity of fenugreek (*Trigonella foenum-graecum* L.) powder seeds in vivo." *Annals of Oncology* 26 (Suppl 7):vii123.1–vii123. doi: 10.1093/annonc/mdv472.78.

Berhe, B., F. Mardu, K. Tesfay, H. Legese, G. Adhanom, H. Haileslasie, G. Gebremichail, B. Tesfanchal, N. Shishay, and H. Negash. 2020. "More than half prevalence of protozoan parasitic infections among diarrheic outpatients in eastern Tigrai, Ethiopia, 2019; A cross-sectional study." *Infect Drug Resist* 13:27–34. doi: 10.2147/IDR.S238493.

Bertuccio, P., F. Turati, G. Carioli, T. Rodriguez, C. La Vecchia, M. Malvezzi, and E. Negri. 2017. "Global trends and predictions in hepatocellular carcinoma mortality." *J Hepatol* 67 (2):302–9. doi: 10.1016/j.jhep.2017.03.011.

Bhalke, R. D., S. J. Anarthe, K. Sasane, S. N. Satpute, S. N. Shinde, and V. S. Sangle. 2008. "In- vitro anthelmintic activity of *Trigonella foenum-graecum* leaves and seeds (Fabaceae)." *Natural Products: An Indian Journal* 4 (1):85–7.

Bhuvanalakshmi, G., Basappa, K. S. Rangappa, A. Dharmarajan, G. Sethi, A. P. Kumar, and S. Warrier. 2017. "Breast cancer stem-like cells are inhibited by diosgenin, a steroidal saponin, by the attenuation of the Wnt β-Catenin Signaling via the Wnt Antagonist Secreted Frizzled Related Protein-4." *Front Pharmacol* 8:124. doi: 10.3389/fphar.2017.00124.

Bin-Hafeez, B., R. Haque, S. Parvez, S. Pandey, I. Sayeed, and S. Raisuddin. 2003. "Immunomodulatory effects of fenugreek (*Trigonella foenum-graecum* L.) extract in mice." *Int Immunopharmacol* 3 (2):257–65. doi: 10.1016/s1567-5769(02)00292-8.

Blumenthal, K. G., J. G. Peter, J. A. Trubiano, and E. J. Phillips. 2019. "Antibiotic allergy." *Lancet* 393 (10167):183–98. doi: 10.1016/s0140-6736(18)32218-9.

Bonilla, F. A., and H. C. Oettgen. 2010. "Adaptive immunity." *J Allergy Clin Immunol* 125 (2 Suppl 2):S33–40. doi: 10.1016/j.jaci.2009.09.017.

Bosnjak, B., B. Stelzmueller, K. J. Erb, and M. M. Epstein. 2011. "Treatment of allergic asthma: Modulation of Th2 cells and their responses." *Respiratory Res* 12 (1):1–17.

Bourke, C. D., J. A. Berkley, and A. J. Prendergast. 2016. "Immune dysfunction as a cause and consequence of malnutrition." *Trends Immunol* 37 (6):386–98. doi: 10.1016/j.it.2016.04.003.

Branca, F., and S. Lorenzetti. 2005. "Health effects of phytoestrogens." *Forum Nutr*:100–11. doi: 10.1159/000083773.

Brillante, C., and J. B. Mantaring. 2014. "O-203 Triple-blind, randomised controlled trial on the use of fenugreek (*Trigonella foenum-graecum* L.) for augmentation of breastmilk volume among postpartum mothers." *Archives of Disease in Childhood* 99 (Suppl 2):A101. doi: 10.1136/archdischild-2014-307384.271.

Brogi, H., H. Elbachir, N. El Amrani, S. Amsaguine, and D. Radallah. 2019. "Fenugreek seeds estrogenic activity in ovariectomized female rats." *Curr Issues Pharm Med Sci* 32 (3):138–45. doi: 10.2478/cipms-2019-0026.

Bui, J. D., and R. D. Schreiber. 2007. "Cancer immunosurveillance, immunoediting and inflammation: Independent or interdependent processes?" *Current Opinion in Immunology* 19 (2):203–8. doi: 10.1016/j.coi.2007.02.001.

Butcher, L. D., G. den Hartog, P. B. Ernst, and S. E. Crowe. 2017. "Oxidative stress resulting from *Helicobacter pylori* infection contributes to gastric carcinogenesis." *Cell Mol Gastroenterol Hepatol* 3 (3):316–22. doi: 10.1016/j.jcmgh.2017.02.002.

Cao, W., H. D. Chen, Y. W. Yu, N. Li, and W. Q. Chen. 2021. "Changing profiles of cancer burden worldwide and in China: A secondary analysis of the global cancer statistics 2020." *Chin Med J (Engl)* 134 (7):783–91. doi: 10.1097/cm9.0000000000001474.

Castro, C., and M. Gourley. 2010. "Diagnostic testing and interpretation of tests for autoimmunity." *J Allergy Clin Immunol* 125 (2 Suppl 2):S238–47. doi: 10.1016/j.jaci.2009.09.041.

Chalghoumi, R., and A. Hedi. 2016. "Antibacterial activity of fenugreek seeds (Trigonella foenum-graecum) crude extracts against a rabbit Escherichia coli isolate." *Academia J Microbiol Res* 4. doi: 10.15413/ajmr.2016.0117.

Chanput, W., J. J. Mes, and H. J. Wichers. 2014. "THP-1 cell line: An in vitro cell model for immune modulation approach." *Int Immunopharmacol* 23 (1):37–45. doi: 10.1016/j.intimp.2014.08.002.

Chatterjee, S., M. Kumar, and A. Kumar. 2012. "Chemomodulatory effect of *Trigonella foenum-graecum* (L.) seed extract on two stage mouse skin carcinogenesis." *Toxicol Int* 19 (3):287–94. doi: 10.4103/0971-6580.103670.

Chen, F. P., and M. H. Chien. 2019. "Effects of phytoestrogens on the activity and growth of primary breast cancer cells ex vivo." *J Obstet Gynaecol* 45 (7):1352–62. doi: 10.1111/jog.13982.

Chen, J. 2016. "The cell-cycle arrest and apoptotic functions of p53 in tumor initiation and progression." *Cold Spring Harb Perspect Med* 6 (3):a026104. doi: 10.1101/cshperspect.a026104.

Chen, P.-S., Y.-W. Shih, H.-C. Huang, and H.-W. Cheng. 2011. "Diosgenin, a steroidal saponin, inhibits migration and invasion of human prostate cancer PC-3 cells by reducing matrix metalloproteinases expression." *PLoS One* 6 (5):e20164. doi: 10.1371/journal.pone.0020164.

Chen, Z., J. Xu, Y. Wu, S. Lei, H. Liu, Q. Meng, and Z. Xia. 2018. "Diosgenin inhibited the expression of TAZ in hepatocellular carcinoma." *Biochem Biophys Res Commun* 503 (3):1181–5. doi: 10.1016/j.bbrc.2018.07.022.

Chiang, C.-T., T.-D. Way, S.-J. Tsai, and J.-K. Lin. 2007. "Diosgenin, a naturally occurring steroid, suppresses fatty acid synthase expression in HER2-overexpressing breast cancer cells through modulating Akt, mTOR and JNK phosphorylation." *FEBS Letters* 581 (30):5735–42. doi: 10.1016/j.febslet.2007.11.021.

Childs, C. E., P. C. Calder, and E. A. Miles. 2019. "Diet and immune function." *Nutrients* 11 (8):1933. doi: 10.3390/nu11081933.

Chinen, J., and W. T. Shearer. 2010. "Secondary immunodeficiencies, including HIV infection." *J Allergy Clin Immunol* 125 (Suppl 2):S195–203. doi: 10.1016/j.jaci.2009.08.040.

Chopra, A., M. Saluja, and G. Tillu. 2010. "Ayurveda-modern medicine interface: A critical appraisal of studies of Ayurvedic medicines to treat osteoarthritis and rheumatoid arthritis." *J Ayurveda Integr Med* 1 (3):190–8. doi: 10.4103/0975-9476.72620.

Choudhary, M., V. Kumar, H. Malhotra, and S. Singh. 2015. "Medicinal plants with potential anti-arthritic activity." *J Int Ethnopharmacol* 4 (2):147–79. doi: 10.5455/jice.20150313021918.

Chow, M. T., A. Möller, and M. J. Smyth. 2012. "Inflammation and immune surveillance in cancer." *Semin Cancer Biol* 22 (1):23–32. doi: 10.1016/j.semcancer.2011.12.004.

Clemons, M., and P. Goss. 2001. "Estrogen and the risk of breast cancer." *N Engl J Med* 344:276–85.

Corbiere, C., B. Liagre, A. Bianchi, K. Bordji, M. Dauca, P. Netter, and J. L. Beneytout. 2003. "Different contribution of apoptosis to the antiproliferative effects of diosgenin and other plant steroids, hecogenin and tigogenin, on human 1547 osteosarcoma cells." *Int J Oncol* 22 (4):899–905.

Corbiere, C., B. Liagre, F. Terro, and J. L. Beneytout. 2004. "Induction of antiproliferative effect by diosgenin through activation of p53, release of apoptosis-inducing factor (AIF) and modulation of caspase-3 activity in different human cancer cells." *Cell Res* 14 (3):188–96. doi: 10.1038/sj.cr.7290219.

Das, S., K. K. Dey, G. Dey, I. Pal, A. Majumder, S. MaitiChoudhury, S. C. kundu, and M. Mandal. 2012. "Antineoplastic and apoptotic potential of traditional medicines thymoquinone and diosgenin in squamous cell carcinoma." *PLoS One* 7 (10):e46641. doi: 10.1371/journal.pone.0046641.

Dash, B. K., S. Sultana, and N. Sultana. 2011. "Antibacterial activities of methanol and acetone extracts of fenugreek (*Trigonella foenum*) and coriander (*Coriandrum sativum*)." *Life Sci Med Res* 27:1–8.

Demain, A. L., and P. Vaishnav. 2011. "Natural products for cancer chemotherapy." *Microbial Biotechnol* 4 (6):687–99. doi: 10.1111/j.1751-7915.2010.00221.x.

Denning, D. W., M. Kneale, J. D. Sobel, and R. Rautemaa-Richardson. 2018. "Global burden of recurrent vulvovaginal candidiasis: A systematic review." *The Lancet Infectious Diseases* 18 (11):e339–e47. doi: 10.1016/S1473-3099(18)30103-8.

Deshmukh, A. R., A. Gupta, and B. S. Kim. 2019. "Ultrasound assisted green synthesis of silver and iron oxide nanoparticles using fenugreek seed extract and their enhanced antibacterial and antioxidant activities." *Biomed Res Int* 2019:1–14. doi: 10.1155/2019/1714358.

Devasena, T., and V. P. Menon. 2002. "Enhancement of circulatory antioxidants by fenugreek during 1,2-dimethylhydrazine-induced rat colon carcinogenesis." *Journal of Biochemistry, Molecular Biology, and Biophysics: JBMBB: The Official Journal of the Federation of Asian and Oceanian Biochemists and Molecular Biologists (FAOBMB)* 6 (4):289–92. doi: 10.1080/10258140290030915.

Dharajiya, D., H. Jasani, T. Khatrani, M. Kapuria, K. Pachchigar, and P. Patel. 2016. "Evaluation of antibacterial and antifungal activity of fenugreek (*Trigonella foenum-graecum*) extracts." *Int J Pharm Pharm Sci* 8 (4):212–7.

Dog, T. L. 2009. "The use of botanicals during pregnancy and lactation." *Alternative Therapies in Health and Medicine* 15 (1):54–8.

Dong, M., Z. Meng, K. Kuerban, F. Qi, J. Liu, Y. Wei, Q. Wang, S. Jiang, M. Feng, and L. Ye. 2018. Diosgenin promotes antitumor immunity and PD-1 antibody efficacy against melanoma by regulating intestinal microbiota. *Cell Death & Disease* 9 (10): 1039. Accessed 2018/10//. doi:10.1038/s41419-018-1099-3.

Dörner, T., K. Egerer, E. Feist, and G. R. Burmester. 2004. "Rheumatoid factor revisited." *Current Opinion in Rheumatology* 16 (3):246–53. doi: 10.1097/00002281-200405000-00013.

Du, T., L. Xu, M. Lei, N. Wang, D. Eidelman, H. Ghezzo, and J. Martin. 1992. "Morphometric changes during the early airway response to allergen challenge in the rat." *Am Rev Respiratory Disease* 146 (4):1037–41.

Duru, M., Z. Erdogan, A. Duru, A. Kucukgul, V. Duzguner, A. Kaya, and A. Sahin. 2013. "Effect of seed powder of a herbal legume fenugreek (*Trigonella foenum-graecum* L.) on growth performance, body components, digestive parts, and blood parameters of broiler chicks." *Pakistan Journal of Zoology* 45 (4):1007–14.

El Bairi, K., M. Ouzir, N. Agnieszka, and L. Khalki. 2017. "Anticancer potential of *Trigonella foenum graecum*: Cellular and molecular targets." *Biomed Pharmacother* 90:479–91. doi: 10.1016/j.biopha.2017.03.071.

El-Serag, H. B. 2020. "Epidemiology of hepatocellular carcinoma." In *The Liver*, edited by Arias, I. M., Alter, H. J., Boyer, J. L., Cohen, D. E., Shafritz, D. A., Thorgeirsson, S. S. and Wolkoff, A. W., 758–772. Hoboken, NJ: Wiley.

Elnour, M. E., M., A. Ali, B. Eldin, and A. E. Saeed. 2015. "Antimicrobial activities and phytochemical screening of callus and seeds extracts of fenugreek (*Trigonella foenum-graecum*)." *Int J Curr Microbiol Appl Sci* 4 (2):147–57.

Elwood, W., J. O. Lötvall, P. J. Barnes, and K. F. Chung. 1991. "Characterization of allergen-induced bronchial hyperresponsiveness and airway inflammation in actively sensitized brown-Norway rats." *J Allergy Clin Immunol* 88 (6):951–60.

Emtiazy, M., L. Oveidzadeh, M. Habibi, L. Molaeipour, D. Talei, Z. Jafari, M. Parvin, and M. Kamalinejad. 2018. "Investigating the effectiveness of the Trigonella foenum-graecum L. (fenugreek) seeds in mild asthma: A randomized controlled trial." *Allergy Asthma Clin Immunol* 14:19. doi: 10.1186/s13223-018-0238-9.

Fair, R. J., and Y. Tor. 2014. "Antibiotics and bacterial resistance in the 21st century." *Perspect Medicin Chem* 6:25–64. doi: 10.4137/pmc.S14459.

Fallatah, S. A. 2010. "Histopathological effects of fenugreek (*Trigonella foenum- graceum*) extracts on the larvae of the mosquito *Culex quinquefasciatus*." *J Arab Soc Med Res* 5 (2):123–30.

Felson, D. T., J. J. Anderson, M. Boers, C. Bombardier, D. Furst, C. Goldsmith, L. M. Katz, R. Lightfoot, Jr., H. Paulus, V. Strand, and et al. 1995. "American college of rheumatology. Preliminary definition of improvement in rheumatoid arthritis." *Arthritis Rheum* 38 (6):727–35. doi: 10.1002/art.1780380602.

Ferlay, J., M. Colombet, I. Soerjomataram, D. M. Parkin, M. Piñeros, A. Znaor, and F. Bray. "Cancer statistics for the year 2020: An overview." *Int J Cancer* n/a (n/a). doi: 10.1002/ijc.33588.

Fernandes, J. V., T. A. A. D. M. Fernandes, J. C. V. De Azevedo, R. N. O. Cobucci, M. G. F. De Carvalho, V. S. Andrade, and J. M. G. De Araujo. 2015. "Link between chronic inflammation and human papillomavirus-induced carcinogenesis." *Oncology Letters* 9 (3):1015–26. doi: 10.3892/ol.2015.2884.

Fernández, L., M. D. Cima-Cabal, A. C. Duarte, A. Rodriguez, P. García, and M. D. García-Suárez. 2020. "Developing diagnostic and therapeutic approaches to bacterial infections for a new era: Implications of globalization." *Antibiotics* 9 (12). doi: 10.3390/antibiotics9120916.

Ferreira, S. S., C. P. Passos, P. Madureira, M. Vilanova, and M. A. Coimbra. 2015. "Structure—function relationships of immunostimulatory polysaccharides: A review." *Carbohydrate Polymers* 132:378–96. doi: 10.1016/j.carbpol.2015.05.079.

Firestein, G. S., and I. B. McInnes. 2017. "Immunopathogenesis of rheumatoid arthritis." *Immunity* 46 (2):183–96. doi: 10.1016/j.immuni.2017.02.006.

Fletcher, S. M., D. Stark, J. Harkness, and J. Ellis. 2012. "Enteric protozoa in the developed world: A public health perspective." *Clin Microbiol Rev* 25 (3):420–49. doi: 10.1128/cmr.05038-11.

Folwarczna, J., M. Zych, B. Nowinska, and M. Pytlik. 2014. "Effects of fenugreek (Trigonella foenum-graecum L.) seed on bone mechanical properties in rats." *Eur Rev Med Pharmacol Sci* 18 (13):1937–47.

Forinash, A. B., A. M. Yancey, K. N. Barnes, and T. D. Myles. 2012. "The use of galactogogues in the breast-feeding mother." *Annals of Pharmacotherapy* 46 (10):1392–404. doi: 10.1345/aph.1R167.

França, T., L. Ishikawa, S. Zorzella-Pezavento, F. Chiuso-Minicucci, M. da Cunha, and A. Sartori. 2009. "Impact of malnutrition on immunity and infection." *Journal of Venomous Animals and Toxins Including Tropical Diseases* 15:374–90. doi: 10.1590/S1678-91992009000300003.

Giancane, G., J. F. Swart, E. Castagnola, A. H. Groll, G. Horneff, H.-I. Huppertz, D. J. Lovell, T. Wolfs, T. Herlin, and P. Dolezalova. 2020. "Opportunistic infections in immunosuppressed patients with juvenile idiopathic arthritis: Analysis by the pharmachild safety adjudication committee." *Arthritis Res Ther* 22:1–15. doi: 10.1186/s13075-020-02167-2.

Gonzalez, H., C. Hagerling, and Z. Werb. 2018. "Roles of the immune system in cancer: From tumor initiation to metastatic progression." *Genes Dev* 32 (19–20):1267–84. doi: 10.1101/gad.314617.118.

Goyal, S., N. Gupta, and S. Chatterjee. 2016. "Investigating therapeutic potential of *Trigonella foenum-graecum* L. As our defense mechanism against several human diseases." *J Toxicol* 2016:1250387. doi: 10.1155/2016/1250387.

Grant, S. S., and D. T. Hung. 2013. "Persistent bacterial infections, antibiotic tolerance, and the oxidative stress response." *Virulence* 4 (4):273–83. doi: 10.4161/viru.23987.

Gruden, M. A., K. Yanamandra, V. G. Kucheryanu, O. R. Bocharova, V. V. Sherstnev, L. A. Morozova-Roche, and R. D. E. Sewell. 2012. "Correlation between protective immunity to α-synuclein aggregates, oxidative stress and inflammation." *Neuroimmunomodulation* 19 (6):334–42. doi: 10.1159/000341400.

Guardiola, F. A., A. Bahi, and M. A. Esteban. 2018. "Effects of dietary administration of fenugreek seeds on metabolic parameters and immune status of gilthead seabream (Sparus aurata L.)." *Fish Shellfish Immunol* 74:372–9. doi: 10.1016/j.fsi.2018.01.010.

Guardiola, F. A., A. Bahi, A. M. Jimenez-Monreal, M. Martinez-Tome, M. A. Murcia, and M. A. Esteban. 2018. "Dietary administration effects of fenugreek seeds on skin mucosal antioxidant and immunity status of gilthead seabream (Sparus aurata L.)." *Fish Shellfish Immunol* 75:357–64. doi: 10.1016/j.fsi.2018.02.025.

Guo, Q., Y. Wang, D. Xu, J. Nossent, N. J. Pavlos, and J. Xu. 2018. "Rheumatoid arthritis: Pathological mechanisms and modern pharmacologic therapies." *Bone Res* 6:15. doi: 10.1038/s41413-018-0016-9.

Guo, X., and X. Ding. 2018. "Dioscin suppresses the viability of ovarian cancer cells by regulating the VEGFR2 and PI3K/AKT/MAPK signaling pathways." *Oncol Lett* 15 (6):9537–42. doi: 10.3892/ol.2018.8454.

Hamden, K., H. Masmoudi, S. Carreau, and A. Elfeki. 2010. "Immunomodulatory, β-cell, and neuroprotective actions of fenugreek oil from alloxan-induced diabetes." *Immunopharmacol Immunotoxicol* 32:437–45. doi: 10.3109/08923970903490486.

Haouala, R., S. Hawala, A. El-Ayeb, R. Khanfir, and N. Boughanmi. 2008. "Aqueous and organic extracts of *Trigonella foenum-graecum* L. inhibit the mycelia growth of fungi." *J Environ Sci* 20 (12):1453–57. doi: 10.1016/S1001-0742(08)62548-6.

Hart, P. H., and M. Norval. 2018. "Ultraviolet radiation-induced immunosuppression and its relevance for skin carcinogenesis." *Photochem Photobiol Sci* 17 (12):1872–84. doi: 10.1039/c7pp00312a.

He, Z., H. Chen, G. Li, H. Zhu, Y. Gao, L. Zhang, and J. Sun. 2014. "Diosgenin inhibits the migration of human breast cancer MDA-MB-231 cells by suppressing Vav2 activity." *Phytomedicine* 21 (6):871–6. doi: 10.1016/j.phymed.2014.02.002.

He, Z., Y. Tian, X. Zhang, B. Bing, L. Zhang, H. Wang, and W. Zhao. 2012. "Anti-tumour and immunomodulating activities of diosgenin, a naturally occurring steroidal saponin." *Nat Prod Res* 26 (23):2243–6. doi: 10.1080/14786419.2011.648192.

Heer, E., A. Harper, N. Escandor, H. Sung, V. McCormack, and M. M. Fidler-Benaoudia. 2020. "Global burden and trends in premenopausal and postmenopausal breast cancer: A population-based study." *The Lancet Global Health* 8 (8):e1027–e37. doi: 10.1016/S2214-109X(20)30215-1.

Hibasami, H., H. Moteki, K. Ishikawa, H. Katsuzaki, K. Imai, K. Yoshioka, Y. Ishii, and T. Komiya. 2003. "Protodioscin isolated from fenugreek (*Trigonella foenum-graecum* L.) induces cell death and morphological change indicative of apoptosis in leukemic cell line H-60, but not in gastric cancer cell line KATO III." *Int J Mol Med* 11 (1):23–6.

Hou, R., Q. L. Zhou, B. X. Wang, S. Tashiro, S. Onodera, and T. Ikejima. 2004. "Diosgenin induces apoptosis in HeLa cells via activation of caspase pathway." *Acta Pharmacol Sin* 25 (8):1077–82.

Howland, R. E., N. C. Deziel, G. R. Bentley, M. Booth, O. A. Choudhury, J. N. Hofmann, R. N. Hoover, H. A. Katki, B. Trabert, and S. D. Fox. 2020. "Assessing endogenous and exogenous hormone exposures and breast development in a migrant study of Bangladeshi and British girls." *Int J Environ Res Public Health* 17 (4):1185.

Hsieh, C.-J., Y.-L. Hsu, Y.-F. Huang, and E.-M. Tsai. 2018. "Molecular mechanisms of anticancer effects of phytoestrogens in breast cancer." *Curr Protein Pept Sci* 19 (3):323–32. doi: 10.2174/1389203718666170111121255.

Hsieh, M. J., T. L. Tsai, Y. S. Hsieh, C. J. Wang, and H. L. Chiou. 2013. "Dioscin-induced autophagy mitigates cell apoptosis through modulation of PI3K/Akt and ERK and JNK signaling pathways in human lung cancer cell lines." *Arch Toxicol* 87 (11):1927–37. doi: 10.1007/s00204-013-1047-z.

Huang, M., J.-J. Lu, and J. Ding. 2021. "Natural products in cancer therapy: Past, present and future." *Natural Products and Bioprospecting* 11 (1):5–13.

Hussain, S., J. Schwank, F. Staib, X. Wang, and C. Harris. 2007. "TP53 mutations and hepatocellular carcinoma: Insights into the etiology and pathogenesis of liver cancer." *Oncogene* 26 (15):2166–76. doi: 10.1038/sj.onc.1210279.

Hussain, Z., H. E. Thu, A. N. Shuid, P. Kesharwani, S. Khan, and F. Hussain. 2017. "Phytotherapeutic potential of natural herbal medicines for the treatment of mild-to-severe atopic dermatitis: A review of human clinical studies." *Biomedicine & Pharmacotherapy* 93:596–608. doi: 10.1016/j.biopha.2017.06.087.

Inokuma, T., M. Haraguchi, F. Fujita, Y. Tajima, and T. Kanematsu. 2009. "Oxidative stress and tumor progression in colorectal cancer." *Hepato-Gastroenterology* 56 (90):343–7.

Ishnava, K. B., and P. S. Konar. 2020. "In vitro anthelmintic activity and phytochemical characterization of Corallocarpus epigaeus (Rottler) Hook. f. tuber from ethyl acetate extracts." *Bull Natl Res Cent* 44 (1):33. doi: 10.1186/s42269-020-00286-z.

Ivanov, A. V., B. Bartosch, and M. G. Isaguliants. 2017. "Oxidative stress in infection and consequent disease." *Oxid Med Cell Longev* 2017:3496043. doi: 10.1155/2017/3496043.

Iwuagwu, M., P. Etusim, N. Emmanuel, J. Igwe, V. Nwaugo, and R. Onyeagba. 2020. "Exploitation of plant herbs in the control of disease vectors: A review." *Pharma Bio J* 8 (5):7–21.

Jagadeesan, J., N. Nandakumar, T. Rengarajan, and M. P. Balasubramanian. 2012. "Diosgenin, a steroidal saponin, exhibits anticancer activity by attenuating lipid peroxidation via enhancing antioxidant defense system during NMU-induced breast carcinoma." *J Environ Pathol Toxicol Oncol* 31 (2):121–9. doi: 10.1615/jenvironpatholtoxicoloncol.v31.i2.40.

Jain, N., Q. Chen, T. Ayer, S. M. O'Brien, M. Keating, W. Wierda, H. M. Kantarjian, and C. Jagpreet. 2015. "Prevalence and economic burden of chronic lymphocytic leukemia (CLL) in the era of oral targeted therapies." *Blood* 126 (23): 871. doi: 10.1182/blood.V126.23.871.871.

Jatav, P., R. Tenguria, M. Naikoo, and S. Ahirwar. 2019. "Phytochemical analysis and antibacterial activity of fenugreek seed extract against *Escherichia coli* and *Pseudomonas*." *Gobal J Bio Biotechnol* 8 (4):347–52.

Jiang, S., J. Fan, Q. Wang, D. Ju, M. Feng, J. Li, Z. B. Guan, D. An, X. Wang, and L. Ye. 2016. "Diosgenin induces ROS-dependent autophagy and cytotoxicity via mTOR signaling pathway in chronic myeloid leukemia cells." *Phytomedicine* 23 (3):243–52. doi: 10.1016/j.phymed.2016.01.010.

Johnson, P. M., and W. P. Faulk. 1976. "Rheumatoid factor: Its nature, specificity, and production in rheumatoid arthritis." *Clin Immunol Immunopathol* 6 (3):414–30. doi: 10.1016/0090-1229(76)90094-5.

Kahaleq, M. A. A., A. R. Abu-Raghif, and S. R. Kadhim. 2016. "Antibacterial activity of fenugreek essential oil against *Pseudomonas aeroginosa*: In vitro and in vivo studies." *Iraqi JMS* 13 (3):227–34.

Kainz, K., M. A. Bauer, M. Frank, and C.-G. Didac. 2020. "Fungal infections in humans: The silent crisis." *Microbial Cell* 7 (6):143–5.

Kany, S., J. T. Vollrath, and B. Relja. 2019. "Cytokines in inflammatory disease." *Int J Mol Sci* 20 (23). doi: 10.3390/ijms20236008.

Katona, P., and J. Katona-Apte. 2008. "The interaction between nutrition and infection." *Clin Infect Dis* 46 (10):1582–8. doi: 10.1086/587658.

Kaveri, C. M., and J. Sincy. 2019. "Comparative analysis of phytochemical and antibacterial activity of ginger (*Zingiber officinale*, Rosc.) and fenugreek (*Trigonella foenum-graecum*, L.)." *Int J Botany Studies* 4 (2):100–3.

Kawabata, T., M. Y. Cui, T. Hasegawa, F. Takano, and T. Ohta. 2011. "Anti-inflammatory and anti-melanogenic steroidal saponin glycosides from fenugreek (*Trigonella foenum-graecum* L.) seeds." *Planta Med* 77 (7):705–10. doi: 10.1055/s-0030-1250477.

Kaya, Y., A. Baldemir, U. Karaman, N. Ildiz, Y. K. Arici, G. Kacmaz, Z. Koloren, and Y. Konca. 2019. "Amebicidal effects of fenugreek (trigonella foenum-graecum) against acanthamoeba cysts." *Food Sci Nutr* 7 (2):563–71. doi: 10.1002/fsn3.849.

Kayondo, N., A. Jumaa, and A. M. Elnazeer. 2019. "Antigiardial activity of some plant extracts used in traditional medicine in Sudan in comparison with metronidazole." *Microbiol Curr Res* 2 (4):75–84.

Khadse, C. D., and R. Kakde. 2010. "In vitro anthelmintic activity of Fenugreek seeds extract against *Pheritima posthuma*." *Int J Res Pharm Sci* 1 (3):267–9.

Khalil, M. I., M. M. Ibrahim, G. A. El-Gaaly, and A. S. Sultan. 2015. "Trigonella foenum (Fenugreek) induced apoptosis in hepatocellular carcinoma cell line, HepG2, mediated by upregulation of p53 and proliferating cell nuclear antigen." *Biomed Res Int* 2015:914645. doi: 10.1155/2015/914645.

Khan, N., C. Hwa, N. Perveen, and N. Paliwal. 2019. "Phytochemical screening, antimicrobial and antioxidant activity determination of *Trigonella foenum-graecum* seeds." *Pharm Pharmacol Int J* 7 (4):175–86. doi: 10.15406/ppij.2019.07.00249.

Khan, T. M., D. B.-C. Wu, and A. V. Dolzhenko. 2018. "Effectiveness of fenugreek as a galactagogue: A network meta-analysis." *Phytotherapy Res* 32 (3):402–12. doi: 10.1002/ptr.5972.

Khoja, K. K., G. Shaf, T. N. Hasan, N. A. Syed, A. S. Al-Khalifa, A. H. Al-Assaf, and A. A. Alshatwi. 2011. "Fenugreek, a naturally occurring edible spice, kills MCF-7 human breast cancer cells via an apoptotic pathway." *Asian Pac J Cancer Prev* 12 (12):3299–304.

Kim, D. S., B. K. Jeon, Y. E. Lee, W. H. Woo, and Y. J. Mun. 2012. "Diosgenin induces apoptosis in HepG2 cells through generation of reactive oxygen species and mitochondrial pathway." *Evid Based Complement Alternat Med* 2012:981675. doi: 10.1155/2012/981675.

Kontoyiannis, D. P. 2017. "Antifungal Resistance: An Emerging Reality and A Global Challenge." *The Journal of Infectious Diseases* 216 (Suppl 3):S431–S5. doi: 10.1093/infdis/jix179.

Koshak, A. E. 2019. "Prevalence of herbal medicines in patients with chronic allergic disorders in Western Saudi Arabia." *Saudi Medical Journal* 40 (4):391.

Kulkarni, M., V. Hastak, V. Jadhav, and A. Date. 2019. "Fenugreek leaf extract and its gel formulation show activity against *Malassezia furfur*." *ASSAY and Drug Development Technologies* 18. doi: 10.1089/adt.2019.918.

Kumar, A., A. N. Malaviya, A. Pandhi, and R. Singh. 2002. "Validation of an Indian version of the Health Assessment Questionnaire in patients with rheumatoid arthritis." *Rheumatology* 41 (12):1457–9. doi: 10.1093/rheumatology/41.12.1457.

Kumar, A., P. Vasmatkar, P. Baral, S. Agarawal, and A. Mishra. 2017. "Immunomodulatory and growth promoting effect of dietary fenugreek seeds in fingerlings of common carp (Cyprinus carpio Lin.)." *Fish Technol* 54:170–5.

Kumari, O., N. B. Rao, and R. G. Gajula. 2016. "Phytochemical analysis and anti-microbial activity of Trigonella foenum gracum (Methi seeds)." *J Med Plants Studies* 4:278–81.

Lalthanpuii, P. B., and K. Lalchhandama. 2020. "Phytochemical analysis and in vitro anthelmintic activity of Imperata cylindrica underground parts." *BMC Complement Med Therapies* 20 (1):332. doi: 10.1186/s12906-020-03125-w.

Landskron, G., M. De la Fuente, P. Thuwajit, C. Thuwajit, and M. A. Hermoso. 2014. "Chronic inflammation and cytokines in the tumor microenvironment." *J Immunol Res*. 2014:149185. doi: 10.1155/2014/149185.

Lee, J., K. Jung, Y. S. Kim, and D. Park. 2007. "Diosgenin inhibits melanogenesis through the activation of phosphatidylinositol-3-kinase pathway (PI3K) signaling." *Life Sci* 81 (3):249–54. doi: 10.1016/j.lfs.2007.05.009.

Leger, D. Y., B. Liagre, and J. L. Beneytout. 2006. "Role of MAPKs and NF-kappaB in diosgenin-induced megakaryocytic differentiation and subsequent apoptosis in HEL cells." *Int J Oncol* 28 (1):201–7.

Leger, D. Y., B. Liagre, C. Corbiere, J. Cook-Moreau, and J. L. Beneytout. 2004. "Diosgenin induces cell cycle arrest and apoptosis in HEL cells with increase in intracellular calcium level, activation of cPLA2 and COX-2 overexpression." *Int J Oncol* 25 (3):555–62.

Lepage, C., D. Y. Leger, J. Bertrand, F. Martin, J. L. Beneytout, and B. Liagre. 2011. "Diosgenin induces death receptor-5 through activation of p38 pathway and promotes TRAIL-induced apoptosis in colon cancer cells." *Cancer Lett* 301 (2):193–202. doi: 10.1016/j.canlet.2010.12.003.

Lewnard, J. A., and A. L. Reingold. 2019. "Emerging challenges and opportunities in infectious disease epidemiology." *Am J Epidemiol* 188 (5):873–82. doi: 10.1093/aje/kwy264.

Li, F., P. P. Fernandez, P. Rajendran, K. M. Hui, and G. Sethi. 2010. "Diosgenin, a steroidal saponin, inhibits STAT3 signaling pathway leading to suppression of proliferation and chemosensitization of human hepatocellular carcinoma cells." *Cancer Lett* 292 (2):197–207. doi: 10.1016/j.canlet.2009.12.003.

Li, N., Y. Deng, L. Zhou, T. Tian, S. Yang, Y. Wu, Y. Zheng, Z. Zhai, Q. Hao, and D. Song. 2019. "Global burden of breast cancer and attributable risk factors in 195 countries and territories, from 1990 to 2017: Results from the global burden of disease study 2017." *Journal of Hematology & Oncology* 12 (1):1–12.

Li, S., B. Cheng, L. Hou, L. Huang, Y. Cui, D. Xu, X. Shen, and S. Li. 2018. "Dioscin inhibits colon cancer cells' growth by reactive oxygen species-mediated mitochondrial dysfunction and p38 and JNK pathways." *Anticancer Drugs* 29 (3):234–42. doi: 10.1097/cad.0000000000000590.

Li, Y., X. Wang, S. Cheng, J. Du, Z. Deng, Y. Zhang, Q. Liu, J. Gao, B. Cheng, and C. Ling. 2015. "Diosgenin induces G2/M cell cycle arrest and apoptosis in human hepatocellular carcinoma cells." *Oncol Rep* 33 (2):693–8. doi: 10.3892/or.2014.3629.

Liagre, B., J. Bertrand, D. Y. Leger, and J. L. Beneytout. 2005. "Diosgenin, a plant steroid, induces apoptosis in COX-2 deficient K562 cells with activation of the p38 MAP kinase signalling and inhibition of NF-kappaB binding." *Int J Mol Med* 16 (6):1095–101.

Liao, J. C., K. T. Lee, B. J. You, C. L. Lee, W. T. Chang, Y. C. Wu, and H. Z. Lee. 2015. "Raf/ERK/Nrf2 signaling pathway and MMP-7 expression involvement in the trigonelline-mediated inhibition of hepatocarcinoma cell migration." *Food Nutr Res* 59:29884. doi: 10.3402/fnr.v59.29884.

Liao, W. L., J. Y. Lin, J. C. Shieh, H. F. Yeh, Y. H. Hsieh, Y. C. Cheng, H. J. Lee, C. Y. Shen, and C. W. Cheng. 2019. "Induction of G2/M phase arrest by diosgenin via activation of Chk1 kinase and Cdc25C regulatory pathways to promote apoptosis in human breast cancer cells." *Int J Mol Sci* 21 (1). doi: 10.3390/ijms21010172.

Liu, M. J., Z. Wang, Y. Ju, J. B. Zhou, Y. Wang, and R. N. Wong. 2004. "The mitotic-arresting and apoptosis-inducing effects of diosgenyl saponins on human leukemia cell lines." *Biol Pharm Bull* 27 (7):1059–65. doi: 10.1248/bpb.27.1059.

Liu, M. J., Z. Wang, Y. Ju, R. N. Wong, and Q. Y. Wu. 2005. "Diosgenin induces cell cycle arrest and apoptosis in human leukemia K562 cells with the disruption of Ca2+ homeostasis." *Cancer Chemother Pharmacol* 55 (1):79–90. doi: 10.1007/s00280-004-0849-3.

Liu, S., G. Rong, X. Li, L. Geng, Z. Zeng, D. Jiang, J. Yang, and Y. Wei. 2020. "Diosgenin and GSK126 produce synergistic effects on epithelial-mesenchymal transition in gastric cancer cells by mediating EZH2 via the Rho/ROCK signaling pathway." *Onco Targets Ther* 13:5057–67. doi: 10.2147/ott.s237474.

Lopes, F. C. P. S., M. G. Sleiman, K. Sebastian, R. Bogucka, E. A. Jacobs, and A. S. Adamson. 2021. "UV exposure and the risk of cutaneous melanoma in skin of color: A systematic review." *JAMA Dermatology* 157 (2):213–19. doi: 10.1001/jamadermatol.2020.4616.

Mabrouki, S., R. Chalghoumi, and H. Abdouli. 2017. "Effects of pre-germinated fenugreek seeds inclusion in low-fiber diets on post-weaned rabbits' health status, growth performances, carcass characteristics, and meat chemical composition." *Trop Anim Health Prod* 49 (3):459–65. doi: 10.1007/s11250-016-1214-3.

Maeda, K., M. J. Caldez, and S. Akira. 2019. "Innate immunity in allergy." *Allergy* 74 (9):1660–74. doi: 10.1111/all.13788.

Magdoleen, O. G., H. M. Daffalla, M. A. M. M., K. S. e.-k. Ali, S. A. Saleh, and H. A. Abdelhalim. 2020. "Total phenolic content, antioxidant and antimicrobial activities of seeds and callus of *Trigonella foenum-graecum* Linn." *GSC Biol Pharm Sci* 10 (3):1–9. doi: 10.30574/gscbps.2020.10.3.0033.

Maggini, S., A. Pierre, and P. C. Calder. 2018. "Immune function and micronutrient requirements change over the life course." *Nutrients* 10 (10). doi: 10.3390/nu10101531.

Mahady, G. 2005. "Medicinal plants for the prevention and treatment of bacterial infections." *Current Pharmaceutical Design* 11:2405–27. doi: 10.2174/1381612054367481.

Mali, R. G., and A. A. Mehta. 2008. "A review on anthelmintic plants." *Natural Product Radiance* 7 (5):466–475.

Malik, S., A. Sawhney, M. Vyas, A. Kakroo, and P. Aggarwal. 2013. "Evaluation of antibacterial activity of methanol and acetone extracts of *Trigonella foenum*, *Coriandrum sativum* and *Brassica nigra*." *Int J Drug Dev & Res* 5 (3):316–21.

Mandal, P. 2017. "Potential biomarkers associated with oxidative stress for risk assessment of colorectal cancer." *Naunyn-Schmiedeberg's Archives of Pharmacology* 390 (6):557–65. doi: 10.1007/s00210-017-1352-9.

Mann, J. 2002. "Natural products in cancer chemotherapy: Past, present and future." *Nature Reviews Cancer* 2 (2):143–8. doi: 10.1038/nrc723.

Mao, Z. J., Q. J. Tang, C. A. Zhang, Z. F. Qin, B. Pang, P. K. Wei, B. Liu, and Y. N. Chou. 2012. "Antiproliferation and anti-invasion effects of diosgenin on gastric cancer BGC-823 cells with HIF-1alpha shRNAs." *Int J Mol Sci* 13 (5):6521–33. doi: 10.3390/ijms13056521.

Marshall, J. S., R. Warrington, W. Watson, and H. L. Kim. 2018. "An introduction to immunology and immunopathology." *Allergy, Asthma & Clin Immunol* 14 (2):49. doi: 10.1186/s13223-018-0278-1.

Mathers, C. D. 2020. "History of global burden of disease assessment at the world health organization." *Archives of Public Health* 78 (1):1–13.

Mattsson, M., F. Sandin, E. Kimby, M. Höglund, and I. Glimelius. 2020. "Increasing prevalence of chronic lymphocytic leukemia with an estimated future rise: A nationwide population-based study." *American journal of hematology* 95 (2):E36–E8.

Meng, X., H. Dong, Y. Pan, L. Ma, C. Liu, S. Man, and W. Gao. 2019. "Diosgenyl saponin inducing endoplasmic reticulum stress and mitochondria-mediated apoptotic pathways in liver cancer cells." *J Agric Food Chem* 67 (41):11428–35. doi: 10.1021/acs.jafc.9b05131.

Meyer, K. C., C. Decker, and R. Baughman. 2010. "Toxicity and monitoring of immunosuppressive therapy used in systemic autoimmune diseases." *Clin Chest Med* 31 (3):565–88. doi: 10.1016/j.ccm.2010.05.006.

Michaud, D. S., E. A. Houseman, C. J. Marsit, H. H. Nelson, J. K. Wiencke, and K. T. Kelsey. 2015. "Understanding the role of the immune system in the development of cancer: New opportunities for population-based research." *Cancer Epidemiol Biomarkers Prev* 24 (12):1811–9. doi: 10.1158/1055-9965.EPI-15-0681.

Misra, A., N. Singhal, B. Sivakumar, N. Bhagat, A. Jaiswal, and L. Khurana. 2011. "Nutrition transition in India: Secular trends in dietary intake and their relationship to diet-related non-communicable diseases." *J Diabetes* 3 (4):278–92. doi: 10.1111/j.1753-0407.2011.00139.x.

Mlynska, A. 2018. "The role of systemic and local immunity in tumor development and response to treatment." Ph.D. Thesis, Vilniaus universitetas. Vilnius: Vilniaus universitetas.

Moalic, S., B. Liagre, C. Corbiere, A. Bianchi, M. Dauca, K. Bordji, and J. L. Beneytout. 2001. "A plant steroid, diosgenin, induces apoptosis, cell cycle arrest and COX activity in osteosarcoma cells." *FEBS Lett* 506 (3):225–30. doi: 10.1016/s0014-5793(01)02924-6.

Mohammad, R. Y., G. Somayyeh, H. Gholamreza, M. Majid, and R. Yousef. 2013. "Diosgenin inhibits hTERT gene expression in the A549 lung cancer cell line." *Asian Pac J Cancer Prev* 14 (11):6945–8. doi: 10.7314/apjcp.2013.14.11.6945.

Moradi-Kor, N., M. B. Didarshetaban, and H. R. Saeid Pour. 2013. "Fenugreek (Trigonella foenum-graecum L.) as a Valuable Medicinal Plant." *Int J Adv Biol Bio Res* 1 (8):922–31.

Moreno, C., K. Hodgson, G. Ferrer, M. Elena, X. Filella, A. Pereira, T. Baumann, and E. Montserrat. 2010. "Autoimmune cytopenia in chronic lymphocytic leukemia: Prevalence, clinical associations, and prognostic significance." *Blood, the J Am Soc Hematol* 116 (23):4771–6.

Mortaz, E., P. Tabarsi, D. Mansouri, A. Khosravi, J. Garssen, A. Velayati, and I. M. Adcock. 2016. "Cancers related to immunodeficiencies: Update and perspectives." *Frontiers in Immunology* 7:365–5. doi: 10.3389/fimmu.2016.00365.

Moustafa, E. M., M. A. O. Dawood, D. H. Assar, A. A. Omar, Z. I. Elbialy, F. A. Farrag, M. Shukry, and M. M. Zayed. 2020. "Modulatory effects of fenugreek seeds powder on the histopathology, oxidative status, and immune related gene expression in Nile tilapia (Oreochromis niloticus) infected with Aeromonas hydrophila." *Aquaculture* 515:734589. doi: 10.1016/j.aquaculture.2019.734589.

Moxley, A. H., and D. Reisman. 2021. "Context is key: Understanding the regulation, functional control, and activities of the p53 tumour suppressor." *Cell Biochemistry and Function* 39 (2):235–47.

Multhoff, G., M. Molls, and J. Radons. 2011. "Chronic inflammation in cancer development." *Front Immunol* 2:98. doi: 10.3389/fimmu.2011.00098.

Murugamani, V., L. Raju, V. B. Anand Raj, M. Sarma Kataki, and G. G. Sankar. 2012. "The new method developed for evaluation of anthelmintic activity by housefly worms and compared with conventional earthworm method." *ISRN Pharmacology* 2012:709860. doi: 10.5402/2012/709860.

Nagai, H., and Y. H. Kim. 2017. "Cancer prevention from the perspective of global cancer burden patterns." *J Thorac Dis* 9 (3):448–51. doi: 10.21037/jtd.2017.02.75.

Nandagopal, S., D. P. Dhanalakshmi, G. K. Anbazhagan, and D. Sujitha. 2018. "Phytochemical and antibacterial studies of fenugreek *Trigonella foenum-graecum* L.—A multipurpose medicinal plant." *J Pharm Res* 5 (1):413–5.

Neagu, M., C. Caruntu, C. Constantin, D. Boda, S. Zurac, D. A. Spandidos, and A. M. Tsatsakis. 2016. "Chemically induced skin carcinogenesis: Updates in experimental models." *Oncology Reports* 35 (5):2516–28.

Nessa, M. U., M. A. Rahman, and Y. Kabir. 2020. "Plant-Produced Monoclonal Antibody as Immunotherapy for Cancer." *BioMed Res Int* 2020:3038564. doi: 10.1155/2020/3038564.

Newaj-Fyzul, A., and B. Austin. 2015. "Probiotics, immunostimulants, plant products and oral vaccines, and their role as feed supplements in the control of bacterial fish diseases." *J Fish Dis* 38 (11):937–55. doi: 10.1111/jfd.12313.

Nie, C., J. Zhou, X. Qin, X. Shi, Q. Zeng, J. Liu, S. Yan, and L. Zhang. 2016. "Diosgenin-induced autophagy and apoptosis in a human prostate cancer cell line." *Mol Med Rep* 14 (5):4349–59. doi: 10.3892/mmr.2016.5750.

Nimbalkar, V., U. Kadu, R. Shelke, S. Shendge, P. Tupe, and P. Gaikwad. 2018. "Evaluation of immunomodulatory activity of diosgenin in rats." *Int J Clin Bio Res* 4:70–5. doi: 10.31878/ijcbr.2018.43.15.

O'Connor, C., and A. Nichol. 2015. "Inflammation, immunity and allergy." *Anaesthesia & Intensive Care Medicine* 16 (7):328–33. doi: 10.1016/j.mpaic.2015.05.001.

Oddepally, R., and L. Guruprasad. 2015. "Isolation, purification, and characterization of a stable defensin-like antifungal peptide from Trigonella foenum-graecum (fenugreek) seeds." *Biochemistry (Mosc)* 80 (3):332–42. doi: 10.1134/s0006297915030086.

Over, K. F., N. Hettiarachchy, M. G. Johnson, and B. Davis. 2009. "Effect of organic acids and plant extracts on Escherichia coli O157:H7, *Listeria monocytogenes*, and *Salmonella typhimurium* in broth culture model and chicken meat systems." *J Food Sci* 74 (9):M515–21. doi: 10.1111/j.1750-3841.2009.01375.x.

Ozaslan, M., I. D. Karagoz, I. H. Kilic, and M. E. Guldur. 2011. "Ehrlich ascites carcinoma." *African J Biotechnol* 10 (13):2375–8.

Pai, V., A. Ganavalli, and N. N. Kikkeri. 2018. "Antifungal resistance in dermatology." *Indian J Dermatol* 63 (5):361–8. doi: 10.4103/ijd.IJD_131_17.

Palaniswamy, M., B. V. Pradeep, R. Sathya, and J. Angayarkanni. 2010. "In Vitro Anti-plasmodial activity of Trigonella foenum-graecum L." *Evid Based Complement Alternat Med* 7 (4):441–5. doi: 10.1093/ecam/nen030.

Park, Y. J., G. H. Kwon, J. O. Kim, N. K. Kim, W. S. Ryu, and K. S. Lee. 2020. "A retrospective study of changes in skin cancer characteristics over 11 years." *Arch Craniofac Surg* 21 (2):87–91. doi: 10.7181/acfs.2020.00024.

Patel, A., Y. Pathak, J. Patel, and V. Sutariya. 2018. "Role of nutritional factors in pathogenesis of cancer." *Food Quality and Safety* 2 (1):27–36. doi: 10.1093/fqsafe/fyx033.

Perše, M. 2013. "Oxidative stress in the pathogenesis of colorectal cancer: Cause or consequence?" *BioMed Res Int* 2013:725710. doi: 10.1155/2013/725710.

Piao, C. H., T. T. Bui, C. H. Song, H. S. Shin, D. H. Shon, and O. H. Chai. 2017. "Trigonella foenum-graecum alleviates airway inflammation of allergic asthma in ovalbumin-induced mouse model." *Biochem Biophys Res Commun* 482 (4):1284–8. doi: 10.1016/j.bbrc.2016.12.029.

Platts-Mills, T. A. E. 2001. "The role of immunoglobulin E in allergy and asthma." *Am J Respiratory and Critical Care Med* 164 (Suppl 1):S1–S5. doi: 10.1164/ajrccm.164.supplement_1.2103024.

Pons-Fuster López, E., Q.-T. Wang, W. Wei, and P. López Jornet. 2017. "Potential chemotherapeutic effects of diosgenin, zoledronic acid and epigallocatechin-3-gallate on PE/CA-PJ15 oral squamous cancer cell line." *Archives of Oral Biology* 82:141–6. doi: 10.1016/j.archoralbio.2017.05.023.

Prabhu, A., and M. Krishnamoorthy. 2010. "Anticancer activity of Trigonella foenum-graecum on Ehrlich Ascites carcinoma in Mus musculus system." *J Pharm Res* 3:1181–3.

Premnath, R., N. Lakshmidevi, and S. M. Aradhya. 2011. "Antibacterial and anti-oxidant activities of fenugreek (*Trigonella foenum-graecum* L.) leaves." *Res J Med Plant* 5 (6):695–705. doi: 10.3923/rjmp.2011.695.705.

Pundarikakshudu, K. 2019. "Chapter 4 — antiinflammatory and antiarthritic activities of some foods and spices." In *Bioactive Food as Dietary Interventions for Arthritis and Related Inflammatory Diseases* (Second Edition), edited by Watson, R.R. and Preedy, V.R., 51–68. Cambridge: Academic Press.

Putta, P. 2018. "Helmintholytic activity of leaves of fenugreek plant (*Trigonella foenum-graecum*)." *World Journal of Pharmaceutical Research* 7 (12):828–35. doi: 10.20959/wjpr201812-12658.

Rahmati-Yamchi, M., S. Ghareghomi, G. Haddadchi, M. Milani, M. Aghazadeh, and H. Daroushnejad. 2014. "Fenugreek extract diosgenin and pure diosgenin inhibit the hTERT gene expression in A549 lung cancer cell line." *Mol Biol Rep* 41 (9):6247–52. doi: 10.1007/s11033-014-3505-y.

Rainsford, K. D. 1982. "Adjuvant polyarthritis in rats: Is this a satisfactory model for screening anti-arthritic drugs?" *Agents Actions* 12 (4):452–8. doi: 10.1007/bf01965926.

Raju, J., and R. P. Bird. 2007. "Diosgenin, a naturally occurring steroid [corrected] saponin suppresses 3-hydroxy-3-methylglutaryl CoA reductase expression and induces apoptosis in HCT-116 human colon carcinoma cells." *Cancer Lett* 255 (2):194–204. doi: 10.1016/j.canlet.2007.04.011.

Raju, J., J. M. Patlolla, M. V. Swamy, and C. V. Rao. 2004. "Diosgenin, a steroid saponin of *Trigonella foenum-graecum* (fenugreek), inhibits azoxymethane-induced aberrant crypt foci formation in F344 rats and induces apoptosis in HT-29 human colon cancer cells." *Cancer Epidemiol Biomarkers Prev* 13 (8):1392–8.

Ramadan, G., N. M. El-Beih, and H. F. Abd El-Kareem. 2011. "Anti-metabolic syndrome and immunostimulant activities of Egyptian fenugreek seeds in diabetic/obese and immunosuppressive rat models." *Br J Nutr* 105 (7):995–1004. doi: 10.1017/s0007114510004708.

Ramesh, H., K. Yamaki, and T. Tsushida. 2002. "Effect of fenugreek (Trigonella foenum-graecum L.) galactomannan fractions on phagocytosis in rat macrophages and on proliferation and IgM secretion in HB4C5 cells." *Carbohydrate Polymers* 50 (1):79–83.

Rana, R. H. 2016. "Experimental study for the effect of *Trigonella foenum-graecum* (fenugreek) seeds extract on some biochemical and histo pathological study in induced diarrhea in mice by *Klebsiella pneumoniae*." *IOSR Journal of Pharmacy* 6 (2):4–13.

Randhir, R., Y.-T. Lin, and K. Shetty. 2004. "Phenolics, their antioxidant and antimicrobial activity in dark germinated fenugreek sprouts in response to peptide and phytochemical elicitors." *Asian Pacific J Clin Nutr* 13 (3):295–307.

Rivlin, N., R. Brosh, M. Oren, and V. Rotter. 2011. "Mutations in the p53 tumor suppressor gene: Important milestones at the various steps of tumorigenesis." *Genes Cancer* 2 (4):466–74. doi: 10.1177/1947601911408889.

Rosenblum, M. D., K. A. Remedios, and A. K. Abbas. 2015. "Mechanisms of human autoimmunity." *J Clin Invest* 125 (6):2228–33. doi: 10.1172/jci78088.

Salah, S. Z. A.-a., M. F. Bestoon, and J. N. Osman. 2010. "Antibacterial effects of fenugreek (*Trigonella foenum-graecum*)." *Basrah Journal of Veterinary Research* 10 (2):133–40. doi: 10.33762/bvetr.2010.55057.

Saleem, F., M. T. J. Khan, M. A. Mahmood, K. I. Khan, S. Bashir, and M. Jamshaid. 2008. "Antimicrobial activity of the extracts of seeds of Trigonella foenum-graecum." *Pakistan Journal of Zoology* 40:385–7.

Samavat, H., and M. S. Kurzer. 2015. "Estrogen metabolism and breast cancer." *Cancer Letters* 356 (2):231–43.

Sarwar, S., M. A. Hanif, M. A. Ayub, Y. D. Boakye, and C. Agyare. 2020. "Chapter 20 — Fenugreek." In *Medicinal Plants of South Asia: Novel Sources for Drug Discovery*. edited by Hanif, M.A., Nawaz, H., Khan, M.M. and Byrne, H.J., 257–71. Amsterdam: Elsevier.

Saunte, D. M. L., G. Gaitanis, and R. J. Hay. 2020. "Malassezia-Associated Skin Diseases, the Use of Diagnostics and Treatment." *Front Cell Infect Microbiol* 10:112. doi: 10.3389/fcimb.2020.00112.

Schneider, C., W.-D. F. Döcke, T. M. Zollner, and L. Röse. 2009. "Chronic mouse model of TMA-induced contact hypersensitivity." *J Invest Dermatol* 129 (4):899–907. doi: 10.1038/jid.2008.307.

Sebastian, K. S., and R. V. Thampan. 2007. "Differential effects of soybean and fenugreek extracts on the growth of MCF-7 cells." *Chem Biol Interact* 170 (2):135–43. doi: 10.1016/j.cbi.2007.07.011.

Selim, S., and S. Al Jaouni. 2016. "Anti-inflammatory, antioxidant and antiangiogenic activities of diosgenin isolated from traditional medicinal plant, *Costus speciosus* (Koen ex.Retz.) Sm." *Natural Prod Res* 30 (16):1830–3. doi: 10.1080/14786419.2015.1065493.

Sepahvand, A., H. Eliasy, M. Mohammadi, A. Safarzadeh, K. Azarbaijani, S. Shahsavari, M. Alizadeh, and F. Beyranvand. 2018. "A review of the most effective medicinal plants for dermatophytosis in traditional medicine." *Bio Res Therapy* 5 (6):2378–88.

Sethi, G., M. K. Shanmugam, S. Warrier, M. Merarchi, F. Arfuso, A. P. Kumar, and A. Bishayee. 2018. "Pro-apoptotic and anti-cancer properties of diosgenin: A comprehensive and critical review." *Nutrients* 10 (5). doi: 10.3390/nu10050645.

Shabbeer, S., M. Sobolewski, R. K. Anchoori, S. Kachhap, M. Hidalgo, A. Jimeno, N. Davidson, M. A. Carducci, and S. R. Khan. 2009. "Fenugreek: A naturally occurring edible spice as an anticancer agent." *Cancer Biol Ther* 8 (3):272–8. doi: 10.4161/cbt.8.3.7443.

Shamshad Begum, S., H. K. Jayalakshmi, H. G. Vidyavathi, G. Gopakumar, I. Abin, M. Balu, K. Geetha, S. V. Suresha, M. Vasundhara, and I. M. Krishnakumar. 2016. "A novel extract of fenugreek husk (FenuSMART) alleviates postmenopausal symptoms and helps to establish the hormonal balance: A randomized, double-blind, placebo-controlled study." *Phytother Res* 30 (11):1775–84. doi: 10.1002/ptr.5680.

Sharma, S., and K. R. Rai. 2019. "Chronic lymphocytic leukemia (CLL) treatment: So many choices, such great options." *Cancer* 125 (9):1432–40.

Sharma, V., P. Singh, and A. Rani. 2017. "Antimicrobial activity of *Trigonella foenum-graecum* L. (fenugreek)." *Eur J Exper Biol* 07 (1(4)):1–4. doi: 10.21767/2248-9215.100004.

Shibeshi, M. A., Z. D. Kifle, and S. A. Atnafie. 2020. "Antimalarial Drug Resistance and Novel Targets for Antimalarial Drug Discovery." *Infect Drug Resist* 13:4047–60. doi: 10.2147/idr.S279433.

Shilovskiy, I. P., M. S. Sundukova, Babakhin A. A., A. R. Gaisina, A. V. Maerle, I. V. Sergeev, A. A. Nikolskiy, E. D. Barvinckaya, V. I. Kovchina, D. A. Kudlay, A. A. Nikonova, and M. R. Khaitov. 2019. "Experimental protocol for development of adjuvant-free murine chronic model of allergic asthma." *J Immunol Methods* 468:10–19. doi: 10.1016/j.jim.2019.03.002.

Shim, S. H., E. J. Lee, J. S. Kim, S. S. Kang, H. Ha, H. Y. Lee, C. Kim, J. H. Lee, and K. H. Son. 2008. "Rat growth-hormone release stimulators from fenugreek seeds." *Chem Biodivers* 5 (9):1753–61. doi: 10.1002/cbdv.200890164.

Shishodia, S., and B. B. Aggarwal. 2006. "Diosgenin inhibits osteoclastogenesis, invasion, and proliferation through the downregulation of Akt, I kappa B kinase activation and NF-kappa B-regulated gene expression." *Oncogene* 25 (10):1463–73. doi: 10.1038/sj.onc.1209194.

Shoemaker, T., C. Boulianne, M. J. Vincent, L. Pezzanite, M. M. Al-Qahtani, Y. Al-Mazrou, A. S. Khan, P. E. Rollin, R. Swanepoel, T. G. Ksiazek, and S. T. Nichol. 2002. "Genetic analysis of viruses associated with emergence of Rift Valley fever in Saudi Arabia and Yemen, 2000–01." *Emerg. Infect. Dis.* 8 (12):1415–20.

Shum, B. O., M. S. Rolph, and W. A. Sewell. 2008. "Mechanisms in allergic airway inflammation—lessons from studies in the mouse." *Expert Rev Mol Med* 10.

Siciliano, V., E. C. Nista, T. Rosà, M. Brigida, and F. Franceschi. 2020. "Clinical management of infectious diarrhea." *Reviews on Recent Clinical Trials* 15 (4):298–308.

Siddiqui, F. J., G. Belayneh, and Z. A. Bhutta. 2021. "Nutrition and diarrheal disease and enteric pathogens." In *Nutrition and Infectious Diseases: Shifting the Clinical Paradigm*, edited by Humphries, D.L., Scott, M.E. and Vermund, S.H., 219–41. Cham: Springer International Publishing.

Sies, H., C. Berndt, and D. P. Jones. 2017. "Oxidative stress." *Ann Rev Biochem* 86:715–48.

Sim, T. F., J. Sherriff, H. L. Hattingh, R. Parsons, and L. B. Tee. 2013. "The use of herbal medicines during breastfeeding: A population-based survey in Western Australia." *BMC Complement Altern Med* 13:317. doi: 10.1186/1472-6882-13-317.

Sindhu, G., M. Ratheesh, G. L. Shyni, B. Nambisan, and A. Helen. 2012. "Anti-inflammatory and antioxidative effects of mucilage of Trigonella foenum-graecum (Fenugreek) on adjuvant induced arthritic rats." *Int Immunopharmacol* 12 (1):205–11. doi: 10.1016/j.intimp.2011.11.012.

Sindhu, G., M. Ratheesh, G. L. Shyni, B. Nambisan, and A. Helen. 2018. "Evaluation of anti-arthritic potential of Trigonella foenum-graecum L. (Fenugreek) mucilage against rheumatoid arthritis." *Prostaglandins Other Lipid Mediat* 138:48–53. doi: 10.1016/j.prostaglandins.2018.08.002.

Singh, J., T. Dutta, K.-H. Kim, M. Rawat, P. Samddar, and P. Kumar. 2018. "'Green' synthesis of metals and their oxide nanoparticles: Applications for environmental remediation." *J Nanobiotechnol* 16 (1):84. doi: 10.1186/s12951-018-0408-4.

Singh, S. R., J. S. Singh, K. Anand, L. Singh, and V. P. Shahi. 2008. "Performance of medicinal plant-based cropping system and changes in soil fertility status of aquic hapludoll of Uttarakhand." *J Indian Soc Soil Sci* 56 (4):442–7.

Slichenmyer, W. J., and D. D. Von Hoff. 1990. "New natural products in cancer chemotherapy." *The J Clin Pharmacol* 30 (9):770–88. doi: 10.1002/j.1552-4604.1990.tb01873.x.

Sobhy, M. Y., S. Abd-Alrahman, M. Salem-Bekhet, and A. A. Mostafa. 2013. "Chemical characterization and antibacterial activity of ethanolic extract of *Trigonella foenum-graecum* (fenugreek) seeds." *J Pure and Appl Microbiol* 7:1373–8.

Song, X., Z. Wang, H. Liang, W. Zhang, Y. Ye, H. Li, Y. Hu, Y. Zhang, H. Weng, J. Lu, X. Wang, M. Li, Y. Liu, and J. Gu. 2017. "Dioscin Induces Gallbladder Cancer Apoptosis by Inhibiting ROS-Mediated PI3K/AKT Signalling." *Int J Bio Sci* 13 (6):782–93. doi: 10.7150/ijbs.18732.

Sreeja, S., V. S. Anju, and S. Sreeja. 2010. "In vitro estrogenic activities of fenugreek Trigonella foenum-graecum seeds." *Indian J Med Res* 131:814–9.

Srinivasan, K. 2019. "Chapter 3.15 — Fenugreek (Trigonella foenum-graecum L.) seeds used as functional food supplements to derive diverse health benefits." In *Nonvitamin and Nonmineral Nutritional Supplements*, edited by Nabavi, S.M. and Silva, A.S., 217–21. Cambridge: Academic Press.

Srinivasan, S., S. Koduru, R. Kumar, G. Venguswamy, N. Kyprianou, and C. Damodaran. 2009. "Diosgenin targets Akt-mediated prosurvival signaling in human breast cancer cells." *Int J Cancer* 125 (4):961–7. doi: 10.1002/ijc.24419.

Srivastava, A., Z. Singh, V. Verma, and T. Choedon. 2020. "Potential health benefits of fenugreek with multiple pharmacological properties." In *Ethnopharmacological Investigation of Indian Spices*, 137–53. Hershey: IGI Global.

Strati, P., J. H. Uhm, T. J. Kaufmann, C. Nabhan, S. A. Parikh, C. A. Hanson, K. G. Chaffee, T. G. Call, and T. D. Shanafelt. 2016. "Prevalence and characteristics of central nervous system involvement by chronic lymphocytic leukemia." *Haematologica* 101 (4):458.

Sudan, P., M. Goswami, and J. Singh. 2020. "Antifungal potential of Fenugreek Seeds (Trigonella foenum-graecum) Crude Extracts against Microsporum gypseum." *Int J Res Pharm Sci* 11:646–9. doi: 10.26452/ijrps.v11i1.1870.

Sur, P., M. Das, A. Gomes, J. R. Vedasiromoni, N. P. Sahu, S. Banerjee, R. M. Sharma, and D. K. Ganguly. 2001. "*Trigonella foenum-graecum* (fenugreek) seed extract as an antineoplastic agent." *Phytother Res* 15 (3):257–9. doi: 10.1002/ptr.718.

Suresh, P., N. Kavitha Ch, S. M. Babu, V. P. Reddy, and A. K. Latha. 2012. "Effect of ethanol extract of Trigonella foenum-graecum (Fenugreek) seeds on Freund's adjuvant-induced arthritis in albino rats." *Inflammation* 35 (4):1314–21. doi: 10.1007/s10753-012-9444-7.

Tang, T., K. Tay, R. Q. Lee, M. Tao, S.-Y. Tan, L. Tan, and S. Lim. 2010. "Peripheral T-Cell Lymphoma: Review and Updates of Current Management Strategies." *Adv Hematol* 2010:624040. doi: 10.1155/2010/624040.

Tang, Y.-Q., Q. Ye, H. Huang, and W.-Y. Zheng. 2020. "An Overview of Available Antimalarials: Discovery, Mode of Action and Drug Resistance." *Curr Mol Med* 20 (8):583–92. doi: 10.2174/1566524020666200207123253.

Tanwar, A. K., N. Dhiman, A. Kumar, and V. Jaitak. 2020. "Engagement of Phytoestrogens in Breast Cancer suppression: Structural classification and mechanistic approach." *Eur J Med Chem*:113037.

Taudorf, E. H., G. B. E. Jemec, R. J. Hay, and D. M. L. Saunte. 2019. "Cutaneous candidiasis—an evidence-based review of topical and systemic treatments to inform clinical practice." *J Eur Acad Dermatol Venereol* 33 (10):1863–73. doi: 10.1111/jdv.15782.

Tenali, R., N. Badri, J. Kandati, and M. Ponugoti. 2018. "Prevalence of intestinal parasitic infections in cases of diarrhea among school children attending a tertiary care hospital: A two-year study." *Int J Contemp Pediatr*. 5 (3):873–8.

Thomas, J. V., J. Rao, F. John, S. Begum, B. Maliakel, K. Im, and A. Khanna. 2020. "Phytoestrogenic effect of fenugreek seed extract helps in ameliorating the leg pain and vasomotor symptoms in postmenopausal women: A randomized, double-blinded, placebo-controlled study." *PharmaNutrition* 14:100209. doi: 10.1016/j.phanu.2020.100209.

Tin, S. T., G. K. Reeves, and T. J. Key. 2021. "Endogenous hormones and risk of invasive breast cancer in pre- and post-menopausal women: Findings from the UK Biobank." *Br J Cancer*:1–9.

Tripathi, S., A. Maurya, M. Kahrana, A. Kaul, and D. R. Sahu. 2012. "Immunomodulatory property of ethanolic extract of Trigonella Foenum-Graeceum leaves on mice." *Der Pharmacia Lettre* 4 (2):708–13.

Turkyılmaz, C., E. Onal, I. M. Hirfanoglu, O. Turan, E. Koç, E. Ergenekon, and Y. Atalay. 2011. "The effect of galactagogue herbal tea on breast milk production and short-term catch-up of birth weight in the first week of life." *The J Alternat Complement Med* 17 (2):139–42.

Turvey, S. E., and D. H. Broide. 2010. "Innate immunity." *J Allergy Clin Immunol* 125 (Suppl 2):S24–32. doi: 10.1016/j.jaci.2009.07.016.

Upadhyay, R. K., S. Ahmad, G. Jaiswal, P. Dwivedi, and R. Tripathi. 2008. "Antimicrobial effects of *Cleome viscosa* and *Trigonella foenum-graecum* seed extracts." *Journal of Cell and Tissue Research* 8 (2):1355–60.

Uzor, P. F., V. D. Prasasty, and C. O. Agubata. 2020. "Natural products as sources of antimalarial drugs." *Evid Based Complement Alternat Med* 2020:9385125. doi: 10.1155/2020/9385125.

Valdes-Ramos, R., and A. Benitez-Arciniega. 2007. "Nutrition and immunity in cancer." *The Br J Nutr* 98 (Suppl 1):S127–32. doi: 10.1017/S0007114507833009.

Varadarajan, S., M. Narasimhan, M. Malaisamy, and C. Duraipandian. 2015. "Invitro anti-mycotic activity of hydro alcoholic extracts of some Indian medicinal plants against fluconazole resistant *Candida albicans*." *J Clin Diagn Res* 9 (8):Zc07–10. doi: 10.7860/jcdr/2015/14178.6273.

Venkatachalam, K., R. Vinayagam, M. Arokia Vijaya Anand, N. M. Isa, and R. Ponnaiyan. 2020. "Biochemical and molecular aspects of 1,2-dimethylhydrazine (DMH)-induced colon carcinogenesis: A review." *Toxicology Research* 9 (1):2–18. doi: 10.1093/toxres/tfaa004.

Ventola, C. L. 2015. "The antibiotic resistance crisis: Part 2: Management strategies and new agents." *Pharmacy and Therapeutics* 40 (5):344.

Vinothini, G., R. S. Murugan, and S. Nagini. 2009. "Evaluation of molecular markers in a rat model of mammary carcinogenesis." *Oncology Research Featuring Preclinical and Clinical Cancer Therapeutics* 17 (10):483–93.

Walaa, F., W. Emeish, and A. El-Deen. 2016. "Immunomodulatory effects of thyme and fenugreek in *Sharptooth catfish, Clarias gariepinus*." *Assiut Veterinary Medical Journal* 62 (150):1–7.

Walli, R., R. Al-Musrati, H. Esthtewi, and F. M. Sherif. 2015. "Screening of antimicrobial activity of fenugreek seeds." *Pharm Pharmacol Int J* 2:00028. doi: 10.15406/ppij.2015.02.00028.

Wang, C., X. Huo, L. Wang, Q. Meng, Z. Liu, Q. Liu, H. Sun, P. Sun, J. Peng, and K. Liu. 2016. "Dioscin strengthens the efficiency of adriamycin in MCF-7 and MCF-7/ADR cells through autophagy induction: More than just down-regulation of MDR1." *Scientific Rep* 6 (1):28403. doi: 10.1038/srep28403.

Wang, J.-Z., R.-Y. Zhang, and J. Bai. 2020. "An anti-oxidative therapy for ameliorating cardiac injuries of critically ill COVID-19-infected patients." *Int J Cardiol* 312:137.

Wang, L., Q. Meng, C. Wang, Q. Liu, J. Peng, X. Huo, H. Sun, X. Ma, and K. Liu. 2013. "Dioscin restores the activity of the anticancer agent Adriamycin in multidrug-resistant human leukemia K562/Adriamycin cells by down-regulating MDR1 via a mechanism involving NF-κB signaling inhibition." *J Natur Products* 76 (5):909–14. doi: 10.1021/np400071c.

Wang, L., C. Wang, J. Peng, Q. Liu, Q. Meng, H. Sun, X. Huo, P. Sun, X. Yang, Y. Zhen, and K. Liu. 2014. "Dioscin enhances methotrexate absorption by down-regulating MDR1 in vitro and in vivo." *Toxicol Appl Pharmacol* 277 (2):146–54. doi: 10.1016/j.taap.2014.03.013.

Wang, S. L., B. Cai, C. B. Cui, H. W. Liu, C. F. Wu, and X. S. Yao. 2004. "Diosgenin-3-O-alpha-L-rhamnopyranosyl-(1→4)-beta-D-glucopyranoside obtained as a new anticancer agent from Dioscorea futschauensis induces apoptosis on human colon carcinoma HCT-15 cells via mitochondria-controlled apoptotic pathway." *J Asian Nat Prod Res* 6 (2):115–25. doi: 10.1080/1028602031000147357.

Warpeha, K. M., V. Munster, C. Mullié, and S. H. Chen. 2020. "Editorial: Emerging infectious and vector-borne diseases: A global challenge." *Frontiers in Public Health* 8 (214). doi: 10.3389/fpubh.2020.00214.

Weledji, E., Weledji, and J. Assob. 2017. "Pros, Cons and future of antibiotics." *New Horizons in Translational Medicine* 4 (1):9–14. doi: https://doi.org/10.1016/j.nhtm.2017.08.001.

Wiederhold, N. P. 2017. "Antifungal resistance: Current trends and future strategies to combat." *Infect Drug Resist* 10:249–59. doi: 10.2147/idr.S124918.

Wiseman, A. C. 2016. "Immunosuppressive medications." *Clin J Am Soc Nephrol* 11 (2):332–43. doi: 10.2215/cjn.08570814.

Wooley, P. H. 1991. "Animal models of rheumatoid arthritis." *Curr Opin Rheumatol* 3 (3):407–20. doi: 10.1097/00002281-199106000-00013.

Wu, Y., W. Li, W. Cui, N. A. M. Eskin, and H. D. Goff. 2012. "A molecular modeling approach to understand conformation—functionality relationships of galactomannans with different mannose/galactose ratios." *Food Hydrocolloids* 26 (2):359–64. doi: 10.1016/j.foodhyd.2011.02.029.

Wu, Z., X. Han, G. Tan, Q. Zhu, H. Chen, Y. Xia, J. Gong, Z. Wang, Y. Wang, and J. Yan. 2020. "Dioscin inhibited glycolysis and induced cell apoptosis in colorectal cancer via promoting c-myc Ubiquitination and subsequent hexokinase-2 suppression." *Onco Targets Ther* 13:31–44. doi: 10.2147/ott.S224062.

Xian, D., R. Lai, J. Song, X. Xiong, and J. Zhong. 2019. "Emerging Perspective: Role of Increased ROS and Redox Imbalance in Skin Carcinogenesis." *Oxidative Medicine and Cellular Longevity* 2019:8127362. doi: 10.1155/2019/8127362.

Xiao, J., G. Zhang, B. Li, Y. Wu, X. Liu, Y. Tan, and B. Du. 2016. "Dioscin augments HSV-tk-mediated suicide gene therapy for melanoma by promoting connexin-based intercellular communication." *Oncotarget* 8 (1).

Xu, X.-H., T. Li, C. M. V. Fong, X. Chen, X.-J. Chen, Y.-T. Wang, M.-Q. Huang, and J.-J. Lu. 2016. "Saponins from Chinese medicines as anticancer agents." *Molecules* 21 (10):1326.

Yang, H.-P., L. Yue, W.-W. Jiang, Q. Liu, J.-P. Kou, and B.-Y. Yu. 2013. "Diosgenin inhibits tumor necrosis factor-induced tissue factor activity and expression in THP-1 cells via down-regulation of the NF-κB, Akt, and MAPK signaling pathways." *Chinese J Natur Med* 11 (6):608–15. doi: 10.1016/S1875-5364(13)60070-9.

Yang, L., S. Ren, F. Xu, Z. Ma, X. Liu, and L. Wang. 2019. "Recent advances in the pharmacological activities of dioscin." *BioMed Res Int* 2019:5763602. doi: 10.1155/2019/5763602.

Yao, D., B. Zhang, J. Zhu, Q. Zhang, Y. Hu, S. Wang, Y. Wang, H. Cao, and J. Xiao. 2019. "Advances on application of fenugreek seeds as functional foods: Pharmacology, clinical application, products, patents and market." *Crit Rev Food Sci Nutr*:1–11. doi: 10.1080/10408398.2019.1635567.

Yap, H. Y., S. Z. Tee, M. M. Wong, S. K. Chow, S. C. Peh, and S. Y. Teow. 2018. "Pathogenic role of immune cells in rheumatoid arthritis: Implications in clinical treatment and biomarker development." *Cells* 7 (10). doi: 10.3390/cells7100161.

Yu, H., H. Liang, M. Ren, K. Ji, Q. Yang, X. Ge, B. Xi, and L. Pan. 2019. "Effects of dietary fenugreek seed extracts on growth performance, plasma biochemical parameters, lipid metabolism, Nrf2 antioxidant capacity and immune response of juvenile blunt snout bream (Megalobrama amblycephala)." *Fish Shellfish Immunol* 94:211–19. doi: 10.1016/j.fsi.2019.09.018.

Yu, H., Y. Liu, C. Niu, and Y. Cheng. 2018. "Diosgenin increased DDX3 expression in hepatocellular carcinoma." *Am J Transl Res* 10 (11):3590–99.

Yum, C. H., H. J. You, and G. E. Ji. 2010. "Cytotoxicity of dioscin and biotransformed fenugreek." *J the Korean Soc Appl Biological Chem* 53 (4):470–7. doi: 10.3839/jksabc.2010.072.

Zailei, R., S. Ali Abdulhadi, F. Khorshid, H. Albar, and G. Karrouf. 2016. "Ameliorative effect of *Trigonella foenum-graecum* (Fenugreek) seeds infusion on mouse lymphocytic leukemia cells induced ascities in mice." *Global J Pharmacol* 10:74–81.

Zentek, J., S. Gartner, L. Tedin, K. Manner, A. Mader, and W. Vahjen. 2013. "Fenugreek seed affects intestinal microbiota and immunological variables in piglets after weaning." *Br J Nutr* 109 (5):859–66. doi: 10.1017/S000711451200219X.

Zhang, J., J.-J. Xie, S.-J. Zhou, J. Chen, Q. Hu, J.-X. Pu, and J.-L. Lu. 2019. Diosgenin inhibits the expression of NEDD4 in prostate cancer cells. *Am J Transl Res* 11 (6): 3461–71. Accessed 2019.

Zhu, X., X. Chen Z Fau—Li, and X. Li. "Diosgenin inhibits the proliferation, migration and invasion of the optic nerve sheath meningioma cells via induction of mitochondrial-mediated apoptosis, autophagy and G0/G1 cell cycle arrest." *J. BUON* 25 (1):508–513.

Zitvogel, L., F. Pietrocola, and G. Kroemer. 2017. "Nutrition, inflammation and cancer." *Nature Immunol* 18 (8):843–50. doi: 10.1038/ni.3754.

Part V Formulations and Food Preparations

20 Fenugreek
Novel Delivery Technologies and Versatile Formulation Excipients

Ujjwala Kandekar, Sunil Ramdasi and Prasad Thakurdesai

20.1 Introduction 362
20.2 Physicochemical Properties of Fenugreek Phytoconstituents 362
    20.2.1 Galactomannan (Gum/Mucilage) 362
        20.2.1.1 Solubility 363
        20.2.1.2 Molecular Weight and Intrinsic Viscosity 363
        20.2.1.3 Rheological Properties 363
        20.2.1.4 Surface and Interfacial Tension 363
        20.2.1.5 Water-Holding Capacity 364
        20.2.1.6 Swelling Index 364
    20.2.2 Trigonelline 364
    20.2.3 Diosgenin 364
    20.2.4 4-Hydroxyisoleucine 364
20.3 Pharmacokinetic Properties of Fenugreek and Constituents 365
    20.3.1 Trigonelline (Oral and Intravenous Route) 365
    20.3.2 Diosgenin and Trigonelline (Intravenous and Oral Route) 365
    20.3.3 Lutein in Fenugreek Leaves (Oral Route) 366
    20.3.4 Saponin and Sapogenins (Oral Route) 366
    20.3.5 Vicenin-1 from Fenugreek Seeds (Oral Route) 366
    20.3.6 Furostenol Glycoside from Fenugreek Seeds (Oral Route) 366
20.4 Formulations—Fenugreek Constituents as a Bioactive Ingredient 367
    20.4.1 Fenugreek Formulations for Skin and Hair Care 371
    20.4.2 Fenugreek Topical Formulations against Inflammation and Pain 372
    20.4.3 Fenugreek Nano-Formulations as Antifungal Medications 372
    20.4.4 Fenugreek Nano-Formulations for Complex Wound Dressings 373
20.5 Formulations—Fenugreek Constituents as Excipients 374
    20.5.1 Fenugreek Gum as Natural Super-Disintegrant 375
    20.5.2 Fenugreek Gum as Binding Agent 375
    20.5.3 Fenugreek Gum as Emulsion Stabilizer 376
    20.5.4 Fenugreek Gum as a Suspending Agent 377
    20.5.5 Fenugreek Fiber as a Bioavailability Enhancer 377
    20.5.6 Fenugreek Galactomannan for Sustained and Controlled Release Tablets 378
    20.5.7 Fenugreek Gum Micelles for Liver-Targeted Oral Delivery 379
    20.5.8 Fenugreek Gum for Colon-Targeted Drug Delivery 379
    20.5.9 Fenugreek Gum for Microencapsulation of Probiotics 380
    20.5.10 Fenugreek Gum-Based Silica Lipid Drug Delivery System 380
    20.5.11 Fenugreek Mucilage for Controlled Release Beads 381
    20.5.12 Nanocomposite for Lung Delivery 381

20.5.13	Fenugreek Gum as Mucoadhesive Nasal Gel	382
20.5.14	Fenugreek Seed Mucilage for Mucoadhesive Buccal Patch	382
20.5.15	Nanoparticulate Formulations Ocular Delivery	382
20.5.16	Vaginal Applications of Bio-Adhesive Film and Cream	383
20.5.17	Aerogel	383
20.5.18	Biodegradable Graft Copolymer	384
20.5.19	Fenugreek in Formulations of Nanoparticles	384
20.6 Conclusion		386
References		387

ABBREVIATIONS

AO/EB: Acridine orange/Ethidium bromide; ASGP-R: Asialoglycoprotein receptor; Cmax: Maximum serum concentration; CMC: Critical micelle concentration; HPLC: High performance liquid chromatography; HepG2: Liver hepatocellular cells; HPLC-DAD: High performance liquid chromatography with diode-array detector; HPLC-ESI-MS/MS: high-performance liquid chromatography/electrospray ionization tandem mass spectrometry; HPMC: Hydroxy propyl methyl cellulose; MCC: Microcrystalline cellulose; MCF-7: Michigan cancer foundation-7; MRT: Mean residence time; PEG: Polyethylene glycol; PK: Pharmacokinetic; PLGA: Polylactic-co-glycolic acid; PVP: Polyvinyl pyrrolidone; SHRs: Spontaneously hypertensive rats; T1/2: Half-life (time required for a quantity to reduce to half of its initial value); Tmax: Time to achieve maximum concentration

20.1 INTRODUCTION

Fenugreek leaves and seeds have been used universally as spices, an edible vegetable source, nutritional supplement, and medicine. Many formulas and formulations containing whole or part of fenugreek are documented in ancient literature of traditional medicinal systems to prevent and treat numerous disorders (Rose, Patel, and Asha 2020). Recently, fenugreek constituents have gained importance in the modern pharmaceutical industry as an active ingredient or as functional excipients. The present chapter describes the physicochemical properties of fenugreek and its phytoconstituents and their use in various modern pharmaceutical formulation applications.

20.2 PHYSICOCHEMICAL PROPERTIES OF FENUGREEK PHYTOCONSTITUENTS

Fenugreek comprises coumarin, amino acids, saponins, flavonoids, fibers, vitamin A and C, minerals, lipids, mucilage, carbohydrates and protein, and volatile aromatic oil (Sowmya and Rajyalakshmi 1999; Wani and Kumar 2018). Each of these constituents' presence and concentration impart characteristic physicochemical properties to the different parts of the fenugreek. These physicochemical properties are described next.

20.2.1 Galactomannan (Gum/Mucilage)

Fenugreek gum is a storage polysaccharide present in the seed endosperm; it is also called galactomannan. Fenugreek contains both high and low molecular weight galactomannans. It has been reported as a linear mannan backbone, made up of β (1→4) linked mannopyranosyl units to which galactopyranosyl group attaches via α (1→6) linkage. The molar ratio of galactose to mannose is reported as 1:1, 1:1.08, 1:1.5, 1:2.5, (Liu et al. 2020; Madar and Shomer 1990; Maier et al. 1993a,

1993b; Prajapati, Jani, Moradiya, and Randeria 2013; Prajapati, Jani, Moradiya, Randeria et al. 2013). The yield of gum varies from 13.6 to 38%. Approximately 48% galactose is reported in fenugreek gum, which provides higher solubility due to presence of hydroxyl groups (Mathur and Mathur 2005; Meghwal and Goswami 2012). Because of suitable viscosity and gelling properties, fenugreek galactomannan was explored for various pharmaceutical applications such as emulsion stabilizer, sustained release agent, and gelling agent (Kumar et al. 2017).

20.2.1.1 Solubility

Solubility is an essential parameter for the dissolution and bioavailability of drugs (Patel and Jain 2014). The solubility of galactomannan depends on the pattern and degree of substitution of the galactosyl unit. The higher degree of galactose substitution in fenugreek offers a significantly higher number of hydroxyl groups resulting in enhanced binding with the water molecules. Similarly, mannose is a linear unit that shares properties like cellulose and insolubility in water (Mittal, Mattu, and Kaur 2016; Pollard and Fischer 2006; Prajapati, Jani, Moradiya, and Randeria 2013). Fenugreek gum exhibits an almost 1:1 ratio galactose to mannose with more galactose units. Therefore, fenugreek galactomannan's solubility is higher than other galactomannans such as guar gum and locust bean gum (Izydorczyk, Cui, and Wang 2005). The solubility of the gums has a positive correlation with temperature. The higher molecular weight of galactomannan makes it solubilize at a higher temperature than the lower molecular weight galactomannans. Other factors affecting galactomannan solubility are particle size, pH, impurities, time spent, and dissolution media (Rodriguez-Canto et al. 2019).

20.2.1.2 Molecular Weight and Intrinsic Viscosity

The molecular weight and intrinsic viscosity of galactomannan were found to be 1,418,000 and 9.61(dl/g), respectively, determined by high-performance size exclusion chromatography (Brummer, Cui, and Wang 2003). Despite the high molecular weight, the intrinsic viscosity is lower due to the lower chain dimensions by galactosyl residues on the mannose backbone (Liu et al. 2020; Petkowicz, Reicher, and Mazeau 1998; Petkowicz et al. 1999). The inverse correlation was observed with the extent of aggregation and degree of substitution of mannan backbone by galactosyl residues. The higher degree of substitution resulted in a reduction in intermolecular interaction, which leads to a reduction in viscosity (Brummer, Cui, and Wang 2003).

20.2.1.3 Rheological Properties

Fenugreek galactomannan shows Newtonian behavior at a lower concentration. At higher concentrations, it follows pseudoplastic flow where the viscosity decreases with a higher shear rate and then set at a minimum value depending on concentration. The lower viscosity of fenugreek gum could be correlated with a lower intrinsic viscosity (Brummer, Cui, and Wang 2003). The viscosity of fenugreek gum decreases with rising temperature and exhibited non-Newtonian flow except a very dilute solution. If fully hydrated gum solutions were subjected to gradually raised temperatures from 30 °C to 90 °C, the viscosity was reduced. When the time of hydration was less than 12 h, the viscosity increased with time.

Conversely, after 12 h, gum begins to be degraded and lose its viscosity (Jiang et al. 2007). It does not interact with other polysaccharides to impart synergistic viscosity owing to a fully substituted backbone (Mathur and Mathur 2005). Due to its rheological attributes, it has been explored as an emulsifying agent, mucoadhesive, and surfactant in pharmaceutical formulation (Gharibzahedi, Razavi, and Mousavi 2012; Hefnawy and Ramadan 2011; Huang, Kakuda, and Ciu 2001; Jiang et al. 2007; Nayak et al. 2012, 2013; Nayak and Pal 2014; Siew and Williams 2008).

20.2.1.4 Surface and Interfacial Tension

Surface tension is a force acting at a right angle to a line of unit length present at the surface. Along with interfacial tension, surface tension plays a vital role in the stability of emulsion and suspension.

The interfacial tension is the force acting per unit length existing at the interface between two immiscible liquids. Fenugreek gum exhibits the ability to reduce the surface tension of water (Garti et al. 1997). The oil and water interfacial tension was reduced remarkably with fenugreek gum (Akiladevi et al. 2019). Some researchers reported that removing protein does not alter the surface activity of fenugreek gum though other galactomannans like guar gum lose the surface activity upon removal of protein (Garti et al. 1997).

20.2.1.5 Water-Holding Capacity

The ability of gum to hold the water is water-holding capacity. Residual proteins and an ample number of hydroxyl groups majorly contribute to moisture absorption in polysaccharides (Naqvi et al. 2011). Pore size, capillary, and the charges on the protein molecules are the determining factors of water-holding capacity (Rashid, Hussain, and Ahmed 2018). The presence of protein and mineral content in fenugreek galactomannan contributes to superior water-holding capacity (Rashid, Hussain, and Ahmed 2018). The water-holding capacity of galactomannan plays a vital role in the preparation of controlled release formulations and hydrogels.

20.2.1.6 Swelling Index

The swelling index is the volume (mL) taken up by the swelling of 1 g of material. Swelling index of fenugreek galactomannan is directly proportional to the concentration of mucilage (Kumar et al. 2014). Higher swelling index is attributed to increased viscosity due to water absorption. Therefore, fenugreek mucilage or galactomannan can be used as a thickening and controlled release agent. The high viscosity results in a gel-like three-dimensional network with aqueous channels that might hinder the drug's release (Kulkarni et al. 2005).

20.2.2 Trigonelline

Trigonelline is a pyridine alkaloid, vitamin B6 derivative, and bitter. It is soluble in water and warm alcohol. Upon heating, it is converted into nicotinic acid (Zhou, Chan, and Zhou 2012). The content of trigonelline in fenugreek seeds is about 0.1–0.15% of seed weight. The maximum content of trigonelline across all varieties is around 2000 ppm (Mohan and Hari 2016). Germination of fenugreek seeds for 12 days in the presence of 20 µM arginine and 75 mM NaCl resulted in the highest trigonelline (Naghdi Badi et al. 2018). It exists as a solid powder in a physical state. It is very soluble in water and alcohol; practically insoluble in ether and chloroform (PubChem 2004).

20.2.3 Diosgenin

Diosgenin is found in oily embryos of fenugreek seed in the range of 0.1–0.9%. It is spirostanol saponin [(25R)-spirost-5-en-3b-ol], composed of a hydrophilic sugar moiety linked to a hydrophobic steroid aglycone (Wani and Kumar 2018). The content of diosgenin and plant growth was found to be enhanced when fenugreek seeds were treated with silver nanoparticles (Jasim et al. 2017). It is available as white needle crystal or light amorphous solid powder. It is strongly hydrophobic (Log P- 5.7) and insoluble in water. The diosgenin is soluble in acetic ether, propyl acetate, acetone, and alcohol (Chen et al. 2012). The solubility in acetic ether, propyl acetate, is found to increase at higher temperatures (Chen et al. 2012). Diosgenin shows good stability under exposure to light or high temperature but is unstable in the presence of hydrochloric acid (Cai et al. 2020).

20.2.4 4-Hydroxyisoleucine

It is the free amino acid found specifically in selected plants of the *Trigonella* genus. It comprises nearly up to 80% of the total amino acids (0.015–0.4%) of fenugreek seeds (Hajimehdipoor et al. 2010). It is derived from isoleucine hydroxylation and exists in two isomeric forms, and major fraction constitutes 2S, 3R, 4S while the minor exhibits 2R, 3R, 4S configuration. Optical rotation yields a value of [αD] + 30.5 (c = 1, H2O), [αD] + 31 (c = 1.1, H2O). It is available as white powder or flakes. The melting point is 223.5 °C. The molecular weight is 147.1 as per the mass spectrometric evaluation (Hari and Mohan 2014). Under acidic conditions, the linear form may convert to lactone form. The linear form of isoleucine is biologically active. In S-confirmation, full methylation at carbon-α and carbon-γ hydroxylation are essential for insulinotropic activity (Broca et al. 2000).

20.3 PHARMACOKINETIC PROPERTIES OF FENUGREEK AND CONSTITUENTS

20.3.1 TRIGONELLINE (ORAL AND INTRAVENOUS ROUTE)

The major alkaloid of fenugreek seeds, trigonelline, is known for its efficacy against diabetes melilotus (Shah et al. 2006; Subramanian and Prasath 2014; Upaganlawar, Badole, and Bodhankar 2013; Zhou, Chan, and Zhou 2012). The pharmacokinetics evaluation using specially developed and validated HPLC methods suggested slow absorption, longer half life, and high resident time of trigonelline in diabetic rats than normal rats (Manwatkar et al. 2008). Single oral administration of trigonelline (75 mg/kg) to rats showed significantly different pharmacokinetic parameters, namely Tmax (120 min v/s 60 min), Cmax (28.52 µg/ml v/s 42.26 µg/ml), $AUC_{0 \to \infty}$ (459.9713 µg/ml v/s 190.0253 µg/ml), $t_{1/2}$/Half-life (24 h v/s 5.13 h), and mean resident time (MRT) (10.6 h v/s 6.29 h) (Manwatkar et al. 2008).

In another set of studies, development and validation of HPLC-DAD assay (Makowska et al. 2014) and HPLC-ESI-MS/MS method (Szczesny et al. 2018) for determination of trigonelline concentration-time profile in mice blood was reported, but pharmacokinetic properties were not calculated in these reports (Makowska et al. 2014; Szczesny et al. 2018).

Recently, the development and validation of bioanalytical methods for trigonelline determination in human plasma with bio extraction using a magnetic nanocomposite have been reported (Mohamadi et al. 2021). Single oral administration of trigonelline to humans showed a rapid absorption, middle rate of elimination with the fitted two-compartment model of pharmacokinetics (Tmax = 1 h; Cmax = 0.115 µg/ml, T1/2α = 0.79 h and t1/2β = 13.68 h, MRT ~24 h).

20.3.2 DIOSGENIN AND TRIGONELLINE (INTRAVENOUS AND ORAL ROUTE)

The development and validation of the simultaneous spectrophotometric determination of trigonelline, diosgenin, and nicotinic acid in tablet, capsule, and thin films prepared from fenugreek seed extract, using the mathematical algorithm based UV-spectrophotometric method is reported (Mohamadi, Pournamdari et al. 2020). The maximum absorbance (λmax) was determined for trigonelline, diosgenin, and nicotinic acid to be 232.65 nm, 296.23 nm, and 262.60 nm, respectively (Mohamadi, Pournamdari et al. 2020).

The pharmacokinetics profile of diosgenin and trigonelline extracted from fenugreek seeds extract can be best fitted in a two-compartment model as per the recently reported study using the ion-pairing reversed-phase HPLC method in rabbit plasma (Mohamadi, Pournamdari et al. 2020; Mohamadi, Sharififar et al. 2020). To determine diosgenin and trigonelline in rabbits, a plasma micro-extraction method using bentonite/β cyclodextrin-coated magnetic nanoparticles were coupled with iron-pairing reverse phase HPLC. The pure diosgenin: Cmax = 0.48mg/ml; Tmax = 1 h; $T_{1/2}$ = 6.23 h. The trigonelline PK parameters: Cmax = 0.121 mg/ml; Tmax= 1.30 h and T1/2 = 20.06 h. The diosgenin was rapidly absorbed after single oral administration with short residence time and metabolism in the liver before attaining the plasma concentration resulting in poor bioavailability. These results might be attributed to low lipid solubility and poor distribution in lipophilic cells. Trigonelline was also rapidly absorbed with moderate residence time, resulting in high bioavailability (Mohamadi, Sharififar et al. 2020).

20.3.3 Lutein in Fenugreek Leaves (Oral Route)

Fenugreek leaves are a good source of lutein, a carotenoid with many health benefits (Abdel-Aal el et al. 2013; Bartlett and Eperjesi 2003; Thammanna et al. 2010). Fatty substances such as cooking oils such as olive, sunflower, or groundnut oil are reported to increase lutein bioavailability from fenugreek leaves (Lakshminarayana et al. 2009). Oral supplementation of an isocratic diet composed of powdered fenugreek leaf (0.421%) as the source of lutein (2.69 mg/kg diet), with 10% of either sunflower oil (linoleic acid source) or olive oil (oleic acid source) or groundnut oil, or 10% soy lecithin (phospholipids source), vitamin mix (1%), mineral mix (3.5%), DL-methionine (0.3%), sucrose (59.57%), casein (20%), cellulose (5%), and choline bitartrate (0.2%) was fed to rats for 4 weeks. The lutein levels in plasma and eyes were higher (20% and 31.3%, respectively) with oil combinations, which suggest higher lutein absorption (Lakshminarayana et al. 2009).

20.3.4 Saponin and Sapogenins (Oral Route)

The recent report demonstrated the gastrointestinal digestion process enhances the permeability of saponins and sapogenins from fenugreek extract (Navarro del Hierro et al. 2020). The parallel artificial membrane permeability assay was used to assess the permeability as an indicator of gastrointestinal absorption and bioavailability. The fenugreek seed extract was digested with the gastric solution and gastric enzymes containing gastric lipase and pepsin. Thereafter, simulated biliary secretion and the pancreatic extract were added to mimic intestinal digestion. Finally, a parallel artificial membrane permeability assay was performed to compare the permeability of saponin and sapogenins digested with the un-digested extract. The digested extract demonstrated significantly higher permeability than un-digested extract (Navarro del Hierro et al. 2020). The presence of bile salts in digested extract might have formed the emulsion, lamellar and micellar structure to remove larger micelles' hinderance in un-digested extract and resulted in higher dispersion, solubility, and permeability (Navarro del Hierro et al. 2020).

20.3.5 Vicenin-1 from Fenugreek Seeds (Oral Route)

The flavonoids are often reported with poor solubility, affecting bioavailability due to their pre-systemic metabolism (Gee et al. 2000; Peng et al. 2018; Zhao, Yang, and Xie 2019). The pharmacokinetic study on Vicenin-1 (a flavonoid 8-C glycoside from fenugreek seeds) reported the highest tissue distribution of Vicenin-1 in the liver followed by the lungs suggesting its potential use in the management of hepatic and respiratory disorders, by oral administration (Kandhare et al. 2016). The rapid and sensitive high-performance liquid chromatography (HPLC) method for the determination of Vicenin-1 in the rat by plasma using solid-liquid extraction (SLE) was developed and validated. The Vicenin-1 on single oral administration (60 mg/kg) is reported to have Cmax = 7.039 lg/mL, Tmax = 2 h, half-life (T1/2) = 11.2 h, MRT = 17.97 h, and excretion (24.2%) in urine, and fits the two-compartment model (Kandhare et al. 2016). The low availability, delayed absorption, and long resident time of Vicenin-1 during this study can be attributed to the variation of gastric emptying and gastrointestinal tract motility, reducing sugars in structure (reduced intestinal absorption), and higher first-pass metabolism (Kandhare et al. 2016).

20.3.6 Furostenol Glycoside from Fenugreek Seeds (Oral Route)

The oral supplementation of furostenol glycoside isolated from fenugreek seed (SFSE-G) is reported with many pharmacological activities (Aswar et al. 2010; Kim et al. 2004; Lee et al. 1993). However, the pharmacokinetic profile of SFSE-G was reported after the development of the rapid, sensitive, and reproducible HPLC method and subsequent evaluation in plasma and urine concentration profiles in rats (Kandhare et al. 2015). Single oral dosing of SFSE-G (200 mg/kg) showed slow distribution and elimination in urine and feces after 24 h (Cmax = 3.03 µg/mL, Tmax = 72 h, MRT = 52 h, and half-life (T1/2) = 40.10 h). The SFSE-G was detected in the lung and brain, indicating its capacity to cross the blood-brain barrier (Kandhare et al. 2015).

20.4 FORMULATIONS—FENUGREEK CONSTITUENTS AS A BIOACTIVE INGREDIENT

Since ancient times, fenugreek has been used by oral route or topical applications (Morcos, Elhaway, and Gabrial 1981). Many oral formulations containing fenugreek and its bioactive constituents as active ingredients in conventional dosage forms such as such as capsules, tablets, packets, solutions, and syrups have been explored for clinical studies during the past few years and are presented as Table 20.1. This section of the chapter reviews the publications involving formulation studies towards non-oral dosage forms (e.g. creams, ointments, transdermal patch, gel and wound dressings) where fenugreek or constituents were used as active ingredient. However, any publications involving multi-herbal formulations containing fenugreek as one of the many ingredients are excluded.

TABLE 20.1
Clinical Trials of Fenugreek as Active Ingredient in Conversional Dosage Forms

Dosage Form	Fenugreek Active	Dosage, Frequency, and Duration	Design, Population/ Indication (Comparator)	Reference
Capsule	Fenugreek seeds	2.5 mg, BID, 3 months	RCT in 60 patients with coronary artery disease (v/s placebo)	Bordia, Verma, and Srivastava (1997)
Capsule	Fenugreek extract rich in 4-hydroxyisoleucine	Extract equivalent to 2 mg/kg 4-HI, single dose	Double-blind, crossover RCT in 6 trained male cyclists (v/s Placebo—microcrystalline cellulose)	Ruby et al. (2005)
Capsule	Fenugreek seed (20%)	1 capsule for 60 days	RCT in 75 postpartum mothers (v/s Molocco + B12 tablets or *Coleus amboinicus* Lour soup)	Damanik, Wahlqvist, and Wattanapenpaiboon (2006)
Capsule	Total saponins from fenugreek	0.35 g/capsule, 6 capsules at a time, TID for 12 weeks	Double-blind RCT in 69, diabetes patients on oral sulphonyura treatment (v/s placebo—Chinese yam)	Lu et al. (2008)
Capsule	Standardized fenugreek seed extract (Torabolic)	500 mg for 8 weeks	Double-blind RCT in 49 resistance trained men (v/s placebo)	Poole et al. (2010)
Capsule	Hydroalcoholic extract of fenugreek seeds	500 mg, BID, for 8 weeks	Double-blind RCT in 58 women with polycystic ovary syndrome (v/s 500 mg metformin)	Bashtian et al. (2013)
Capsule	Fenugreek powder	575 mg per capsule X 3 capsules TID for 21 days	Double-blind RCT in 26 postpartum mothers (v/s placebo— starch)	Reeder, Legrand, and O'connor-Von (2013)
Capsule	Glycosides-based standardized fenugreek seed extract (IND09)	300 mg X 2 capsules, Single dose, 12 h duration with 7 days washout	Double-blind crossover RCT in 60 healthy sedentary male subjects (v/s placebo—lactose)	Mokashi et al. (2014)

(Continued)

TABLE 20.1 (Continued)

Dosage Form	Fenugreek Active	Dosage, Frequency, and Duration	Design, Population/ Indication (Comparator)	Reference
Capsule	Fenugreek seed powder	900 mg, TID, first 3 days of 2 consecutive menstrual cycles	Double-blind RCT in 101 women with primary dysmenorrhea (placebo—potato starch)	Younesy et al. (2014)
Capsule	Fenugreek seeds	610 mg per capsule, TID for 5 days	Triple-blind RCT in 60 postpartum mothers	Brillante and Mantaring (2014)
Capsule	Standardized hydroalcoholic fenugreek seeds extract with trigonelline as a marker compound (IBHB)	300 mg, BID for 6 months	Double-blind RCT in 50 patients of Parkinson's Disease with stable dose of L-Dopa (v/s placebo—di-calcium phosphate)	Nathan et al. (2014)
Capsule	Standardized fenugreek seed extract (Libifem)	600 mg, OD over 8 weeks/two menstrual cycles	Double-blind RCT in 80 healthy menstruating women who reported low sexual drive (v/s placebo—330 mg maltodextrin)	Rao et al. (2015)
Capsule	Fenugreek extract enriched in furostanolic saponins (Furocyst)	500 mg, BID for 90 days	Open-label RCT in 50 women with Polycystic ovary syndrome	Sankhwar et al. (2018), Swaroop et al. (2015)
Capsule	Fenugreek seed hydroalcoholic extract	500 mg, BID/500 mg, OD + glipizide (2.5 mg, OD)—for 12 weeks	Open-label RCT in 60, diabetes patients (v/s glipizide 5 mg, OD)	Singh, Rai, and Mahajan (2016)
Capsule	Standardized fenugreek seed extract	600 mg OD, for 12 weeks.	Double-blind RCT in 120 healthy subjects	Rao et al. (2016)
Capsule	Furostenol glycosides-based fenugreek seeds extract (Fenu-FG)	300 mg, BID for 8 weeks	Double-blind RCT in 60 subjects (v/s placebo—di-calcium phosphate)	Wankhede, Mohan, and Thakurdesai (2016)
Capsule	Fenugreek husk extract (FenuSMART)	500 mg, OD 45 days, 1000 mg OD for day 45 to day 90	Double-blind RCT in 88 women with postmenopausal discomforts (v/s placebo—microcrystalline cellulose)	Shamshad Begum et al. (2016)
Capsule	Fenugreek seed powder	3 g (3 X 1g) BID from day 1 to 3 of menstrual cycle	Open-label RCT in 60 primary dysmenorrhea patients (v/s 500 mg mefenamic acid)	Inanmdar et al. (2016)
Capsule	Fenugreek seed extract enriched in 20% protodioscin (Furosap)	500 mg, OD for 12 weeks	Open-label RCT in 50 male subjects	Maheshwari et al. (2017), Swaroop et al. (2017)
Capsule	Blend containing glycoside-based standardized fenugreek seed extract (IBPR)	450 mg IBPR (contains 400 mg of glycoside-based fenugreek seed extract) for 14 days	Double-blind RCT in 20 non-resistance trained young male and female subjects during eccentric exercise bout (v/s placebo)	Wilborn et al. (2017)

Dosage Form	Fenugreek Active	Dosage, Frequency, and Duration	Design, Population/ Indication (Comparator)	Reference
Capsule	Fenugreek extract	1000 mg, TID, for 10 days	Double-blind, RCT in 13 patients with diabetes (v/s placebo—ground corn)	Kiss et al. (2018)
Capsule	Fenugreek seed extract enriched in 20% protodioscin (Furosap)	250 mg, BID for 12 weeks	Double-blind RCT in 40 healthy male athletes (v/s placebo)	Guo et al. (2018)
Capsule	Fenugreek seed powder	2 g, in three divided doses (500 mg after breakfast, 1000 mg after lunch, 500 mg after dinner) for 12 weeks	Open-label RCT in 12 patients with diabetes (v/s 5 mg glibenclimide)	Najdi et al. (2019)
Capsule	Fenugreek seed galactomannan	1 g, OD, 12 weeks	Single-blind RCT in 70 diabetes patients (v/s placebo)	Rashid et al. (2019)
Capsule	Fenugreek seed hydroalcoholic extract	500 mg, BID for 60 days	Triple-blind RCT in 20 acne vulgaris patients (v/s control—125 mg Azithromycin)	Mozhdeh et al. (2019)
Capsule	Low molecular weight galactomannans-based standardized fenugreek seeds extract (LMWGAl-TF, Torabolic)	500 mg, OD for 8 weeks	Double-blind RCT in 24 subjects with high fat (v/s placebo— di-calcium phosphate)	Deshpande et al. (2020)
Capsule	Fenugreek extract with 50% saponin glycosides (Testofen)	300 mg or 600 mg, OD for 12 weeks	Double-blind RCT in 98 healthy exercising males (v/s placebo—maltodextrin)	Rao, Mallard, and Grant (2020)
Capsule	Fenugreek extract (AlphaFen)	400 mg or 500 mg, OD for 60 days	Double-blind RCT in 57 healthy males (v/s placebo)	Hausenblas et al. (2020)
Capsule	Fenugreek husk extract (FenuSMART)	250 mg, BID for 42 days	Double-blind RCT in 48 women with postmenopausal discomforts (v/s placebo—microcrystalline cellulose)	Khanna et al. (2020), Thomas et al. (2020)
Capsule	Hydroalcoholic extract of Fenugreek seeds	500 mg, BID for 3 months	Triple-blind RCT in patients with nonalcoholic fatty liver disease (v/s placebo—rice)	Babaei et al. (2020)
Capsule	Standardized fenugreek seed extract (Testofen)	300 mg, BID for 8 weeks	Double-blind RCT in 100 male with benign prostate hyperplasia (v/s placebo—maltodextrin)	Rao and Grant (2020)
Capsules	Fenugreek seed extract (Fenfuro) enriched in furostanolic saponins	500 mg, BID for 30 days	Double-blind RCT in 154 patients with diabetes (v/s placebo—di-calcium phosphate)	Verma et al. (2016)
Packets	Germinated fenugreek seed powder	12.5 g and 18.0 g in routine diet, OD, 1 month	20 subjects with old age hypercholesterolemia	Sowmya and Rajyalakshmi (1999)

(Continued)

TABLE 20.1 (Continued)

Dosage Form	Fenugreek Active	Dosage, Frequency, and Duration	Design, Population/Indication (Comparator)	Reference
Packets	Fenugreek seed powder	25 g and 50 g, BID, 20 days	Double-blind RCT in 18 subjects with hyperlipidemia (v/s rice and bengal gram powder)	Prasanna (2000)
Packets	Fenugreek seed powder	60 g, for 12 weeks	Double-blind RCT in 61 subjects with hyperlipidemia (v/s placebo—3 g cellulose)	Sajty et al. (2014) Fedacko et al. (2016)
Packets	Fenugreek extract	8 g, OD for 8 weeks	Double-blind RCT in 56 subjects with borderline hyperlipidemia (v/s placebo—starch powder)	Yousefi et al. (2017)
Packets	Fenugreek seeds	5 g, TID for 8 weeks with and without exercise	Open-label RCT in diabetes patients (v/s exercise)	Gholaman (2018)
Packets	Fenugreek powder	10 g BID, with or without nutrition training, 2 months	Double-blind RCT in 12 diabetes patients (v/s placebo—wheat flour)	Hassani et al. (2019)
Packets	Fenugreek seed powder	5 g, TID for 8 weeks	Open-label RCT, 50 patients of diabetes with anti-diabetic medications	Hadi et al. (2020)
Powder	Fenugreek seed powder	25 g in diet for 24 weeks	RCT in 60 diabetic patients on oral hypoglycemic agents (v/s prescribed diet)	Sharma et al. (1996)
Powder	Aqueous extract powder in 10 ml water	40 mg/kg, single dose	Double-blind RCT in 20 male healthy volunteers (v/s placebo— diluted coffee)	Abdel-Barry et al. (2000)
Powder	Fenugreek extract with >75% galactomannans (FenuLife®)	4 g and 8 g, single dose in low-calorie beverage	Single-blind, 3-period, crossover RCT, in 18 obese subjects (v/s control beverage)	Mathern et al. (2009)
Powder	Debitterized fenugreek fiber	5 g, BID for 3 years	Single-blinded RCT in 140 prediabetes subjects (v/s Placebo	Gaddam et al. (2015)
Powder	Fenugreek seed powder	5 g, QID for 8 weeks	Open-label, RCT in 50 patients of diabetes with dyslipidemia on diet and exercise (v/s oral hypoglycemic and hypolipidemic agents)	Kumar et al. (2015)
Powder	Fenugreek seed powder	5 g, TID for 8 weeks	Open-label RCT in 48 patients of diabetes with routine anti-diabetic drugs	Tavakoly et al. (2018)
Powder	Fenugreek flakes (FenuLean™)	5g and 10 g for 45 days	Double blind, RCT in 100 healthy subjects (v/s Control—No intervention)	Mehkri et al. (2019)
Solution	Fenugreek seeds extract	2.5 g and 5 g, soaked in dextrose solution (1:5 (wt:vol)), single dose	Open-label RCT in 166 diabetes patients (v/s placebo drink—dextrose solution, 0.8 g in 25 ml of warm water)	Bawadi et al. (2009)

Dosage Form	Fenugreek Active	Dosage, Frequency, and Duration	Design, Population/ Indication (Comparator)	Reference
Solution	Fenugreek seed extract solution	25 mg, BID, 1 month	RCT in 114 diabetic and hyperlipidemic patients (v/s metformin)	Geberemeskel, Debebe, and Nguse (2019)
Syrup	Fenugreek seed syrup (50 g in 100 ml of syrup)	10 ml, BID for 4 weeks	Double-blind RCT in 79 asthma patients (v/s honey syrup v/s placebo-sugar syrup)	Emtiazy et al. (2018)
Tablet	Fenugreek seed hydroalcoholic extract	1176 mg TID for 6 weeks	Double-blind RCT in 39 overweight subjects (v/s placebo)	Chevassus et al. (2010)
Tablet	Fenugreek extract	300 mg, BID, for 6 weeks	Double-blind RCT in 60 male subjects (v/s placebo—50 mg rice bran)	Steels, Rao, and Vitetta (2011)

RCT—Randomized controlled trial, OD—Once a day, BID—Twice a day, TID—Three times a day, QID—four times a day.

20.4.1 Fenugreek Formulations for Skin and Hair Care

Skin is a protective organ, interfaces with the environment, and is highly vulnerable to infections and climatic changes. The structure and functions of the skin are highly complicated (Monteiro-Riviere 2006). Therefore, the skin barrier and mechanical properties of the formulation and compatibility of the formulation with skin are essential parameters. Numerous herbal formulations have been applied to the skin for cosmetic and therapeutic purposes (Ghafari, Fahimi, and Sahranavard 2017). Many natural gums and mucilage are explored as safe excipients for skin (Bhasha et al. 2013; Djiobie Tchienou et al. 2018).

Melanin pigments impart color to skin and hair, and excess of melanin secretion leads to hyperpigmentation and melasma (Ogbechie-Godec and Elbuluk 2017). Stable water-in-oil emulsion composed of methanolic extract of fenugreek (4%) was found to reduce the skin melanin and insignificant transepidermal water loss when evaluated for 6 weeks in human volunteers (Waqas et al. 2010). Fenugreek seed extract was incorporated in the dispersed water phase and liquid paraffin was used as an oil phase (external phase). The formulation was tested and found stable regarding color, liquefaction, and phase separation (Waqas et al. 2010). Subsequently, a cream formulation containing a stable water-in-oil emulsion of methanol extract of fenugreek seeds and olive oil was reported to be homogeneous, non-greasy with good consistency, spreadability, and stability at the accelerated conditions (Gade, More, and Bhalerao 2015). The formulation was reported to improve skin hydration and reduction in melanin and was devoid of skin irritation or allergic reactions (Gade, More, and Bhalerao 2015).

The water-in-oil cream formulation containing 4% methanolic extract of fenugreek seeds in the internal phase was found to improve the mechanical parameters such as hydration, elasticity, and resistance to photo-aging during 28 days of applications on the cheeks of 10 male human volunteers (Akhtar et al. 2010). Researchers attributed the skin hydration properties of galactomannan as the possible cause of improved elasticity and skin recovery and a major contribution of the presence of antioxidants (flavones and polyphenols) for reduction of photo-aging (Akhtar et al. 2010).

The six months of oral supplementation (2 capsules, containing 300 mg extract, equivalent to 1200 mg fenugreek seeds) was reported to have favorable effects as a conditioner and reduced hair loss in 60 men and women with low to moderate hair loss during a mono-centric, randomized, double-blind, placebo-controlled clinical study (Medizin et al. 2006). Shampoos are preparations used in hair care for cleansing or therapeutic purposes to impart conditioning and shine to the hair (Trüeb 2001). The shampoo formulation of hydroalcoholic extract of fenugreek seeds was reported to have good thixotropy properties, pH, foam formation, viscosity, conditioning and wettability, and user satisfaction (Noudeh et al. 2011). The shampoo was formulated using sodium lauryl sulfate, sodium sulfosuccinate, N-alkyl betaine, coconut fatty acid diethanol, and aqueous demineralized water solution containing propyl and butyl paraben (1%) (Noudeh et al. 2011). Furthermore, the effectiveness and acceptability of shampoo was evaluated as compared with standard marketed conditioner shampoo, during a double-blind, crossover study in 24 healthy female subjects for 2 weeks (use of shampoo a minimum of three times weekly), after a 1-week washout period (Noudeh et al. 2011). Fenugreek-containing shampoo showed excellent acceptability with high foam production and shiny hair, which is attributed to the presence of saponins and trigonelline content, respectively. Researchers also suggested the contribution of protein, lecithin, and amino acid content of fenugreek seed imparts hair strength and softness (Noudeh et al. 2011).

20.4.2 Fenugreek Topical Formulations against Inflammation and Pain

A transdermal patch is designed to release a controlled dose of a drug through the skin for a prolonged time (Saroha, Yadav, and Sharma 2011). The transdermal patch containing trigonelline (10%)-based standardized fenugreek extract (carbomer-940 and polyethylene glycol) was reported to decrease pain score, post-surgery demand for subcutaneous morphine in comparison to diclofenac dermal patch (1%) in patients of inguinal hernia post-operative pain during a double-blind placebo-controlled clinical study (Ansari et al. 2019). Gradual reduction in pain during 6 h and after 24 h of surgery was reported on patients treated with fenugreek patch as measured by a visual analogue scale (the pain sensation in the range of 0 to 10) with reduced side effects of morphine and no allergic side effects (Ansari et al. 2019).

Cream formulations are semisolid dosage forms containing one or more drug substances dissolved or dispersed in a suitable base (Buhse et al. 2005). The natural polymers are suggested to enhance texture, viscosity of external phase, rheological behavior, and storage stability of creams (Gade, More, and Bhalerao 2015). The cream formulation of fenugreek methanolic extract (2–5% concentrations) reported significant anti-inflammatory activity similar to 1% hydrocortisone cream against inflammation during carrageenan induce rat paw edema (Sharififar, Khazaeli, and Alli 2009).

Ointments are semisolid formulations designed for external application to skin or mucus membrane. Active ingredients are either dissolved or dispersed in various ointment bases as per the need (De Villiers 2009a). Many fenugreek seed constituents such as flavonoids, alkaloids, and saponins are reported with anti-inflammatory activity (Ahmadiani et al. 2001) through inhibition of proinflammatory mediators such as cyclooxygenase, nitric acid synthase, and lipoxygenase (Belguith-Hadriche et al. 2013). At the same time, a ointment containing ethanolic extract of fenugreek seeds (5% w/w) was found effective as an anti-inflammatory agent against carrageenan-induced paw edema in rats (Jyothi, Koland, and Priya 2014). The formulation had a brown color, an aromatic odor, good homogeneity, optimum pH, viscosity, and spreadability (Jyothi, Koland, and Priya 2014). A subsequent study reported anti-inflammatory and analgesic efficacy of gel formulations of ethanolic extract of fenugreek seeds against carrageenan-induced paw edema in rats (Jyothi and Koland 2016). A gel is a three-dimensional cross-linked polymer network swollen in a liquid medium (water) (Almdal et al. 1993). Three variations of fenugreek extract gels were prepared by dispersing ethanolic extract fenugreek seeds with carbopol-934 or HPMC K4M or their combination as a gelling agent. All gels exhibited good viscosity, spreadability, and homogeneity, but gel containing a combination of carbopol-934 and HPMC K4M showed better release profiles of polyphenol content (alleged active constituent) (Jyothi and Koland 2016).

20.4.3 Fenugreek Nano-Formulations as Antifungal Medications

The fungi, *Malassezia furfur*, *Candida albicans*, and *Aspergillus niger* are the most common causes of fungal pathogenic infections to the human body such as dandruff and seborrheic dermatitis to produce an itchy scalp and skin, and flaky scales (Gubitosa et al. 2019; Trueb 2007). Recently, gel formulation containing aqueous fenugreek leaf extract (30%) with sodium alginate as a gelling agent, was reported to have significant antifungal and anti-dandruff activity against *Malassezia furfur*, *Candida albicans*, and *Aspergillus niger* during 3-month accelerated stability studies without irritation or sensitization in New Zealand rabbits (Kulkarni et al. 2020). Furthermore, the gel was found safe in rabbits, which were devoid of skin irritation (Kulkarni et al. 2020). Fenugreek leaf extract could be an effective and safe alternative to the synthetic anti-dandruff agents.

Another recent study on hydroalcoholic extract of fenugreek leaves containing silver-zinc oxide nanocomposite reported its antioxidant, antimicrobial, and antifungal potential (Noohpisheh et al. 2020). Nanocomposites were synthesized by simple hydrothermal method (Baruwati and Varma 2009). The fenugreek silver-zinc oxide nanocomposite showed antimicrobial activity against gram-negative bacteria (*Escherichia coli*), gram-positive bacteria (*Staphylococcus aureus*), and fungi (*Candida albicans*), and free radical scavenging activity *in vitro* (Noohpisheh et al. 2020).

20.4.4 FENUGREEK NANO-FORMULATIONS FOR COMPLEX WOUND DRESSINGS

Complex wounds are well-known difficult wounds, such as pressure sores, post-infection soft-tissue gangrenes, or chronic or acute venous ulcers, that challenge medical and nursing teams (Ferreira et al. 2006). These wounds do not heal with conventional dressings and demand specialized care (Ferreira et al. 2006). In the last few years, nanotechnological research has found it useful to develop strategies using different biocompatible nanoparticles and nano-formulations with specific properties as wound dressing materials.

The nanofiber (fibers in the nanometer range) is known to have great potential in drug delivery owing to its unique characteristics such as its 3D network structure, high surface area, porosity, and tunable pore size (Barhoum et al. 2019). Many researchers explored have the potential of fenugreek extract and its constituents such as gum and fibers as nanoparticles (Bakthavatchalam et al. 2017; Muniyan et al. 2017; Pooloth 2013; Singh, Magesh, and Rakkiyappan 2011; Ying et al. 2019), nanofibers (Selvaraj et al. 2018), and nanocomposites (Deng et al. 2020; Memis et al. 2017) as novel wound-healing materials.

Silver nanoparticles conjugated with fenugreek saponins and *quillia saponaria* extract showed antimicrobial potential against multidrug-resistant bacteria that cause severe burn wound infections (Muniyan et al. 2017). Saponin solution was incorporated in silver nitrate at a concentration of 30, 60, 120, 240, and 500 mg/mL to prepare silver nanoparticles. The saponin content of fenugreek is used to produce a reduction of silver nitrate for enhanced antibacterial activity with a positive concentration-activity correlation. Silver ions and related compounds are toxic to microorganisms and used to treat various infections (Sondi and Salopek-Sondi 2004). Saponins are known to act on lipopolysaccharides of the bacterial cell wall and perhaps increase the permeability of the cell wall to silver to produce synergistic effects (Arabski et al. 2012).

The preparation and evaluation of ethanolic extract of fenugreek leaves containing silver nanoparticles (highly stable, monodispersed, spherical shaped with the size ranging from 20 to 30 nm) with antibacterial activities against pathogenic gram-positive (*Staphylococcus aureus*) and gram-negative (*Escherichia coli*) bacteria was also reported (Bakthavatchalam et al. 2017). The mechanism of nucleation of a silver nanoparticle, subsequent adsorption of molecules from the phytochemicals of the extract, and forming the layers to grow its size under influence of coulombic forces' circular canal-like structure with serrated boundary around the silver nanoparticles were suggested to be correlated to the presence of poorly soluble and insoluble molecules of the fenugreek phytochemicals (Singh, Sadasivam, and Rakkiyappan 2011). A similar study on fenugreek seed extract containing silver nitrate nanoparticles and their antimicrobial activity is reported (Pooloth 2013), further substantiating the antimicrobial potential of fenugreek extract-incorporated silver nanoparticles.

The incorporation of fenugreek seed absolute in collagen-silk fibroin was reported for better morphology, more tensile strength, biocompatibility, and cell viability as compared to collagen fibers for bioactive wound dressing (Selvaraj et al. 2018). Collagen and silk fibroin was dissolved in 1, 1, 1, 3, 3, 3-Hexafluoro 2 propanol in various ratios, mainly 100:0, 75:25, 50:50, 25:75, 0:100, and 1.5% of fenugreek was added to the solution, which was electrospun to obtain nanofibers. During *in vivo* experiment in rats, the topical application of collagen-silk fibroin-fenugreek (50:50:50) nanofibers for 16 days was reported to enhance wound closure, early epithelialization, and complete wound closure with minimal inflammation to skin (Selvaraj et al. 2018). The nanofibers also demonstrated antioxidant potential (*in vitro*) probably due to presence of fenugreek constituents such as trigonelline, nicotinic acid, saponins, and polyphenols in fenugreek (Selvaraj et al. 2018).

In another study, the design, characterization, and evaluation of the wound dressing system of silver nanoparticles fabricated with fenugreek aqueous leaf extract was reported for effective wound healing in rats (Ying et al. 2019). The fenugreek extract containing nanoparticles were found to be stable, well-dispersed, spherical, and small well-formed crystals with uniform and distinct lattice fringes with good stability (6 months). The application of fenugreek-sliver nanoparticles to the rats' wounds showed good biocompatibility tensile strength with complete, more uniform epithelialization within 17 days, while the control group required 25 days (Ying et al. 2019).

Nanocomposites are nanosized high-performance combinations of two or more different materials and mixed to obtain the best advantages of both (Camargo, Satyanarayana, and Wypych 2009). Nanocomposites are based on natural polymers developed owing to their biodegradable properties and other benefits (Adeosun et al. 2012). Fenugreek seed gum and nanoclay composite films exhibited better mechanical strength and showed antimicrobial effects (Memis et al. 2017). Nanocomposite films were composed of 1.5% w/v of fenugreek and nanoclays were made of mainly montmorillonite, halloysite, and nanomer in the concentration of 2.5, 5, and 7.5% w/v. The fenugreek seed gum-containing nanocomposite film had higher moisture content than the films reinforced with nanoclays, perhaps because of the water-holding ability of fenugreek gum. Significant reduction in oxygen permeability was found in films reinforced with fenugreek gum. The antimicrobial activity of fenugreek-based nanocomposite might be the result of the inherent bactericidal effect of a few residual components of fenugreek (Memis et al. 2017).

Because of the high water-holding capacity, fenugreek gum could be effectively used as antimicrobial packing material for hydrogel formulations for wound-healing effects. Hydrogels are hydrophilic three-dimensional structures having the ability to hold a large amount of water (Ahmed 2015). Composite hydrogel from fenugreek gum and cellulose was reported as porous, thermally stable with good water retention capability, with sustained effects that are desirable for good wound-healing formulation (Deng et al. 2020). Furthermore, the fenugreek gum hydrogel formulation (20% and 40%) prevented loss of blood and better wound-healing rate with excellent biocompatibility during 10-day application to wounds in mice (Deng et al. 2020). Researchers suggested the fenugreek gum hydrogel's property of increased expression of factors such as vascular endothelial growth factor (VEGF) and transforming growth factor beta (TGF-β) (Cheng et al. 2018; Miscianinov et al. 2018) might be responsible for promoting neovascularization, tissue regeneration, and stimulated collagen synthesis and accelerated wound healing (Deng et al. 2020).

20.5 FORMULATIONS—FENUGREEK CONSTITUENTS AS EXCIPIENTS

Natural excipients, especially plant gums and mucilage, are attractive alternatives to the synthetic excipients because of their easy availability, biocompatibility, biodegradability, low toxicity, environmental friendliness, and low price (Avachat, Dash, and Shrotriya 2011; Satturwar, Fulzele, and Dorle 2003). Most of the uses of excipients occur in oral dosage forms, which are most convenient, cost effective, and preferred for patient acceptance (Homayun, Lin, and Choi 2019; Liu et al. 2003).

Fenugreek gum's mucilage and husk are the most prominent amongst them, and have been explored as excipients during manufacturing of synthetic medicines in various dosage forms. Fenugreek polysaccharide of galactomannan, in the firm of gum, or mucilage, had been widely reported as a natural pharmaceutical excipient in oral dosage forms (Adhikari et al. 2010; Bahadur et al. 2016; Kaltsa et al. 2016; Kaltsa, Yanniotis, and Mandala 2016; Kumar et al. 2009, 2017; Nayak, Pal, and Das 2013; Nayak et al. 2013; Tavakoli et al. 2012). Along with conventional dosage forms (oral, topical, external, and intravenous route), fenugreek constituents are reported for design and development of novel delivery systems such as nanoparticles, aerogels, nanocomposites, and graft copolymers. The following section reviews all excipient applications of fenugreek constituents.

20.5.1 Fenugreek Gum as Natural Super-Disintegrant

The active ingredients in solid oral dosage forms such as tablets or capsules need to be dissolved and disintegrated to initiate its medicinal actions. The disintegrants are used to facilitate the breakup or disintegration of the tablet or capsule into smaller particles, eventually resulting in the faster dissolution of the tablet. The super-disintegrant are effective in lower concentrations and have higher disintegration efficiency and mechanical strength than normal disintegrants (Mohanachandran, Sindhumol, and Kiran 2011).

Fenugreek gum demonstrated better super-disintegrant properties than synthetic alternatives during a fast-dissolving tablet formulation study of diclofenac sodium (50 mg) (Kumar and Babu 2014). The fenugreek gum (6%)-incorporated diclofenac sodium tablets showed mechanical strength, wetting time (indicative of high fluid absorption ability) with fast disintegration time (21 sec), higher release rate (93.74% at 25 min) with additive anti-inflammatory activity *in vivo* against carrageenan-induced rat paw edema in rats (Kumar and Babu 2014).

Fenugreek seed mucilage (1%–6%) matrix-forming agent demonstrated super-disintegrant properties (better fast disintegration and high crushing strength) during a formulation study of orodispersible pharmaceutical lyophilizates of meloxicam (Iurian et al. 2017). The results tablet was dry and elegant, with highly porous structure, with a longer disintegration time as compared with control formulation using gelatin (Iurian et al. 2017).

20.5.2 Fenugreek Gum as Binding Agent

Binding agents or binders are common excipients in tablet formulation to enhance the bonding strength of drug particles or granules by imparting plasticity, providing the cohesiveness that is required under compression to form a tablet (Tavakoli et al. 2012).

Dried and powdered fenugreek raw husk (4–5% concentration, particle size 180–250 μ), was reported as a better granulating agent than starch during tablet formulation of paracetamol and diclofenac with better friability, hardness, disintegration, weight variation, and dissolution parameters (Avachat et al. 2007). The raw husk showed excellent physicochemical properties swelling index, flow properties, and particle size distribution. The fenugreek husk and gunda glue-containing tablets showed excellent binding with minimum capping tendencies which are important for therapeutic efficacy (Avachat et al. 2007). A similar study was performed using starch extracted from fenugreek seed husk for formulation of paracetamol tablets for better binder properties than that of gelatin (Doharey and Sharma 2010). The dried fenugreek husk was separated from the 'core and oily portion' of seeds to yield about 40% w/w of the dry husk (particle size 180–250 μ). Two separate batches of tablets containing 400 mg of paracetamol were made by wet granulation method, each with fenugreek husk and gelatin and a common disintegrant and lubricant. Fenugreek husk-containing tablets showed comparative effectiveness and tablet properties as that of gelatin with respect to hardness, friability, weight uniformity, disintegration, and dissolution profiles (Doharey and Sharma 2010).

Fenugreek gum (mucilage) as a natural tablet binder in concentration of 2.5% w/w during a formulation study of three model drugs of diverse solubility characteristics (freely soluble (Calcium acetate), slightly soluble (theophylline), and practically water insoluble (ibuprofen)) showed comparable tablet characteristics as that of the polyvinyl pyrrolidone (PVP) and corn starch (Tavakoli et al. 2012). A fenugreek gum spreads more readily than starch paste in powder mass. Granules prepared from fenugreek gum (0.1–2.5%) exhibited good flow characteristics. The increased concentration of gum in the range of 0.1–0.5% w/w enhanced the disintegration time, hardness, and reduced the friability. These findings might be attributed to the gel-forming potential of gum (Tavakoli et al. 2012).

Use of fenugreek seed mucilage (FSM) was reported to be not only useful as a binder but also provided additional anti-diabetic efficacy to glipizide tablets (Bahadur et al. 2016). The study utilized nine formulations (F1 to F7) of the glipizide tablets with fenugreek seed mucilage (40, 80, and 120 mg) and microcrystalline cellulose, MCC, (150, 200, and 250 mg) as independent variables. The dependent (response) variables were % drug release at 1 h and 12 h and time required for 50% release (T50). All formulated tablets showed compliance with the requirement of United States Pharmacopeia NF 24/19 for hardness and friability, no drug-polymer compatibility, and positive effects during *in vitro* release studies. The oral administration of optimum selected formulation—F7 (FSM (120 mg) + MCC (150 mg)) showed better reduction of blood glucose level against streptozotocin-induced hyperglycemia as compared with glipizide alone) (Bahadur et al. 2016).

20.5.3 Fenugreek Gum as Emulsion Stabilizer

The emulsion is a bi-phasic dispersed system containing oil and water phases stabilized by an emulsifying agent. As emulsions are highly unstable systems, emulsifying agents play a vital role in emulsion formulation (Akbari 2018). Many natural gums and mucilage including fenugreek gum (mucilage, galacromenan) were reviewed for their emulsifying, surfactant, and thickening properties (Prajapati, Jani, Moradiya, and Randeria 2013), with fenugreek gum having the highest stabilizing properties (Prajapati, Jani, Moradiya, and Randeria 2013).

Fenugreek galactomannan is a natural polysaccharide, covalently or non-covalently attached to hydrophobic proteins and useful as an emulsion stabilizer that adsorbs at the oil-water interface and stabilizes the oil layer (Rashid, Hussain, and Ahmed 2018; Siew and Williams 2008). The concentration of 0.5% w/v is reported as optimum for emulsion stabilization (Gharibzahedi, Razavi, and Mousavi 2012; Huang, Kakuda, and Ciu 2001). An exhaustive comparative study of emulsions using 14 hydrocolloidal gums including fenugreek gum demonstrated better efficiency than others gums (propylene glycol alginate, gellan, carrageenan, pectin, methylcellulose, microcrystalline cellulose, gum arabic, locust bean gum, guar, xanthan, mustard, flaxseed, and oat) (Huang, Kakuda, and Ciu 2001). During this study, fenugreek gum produced the most stable emulsion with desirable properties such as lower interfacial free energy, capacity to develop very small and uniform oil droplets (70% < 1 μm), and narrow particle size distribution (Huang, Kakuda, and Ciu 2001).

The emulsifying ability of fenugreek gum in combination with soy protein isolate was found to increase by 4-fold along with improvement in solubility (Hefnawy and Ramadan 2011). Various blends of soy protein isolate and fenugreek gums (3:1, 2:1, 1:1, 1:2, 1:3, and 1:4) were reported with a positive correlation between concentration and emulsifying activity. The combination of 1:2 showed optimum emulsifying stability in the pH range of 3–9, in various salt concentrations and under heat treatments, which was 3 and 2 times higher than soy protein isolate and standard bovine serum albumin respectively (Hefnawy and Ramadan 2011).

Fenugreek gum–soy whey protein isolate conjugate showed better protein solubility and emulsifying properties than non-conjugated mixture, especially in the pH range 3–8 at 22 °C at isoelectric point after undergoing a Maillard-type reaction in a controlled dry state condition for 3 days (Kasran, Cui, and Goff 2013b). This property is important as proteins are prone to aggregate near the isoelectric pH of protein and could destabilize the emulsion. Heating of the solutions and the presence of high salt concentration, conjugate shows improved emulsification properties more than soy whey protein isolate alone (Kasran, Cui, and Goff 2013b). A subsequent study confirmed unhydrolyzed fenugreek gum has better emulsification properties as compared to its partially hydrolyzed counterpart when conjugating with soy whey protein isolate at 1:3 and 1:5 ratios (Kasran, Cui,

Fenugreek: Novel Delivery Technologies

and Goff 2013a). Olive oil–water micro emulation (with fenugreek gum) showed better viscoelastic properties, consistency, and stability than the nanoemulsion two-step sequential ultrasound process with whey-protein isolate (as emulsifier) followed by fenugreek gum (as stabilizer) in 5 months of storage period (Kaltsa et al. 2016). The preparation, formulation, and stability study of emulsions prepared with high-shear and ultrasonic methods using fenugreek gum with different concentrations of galactomannans and soy protein isolates as emulsifying agents was reported (Kaltsa, Yanniotis, and Mandala 2016). The 0.5% concentration of fenugreek gum was found to have optimum viscosity and stability of emulsion with reduced droplet size. Formulations with fenugreek gum were more found more stable than other formulations with guar gum and locust bean gum (Kaltsa, Yanniotis, and Mandala 2016).

Recently, a detailed report on the extraction, purification, and detailed characterization of fenugreek galactomannans revealed excellent functional attributes such as lower water-holding capacity (compared to guar gum), good oil-holding capacity, foaming capacity, swelling index and foaming stability, and excellent emulsifying activity and emulsion stability (Rashid, Hussain, and Ahmed 2018). In addition, a wide melting range, large degree of crystallinity, and a glittering surface makes fenugreek galactomannan an excellent candidate for an industrial purposes study (Rashid, Hussain, and Ahmed 2018).

20.5.4 FENUGREEK GUM AS A SUSPENDING AGENT

The suspension is a bi-phasic dispersed system containing solids dispersed in a liquid phase. The suspensions are a highly unstable formulation, so the addition of suspending agent is required (Yarnykh et al. 2018). Fenugreek gum was reported as an effective suspending agent during many scientific studies (Akiladevi et al. 2019; Nayak et al. 2012; Senthil and Sripreethi 2011).

The formulations of Cefpodoxime proxetil suspension with fenugreek mucilage (as suspending agent) and varying proportions of acacia showed superior stability, greater flocculation, good flow rate with easy dispersibility characteristics than similar formulations with different suspending agent (xanthan gum) (Akiladevi et al. 2019). Powdered fenugreek seed mucilage, extracted from fenugreek seed, showed better suspending agent properties than other natural suspending agents such as acacia and tragacanth, during formulation of paracetamol suspension (Senthil and Sripreethi 2011) and bentonite, during zinc oxide suspensions (Nayak et al. 2012).

20.5.5 FENUGREEK FIBER AS A BIOAVAILABILITY ENHANCER

When administered by the oral route, the drug needs to be solubilized and absorbed, and bioavailable to exert the pharmacological action. Curcuminoids, the bioactive constituents of turmeric (*Curcuma longa* L.) have many medicinal applications. However, poor solubility and permeability limit the oral bioavailability of curcuminoids and become a major for obstacle optimum usage (Anand et al. 2007; Lopresti 2018). Many approaches have been explored to enhance the solubility (Aditya et al. 2015; Kumar, Kesharwani et al. 2016; Li et al. 2016; Mohammadian et al. 2019; Peng et al. 2018; Rezaei and Nasirpour 2018; Sarika et al. 2015), but systemic oral bioavailability of unconjugated curcuminoids remains a challenge (Olotu et al. 2020; Sanidad et al. 2019). A water dispersible formulation of curcuminoids with fenugreek galactomannans ('curcumagalactomannosides' (CGM)) has reported improved blood-brain barrier permeability and tissue distribution of free curcuminoids following its oral administration to rats (Krishnakumar et al. 2015). A subsequent double-blind randomized controlled crossover study in 50 healthy volunteers demonstrated high relative bioavailability (45.5 fold) and pharmacokinetics of free curcuminoids, when formulated with fenugreek dietary fiber as CGM, when compared to unformulated standard curcumin (Kumar, Jacob et al. 2016). These results were attributed to the strong interaction of curcumin with mucoadhesive fenugreek galactomannans to generate sub-micronized colloidal particles of CGM complexes that were resistant to the enzymatic biotransformation (Kumar, Jacob et al. 2016).

20.5.6 FENUGREEK GALACTOMANNAN FOR SUSTAINED AND CONTROLLED RELEASE TABLETS

Sustained and controlled release tablets are intended to modify the drug release rate for a prolonged period with the help of customized polymers (Ainurofiq and Choiri 2015; Cazorla-Luna et al. 2019; Mylangam et al. 2016; Ofori-Kwakye et al. 2016). The sustained release tablets help to maintain the constant plasma of drug concentration and retards the release rate of the drug, thereby extending the duration of action (Sharma et al. 2019).

Many research reports demonstrated the release-retardant potential of fenugreek galactomannan (Kumar and Sinha 2012; Nokhodchi et al. 2008; Sav et al. 2013a; Shukla et al. 2019).

The fenugreek gum demonstrated better controlled release characteristics as compared to hypromellose as a controlled release agent for formulation of propranolol hydrochloride matrix tablets (Nokhodchi et al. 2008). The matrix tablets incorporating the drug with fenugreek mucilage in the ratio of 1:1, 1:1.25, 1:1.5, and 1:2 were prepared and compared to tablets with hypromellose as an alternative polymer. Because of fenugreek mucilage's compactness, enhanced, better friability characteristics and crushing strength were obtained. The dissolution rate of formulation containing fenugreek mucilage was higher at lower concentration (100 mg). However, beyond 200 mg of fenugreek mucilage addition, the release rate of the drug was slower than hypromellose-containing formulation (Nokhodchi et al. 2008).

Sustained release matrix tablets of acarbose (100 mg) with various proportions of fenugreek gum (AF40–60) was able to control drug release for a longer duration of approximately 10–12 h, which is much more than boswellia gum and locust bean gum (Nayak et al. 2012). The fenugreek gum-containing tablets showed first-order release kinetics; faster drug release was observed for the initial 3 h followed by constant drug release, owing to the highly soluble nature of the drug during *in vitro* release studies. Furthermore, the oral administration of formulation AF40 (120 mg of fenugreek gum and 71 mg MCC PH101 and 100 mg acarbose) resulted in a significant reduction in the postprandial blood glucose and triglyceride levels in the diabetic rats (Kumar and Sinha 2012).

Octenyl succinate anhydride derivative of fenugreek was more hydrophobic and exhibited extended release potential (Sav et al. 2013a). Esterification of fenugreek gum was carried out by octenyl succinate anhydride and sodium bicarbonate (catalyst). Swelling behavior of modified gum was found higher at acidic pH, hence it can be employed as a gastro retentive polymer. Tablets prepared with untreated fenugreek gum showed better swelling owing to its hydrophilic nature and released more than 70% of the drug within 1 h. The release profile of the drug was also not consistent. Modified gum alone found to release 25% of drug within 1 h at high concentration but it still could not be extended for a prolonged time. Modified gum in combination with untreated gum demonstrated extended drug release up to 20 h. An optimum combination polymer with both hydrophilic and hydrophobic properties can provide a desired drug release profile (Sav et al. 2013b).

Bioadhesive sustained release tablets of venalflaxin hydrochloride using fenugreek gum in combination with other bioadhesive polymers reported to extend release from 12 h as compared to immediate release (1 h) (Momin, Kane, and Abhang 2015). Bilayer tablets were prepared, composed of sustained and immediate release layers (containing different proportions of fenugreek gum, HPMC K 100M, Carbopol 934P, Xanthan gum, and the drug). The combination of other polymers with fenugreek gum exhibited good bioadhesion, and were stable at different stability conditions. The combination of fenugreek gum with HPMC resulted in the polymer mixture for better bioadhesion and sustained drug release (Momin, Kane, and Abhang 2015).

Matrix tablets using alginate-fenugreek gum gel membrane in formulation of intragastric quetiapine fumarate delivery by combining floating and swelling mechanisms showed better buoyancy, swelling ability, and slower release (Bera et al. 2016). Matrix tablets were prepared by wet granulation and coated with alginate-fenugreek gum and hardened by calcium chloride solution with various combinations using factorial design. The resulting optimized tablets exhibited superior release characteristics such as time taken for 50% drug release (T50% = 24.67 min) with 71.11% drug release at 8 h. The increased floating ability and swelling rate of the tablet was attributed to the tight barrier provided by the alginate-fenugreek layer that hindered the imbibition of fluid (Bera et al. 2016).

Tablets coated with modified fenugreek gum (complex with sodium trimetaphosphate) in combination with sodium alginate was found to produce better retardation of release of nifedipine from tablets as compared to natural fenugreek gum (Shukla et al. 2019). The film-coated tablet of nifedipine using natural and modified fenugreek gum at 2% w/v in combination with 1% sodium alginate showed prolonged retard rate (9% and 11% respectively) as compared with 15% rate shown by tablets using HPMC during dissolution testing (Shukla et al. 2019).

20.5.7 Fenugreek Gum Micelles for Liver-Targeted Oral Delivery

Despite excellent medicinal applications, poor bioavailability to the target organs can be a dose-limiting factor for many drugs (e.g. coumarins) for oral route of administration (Van der Merwe et al. 2020). Polymeric micelles (PMs) are one of the modern approaches to overcoming the limitations of poorly water-soluble drugs (Kapare and Metkar 2020). Micelles are self-assembled, core-shell nanostructures of amphiphilic polymers (both hydrophilic and lipophilic) formed above a certain concentration, critical micelle concentration (CMC) in an aqueous solution (Xu, Ling, and Zhang 2013). Natural polysaccharides have an abundance of hydroxyl, amino, and carboxylic acid functional groups that can be partially grafted by hydrophobic segments to acquire amphiphilic molecules for better solubility (Wang et al. 2017).

The technology of self-assembled nanomicelles was explored to prepare hydrophobically modified fenugreek gum with stearic acid conjugation by esterification reaction and ultrasonication process for a liver-targeted drug delivery system (Zhou et al. 2019). The resulting micelles were spherical, negatively charged, physical stable particles with narrow size distribution. Furthermore, the micelles showed low toxicity on both HepG2 and MCF-7 cells (at a range of 0.1–100 mg/ml) during *in vitro* cytotoxicity assay, low hemolysis during hemolysis assay on rabbit RBCs (*ex vivo*) and enhanced cellular uptake from coumarin-loaded nanomicelles into HepG2 cells (*in vitro*, as compared to free coumarin-6). These results were attributed to galactose residues which can recognize ASGP-R receptors on HepG2 cell surfaces and enhance cellular uptake (Zhou et al. 2019).

20.5.8 Fenugreek Gum for Colon-Targeted Drug Delivery

Colon-targeted drug delivery aims to release the drug in the colon for a local and systemic effect. It is desirable to target the drug into the colon for gastrointestinal diseases such as inflammatory bowel diseases, ulcerative colitis, colonic cancer, and amebiasis (Amidon, Brown, and Dave 2015; Vass et al. 2019). The colon is a suitable absorption site for certain drugs with high molecular weight such as protein and peptides due to the absence of stomach enzymes and less proteolytic activity (Philip and Philip 2010).

Fenugreek gum was studied as an effective and safe excipient for colon-specific drug delivery (Randhawa, Bassi, and Kaur 2012; Sharma et al. 2018).

An inter-polymer complex between carboxymethyl cellulose + fenugreek gum with chitosan protected the tamoxifen matrix tablet to release in stomach and small intestine, and released in the colon for the treatment of colorectal cancer (Randhawa, Bassi, and Kaur 2012). The colon-targeted tablets were prepared by coating tamoxifen (25 mg) core tablets with aqueous solution of carboxymethyl fenugreek gum and chitosan in the ratio of 40:60; this showed prevention of drug release in pH 1.2 (stomach) and pH 7.4 (intestine) and released 90% of the tamoxifen in pH 6.8 (*in vitro* dissolution test) and 94% in rat cecal content within 19 h with Tmax delayed by 10 h (Randhawa, Bassi, and Kaur 2012). Another study of colon-targeted tablet formulation using fenugreek gum resulted in the metronidazole release from tablets in the stomach and reaching the colon within 6–8 h for localized treatment of amoebiasis (Sharma et al. 2018). Tablets of metronidazole (100 mg) were prepared in two stages; core and compression coated by the direct compression method. A mixture of fenugreek polysaccharide and MCC was used as an excipient. The resultant tablets demonstrated higher swelling in simulated intestinal fluid as compared to simulated gastric fluid during *in vitro* dissolution tests. Furthermore, colon-targeted tablets demonstrated pH-dependent drug release and prevented release into the stomach, allowing traces into the intestine and the highest availability into the colon. Release of the drug was found to be hindered in gastric pH and higher in intestinal pH. The increased level of coating resulted in increased lag time of release. The *in vivo* study on human volunteers indicated smaller intestinal transit time (i.e. 3–5 h) with tablets reaching the colon within 6–8 h (Sharma et al. 2018). Researchers attributed these properties to the cleavage of the glycosidic bond from polysaccharides and the migration of the drug into the dissolution media (Sharma et al. 2018). Taken together, the results were suggestive of the suitability of fenugreek gum for colon-targeting formulations (Randhawa, Bassi, and Kaur 2012; Sharma et al. 2018).

20.5.9 Fenugreek Gum for Microencapsulation of Probiotics

Probiotics are live microbial feed supplements, which contribute to maintain a healthy microbial balance in the gut (Kechagia et al. 2013). The viability of probiotics is impeded in the harsh acidic gastric environment. Therefore, some kind of protective mechanisms are necessary to protect bacteria for successful probiotic formulation (Prakash et al. 2011). Microencapsulation is the process to shield the active material from the surrounding environment, where active material is encapsulated in a polymeric shell (Dubey, Shami, and Bhasker Rao 2009).

Fenugreek gum is explored as a polymer for microencapsulation during the herbal gel formulation of probiotic *Lactobacillus plantarum* 15HN, which enhanced oral delivery, facilitated release in colonic conditions for 12 h, and improved survival rate of these probiotic cultures in the gastrointestinal environment (Haghshenas et al. 2015). The encapsulation formulation using Alginate + fenugreek gum (1.5% + 0.5%) blend showed 80% survival rate of viable probiotic cells under both low pH and high bile salt conditions, which is better than non-encapsulated formulations. Furthermore, encapsulated strains exhibited favorable anti-pathogen activity and acceptable antibiotic susceptibility. Non-encapsulated strains were sensitive to low pH condition and lost viability (Haghshenas et al. 2015).

In another study, *Pediococcus pentosaceus KID7*, *Lactobacillus plantarum KII2*, *Lactobacillus fermentum KLAB6*, and *Lactobacillus helveticus KII13* probiotics were encapsulated in an alginate-fenugreek gum-locust bean gum matrix, and the viability of the bacteria were found to be retained in different storage conditions (Damodharan et al. 2017). Polymeric matrices were prepared by varying the ratio of sodium alginate (0.3 to 1.2%), fenugreek gum (0.5 to 1%), and locust bean gum (0.3 to 1%). The combination of sodium alginate (1%), fenugreek gum (0.5%), and locust bean gum (0.5%) showed improved dissolution on colonic fluid, reduced water activity, and improved stability gums (Damodharan et al. 2017).

20.5.10 Fenugreek Gum-Based Silica Lipid Drug Delivery System

Lipid-based formulations are used to enhance the solubility of poorly water-soluble (lipophilic) drugs. These exist in a solid state (more stable and flexible formulation) or liquid state (Schultz et al. 2018). Highly purified solid state formulation using fenugreek gum-based, silica-based lipid systems as hydrophilic solid carriers is reported to enhance the solubility for simvastatin formulation (Sav et al. 2013b). The silica-based drug delivery composed of medium-chain triglycerides as a lipid phase and different grades of colloidal silica were prepared by an emulsion composed of solubilized drug followed by adsorption on hydrophilic solid carriers with 0.6% fenugreek gum as an emulsifier in the first step, to produce a free-flowing powder of simvastatin lipid system. The formulation showed better aqueous solubility, re-dispersibility, excellent flow property, and

3–4-fold increase in dissolution rate in comparison with the pure drug and the marketed formulation. Enhanced solubility and dissolution rate might be due to conversion of the drug into an amorphous state, improved wetting, and facilitated diffusion from the lipid-based system (Sav et al. 2013b). A highly purified fenugreek gum-based silica lipid system offers an alternative to synthetic emulsifier for enhancement of solubility, dissolution, and stability in the oral solid dosage form.

20.5.11 Fenugreek Mucilage for Controlled Release Beads

Beads are spherical particles of size 50 nm to 2 mm, mostly designed for sustained release of drugs. Versatile drug delivery could be attained by the use of a variety of polymers in bead formulation. In recent years, a great deal of attention has been paid to the development of mucoadhesive hydrogel beads using plant-derived mucilages for controlled release drug delivery (Ahmad et al. 2019; Mishra et al. 2021). Many natural polymers exhibited good mucoadhesive properties and are effectively used for the controlled release of the drug in the formulation of beads (Kandekar et al. 2019; Li et al. 2019; Vakili et al. 2019).

Preparation, optimization, and evaluation of mucoadhesive beads of calcium pectinate + fenugreek seed mucilage with metformin hydrochloride demonstrated good mucoadhesive and swelling properties with significant hypoglycemic effects in alloxan-induced diabetic rats over a prolonged period after oral administration (Nayak, Pal, and Das 2013). Similar results were reported in the next series of experiment on fenugreek seed mucilage with alginate (Nayak et al. 2013) and gellan gum (Nayak and Pal 2014).

Floating and bioadhesive diethonolamine-modified, high-methoxyl pectin-alginate beads for controlled intragastric delivery of metformin hydrochloride exhibited excellent drug encapsulation efficiency and sustained release (Bera and Kumar 2018). High-methoxyl pectin, modified by amino de-alkoxylation, and combined with alginate and added zinc acetate was used for metformin hydrochloride entrapment resulting in rigid beads. The prepared formulation was coated with fenugreek gum-alginate composite by diffusion-controlled interfacial complexation technique. The formulation showed less than 5 min of floating lag time, and higher mucoadhesion, and good entrapment efficiency. The researcher attributed restricted and delayed drug release to interaction of hydrophilic polymer with mucin, conversion of pectin into pectinic acid in an acidic environment and formation of a diffusion barrier by coating the membrane to retard metformin release (Bera and Kumar 2018). A similar release profile of glimepiride in the colon was achieved using carboxymethyl fenugreek galactomannan-gellan gum-calcium silicate composite beads (Bera et al. 2018). A strong interaction was noted between carboxylate ion polymers and $Ca^{+2}/Zn^{+2}/Al^{+3}$ ions that were directed to the generation of a cross-linked network. Similarly, all the formulations showed slower drug release in gastric pH and faster in intestinal pH. This might be attributed to ionization of carboxyl moieties and increased hydrophilic characteristics (Bera et al. 2018).

20.5.12 Nanocomposite for Lung Delivery

The nanocomposite of carboxymethyl fenugreek galactomannan-g-poly (N-isopropylacrylamide-co-N, N'-methylene-bis-acrylamide)-bentonite was used for erlotinib (an anticancer agent) formulation, which was reported for its sustained release and greater efficacy than the pristine drug, as detected by cellular uptake analysis, cytotoxicity test, and AO/EB staining assay (to detect morphological changes in carcinoma cells) (Bera et al. 2020). Carboxymethylation of fenugreek gum was carried out by Williamson ether synthesis and grafted with other polymers (N-isopropylacrylamide, N'-methylene-bis-acrylamide, bentonite). The resultant nanocomposites showed pH (higher in neutral and lower in acidic medium) and temperature-dependent swelling (reduced with higher temperature) within 3–5 h, owing to osmotically driven water molecules with biodegradability potential due to enzymes-induced degradation of nanoscaffold. The Erlotinib loaded composites formulation showed an initial burst release and was sustained for 8 h and enhanced efficacy to suppress the proliferation of A549 (human alveolar adenocarcinomic cells) and promoted apoptosis as compared with formulation without nanocomposite (Bera et al. 2020).

20.5.13 Fenugreek Gum as Mucoadhesive Nasal Gel

The nasal drug delivery system is a reliable alternative to oral and more convenient than parenteral drug delivery (Saxena, Arora, and Chaurasia 2019). Nasal gel offers the advantage of prolonged drug release and improves bioavailability along with the reduction in side effects (Vibha 2014).

Use of fenugreek gum (seed mucilage) as a gelling agent for diclofenac potassium nasal gel formulation was found to be well-tolerated without any dermatological reaction (Mundhe, Pagore, and Biyani 2012). Diclofenac potassium at the level of 1% w/v was incorporated in different concentrations of fenugreek seed mucilage, 4.0%, 5.0%, 6.5%, 8.0%, 10% w/v to produce gel formulation. The resulting gel formulation was compatible at nasal pH, had a good swelling index, spreadability, optimum viscosity, satisfactory drug content release (permeability), stability (up to 3 weeks), and biocompatibility. The optimum gel formulation with 8% w/v fenugreek seed mucilage was found to be comparable to anti-inflammatory efficacy with a commercial product (carrageenan-induced rat paw edema) and devoid of dermatological reactions in rats when applied twice a day for 7 days (Mundhe, Pagore, and Biyani 2012).

More recently, the mucoadhesive properties of fenugreek seed mucilage was explored to formulate the domperidone nasal gel using mixed solvency concept (Kumar 2019). The extraction and characterization of fenugreek seed mucilage from fenugreek seed showed higher viscosity (at 1% solution) and mucoadhesive properties (0.5% solution) than synthetic polymers (HPMC and Carbopol 934) (Kumar 2019). Domperodone (0.3%) nasal gel formulation using PVP K-30 + PEG 600 combination, and fenugreek seed mucilage, showed nasal compatible pH, optimum viscosity, and good mucoadhesive properties (Kumar 2019).

20.5.14 Fenugreek Seed Mucilage for Mucoadhesive Buccal Patch

Buccal patches are the formulations are targeted to buccal mucosa for the benefits such as prevention of first pass metabolism and faster delivery (Guo and Pratap-Singh 2019). The use of increasing concentrations of fenugreek seed mucilage (500 mg to 800 mg) with decreasing proportion of hydroxypropyl methyl cellulose (HPMC K4M, 500 mg to 200 mg) and a backing membrane of ethyl cellulose (1%) in buccal patch formulations of atenolol (anti-hypertensive drug) were found satisfactory in terms of average weight, thickness, drug content, folding endurance, and moisture content. The formulation with combination of 800 mg of fenugreek seed mucilage and 200 mg of HPMC K4M produced buccal patch with optimum mucoadhesive strength, force of adhesion, bond strength, first order drug permeation kinetics across the excised porcine buccal mucosa over 12 h (ex vivo) and maximum permeation flux (Adhikari et al. 2010).

20.5.15 Nanoparticulate Formulations Ocular Delivery

Nanoparticles are particles with characteristic dimension from 1 nm to 100 nm and have been widely utilized for various pharmaceutical applications (Karuppusamy and Venkatesan 2017). Nanoparticles of water-soluble fenugreek seed mucilage have been used in combination with chitosan as an alternative for synthetic anionic polymers for ocular targeting formulation systems (Pathak et al. 2014). The fenugreek seed mucilage was isolated, purified, sterilized, and characterized with modern instrumentations. The fenugreek seed mucilage was identified to be a galactomannan chain consisting of 4 units of galactose attached to the backbone of 6 mannose units in a 1:1.5 ratio. The nano-formulation, ocular suspension, with purified fenugreek seed mucilage (0.04% w/v) and chitosan (0.02% w/v), was found to have negative surface charges, having a spherical shape and being sterile, a non-irritant (*in vivo*), and hemocompatible (*ex vivo*) (Pathak et al. 2014).

20.5.16 VAGINAL APPLICATIONS OF BIO-ADHESIVE FILM AND CREAM

Bio-adhesive films are thin films that exhibit the ability to attach to biological surfaces of tissue and extend the drug release. These are mostly made up of natural polymers and composed of carbohydrates and protein (Huanbutta and Sangnim 2020). Amine derivatives of fenugreek gum in bioadhesive films of nystatin were found to exhibit higher bioadhesion as compared to native gum films during antifungal topical formulation, against vaginal candidiasis (Bassi and Kaur 2015). Vaginal candidiasis is an acute inflammatory condition and the most frequent reason for gynecological consultation (Cassone 2015). Antifungal agents such as nystatin are the priority drugs used to treat vaginal candidiasis (Willems et al. 2020). Bioadhesive film of nystatin comprised of aminated gum exhibited higher bioadhesion, extended release (8 h), and better antifungal activity at lower concentrations (Bassi and Kaur 2015), perhaps because of stronger interaction of amino groups with negatively charged mucin chains via electrostatic interaction (Kaur, Mahajan, and Bassi 2013). The better interfacial and dissolution characteristics of a fenugreek-containing formulation might have contributed to better antifungal activity (Bassi and Kaur 2015).

Suppositories are solid dosage formulations with various shapes and weights, developed for drug administration in the vagina, rectum, or urethra of the human body, mostly to exert local effect. These are made up of cream bases that melt at body temperature (De Villiers 2009b). A vaginal application containing 5% of fenugreek extract cream (0.5 g, with the applicator filled to the half-full mark, twice a week for 12 weeks) was reported to reduce atopic vaginitis score and Vaginal Maturation Index (VMI) in 60 postmenopausal Iranian women during a randomized controlled study (Hakimi et al. 2020). However, the efficacy was less than a comparator cream with 0.625 mg of synthetic conjugated estrogens, with applicator containing 0.3 mg of conjugated estrogens). Both formulations were well tolerated, maintained vaginal pH without adverse effects (Hakimi et al. 2020). The researcher attributed the efficacy of the fenugreek extract-containing formulation to constituents such as sapogenin and diosgenin, which can mimic progesterone- and estrogen-like effects respectively (Hakimi et al. 2020).

20.5.17 AEROGEL

Aerogels are solids with lowermost density and very high porosities. The polysaccharide-based aerogels, called organic hydrogels, have a highly porous structure, high surface area, are biocompatible, and can be tuned for numerous pharmaceutical applications (Ulker and Erkey 2014).

Lyophilized TEMPO (an oxidation catalyst) and laccase (an enzyme)-oxidized galactomannan obtained from fenugreek seeds was found to be a versatile drug delivery system for the antibiotic polymyxin B, antimicrobial peptide nisin, and muramidase lysozymes against different bacterial strains (Campia et al. 2017). Oxidized fenugreek-generated water-soluble, structured, elastic, stable hydrogels showed 15-fold viscosity increase, presumably due to establishment of hemiacetalic bonds between newly formed carbonyl groups and free hydroxyl groups (Campia et al. 2017). Upon lyophilization of these hydrogels, water-insoluble aerogels were generated, capable of aqueous solution uptake several times their own weight (Rossi et al. 2017).

Further characterization of chemo-enzymatically oxidized galactomannans aerogels with electrospray ionization mass spectrometry (ESI-MS) after enzymatic depolymerization, scanning electron microscopy, thermal and X-ray analyses confirmed they are more structured and stable as compared to guar and sesbania aerogels (Rossi et al. 2017). Fenugreek gum undergoes more cross-linking with hemiacetalic bonds, and upon oxidation gains the highest viscosity and is found to be stable after the lyophilization (Rossi et al. 2017). The researchers attributed better stability of fenugreek galactomannan aerogels to its higher amount of oxidizable galactose units linked to the mannose backbone (i.e. Gal: Man = 1:1) and, subsequent extensive cross-linking of the resulting high viscosity, elastic gel (Ponzini et al. 2019; Rossi et al. 2017). The follow-up study with detailed characterization confirmed the potential of oxidized galactomannans as an attractive polymers backbone to develop targeted delivery systems with flexible release profile, uptake, and stability characteristics (Ponzini et al. 2019).

20.5.18 BIODEGRADABLE GRAFT COPOLYMER

Biodegradable grafts have versatile drug delivery (Sakhare and Rajput 2017) and medical applications for skin reconstruction and tissue repair (Chan et al. 2000; Chung et al. 2017; Jadoun, Riaz, and Budhiraja 2020). A variety of functional groups can be attached to the polymer to prepare biodegradable grafts (Verbeek and Gavin 2020). Graft copolymers contain at least two different kinds of monomers which are structurally distinct from the main backbone (Sherazi 2016). In the earliest study to produce biodegradable graft, acrylamide was grafted onto fenugreek seed mucilage to produce pure biodegradable copolymer (Mishra et al. 2006). In this reaction, oxidation was proceeded under the nitrogen flushing by single electron transfer and free radical formation on fenugreek mucilage in the presence of acrylamide to form a grafted copolymer with improved characteristics without affecting biodegradability, with partial solubility in water and complete solubility in a solution of basic pH (Mishra et al. 2006).

In separate study, fenugreek galactomannan was grafted with collagen to convert a disordered, unbound state collagen-water network to a structured bound state reported to improve the physicochemical properties of collagen as a smart biomaterial for cosmetics (Kanungo et al. 2014). Collagen-fenugreek galactomannan composites were prepared by mixing in the ratios of collagen and fenugreek 2:1, 1:1, 2:3, and 1:2, with the concentration of collagen fixed at 0.5 mM. The molecular motion of collagen was modulated by fenugreek, that causing the changed solvation dynamics of collagen. The concentration of fenugreek plays a vital role in improved interfacial behavior, clustering of bound water, structure, folding, and stability of collagen (Kanungo et al. 2014).

In the more recent study, the fenugreek seed mucilage grafted to polyvinyl alcohol with acrylamide, free radical polymerization was reported as a controlled release polymer for 95% release of loaded drug Enalapril maleate for 16 h (Bal and Swain 2019). The graft copolymer was prepared by microwave synthesis and varying concentration of ammonium per-sulphate as redox initiator. The resultant graft was a stable and well-formed copolymer with prominent presence of amide and the hydroxyl groups indicating that a grafting mechanism had efficiently taken place. The optimized sample showed higher intrinsic viscosity owing to formation of long polymeric chains that led to higher swelling and delayed drug release (Bal and Swain 2019). Assessment of graft copolymer as a tissue-engineered scaffold using subcutaneous tissue implants of cotton balls to back skin of mice for 15 days revealed rapid reepithelization, complete wound healing, and good biodegradability without infection (Bal and Swain 2019).

Taken together, the use of fenugreek seed mucilage as grafted copolymer is suggested for controlled drug release and in tissue engineering applications (Bal and Swain 2019; Kanungo et al. 2014; Mishra et al. 2006).

20.5.19 FENUGREEK IN FORMULATIONS OF NANOPARTICLES

Nanotechnology has gained huge attention over time due to numerous applications in diverse variety of fields including drug delivery (Anu Mary Ealia and Saravanakumar 2017; Christian et al. 2008; De Jong and Borm 2008; Ghaffari and Dolatabadi 2019; Patra et al. 2018) (Pallotta et al. 2019). The fundamental component of nanotechnology is the nanoparticles (in the size range of 1–100 nm). Because of their small size, nanoparticles impart enhanced properties to drug formulations, such as high reactivity, strength, surface area, sensitivity, and stability (Anu Mary Ealia and Saravanakumar 2017). This section of the chapter reviews the reports of applications of the nanoparticle systems to enhance the medicinal efficacy and optimize the targeted delivery of fenugreek extracts and constituents for management of many diseases and disorders (Aswathy Aromal and Philip 2012; El-Batal, Mosalam et al. 2018; El-Batal et al. 2020; Ghosh et al. 2014; Kestwal, Bagal-Kestwal, and Chiang 2015; Mallikarjuna et al. 2017; Mallikarjuna et al. 2019; Pol 2014; Seetharaman, Balya, and Kuppusamy 2016).

The preparation, characterization, and evaluation of fenugreek extract entrapped in polylactic-co-glycolic acid (PLGA) nanoparticles for the treatment of diabetes mellitus is reported in Pol (2014). Ethanolic fenugreek seed extract was loaded into PLGA to formulate nanoparticles (20 mg of fenugreek extract in 200 mg of PLGA) by single emulsion solvent evaporation process resulting in spherical and well-formed nanoparticles. The subcutaneous injection of nano-formulation (15 mg/kg, once a day, 15 days) to alloxan-induced diabetic mice showed anti-hyperglycemic effects with enhanced liver glycogen levels (Pol 2014).

Fenugreek seed mucilage (0.04% w/v) and chitosan (0.02%)-based cefixime nanoparticles were reported as controlled and sustained bi-phasic release pattern with initial burst release followed by the sustained release of the drug up to 24 h (Seetharaman, Balya, and Kuppusamy 2016). The nanoparticles were spherical and physically stable, with good surface morphology, particle size distribution, entrapment efficiency, yield and drug content. The nanoparticle formulation showed a better antimicrobial activity during a disc agar diffusion test as revealed by a higher zone of inhibition as compared to pure cefixime. The fenugreek seed mucilage, a natural polymer, was attributed to increased bacterial adhesion in the nanoparticulate formulation. The enhanced antimicrobial properties and sustained release for 24 h will help to decrease dose and frequency of administration of cefixime for better patient compliance (Seetharaman, Balya, and Kuppusamy 2016).

Gold nanoparticles are being explored for numerous applications such as cancer treatment, diabetes, Parkinson's disease, and coronary disease (Pallotta et al. 2019). However, the gradual degradation of catalytic activity on multiple usage limits the monometallic nanoparticles to industrial applications (Mallikarjuna et al. 2019). The incorporation of aqueous extract of fenugreek seeds (0.5 to 3 ml) in the gold nanoparticles formulation was effectively used as a reducing and protecting agent in the catalytic formulation of stable and spherical nanoparticles (Aswathy Aromal and Philip 2012). The presence of flavonoids in fenugreek extract is suggested as a reducing agent and capping material, whereas protein can contribute to the stabilization of nanoparticles (Aswathy Aromal and Philip 2012).

Due to its highly porous nature, the fenugreek polysaccharide (seed mucilage) was reported to be useful as a catalyst in the development of an environmentally friendly, simple, and low-cost procedure for palladium (Mallikarjuna et al. 2017) and gold-palladium (Mallikarjuna et al. 2019) nanoparticles. Furthermore, fenugreek (2%)-agarose (2%) hydrogel matrix composite entrapped gold nanoparticles with acetylcholinesterase-based biosensor for detection of carbamates, the group of pesticides on food samples (Kestwal, Bagal-Kestwal, and Chiang 2015). The resultant composite biosensor was a transparent clear film with good mechanical properties, stability, biocompatibility, superior enzyme retention efficiency, and prolonged shelf life (Kestwal, Bagal-Kestwal, and Chiang 2015).

Aqueous fenugreek seed extract demonstrated its potential as a reducing agent, stability enhancer, hematopoietic with antioxidant peripheries when incorporated in gold nanoparticles formulation (Ghosh et al. 2014). Fenugreek extract (200 mg) containing gold nanoparticles were pinkish red in color, spherical, monodispersed in nature, stable at room temperature (30 days), and 2–8 °C (90 days). Ten days, once daily, with oral treatment of fenugreek seed extract nanoparticles (0.25 mg/kg) and standard drug iron-sucrose (1.5 mg/kg, intraperitoneal injection) showed improved hematological (serum iron concentration, total RBC count, blood hemoglobin, and hematocrit) and antioxidant parameters in iron-deficient anemic mice (Ghosh et al. 2014). However, normal aqueous fenugreek seed extract (100 mg/kg) did not show such improvement to anemic mice (Ghosh et al. 2014).

In a similar manner, aqueous dispersed zinc nanoparticles incorporating aqueous fermented fenugreek powder (FFP) by *Pleurotus ostreatus* as a reducing and stabilizing agent demonstrated excellent anticancer (against Ehrlich Ascites Carcinoma and human Colon Adenocarcinoma) and bactericidal (against gram-positive and negative bacteria and yeast) activity (El-Batal, Mosalam et al. 2018) without affecting normal cells (El-Batal, Mosalam et al. 2018). Furthermore, fermented FFP incorporated copper nanoparticles demonstrated excellent antioxidant activity and antimicrobial potential against microbes causing burn skin infection such as *Klebsiella pneumoniae*, *Staphylococcus aureus*, and *Candida albicans* (El-Batal, Al-Hazmi et al. 2018).

The most recent application of *Pleurotus ostreatus* fermented aqueous extract of fenugreek powder (FPP) is a factorial design-optimized and gamma irradiation-assisted fabrication of selenium nanoparticles for enhanced anticancer and antioxidant activity (El-Batal et al. 2020). Selenium is an exceptional and interesting material with photoreactive, biocidal, catalytic, antioxidant, and anticancer properties, for applications in diverse areas such as nanotherapeutics, nutritional supplements, antimicrobial coatings, and diagnostics (Vrček 2018). FPP-incorporated selenium nanoparticles demonstrated significant *in vitro* antitumor effects against Ehrlich Ascites Carcinoma and human Colon Adenocarcinoma (El-Batal et al. 2020).

Taken together, fenugreek seed extract-incorporated metal nanoparticle formulations can gain importance in biomedical and pharmaceutical applications because of their unique benefits such as environmental friendliness, non-toxic nature, cost-effectiveness, consistent stabilization, and added bioactivity.

20.6 CONCLUSION

The available evidence indicates the ever-growing and diverse applications of fenugreek extract and constituents in formulations of conventional and novel dosage forms. The bioactive constituents of fenugreek seed extracts have been formulated for oral and topical dosage forms for skin and hair

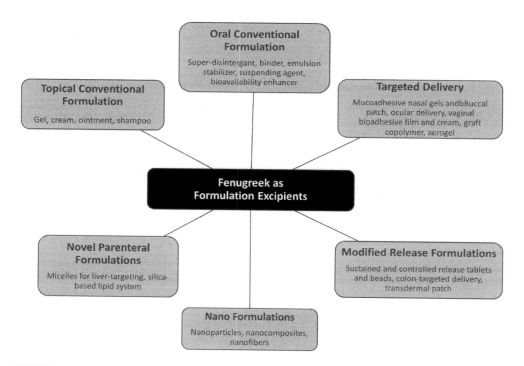

FIGURE 20.1 Fenugreek as formulation excipients.

care, inflammation and pain relief, antifungal properties, and wound dressings. However, fenugreek extract and constituents had been explored as excipients for improved formulations of established synthetic medicines in conventional or novel delivery systems. Fenugreek seed and leaves extract and constituents (especially seed mucilage) can provide excellent value during formulations of novel drug delivery systems because of their eco-friendliness, cost effectiveness, and safe nature with additive bioactivity potential.

REFERENCES

Abdel-Aal el, S. M., H. Akhtar, K. Zaheer, and R. Ali. 2013. "Dietary sources of lutein and zeaxanthin carotenoids and their role in eye health." *Nutrients* 5 (4):1169–85. doi: 10.3390/nu5041169.

Abdel-Barry, J. A., I. A. Abdel-Hassan, A. M. Jawad, and M. H. al-Hakiem. 2000. "Hypoglycaemic effect of aqueous extract of the leaves of *Trigonella foenum-graecum* in healthy volunteers." *East Mediterr Health J* 6 (1):83–8.

Adeosun, S. O., G. I. Lawal, S. A. Balogun, and E. I. Akpan. 2012. "Review of green polymer nanocomposites." *Journal of Minerals and Materials Characterization and Engineering* 11 (04):385. doi: 10.4236/jmmce.2012.114028.

Adhikari, S. N. R., B. S. Nayak, A. K. Nayak, and B. Mohanty. 2010. "Formulation and evaluation of buccal patches for delivery of atenolol." *AAPS PharmSciTech* 11 (3):1038–44. doi: 10.1208/s12249-010-9459-z.

Aditya, N. P., H. Yang, S. Kim, and S. Ko. 2015. "Fabrication of amorphous curcumin nanosuspensions using beta-lactoglobulin to enhance solubility, stability, and bioavailability." *Colloids Surf B Biointerfaces* 127:114–21. doi: 10.1016/j.colsurfb.2015.01.027.

Ahmad, S., M. Ahmad, K. Manzoor, R. Purwar, and S. Ikram. 2019. "A review on latest innovations in natural gums based hydrogels: Preparations & applications." *Int J Biol Macromol* 136:870–90. doi: 10.1016/j.ijbiomac.2019.06.113.

Ahmadiani, A., M. Javan, S. Semnanian, E. Barat, and M. Kamalinejad. 2001. "Anti-inflammatory and antipyretic effects of *Trigonella foenum-graecum* leaves extract in the rat." *J Ethnopharmacol* 75 (2–3):283–6. doi: 10.1016/s0378-8741(01)00187-8.

Ahmed, E. M. 2015. "Hydrogel: Preparation, characterization, and applications: A review." *J Adv Res* 6 (2):105–21. doi: 10.1016/j.jare.2013.07.006.

Ainurofiq, A., and S. Choiri. 2015. "Drug release model and kinetics of natural polymers-based sustained release tablet." *Latin American Journal of Pharmacy* 34 (7):1328–37.

Akbari, S. 2018. "Emulsion types, stability mechanisms and rheology: A review." *International Journal of Innovative Research and Scientific Studies* 1 (1):14–21.

Akhtar, N., M. K. Waqas, M. Ahmed, A. Ali, T. Saeed, G. Murtaza, A. Rasool, M. N. Aamir, S. A. Khan, and N. S. Bhatti. 2010. "Effect of cream formulation of fenugreek seed extract on some mechanical parameters of human skin." *Tropical Journal of Pharmaceutical Research* 9 (4). doi: 10.4314/tjpr.v9i4.58922.

Akiladevi, D., S. Umadevi, R. S. Kumar, and N. Arunkumar. 2019. "Formulation and evaluation of reconstitutable oral suspension of cefpodoxime proxetil using natural suspending agents." *Int Res J Pharm* 10 (2):126–33. doi: 10.7897/2230-8407.100256.

Almdal, K., J. Dyre, S. Hvidt, and O. Kramer. 1993. "Towards a phenomenological definition of the term 'gel'." *Polymer Gels and Networks* 1 (1):5–17. doi: 10.1016/0966-7822(93)90020-I.

Amidon, S., J. E. Brown, and V. S. Dave. 2015. "Colon-targeted oral drug delivery systems: Design trends and approaches." *AAPS PharmSciTech* 16 (4):731–41. doi: 10.1208/s12249-015-0350-9.

Anand, P., A. B. Kunnumakkara, R. A. Newman, and B. B. Aggarwal. 2007. "Bioavailability of curcumin: Problems and promises." *Mol Pharm* 4 (6):807–18. doi: 10.1021/mp700113r.

Ansari, M., P. Sadeghi, H. Mahdavi, M. Fattahi-Dolatabadi, N. Mohamadi, A. Asadi, and F. Sharififar. 2019. "Fenugreek dermal patch, a new natural topical antinociceptive medication for relieving the postherniotomy pain, a double-blind placebo controlled trial." *J Complement Integr Med* 16 (3):1–8. doi: 10.1515/jcim-2018-0082.

Anu Mary Ealia, S., and M. P. Saravanakumar. 2017. "A review on the classification, characterisation, synthesis of nanoparticles and their application." *IOP Conference Series: Materials Science and Engineering* 263:032019. doi: 10.1088/1757-899x/263/3/032019.

Arabski, M., A. Wegierek-Ciuk, G. Czerwonka, A. Lankoff, and W. Kaca. 2012. "Effects of saponins against clinical *E. coli* strains and eukaryotic cell line." *J Biomed Biotechnol* 2012:286216. doi: 10.1155/2012/286216.

Aswar, U., S. L. Bodhankar, V. Mohan, and P. A. Thakurdesai. 2010. "Effect of furostanol glycosides from Trigonella foenum-graecum on the reproductive system of male albino rats." *Phytother Res* 24 (10):1482–8. doi: 10.1002/ptr.3129.

Aswathy Aromal, S., and D. Philip. 2012. "Green synthesis of gold nanoparticles using Trigonella foenum-graecum and its size-dependent catalytic activity." *Spectrochim Acta A Mol Biomol Spectrosc* 97:1–5. doi: 10.1016/j.saa.2012.05.083.

Avachat, A. M., R. R. Dash, and S. N. Shrotriya. 2011. "Recent investigations of plant based natural gums, mucilages and resins in novel drug delivery systems." *Ind J Pharm Edu Res* 45 (1):86–99.

Avachat, A. M., K. N. Gujar, V. B. Kotwal, and S. Patil. 2007. "Isolation and evaluation of fenugreek seed husk as a granulating agent." *Indian J Pharmaceutical Sci* 69 (5):676–9. doi: 10.4103/0250-474X.38475.

Babaei, A., S. A. Taghavi, A. Mohammadi, M. A. Mahdiyar, P. Iranpour, F. Ejtehadi, and A. Mohagheghzadeh. 2020. "Comparison of the efficacy of oral fenugreek seeds hydroalcoholic extract versus placebo in nonalcoholic fatty liver disease; A randomized, triple-blind controlled pilot clinical trial." *Indian J Pharmacol* 52 (2):86–93. doi: 10.4103/ijp.IJP_17_19.

Bahadur, S., A. Roy, P. Baghel, and R. Chanda. 2016. "Formulation of glipizide tablets using fenugreek seed mucilage: Optimization by factorial design." *Asian Journal of Pharmaceutics* 10 (4):S662–S8. doi: 10.13140/RG.2.2.16112.97289.

Bakthavatchalam, S., T. Devasena, B. Prakash, and R. A. 2017. "Non-cytotoxic effect of green synthesized silver nanoparticles and its antibacterial activity." *J Photochem Photobiol B: Biology* 177. doi: 10.1016/j.jphotobiol.2017.10.010.

Bal, T., and S. Swain. 2019. "Microwave assisted synthesis of polyacrylamide grafted polymeric blend of fenugreek seed mucilage-Polyvinyl alcohol (FSM-PVA-g-PAM) and its characterizations as tissue engineered scaffold and as a drug delivery device." *DARU Journal of Pharmaceutical Sciences*:1–12. doi: 10.1007/s40199-019-00237-8.

Barhoum, A., R. Rasouli, M. Yousefzadeh, H. Rahier, and M. Bechelany. 2019. "Nanofiber technologies: History and development." In *Handbook of Nanofibers*, edited by Barhoum, A., Bechelany, M. and Makhlouf, A.S.H., 3–43. Cham: Springer International Publishing.

Bartlett, H., and F. Eperjesi. 2003. "Age-related macular degeneration and nutritional supplementation: A review of randomised controlled trials." *Ophthalmic Physiol Opt* 23 (5):383–99. doi: 10.1046/j.1475-1313.2003.00130.x.

Baruwati, B., and R. Varma. 2009. "High value products from waste: Grape pomace extract-a three-in-one package for the synthesis of metal nanoparticles." *ChemSusChem* 2 (11):1041–4. doi: 10.1002/cssc.200900220.

Bashtian, M. H., S. A. Emami, N. Mousavifar, H. A. Esmaily, M. Mahmoudi, and A. H. M. Poor. 2013. "Evaluation of fenugreek (*Trigonella Foenum-graceum* L.), effects seeds extract on insulin resistance in women with polycystic ovarian syndrome." *Iran J Pharm Res* 12 (2):475.

Bassi, P., and G. Kaur. 2015. "Fenugreek gum derivatives with improved bioadhesion and controlled drug release: In vitro and in vivo characterization." *Journal of Drug Delivery Science and Technology* 29:42–54. doi: 10.1016/j.jddst.2015.06.006.

Bawadi, H. A., S. N. Maghaydah, R. F. Tayyem, and R. F. Tayyem. 2009. "The postprandial hypoglycemic activity of fenugreek seed and seeds' extract in type 2 diabetics: A pilot study." *Pharmacognosy Magazine* 5 (18):134.

Belguith-Hadriche, O., M. Bouaziz, K. Jamoussi, M. S. J. Simmonds, A. El Feki, and F. Makni-Ayedi. 2013. "Comparative study on hypocholesterolemic and antioxidant activities of various extracts of fenugreek seeds." *Food Chemistry* 138 (2–3):1448–53. doi: 10.1016/j.foodchem.2012.11.003.

Bera, H., Y. F. Abbasi, V. Gajbhiye, K. F. Liew, P. Kumar, P. Tambe, A. K. Azad, D. Cun, and M. Yang. 2020. "Carboxymethyl fenugreek galactomannan-g-poly(N-isopropylacrylamide-co-N,N′-methylene-bis-acrylamide)-clay based pH/temperature-responsive nanocomposites as drug-carriers." *Materials Science and Engineering: C* 110:110628. doi: 10.1016/j.msec.2020.110628.

Bera, H., C. Gaini, S. Kumar, S. Sarkar, S. Boddupalli, and S. R. Ippagunta. 2016. "HPMC-based gastroretentive dual working matrices coated with Ca(+2) ion crosslinked alginate-fenugreek gum gel membrane." *Mater Sci Eng C Mater Biol Appl* 67:170–81. doi: 10.1016/j.msec.2016.05.016.

Bera, H., and S. Kumar. 2018. "Diethanolamine-modified pectin based core-shell composites as dual working gastroretentive drug-cargo." *Int J Biol Macromol* 108:1053–62. doi: 10.1016/j.ijbiomac.2017.11.019.

Bera, H., S. Mothe, S. Maiti, and S. Vanga. 2018. "Carboxymethyl fenugreek galactomannan-gellan gum-calcium silicate composite beads for glimepiride delivery." *Int J Biol Macromol* 107 (Pt A):604–14. doi: 10.1016/j.ijbiomac.2017.09.027.

Bhasha, S. A., S. A. Khalid, S. Duraivel, D. Bhowmik, and K. S. Kumar. 2013. "Recent trends in usage of polymers in the formulation of dermatological gels." *Indian J Res Pharm Biotechnol* 1 (2):161–8.

Bordia, A., S. K. Verma, and K. C. Srivastava. 1997. "Effect of ginger (Zingiber officinale Rosc.) and fenugreek (Trigonella foenumgraecum L.) on blood lipids, blood sugar and platelet aggregation in patients with coronary artery disease." *Prostaglandins Leukot Essent Fatty Acids* 56 (5):379–84. doi: 10.1016/s0952-3278(97)90587-1.

Brillante, C., and J. B. Mantaring. 2014. "O-203 Triple-blind, randomised controlled trial on the use of fenugreek (*Trigonella foenum-graecum* L.) for augmentation of breastmilk volume among postpartum mothers." *Archives of Disease in Childhood* 99 (Suppl 2):A101.

Broca, C., M. Manteghetti, R. Gross, Y. Baissac, M. Jacob, P. Petit, Y. Sauvaire, and G. Ribes. 2000. "4-Hydroxyisoleucine: Effects of synthetic and natural analogues on insulin secretion." *Eur J Pharmacol* 390 (3):339–45. doi: 10.1016/s0014-2999(00)00030-3.

Brummer, Y., W. Cui, and Q. Wang. 2003. "Extraction, purification and physicochemical characterization of fenugreek gum." *Food Hydrocolloids* 17 (3):229–36. doi: 10.1016/s0268-005x(02)00054-1.

Buhse, L., R. Kolinski, B. Westenberger, A. Wokovich, J. Spencer, C. W. Chen, S. Turujman, M. Gautam-Basak, G. J. Kang, A. Kibbe, B. Heintzelman, and E. Wolfgang. 2005. "Topical drug classification." *Int J Pharm* 295 (1–2):101–12. doi: 10.1016/j.ijpharm.2005.01.032.

Cai, B., Y. Zhang, Z. Wang, D. Xu, Y. Jia, Y. Guan, A. Liao, G. Liu, C. Chun, and J. Li. 2020. "Therapeutic potential of diosgenin and its major derivatives against neurological diseases: Recent advances." *Oxid Med Cell Longev* 2020:3153082. doi: 10.1155/2020/3153082.

Camargo, P., K. G. Satyanarayana, and F. Wypych. 2009. "Nanocomposites: Synthesis, structure, properties and new application opportunities." *Materials Research-IBERO-American Journal of Materials* 12 (1):1–39. doi: 10.1590/S1516-14392009000100002.

Campia, P., E. Ponzini, B. Rossi, S. Farris, T. Silvetti, L. Merlini, M. Brasca, R. Grandori, and Y. M. Galante. 2017. ""Aerogels of enzymatically oxidized galactomannans from leguminous plants: Versatile delivery systems of antimicrobial peptides and enzymes"." *Carbohydr Polym* 158:102–11. doi: 10.1016/j.carbpol.2016.11.089.

Cassone, A. 2015. "Vulvovaginal Candida albicans infections: Pathogenesis, immunity and vaccine prospects." *BJOG* 122 (6):785–94. doi: 10.1111/1471-0528.12994.

Cazorla-Luna, R., A. Martín-Illana, F. Notario-Pérez, L.-M. Bedoya, P. Bermejo, R. Ruiz-Caro, and M.-D. Veiga. 2019. "Dapivirine bioadhesive vaginal tablets based on natural polymers for the prevention of sexual transmission of HIV." *Polymers* 11 (3):483. doi: 10.3390/polym11030483.

Chan, E. S., P. K. Lam, C. T. Liew, R. S. Yen, and J. W. Lau. 2000. "The use of composite biodegradable skin graft and artificial skin for burn reconstruction." *Plast Reconstr Surg* 105 (2):807–8. doi: 10.1097/00006534-200002000-00067.

Chen, F.-x., M.-r. Zhao, B.-z. Ren, C.-r. Zhou, and F.-f. Peng. 2012. "Solubility of diosgenin in different solvents." *The J Chem Thermodyn* 47:341–6. doi: 10.1016/j.jct.2011.11.009.

Cheng, T.-L., P.-K. Chen, W.-K. Huang, C.-H. Kuo, C.-F. Cho, K.-C. Wang, G.-Y. Shi, H.-L. Wu, and C.-H. Lai. 2018. "Plasminogen/thrombomodulin signaling enhances VEGF expression to promote cutaneous wound healing." *J Mol Med* 96 (12):1333–44. doi: 10.1007/s00109-018-1702-1.

Chevassus, H., J. B. Gaillard, A. Farret, F. Costa, I. Gabillaud, E. Mas, A. M. Dupuy, F. Michel, C. Cantie, E. Renard, F. Galtier, and P. Petit. 2010. "A fenugreek seed extract selectively reduces spontaneous fat intake in overweight subjects." *Eur J Clin Pharmacol* 66 (5):449–55. doi: 10.1007/s00228-009-0770-0.

Christian, P., F. Von der Kammer, M. Baalousha, and T. Hofmann. 2008. "Nanoparticles: Structure, properties, preparation and behaviour in environmental media." *Ecotoxicology* 17 (5):326–43. doi: 10.1007/s10646-008-0213-1.

Chung, E. J., M. J. Sugimoto, J. L. Koh, and G. A. Ameer. 2017. "A biodegradable tri-component graft for anterior cruciate ligament reconstruction." *Journal of Tissue Engineering and Regenerative Medicine* 11 (3):704–12. doi: 10.1002/term.1966.

Damanik, R., M. L. Wahlqvist, and N. Wattanapenpaiboon. 2006. "Lactagogue effects of Torbangun, a Bataknese traditional cuisine." *Asia Pac J Clin Nutr* 15 (2):267–74.

Damodharan, K., S. A. Palaniyandi, S. H. Yang, and J. W. Suh. 2017. "Co-encapsulation of lactic acid bacteria and prebiotic with alginate-fenugreek gum-locust bean gum matrix: Viability of encapsulated bacteria under simulated gastrointestinal condition and during storage time." *Biotechnology and Bioprocess Engineering* 22 (3):265–71. doi: 10.1007/s12257-017-0096-1.

De Jong, W. H., and P. J. A. Borm. 2008. "Drug delivery and nanoparticles: Applications and hazards." *International Journal of Nanomedicine* 3 (2):133. doi: 10.2147/ijn.s596.

De Villiers, M. 2009a. "Ointment bases." In *A Practical Guide to Contemporary Pharmacy Practice*, edited by Thompson, J.E., 277–90. Philadelphia: Lippincott Williams & Wilkins.

De Villiers, M. 2009b. "Suppository bases." In *A Practical Guide to Contemporary Pharmacy Practice*, edited by Thompson, J.E. and Davidow, L.W., 291–7. Philadelphia: Lippincott Williams & Wilkins.

Deng, Y., X. Yang, X. Zhang, H. Cao, L. Mao, M. Yuan, and W. Liao. 2020. "Novel fenugreek gum-cellulose composite hydrogel with wound healing synergism: Facile preparation, characterization and wound healing activity evaluation." *Int J Biol Macromol* 160:1242–51. doi: 10.1016/j.ijbiomac.2020.05.220.

Deshpande, P. O., V. Bele, K. Joshi, and P. A. Thakurdesai. 2020. "Effects of low molecular weight galactomannans based standardized fenugreek seed extract in subjects with high fat mass: A randomized, double-blind, placebo-controlled clinical study." *J Appl Pharmaceutical Sci* 10 (1):62–9. doi: 10.7324/japs.2020.101008.

Djiobie Tchienou, G. E., R. K. Tsatsop TsaguE, T. F. Mbam Pega, V. Bama, A. Bamseck, S. Dongmo Sokeng, and M. B. Ngassoum. 2018. "Multi-response optimization in the formulation of a topical cream from natural ingredients." *Cosmetics* 5 (1):7. doi: 10.3390/cosmetics5010007.

Doharey, V., and N. Sharma. 2010. "The permutation role of fenugreek seeds starch and Gunda glue as a binder in Paracetamol tablets." *J Pharmaceutical Sci Res* 2 (2):64.

Dubey, R., T. C. Shami, and K. U. Bhasker Rao. 2009. "Microencapsulation technology and applications." *Defence Science Journal* 59 (1):82. doi: 10.14429/dsj.59.1489.

El-Batal, A. I., N. E. Al-Hazmi, F. M. Mosallam, and G. S. El-Sayyad. 2018. "Biogenic synthesis of copper nanoparticles by natural polysaccharides and Pleurotus ostreatus fermented fenugreek using gamma rays with antioxidant and antimicrobial potential towards some wound pathogens." *Microb Pathog* 118:159–69. doi: 10.1016/j.micpath.2018.03.013.

El-Batal, A. I., F. M. Mosalam, M. M. Ghorab, A. Hanora, and A. M. Elbarbary. 2018. "Antimicrobial, antioxidant and anticancer activities of zinc nanoparticles prepared by natural polysaccharides and gamma radiation." *Int J Biol Macromol* 107 (Pt B):2298–311. doi: 10.1016/j.ijbiomac.2017.10.121.

El-Batal, A. I., F. M. Mosallam, M. M. Ghorab, A. Hanora, M. Gobara, A. Baraka, M. A. Elsayed, K. Pal, R. M. Fathy, M. Abd Elkodous, and G. S. El-Sayyad. 2020. "Factorial design-optimized and gamma irradiation-assisted fabrication of selenium nanoparticles by chitosan and Pleurotus ostreatus fermented fenugreek for a vigorous in vitro effect against carcinoma cells." *Int J Biol Macromol* 156:1584–99. doi: 10.1016/j.ijbiomac.2019.11.210.

Emtiazy, M., L. Oveidzadeh, M. Habibi, L. Molaeipour, D. Talei, Z. Jafari, M. Parvin, and M. Kamalinejad. 2018. "Investigating the effectiveness of the Trigonella foenum-graecum L. (fenugreek) seeds in mild asthma: A randomized controlled trial." *Allergy Asthma Clin Immunol* 14 (1):19. doi: 10.1186/s13223-018-0238-9.

Fedacko, J., R. B. Singh, M. A. Niaz, S. Ghosh, P. Fedackova, A. D. Tripathi, A. Etharat, E. Onsaard, V. K. Singh, and S. Shastun. 2016. "Fenugreeg seeds decrease blood cholesterol and blood glucose as adjunct to diet therapy in patients with hypercholesterolemia." *World Heart Journal* 8 (3):239.

Ferreira, M. C., P. Tuma, Jr., V. F. Carvalho, and F. Kamamoto. 2006. "Complex wounds." *Clinics (Sao Paulo)* 61 (6):571–8. doi: 10.1590/s1807-59322006000600014.

Gaddam, A., C. Galla, S. Thummisetti, R. K. Marikanty, U. D. Palanisamy, and P. V. Rao. 2015. "Role of Fenugreek in the prevention of type 2 diabetes mellitus in prediabetes." *J Diabetes Metab Disord* 14 (1):74. doi: 10.1186/s40200-015-0208-4.

Gade, J., S. More, and N. Bhalerao. 2015. "Formulation and characterization of herbal cream containing fenugreek seed extracts." *Int J Sci Res Publ* 5 (10):284–7.

Garti, N., Z. Madar, A. Aserin, and B. Sternheim. 1997. "Fenugreek galactomannans as food emulsifiers." *LWT—Food Science and Technology* 30 (3):305–11. doi: 10.1006/fstl.1996.0179.

Geberemeskel, G. A., Y. G. Debebe, and N. A. Nguse. 2019. "Antidiabetic effect of Fenugreek seed powder solution (*Trigonella foenum-graecum* L.) on hyperlipidemia in diabetic patients." *J Diabetes Res* 2019:8507453. doi: 10.1155/2019/8507453.

Gee, J. M., M. S. DuPont, A. J. Day, G. W. Plumb, G. Williamson, and I. T. Johnson. 2000. "Intestinal transport of quercetin glycosides in rats involves both deglycosylation and interaction with the hexose transport pathway." *J Nutr* 130 (11):2765–71. doi: 10.1093/jn/130.11.2765.

Ghafari, S., S. Fahimi, and S. Sahranavard. 2017. "Plants used to treat hyperpigmentation in Iranian traditional medicine: A review." *Research Journal of Pharmacognosy* 4 (4):71–85.

Ghaffari, M., and J. E. N. Dolatabadi. 2019. "Nanotechnology for pharmaceuticals." In *Industrial Applications of Nanomaterials*, 475–502. Amsterdam: Elsevier.

Gharibzahedi, S. M., S. H. Razavi, and S. M. Mousavi. 2012. "Developing an emulsion model system containing canthaxanthin biosynthesized by Dietzia natronolimnaea HS-1." *Int J Biol Macromol* 51 (4):618–26. doi: 10.1016/j.ijbiomac.2012.06.030.

Gholaman, M. 2018. "Effect of eight weeks' endurance training along with fenugreek ingestion on lipid profile, body composition, insulin resistance and vo2max in obese women's with type2 diabetes." *J Med Plants* 1 (65):83–92.

Ghosh, S., J. Sengupta, P. Datta, and A. Gomes. 2014. "Hematopoietic and antioxidant activities of gold nanoparticles synthesized by aqueous extract of Fenugreek (Trigonella foenum-graecum) seed." *Adv Sci Eng Med* 6 (5):546–52. doi: 10.1166/asem.2014.1511.

Gubitosa, J., V. Rizzi, P. Fini, and P. Cosma. 2019. "Hair care cosmetics: From traditional shampoo to solid clay and herbal shampoo, a review." *Cosmetics* 6 (1):13. doi: 10.3390/cosmetics6010013.

Guo, R., Q. Wang, R. P. Nair, S. L. Barnes, D. T. Smith, B. Dai, T. J. Robinson, and S. Nair. 2018. "Furosap, a novel Fenugreek seed extract improves lean body mass and serum testosterone in a randomized, placebo-controlled, double-blind clinical investigation." *Functional Foods in Health and Disease* 8 (11):519. doi: 10.31989/ffhd.v8i11.565.

Guo, Y., and A. Pratap-Singh. 2019. "Emerging technologies in buccal and sublingual drug administration." *Journal of Drug Delivery Science and Technology* 52:440–51. doi: 10.1016/j.jddst.2019.05.014.

Hadi, A., A. Arab, H. Hajianfar, B. Talaei, M. Miraghajani, S. Babajafari, W. Marx, and R. Tavakoly. 2020. "The effect of fenugreek seed supplementation on serum irisin levels, blood pressure, and liver and kidney function in patients with type 2 diabetes mellitus: A parallel randomized clinical trial." *Complement Ther Med* 49:102315. doi: 10.1016/j.ctim.2020.102315.

Haghshenas, B., N. Abdullah, Y. Nami, D. Radiah, R. Rosli, and A. Yari Khosroushahi. 2015. "Microencapsulation of probiotic bacteria *Lactobacillus plantarum* 15HN using alginate-psyllium-fenugreek polymeric blends." *J Appl Microbiol* 118 (4):1048–57. doi: 10.1111/jam.12762.

Hajimehdipoor, H., S. E. Sadat-Ebrahimi, Y. Amanzadeh, M. Izaddoost, and E. Givi. 2010. "Identification and Quantitative Determination of 4-Hydroxyisoleucine in *Trigonella foenum-graecum* L. from Iran." *J Med Plants* 9:29–34.

Hakimi, S., A. Delazar, N. Mobaraki-Asl, P. Amiri, H. Tavassoli, and M. Safari. 2020. "Comparison of the effects of fenugreek vaginal cream and ultra low-dose estrogen on atrophic vaginitis." *Current Drug Delivery* 17 (9):1–9. doi: 10.2174/1567201817666200708112655.

Hari, N., and V. Mohan. 2014. "Isolation of (2S,3R,4S)-4-Hydroxyisoleucine from Trigonella foenum-graecum Seeds." *Asian J Chem* 26 (10):3082–4. doi: 10.14233/ajchem.2014.16831.

Hassani, S. S., A. Fallahi, S. S. Esmaeili, and M. GholamiFesharaki. 2019. "The effect of combined therapy with fenugreek and nutrition training based on Iranian traditional medicine on fbs, hga1c, bmi, and waist circumference in type 2 diabetic patients: A randomized double blinded clinical trial." *J Adv Med Bio Res* 27 (120):37–42.

Hausenblas, H. A., K. L. Conway, K. R. M. Coyle, E. Barton, L. D. Smith, M. Esposito, C. Harvey, D. Oakes, and D. R. Hooper. 2020. "Efficacy of fenugreek seed extract on men's psychological and physical health: A randomized placebo-controlled double-blind clinical trial." *J Complement Integr Med* 1 (ahead-of-print). doi: 10.1515/jcim-2019-0101.

Hefnawy, H. T., and M. F. Ramadan. 2011. "Physicochemical characteristics of soy protein isolate and fenugreek gum dispersed systems." *J Food Sci Technol* 48 (3):371–7. doi: 10.1007/s13197-010-0203-1.

Homayun, B., X. Lin, and H.-J. Choi. 2019. "Challenges and recent progress in oral drug delivery systems for biopharmaceuticals." *Pharmaceutics* 11 (3):129. doi: 10.3390/pharmaceutics11030129.

Huanbutta, K., and T. Sangnim. 2020. "Bioadhesive films for drug delivery systems." In *Bioadhesives in Drug Delivery*, edited by Mittal, K.L., S., B.I. and Narang, J.K., 99–122. Beverly: Scrivener Publishing LLC.

Huang, X., Y. Kakuda, and W. Ciu. 2001. "Hydrocolloids in emulsions: Particle size distribtion and interfacia activity." *Food Hydrocolloids* 15:533–42. doi: 10.1016/S0268-005X(01)00091-1.

Inanmdar, W., A. Sultana, U. Mubeen, and K. Rahman. 2016. "Clinical efficacy of *Trigonella foenum-graecum* (Fenugreek) and dry cupping therapy on intensity of pain in patients with primary dysmenorrhea." *Chin J Integr Med*:1–8. doi: 10.1007/s11655-016-2259-x.

Iurian, S., E. Dinte, C. Iuga, C. Bogdan, I. Spiridon, L. Barbu-Tudoran, A. Bodoki, I. Tomuta, and S. E. Leucuta. 2017. "The pharmaceutical applications of a biopolymer isolated from *Trigonella foenum-graecum* seeds: Focus on the freeze-dried matrix forming capacity." *Saudi Pharm J* 25 (8):1217–25. doi: 10.1016/j.jsps.2017.09.006.

Izydorczyk, M., S. W. Cui, and Q. Wang. 2005. "Polysaccharide gums: Structures, functional properties, and applications." In *Food Carbohydrates: Chemistry, Physical Properties, and Applications*, edited by Cui, S.W., 299. Boca Raton: CRC Press.

Jadoun, S., U. Riaz, and V. Budhiraja. 2020. "Biodegradable conducting polymeric materials for biomedical applications: A review." *Medical Devices & Sensors*:e10141. doi: 10.1002/mds3.10141.

Jasim, B., R. Thomas, J. Mathew, and E. K. Radhakrishnan. 2017. "Plant growth and diosgenin enhancement effect of silver nanoparticles in Fenugreek (*Trigonella foenum-graecum* L.)." *Saudi Pharm J* 25 (3):443–7. doi: 10.1016/j.jsps.2016.09.012.

Jiang, J. X., L. W. Zhu, W. M. Zhang, and R. C. Sun. 2007. "Characterization of Galactomannan Gum from Fenugreek (Trigonella foenum-graecum) Seeds and Its Rheological Properties." *International Journal of Polymeric Materials* 56 (12):1145–54. doi: 10.1080/00914030701323745.

Jyothi, D., and M. Koland. 2016. "Formulation and evaluation of an herbal anti-inflammatory gel containing *trigonella foenum greacum* seed extract." *International Journal of Pharmacy and Pharmaceutical Sciences* 8 (1):41–4.

Jyothi, D., M. Koland, and S. Priya. 2014. "Investigation of anti-inflammatory activity of ointments containing fenugreek extract." *Asian J Pharm Clin Res* 7:66–9.

Kaltsa, O., N. Spiliopoulou, S. Yanniotis, and I. Mandala. 2016. "Stability and physical properties of model macro- and nano/submicron emulsions containing fenugreek gum." *Food Hydrocolloids* 61:625–32. doi: 10.1016/j.foodhyd.2016.06.025.

Kaltsa, O., S. Yanniotis, and I. Mandala. 2016. "Stability properties of different fenugreek galactomannans in emulsions prepared by high-shear and ultrasonic method." *Food Hydrocolloids* 52:487–96. doi: 10.1016/j.foodhyd.2015.07.024.

Kandekar, U. Y., P. D. Chaudhari, K. B. Chandrasekhar, and R. R. Pujari. 2019. "Exploration of mucoadhesive microparticles by using *Linum usitatissimum* mucilage." *Latin American Journal of Pharmacy* 38 (12):2463–72.

Kandhare, A. D., S. L. Bodhankar, V. Mohan, and P. A. Thakurdesai. 2015. "Pharmacokinetics, tissue distribution and excretion study of a furostanol glycoside-based standardized fenugreek seed extract in rats." *Ren Fail* 37 (7):1208–18. doi: 10.3109/0886022X.2015.1057472.

Kandhare, A. D., S. L. Bodhankar, V. Mohan, and P. A. Thakurdesai. 2016. "Development and validation of HPLC method for vicenin-1 isolated from fenugreek seeds in rat plasma: Application to pharmacokinetic, tissue distribution and excretion studies." *Pharm Biol* 54 (11):2575–83. doi: 10.3109/13880209.2016.1172245.

Kanungo, I., N. N. Fathima, R. Rao, J. Balachandran, and U. Nair. 2014. "Go natural and smarter: Fenugreek as a hydration designer of collagen based biomaterials." *Phys Chem Chem Phys* 17 (4):2778–93. doi: 10.1039/C4CP04363D.

Kapare, H. S., and S. R. Metkar. 2020. "Micellar drug delivery system: A review." *Pharmaceutical Resonance* 2 (2):21–6.

Karuppusamy, C., and P. Venkatesan. 2017. "Role of nanoparticles in drug delivery system: A comprehensive review." *J Pharm Sci Res* 9 (3):318.

Kasran, M., S. W. Cui, and H. D. Goff. 2013a. "Covalent attachment of fenugreek gum to soy whey protein isolate through natural Maillard reaction for improved emulsion stability." *Food Hydrocolloids* 30 (2):552–8. doi: 10.1016/j.foodhyd.2012.08.004.

Kasran, M., S. W. Cui, and H. D. Goff. 2013b. "Emulsifying properties of soy whey protein isolate—fenugreek gum conjugates in oil-in-water emulsion model system." *Food Hydrocolloids* 30 (2):691–7. doi: 10.1016/j.foodhyd.2012.09.002.

Kaur, G., M. Mahajan, and P. Bassi. 2013. "Derivatized polysaccharides: Preparation, characterization, and application as bioadhesive polymer for drug delivery." *Int J Polym Mat* 62 (9):475–81. doi: 10.1080/00914037.2012.734348.

Kechagia, M., D. Basoulis, S. Konstantopoulou, D. Dimitriadi, K. Gyftopoulou, N. Skarmoutsou, and E. M. Fakiri. 2013. "Health benefits of probiotics: A review." *ISRN Nutr* 2013:481651. doi: 10.5402/2013/481651.

Kestwal, R. M., D. Bagal-Kestwal, and B. H. Chiang. 2015. "Fenugreek hydrogel-agarose composite entrapped gold nanoparticles for acetylcholinesterase based biosensor for carbamates detection." *Anal Chim Acta* 886:143–50. doi: 10.1016/j.aca.2015.06.004.

Khanna, A., F. John, S. Das, J. Thomas, J. Rao, B. Maliakel, and K. Im. 2020. "Efficacy of a novel extract of fenugreek seeds in alleviating vasomotor symptoms and depression in perimenopausal women: A randomized, double-blinded, placebo-controlled study." *J Food Biochem* 44 (12):e13507. doi: 10.1111/jfbc.13507.

Kim, H. P., K. H. Son, H. W. Chang, and S. S. Kang. 2004. "Anti-inflammatory plant flavonoids and cellular action mechanisms." *J Pharmacol Sci* 96 (3):229–45. doi: 10.1254/jphs.crj04003x.

Kiss, R., K. Szabo, K. Gesztelyi, S. Somodi, P. Kovacs, Z. Szabo, J. Nemeth, D. Priksz, A. Kurucz, B. Juhasz, and Z. Szilvassy. 2018. "Insulin-sensitizer effects of fenugreek seeds in parallel with changes in plasma MCH levels in healthy volunteers." *Int J Mol Sci* 19 (3):771. doi: 10.3390/ijms19030771.

Krishnakumar, I., A. Maliakel, G. Gopakumar, D. Kumar, B. Maliakel, and R. Kuttan. 2015. "Improved blood—brain-barrier permeability and tissue distribution following the oral administration of a food-grade formulation of curcumin with fenugreek fibre." *J Funct Foods* 14:215–25. https://doi.org/10.1016/j.jff.2015.01.049.

Kulkarni, G. T., K. Gowthamarajan, R. R. Dhobe, F. Yohanan, and B. Suresh. 2005. "Development of controlled release spheroids using natural polysaccharide as release modifier." *Drug Delivery* 12 (4):201–6. doi: 10.1080/10717540590952537.

Kulkarni, M., V. Hastak, V. Jadhav, and A. A. Date. 2020. "Fenugreek leaf extract and its gel formulation show activity against malassezia furfur." *ASSAY and Drug Development Technologies* 18 (1):45–55. doi: 10.1089/adt.2019.918.

Kumar, D. 2019. "Domperidone nasal gel using fenugreek seed as mucoadhesive agent: Water solubility problem sort-out." *Trends in Pharm Nanotechnol* 2 (1):1–11. doi: 10.5281/zenodo.3477840.

Kumar, D., D. Jacob, S. Ps, A. Maliakkal, J. Nm, R. Kuttan, B. Maliakel, V. Konda, and K. Im. 2016. "Enhanced bioavailability and relative distribution of free (unconjugated) curcuminoids following the oral administration of a food-grade formulation with fenugreek dietary fibre: A randomised double-blind crossover study." *J Funct Foods* 22:578–87. doi: 10.1016/j.jff.2016.01.039.

Kumar, D., A. Singhal, S. Bansal, and S. K. Gupta. 2014. "Extraction, isolation and evaluation *Trigonella foenum-graecum* as mucoadhesive agent for nasal gel drug delivery." *J Nepal Pharm Assoc* 27 (1):40–5. doi: 10.3126/jnpa.v27i1.12149.

Kumar, K., S. Kumar, A. Datta, and A. Bandyopadhyay. 2015. "Effect of fenugreek seeds on glycemia and dyslipidemia in patients with type 2 diabetes mellitus." *Int J Med Sci Public Health* 4 (7):997–1000. doi: 10.5455/ijmsph.2015.11032015202.

Kumar, M. U., and M. K. Babu. 2014. "Design and evaluation of fast dissolving tablets containing diclofenac sodium using fenugreek gum as a natural superdisintegrant." *Asian Pac J Trop Biomed* 4 (Suppl 1):S329–34. doi: 10.12980/APJTB.4.2014B672.

Kumar, M. U., A. K. Shukla, R. S. Bishnoi, and C. P. Jain. 2017. "Application of fenugreek seed gum: In novel drug delivery." *Asian J Biomat Res 2017* 3 (6):1–10.

Kumar, R. V., S. Patil, M. B. Patil, S. R. Patil, and M. S. Paschapur. 2009. "Isolation and evaluation of disintegrant properties of fenugreek seed mucilage." *Int J PharmTech Res* 1 (4):982–96.

Kumar, R. V., and V. R. Sinha. 2012. "A novel synergistic galactomannan-based unit dosage form for sustained release of acarbose." *AAPS PharmSciTech* 13 (1):262–75. doi: 10.1208/s12249-011-9724-9.

Kumar, S., S. S. Kesharwani, H. Mathur, M. Tyagi, G. J. Bhat, and H. Tummala. 2016. "Molecular complexation of curcumin with pH sensitive cationic copolymer enhances the aqueous solubility, stability and bioavailability of curcumin." *Eur J Pharm Sci* 82:86–96. https://doi.org/10.1016/j.ejps.2015.11.010.

Lakshminarayana, R., M. Raju, M. N. Keshava Prakash, and V. Baskaran. 2009. "Phospholipid, oleic acid micelles and dietary olive oil influence the lutein absorption and activity of antioxidant enzymes in rats." *Lipids* 44 (9):799–806. doi: 10.1007/s11745-009-3328-0.

Lee, S. J., K. H. Son, H. W. Chang, J. C. Do, K. Y. Jung, S. S. Kang, and H. P. Kim. 1993. "- Antiinflammatory activity of naturally occurring flavone and flavonol glycosides." *Archives of Pharmacal Research* 16 (1):25–8. https://doi.org/10.1007/BF02974123.

Li, J., G. H. Shin, I. W. Lee, X. Chen, and H. J. Park. 2016. "Soluble starch formulated nanocomposite increases water solubility and stability of curcumin." *Food Hydrocolloids* 56:41–9. https://doi.org/10.1016/j.foodhyd.2015.11.024.

Li, L., J. Zhao, Y. Sun, F. Yu, and J. Ma. 2019. "Ionically cross-linked sodium alginate/κ-carrageenan double-network gel beads with low-swelling, enhanced mechanical properties, and excellent adsorption performance." *Chemical Engineering Journal* 372:1091–3. https://doi.org/10.1016/j.cej.2019.05.007.

Liu, L., M. L. Fishman, J. Kost, and K. B. Hicks. 2003. "Pectin-based systems for colon-specific drug delivery via oral route." *Biomaterials* 24 (19):3333–43. doi: 10.1016/S0142-9612(03)00213-8.

Liu, Y., F. Lei, L. He, W. Xu, and J. Jiang. 2020. "Physicochemical characterization of galactomannans extracted from seeds of Gleditsia sinensis Lam and fenugreek. Comparison with commercial guar gum." *Int J Biol Macromol* 158:1047–54. doi: 10.1016/j.ijbiomac.2020.04.208.

Lopresti, A. L. 2018. "The problem of curcumin and its bioavailability: Could its gastrointestinal influence contribute to its overall health-enhancing effects?" *Advances in Nutrition* 9 (1):41–50. doi: 10.1093/advances/nmx011.

Lu, F. R., L. Shen, Y. Qin, L. Gao, H. Li, and Y. Dai. 2008. "Clinical observation on trigonella foenum-graecum L. total saponins in combination with sulfonylureas in the treatment of type 2 diabetes mellitus." *Chin J Integr Med* 14 (1):56–60. doi: 10.1007/s11655-007-9005.

Madar, Z., and I. Shomer. 1990. "Polysaccharide composition of a gel fraction derived from fenugreek and its effect on starch digestion and bile acid absorption in rats." *J Agr Food Chem* 38 (7):1535–9. doi: 10.1021/jf00097a023.

Maheshwari, A., N. Verma, A. Swaroop, M. Bagchi, H. G. Preuss, K. Tiwari, and D. Bagchi. 2017. "Efficacy of FurosapTM, a novel *Trigonella foenum-graecum* seed extract, in enhancing testosterone level and improving sperm profile in male volunteers." *Int J Med Sci* 14 (1):58. doi: 10.7150/ijms.17256.

Maier, H., M. Anderson, C. Karl, K. Magnuson, and R. L. Whistler. 1993a. "Chapter 8 — Guar, locust bean, Tara, and fenugreek gums." In *Industrial Gums* (Third Edition), edited by Whistler, R.L. and Bemiller, J.N., 181–226. London: Academic Press.

Maier, H., M. Anderson, C. Karl, K. Magnuson, and R. L. Whistler. 1993b. "Guar, locust bean, tara, and fenugreek gums." In *Industrial Gums: Polysaccharides and Their Derivatives*, edited by Bemiller, R.l.W.A.J.N., 181–226. Amsterdam: Elsevier.

Makowska, J., D. Szczesny, A. Lichucka, A. Gieldon, L. Chmurzynski, and R. Kaliszan. 2014. "Preliminary studies on trigonelline as potential anti-Alzheimer disease agent: Determination by hydrophilic interaction liquid chromatography and modeling of interactions with beta-amyloid." *J Chromatogr B Analyt Technol Biomed Life Sci* 968:101–4. doi: 10.1016/j.jchromb.2013.12.001.

Mallikarjuna, K., C. Bathula, K. Buruga, N. K. Shrestha, Y.-Y. Noh, and H. Kim. 2017. "Green synthesis of palladium nanoparticles using fenugreek tea and their catalytic applications in organic reactions." *Materials Letters* 205:38–141. doi: 10.1016/j.matlet.2017.06.081.

Mallikarjuna, K., C. Bathula, G. Dinneswara Reddy, N. K. Shrestha, H. Kim, and Y. Y. Noh. 2019. "Au-Pd bimetallic nanoparticles embedded highly porous Fenugreek polysaccharide based micro networks for catalytic applications." *Int J Biol Macromol* 126:352–8. doi: 10.1016/j.ijbiomac.2018.12.137.

Manwatkar, S., P. A. Thakurdesai, S. L. Bodhankar, V. Mohan, and G. N. Zambare. 2008. "Pharmacokinetic and pharmacodynamic study of trigonelline in normal and diabetic rats." In *IBRO sponsored International Structural Neuroscience Conference on Peptides*, Nagpur, p. 10, Feb 2–3, 2008. Nagpur: RTM Nagpur University.

Mathern, J. R., S. K. Raatz, W. Thomas, and J. L. Slavin. 2009. "Effect of fenugreek fiber on satiety, blood glucose and insulin response and energy intake in obese subjects." *Phytotherapy Research* 23 (11):1543–8. doi: 10.1002/ptr.2795.

Mathur, V., and N. K. Mathur. 2005. "Fenugreek and other lesser known legume galactomannan-polysaccharides: Scope for developments." *J Sci Ind Res* 64:475–81.

Medizin, K., C. Schoen, S. Bielfeldt, and J. Reimann. 2006. "Fenugreek+micronutrients: Efficacy of a food supplement against hair loss." *Kosmetische Medizin* 27.

Meghwal, M., and T. K. Goswami. 2012. "A review on the functional properties, nutritional content, medicinal utilization and potential application of fenugreek." *Journal of Food Processing and Technology* 3 (9):(1–10) 1000181. doi: 10.4172/2157-7110.1000181.

Mehkri, S., C. Ambarish, B. J. Madankumar, P. Shivayogi, and K. Bopanna. 2019. "Effect of two doses of Fenugreek Flakes (FenuLean TM) on appetite, body-weight and blood glucose homeostasis: A randomized, double-blind, multicenter, three-arm, long-term, control study in 100 healthy subjects." *Int J Med Sci Clin Res* 1 (4):7–14.

Memis, S., F. Tornuk, F. Bozkurt, and M. Z. Durak. 2017. "Production and characterization of a new biodegradable fenugreek seed gum based active nanocomposite film reinforced with nanoclays." *Int J Biol Macromol* 103:669–75. doi: 10.1016/j.ijbiomac.2017.05.090.

Miscianinov, V., A. Martello, L. Rose, E. Parish, B. Cathcart, T. Mitić, G. A. Gray, M. Meloni, A. Al Haj Zen, and A. Caporali. 2018. "MicroRNA-148b targets the TGF-β pathway to regulate angiogenesis and endothelial-to-mesenchymal transition during skin wound healing." *Mol Ther* 26 (8):1996–2007. doi: 10.1016/j.ymthe.2018.05.002.

Mishra, A., A. Yadav, S. Pal, and A. Singh. 2006. "Biodegradable graft copolymers of fenugreek mucilage and polyacrylamide: A renewable reservoir to biomaterials." *Carbohydrate Polymers* 65:58–63. doi: 10.1016/j.carbpol.2005.12.015.

Mishra, P., A. K. Srivastava, T. C. Yadav, V. Pruthi, and R. Prasad. 2021. "Pharmaceutical and Therapeutic Applications of Fenugreek Gum." *Bioactive Natural Products for Pharmaceutical Applications*:379–408. doi: 10.1007/978-3-030-54027-2_11.

Mittal, N., P. Mattu, and G. Kaur. 2016. "Extraction and derivatization of *Leucaena leucocephala* (Lam.) galactomannan: Optimization and characterization." *Int J Biol Macromol* 92:831–41. doi: 10.1016/j.ijbiomac.2016.07.046.

Mohamadi, N., M. Pournamdari, F. Sharififar, and M. Ansari. 2020. "Simultaneous spectrophotometric determination of trigonelline, diosgenin and nicotinic acid in dosage forms prepared from fenugreek seed extract." *Iran J Pharm Sci* 19 (2):153–9. doi: 10.22037/ijpr.2019.1100790.

Mohamadi, N., F. Sharififar, M. Ansari, M. Pournamdari, M. Rezaei, and N. Hassanabadi. 2020. "Pharmacokinetic profile of diosgenin and trigonelline following intravenous and oral administration of fenugreek seed extract and pure compound in rabbit." *J Asian Nat Prod Res*:1–12. doi: 10.1080/10286020.2020.1769609.

Mohamadi, N., F. Sharififar, M. Pournamdari, and M. Ansari. 2021. "Determination of trigonelline in human plasma by magnetic solid-phase extraction: A pharmacokinetic study." *Nanomedicine (Lond)*. doi: 10.2217/nnm-2020-0365.

Mohammadian, M., M. Salami, S. Momen, F. Alavi, Z. Emam-Djomeh, and A. A. Moosavi-Movahedi. 2019. "Enhancing the aqueous solubility of curcumin at acidic condition through the complexation with whey protein nanofibrils." *Food Hydrocolloids* 87:902–14. https://doi.org/10.1016/j.foodhyd.2018.09.001.

Mohan, V., and N. Hari. 2016. "Novel isolation and characterization of trigonelline monohydrate." Phytocongress—2016, in Thanjavuar, India, 21 /07/2016–22/07/2016, p. 21.

Mohanachandran, P. S., P. G. Sindhumol, and T. S. Kiran. 2011. "Superdisintegrants: An overview." *Int J Pharm Sci Rev Res* 6 (1):105–9.

Mokashi, M., R. Singh-Mokashi, V. Mohan, and P. Thakurdesai. 2014. "Effects of glycosides based fenugreek seed extract on serum testosterone levels of healthy sedentary male subjects: A exploratory double blind, placebo controlled, crossover study." *Asian J Clin Res* 7:177–81.

Momin, M. M., S. Kane, and P. Abhang. 2015. "Formulation and evaluation of bilayer tablet for bimodal release of venlafaxine hydrochloride." *Front Pharmacol* 6 (144):144. doi: 10.3389/fphar.2015.00144.

Monteiro-Riviere, N. 2006. "Structure and function of skin." In *Toxicology of the Skin*, edited by Roberts, M.S., 1–19. Boca Raton: CRC Press.

Morcos, S. R., Z. Elhawary, and G. N. Gabrial. 1981. "Protein-rich food mixtures for feeding the young in Egypt. 1. formulation." *Z Ernahrungswiss* 20 (4):275–82. doi: 10.1007/bf02021639.

Mozhdeh, S., M. Ali, N. Sara Sadat, F. Pouya, and B. Amir Hossein. 2019. "Comparison of the efficacy of oral fenugreek seed extract and azithromycin in the treatment of acne vulgaris: A randomized, triple-blind controlled pilot clinical trial." *Iran J Dermatol* 22 (2):58–64. doi: 10.22034/IJD.2019.98373.

Mundhe, M. R., R. R. Pagore, and K. R. Biyani. 2012. "Isolation and evaluation of *Trigonella Foenum-graecum* mucilage as gelling agent in diclofenac potassium gel." *International Journal of Ayurvedic and Herbal Medicine* 2 (2):300–6.

Muniyan, A., K. Ravi, U. Mohan, and R. Panchamoorthy. 2017. "Characterization and in vitro antibacterial activity of saponin-conjugated silver nanoparticles against bacteria that cause burn wound infection." *World Journal of Microbiology and Biotechnology* 33 (7):147. doi: 10.1007/s11274-017-2309-3.

Mylangam, C. K., S. Beeravelli, J. Medikonda, J. S. Pidaparthi, and V. R. M. Kolapalli. 2016. "Badam gum: A natural polymer in mucoadhesive drug delivery. Design, optimization, and biopharmaceutical evaluation of badam gum-based metoprolol succinate buccoadhesive tablets." *Drug delivery* 23 (1):195–206. doi: 10.3109/10717544.2014.908979.

Naghdi Badi, H., A. Mehrafarin, S. H. Mustafavi, and M. Labbafi. 2018. "Exogenous arginine improved fenugreek sprouts growth and trigonelline production under salinity condition." *Industrial Crops and Products* 122:609–16. https://doi.org/10.1016/j.indcrop.2018.06.042.

Najdi, R. A., M. M. Hagras, F. O. Kamel, and R. M. Magadmi. 2019. "A randomized controlled clinical trial evaluating the effect of *Trigonella foenum-graecum* (fenugreek) versus glibenclamide in patients with diabetes." *African Health Sciences* 19 (1):1594–601.

Naqvi, S. A., M. M. Khan, M. Shahid, M. J. Jaskani, I. A. Khan, M. Zuber, and K. M. Zia. 2011. "Biochemical profiling of mucilage extracted from seeds of different citrus rootstocks." *Carbohydrate Polymers* 83 (2):623–8. doi: 10.1016/j.carbpol.2010.08.031.

Nathan, J., S. Panjwani, V. Mohan, V. Joshi, and P. A. Thakurdesai. 2014. "Efficacy and safety of standardized extract of Trigonella foenum-graecum L seeds as an adjuvant to L-Dopa in the management of patients with Parkinson's disease." *Phytother Res* 28 (2):172–8. doi: 10.1002/ptr.4969.

Navarro del Hierro, J., V. Piazzini, G. Reglero, D. Martin, and M. C. Bergonzi. 2020. "In vitro permeability of saponins and sapogenins from seed extracts by the parallel artificial membrane permeability assay: Effect of in vitro gastrointestinal digestion." *J Agric Food Chem* 68 (5):1297–305. doi: 10.1021/acs.jafc.9b07182.

Nayak, A. K., D. Pal, J. Pradhan, and T. Ghora. 2012. "The potential of *Trigonella foenum-graecum* L. Seed mucilage as suspending agent." *Indian J Pharm Educ Res* 45 (4):312–17.

Nayak, A. K., and D. Pal. 2014. "*Trigonella foenum-graecum* L. seed mucilage-gellan mucoadhesive beads for controlled release of metformin HCl." *Carbohydrate Polymers* 107:31–40. doi: 10.1016/j.carbpol.2014.02.031.

Nayak, A. K., D. Pal, and S. Das. 2013. "Calcium pectinate-fenugreek seed mucilage mucoadhesive beads for controlled delivery of metformin HCl." *Carbohydrate Polymers* 96 (1):349–57. doi: 10.1016/j.carbpol.2013.03.088.

Nayak, A. K., D. Pal, J. Pradhan, and M. S. Hasnain. 2013. "Fenugreek seed mucilage-alginate mucoadhesive beads of metformin HCl: Design, optimization and evaluation." *Int J Biol Macromol* 54:144–54. doi: 10.1016/j.ijbiomac.2012.12.008.

Nokhodchi, A., H. Nazemiyeh, A. Khodaparast, T. Sorkh-Shahan, H. Valizadeh, and J. L. Ford. 2008. "An in vitro evaluation of fenugreek mucilage as a potential excipient for oral controlled-release matrix tablet." *Drug Dev Ind Pharm* 34 (3):323–9. doi: 10.1080/03639040701662594.

Noohpisheh, Z., H. Amiri, S. Farhadi, and A. Mohammadi-gholami. 2020. "Green synthesis of Ag-ZnO nanocomposites using Trigonella foenum-graecum leaf extract and their antibacterial, antifungal, antioxidant and photocatalytic properties." *Spectrochimica Acta Part A: Molecular and Biomolecular Spectroscopy* 240:118595. doi: 10.1016/j.saa.2020.118595.

Noudeh, G. D., F. Sharififar, P. Khazaeli, E. Mohajeri, and J. Jahanbakhsh. 2011. "Formulation of herbal conditioner shampoo by using extract of fenugreek seeds and evaluation of its physicochemical parameters." *Afr J Pharmacy Pharmacol* 5 (22):2420–7. doi: 10.5897/AJPP11.121.

Ofori-Kwakye, K., K. A. Mfoafo, S. L. Kipo, N. Kuntworbe, and M. E. Boakye-Gyasi. 2016. "Development and evaluation of natural gum-based extended release matrix tablets of two model drugs of different water solubilities by direct compression." *Saudi Pharmaceutical Journal* 24 (1):82–91. https://doi.org/10.1016/j.jsps.2015.03.005.

Ogbechie-Godec, O. A., and N. Elbuluk. 2017. "Melasma: An up-to-date comprehensive review." *Dermatology and Therapy* 7 (3):305–18. doi: 10.1007/s13555-017-0194-1.

Olotu, F., C. Agoni, O. Soremekun, and M. E. Soliman. 2020. "An update on the pharmacological usage of curcumin: Has it failed in the drug discovery pipeline?" *Cell Biochem Biophy* 78:267–89. doi: 10.1007/s12013-020-00922-5.

Pallotta, A., A. Boudier, B. Creusot, E. Brun, C. Sicard-Roselli, R. Bazzi, S. Roux, and I. Clarot. 2019. "Quality control of gold nanoparticles as pharmaceutical ingredients." *Int J Pharm* 569:118583. doi: 10.1016/j.ijpharm.2019.118583.

Patel, N., and S. Jain. 2014. "Solubility and dissolution enhancement strategies: Current understanding and recent trends." *Drug Dev Ind Pharm* 41:875–87. doi: 10.3109/03639045.2014.971027.

Pathak, D., P. Kumar, G. Kuppusamy, A. Gupta, B. Kamble, and A. Wadhwani. 2014. "Physicochemical characterization and toxicological evaluation of plant-based anionic polymers and their nanoparticulated system for ocular delivery." *Nanotoxicology* 8 (8):843–55. doi: 10.3109/17435390.2013.834996.

Patra, J. K., G. Das, L. F. Fraceto, E. V. R. Campos, M. d. P. Rodriguez-Torres, L. S. Acosta-Torres, L. A. Diaz-Torres, R. Grillo, M. K. Swamy, S. Sharma, S. Habtemariam, and H.-S. Shin. 2018. "Nano based drug delivery systems: Recent developments and future prospects." *Journal of Nanobiotechnology* 16 (1):71. doi: 10.1186/s12951-018-0392-8.

Peng, S., Z. Li, L. Zou, W. Liu, C. Liu, and D. J. McClements. 2018. "Improving curcumin solubility and bioavailability by encapsulation in saponin-coated curcumin nanoparticles prepared using a simple pH-driven loading method." *Food & Function* 9 (3):1829–39. doi: 10.1039/c7fo01814b.

Petkowicz, C. L. O., M. Milas, K. Mazeau, T. Bresolin, F. Reicher, J. L. M. S. Ganter, and M. Rinaudo. 1999. "Conformation of galactomannan: Experimental and modelling approaches." *Food Hydrocolloids* 13 (3):263–6. doi: 10.1016/S0268-005X(99)00008-9.

Petkowicz, C. L. O., F. Reicher, and K. Mazeau. 1998. "Conformational analysis of galactomannans: From oligomeric segments to polymeric chains." *Carbohydrate Polymers* 37 (1):25–39. doi: 10.1016/S0144-8617(98)00051-4.

Philip, A. K., and B. Philip. 2010. "Colon targeted drug delivery systems: A review on primary and novel approaches." *Oman Med J* 25 (2):70–8. doi: 10.5001/omj.2010.24.

Pol, S. B. 2014. "Effect of *Trigonella foenum-graecum* fenugreek nanoparticles on islets of langerhans of alloxan induced diabetic mice." PHD, Zoology, Shivaji University, Maharashtra.

Pollard, M. A., and P. Fischer. 2006. "Partial aqueous solubility of low-galactose-content galactomannans—What is the quantitative basis?" *Current Opinion in Colloid & Interface Science* 11 (2–3):184–90. doi: 10.1016/j.cocis.2005.12.001.

Ponzini, E., A. Natalello, F. Usai, M. Bechmann, F. Peri, N. Muller, and R. Grandori. 2019. "Structural characterization of aerogels derived from enzymatically oxidized galactomannans of fenugreek, sesbania and guar gums." *Carbohydrate Polymers* 207 (1):510–20. doi: 10.1016/j.carbpol.2018.11.100.

Poole, C., B. Bushey, C. Foster, B. Campbell, D. Willoughby, R. Kreider, L. Taylor, and C. Wilborn. 2010. "The effects of a commercially available botanical supplement on strength, body composition, power output, and hormonal profiles in resistance-trained males." *J Int Soc Sports Nutr* 7 (1):34. doi: 10.1186/1550-2783-7-34.

Pooloth, J. 2013. "Biosynthesis of silver nanoparticles using *Trigonella Foenum-graecum* and the determination of their antimicrobial activity." *Int J Sci Res* 2 (5):287–90.

Prajapati, V. D., G. K. Jani, N. G. Moradiya, and N. P. Randeria. 2013. "Pharmaceutical applications of various natural gums, mucilages and their modified forms." *Carbohydrate Polymers* 92 (2):1685–99. doi: 10.1016/j.carbpol.2012.11.021.

Prajapati, V. D., G. K. Jani, N. G. Moradiya, N. P. Randeria, B. J. Nagar, N. N. Naikwadi, and B. C. Variya. 2013. "Galactomannan: A versatile biodegradable seed polysaccharide." *Int J Biol Macromol* 60:83–92. doi: 10.1016/j.ijbiomac.2013.05.017.

Prakash, S., C. Tomaro-Duchesneau, S. Saha, and A. Cantor. 2011. "The gut microbiota and human health with an emphasis on the use of microencapsulated bacterial cells." *J Biomed Biotechnol* 2011:981214. doi: 10.1155/2011/981214.

Prasanna, M. 2000. "Hypolipidemic effect of fenugreek: A clinical study." *Indian J Pharm* 32 (1):34–6.

PubChem. 2004. *Trigonelline (PubChem Compound Summary for CID 5570)*. Bethesda, MD: National Center for Biotechnology Information, National Library of Medicine.

Randhawa, R., P. Bassi, and G. Kaur. 2012. "In vitro, in vivo evaluation of inter polymer complexes between carboxymethyl fenugreek gum and chitosan or carboxymethyl guar gum and chitosan for colon delivery of tamoxifen." *Asian Pacific Journal of Tropical Disease* 2:S202–S7. doi: 10.1016/s2222-1808(12)60152-2.

Rao, A., and R. Grant. 2020. "The effect of Trigonella foenum-graecum extract on prostate-specific antigen, and prostate function in otherwise healthy men with benign prostate hyperplasia." *Phytother Res* 34 (3):634–9. doi: 10.1002/ptr.6554.

Rao, A. J., A. R. Mallard, and R. Grant. 2020. "Testofen (Fenugreek extract) increases strength and muscle mass compared to placebo in response to calisthenics. A randomized control trial." *Translational Sports Medicine* 3 (4):374–80. doi: 10.1002/tsm2.153.

Rao, A., E. Steels, G. Beccaria, W. J. Inder, and L. Vitetta. 2015. "Influence of a Specialized Trigonella foenum-graecum Seed Extract (Libifem), on Testosterone, Estradiol and Sexual Function in Healthy Menstruating Women, a Randomised Placebo Controlled Study." *Phytother Res* 29 (8):1123–30. doi: 10.1002/ptr.5355.

Rao, A., E. Steels, W. J. Inder, S. Abraham, and L. Vitetta. 2016. "Testofen, a specialised Trigonella foenum-graecum seed extract reduces age-related symptoms of androgen decrease, increases testosterone levels and improves sexual function in healthy aging males in a double-blind randomised clinical study." *Aging Male* 19 (2):134–42. doi: 10.3109/13685538.2015.1135323.

Rashid, F., S. Hussain, and Z. Ahmed. 2018. "Extraction purification and characterization of galactomannan from fenugreek for industrial utilization." *Carbohydr Polym* 180:88–95. doi: 10.1016/j.carbpol.2017.10.025.

Rashid, R., H. Ahmad, Z. Ahmed, F. Rashid, and N. Khalid. 2019. "Clinical investigation to modulate the effect of fenugreek polysaccharides on type-2 diabetes." *Bioactive Carbohydrates and Dietary Fibre* 19:100194. doi: 10.1016/j.bcdf.2019.100194.

Reeder, C., A. Legrand, and S. K. O'connor-Von. 2013. "The effect of fenugreek on milk production and prolactin levels in mothers of preterm infants." *Clinical Lactation* 4 (4):159–65. doi: 10.1891/2158-0782.4.4.159.

Rezaei, A., and A. Nasirpour. 2018. "Encapsulation of curcumin using electrospun almond gum nanofibers: Fabrication and characterization." *Int J Food Prop* 21 (1):1608–18. doi: 10.1080/10942912.2018.1503300.

Rodriguez-Canto, W., L. Chel-Guerrero, V. V. A. Fernandez, and M. Aguilar-Vega. 2019. "Delonix regia galactomannan hydrolysates: Rheological behavior and physicochemical characterization." *Carbohydr Polym* 206:573–82. doi: 10.1016/j.carbpol.2018.11.028.

Rose, B. P. K., A. K. Patel, and K. V. Asha. 2020. "A review on methika (*Trigonella Foenum-graecum Linn*): A widely used drug in postnatal care in Kerala, India." *Int J Ayurveda Res* 11 (3):61–3. doi: 10.7897/2277-4343.110361.

Rossi, B., E. Ponzini, L. Merlini, R. Grandori, and Y. M. Galante. 2017. "Characterization of aerogels from chemo-enzymatically oxidized galactomannans as novel polymeric biomaterials." *European Polymer Journal* 93:347–57. doi: 10.1016/j.eurpolymj.2017.06.016.

Ruby, B. C., S. E. Gaskill, D. Slivka, and S. G. Harger. 2005. "The addition of fenugreek extract (T*rigonella foenum-graecum*) to glucose feeding increases muscle glycogen resynthesis after exercise." *Amino Acids* 28 (1):71–6. doi: 10.1007/s00726-004-0143-z.

Sajty, M., L. Jackova, L. Merkovska, L. Jedlickova, J. Fedacko, M. Janicko, M. Vachalcova, and R. B. Singh. 2014. "Fenugreek seeds decrease oxidative stress and blood lipids and increase nitric oxide in patients with hyperlipidemia." *Atherosclerosis* 235 (2):e113. doi: 10.1016/j.atherosclerosis.2014.05.306.

Sakhare, M., and H. Rajput. 2017. "Polymer grafting and applications in pharmaceutical drug delivery systems—a brief review." *Asian J Pharm Clin Res* 10 (6):59. doi: 10.22159/ajpcr.2017.v10i6.18072.

Sanidad, K. Z., E. Sukamtoh, H. Xiao, D. J. McClements, and G. Zhang. 2019. "Curcumin: Recent advances in the development of strategies to improve oral bioavailability." *Annu Rev Food Sci Technol* 10:597–617. doi: 10.1146/annurev-food-032818-121738.

Sankhwar, P., S. P. Jaiswar, A. Goel, and K. Tiwari. 2018. "Clinical evaluation of furostanolic saponins and flavanoids in polycystic ovarian syndrome (PCOS) patients." *Journal of Endocrinology and Diabetes Care* 1 (1):103. doi: 10.33425/2639-9342.1049.

Sarika, P. R., N. R. James, P. R. A. Kumar, D. K. Raj, and T. V. Kumary. 2015. "Gum arabic-curcumin conjugate micelles with enhanced loading for curcumin delivery to hepatocarcinoma cells." *Carbohydrate Polymers* 134:167–74. https://doi.org/10.1016/j.carbpol.2015.07.068.

Saroha, K., B. Yadav, and B. Sharma. 2011. "Transdermal patch: A discrete dosage form." *Int J Curr Pharm Res* 3 (3):98–108.

Satturwar, P. M., S. V. Fulzele, and A. K. Dorle. 2003. "Biodegradation and in vivo biocompatibility of rosin: A natural film-forming polymer." *Aaps Pharmscitech* 4 (4):434–9. doi: 10.1208/pt040455.

Sav, A. K., M. T. Ali, R. A. Fule, and P. D. Amin. 2013a. "Synthesis and evaluation of octenyl succinate anhydride derivative of fenugreek gum as extended release polymer." *Int J. Pharm Investig* 43 (5):417–29. doi. org/10.1007/s40005-013-0088-x.

Sav, A. K., M. T. Ali, R. A. Fule, and P. D. Amin. 2013b. "Formulation of highly purified fenugreek gum based silica lipid drug delivery system for simvastatin with enhanced dissolution rate and *in-vitro* characterization." *J. Pharm Investig* 43 (5):363–73. doi: 10.1007/s40005-013-0081-4.

Saxena, C., K. Arora, and L. Chaurasia. 2019. "Importance of different novel nasal drug delivery system—A review." *Int J Pharm Clin Res* 11 (1):13–19.

Schultz, H. B., N. Thomas, S. Rao, and C. A. Prestidge. 2018. "Supersaturated silica-lipid hybrids (super-SLH): An improved solid-state lipid-based oral drug delivery system with enhanced drug loading." *Eur J Pharm Biopharm* 125:13–20. doi: 10.1016/j.ejpb.2017.12.012.

Seetharaman, S., H. Balya, and G. Kuppusamy. 2016. "Preparation and evaluation of cefixime nanoparticles prepared using fenugreek seed mucilage and chitosan as natural polymers." *Int J Pharm Clin Res* 8 (3):179–88.

Selvaraj, S., N. Duraipandy, M. S. Kiran, and N. N. Fathima. 2018. "Anti-oxidant enriched hybrid nanofibers: Effect on mechanical stability and biocompatibility." *Int J Biol Macromol* 117:209–17. doi: 10.1016/j.ijbiomac.2018.05.152.

Senthil, V., and D. Sripreethi. 2011. "Formulation and evaluation of paracetamol suspension from *Trigonella Foenum-graecum* mucilage." *J Adv Pharm Ed Res* 1 (5):225–33.

Shah, S., S. L. Bodhankar, S. L. Badole, H. V. Kamble, and V. Mohan. 2006. "Effect of trigonelline: An active compound from *Trigonella foenum-graecum* linn. in alloxan induced diabetes in mice." *Journal of Cell and Tissue Research* 6 (1):585–90.

Shamshad Begum, S., H. K. Jayalakshmi, H. G. Vidyavathi, G. Gopakumar, I. Abin, M. Balu, K. Geetha, S. V. Suresha, M. Vasundhara, and I. M. Krishnakumar. 2016. "A novel extract of Fenugreek Husk (FenuSMART) alleviates postmenopausal symptoms and helps to establish the hormonal balance: A randomized, double-blind, placebo-controlled study." *Phytother Res* 30 (11):1775–84. doi: 10.1002/ptr.5680.

Sharififar, F., P. Khazaeli, and N. Alli. 2009. "*In Vivo* evaluation of anti-inflammatory activity of topical preparations from fenugreek (*Trigonella foenum-graecum* L.) seeds in a cream base." *Iran J Pharm Sci* 5 (3):157–62.

Sharma, D., D. Dev, D. Prasad, and M. Hans. 2019. "Sustained release drug delivery system with the role of natural polymers: A review." *J Drug Deliv Ther* 9 (3):913–23.

Sharma, N., A. Sharma, D. K. Nishad, K. Khanna, B. G. Sharma, D. Kakkar, and A. Bhatnagar. 2018. "Development and Gamma scintigraphy study of Trigonella foenum-graecum (Fenugreek) polysaccharide-based colon tablet." *AAPS PharmSciTech* 19 (6):2564–71. doi: 10.1208/s12249-018-1066-4.

Sharma, R. D., A. Sarkar, D. K. Hazara, B. Mishra, J. B. Singh, S. K. Sharma, B. B. Maheshwari, and P. K. Maheshwari. 1996. "Use of Fenugreek seed powder in the management of non-insulin dependent diabetes mellitus." *Nutrition Research* 16 (8):1331–9. doi: 10.1016/0271-5317(96)00141-8.

Sherazi, T. A. 2016. "Graft polymerization." In *Encyclopedia of Membranes*, edited by Drioli, E. and Giorno, L., 886–7. Berlin, Heidelberg: Springer Berlin Heidelberg.

Shukla, A. K., R. S. Bishnoi, M. Kumar, and C. P. Jain. 2019. "Development of natural and modified gum based sustained-release film-coated tablets containing poorly water-soluble drug." *Asian J Pharm Clin Res* 12 (3):266–71. doi: 10.22159/ajpcr.2019.v12i3.30296.

Siew, C. K., and P. A. Williams. 2008. "Characterization of the surface-active components of sugar beet pectin and the hydrodynamic thickness of the adsorbed pectin layer." *J Agr Food Chem* 56 (17):8111–20. doi: 10.1021/jf801588a.

Singh, A., J. Rai, and D. Mahajan. 2016. "Comparative evaluation of glipizide and fenugreek (*Trigonella foenum-graecum*) seeds as monotherapy and combination therapy on glycaemic control and lipid profile in patients with type 2 diabetes mellitus." *Int J Basic Clin Pharmacol* 5 (3):942–50. doi: 10.18203/2319-2003.ijbcp20161549.

Singh, R., S. Magesh, and C. Rakkiyappan. 2011. "Formation of fenugreek (*Trigonella foenum-graecum*) extract mediated Ag nanoparticles: Mechanism and applications." *Int J Bioeng Sci Technol* 2 (3):64–73.

Sondi, I., and B. Salopek-Sondi. 2004. "Silver nanoparticles as antimicrobial agent: A case study on *E. coli* as a model for Gram-negative bacteria." *Journal of Colloid And Interface Science* 275 (1):177–82. doi: 10.1016/j.jcis.2004.02.012.

Sowmya, P., and P. Rajyalakshmi. 1999. "Hypocholesterolemic effect of germinated fenugreek seeds in human subjects." *Plant Foods Hum Nutr* 53 (4):359–65. doi: 10.1023/a:1008021618733.

Steels, E., A. Rao, and L. Vitetta. 2011. "Physiological aspects of male libido enhanced by standardized Trigonella foenum-graecum extract and mineral formulation." *Phytother Res* 25 (9):1294–300. doi: 10.1002/ptr.3360.

Subramanian, S. P., and G. S. Prasath. 2014. "Antidiabetic and antidyslipidemic nature of trigonelline, a major alkaloid of fenugreek seeds studied in high-fat-fed and low-dose streptozotocin-induced experimental diabetic rats." *Biomedicine & Preventive Nutrition* 4 (4):475–80. doi: 10.1016/j.bionut.2014.07.001.

Swaroop, A., A. S. Jaipuriar, S. K. Gupta, M. Bagchi, P. Kumar, H. G. Preuss, and D. Bagchi. 2015. "Efficacy of a Novel Fenugreek Seed Extract (Trigonella foenum-graecum, Furocyst) in Polycystic Ovary Syndrome (PCOS)." *Int J Med Sci* 12 (10):825–31. doi: 10.7150/ijms.13024.

Swaroop, A., A. Maheshwari, N. Verma, K. Tiwari, P. Kumar, M. Bagchi, H. G. Preuss, and D. Bagchi. 2017. "A novel protodioscin-enriched fenugreek seed extract (*Trigonella foenum-graecum*, family Fabaceae) improves free testosterone level and sperm profile in healthy volunteers." *Functional Foods in Health and Disease* 7 (4):235–45. doi: 10.31989/ffhd.v7i4.326.

Szczesny, D., E. Bartosinska, J. Jacyna, M. Patejko, D. Siluk, and R. Kaliszan. 2018. "Quantitative determination of trigonelline in mouse serum by means of hydrophilic interaction liquid chromatography-MS/MS analysis: Application to a pharmacokinetic study." *Biomed Chromatogr* 32 (2):e4054. doi: 10.1002/bmc.4054.

Tavakoli, N., J. Varshosaz, A. Ghannadi, and N. Bavarsad. 2012. "Evaluation of *Trigonella Foenum- graecum* seeds gum as a novel tablet binder." *Int J Pharm Pharm Sci* 4 (1):97–101.

Tavakoly, R., M. R. Maracy, M. Karimifar, and M. H. Entezari. 2018. "Does fenugreek (Trigonella foenum-graecum) seed improve inflammation, and oxidative stress in patients with type 2 diabetes mellitus? A parallel group randomized clinical trial." *Eur J Integrative Med* 18:13–17. doi: 10.1016/j.eujim.2018.01.005.

Thammanna, G. S. S., D. D. Ramadas, H. ramakrishna, and L. Srinivas. 2010. "Free radical scavenging activity of lutein—isolated from methi leaves (*Trigonella foenum graecum*)." *Int J Pharm Pharm Sci* 2:113–17.

Thomas, J. V., J. Rao, F. John, S. Begum, B. Maliakel, K. Im, and A. Khanna. 2020. "Phytoestrogenic effect of fenugreek seed extract helps in ameliorating the leg pain and vasomotor symptoms in postmenopausal women: A randomized, double-blinded, placebo-controlled study." *PharmaNutrition* 14:100209. doi: 10.1016/j.phanu.2020.100209.

Trüeb, R. M. 2001. "The value of hair cosmetics and pharmaceuticals." *Dermatology* 202 (4):275–82. doi: 10.1159/000051658.

Trueb, R. M. 2007. "Shampoos: Ingredients, efficacy and adverse effects." *J Dtsch Dermatol Ges* 5 (5):356–65. doi: 10.1111/j.1610-0387.2007.06304.x.

Ulker, Z., and C. Erkey. 2014. "An emerging platform for drug delivery: Aerogel based systems." *J Control Release* 177:51–63. doi: 10.1016/j.jconrel.2013.12.033.

Upaganlawar, A. B., S. L. Badole, and S. L. Bodhankar. 2013. "Chapter 6 — antidiabetic potential of trigonelline and 4-hydroxyisoleucine in Fenugreek." In *Bioactive Food as Dietary Interventions for Diabetes*, edited by Watson, R.R. and Preedy, V.R., 59–64. San Diego: Academic Press.

Vakili, M., P. Amouzgar, G. Cagnetta, B. Wang, X. Guo, A. Mojiri, E. Zeimaran, and B. Salamatinia. 2019. "Ultrasound-assisted preparation of chitosan/nano-activated carbon composite beads aminated with (3-Aminopropyl)Triethoxysilane for adsorption of acetaminophen from aqueous solutions." *Polymers (Basel)* 11 (10):1701. doi: 10.3390/polym11101701.

Van der Merwe, J., J. Steenekamp, D. Steyn, and J. Hamman. 2020. "The role of functional excipients in solid oral dosage forms to overcome poor drug dissolution and bioavailability." *Pharmaceutics* 12 (5):393. doi: 10.3390/pharmaceutics12050393.

Vass, P., B. Demuth, E. Hirsch, B. Nagy, S. K. Andersen, T. Vigh, G. Verreck, I. Csontos, Z. K. Nagy, and G. Marosi. 2019. "Drying technology strategies for colon-targeted oral delivery of biopharmaceuticals." *J Control Release* 296:162–78. doi: 10.1016/j.jconrel.2019.01.023.

Verbeek, C. J., and C. Gavin. 2020. "Grafting functional groups onto biodegradable thermoplastic polyesters." In *Reactive and Functional Polymers*, 245–81. New York: Springer.

Verma, N., K. Usman, N. Patel, A. Jain, S. Dhakre, A. Swaroop, M. Bagchi, P. Kumar, H. G. Preuss, and D. Bagchi. 2016. "A multicenter clinical study to determine the efficacy of a novel fenugreek seed (Trigonella foenum-graecum) extract (Fenfuro) in patients with type 2 diabetes." *Food Nutr Res* 60 (1):32382. doi: 10.3402/fnr.v60.32382.

Vibha, B. 2014. "*In-situ* gel nasal drug delivery system-a review." *Int J Pharm Sci* 4 (3):577–80.

Vrček, I. V. 2018. "Selenium nanoparticles: Biomedical applications." In *Selenium*, edited by Michalke, B., 393–412. New York: Springer.

Wang, W., S. He, T. Hong, Y. Zhang, H. Sui, X. Zhang, and Y. Ma. 2017. "Synthesis, self-assembly, and *in vitro* toxicity of fatty acids-modified *Bletilla striata* polysaccharide." *Artif Cells Nanomed Biotechnol* 45 (1):69–75. doi: 10.3109/21691401.2015.1129621.

Wani, S. A., and P. Kumar. 2018. "Fenugreek: A review on its nutraceutical properties and utilization in various food products." *Journal of the Saudi Society of Agricultural Sciences* 17 (2):97–106. doi: 10.1016/j.jssas.2016.01.007.

Wankhede, S., V. Mohan, and P. Thakurdesai. 2016. "Beneficial effects of fenugreek glycoside supplementation in male subjects during resistance training: A randomized controlled pilot study." *J Sport Health Sci* 5 (2):176–82. doi: 10.1016/j.jshs.2014.09.005.

Waqas, M. K., N. Akhtar, M. Ahmad, G. Murtaza, H. M. Khan, M. Iqbal, A. Rasul, and N. S. Bhatti. 2010. "Formulation and characterization of a cream containing extract of fenugreek seeds." *Acta Pol Pharm* 67 (2):173–8.

Wilborn, C., S. Hayward, L. Taylor, S. Urbina, C. Foster, P. Deshpande, V. Mohan, and P. Thakurdesai. 2017. "Effects of a proprietary blend rich in glycoside based standardized fenugreek seed extract (Ibpr) on inflammatory markers during acute eccentric resistance exercise in young subjects." *Asian J Pharm Clin Res* 10 (10):99. doi: 10.22159/ajpcr.2017.v10i10.18811.

Willems, H. M. E., S. S. Ahmed, J. Liu, Z. Xu, and B. M. Peters. 2020. "Vulvovaginal candidiasis: A current understanding and burning questions." *J Fungi (Basel)* 6 (1):1–20. doi: 10.3390/jof6010027.

Xu, W., P. Ling, and T. Zhang. 2013. "Polymeric micelles, a promising drug delivery system to enhance bioavailability of poorly water-soluble drugs." *J Drug Deliv* 2013:340315. doi: 10.1155/2013/340315.

Yarnykh, T. G., O. I. Tykhonov, G. M. Melnyk, and G. B. Yuryeva. 2018. "Pharmacopoeian aspects of suspensions preparation in pharmacy conditions." *Asian J Pharm (AJP): Free Full Text Articles from Asian J Pharm* 11 (04):S859–S64. doi: 10.22377/ajp.v11i04.1727.

Ying, W., J. Tan, C. Chen, T. Sun, S. Wang, and M. Zhang. 2019. "Biofabrication of silver nanoparticles and its application for development of wound dressing system in nursing care for burn injuries in children." *J Drug Deliv Ther* 54:101236. doi: 10.1016/j.jddst.2019.101236.

Younesy, S., S. Amiraliakbari, S. Esmaeili, H. Alavimajd, and S. Nouraei. 2014. "Effects of fenugreek seed on the severity and systemic symptoms of dysmenorrhea." *J Reprod Infertil* 15 (1):41–8.

Yousefi, E., S. Zareiy, R. Zavoshy, M. Noroozi, H. Jahanihashemi, and H. Ardalani. 2017. "Fenugreek: A therapeutic complement for patients with borderline hyperlipidemia: A randomised, double-blind, placebo-controlled, clinical trial." *Adv Integr Med* 4 (1):31–5. doi: 10.1016/j.aimed.2016.12.002.

Zhao, J., J. Yang, and Y. Xie. 2019. "Improvement strategies for the oral bioavailability of poorly water-soluble flavonoids: An overview." *Int J Pharm* 570:118642. doi: 10.1016/j.ijpharm.2019.118642.

Zhou, J., L. Chan, and S. Zhou. 2012. "Trigonelline: A plant alkaloid with therapeutic potential for diabetes and central nervous system disease." *Curr Med Chem* 19 (21):3523–31. doi: 10.2174/092986712801323171.

Zhou, M., B. Li, X. Zhang, S. He, Y. Ma, and W. Wang. 2019. "Synthesis and evaluation of hydrophobically modified fenugreek gum for potential hepatic drug delivery." *Artif Cells Nanomed Biotechnol* 47 (1):1702–9. doi: 10.1080/21691401.2019.1606009.

21 Applications of Fenugreek in Nutritional and Functional Food Preparations

Ujjwala Kandekar, Rohini Pujari and Prasad Thakurdesai

CONTENTS

21.1 Introduction 402
21.2 Sensory and Organoleptic Properties of Fenugreek...403
21.3 Nutritional Properties of Fenugreek..403
21.4 Effects of the Cooking Process on Fenugreek 404
    21.4.1 Effect of the Germination Process..404
    21.4.2 Effect of the Soaking Process..405
    21.4.3 Effect of the Roasting Process...405
    21.4.4 Effect of the Boiling Process 405
    21.4.5 Effect of the Microwave Drying Process..406
    21.4.6 Effects of the Refrigeration Process 406
    21.4.7 Effects of Gamma Irradiation Treatment 406
    21.4.8 Effect of the Blanching Process...407
    21.4.9 Effect of Storage Conditions on Nutritional Content in the Foods.....................407
21.5 Fenugreek in Bakery Products Preparations..407
    21.5.1 Fenugreek in Bread..407
    21.5.2 Fenugreek Seed Husk in Muffins..408
    21.5.3 Fenugreek Seed Gum in Barbari and Lavash Flatbreads 408
    21.5.4 Fenugreek Amylase in Cake 408
    21.5.5 Fenugreek Leaves in Pizza 410
    21.5.6 Fenugreek in Biscuits...410
    21.5.7 Fenugreek in Cookies 410
    21.5.8 Debittered Fenugreek Flour in Rusk...410
    21.5.9 Fenugreek in Puff Pastry 412
    21.5.10 Fenugreek Leaves in Waffles..412
21.6 Fenugreek in Snacks Preparations 412
    21.6.1 Fenugreek in Indian Snack Preparations 412
        21.6.1.1 Fenugreek in Matthi and Seviyan..412
        21.6.1.2 Fenugreek in Kachori...413
        21.6.1.3 Fenugreek in Masala Parwal 413
        21.6.1.4 Fenugreek Flour in Pakoras...413
        21.6.1.5 Fenugreek in Upma..413
        21.6.1.6 Fenugreek in Idlis Flour...413
    21.6.2 Fenugreek in Extruded Snacks..413
    21.6.3 Fenugreek Leaves in Pasta..414
    21.6.4 Fenugreek in Noodles..414
    21.6.5 Fenugreek in Spent Hen Patties..414
    21.6.6 Fenugreek Polysaccharide in Sausage 415

DOI: 10.1201/9781003082767-26

	21.6.7	Fenugreek in Meat Patties and Burgers	415
	21.6.8	Fenugreeks Paste in Pastirma	415
21.7	Fenugreek in the Main Course		416
	21.7.1	Fenugreek Leaves as Vegetables	416
	21.7.2	Fenugreek Seed Flour in Indian Flatbread Preparations	416
		21.7.2.1 Fenugreek in Injera	416
		21.7.2.2 Fenugreek in Chapattis	416
		21.7.2.3 Fenugreek Leaves in Thepla	416
		21.7.2.4 Fenugreek Dried Leaves in Missi Roti	417
		21.7.2.5 Fenugreek Leaves in Bhakri	417
		21.7.2.6 Fenugreek Leaves in Paratha	417
		21.7.2.7 Fenugreek Leaves in Doli Ki Roti	417
21.8	Fenugreek as Spices and Pickles		417
21.9	Fenugreek in Animal Feed		419
	21.9.1	Fenugreek Seeds for Improving Animal Milk Characteristics	419
	21.9.2	Fenugreek as a Growth Promoter in Poultry and Pigs Feed	420
	21.9.3	Fenugreek Seeds as High Fiber Diet in Rabbit Feed	420
	21.9.4	Fenugreek Seeds in Fish Feed	420
21.10	Fenugreek Preparations as a Functional Food		420
	21.10.1	Fenugreek Food Preparations and Diabetes Mellitus	420
	21.10.2	Fenugreek Herbal Tea as a Galactagogue	422
	21.10.3	Fenugreek Sprouts and Biscuits for Anemia	422
21.11	Conclusion		423
References			424

ABBREVIATIONS

Caco-2 cells—Cancer coli-2 cells; CRS—Cardiovascular risk score; GI—Glycemic index; Gy—Gray; HbA1c—Hemoglobin A1c; HDL—High density lipoprotein; HSI—Hepatic steatosis index; IgG—Immunoglobulin G; IgM—Immunoglobulin M; KMS—Potassium metabisulphite; LDL—Low-density lipoprotein; RBC—Red blood cells; TC—Total cholesterol; TG—Triglycerides; VLDL—Very-low-density lipoprotein

21.1 INTRODUCTION

Food is meant to provide essential nutrition for the health and wellbeing of humans. The primary concern for food products is maintaining the balanced intake of nutrients by fibers, amino acids, proteins, fats, vitamins, and minerals (Dey 2018). It is always desirable to consume the nutrients in food and food-based products to attain optimum nutritional status (Ruel and Levin 2002).

Fenugreek has been used as multipurpose food since the ancient period. For example, fenugreek health and medicinal benefits have been documented in the Ebers Papyrus around 1500 BC. The use of fenugreek as a spice was explored in several traditional recipes. Dried leaves and seeds of fenugreek have extensive applications as a flavoring agent in spices. In traditional knowledge, fenugreek seeds are suggested to enhance digestion and nutrient absorption, promote muscle growth, reduce fat, and exhibit strengthening and nourishing properties (Roberts et al. 2015). Fenugreek was generally regarded as a safe herb, even in large amounts (Alfarisi et al. 2017).

Fenugreek is one of the most popular sources of nutrients. Fenugreek leaves are composed of many vital nutrients, such as proteins, fats, fibers, and carbohydrates. The chemical composition of fenugreek seeds has been documented in the past scientific literature (Alsebaeai et al. 2017;

Bienkowski et al. 2017; Kochhar, Nagi, and Sachdeva 2005). The predominant constituents present in fenugreek are carbohydrate, gum, mucilage, crude fiber, protein, and amino acid (Naidu et al. 2011). Other minor constituents include essential oils; saponins (mainly yamogenin and diosgenin) (Acharya et al. 2006; Taylor et al. 2002); alkaloids (mainly trigonelline); choline, gentianine, and carpine (Mehrafarin et al. 2010); minerals such as calcium, phosphorus, iron, sodium, potassium, and vitamins (mainly A, B1, B2, B3, and C) (Bienkowski et al. 2017; Wani and Kumar 2018). Young seeds contain a small amount of sucrose, fructose, glucose, galactinol, stachyose, and raffinose traces (Leela and Shafeekh 2008). Leaves are a rich source of lysine and other essential amino acids, vitamin C, ß-carotene, and folic acid (Yadav and Sehgal 1997). Galactomannan, a water-soluble polysaccharide present in the seeds, has attracted attention towards its application as an excipient in pharmaceutical products. Almost 14–15% galactomannan was present in the seeds' endosperm (Majeed et al. 2018). The presence of these vital constituents thus enhances the nutritional and medicinal applications of fenugreek seeds.

A combination of a sweet and bitter flavor and richness of essential nutrients has rendered fenugreek indispensable in preparing various foods (Sarwar et al. 2020). In India, ground seeds constitute a vital ingredient of spices and condiments, and leaves are routinely consumed as a vegetable (Ahmad et al. 2016). Fenugreek seeds and dried leaves are widely used as flavoring agents, taste enhancers, and coloring agents in various food preparations (Meghwal and Goswami 2012). The present chapter discusses scientific publications focused on the use of fenugreek in any form of food preparations and its influence on sensory, nutritional, physical, and medicinal properties for value addition.

21.2 SENSORY AND ORGANOLEPTIC PROPERTIES OF FENUGREEK

Fenugreek seeds and leaves are extensively used as spices for characteristic aroma and taste. The volatile oil present in the seeds contains hemiterpenoid-glactone, sotolon (3-hydroxy-4,5-dimethyl-2(5H)-furanone), and other furans contribute to aroma (Girardon et al. 1985). Dried fenugreek leaves have been used widely in various food preparations for aroma (Ahmad et al. 2016; Nagulapalli Venkata et al. 2017). Fenugreek seed oil is used to flavor vanilla, butterscotch, maple syrup, rum, and cheese (Ahmad et al. 2016).

Fenugreek leaves and seeds smell like caramel and impart a bitter taste upon heating. The nonvolatile furostenol glycosides present in the seed might be contributing to the bitter taste (Leela and Shafeekh 2008). When seeds are roasted, trigonelline (an alkaloid present in seeds) gets degraded into different chemical moieties such as nicotinic acid, pyridines, and pyrroles, which result in the characteristic flavor (Acharya et al. 2006; Srinivasan 2006; Yadav, Biyani, and Umekar 2019). The bitterness of seeds could be modified by roasting, soaking, and germination (Dwivedi, Singh, and Giri 2019; Hooda and Jood 2003).

Different constituents from fenugreek have been added to foods for modification of physical properties and preservation. Fenugreek seeds and their phytoconstituents, especially polysaccharides and fiber extracted from fenugreek, have been studied as texture modifiers (Huang et al. 2016; Khemakhem et al. 2018; Koh et al. 2011; Srivastava et al. 2012).

21.3 NUTRITIONAL PROPERTIES OF FENUGREEK

The fiber component of food is known to improve health by enhancing chewing time, replacing calories, suppressing appetite, and reducing overeating, ultimately helping people maintain an appropriate weight. Dietary fiber also induces a feeling of satiety, delays gastric emptying, and enhances mouth-to-caecum food transit time (Dhingra et al. 2012). Fenugreek seed contains edible dietary fibers almost nearly 50% of its weight, out of which 30% is composed of soluble gel-forming galactomannan, and 20% is bulk-forming fiber. Dietary fiber of fenugreek has numerous applications in various food products because of its stability and long shelf life (Srinivasan 2006).

Fenugreek contains proteins with a high content of essential amino acids and total amino acids, especially methionine, histidine, isoleucine, lysine, glutamine, leucine, arginine, threonine, asparagine, and cysteine (Feyzi et al. 2015; Mahfouz, Elaby, and Hassouna 2012; Mandal and DebMandal 2016). The major amino acid of fenugreek seeds, 4-hydroxyisoleucine (4HI), is documented to stimulate insulin secretion and reduce serum cholesterol levels (Avalos-Soriano et al. 2016). The seeds can enhance the nutritional value of cereals and snacks (Anderson and Bush 2011; Işıklı and Karababa 2005). Hence fenugreek seeds are consumed as a dietary supplement (Srinivasan 2006, 2019).

Fenugreek seeds and leaves contain a substantial amount of varied minerals such as phosphorus, iron, calcium, sulfur, potassium, and sodium (Akbari, Rasouli, and Bahdor 2012; Al-Jasass and Al-Jasser 2012; El Nasri and El Tinay 2006). Curry containing fenugreek seeds is a rich source of zinc, calcium, and iron (Jani, Udipi, and Ghugre 2009). Selenium present in fenugreek regulates thyroid function. It also possesses antioxidant and anti-carcinogenic potential (Uras Güngör et al. 2014). Researchers have reported that the number of minerals present in the fenugreek leaves are sufficient for nutritional purposes (Gharneh and Davodalhosseini 2015; Khorshidian et al. 2016; Żuk-Gołaszewska and Wierzbowska 2017). Fenugreek leaves are enriched with proteins, fiber, iron, and vitamin K, and insoluble fibers reduce the appetite and provide early satiety (Samra and Anderson 2007) and fulfill the nutritional needs of overweight and obese individuals (Porecha and Sengupta 2013).

Fenugreek seed lipids are comprised of neutral fats such as glycolipids, phospholipids, triglycerides, diglycerides, small amounts of free fatty acids, and sterols monoglycerides (Hemavathy and Prabhakar 1989). The fatty acids such as oleic acid, linoleic acid, palmitic acid, stearic acid, and arachidic acid are found in the seeds (Al-Jasass and Al-Jasser 2012).

21.4 EFFECTS OF THE COOKING PROCESS ON FENUGREEK

Food preparation involves many processes and treatments (e.g., germination, soaking, roasting, boiling, microwave cooking) and preservation methods (e.g., refrigeration, gamma radiation, blanching). Therefore, these processes' effects on the nutritional content of food preparations are crucial to ensure optimum nutritional properties and food preparation stability.

This section describes the research reports on the effects of diverse processing and preservation techniques and storage conditions on the sensory or nutritional values of fenugreek-containing food preparations (Abdel-Nabey and Damir 1990; Hooda and Jood 2003; Mansour and El-Adawy 1994; Saini et al. 2016).

21.4.1 Effect of the Germination Process

Germination is the process of resumption of the plant embryo inside the seeds under specific temperature and humidity conditions (Luna, Wilkinson, and Dumroese 2014). The germination process was carried out by soaking the seeds in distilled water, followed by enclosing them in cotton cloth. Germination of fenugreek seeds improves the amount of vitamin C, protein, and amino acids and reduces the fat content (Dwivedi, Singh, and Giri 2019; Hooda and Jood 2003; Hussain, Murtaza, and Khan 2012). The germination showed a reduced amount of sugars (raffinose, stachyose, and verbascose) (Mansour and El-Adawy 1994). The level of crude fiber content was elevated owing to enhanced structural carbohydrates, such as celluloses and hemicellulose mobilization and hydrolysis of seed polysaccharides during sprouting, and enriched carbohydrate contents (El-Mahdy and El-Sebaiy 1983). Enzymatic breakdown of galactomannan by α-galactosidase during sprouting resulted in a marked decrease in total dietary fiber, soluble and insoluble dietary fiber (Sathyanarayana et al. 2011; Spyropoulos and Reid 1988).

The germination process resulted in the mobilization of protein, leading to peptides formation, oligopeptides, and free amino acids with protease enzyme help (Jood and Kapoor 1997). Increased protein composition was primarily due to reduced food nitrates or ammonium compounds with improved lysine content. The enzymatic hydrolysis of phytate phosphorus occurs during sprouting, which reduces phytic acid content responsible for protein and starch digestibility (Hooda and Jood 2003; Raju et al. 2001). Phytic acid is a distinctive natural substance found in many plant seeds. Phytic acid impairs iron, zinc, and calcium absorption and may cause mineral deficiencies, hence referred to as an anti-nutrient (Schlemmer et al. 2009). Minerals like calcium and phosphorus increased while zinc and iron decreased significantly in the process of sprouting. Increased calcium and phosphorus might be due to reduced tannins, phytates, and other anti-nutritional factors that bind to the minerals. The reduction in zinc and iron leaked the minerals in the soaking medium (El-Mahdy and El-Sebaiy 1982; El-Shimi, Damir, and Ragab 1984).

Substantial enhancement in antioxidant activity was observed in germinated seeds as compared to raw seeds due to increased phenolic content (Cevallos-Casals and Cisneros-Zevallos 2010; Naidu et al. 2011; Randhir, Lin, and Shetty 2004). These results support the benefits of germinated fenugreeks seeds to provide health and nutritional benefits.

21.4.2 Effect of the Soaking Process

The soaking process resulted in enhanced protein content and reduction in fat content (Pandey and Awasthi 2015a), total carbohydrate, total dietary fiber, insoluble dietary, and soluble dietary fiber (Hooda and Jood 2003). Increased phosphorus and decreased calcium, iron, and zinc content were also observed, probably due to the leaching of minerals in soaking water. The reduced phytic acid content might increase starch and protein digestibility (Pandey and Awasthi 2015a). On the other hand, the soaking process increased the phenolic content and antioxidant activity of fenugreek seeds (Hooda and Jood 2003). When fenugreek seeds are soaked for 16–24 hours, the α-galactosidase enzyme acts on α-galacto-oligosaccharide to remove terminal galactose α-galacto-oligosaccharide to show increased levels of galactose with an increase in glucose and fructose content (due to sucrose hydrolysis) (Njoumi et al. 2019).

21.4.3 Effect of the Roasting Process

The roasting process enhances the crunchiness, flavor, and nutrient levels (Kahyaoglu and Kaya 2006) and antioxidant activity (Açar et al. 2009; Rizki et al. 2015). Fenugreek seed roasting was reported to decrease fat, phytic acid, tannins, and oxalates and significantly increase proteins and phenolic content (Pandey and Awasthi 2015a). The reduced fat content was attributed to the shrinkage of seeds, evaporation of volatile fatty acids, and fats' breakdown, leading to a higher percentage of free fatty acids, esters, formic acid, and acrolein, with retrogradation of starch causing a reduction in total dietary fiber, soluble and insoluble dietary fiber (Mathur and Choudhry 2009a). The decreased phytates, oxalates, and tannins tend to improve *in vitro* digestibility of starch and proteins. Calcium, phosphorus, iron, and zinc levels increase slightly upon roasting (Pandey and Awasthi 2015a). Reported studies demonstrated that roasting of fenugreek seeds increases nutrients and decreases anti-nutrient levels to impact overall health positively.

21.4.4 Effect of the Boiling Process

During the boiling process, the cell membrane weakens and releases content, and helps solubilize individual constituents. The weight and volume of typical seeds are known to increase during the boiling process (Meghwal and Goswami 2011; Shreelalitha and Sridhar 2018). During the scientific studies that involve boiling fenugreek seeds, total fiber content was not altered. However, a noteworthy reduction in phenolic, total sugars, calcium, magnesium, total phosphorus, phytate phosphorus, phytic acid protein, and flavonoid was reported (Azizah et al. 2009; Nguekouo et al. 2018; Ogunmoyole et al. 2012).

The boiling process can affect fenugreek seeds' nutrients within 15 mins (Abdel-Nabey and Damir 1990; Mansour and El-Adawy 1994). Data indicated the increase in both weight and volume and the *in vitro* digestibility of fenugreek seeds after the first 5 minutes of boiling (Abdel-Nabey and Damir 1990). However, at 15 mins of the boiling process of fenugreek seeds, the reduction in phosphorus content, phytic acid, and phytate phosphorus and higher content of minerals, such as iron, zinc, and inorganic phosphorus, was observed. The boiling process increased valine and alanine contents, decreased aspartic and glutamic acids, proline, threonine, isoleucine, glycine, histidine, methionine, tyrosine, tryptophan, and arginine, with no effect on phenylalanine and serine content (Abdel-Nabey and Damir 1990). These findings were supported by a subsequent report of the destruction of trypsin inhibitor activity and verbascose and reduced raffinose, stachyose, phytic acid, and tannic acid levels in fenugreek seeds after heat treatment (boiling for 20 min) (Mansour and El-Adawy 1994). In this study, heat treatment decreased the content of the sulfur-containing amino acids and tryptophan contents, improved *in vitro* protein digestibility, decreased the mineral contents, decreased the nitrogen solubility index, improved the water absorption capacity without affecting emulsion stability (Mansour and El-Adawy 1994).

21.4.5 Effect of the Microwave Drying Process

Drying is one of the traditional preservation methods, which increases the shelf life of a vegetable and reduces the weight, storage space, and cost of transportation (Khatoniar, Barooah, and Das 2018). Green leafy vegetables could be dried during a specific season and stored for further use. The effect of dehydration with microwave drying on nutrients was investigated on fenugreek leaves at the output power of 135, 270, 405, 540, and 675 W (Patil, Pardeshi, and Shinde 2015). Microwave-dried fenugreek leaves indicated that at low power irradiation, color, quality, chlorophyll, protein, and calcium content were not affected as compared to high power irradiation. Dehydration time was reduced with increased power, but with the degradation of the heat-liable constituents. The microwave drying process enhanced nutrients and protected the fenugreek leaves from degradation (Patil, Pardeshi, and Shinde 2015).

21.4.6 Effects of the Refrigeration Process

Refrigeration is the most prevalent food preservation method (Aste, Del Pero, and Leonforte 2017; Fernandes et al. 2015; Lorentzen 1978). Understanding changes in phytonutrients during refrigeration are needed in the food and health industry. The knowledge of the nutrient content of health-promoting products after storage at low temperatures is crucial (Blessington et al. 2010; Grace et al. 2014). The refrigeration of fruits and vegetables was found to increase phenolic acid, and anthocyanin (Prabhu and Barrett 2009). On storage at 4 °C for 15 days, fenugreek showed enhanced phenolic acid content such as caffeic acid, vanillic acid, syringic acid, and ferulic acid, whereas vitamin C content was reduced (Galani et al. 2017).

21.4.7 Effects of Gamma Irradiation Treatment

The gamma irradiation treatment restricts food's microbial load and increases its nutritional composition (Hong et al. 2008; Lacroix 2005). A positive correlation between the radiation process, total phenolic content, and antioxidant activity was reported for many medicinal plants (Rajurkar, Gaikwad, and Razavi 2012; Topuz and Ozdemir 2004; Variyar, Limaye, and Sharma 2004). Fenugreek leaves after exposure to gamma irradiation in the applied dose range of 0.25–1.5 kGy and cobalt 60 as the radiation source at a temperature of 10 ± 2 °C is reported (Hussain et al. 2016). The Gy (gray) is the unit for the absorbed dose of the ionizing radiation in the International System of Units, where 1 gray is equivalent to the ionizing energy dose absorbed by 1 kg of irradiated material (Amit et al. 2017). Gamma irradiated fenugreek leaves significantly enhanced the

Nutritional, Functional Food Preparations

antioxidant activity (higher phenolic and flavonoid content and found ferric reducing power) without causing any deleterious effects on the color and visual quality of the fenugreek (Hussain et al. 2016). These reports suggested that postharvest gamma radiation treatment, besides phytosanitary action, enhances the antioxidant potential of fenugreek.

21.4.8 Effect of the Blanching Process

Blanching is a primary step in food preservation to inactivate the peroxidase enzyme. However, the blanching process can cause adverse effects on color, flavor, texture, and nutrient composition (Lin and Schyvens 2007; Richter Reis 2016; Xiao et al. 2017). Ascorbic acid, ß-carotene, vitamins, and chlorophyll were sensitive to oxidation in the blanching process (Gupta, Lakshmi A, and Prakash 2008; Oboh 2005; Selman 1994; Xiao et al. 2017).

The contents of ß carotene, ascorbic acid, and chlorophyll in fenugreek leaves were evaluated after blanching and drying treatment with plain water, water with potassium metabisulphite (5 g/L in water) for 1 min, salt solution (20 g/L in water), salt solution followed by potassium metabisulphite dip for 1 min, sodium bicarbonate (1 g/L in water), mixture of magnesium oxide (5 g/L in water) and potassium metabisulphite at 95 ± 3 °C for 30 sec to 180 sec (Negi and Roy 2000). The blanching process of dipping in water followed by potassium metabisulphite (KMS) dip and drying process with low temperature provided maximum protection from drastic effects on contents of ß carotene, ascorbic acid, and chlorophyll with no effects on antioxidant potential (Negi and Roy 2000). The refrigeration conditions with the blanching process for 15 min were found to enhance iron bioavailability and reduce anti-nutrient (phytic acid) contents without loss of antioxidant potential during the subsequent study (Yadav and Sehgal 2003). Taken together, available reports on blanching highlight the use of optimum blanching chemicals and drying conditions in fenugreek preservation.

21.4.9 Effect of Storage Conditions on Nutritional Content in the Foods

Food packages are an integral part of the food to hold and protect them from the external environment. Two food preparations of fenugreek leaves, matthi, and seviyan, were evaluated for nutritional content when stored in different packing materials at a temperature of 30 ± 2 °C for 90 days (Dhanesh, Kochhar, and Javed 2018). Matthi is a northwest Indian flaky biscuit, and seviyan is crunchy noodles with varied thickness. Although all the products' proximate and mineral composition was reduced in all three packaging materials, the overall composition was maintained in glass jars, aluminum zip-pouch + glass jar compared with storage in plastic zip-lock pouch + glass jar. The good microbial quality of products was maintained with no aflatoxin or pathogenic bacteria, except fungal count was higher in plastic zip-lock pouch + glass jar storage after 60 days (Dhanesh, Kochhar, and Javed 2018).

21.5 FENUGREEK IN BAKERY PRODUCTS PREPARATIONS

Bakery products are the most popular food products in all age groups. Though they lack nutritional content as such, the value addition can be done by adding nutritional rich compounds (Angioloni and Collar 2012; Haruna, Udobi, and Ndife 2011; Sęczyk et al. 2017; Turfani et al. 2017). Fenugreek has been studied in various bakery products for enhancement of flavor and nutrition, stability, texture, and other physical properties of bakery food preparations as described in this section.

21.5.1 Fenugreek in Bread

Bread is the most popular food prepared from wheat dough. Wheat flour composite enriched with dietary fiber, total protein, lysine, and minerals with fenugreek seeds (raw seed, soaked seed, and germinated seed) is reported (Hooda and Jood 2004). The resultant flour can be incorporated into bakery preparations, mainly bread, macaroni, noodles, and biscuits. Seeds powder of fenugreek (13.6–15.3%) in raw, soaked, and germinated forms, when blended in combination with wheat flour (5, 10, 15, and 20%), increased the protein, fat, lysine, and dietary fibers and antioxidant activity. The composite flour containing germinated fenugreek seeds was superior in nutritional content than others with good organoleptic properties and up to 20% of fenugreek flour (Hooda and Jood 2004). Similar nutritional benefits and organoleptic properties with fenugreek flour in bread preparations were observed during many studies and publications, as summarized in Table 21.1.

The addition of fenugreek fiber (0, 3, 6, 9, and 12 g/100g) to wheat flour is reported to enhance the stability, strength, and softness, increase water absorption, retain good texture but delay the development time of dough of home-style bread (Huang et al. 2016). These results might be attributed to the high water-holding capacity of fenugreek fiber content and interaction between fenugreek fiber and starch (Huang et al. 2016).

21.5.2 Fenugreek Seed Husk in Muffins

The husk is the outer coating of seeds and provides better structural characteristics to food (Bashir et al. 2017; Buamard and Benjakul 2015). Fenugreek husk is a rich source of soluble and insoluble dietary fibers and polyphenols (Naidu et al. 2011; Sakhare, Inamdar, and Prabhasankar 2015). The effect of incorporating fenugreek seed husk in muffin batter showed increased viscosity of muffin batter and muffin volumes (Srivastava et al. 2012). In the study, the addition of fenugreek seed husk (5% and 10%) to wheat flour resulted in white creamy colored muffins with enhanced protein and dietary fiber contents, reduced hardness, softer crumb texture with good sensory properties. The addition of 15% fenugreek husk resulted in sticky, yellow-colored, more fragile muffins (Srivastava et al. 2012).

21.5.3 Fenugreek Seed Gum in Barbari and Lavash Flatbreads

Barbari, thick flatbread, is known as Iranian or Persian flatbread. Lavash is thin flatbread leavened and baked in a tandoor (Hejrani et al. 2017, 2019; Karizaki 2017; Pontonio et al. 2015; Pourafshar, Rosentrater, and Krishnan 2018). Fenugreek gum's hydrocolloids nature was explored as a texture modifier in bread (Rahnama, Milani, and Ardabili 2017). The fenugreek gum concentration (1.46 to 8.54) was partially incorporated in wheat flour for better color, textural and sensory attributes and surface methodology response during barbari and lavash flatbread preparation (Rahnama, Milani, and Ardabili 2017). The enhanced water absorption (with dough development time) with increased gum concentration was observed due to interactions between hydroxyl groups (Collar et al. 1999) and flour protein with fenugreek gum (Rahnama, Milani, and Ardabili 2017).

Water absorption, dough development time, and stability time were selected as independent variables. The fenugreek gum levels (4.93% and 5.31%) and leavening time (2.38 h and 2.33 h) were found to be optimal for barberi and lavash, respectively. The 7% fenugreek gum level in dough provided optimum stability time, density and sensory characteristics, and moisture content of the dough's fresh bread (Rahnama, Milani, and Ardabili 2017).

21.5.4 Fenugreek Amylase in Cake

The cake is a delicious, sweet bakery food made from flour, sugar, and other ingredients. Amylases enzymes are commercially used to impart quality to baked products, optimize baking properties, increase the volume of bread, reduce the rate of staling of crumbs, and improve color and texture of the cake (Bueno, Thys, and Rodrigues 2016; Giannone et al. 2016; Pérez-Quirce et al. 2017; Tsatsaragkou et al. 2017). Recently, the addition of purified amylase from fenugreek is reported to enhance the physical, sensory, and textural properties of cake (Khemakhem et al. 2018). The dough

TABLE 21.1
Use of Fenugreek in Preparation of Bread

Content	Nutritional	Organoleptic - Color	Organoleptic - Odor	Organoleptic - Taste	Texture	Overall	Reference
Seed powder (5%, 10%, 15%)	↑ protein, fibers, flavonoid, antioxidant	-	-	-	-	↑	Afzal et al. (2016)
Seed powder (50 g/kg flour)	↑	↔	↔	↓ bitter	↔	↔	Makowska et al. (2017)
Fenugreek flour (2%, 5% and 8%)	↑ protein, fibers, antioxidant	↔	↔	↔ to ↓ bitter	↔	↔	Man et al. (2019)
Fenugreek flour (3%, 7.5% and 9%)	↑ protein, lysine and dietary fiber contents	-	-	-	-	↑ baking absorption	Sharma and Chauhan (2000)
Debittered and germinated fenugreek seed flour (10%)	↑ fibers, protein, total polyphenols, flavonoid, antioxidant	↔	↔	↔	↔	↑ bulk density and crumb firmness ↓ loaf volume, and	Chaubey et al. (2018)
4% fenugreek flour. Mixed with other flours	↑ protein, calorie, iron	-	↓	-	-	↑ water absorption,	Bakr (1997)
Fenugreek (3%, 6%, 9%)	-	↔	↔	↔	↑ water absorption, extensibility	↑ stability and quality of dough.	Bushuty (2015)
Raw and germinated seed flour (2%, 4% and 6%)	↑ amino acids (lysine and threonine)	-	-	↓ bitter	↔	↔ dough thickness and volume	Sidahmmed and Mustafa (2016)
Fenugreek flour (5%)	↑ protein, fiber, calcium, magnesium, iron, zinc	↔	↔	↔	↔	↑ stability and quality of dough	Kasaye and Jha (2015)
Fenugreek flour (5% to 10%)	↑ Vitamin B1	-	-	-	-	-	El-Arab, Ali, and Hussein (2004)
Fenugreek flour (1% to 3%) mixed with corn floor	↑ (calcium, iron, cobalt, Rb)	-	-	-	-	-	Iskander and Davis (1992)
Flour (5%, 10,%, 15% and 20%) germinated fenugreek seed	↑ protein, fat, fibers	↔	↔	↔	↔	acceptable sensory attributes up to 10%, flour	Kasaye (2015)

with fenugreek-derived amylase has a higher gas retention profile, improved cake texture, decreased hardness, increased cohesion, improved microstructure, good crust properties, and qualities affecting density and water retention ability (Khemakhem et al. 2018). Overall, fenugreek-derived amylase addition can be a promising method to improve the quality of the cake.

21.5.5 Fenugreek Leaves in Pizza

Pizza is one of the bread-containing foods, which is popular because of its good taste, easy preparation and consumption. It contains high white flour contents, refined sugar, polyunsaturated fats, and other additives, but it is devoid of proteins, vitamins, and fibers. The inclusion of fenugreek and lotus stem flour in pizza is reported to increase the dietary fibers' content along with acceptable sensory characteristics (Kanaujiya and Singh 2017). Pizza bases prepared with the inclusion of levels of fenugreek leaves in the range of 10–15% content were evaluated. The resultant product exhibited acceptable sensory properties in terms of flavor and taste, body and texture, color and appearance, and overall acceptability, with higher dietary fiber content as compared to refined wheat flour pizza (Kanaujiya and Singh 2017). A second study reported enhanced contents of protein, fat, dietary fibers, vitamin C, and calcium in a pizza base with fenugreek leaves (Kanaujiya and Singh 2017).

21.5.6 Fenugreek in Biscuits

Biscuit is an edible flour-based sweet baked product. Wheat is the most used flour to prepare biscuits due to the formation of dough and gas retention. Fenugreek incorporation (5, 10, and 15%) to wheat flour is reported to enhance the nutritional and sensory properties of salted biscuits (Kumar et al. 2016). The increased fenugreek content shows a gradual reduction in diameter, increased thickness, and more protein content (Kumar et al. 2016). Other researchers observed similar benefits in biscuits using fenugreek-enriched flour, as summarized in Table 21.2.

21.5.7 Fenugreek in Cookies

The cookies made from wheat flour incorporated with fenugreek and oat flours exhibited significantly higher energy contents, proteins, minerals, fats, crude fibers, and reduced phytic acid and carbohydrates (Negu, Zegeye, and Astatkie 2020). Fenugreek seed flour containing cookies contained the highest amount of fat, protein, crude fiber, minerals, energy value, high water, and oil absorption capacity, and the lowest moisture, ash, and carbohydrate. The anti-nutrient contents, such as phytic acid and condensed tannins, were increased moderately. In another study, the replacement of wheat flour with germinated fenugreek flour (10%) during cookie preparation is reported to be a considerable increase in the protein, fiber, and ash contents and decrease in the carbohydrate and fat content (Agarwal and Syed 2017). The weight and thickness of cookies and sensory score were reduced, whereas hardness, brittleness, protein content, and fiber were found to increase (Agarwal and Syed 2017). Overall, fenugreek (10%) seed flour-containing baked cookies showed improved nutritional content with acceptable sensory qualities (Agarwal and Syed 2017; Negu, Zegeye, and Astatkie 2020).

The use of fenugreek seed powder extracted and encapsulated in gum arabic during the cookie's preparation is reported to have higher fracture strength, spread ratio, sensory acceptability, and an average value for fracture strength (Wani and Kumar 2015). These results were supported by the report of improved physicochemical, sensory, and nutritional attributes of cookies using fenugreek seed or leaf powder with other millet powders (Paul et al. 2018).

21.5.8 Debittered Fenugreek Flour in Rusk

Rusk is a dry hard-baked biscuit or bread. The incorporation of debittered fenugreek flour in wheat flour is reported to provide an excellent nutritional and antioxidant profile without compromising

TABLE 21.2
Use of Fenugreek in Preparation of Biscuits

Content	Nutritional	Organoleptic				Overall	Reference
		Color	Odor	Taste	Texture		
Soaked fenugreek (10%) and germinated fenugreek flours	↑ protein, fibers, minerals, antioxidant anti-hypercholesterolemic	-	-	-	↑ viscosity	Acceptable baking quality and sensory attributes	Hussein et al. (2011)
Flour (10%, 20%, and 30% germinated fenugreek seed)	↑ protein, fibers	↔	↔	↔	↔	Acceptable sensory attributes	Al-Gemeai (2016)
Flour (5%, 10%, 15%, and 20% germinated fenugreek seed)	↑ protein, fat, fibers, amino acids	↑ (5–10%) ↓ (15% and 20%)	↑ (5–10%) ↓ (15% and 20%)	↑ (5–10%) ↓ (15% and 20%)	↑ (5–10%) ↓ (15% and 20%)	Improves protein efficacy ratio (PER), net protein ratio (NPR), net protein utilization (NPU) ratios in rats	Hegazy and Ibrahium (2009)
Flour (5%, 10%, 15%, and 20% germinated fenugreek seed)	↑ protein, fat, fibers, amino acids, minerals	↔	↔	↔	↑ dough development time, water absorption, and mixing tolerance indices		El-Naggar (2019)
Flour (5%, 10%, 15%, and 20% germinated fenugreek seed)	-	↔	↔	↔	↔	Acceptable sensory attributes up to 10% flour	Kasaye (2015)
Flour (5%, 10%, 15%, and 20% germinated fenugreek seed)	↑ protein, fat, fiber, and minerals (Fe, Ca, and Zn), polyphenols, vitamin B2, and carotene	↔	↔	↔	↔	Acceptable sensory attributes up to 5% flour ↑ body weight and feed efficiency ratio in anemic rats	Mahmoud, Salem, and Mater (2012)

sensory properties (Dhull et al. 2020). With increasing concentration of fenugreek, loaf volume, specific loaf volume, and pasting properties decreased (Dhull et al. 2020). In contrast, properties such as loaf weight, hardness, antioxidant potential, water absorption, and color were increased (Dhull et al. 2020).

The debittered fenugreek seeds, extracts, or formulation were claimed to be a rich source of fibers with health benefits such as reduced body weight gain, food intake, epididymal white adipose tissue, weight of soleus muscle, plasma and hepatic lipid levels, and enhanced fecal lipid levels (Muraki et al. 2012).

21.5.9 Fenugreek in Puff Pastry

Puff pastry (khari) is distinctly layered non-yeasted, flaky textured pastry. It is a laminated bakery product made up of many thin layers of dough separated by alternate fat layers (Wickramarachchi, Sissons, and Cauvain 2015). The incorporation of dried fenugreek leaves imparts organoleptic characteristics to puff pastry and showed fat, carbohydrates, ash, and protein (Patil and Nikam 2019). The fenugreek enriched puff pastry showed better physicochemical properties and sensory qualities on a 9-point Hedonic scale. The Hedonic scale (Jones, Peryam, and Thurstone 1955; Peryam and Girardot 1952) is the most widely used for measuring acceptability of food, personal care products, household, and cosmetics products (Xia et al. 2020). The rating was allocated using a nine-point hedonic scale (9 = 'liked extremely' down to 1 = 'disliked extremely') for each of the parameters such as color, appearance, texture, taste, flavor, and overall acceptability. The fenugreek leaves-containing puff pastry were found more acceptable (score 9) than puff pastry with ajwain (carom) and cumin with a score of 8 (Patil and Nikam 2019).

21.5.10 Fenugreek Leaves in Waffles

Waffles are made from leavened dough or batter cooked between two plates that impart specific size, shape, and impression. Fenugreek leaves-incorporated waffles were reported with protein, fiber, iron, and vitamins (Porecha and Sengupta 2013).

21.6 FENUGREEK IN SNACKS PREPARATIONS

Snacks are ready-to-eat food products and popular owing to organoleptic attractions. Recently, because of people's changing lifestyles, the need and expectations from snacks preparations go beyond refreshment. The snacks are now expected to provide nutritional and health benefits. As a spice enriched with nutrients, fenugreek is widely used in various snack preparations (Chakraborty et al. 2016; Dhanesh, Kochhar, and Javed 2018; Llavata, Albors, and Martin-Esparza 2020). This section reviews some of the scientific publications evaluating nutritional and other benefits provided by fenugreek addition in snack preparations.

21.6.1 Fenugreek in Indian Snack Preparations

21.6.1.1 Fenugreek in Matthi and Seviyan

Matthi and seviyan are salted snacks made up of wheat and chickpea flour, respectively. Dried fenugreek leaves (1%), when incorporated in wheat-chickpea flour and chickpea flour to develop matthi and seviyan, exhibited a noteworthy enhancement in protein, amino acid, minerals, fat, and fibers (Dhanesh, Kochhar, and Javed 2018). Matthi was prepared by incorporating 1% dried fenugreek leaves and 10% peanut cake in wheat-chickpea composite flour. To this mixture, carom seeds, salt, and dalda (ghee) were added. The stiff dough was prepared by adding a little amount of water. Small balls were prepared, flattened, and pricked with a fork, finally deep-fried in oil. Seviyan was prepared by incorporating 1% dried fenugreek leaves and 10% of peanut cake in chickpea flour.

To this mixture, salt, oil, and water were added to form a soft dough. This mass was then extruded by seviyan machine into hot oil and deep-fried. Both the products were enriched with protein and dietary fibers and mineral contents (calcium, iron, and zinc) (Dhanesh, Kochhar, and Javed 2018).

21.6.1.2 Fenugreek in Kachori

Kachori is a popular snack all over India and a few Asian countries. It is a round flattened ball made up of fine flour and stuffed with a baked mixture of various spices and split and de-husked green or black gram lentils. Modified kachori stuffed with fenugreek leaves was found to be enriched with fiber and proteins (Agrawal and Sengupta 2014). Nutritional analysis was indicative of the higher levels of protein, fat, dietary fibers, carbohydrates, sodium, and potassium with good sensory properties (Agrawal and Sengupta 2014).

21.6.1.3 Fenugreek in Masala Parwal

Masala parwal (pointed gourd) is a salty snack, originated from the Bihar state in India. The addition of dried fenugreek leaves to wheat flour during masala parwal preparation showed enhanced nutritive values with acceptable sensory properties (Sharma 2010).

21.6.1.4 Fenugreek Flour in Pakoras

Pakoras, or bhajji, is a popular snack on the Indian subcontinent. The ingredients such as onion, potato, spinach, paneer, fish, and/or chicken are dipped in batter (gram flour) and deep-fried. The fenugreek-containing pakoras with multi-mix pulses and dried spinach were found to possess enhanced nutritional value (enhanced content of crude fiber, protein, fat, and minerals) with optimum sensory characteristics (Ahsan et al. 2019).

21.6.1.5 Fenugreek in Upma

Upma is a traditional southern Indian breakfast food. Upma supplemented with germinated fenugreek seeds results in the enhancement of proteins, dietary fibers, fats, calcium, iron, and moisture with reduced carbohydrate and calorie content (Kumari and Dubey 2013).

21.6.1.6 Fenugreek in Idlis Flour

An idli is a steamed rice cake with a soft and spongy texture (Krishnamoorthy, Kunjithapatham, and Manickam 2013). It is the staple food of south India, easy to prepare and digest, and so consumed in breakfast and the main course. Idlis composed of germinated fenugreek seeds, flour, rice, and lentils were reported to have good nutritional and sensory attributes (Pandey and Awasthi 2015b). The fenugreek-containing idlis were found rich in proteins, insoluble and soluble dietary fibers, low in fat, and acceptable sensory characteristics up to 30% addition of fenugreek seeds flour (Pandey and Awasthi 2015b).

Idlis were prepared from dehulled black gram and rice by the fermentation with the lactic acid bacteria to reduce anti-nutrients such as tannin and phytic acid (Mukherjee et al. 1965; Singh and Raghuvanshi 2012; Sridevi, Halami, and Vijayendra 2010). The addition of fenugreek seeds to flour is reported to favor lactic acid bacteria's growth and imparts better texture and flavor to idlis (Guhan 2008).

21.6.2 Fenugreek in Extruded Snacks

Extruded snacks are among the most commercially successful food products because they are ready to eat or ready to cook. These are mainly manufactured from starches or cereal flour and contain high fat and calories with low fiber and protein content (Korkerd et al. 2016). Extruded snakes are perceived as unhealthy to consumers because of the risk of health problems such as high glycemic index, obesity, and cardiovascular diseases that can develop from regular consumption for a long time (Korkerd et al. 2016). Fenugreek was explored as an ingredient to provide the nutritional value to extruded products in many research reports (Chakraborty et al. 2016; Llavata, Albors, and Martin-Esparza 2020; Ravindran, Carr, and Hardacre 2011).

The fenugreek constituent, galactomannan, enhances fiber content, and provides physiological benefits such as cholesterol and glucose-lowering and -enhancing laxation. Fenugreek galactomannans in extruded snack preparation resulted in improved textural and nutritional properties in extruded snacks (Ravindran, Carr, and Hardacre 2011). The extruded snack prepared with pea and rice flour in the ratio of 70:30 showed a more uniform, harder, crisper product after the addition of galactomannan (up to 15%). The resultant snack showed higher dietary fibers, higher proteins, and lower fat and glycemic index (Ravindran, Carr, and Hardacre 2011).

Pasta, an Italian preparation made from an unleavened dough of wheat flour cooked by boiling or baking, is now consumed worldwide. Fenugreek leaves, in dried and powder form, incorporated in extruded snacks and pasta with rice and corn flour exhibited low fat, high carbohydrate, protein, moderate fiber, and polyphenolic content (Chakraborty et al. 2016). The fenugreek-enriched extruded snacks showed higher oil and water-holding capacity. Both preparations, when enriched with fenugreek leaves powdered blend, showed good sensory characteristics at levels of 2 g and 3 g of leaves in 100 g to produce the most acceptable extruded products and pasta, respectively (Chakraborty et al. 2016).

The partial replacement of tiger nut flour with fenugreek flour for pasta was found to provide enhanced high nutritional (gluten-free, high soluble and insoluble dietary fibers, starch, and minerals) and better sensory characteristics due to galactomannan presence (Llavata, Albors, and Martin-Esparza 2020). The product's color became progressively intense, with increasing fenugreek levels due to carotenoids and other pigments (Llavata, Albors, and Martin-Esparza 2020).

21.6.3 Fenugreek Leaves in Pasta

Microwave dried leaf blends of fenugreek, curry, and coriander were reported to provide an antioxidant potential in extruded products and pasta due to phenolic content (Chakraborty et al. 2016). The pasta was prepared with blending of equal proportion of fine powder of fenugreek leaves with wheat flour and co-extruded with rice or corn flour. Fenugreek powder-containing pasta showed higher carbohydrates, protein, fat, fiber, cellulose, tannins, and phenolic contents with superior free radical scavenging activity (Chakraborty et al. 2016).

21.6.4 Fenugreek in Noodles

Noodles share a long history as an essential dietary component in the Asian region. The noodles are prepared from wheat, buckwheat, rice flour, and have low nutritional properties (Hou 2001). However, the use of fenugreek (up to 10%) in wheat flour was reported to enrich noodles with a higher protein, crude fiber, and fat (Dhull and Sandhu 2018). However, the noodles prepared with 7% fenugreek flour and 93% wheat flour were found to show acceptable sensory characteristics (Dhull and Sandhu 2018).

Gums impart textural characteristics to foodstuffs (Clemens and Pressman 2017; Park et al. 2019; Shahzad et al. 2019). The addition of 0.42% w/w of fenugreek gum was reported as a texture enhancer for gluten-free rice noodles preparation for the harder and stickier product with the same tensile strength, elasticity, and springiness as that of ghatti gum (Koh et al. 2011).

21.6.5 Fenugreek in Spent Hen Patties

Spent hen patties supplemented with fenugreek seed powder contained higher proteins, fat, and free radical scavenging properties (Qureshi et al. 2018). Fenugreek seed powder at 0.5, 1, 1.5, and 2% replaced lean meat in the basic formulation, and the mixture was molded and cooked at 180°C for 30 min. The cooking yield was improved due to the hydration and binding properties of fenugreek (Hegazy 2011).

21.6.6 Fenugreek Polysaccharide in Sausage

The color of meat is contributed by water-soluble heme protein myoglobin, making it more susceptible to oxidative degradation during manufacturing, storage, and distribution (Carlez, Veciana-Nogues, and Cheftel 1995; Faustman et al. 2010; Suman and Joseph 2013). Therefore, the addition of natural polysaccharides is gaining considerable interest for meat preservation in recent years (Laurienzo 2010; Venkatesan et al. 2015; Wang et al. 2016).

Sausage is a slice of minced pork or other meat product in a cylindrical shape, encased in skin and sold raw. Fenugreek galactomannan, a polysaccharide, was useful to prevent the lipid and myoglobin oxidation of beef sausage during refrigeration storage (Ktari et al. 2017). During the study, fenugreek seed galactomannan in concentrations of 0.05, 0.125, and 0.25%, when added as an antioxidant, demonstrated better inhibition of lipid peroxidation and myoglobin oxidation than vitamin C (Ktari et al. 2017). Therefore, fenugreek polysaccharide can be an excellent option as compared to synthetic antioxidants during meat product preservation.

21.6.7 Fenugreek in Meat Patties and Burgers

Lipid and protein components of pork or meat patties and burgers are known to undergo oxidation and microbial spoilage during storage (Ahmad, Saleh, and Sabow 2019; Cheng, Wang, and Ockerman 2007; Lund, Hviid, and Skibsted 2007). Numerous efforts were taken to find a natural alternative for food preservation (Del Nobile et al. 2012; Eissa 2007; Hosny et al. 2011; Jalosinska and Wilczak 2009; Ouattara et al. 2000). Natural sources of phenolic compounds, including fenugreek, are useful antioxidants (Dintcheva et al. 2017; López-Vélez, Martínez-Martínez, and Valle-Ribes 2003).

Patties are round and flattened meat or vegetable products. A burger is a sandwich composed of one or more cooked patties of ground meat. Fenugreek and a few other natural ingredients were found more effective than synthetic antioxidants and vitamin E for 9 days of refrigeration storage of raw and cooked patties (McCarthy et al. 2001). Furthermore, spent hen meat contains higher amounts of unsaturated fatty acid, which readily undergoes oxidation during processing and storage. Incorporating fenugreek seed powder (0.5 to 2%) in spent hen patties is reported with higher antioxidant activity and cooking yield (Qureshi et al. 2018).

The antioxidant and antimicrobial potential of fenugreek seed flour (3, 6, 9, and 12%) in beef burgers' manufacturing is reported in two research reports (Ahmad, Saleh, and Sabow 2019; Hegazy 2011). The burger with fenugreek seed flour is reported with better microbial control, good sensory properties, and more acceptability with an enhanced content of essential amino acids, pH, water-holding capacity, cooking shrinkage, total volatile nitrogen, and thiobarbituric acid content even with refrigeration storage for 3 months (Hegazy 2011). During the second study, the samples of mutton and beef cattle's meat with the addition of fenugreek leaves (up to 1.5%) showed a lower content of malondialdehyde (a marker for oxidation) and better microbial control during storage of 7–10 days or refrigeration as compared to control samples (Ahmad, Saleh, and Sabow 2019). The chelator compounds or flavonoid content in fenugreek leaves might be responsible for preventing the free radicals from reducing thiobarbituric acid in meat (Ahmad, Saleh, and Sabow 2019). Taken together, the addition of fenugreek leaves and seeds into the burger can be useful to burger preservation due to their antimicrobial and antioxidant potential.

21.6.8 Fenugreeks Paste in Pastirma

Microbes' growth is a major concern during the preservation of meat products (Casaburi et al. 2015). Pastirma is a traditional dry-cured meat product produced from beef and pasted with cemen (a Turkish preparation with spices). The cemen paste containing fenugreek and red hot pepper paste demonstrated substantial bacteriostatic effects against pathogenic microbes such as *Escherichia coli*, *Staphylococcus aureus*, and *Yersinia enterocolitica* on 4 days of storage (Yetim et al. 2006).

21.7 FENUGREEK IN THE MAIN COURSE

Apart from use as a spice, fenugreek leaves are commonly used worldwide in the main course as the leafy vegetable and in varieties of wheat preparations. These preparations are reported to provide nutritional and sensory value additions in past scientific literature as summarized in the following sections.

21.7.1 Fenugreek Leaves as Vegetables

Methi aloo is a mixed vegetable with fenugreek leaves and boiled potatoes. It is most regularly consumed as a rich source of carotenoid and other constituents (Khokhar, Roe, and Swan 2012). A survey evaluation of the mid-day meal program in Municipal Corporation of Delhi schools found the use of Kasuri methi (dried fenugreek leaves) in a meal with enhanced nutritional quality (Sharma et al. 2006). Recently, fenugreek was used as an infused herb during the preparation of functional paneer because of enrichment of vitamins (A, C, B6), minerals (iron, magnesium), and antioxidants (Trehan et al. 2019).

21.7.2 Fenugreek Seed Flour in Indian Flatbread Preparations

21.7.2.1 Fenugreek in Injera

Injera is a sour fermented, spongy flatbread of Ethiopian origin (Yetneberk et al. 2004). The raw, germinated, and roasted fenugreek seeds (4%–16%), when added to injera flours of millets such as teff, corn, sorghum, barley, and finger millet, reported to protect from microbes (Leykun, Admasu, and Abera 2020). The base used in the preparation of injera (a sour fermented flatbread) is usually teff flour. However, fenugreek seed flour in injera preparation provided the antioxidant capacity (Boka, Woldegiorgis, and Haki 2013) due to enhanced flavonoid and tannin contents (Godebo, Dessalegn, and Nigusse 2019). Recently, injera supplemented with fenugreek seed flour was found to have high nutritional content and lower anti-nutrients levels (phytic acid) (Leykun, Admasu, and Abera 2020). Raw, germinated, and roasted fenugreek seeds flour incorporated in teff flour at 16% levels showed enhanced contents of zinc and phytochemicals (Leykun, Admasu, and Abera 2020).

Injera, containing raw and germinated fenugreek seeds (in range of 12–16%) showed lower phytic acid, tannins, and phenolic content, with no coliform, yeast, or mold and low aerobic plate (Leykun, Admasu, and Abera 2020), perhaps through the contribution of antimicrobial fenugreek constituents (Hegazy 2011). Therefore, fenugreek seeds flour can be useful to restrict the growth of microbes during food preparations.

21.7.2.2 Fenugreek in Chapattis

Chapatti is an Indian unleavened flatbread made up of wheat flour. Germinated fenugreek seed flour addition is reported to provide better nutritional value with acceptable sensory characteristics to chapattis (Pandey and Awasthi 2015b). Chapattis prepared by incorporating germinated fenugreek seed flour (5% to 30%) into wheat flour showed a direct correlation with water requirement during dough making due to the higher water absorption capacity of fenugreek. However, an inverse correlation was noted for sensory properties and fenugreek content (Pandey and Awasthi 2015b).

21.7.2.3 Fenugreek Leaves in Thepla

Methi thepla (delicious flatbreads) is a well-known travel-friendly food all over India. A standard recipe of thepla preparation involves adding fresh fenugreek leaves with whole wheat flour and gram flour (besan). The thepla preparation using fenugreek leaves (12% w/w) of whole wheat flour and other species in minor quantities exhibited the highest sensory acceptability scores and higher fiber content than chapattis, another common flatbread preparation (Patil and Arya 2012).

21.7.2.4 Fenugreek Dried Leaves in Missi Roti

Missi roti is a north Indian flatbread made up of chickpea and wheat flour. Composite flour containing fenugreek powder is reported to enhance the nutritional value of missi roti for protein, minerals, fat, fiber, carbohydrates, and minerals (Kadam et al. 2012). The fenugreek supplemented blend showed higher Iron content and good sensory characteristics. The composite flours can be stored in polyethylene bags or tin boxes for 3 months without degradation in quality (Kadam et al. 2012).

21.7.2.5 Fenugreek Leaves in Bhakri

Bhakri is a traditional flatbread preparation and an integral part of western and central Indian cuisine. Bhakri is made up of ingredients such as sorghum, pearl millet, and finger millets, and therefore considered as a healthy preparation with excellent nutritious properties (Badgujar et al. 2017). Bhakri containing multigrain, flaxseed, and dried fenugreek leaves were found rich in nutritional contents and organoleptically well-accepted (Bafna, Kalpana, and Ramya 2020).

21.7.2.6 Fenugreek Leaves in Paratha

Paratha, native to the Indian subcontinent, is a layered flatbread made with wheat flour and different vegetables and grains. A composite premix formulation for parathas had high nutritional content due to proteins, fats, and crude fibers (Pathania, Kaur, and Sachdev 2017). In the study, paratha was made with fenugreek seed powder to wheat flour, and chickpea flour with other ingredients. Developed parathas were composed of 16.52% protein, 7.11% fat, 4.05% ash, 72.63% carbohydrates, 420.59 Kcal/100 g calories. The quantity and quality of proteins were found better than the respective control parathas (Pathania, Kaur, and Sachdev 2017). In another study, parathas of fresh and dried fenugreek leaves were reported to contain higher dietary fibers, proteins, and fats with enhanced green color and pleasant flavor (Sudha et al. 2015). The content of nutrients such as ascorbic acid, chlorophyll, and ß-carotene was higher in fresh leaves-containing parathas than that of dehydrated fenugreek leaves (Sudha et al. 2015).

The fenugreek seed containing multigrain composition resulted in parathas with better rheological, structural, and quality characteristics such as increased water absorption and extensibility (Indrani et al. 2011). The addition of fenugreek seeds to multigrain flour increased the water absorption and dough development time, perhaps due to fibers' presence. However, the strength, elasticity of dough was adversely affected due to a reduction in wheat gluten. The dough's microstructure with respect to the continuity of the protein matrix was disrupted with the progressive addition of multigrain (Indrani et al. 2011).

21.7.2.7 Fenugreek Leaves in Doli Ki Roti

Doli ki roti (multani flatbread) is a preparation that originated from the Panjab region of India and Pakistan. It is a fermented, stuffed, and fried preparation containing homemade yeast (wheat flour) in an earthen pot called 'doli.' Doli ki roti stuffed with fenugreek leaves was found enriched with calcium and iron, with excellent availability of minerals and reduced anti-nutrient phytic acid (Bhatia and Khetarpaul 2002). The doli ki roti preparation was initiated by boiling the spice mixture of poppy seeds, cloves, cardamom, cinnamon, and jaggery in water for 5 min and then fermenting it in an incubator before the addition of wheat flour. The mass was again fermented, the dough was prepared, divided into small balls, and rolled flat in a small round shape (called puri). These puris were then stuffed with dried fenugreek leaves, closed in a small ball shape, flattened with the hand, and deep-fried (Bhatia and Khetarpaul 2002).

21.8 FENUGREEK AS SPICES AND PICKLES

Spices such as fenugreek seeds have been an integral part of foods worldwide since ancient times. They enhance the taste, color, and flavor of foodstuffs. Blends of numerous natural resources are being used to prepare spices (Yanishlieva, Marinova, and Pokorný 2006). Dried fenugreek leaves, mainly known as 'Kasuri methi,' and seeds are mostly used as spices in plentiful foods. The popular Indian spice composition of 'Garam Masala' (hot Spice) also contains fenugreek seeds. Optimum roasting of fenugreek seeds is essential to generate proper aroma intensity and darker color (Pandey and Awasthi 2015a). The research publications on the use of fenugreek as a spice preparation is summarized in Table 21.3. Apart from imparting taste and flavor, the use of fenugreek as a spice during food preparations were reported to provide essential nutrients, antioxidant potential, other medicinal benefits (Embuscado 2015; Maheshwari et al. 2014; Rathore and Shekhawat 2008; Yanishlieva, Marinova, and Pokorný 2006; Yashin et al. 2017).

Pickling is one of the oldest food preservation methods of numerous foodstuffs such as fruits, vegetables, fish, and meat (Behera et al. 2020; Dwivedi et al. 2017; Joardder and Masud 2019).

TABLE 21.3
Use of Fenugreek as Spice in Various Preparations

Part of Fenugreek	Spice Preparations	Details	References
Dried leaves (Kasuri methi)	Crackers (South Indian)	Crispiness, good sensory acceptability	Karad, Jangale, and Karad (2016)
Seeds	Sepubari/Mukandbari—pulse-based dish in marriage feast (Himachal Pradesh)	Spices containing fenugreek seeds and fenugreek powder	Sharma and Singh (2012)
Seeds	Kadi—milk-based liquid dish (Himachal Pradesh)	It is prepared by simmering the mixture of gram flour and buttermilk along with spices and seed powder	Sharma and Singh (2012)
Seeds	Kaandal (Himachal Pradesh)	Spilt black gram was soaked and converted into the paste, and spices were added. Colocasia stems are coated with the paste, sundried, and used	Sharma and Singh (2012)
Seeds	Papad methi curry, besan methi curry (Rajsthan)	Sweet and sour methi- fenugreek enhances nutrition value, aroma, and medicinal value	Mathur and Choudhry (2009b)
Seeds	Pulihara—tamarind rice (South Indian)	Tamarind, fried cashew nuts, and spices- incorporated rice	Rajni, Thomas, and Kapur (2018)
	Menthula dosa—rice pancake (South Indian)	Dish from a batter of rice and spilt black gram incorporated with fenugreek seeds	
	Masar anja (Assamese)	Fish curries containing various spices	
	Masar tenga/dhekia tenga (Assamese)	Sour tomato dish with fenugreek as a spice in tarka (hot oil)	
	Bhindi sabji—dry dish (all over India)	Lady fingers with fenugreek seeds and other spices.	
	Adrak ki sabji (Punjab)	Dry preparation containing ginger, fenugreek seeds, and other spices	
	Alugedde kut— potato savory (all over India)	Fried potatoes with spices	
	Kaalen—dry dish (South Indian)	Raw banana with spices	
	Theeyal—soupy consistency (South Indian)	Ridge gourd or brinjal curry with coconut and other spices.	
	Dana methi	Curry containing fenugreek seeds	
	Muthia	A stiff dough with butter mil, cut into small oblong pieces, steamed and shallow fried.	
	Kodubade	Preparation of fried rice flour, fenugreek leaves, fat, and coconut oil	

Part of Fenugreek	Spice Preparations	Details	References
Seeds	Extruded snacks (all over India)	Extruded snacks from sweet potato, rice, and barley flour with added fenugreek	Yadav, Singh, and Chatterjee (2016)
Dried leaves	Tandoori chicken—roasted chicken (Indian subcontinent)	The deep cut of chicken marinated with spices.	Srinivas (2011)
Seeds	Chicken bhuna— super concentrated curry of chicken and spices (Indian subcontinent)	Fried bay leaf, ginger, garlic, and fenugreek in oil and added to other spices, onions, salt, and chicken and cooked	Chen, Gilbert, and Khokhar (2009)
Dried leaves	Lamb kebab—thin sliced roasted lamb meat (Middle East)	Minced lamb, coriander chilies, onions, and fenugreek leaves shaped into a round ball, then flattened and fried.	Chen, Gilbert, and Khokhar (2009)
Seeds	Beef sausages (The USA)	Lean beef and different spices-based recipes with garlic to form beef garlic sausages	Savić (1985)
Seeds	Pepperoni—salami of pork and beef (The USA)	The spices are added to the mixture of pork and beef	Savić (1985)
Roasted seeds	Meat, pork curry	Spices incorporated to meat and pork curries	Xing and Ng (2016)

Ancient civilizations such as Indians, Chinese, Egyptians carried out the pickling tradition (El Sheikha and Hu 2018). Food preservation by fermentation involves preserving foodstuffs under high acid concentration for preservation (Behera et al. 2020). This process improves the taste, texture, and flavor of the food.

Fenugreek is added to a variety of pickles for preservation and imparting flavor to foodstuffs. Traditional pickle preparation recipes of the state of Himachal Pradesh in India, such as galgal, lingri, aaroo, plum, lasura, dehu, kachnar, beedana (Savitri et al. 2016), and fresh bamboo shoot pickles (Shinde et al. 2019), use fenugreek powder as a standard and essential ingredient.

21.9 FENUGREEK IN ANIMAL FEED

Apart from use in human food preparations, fenugreek has also been explored to enhance animal feed's nutritional properties, as described in the following section.

21.9.1 FENUGREEK SEEDS FOR IMPROVING ANIMAL MILK CHARACTERISTICS

Though the milk has nutritional value, the levels of cholesterol in milk lipids can limit their applications. Therefore, efforts are being made to reduce milk cholesterol in cow milk for health benefits (Bjermo et al. 2012; Iggman et al. 2011; Rosqvist et al. 2015). Fenugreek seed at 20% of dry matter of feed was supplemented to impact cow's milk characteristics (functional fatty acid profile and reduced blood cholesterol) without affecting milk's flavor and taste (Shah and Mir 2004). Subsequent studies supported the benefits of the addition of 200 g in basic diet to lactating buffaloes, 4 weeks before calving for the next 12 weeks (Sah et al. 2007). The buffaloes fed with fenugreek seed supplementing diet showed a significant increase in dry matter intake, nutrient digestibility, milk yield, fat, proteins, lactose glucose, albumin, and creatinine with non-significant lower values of cholesterol and total lipids as compared to control diet with normal liver function parameters (Sah et al. 2007). Besides, fenugreek seed diet was reported to enhance the daily feed consumption and milk yield in Anatolian water buffaloes' feed during a randomized controlled study (Degirmencioglu et al. 2016). The 50 g/kg of ground fenugreek seed-containing feed consumption produced significantly improved total dry matter, daily concentrated feed, and increased milk production compared with the control diet (Degirmencioglu et al. 2016). These effects can be attributed to fenugreek's effect on the hypothalamus to stimulate the brain's hunger center, as reported earlier (Petit et al. 1993). These reports suggested the excellent befits of fenugreek supplementation to cow and buffalo diets for higher and better milk production.

21.9.2 Fenugreek as a Growth Promoter in Poultry and Pigs Feed

The broiler is a chicken raised for meat production, where higher weights of meat are desirable. Broiler chicks fed on fenugreek supplemented diet (3 g/kg, up to 6 weeks) showed significant improvement in chicks' body weight (Alloui et al. 2012). Recently inclusion of fenugreek powder and extract flour (1%) in a corn-soybean meal-based diet of laying hens revealed enhanced feed intake, feed conversion ratio, and intense color of egg yolk without affecting serum biochemical parameters and immune markers (Samani et al. 2020).

The dietary fenugreek seed extract incorporated in weaning pigs' diet for 42 days reported the average daily weight gain, energy digestibility, and growth efficiency compared to the control (corn-soybean meal-based) diet (Begum, Hossain, and Kim 2016). On day 42, fenugreek seeds-supplemented diet showed increased levels of IgG, serum HDL, cholesterol, and RBCs, with decreased fecal gas emission (Begum, Hossain, and Kim 2016).

21.9.3 Fenugreek Seeds as High Fiber Diet in Rabbit Feed

When fed a low fiber diet, weaned rabbits suffer from digestive disorders (Gidenne et al. 2004). Therefore, dietary fiber-rich food such as fenugreek seed would be a suitable solution. One of the reported studies demonstrated beneficial effects of seven days of pre-germinated fenugreek seeds supplementation in the diet (5%) on rabbits' improved appetite, digestibility, growth performance, weight gain, higher feed conversion ratio without mortality, sickness, or clinical signs of diarrhea and constipation (Mabrouki, Chalghoumi, and Abdouli 2016). These effects were attributed to the gelatinous structure formation due to soluble fiber in fenugreek seeds.

21.9.4 Fenugreek Seeds in Fish Feed

The dietary fenugreek seed extract (0.01%, 0.02%, 0.04%, 0.08%, and 0.16%) added to feed of fish species (juvenile blunt snout) for 8 weeks, did not show weight gain, feed conversion ratio, specific growth rate, and glucose content (Yu et al. 2019). However, a significant reduction in lipid content and enhancement in IgM, albumin, total protein, and antioxidant activity was observed (Yu et al. 2019).

21.10 FENUGREEK PREPARATIONS AS A FUNCTIONAL FOOD

Apart from the nutritional role of fenugreek as a food, its medicinal benefits have been used to treat various ailments since the ancient era. Many functional and medicinal properties of fenugreek-containing food preparations and recipes are reported in modern scientific publications as described in this section.

21.10.1 Fenugreek Food Preparations and Diabetes Mellitus

Several research investigations have documented that the addition of fenugreek to the diets of animals or diabetic patients resulted in a significant fall in blood glucose and improvement in glucose tolerance (Gopalpura, Jayanthi, and Dubey 2007; Lalit and Kochhar 2018; Losso et al. 2009; Pathak, Srivastava, and Grover 2000; Rajamani and Raajeswari 2016; Ranade and Mudgalkar 2017; Robert, Ismail, and Rosli 2016; Sanlier and Gencer 2020; Shirani and Ganesharanee 2009; Srinivasan 2005).

Three nutritious food products such as laddu (sweet balls), upma (kedgeree), and dhokla (leavened steamed cake) formulated from a judicious combination of fenugreek seeds and legumes millets showed hypoglycemic effects in five diabetes and five non-diabetic subjects (Pathak, Srivastava, and Grover 2000). The flattened postprandial blood glucose response curves were observed in diabetic as well as normal subjects. The results indicated the usefulness of fenugreek-containing food products in dietary management for diabetes (Pathak, Srivastava, and Grover 2000).

A survey on diabetic patients revealed that the patients consumed fenugreek leaves as a hypoglycemic food (Sharma and Choudhry 2006). Fenugreek included in the daily diet in amounts of 25/50 g can be effective supportive therapy in the management of diabetes (Ranade and Mudgalkar 2017; Sanlier and Gencer 2020; Srinivasan 2005). All forms of fenugreek seeds, in several forms such as whole dried, soaked, cooked, gum isolate, extract, suspension, have shown a significant reduction in fasting blood glucose, serum triglyceride, total and low-density lipoprotein cholesterol, 24 h urinary sugar excretion, glycated hemoglobin, serum cholesterol with improved glucose tolerance and high-density lipoprotein cholesterol (Ranade and Mudgalkar 2017; Sanlier and Gencer 2020; Srinivasan 2005).

Chapattis made from fenugreek seed powder and wheat flour in a ratio 1:9 reported to produce a slower and more sustained glucose release with a significant reduction in GI value by 21% when consumed for two consecutive weeks in healthy as well as diabetic volunteers (with discontinued antidiabetic medications) (Gopalpura, Jayanthi, and Dubey 2007). In another study, beneficial effects of chapattis supplemented with *Nigella sativa* and fenugreek seeds (two chapatis twice a day 6 days/week with a daily dose of 5.45 g of an *N. sativa* + fenugreek combination over 12 weeks) to 40 type-2 diabetic subjects with no changes in medications or lifestyle was reported (Rao et al. 2020). The decrease in body mass index, body weight, hip and waist circumferences, fasting blood glucose, HbA1c, index of central obesity, 2-h postprandial blood glucose, average glucose over 12 weeks, total cholesterol (TC), triglycerides, very-low-density lipoprotein (VLDL), (TG) and increased high-density lipoprotein (HDL) cholesterol in all subjects were observed in the fenugreek-containing supplemented group. Reduction in HbA1c was correlated with reduced systolic blood pressure, hepatic enzymes aspartate transaminase and alkaline phosphatase, a number of diagnostic metabolic syndrome criteria, cardiovascular risk score (CRS), hepatic steatosis index (HSI), diabetogenic and atherogenic indexes (TG/HDL, VLDL/HDL, TC/HDL, and low-density lipoprotein/HDL) with improvement in CRS, HSI, fatty liver index, lipid accumulation product being observed (Rao et al. 2020). Thus, chapattis supplemented with *N. sativa* and fenugreek seems to present a good and safe dietary modification to overcome cardiometabolic risk (Rao et al. 2020). The fenugreek leaves added to khakhra (thin chapattis and a popular Gujarati snack) resulted in lower glycemic index and are suggested as a wholesome snack for kids, diabetics, and the normal population (Rajamani and Raajeswari 2016).

Similar benefits on postprandial glucose metabolism from fenugreek bread (prepared using a proprietary process) in type 2 diabetes mellitus patients in a double-blind, randomized, crossover study is reported (Losso et al. 2009). Eight diet-controlled diabetic subjects served with two slices of bread (56 g) with 5% fenugreek showed improved insulin sensitivity without significant changes in glucose levels when tested periodically over 4 h after consumption on two occasions with a 1 week washout period (Losso et al. 2009).

Another randomized, placebo-controlled, crossover study demonstrated the usefulness of fenugreek in reducing postprandial glycemia in 10 healthy human subjects (five men, five women) fed with flatbreads and buns, each in the absence or presence of 10% fenugreek seed powder, on six different occasions (Robert, Ismail, and Rosli 2016). The results showed that adding 10% fenugreek seed powder cause a significant reduction in the GI of buns (from 82 to 51) and flatbread (from 63 to 43) (Robert, Ismail, and Rosli 2016).

Another study on the biscuits with germinated fenugreek seed powder was found to have higher proteins (5.8% increase) and fibers (30.3% increase) as compared to biscuits with barley flour and wheat flour, when consumed at an interval of 15 days for 3 months (Lalit and Kochhar 2018). These biscuits were found to have excellent sensory characteristics on the scorecard of the 9-point Hedonic Rating Scale (Peryam and Girardot 1952) for appearance, color, texture, flavor, taste, and overall acceptability (Lalit and Kochhar 2018).

Acute postprandial glycemic response of pudding containing fenugreek gum (viscosity-matched soluble dietary fiber) in 15 adults at elevated risk of diabetes demonstrated a significant decrease in blood glucose and plasma insulin at peak concentrations and time points as compared to a control pudding is observed during a randomized, double-blind, crossover study (Shirani and Ganesharanee 2009). Furthermore, extruded products containing 15% fenugreek polysaccharide showed a significant reduction in glycemic index (Shirani and Ganesharanee 2009).

Taken together, the food preparations with fenugreek leaves or seeds can be beneficial in lowering glycemic index, increasing proteins and fiber content, reducing insulin resistance to and reducing cardiovascular risk for effective management of diabetes patients. Several mechanisms have been suggested for their effectiveness in diabetes patients. The researchers attributed the beneficial effects to the galactomannan component of fenugreek (Gopalpura, Jayanthi, and Dubey 2007). Fenugreek galactomannan is known to provide improved glucose homeostasis in diabetes by delaying carbohydrate digestion and absorption and improving insulin action (Hannan et al. 2007, 2003; Srichamroen et al. 2009), and reduction in LDL cholesterol levels (Srinivasan 2005).

21.10.2 FENUGREEK HERBAL TEA AS A GALACTAGOGUE

Breast milk has been accepted as the gold standard of infant nutrition. The perception of insufficient milk production is the most common worldwide maternal factor of the early cessation of breastfeeding. As a tradition, women consume fenugreek seeds in many forms of food preparations to facilitate lactation during the postpartum period (Gabay 2002; Gartner et al. 2005; Tiran 2003). Several modern scientific preclinical and clinical studies demonstrated the excellent efficacy of fenugreek-containing herbal tea as a galactagogue to increase prolactin levels in lactating mothers (Abdou and Fathey 2018; El Sakka, Salama, and Salama 2014; Ghasemi, Kheirkhah, and Vahedi 2015; Turkyılmaz et al. 2011).

The earliest randomized placebo-controlled study on fenugreek herbal tea as a galactagogue reported the beneficial effects of fenugreek supplementation to produce more breast milk and higher gain in infant birth weight in the early postnatal period (Turkyılmaz et al. 2011). Similar results were reported in another study, wherein herbal fenugreek enhanced breast milk production in lactating mothers with increased birth weight gain in their infants during the early postpartum period (El Sakka, Salama, and Salama 2014).

In a double-blind, randomized clinical trial, herbal tea containing fenugreek (7.5 g fenugreek seed powder + 3 g of black tea, thrice a day) consumed by lactating mothers produced improved breast milk sufficiency, including infant's growth parameters such as head circumference and weight, the number of wet diapers per day, frequency of defecation and infant's breastfeeding times in Iranian girl infants aged between 0 and 4 months, during four weeks (Ghasemi, Kheirkhah, and Vahedi 2015).

Similarly, consumption of 200 ml of fenugreek tea (50 g of fenugreek seeds) three times daily showed increased breastmilk volume and serum prolactin concentration in an early stage of lactogenesis in 30 lactating mothers indicating its usefulness in mother satisfaction and reassurance in the early stages of lactation (Abdou and Fathey 2018).

Despite its unknown mechanism of action, fenugreek is suggested to increase milk flow due to phytoestrogens and diosgenin content (Abdou and Fathey 2018; Penagos Tabares, Bedoya Jaramillo, and Ruiz-Cortés 2014; Turkyılmaz et al. 2011) and the presence of hormone precursors (Abdou and Fathey 2018; Ravi and Joseph 2020). Another possible mechanism is the stimulating action of fenugreek on sweat production, which may boost milk secretion because the breast is a type of sweat gland (Gabay 2002; Gartner et al. 2005; Tiran 2003; Wani and Kumar 2018).

21.10.3 FENUGREEK SPROUTS AND BISCUITS FOR ANEMIA

Anemia occurs because of the deficiency of iron in food. There is an alarming increase in the incidence of anemia in developing countries (Alarcon and Samantha 2020; Gupta and Gadipudi 2018). Various synthetic and natural product options are available to use as iron supplementation.

Nutritional, Functional Food Preparations

Recently, food-based approaches are being suggested as bioavailable dietary iron sources to prevent iron deficiency (Bouis and Saltzman 2017). Fenugreek is a rich iron source, containing 33 mg/100 g dry weight (Wani and Kumar 2018). The iron content in fenugreek sprouts is twice as much as raw seeds (Khoja et al. 2021). Therefore, the food preparation containing fenugreek sprouts is studied as a possible source of bioavailable iron supplementation to manage anemia in a series of past scientific publications (Khoja et al. 2021; Mahmoud, Salem, and Mater 2012; Prasad et al. 2014).

The earliest study demonstrated good nutritive value and favorable physiological changes by fenugreek biscuit supplementation (5–10%) in anemic rats (Mahmoud, Salem, and Mater 2012). The fenugreek biscuit-supplemented anemic rats showed an increased level of ß-carotene and vitamin B2, increased food consumption, weight gain, feed efficacy ratio, serum total iron-binding capacity, serum proteins and minerals (iron and zinc) levels as compared with control rats (Mahmoud, Salem, and Mater 2012). One of the significant challenges in anemia management is the bioavailability of iron from the source (Aspuru et al. 2011). The fenugreek-containing food preparations—laddoos containing dehydrated bathua, and methi leaf powder—had reported high total iron bioavailability with good sensory characteristic and nutrient concentration during *in vitro* study (Prasad et al. 2014). The good bioavailability of iron from fenugreek preparations is also supported by a recent *in vitro* study on fenugreek sprouts (Khoja et al. 2021). In this study, fenugreek sprouts showed good bioaccessibility and bioavailability of iron as measured *in vitro* simulated peptic-pancreatic digestion, followed by ferritin measurement in Caco-2 cells (Khoja et al. 2021). Taken together, the present evidence strongly suggests the benefits of fenugreek-containing food preparations as a source of bioavailable iron in the management of iron deficiency anemia.

21.11 CONCLUSION

The available scientific evidence suggests the expanding use of fenugreek in various food preparations and recipes worldwide (Figure 21.1). The research results demonstrate the benefits of fenugreek

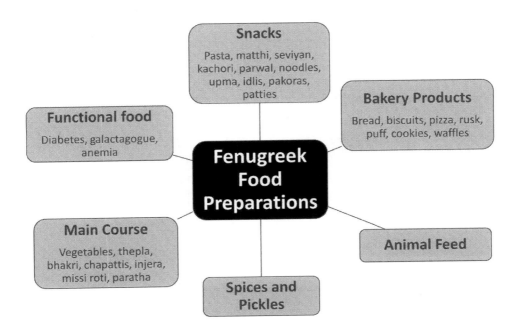

FIGURE 21.1 Application of fenugreek in food recipes.

REFERENCES

Abdel-Nabey, A. A., and A. A. Damir. 1990. "Changes in some nutrients of Fenugreek (*Trigonella foenum-graecum* L.) seeds during water boiling." *Plant Foods for Human Nutrition* 40 (4):267–274.

Abdou, R. M., and M. Fathey. 2018. "Evaluation of early postpartum Fenugreek supplementation on expressed breast milk volume and prolactin levels variation." *Egyptian Pediatric Association Gazette* 66 (3):57–60. doi: 10.1016/j.epag.2018.07.003.

Açar, Ö., V. Gökmen, N. Pellegrini, and V. Fogliano. 2009. "Direct evaluation of the total antioxidant capacity of raw and roasted pulses, nuts and seeds." *European Food Research and Technology* 229 (6):961–969. doi: 10.1007/s00217-009-1131-z.

Acharya, S., B. Ooraikul, A. Srichamroen, S. Basu, and T. Basu. 2006. "Improvement in the nutraceutical properties of fenugreek (*Trigonella foenum-graecum* L.)." *Songklanakarin Journal of Science and Technology* 28:1–9.

Afzal, B., I. Pasha, T. Zahoor, and H. Nawaz. 2016. "Nutritional potential of Fenugreek supplemented bread with special reference to antioxidant profiling." *Pakistan Journal of Agricultural Sciences* 53 (1):217–223. doi: 10.21162/pakjas/16.4664.

Agarwal, R. S., and H. M. Syed. 2017. "Quality evaluation of cookies supplemented with germinated fenugreek seed flour." *The Bioscan* 12 (1):125–128.

Agrawal, V., and R. Sengupta. 2014. "Baked kachori." *International Journal of Food and Nutritional Sciences* 3 (3):175–181.

Ahmad, A., S. S. Alghamdi, K. Mahmood, and M. Afzal. 2016. "Fenugreek a multipurpose crop: Potentialities and improvements." *Saudi Journal of Biological Sciences* 23 (2):300–310. doi: 10.1016/j.sjbs.2015.09.015.

Ahmad, B. H., S. J. Saleh, and A. B. Sabow. 2019. "Role of dried fenugreek (*Trigonella foenum-graecum* L.) leaves as antioxidant and antimicrobial in quality preservation in burgers made of mutton and beef cattle meat during refrigerator storage." *Tikrit Journal for Agricultural Sciences* 19 (2):1–7. doi: 10.25130/tjas.v19i2.373.

Ahsan, F., M. K. Sharif, M. S. Butt, A. Rauf, M. A. Shariati, M. Imran, T. A. Gondal, A. Khan, M. Atif, A. Alsayari, A. B. Muhsinah, H. Algarni, and Y. N. Mabkhot. 2019. "Legumes and leafy vegetables based multi-mix pakoras to alleviate iron and protein deficiency among school aged children." *BioCell* 43 (1–1):8–15.

Akbari, M., H. Rasouli, and T. Bahdor. 2012. "Physiological and pharmaceutical effect of Fenugreek: A review." *IOSR Journal of Pharmacy (IOSRPHR)* 2 (4):49–53.

Al-Gemeai, H. A. 2016. "Effect on organoleptic properties of biscuits fortified with Fenugreek seed germinated until 5 days." *Alexandria Science Exchange Journal* 37 (2):197–205.

Al-Jasass, F. M., and M. S. Al-Jasser. 2012. "Chemical composition and fatty acid content of some spices and herbs under Saudi Arabia conditions." *The Scientific World Journal* 2012 (4):1–5. doi: 10.1100/2012/859892.

Alarcon, B., and G. Samantha. 2020. "Iron-deficiency Anaemia (IDA): Socio-cultural misconceptions intersect the health of vulnerable populations in developing countries." Bachelor of Arts (B.A.) in Public Health Studies, Health Studies, Portland State University, OR, USA.

Alfarisi, H. A. H., A. K. N. Allow, A. H. B. Hamdan, and Z. B. H. Mohamed. 2017. "Acute toxicity of *Trigonella Foenum-Graecum* (fenugreek) seeds aqueous extract on liver in male mice, histopathological study." *International Research Journal of Pharmacy* 8 (4):24–27. doi: 10.7897/2230-8407.080443.

Alloui, N., B. S. Aksa, M. N. Alloui, and F. Ibrir. 2012. "Utilization of fenugreek (*Trigonella foenum-graecum*) as growth promoter for broiler chickens." *Journal of World's Poultry Research* 2 (2):25–27.

Alsebaeai, M., M. Alfawaz, A. Chauhan, A. Al-Farga, and S. Fatma. 2017. "Physicochemical characteristics and nutritional value of fenugreek seeds and seed oil." *International Journal of Food Science and Nutrition* 2 (6):52–55.

Amit, S. K., M. M. Uddin, R. Rahman, S. R. Islam, and M. S. Khan. 2017. "A review on mechanisms and commercial aspects of food preservation and processing." *Agriculture & Food Security* 6 (1):51.

Anderson, J. W., and H. M. Bush. 2011. "Soy protein effects on serum lipoproteins: A quality assessment and meta-analysis of randomized, controlled studies." *Journal of the American College of Nutrition* 30 (2):79–91. doi: 10.1080/07315724.2011.10719947.

Angioloni, A., and C. Collar. 2012. "High legume-wheat matrices: An alternative to promote bread nutritional value meeting dough viscoelastic restrictions." *European Food Research and Technology* 234 (2):273–284. doi: 10.1007/s00217-011-1637-z.

Aspuru, K., C. Villa, F. Bermejo, P. Herrero, and S. G. López. 2011. "Optimal management of iron deficiency anemia due to poor dietary intake." *International Journal of General Medicine* 4:741–50. doi: 10.2147/ijgm.S17788.

Aste, N., C. Del Pero, and F. Leonforte. 2017. "Active refrigeration technologies for food preservation in humanitarian context—A review." *Sustainable Energy Technologies and Assessments* 22:150–160. doi: 10.1016/j.seta.2017.02.014.

Avalos-Soriano, A., R. De la Cruz-Cordero, J. L. Rosado, and T. Garcia-Gasca. 2016. "4-Hydroxyisoleucine from Fenugreek (Trigonella foenum-graecum): Effects on Insulin Resistance Associated with Obesity." *Molecules* 21 (11). doi: 10.3390/molecules21111596.

Azizah, A. H., K. C. Wee, O. Azizah, and M. Azizah. 2009. "Effect of boiling and stir frying on total phenolics, carotenoids and radical scavenging activity of pumpkin (*Cucurbita moschato*)." *International Food Research Journal* 16 (1):45–51.

Badgujar, J., S. Gaikwad, S. K. Sonawane, and S. S. Arya. 2017. "Low glycaemic index bhakri: Indian sorghum unleavened flat bread." *Journal of Food Measurement and Characterization* 11 (2):768–775. doi: 10.1007/s11694-016-9447-4.

Bafna, J., B. Kalpana, and K. G. Ramya. 2020. "Development of nutri-rich bhakri (snack) instant mix." *Journal of Pharmacognosy and Phytochemistry* 9 (3):28–31.

Bakr, A. A. 1997. "Production of iron-fortified bread employing some selected natural iron sources." *Food/Nahrung* 41 (5):293–298.

Bashir, K. M. I., J.-S. Kim, J. H. An, J. H. Sohn, and J.-S. Choi. 2017. "Natural food additives and preservatives for fish-paste products: A review of the past, present, and future states of research." *Journal of Food Quality* 2017:1–31. doi: 10.1155/2017/9675469.

Begum, M., M. M. Hossain, and I. H. Kim. 2016. "Effects of fenugreek seed extract supplementation on growth performance, nutrient digestibility, diarrhoea scores, blood profiles, faecal microflora and faecal noxious gas emission in weanling piglets." *Journal of Animal Physiology and Animal Nutrition* 100 (6):1121–1129. doi: 10.1111/jpn.12496.

Behera, S. S., A. F. El Sheikha, R. Hammami, and A. Kumar. 2020. "Traditionally fermented pickles: How the microbial diversity associated with their nutritional and health benefits?" *Journal of Functional Foods* 70:103971. doi: 10.1016/j.jff.2020.10397.

Bhatia, A., and N. Khetarpaul. 2002. "Effect of fermentation on phytic acid and in vitro availability of calcium and iron of 'Doli ki roti'—an indigenously fermented Indian bread." *Ecology of Food and Nutrition* 41 (3):243–253. doi: 10.1080/03670244.2002.9991685.

Bienkowski, T., K. Zuk-Golaszewska, J. Kaliniewicz, and J. Gołaszewski. 2017. "Content of biogenic elements and fatty acid composition of fenugreek seeds cultivated under different conditions." *Chilean Journal of Agricultural Research* 77 (2):134–141. doi: 10.4067/S0718-58392017000200134.

Bjermo, H., D. Iggman, J. Kullberg, I. Dahlman, L. Johansson, L. Persson, J. Berglund, K. Pulkki, S. Basu, and M. Uusitupa. 2012. "Effects of n-6 PUFAs compared with SFAs on liver fat, lipoproteins, and inflammation in abdominal obesity: A randomized controlled trial." *The American Journal of Clinical Nutrition* 95 (5):1003–1012. doi: 10.3945/ajcn.111.030114.

Blessington, T., N. M. Nzaramba, D. C. Scheuring, A. L. Hale, L. Reddivari, and C. J. Miller. 2010. "Cooking methods and storage treatments of potato: Effects on carotenoids, antioxidant activity, and phenolics." *American Journal of Potato Research* 87 (6):479–491. doi: 10.1016/j.jff.2020.103971.

Boka, B., A. Z. Woldegiorgis, and G. D. Haki. 2013. "Antioxidant properties of Ethiopian traditional bread (*injera*) as affected by processing techniques and tef grain [Eragrostis tef (Zucc.) Trotter] varieties." *Canadian Chemical Transactions* 1 (1):7–24.

Bouis, H. E., and A. Saltzman. 2017. "Improving nutrition through biofortification: A review of evidence from HarvestPlus, 2003 through 2016." *Glob Food Sec* 12:49–58. doi: 10.1016/j.gfs.2017.01.009.

Buamard, N., and S. Benjakul. 2015. "Improvement of gel properties of sardine (*Sardinella albella*) surimi using coconut husk extracts." *Food Hydrocolloids* 51:146–155. doi: 10.1016/j.foodhyd.2015.05.011.

Bueno, M. M., R. C. S. Thys, and R. C. Rodrigues. 2016. "Microbial enzymes as substitutes of chemical additives in baking wheat flour—part i: Individual effects of nine enzymes on flour dough rheology." *Food and Bioprocess Technology* 9 (12):2012–2023. doi: 10.1007/s11947-016-1780-4.

Bushuty, D. H. E. 2015. "Improve the nutritional value of bread by using iron-rich food." *Journal of Home Economics* 31 (31):1–15. doi: 10.21608 / jhe.2015.59383.

Carlez, A., T. Veciana-Nogues, and J.-C. Cheftel. 1995. "Changes in colour and myoglobin of minced beef meat due to high pressure processing." *LWT-Food Science and Technology* 28 (5):528–538. doi: 10.1006/fstl.1995.0088.

Casaburi, A., P. Piombino, G.-J. Nychas, F. Villani, and D. Ercolini. 2015. "Bacterial populations and the Volatilome associated to meat spoilage." *Food Microbiology* 45 (Pt A):83–102. doi: 10.1016/j.fm.2014.02.002.

Cevallos-Casals, B. A., and L. Cisneros-Zevallos. 2010. "Impact of germination on phenolic content and antioxidant activity of 13 edible seed species." *Food Chemistry* 119 (4):1485–1490. doi: 10.1016/j.foodchem.2009.09.030.

Chakraborty, P., A. Bhattacharya, D. K. Bhattacharyya, N. R. Bandyopadhyay, and M. Ghosh. 2016. "Studies of nutrient rich edible leaf blend and its incorporation in extruded food and pasta products." *Materials Today: Proceedings* 3 (10):3473–3483. doi: 10.1016/j.matpr.2016.10.030.

Chaubey, P. S., G. Somani, D. Kanchan, S. Sathaye, S. Varakumar, and R. S. Singhal. 2018. "Evaluation of debittered and germinated fenugreek (*Trigonella foenum-graecum* L.) seed flour on the chemical characteristics, biological activities, and sensory profile of fortified bread." *Journal of Food Processing and Preservation* 42 (1):(1–11) e13395. doi: 10.1111/jfpp.13395.

Chen, A., P. Gilbert, and S. Khokhar. 2009. "Estimating the nutrient composition of South Asian foods using a recipe calculation method." *Food Chemistry* 113 (3):825–831. doi: 10.1016/j.foodchem.2008.05.063.

Cheng, J.-H., S.-T. Wang, and H. W. Ockerman. 2007. "Lipid oxidation and color change of salted pork patties." *Meat Science* 75 (1):71–77. doi: 10.1016/j.meatsci.2006.06.017.

Clemens, R. A., and P. Pressman. 2017. "Food gums: An overview." *Nutrition Today* 52 (1):41–43. doi: 10.1097/nt.0000000000000190.

Collar, C., P. Andreu, J. C. Martı́nez, and E. Armero. 1999. "Optimization of hydrocolloid addition to improve wheat bread dough functionality: A response surface methodology study." *Food Hydrocolloids* 13 (6):467–475. doi: 10.1016/S0268-005X(99)00030-2.

Degirmencioglu, T., H. Unal, S. Ozbilgin, and H. Kuraloglu. 2016. "Effect of ground Fenugreek seeds (*Trigonella foenum-graecum*) on feed consumption and milk performance in Anatolian water buffaloes." *Archives Animal Breeding* 59 (3):345–349. doi: 10.5194/aab-59-345-2016.

Del Nobile, M. A., A. Lucera, C. Costa, and A. Conte. 2012. "Food applications of natural antimicrobial compounds." *Frontiers in Microbiology* 3 (287):287. doi: 10.3389/fmicb.2012.00287.

Dey, A. 2018. "Global consequence of malnutrition and its preventive methodologies." *International Journal of Food Science and Nutrition* 3 (6):263–267.

Dhanesh, T. B., A. Kochhar, and M. Javed. 2018. "Effect of storage on the quality of value added snacks developed using partially defatted peanut cake flour and fenugreek leaves powder." *International Journal of Current Microbiology and Applied Sciences* 7 (2):1127–1135. doi: 10.20546/ijcmas.2018.702.140.

Dhingra, D., M. Michael, H. Rajput, and R. Patil. 2012. "Dietary fibre in foods: A review." *Journal of Food Science and Technology* 49 (3):255–266. doi: 10.1007/s13197-011-0365-5.

Dhull, S. B., S. Punia, K. S. Sandhu, P. Chawla, R. Kaur, and A. Singh. 2020. "Effect of debittered Fenugreek (*Trigonella foenum-graecum* L.) flour addition on physical, nutritional, antioxidant, and sensory properties of wheat flour rusk." *Legume Science* 2 (1):1–9 e21. doi: 10.1002/leg3.21.

Dhull, S. B., and K. S. Sandhu. 2018. "Wheat-fenugreek composite flour noodles: Effect on functional, pasting, cooking and sensory properties." *Current Research in Nutrition and Food Science Journal* 6 (1):174–182. doi: 10.12944/CRNFSJ.6.1.20.

Dintcheva, N. T., R. Arrigo, M. Baiamonte, P. Rizzarelli, and G. Curcuruto. 2017. "Concentration-dependent anti-/pro-oxidant activity of natural phenolic compounds in bio-polyesters." *Polymer Degradation and Stability* 142:21–28. doi: 10.1016/j.polymdegradstab.2017.05.022.

Dwivedi, G., S. Singh, and D. Giri. 2019. "Effect of varieties and processing on nutritional composition of fenugreek seeds." *Pharma Innovation* 8 (12):68–72.

Dwivedi, S., P. Prajapati, N. Vyas, S. Malviya, and A. Kharia. 2017. "A review on food preservation: Methods, harmful effects and better alternatives." *Asian Journal of Pharmacy and Pharmacology* 3 (6):193–199.

Eissa, H. A. A. 2007. "Effect of chitosan coating on shelf life and quality of fresh-cut mushroom." *Journal of Food Quality* 30 (5):623–645. doi: 10.1111/j.1745-4557.2007.00147.x.

El-Arab, A. E., M. Ali, and L. Hussein. 2004. "Vitamin B1 profile of the Egyptian core foods and adequacy of intake." *Journal of Food Composition and Analysis* 17 (1):81–97. doi: 10.1016/s0889-1575(03)00103-0.

El-Mahdy, A. R., and L. A. El-Sebaiy. 1982. "Changes in phytate and minerals during germination and cooking of fenugreek seeds." *Food Chemistry* 9 (3):149–158. doi: 10.1016/0308-8146(82)90092-9.

El-Mahdy, A. R., and L. A. El-Sebaiy. 1983. "Changes in carbohydrates of germinating fenugreek seeds (*Trigonella foenum-graecum* L.)." *Journal of Science of Food and Agricultural* 34:951–956. doi: 10.1002/jsfa.2740340909.

El-Naggar, E. A. 2019. "Influence of fenugreek seeds flour on the rheological characteristics of wheat flour and biscuit quality." *Zagazig Journal of Agricultural Research* 46 (3):721–738.

El Nasri, N. A., and A. H. El Tinay. 2006. "Functional properties of fenugreek (*Trigonella foenum graecum*) protein concentrate." *Food Chemistry* 103 (2):582–589. doi: 10.1016/j.foodchem.2006.09.003.

El Sakka, A., M. Salama, and K. Salama. 2014. "The effect of fenugreek herbal tea and palm dates on breast milk production and infant weight." *The Journal of Pediatrics* 6 (e(202)).

El Sheikha, A. F., and D.-M. Hu. 2018. "Molecular techniques reveal more secrets of fermented foods." *Critical Reviews in Food Science and Nutrition* 60 (1):1–22. doi: 10.1080/10408398.2018.1506906.

El-Shimi, N. M., A. A. Damir, and M. Ragab. 1984. "Changes in some nutrients of fenugreek seeds during germination." *Food Chemistry* 14 (1):11–19. doi: 10.1016/0308-8146(84)90013-X.

Embuscado, M. E. 2015. "Spices and herbs: Natural sources of antioxidants—a mini review." *Journal of Functional Foods* 18:811–819. doi: 10.1016/j.jff.2015.03.005.

Faustman, C., Q. Sun, R. Mancini, and S. P. Suman. 2010. "Myoglobin and lipid oxidation interactions: Mechanistic bases and control." *Meat Science* 86 (1):86–94. doi: 10.1016/j.meatsci.2010.04.025.

Fernandes, P. A. R., S. A. Moreira, L. G. Fidalgo, M. D. Santos, R. P. Queirós, I. Delgadillo, and J. A. Saraiva. 2015. "Food preservation under pressure (hyperbaric storage) as a possible improvement/alternative to refrigeration." *Food Engineering Reviews* 7 (1):1–10.

Feyzi, S., M. Varidi, F. Zare, and M. J. Varidi. 2015. "Fenugreek (Trigonella foenum graecum) seed protein isolate: Extraction optimization, amino acid composition, thermo and functional properties." *Journal of the Science of Food and Agriculture* 95 (15):3165–3176. doi: 10.1002/jsfa.7056.

Gabay, M. P. 2002. "Galactogogues: Medications that induce lactation." *Journal of Human Lactation* 18:274–279. doi: 10.1177/089033440201800311.

Galani, J. H. Y., J. S. Patel, N. J. Patel, and J. G. Talati. 2017. "Storage of fruits and vegetables in refrigerator increases their phenolic acids but decreases the total phenolics, anthocyanins and vitamin C with subsequent loss of their antioxidant capacity." *Antioxidants* 6 (3):(1–19) 59. doi: 10.3390/antiox6030059.

Gartner, L. M., J. Morton, R. A. Lawrence, A. J. Naylor, D. O'Hare, R. J. Schanler, and A. I. Eidelman. 2005. "Breastfeeding and the use of human milk." *Pediatrics* 115 (2):496–506. doi: 10.1542/peds.2004-2491.

Gharneh, H. A. A., and S. Davodalhosseini. 2015. "Evaluation of mineral content in some native Iranian fenugreek (*Trigonella foenum-graceum* L.) genotypes." *Journal of Earth, Environment and Health Sciences* 1 (1):38–41. doi: 10.4103/2423-7752.159926.

Ghasemi, V., M. Kheirkhah, and M. Vahedi. 2015. "The effect of herbal tea containing Fenugreek seed on the signs of breast milk sufficiency in Iranian Girl Infants." *The Iranian Red Crescent Medical Journal* 17 (8):e21848. doi: 10.5812/ircmj.21848.

Giannone, V., M. R. Lauro, A. Spina, A. Pasqualone, L. Auditore, I. Puglisi, and G. Puglisi. 2016. "A novel α-amylase-lipase formulation as anti-staling agent in durum wheat bread." *LWT-Food Science and Technology* 65:381–389. doi: 10.1016/j.lwt.2015.08.020.

Gidenne, T., N. Jehl, A. Lapanouse, and M. Segura. 2004. "Inter-relationship of microbial activity, digestion and gut health in the rabbit: Effect of substituting fibre by starch in diets having a high proportion of rapidly fermentable polysaccharides." *The British Journal of Nutrition* 92 (1):95–104. doi: 10.1079/BJN20041173.

Girardon, P., J. M. Bessiere, J. C. Baccou, and Y. Sauvaire. 1985. "Volatile constituents of Fenugreek seeds." *Planta Medica* 51 (6):533–534. doi: 10.1055/s-2007-969591.

Godebo, D. D., E. Dessalegn, and G. Nigusse. 2019. "Evaluation of phytochemical content and antioxidant capacity of processed Fenugreek (*Trigonella foenum-graecum* L.) flour substituted Injera." *Food Science and Quality Management* 92:1–10. doi: 10.7176/FSQM/92-01.

Gopalpura, P. B., C. Jayanthi, and S. Dubey. 2007. "Effect of Trigonella foenum-graecum seeds on the glycemic index of food: A clinical evaluation." *International Journal of Diabetes in Developing Countries* 27 (2):41–45.

Grace, M. H., G. G. Yousef, S. J. Gustafson, V. D. Truong, G. C. Yencho, and M. A. Lila. 2014. "Phytochemical changes in phenolics, anthocyanins, ascorbic acid, and carotenoids associated with sweet potato storage and impacts on bioactive properties." *Food Chemistry* 145:717–24. doi: 10.1016/j.foodchem.2013.08.107.

Guhan, S. M. 2008. "A behind the scenes look at the idli fermentation." *California State Science Fair, Los Angeles*, May 19–20, 2008.

Gupta, A., and A. Gadipudi. 2018. "Iron deficiency Anaemia in pregnancy: Developed versus developing countries." *Hematology* 6 (1):101–109.

Gupta, S., J. Lakshmi, and J. Prakash. 2008. "Effect of different blanching treatments on ascorbic acid retention in green leafy vegetables." *Natural Product Radiance* 7 (2):111–116.

Hannan, J. M., L. Ali, B. Rokeya, J. Khaleque, M. Akhter, P. R. Flatt, and Y. H. Abdel-Wahab. 2007. "Soluble dietary fibre fraction of *Trigonella foenum-graecum* (fenugreek) seed improves glucose homeostasis in animal models of type 1 and type 2 diabetes by delaying carbohydrate digestion and absorption, and enhancing insulin action." *British Journal of Nutrition* 97 (3):514–521. doi: 10.1017/S0007114507657869.

Hannan, J. M., B. Rokeya, O. Faruque, N. Nahar, M. Mosihuzzaman, A. K. Azad Khan, and L. Ali. 2003. "Effect of soluble dietary fibre fraction of Trigonella foenum-graecum on glycemic, insulinemic, lipidemic and platelet aggregation status of Type 2 diabetic model rats." *Journal of Ethnopharmacology* 88 (1):73–77. doi: 10.1016/s0378-8741(03)00190-9.

Haruna, M., C. E. Udobi, and J. Ndife. 2011. "Effect of added brewers dry grain on the physico-chemical, microbial and sensory quality of wheat bread." *American Journal of Food and Nutrition* 1 (1):39–43. doi: 10.5251/ajfn.2011.1.1.39.43.

Hegazy, A. I. 2011. "Influence of using fenugreek seed flour as antioxidant and antimicrobial agent in the manufacturing of beef burger with emphasis on frozen storage stability." *World Journal of Agricultural Sciences* 7 (4):391–399.

Hegazy, A. I., and M. I. Ibrahium. 2009. "Evaluation of the nutritional protein quality of wheat biscuit supplemented by fenugreek seed flour." *World Journal of Dairy & Food Sciences* 4 (2):129–135.

Hejrani, T., Z. Sheikholeslami, A. S. Mortazavi, and M. G. Davoodi. 2017. "The properties of part baked frozen bread with guar and xanthan gums." *Food Hydrocolloids* 71:252–257. doi: 10.1007/s13197-020-04757-z.

Hejrani, T., Z. Sheikholeslami, A. S. Mortazavi, M. Karimi, and A. H. Elhamirad. 2019. "Impact of the Basil and Balangu gums on physicochemical properties of part baked frozen Barbari bread." *Information Processing in Agriculture* 6 (3):407–413. doi: 10.1016/j.inpa.2018.11.004.

Hemavathy, J., and J. V. Prabhakar. 1989. "Lipid composition of fenugreek (*Trigonella foenum-graecum* L.) seeds." *Food chemistry* 31 (1):1–7. doi: 10.1016/0308-8146(89)90145-3.

Hong, Y.-H., J.-Y. Park, J.-H. Park, M.-S. Chung, K.-S. Kwon, K. Chung, M. Won, and K.-B. Song. 2008. "Inactivation of *Enterobacter sakazakii, Bacillus cereus,* and *Salmonella typhimurium* in powdered weaning food by electron-beam irradiation." *Radiation Physics and Chemistry* 77 (9):1097–1100. doi: 10.1016/j.radphyschem.2008.05.004.

Hooda, S., and S. Jood. 2003. "Effect of soaking and germination on nutrient and antinutrient contents of fenugreek (*Trigonella foenum-graecum* L.)." *Journal of Food Biochemistry* 27 (2):165–176. doi: 10.1111/j.1745-4514.2003.tb00274.x.

Hooda, S., and S. Jood. 2004. "Nutritional evaluation of wheat—fenugreek blends for product making." *Plant Foods for Human Nutrition* 59 (4):149–154. doi: 10.1007/s11130-004-0024-3.

Hosny, I. M., W. I. El Kholy, H. A. Murad, and R. K. El Dairouty. 2011. "Antimicrobial activity of Curcumin upon pathogenic microorganisms during manufacture and storage of a novel style cheese 'Karishcum'." *Journal of American Science* 7 (5):611–618.

Hou, G. 2001. "Oriental noodles." *Advances in Food and Nutrition Research* 43:141–193. doi: 10.1016/s1043-4526(01)43004-x.

Huang, G., Q. Guo, C. Wang, H. H. Ding, and S. W. Cui. 2016. "Fenugreek fibre in bread: Effects on dough development and bread quality." *LWT—Food Science and Technology* 71:274–280. doi: 10.1016/j.lwt.2016.03.040.

Hussain, A., I. Murtaza, and S. Khan. 2012. "Changes in some nutrient contents of fenugreek sprouts with natural food grade additives." *Annals of Horticulture* 5 (2):163–172.

Hussain, P. R., P. Suradkar, S. Javaid, H. Akram, and S. Parvez. 2016. "Influence of postharvest gamma irradiation treatment on the content of bioactive compounds and antioxidant activity of fenugreek (*Trigonella foenum—graeum* L.) and spinach (*Spinacia oleracea* L.) leaves." *Innovative Food Science & Emerging Technologies* 33:268–281. doi: 10.1016/j.ifset.2015.11.017.

Hussein, A. M. S., A. S. Abd El-Azeem, A. M. Hegazy, A. A. Afifi, and G. H. Ragab. 2011. "Physiochemical, sensory and nutritional properties of corn-fenugreek flour composite biscuits." *Australian Journal of Basic and Applied Sciences* 5 (4):84–95.

Iggman, D., I. B. Gustafsson, L. Berglund, B. Vessby, P. Marckmann, and U. Risérus. 2011. "Replacing dairy fat with rapeseed oil causes rapid improvement of hyperlipidaemia: A randomized controlled study." *Journal of Internal Medicine* 270 (4):356–364. doi: 10.1111/j.1365-2796.2011.02383.x.

Indrani, D., P. Swetha, C. Soumya, J. Rajiv, and G. V. Rao. 2011. "Effect of multigrains on rheological, microstructural and quality characteristics of north Indian parotta—An Indian flat bread." *LWT-Food Science and Technology* 44 (3):719–724. doi: 10.1016/j.lwt.2010.11.017.

Işıklı, N. D., and E. Karababa. 2005. "Rheological characterization of fenugreek paste (çemen)." *Journal of Food Engineering* 69 (2):185–190. doi: 10.1016/j.jfoodeng.2004.08.013.

Iskander, F. Y., and K. R. Davis. 1992. "Mineral and trace element contents in bread." *Food Chemistry* 45 (4):269–277. doi: 10.1016/0308-8146(92)90159-Y.

Jalosinska, M., and J. Wilczak. 2009. "Influence of plant extracts on the microbiological shelf life of meat products." *Polish Journal of Food and Nutrition Sciences* 59 (4).

Jani, R., S. A. Udipi, and P. S. Ghugre. 2009. "Mineral content of complementary foods." *Indian Journal of Pediatrics* 76 (1):37–44. doi: 10.1007/s12098-009-0027-z.

Joardder, M. U., and M. H. Masud. 2019. "Food preservation techniques in developing countries." In *Food Preservation in Developing Countries: Challenges and Solutions*, 67–125. Switzerland: Springer International Publishing.

Jones, L. V., D. R. Peryam, and L. Thurstone. 1955. "Development of a scale for measuring soldiers' food preferences." *Food Research* 20:512–520. doi: 10.1111/j.1365-2621.1955.tb16862.x.

Jood, S., and A. C. Kapoor. 1997. "Improvement in bioavailability of minerals of chickpea and blackgram cultivars through processing and cooking methods." *International Journal of Food Sciences and Nutrition* 48 (5):307–312. doi: 10.3109/09637489709028576.

Kadam, M. L., R. V. Salve, Z. M. Mehrajfatema, and S. G. More. 2012. "Development and evaluation of composite flour for *Missi roti*/chapatti." *Journal of Food Processing and Technology* 3 (1):(1–7) 134. doi: 10.4172/2157-7110.1000134.

Kahyaoglu, T., and S. Kaya. 2006. "Modeling of moisture, color and texture changes in sesame seeds during the conventional roasting." *Journal of Food Engineering—J FOOD ENG* 75 (2):167–177. doi: 10.1016/j.jfoodeng.2005.04.011.

Kanaujiya, G., and N. Singh. 2017. "To nutritional profile of dietary fibre pizza base and sensory evaluation of develop product." *International Journal of Advance Research, Ideas and Innovations in Technology* 3 (6):932–936.

Karad, V. A., R. S. Jangale, and K. A. Karad. 2016. "Quality characteristics of crackers made from Kasuri Methi and different flours." *International Journal of Science and Research* 5 (1):734–736.

Karizaki, V. M. 2017. "Ethnic and traditional Iranian breads: Different types, and historical and cultural aspects." *Journal of Ethnic Foods* 4 (1):8–14. doi: 10.1016/j.jef.2017.01.002.

Kasaye, A. T. 2015. "Utilization of fenugreek (*Trigonella foenum-graecum Linn*) to develop value added biscuits and breads." *European Journal of Nutrition & Food Safety* 5 (5):602–603. doi: 10.9734/EJNFS/2015//20986.

Kasaye, A. T., and Y. K. Jha. 2015. "Evaluation of composite blends of fermented fenugreek and wheat flour to assess its suitability for bread and biscuit." *International Journal of Nutrition and Food Sciences* 4 (1):29–35. doi: 10.11648/j.ijnfs.20150401.15.

Khatoniar, S., M. S. Barooah, and M. Das. 2018. "Formulation and evaluation of dehydrated greens incorporated value added products." *International Journal of Current Microbiology and Applied Sciences* 7 (6):94–102.

Khemakhem, B., S. Smaoui, H. El Abed, I. Fendri, H. Hammami, and M. A. Ayadi. 2018. "Improving changes in physical, sensory and texture properties of cake supplemented with purified amylase from fenugreek (*Trigonella foenum graecum*) seeds." *3 Biotech* 8 (3):(1–10) 174. doi: 10.1007/s13205-018-1197-z.

Khoja, K. K., M. F. Aslam, P. A. Sharp, and G. O. Latunde-Dada. 2021. "In vitro bioaccessibility and bioavailability of iron from fenugreek, baobab and moringa." *Food Chemistry* 335:127671. doi: 10.1016/j.foodchem.2020.127671.

Khokhar, S., M. Roe, and G. Swan. 2012. "Carotenoid and retinol composition of South Asian foods commonly consumed in the UK." *Journal of Food Composition and Analysis* 25 (2):166–172. doi: 10.1016/j.jfca.2011.11.005.

Khorshidian, N., M. Y. Asli, M. Arab, A. A. Mirzaie, and A. M. Mortazavian. 2016. "Fenugreek: Potential applications as a functional food and nutraceuticals." *Nutrition and Food Sciences Research* 3 (1):5–16. doi: 10.18869/acadpub.nfsr.3.1.5.

Kochhar, A., M. Nagi, and R. Sachdeva. 2005. "Proximate composition, available carbohydrates, dietary fibre and anti nutritional factors of selected traditional medicinal plants." *Journal of Human Ecology* 19 (3):195–199. doi: 10.1080/09709274.2006.11905878.

Koh, L., B. Jiang, S. Kasapis, and C. W. Foo. 2011. "Structure, sensory and nutritional aspects of soluble-fibre inclusion in processed food products." *Food Hydrocolloids* 25 (2):159–164. doi: 10.1016/j.foodhyd.2010.03.013.

Korkerd, S., S. Wanlapa, C. Puttanlek, D. Uttapap, and V. Rungsardthong. 2016. "Expansion and functional properties of extruded snacks enriched with nutrition sources from food processing by-products." *Journal of food science and technology* 53 (1):561–570. doi: 10.1007/s13197-015-2039-1.

Krishnamoorthy, S., S. Kunjithapatham, and L. Manickam. 2013. "Traditional Indian breakfast (Idli and Dosa) with enhanced nutritional content using millets." *Nutrition & Dietetics* 70:241–246. doi: 10.1111/1747-0080.12020.

Ktari, N., A. Feki, I. Trabelsi, M. Triki, H. Maalej, S. B. Slima, M. Nasri, I. B. Amara, and R. B. Salah. 2017. "Structure, functional and antioxidant properties in Tunisian beef sausage of a novel polysaccharide from *Trigonella foenum-graecum* seeds." *International Journal of Biological Macromolecules* 98:169–181. doi: 10.1016/j.ijbiomac.2017.01.113.

Kumar, S., A. Kumar, A. A. Mishra, R. N. Shukla, and A. K. Gautam. 2016. "Physico-chemical and sensory evaluation of fenugreek enriched salted biscuits." *Food Science Research Journal* 7 (1):89–95. doi: 10.15740/has/fsrj/7.1/89-95.

Kumari, P., and R. P. Dubey. 2013. "Development of semolina based 'upma' incorporated with fenugreek and chickpea seeds flour and evaluation of its sensory and nutritional attributes along with its glycemic index." *Advances in Life Sciences* 2 (1):48–50.

Lacroix, M. 2005. "Combined industrial processes with irradiation to assure innocuity and preservation of fruits and vegetables." Proceedings of the International Nuclear Atlantic Conference (INAC), Santos, Brazil, August 28 to September 2, 2005.

Lalit, H., and A. Kochhar. 2018. "Development and organoleptic evaluation of sweet biscuits formulated by using wheat flour, barley flour and germinated fenugreek seed powder for diabetics." *Current Research in Diabetes & Obesity Journal* 6 (2):1–4. doi: 10.19080/CRDOJ.2018.06.555681.

Laurienzo, P. 2010. "Marine polysaccharides in pharmaceutical applications: An overview." *Marine drugs* 8 (9):2435–2465. doi: 10.3390/md8092435.

Leela, N. K., and K. M. Shafeekh. 2008. "Fenugreek." In *Chemistry of Spices*, edited by V. A. Parthasarathy, B. Chempakam, and T. J. Zachariah, 242–259. Pondicherry: CAB International.

Leykun, T., S. Admasu, and S. Abera. 2020. "Evaluation of the mineral content, phyto-chemicals profile and microbial quality of tef *injera* supplemented by fenugreek flour." *Journal of Food Science and Technology* 57 (7):2480–2489. doi: 10.1007/s13197-020-04283-y.

Lin, Z., and E. Schyvens. 2007. "Influence of blanching on the texture and color of some processed vegetables and fruits." *Journal of Food Processing and Preservation* 19:451–465. doi: 10.1111/j.1745-4549.1995.tb00306.x.

Llavata, B., A. Albors, and M. E. Martin-Esparza. 2020. "High fibre gluten-free fresh pasta with tiger nut, chickpea and fenugreek: Technofunctional, sensory and nutritional properties." *Foods* 9 (11):1–15. doi: 10.3390/foods9010011.

López-Vélez, M., F. Martínez-Martínez, and C. D. Valle-Ribes. 2003. "The study of phenolic compounds as natural antioxidants in wine." *Critical Reviews in Food Science and Nutrition* 43 (2):233–244. doi: 10.1080/10408690390826509.

Lorentzen, G. 1978. "Food preservation by refrigeration, a general introduction." *International Journal of Refrigeration* 1 (1):9–12.

Losso, J. N., D. L. Holliday, J. W. Finley, R. J. Martin, J. C. Rood, Y. Yu, and F. L. Greenway. 2009. "Fenugreek bread: A treatment for diabetes mellitus." *Journal of Medicinal Food* 12 (5):1046–1049. doi: 10.1089/jmf.2008.0199.

Luna, T., K. Wilkinson, and K. R. Dumroese. 2014. "Seed germination and sowing options." In *Nursery Manual for Native Plants: A Guide for Tribal Nurseries*, edited by R. Kasten, Tara Luna Dumroese, and Thomas D. Landis, 162–183. Washington, DC: Department of Agriculture, Forest Service.

Lund, M. N., M. S. Hviid, and L. H. Skibsted. 2007. "The combined effect of antioxidants and modified atmosphere packaging on protein and lipid oxidation in beef patties during chill storage." *Meat Science* 76 (2):226–233. doi: 10.1016/j.meatsci.2006.11.003.

Mabrouki, S., R. Chalghoumi, and H. Abdouli. 2016. "Effects of pre-germinated fenugreek seeds inclusion in low-fiber diets on post-weaned rabbits' health status, growth performances, carcass characteristics, and meat chemical composition." *Tropical Animal Health and Production* 49 (3):459–465. doi: 10.1007/s11250-016-1214-3.

Maheshwari, R. K., A. K. Chauhan, L. Mohan, and M. Maheshwari. 2014. "Spice up for scrumptious tang, cologne & wellbeing." *Journal of Global Biosciences* 3 (1):304–313.

Mahfouz, S. A., S. M. Elaby, and H. Z. Hassouna. 2012. "Effects of some legumes on hypercholesterolemia in rats." *Journal of American Science* 8 (12):1453–1460.

Mahmoud, N. Y., R. H. Salem, and A. A. Mater. 2012. "Nutritional and biological assessment of wheat biscuits supplemented by fenugreek plant to improve diet of anemic rats." *Academic Journal of Nutrition* 1 (1):1–9. doi: 10.5829/idosi.ajn.2012.1.1.63103.

Majeed, M., S. Majeed, K. Nagabhushanam, S. Arumugam, S. Natarajan, K. Beede, and F. Ali. 2018. "Galactomannan from *Trigonella foenum-graecum* L. seed: Prebiotic application and its fermentation by the probiotic *Bacillus coagulans* strain MTCC 5856." *Food Science & Nutrition* 6 (3):666–673. doi: 10.1002/fsn3.606.

Makowska, A., M. Majcher, S. Mildner-Szkudlarz, A. Jedrusek-Golinska, and K. Przygoński. 2017. "Triticale crisp bread enriched with selected bioactive additives: Volatile profile, physical characteristics, sensory and nutritional properties." *Journal of Food Science and Technology* 54 (10):3092–3101. doi: 10.1007/s13197-017-2745-y.

Man, S. M., A. Păucean, I. D. Călian, V. Mureşan, M. S. Chiş, A. Pop, A. E. Mureşan, M. Bota, and S. Muste. 2019. "Influence of fenugreek flour (*Trigonella foenum-graecum* l.) addition on the technofunctional properties of dark wheat flour." *Journal of Food Quality* 2019 (8635806):1–8. doi: 10.1155/2019/8635806.

Mandal, S., and M. DebMandal. 2016. "Fenugreek (*Trigonella foenum-graecum* L.) oils." In *Essential Oils in Food Preservation, Flavor and Safety*, 421–429. Amsterdam: Elsevier Inc.

Mansour, E. H., and T. A. El-Adawy. 1994. "Nutritional potential and functional properties of heat-treated and germinated fenugreek seeds." *Food Science and Technology—Zurich* 27 (6):568–572. doi: 10.1006/fstl.1994.1111.

Mathur, P., and M. Choudhry. 2009a. "Effect of domestic processing on proximate composition of fenugreek seeds." *Journal of Food Science and Technology* 46:255–258.

Mathur, P., and M. Choudhry. 2009b. "Consumption pattern of fenugreek seeds in Rajasthani families." *Journal of Human Ecology* 25 (1):9–12. doi: 10.1080/09709274.2009.11906127.

McCarthy, T. L., J. P. Kerry, J. F. Kerry, P. B. Lynch, and D. J. Buckley. 2001. "Evaluation of the antioxidant potential of natural food/plant extracts as compared with synthetic antioxidants and vitamin E in raw and cooked pork patties." *Meat Science* 57:45–52. doi: 10.1016/s0309-1740(00)00129-7.

Meghwal, M., and T. K. Goswami. 2011. "Effect of moisture content on physical properties of black pepper." *Journal of Agricultural Engineering* 48 (2):8–14.

Meghwal, M., and T. K. Goswami. 2012. "A review on the functional properties, nutritional content, medicinal utilization and potential application of fenugreek." *Journal of Food Processing and Technology* 3 (9):(1–10) 1000181. doi: 10.4172/2157-7110.1000181.

Mehrafarin, A., A. Qaderi, S. Rezazadeh, H. Naghdi Badi, G. Noormohammadi, and E. Zand. 2010. "Bioengineering of important secondary metabolites and metabolic pathways in fenugreek (*Trigonella foenum-graecum* L.)." *Journal of Medicinal Plants* 9 (35):1–18.

Mukherjee, S. K., M. N. Albury, C. S. Pederson, A. G. Vanveen, and K. H. Steinkraus. 1965. "Role of Leuconostoc Mesenteroides in leavening the batter of Idli, a fermented food of India." *Appl Microbiol* 13 (2):227–231.

Muraki, E., H. Chiba, K. Taketani, S. Hoshino, N. Tsuge, N. Tsunoda, and K. Kasono. 2012. "Fenugreek with reduced bitterness prevents diet-induced metabolic disorders in rats." *Lipids in Health and Disease* 11 (1):58.

Nagulapalli Venkata, K. C., A. Swaroop, D. Bagchi, and A. Bishayee. 2017. "A small plant with big benefits: Fenugreek (*Trigonella foenum-graecum* Linn.) for disease prevention and health promotion." *Molecular Nutrition & Food Research* 61 (6):(1–26)1600950. doi: 10.1002/mnfr.201600950.

Naidu, M. M., B. N. Shyamala, P. J. Naik, G. Sulochanamma, and P. Srinivas. 2011. "Chemical composition and antioxidant activity of the husk and endosperm of fenugreek seeds." *LWT-Food Science and Technology* 44 (2):451–456. doi: 10.1016/j.lwt.2010.08.013.

Negi, P. S., and S. K. Roy. 2000. "Effect of blanching and drying methods on β -carotene, ascorbic acid and chlorophyll retention of leafy vegetables." *LWT—Food Science and Technology* 33 (4):295–298. doi: 10.1006/fstl.2000.0659.

Negu, A., A. Zegeye, and T. Astatkie. 2020. "Development and quality evaluation of wheat based cookies enriched with fenugreek and oat flours." *Journal of Food Science and Technology* 57 (4):1–8. doi: 10.1007/s13197-020-04389-3.

Nguekouo, P. T., D. Kuate, A. P. N. Kengne, C. Y. Woumbo, F. A. Tekou, and J. E. Oben. 2018. "Effect of boiling and roasting on the antidiabetic activity of *Abelmoschus esculentus* (Okra) fruits and seeds in type 2 diabetic rats." *Journal of Food Biochemistry* 42 (6):e12669. doi: 10.1111/jfbc.12669.

Njoumi, S., M. Josephe Amiot, I. Rochette, S. Bellagha, and C. Mouquet-Rivier. 2019. "Soaking and cooking modify the alpha-galacto-oligosaccharide and dietary fibre content in five Mediterranean legumes." *International Journal of Food Sciences and Nutrition* 70 (5):551–561. doi: 10.1080/09637486.2018.1544229.

Oboh, G. 2005. "Effect of blanching on the antioxidant properties of some tropical green leafy vegetables." *LWT—Food Science and Technology* 38 (5):513–517. doi: 10.1016/j.lwt.2004.07.007.

Ogunmoyole, T., I. J. Kade, O. D. Johnson, and O. J. Makun. 2012. "Effect of boiling on the phytochemical constituents and antioxidant properties of African pear *Dacryodes edulis* seeds in vitro." *African Journal of Biochemistry Research* 6 (8):105–114. doi: 10.5897/AJBR12.023.

Ouattara, B., R. E. Simard, G. Piette, A. Bégin, and R. A. Holley. 2000. "Inhibition of surface spoilage bacteria in processed meats by application of antimicrobial films prepared with chitosan." *International Journal of Food Microbiology* 62 (1–2):139–148. doi: 10.1016/S0168-1605(00)00407-4.

Pandey, H., and P. Awasthi. 2015a. "Effect of processing techniques on nutritional composition and antioxidant activity of fenugreek (*Trigonella foenum-graecum*) seed flour." *Journal of Food Science and Technology* 52 (2):1054–1060. doi: 10.1007/s13197-013-1057-0.

Pandey, H., and P. Awasthi. 2015b. "Organoleptic evaluation of germinated fenugreek seed flour incorporated recipes: *Chapatti* and *idli*." *Asian Journal of Home Science* 10 (1):41–44. doi: 10.15740/HAS/AJHS/10.1/41-44.

Park, Y., J. Oglesby, S. Aljaloud, R. Gyawali, and S. Ibrahim. 2019. "Impact of different gums on textural and microbial properties of goat milk yogurts during refrigerated storage." *Foods* 8 (5):169. doi: 10.3390/foods8050169.

Pathak, P., S. Srivastava, and S. Grover. 2000. "Development of food products based on millets, legumes and fenugreek seeds and their suitability in the diabetic diet." *International Journal of Food Sciences and Nutrition* 51 (5):409–414. doi: 10.1080/096374800427019.

Pathania, S., A. Kaur, and P. A. Sachdev. 2017. "Chickpea flour supplemented high protein composite formulation for flatbreads: Effect of packaging materials and storage temperature on the ready mix." *Food Packaging and Shelf Life* 11:125–132. doi: 10.1016/j.fpsl.2017.01.006.

Patil, G. D., I. L. Pardeshi, and K. J. Shinde. 2015. "Drying of green leafy vegetables using microwave oven dryer." *Journal Ready to Eat Food* 2 (1):18–26.

Patil, S., and S. S. Arya. 2012. "Ingredients and process standardization of Thepla: An Indian unleavened vegetable flatbread using hierarchical cluster analysis." *Advance Journal of Food Science and Technology* 4 (5):286–293.

Patil, V., and P. Nikam. 2019. "Development and quality evaluation of puff pastry (Fenugreek)." *International Journal of Chemical Studies* 7 (1):2490–2493.

Paul, V., A. Paul, P. Kushwaha, and A. Tungoe. 2018. "Utilization of coarse grains and millet in preparation of ready to eat snack (Cookies)." *Journal of Pharmacognosy and Phytochemistry* 7 (3):3539–3541.

Penagos Tabares, F., J. V. Bedoya Jaramillo, and Z. T. Ruiz-Cortés. 2014. "Pharmacological overview of galactogogues." *Veterinary Medicine International* 2014:602894. doi: 10.1155/2014/602894.

Pérez-Quirce, S., F. Ronda, A. Lazaridou, and C. G. Biliaderis. 2017. "Effect of microwave radiation pretreatment of rice flour on gluten-free breadmaking and molecular size of β-glucans in the fortified breads." *Food and Bioprocess Technology* 10 (8):1412–1421. doi: 10.1007/s11947-017-1910-7.

Peryam, D. R., and N. Girardot. 1952. "Advanced taste-test method." *Food Engineering* 24:58–61.

Petit, P., Y. Sauvaire, G. Ponsin, M. Manteghetti, A. Fave, and G. Ribes. 1993. "Effects of a fenugreek seed extract on feeding behaviour in the rat: Metabolic-endocrine correlates." *Pharmacology Biochemistry and Behavior* 45 (2):369–374. doi: 10.1016/0091-3057(93)90253-p.

Pontonio, E., L. Nionelli, J. A. Curiel, A. Sadeghi, R. Di Cagno, M. Gobbetti, and C. G. Rizzello. 2015. "Iranian wheat flours from rural and industrial mills: Exploitation of the chemical and technology features, and selection of autochthonous sourdough starters for making breads." *Food microbiology* 47:99–110. doi: 10.1016/j.fm.2014.10.011.

Porecha, V. M., and R. Sengupta. 2013. "Nutri waffles with tomato dip." *International Journal of Food and Nutritional Sciences* 2 (2):109.

Pourafshar, S., K. A. Rosentrater, and P. G. Krishnan. 2018. "Production of Barbari bread (traditional Iranian bread) using different levels of distillers dried grains with solubles (DDGS) and sodium Stearoyl lactate (SSL)." *Foods* 7 (3):31. doi: 10.3390/foods7030031.

Prabhu, S., and D. M. Barrett. 2009. "Effects of storage condition and domestic cooking on the quality and nutrient content of African leafy vegetables (*Cassia tora* and *Corchorus tridens*)." *Journal of the Science of Food and Agriculture* 89 (10):1709–1721. doi: 10.1002/jsfa.3644.

Prasad, R., A. Gupta, R. Parihar, and K. Gangwar. 2014. "In vitro method for predicting the bioavailability of iron from Bathua (*Chenopodium album*) and Fenugreek (*Trigonella foenum graecum*) leaves in Indian cookies." *Journal of Applied and Natural Science* 6 (2):701–706. doi: 10.31018/jans.v6i2.521.

Qureshi, A. I., S. R. Ahmad, T. Nazir, M. A. Pal, A. H. Sofi, M. Rovida, and H. Jalal. 2018. "Efficacy of fenugreek seed powder for the development of functional spent hen meat patties." *Journal of Entomology and Zoology Studies* 6 (5):353–356.

Rahnama, F., J. M. Milani, and A. G. Ardabili. 2017. "Improved quality attributes of Brabari and lavash flat breads with wheat doughs incorporated with Fenugreek seed (*Trigonella foenum graecum*) Gum." *Journal of Food Processing and Preservation* 41 (1):e12741. doi: 10.1111/jfpp.12741.

Rajamani, S., and P. A. Raajeswari. 2016. "Development and evaluation of glycemic index of traditional western Indian food *khakhara* using low glycemic ingredients." *International Journal of Science and Research* 5 (11):433–437.

Rajni, B., A. J. Thomas, and D. Kapur. 2018. Practical manual-part-I, section-4, subject: 1-nutrition for the community, course: Certificate in nutrition and child care (CNCC)-01. In *eGyanKosh—IGNOU Self Learning Material*. New Delhi: IGNOU-The People University.

Raju, J., D. Gupta, A. R. Rao, P. K. Yadava, and N. Z. Baquer. 2001. "*Trigonella foenum-graecum* (fenugreek) seed powder improves glucose homeostasis in alloxan diabetic rat tissues by reversing the altered glycolytic, gluconeogenic and lipogenic enzymes." *Molecular and Cellular Biochemistry* 224 (1–2):45–51. doi: 10.1023/a:1011974630828.

Rajurkar, N. S., K. N. Gaikwad, and M. S. Razavi. 2012. "Evaluation of free radical scavenging activity of Justicia adhatoda: A gamma radiation study." *International Journal of Pharmacy and Pharmaceutical Sciences* 4 (Suppl 4):93–96.

Ranade, M., and N. Mudgalkar. 2017. "A simple dietary addition of fenugreek seed leads to the reduction in blood glucose levels: A parallel group, randomized single-blind trial." *AYU (An International Quarterly Journal of Research in Ayurveda)* 38 (1–2):24–27. doi: 10.4103/ayu.AYU_209_15.

Randhir, R., Y.-T. Lin, and K. Shetty. 2004. "Phenolics, their antioxidant and antimicrobial activity in dark germinated fenugreek sprouts in response to peptide and phytochemical elicitors." *Asia Pacific Journal of Clinical Nutrition* 13 (3):295–307.

Rao, A. S., S. Hegde, L. M. Pacioretty, J. DeBenedetto, and J. G. Babish. 2020. "*Nigella sativa* and *Trigonella foenum-graecum* supplemented chapatis safely improve HbA1c, body weight, waist circumference, blood lipids, and fatty liver in overweight and diabetic subjects: A twelve-week safety and efficacy study." *Journal of Medicinal Food* 23 (9):905–919. doi: 10.1089/jmf.2020.0075.

Rathore, M. S., and N. S. Shekhawat. 2008. "Incredible spices of India: From traditions to cuisine." *American-Eurasian Journal of Botany* 1 (3):85–89.

Ravi, R., and J. Joseph. 2020. "Effect of fenugreek on breast milk production and weight gain among Infants in the first week of life." *Clinical Epidemiology and Global Health* 8 (3):656–660. doi: 10.1016/j.cegh.2019.12.021.

Ravindran, G., A. Carr, and A. Hardacre. 2011. "A comparative study of the effects of three galactomannans on the functionality of extruded pea—rice blends." *Food Chemistry* 124 (4):1620–1626. doi: 10.1016/j.foodchem.2010.08.030.

Richter Reis, F. 2016. "Effect of Blanching on Food Physical, Chemical, and Sensory Quality." In *New Perspectives on Food Blanching*, edited by F. Richter Reis, 7–48. Cham: Springer.

Rizki, H., F. Kzaiber, M. Elharfi, S. Ennahli, and H. Hanine. 2015. "Effects of roasting temperature and time on the physicochemical properties of sesame (Sesamum indicum. L) seeds." *International Journal of Innovation and Applied Studies* 11 (1):148–155.

Robert, S. D., A. A.-S. Ismail, and W. I. W. Rosli. 2016. "Reduction of postprandial blood glucose in healthy subjects by buns and flatbreads incorporated with fenugreek seed powder." *European Journal of Nutrition* 55 (7):2275–2280. doi: 10.1007/s00394-015-1037-4.

Roberts, K. T., E. Allen-Vercoe, S. A. Williams, T. Graham, and S. W. Cui. 2015. "Comparative study of the *in vitro* fermentative characteristics of fenugreek gum, white bread and bread with fenugreek gum using human faecal microbes." *Bioactive Carbohydrates and Dietary Fibre* 5 (2):116–124. doi: 10.1016/j.bcdf.2014.09.007.

Rosqvist, F., A. Smedman, H. Lindmark-Månsson, M. Paulsson, P. Petrus, S. Straniero, M. Rudling, I. Dahlman, and U. Risérus. 2015. "Potential role of milk fat globule membrane in modulating plasma lipoproteins, gene expression, and cholesterol metabolism in humans: A randomized study." *The American journal of clinical nutrition* 102 (1):20–30. doi: 10.3945/ajcn.115.107045.

Ruel, M. T., and C. E. Levin. 2002. "Food-based approaches for alleviating micronutrient malnutrition: An overview." *Journal of Crop Production* 6 (1–2):31–53. doi: 10.1300/J144v06n01_05.

Sah, A. E.-N., H. M. Khattab, H. A. Al-Alamy, F. A. Salem, and M. M. Abdou. 2007. "Effect of some medicinal plants seeds in the rations on the productive performance of lactating buffaloes." *Egyptian Journal of Nutrition and Feeds* 3:31–41. doi: 10.3923/ijds.2007.348.355.

Saini, P., N. Dubey, P. Singh, and A. Singh. 2016. "Effect of processing methods on proximate composition and antioxidant activity of fenugreek (*Trigonella foenum-graecum*) seeds." *International Journal of Agricultural and Food Science* 6 (2):82–87.

Sakhare, S. D., A. A. Inamdar, and P. Prabhasankar. 2015. "Roller milling process for fractionation of fenugreek seeds (Trigonella foenumgraecum) and characterization of milled fractions." *Journal of Food Science and Technology* 52 (4):2211–2219. doi: 10.1007/s13197-014-1279-9.

Samani, S. K., M. R. Ghorbani, J. Fayazi, and S. Salari. 2020. "The effect of different levels of Fenugreek (*Trigonella foenum-graecum* L.) powder and extract on performance, egg quality, blood parameters and immune responses of laying hens in second production cycle." In *Veterinary Research Forum*, 53–57. Urmia, Iran: Faculty of Veterinary Medicine, Urmia University.

Samra, R. A., and H. G. Anderson. 2007. "Insoluble cereal fiber reduces appetite and short-term food intake and glycemic response to food consumed 75 min later by healthy men." *The American Journal of Clinical Nutrition* 86:972–979. doi: 10.1093/ajcn/86.4.972.

Sanlier, N., and F. Gencer. 2020. "Role of spices in the treatment of diabetes mellitus: A mini review." *Trends in Food Science & Technology* 99 (May 2020):441–449. doi: 10.1016/j.tifs.2020.03.018.

Sarwar, S., M. A. Hanif, M. A. Ayub, Y. D. Boakye, and C. Agyare. 2020. "Fenugreek." In *Medicinal Plants of South Asia: Novel Sources for Drug Discovery*, edited by Muhammad Asif Hanif, Haq Nawaz, Khan Muhammad Mumtaz, and Byrne Hugh, 257–271. Amsterdam: Elesvier.

Sathyanarayana, S., J. P. Naik, T. Jeyarani, M. M. Naidu, and P. Srinivas. 2011. "Characterisation of germinated fenugreek (*Trigonella foenum-graecum* L.) seed fractions." *International Journal of Food Science & Technology* 46 (11):2337–2343. doi: 10.1111/j.1365-2621.2011.02754.x.

Savić, I. V. 1985. *Small-Scale Sausage Production, FAO Animal Production and Health Paper*. Rome: Food and Agriculture Organization of the United Nations.

Savitri, M., A. Kumari, K. Angmo, and T. C. Bhalla. 2016. "Traditional pickles of Himachal Pradesh." *Indian Journal of Traditional Knowledge* 15 (2):330–336.

Schlemmer, U., W. Frølich, R. M. Prieto, and F. Grases. 2009. "Phytate in foods and significance for humans: Food sources, intake, processing, bioavailability, protective role and analysis." *Molecular Nutrition & Food Research* 53 Suppl 2:S330–375. doi: 10.1002/mnfr.200900099.

Sęczyk, Ł., M. Świeca, D. Dziki, A. Anders, and U. Gawlik-Dziki. 2017. "Antioxidant, nutritional and functional characteristics of wheat bread enriched with ground flaxseed hulls." *Food Chemistry* 214:32–38. doi: 10.1016/j.foodchem.2016.07.068.

Selman, J. D. 1994. "Vitamin retention during blanching of vegetables." *Food Chemistry* 49 (2):137–147. doi: 10.1016/0308-8146(94)90150-3.

Shah, M. A., and P. S. Mir. 2004. "Effect of dietary fenugreek seed on dairy cow performance and milk characteristics." *Canadian Journal of Animal Science* 84 (4):725–729. doi: 10.4141/A04-027.

Shahzad, S. A., S. Hussain, A. A. Mohamed, M. S. Alamri, M. A. Ibraheem, and A. A. A. Qasem. 2019. "Effect of hydrocolloid gums on the pasting, thermal, rheological and textural properties of chickpea starch." *Foods* 8 (12):687. doi: 10.3390/foods8120687.

Sharma, A., and M. Choudhry. 2006. "Food habits and consumption practices of Missi Roti among diabetics." *Journal of Human Ecology* 19 (4):273–276. doi: 10.1080/09709274.2006.11905890.

Sharma, H. R., and G. S. Chauhan. 2000. "Physical, sensory and chemical characteristics of wheat breads supplemented with fenugreek (*Trigonella foenum-graecum L.*)." *Journal of Food Science and Technology (Mysore)* 37 (1):91–94.

Sharma, N. 2010. "Aloe vera leaf powder (AVLP): Nutrient composition, development of ready to eat snacks and its clinical potential." Thesis for PhD in Home Science, Foods and Nutrition, Maharana Pratap University of Agriculture and Technology, Udaipur (Rajasthan).

Sharma, N., and A. Singh. 2012. "An insight into traditional foods of north-western area of Himachal Pradesh." *Indian Journal of Traditional Knowledge* 11 (1):58–65.

Sharma, S., S. P. Jain, S. Thomas, and H. S. Gopalan. 2006. *Evaluation of Mid Day Meal Programme in MCD Schools*. New Delhi: Nutrition Foundation of India.

Shinde, S. T., A. R. Sawate, R. B. Kshirsagar, and S. S. Patangare. 2019. "Effect of pre-treatment on quality attributes of fresh bamboo shoot pickle." *The Pharma Innovation Journal* 8 (3):257–260.

Shirani, G., and R. Ganesharanee. 2009. "Extruded products with Fenugreek (*Trigonella foenum-graecum*) chickpea and rice: Physical properties, sensory acceptability and glycaemic index." *Journal of Food Engineering* 90 (1):44–52. doi: 10.1016/j.jfoodeng.2008.06.004.

Shreelalitha, S. J., and K. R. Sridhar. 2018. "Physical and Cooking Properties of Seeds of Two Wild Legume Landraces of *Sesbania*." *International Journal of Agricultural Technology* 14 (3):363–376.

Sidahmmed, S. S. A., and A. I. Mustafa. 2016. "Effect of fenugreek (*Trigonella foneumgraecium*) seeds flour on amino acids profile, physical and organoleptic properties of Sudanese wheat bread." *IOSR Journal of Engineering* 6 (12):21–28.

Singh, P., and R. S. Raghuvanshi. 2012. "Finger millet for food and nutritional security." *African Journal of Food Science* 6 (4):77–84. doi: 10.5897/AJFSX10.010.

Spyropoulos, C. G., and G. J. S. Reid. 1988. "Water stress and galactomannan breakdown in germinated fenugreek seeds. Stress affects the production and the activities *in vivo* of galactomannan-hydrolysing enzymes "*Planta* 174:473–478. doi: 10.1007/BF00634475.

Srichamroen, A., A. Thomson, C. Field, and T. Basu. 2009. "In vitro intestinal glucose uptake is inhibited by galactomannan from Canadian fenugreek seed (Trigonella foenum-graecum L) in genetically lean and obese rats." *Nutrition Research (New York, N.Y.)* 29:49–54. doi: 10.1016/j.nutres.2008.11.002.

Sridevi, J., P. M. Halami, and S. V. N. Vijayendra. 2010. "Selection of starter cultures for idli batter fermentation and their effect on quality of idlis." *Journal of Food Science and Technology* 47 (5):557–563. doi: 10.1007/s13197-010-0101-6.

Srinivas, T. 2011. "Exploring Indian culture through food." *Education About Asia*:38–41.

Srinivasan, K. 2005. "Plant foods in the management of diabetes mellitus: Spices as beneficial antidiabetic food adjuncts." *International Journal of Food Sciences and Nutrition* 56 (6):399–414. doi: 10.1080/09637480500512872.

Srinivasan, K. 2006. "Fenugreek (*Trigonella foenum-graecum*): A review of health beneficial physiological effects." *Food Reviews International* 22 (2):203–224. doi: 10.1080/87559120600586315.

Srinivasan, K. 2019. "Fenugreek (*Trigonella foenum-graecum L.*) seeds used as functional food supplements to derive diverse health benefits." In *Functional Food Supplements to Derive Diverse Health Benefits*, edited by Seyed Mohammad Nabavi and Ana Sanches Silva, 217–221. Cambridge: Academic Press.

Srivastava, D., J. Rajiv, M. M. Naidu, J. Puranaik, and P. Srinivas. 2012. "Effect of fenugreek seed husk on the rheology and quality characteristics of muffins." *Food and Nutrition Sciences* 3:1473–1479. doi: 10.4236/fns.2012.311191.

Sudha, M. L., S. W. Eipson, H. Khanum, M. M. Naidu, and V. G. Rao. 2015. "Effect of normal/dehydrated greens on the rheological, microstructural, nutritional and quality characteristics of paratha-an Indian flat bread." *Journal of Food Science and Technology* 52 (2):840–848. doi: 10.1007/s13197-013-1062-3.

Suman, S. P., and P. Joseph. 2013. "Myoglobin chemistry and meat color." *Annual Review of Food Science and Technology* 4:79–99. doi: 10.1146/annurev-food-030212-182623.

Taylor, W. G., H. J. Zulyniak, K. W. Richards, S. N. Acharya, S. Bittman, and J. L. Elder. 2002. "Variation in diosgenin levels among 10 accessions of fenugreek seeds produced in western canada." *Journal of Agricultural and Food Chemistry* 50 (21):5994–5997. doi: 10.1021/jf020486y.

Tiran, D. 2003. "The use of fenugreek for breast feeding women." *Complement Ther Nurs Midwifery* 9 (3):155–6. doi: 10.1016/s1353-6117(03)00044-1.

Topuz, A., and F. Ozdemir. 2004. "Influences of gamma irradiation and storage on the capsaicinoids of sun-dried and dehydrated paprika." *Food Chemistry* 86 (4):509–515. doi: 10.1016/j.foodchem.2003.09.003.

Trehan, A., A. Sharma, B. Naik, A. Singh, and V. Kumar. 2019. "Infusion of herbs for development of functional paneer: A mini review." *Journal of Pharmacognosy and Phytochemistry* 8 (3):1969–1972.

Tsatsaragkou, K., T. Kara, C. Ritzoulis, I. Mandala, and C. M. Rosell. 2017. "Improving carob flour performance for making gluten-free breads by particle size fractionation and jet milling." *Food and Bioprocess Technology* 10 (5):831–841. doi: 10.1007/s11947-017-1863-x.

Turfani, V., V. Narducci, A. Durazzo, V. Galli, and M. Carcea. 2017. "Technological, nutritional and functional properties of wheat bread enriched with lentil or carob flours." *LWT* 78:361–366. doi: 10.1016/j.lwt.2016.12.030.

Turkyılmaz, C., E. Onal, I. M. Hirfanoglu, O. Turan, E. Koç, E. Ergenekon, and Y. Atalay. 2011. "The effect of galactagogue herbal tea on breast milk production and short-term catch-up of birth weight in the first week of life." *Journal of Alternative and Complementary Medicine* 17 (2):139–42. doi: 10.1089/acm.2010.0090.

Uras Güngör, Ş. S., S. Güzel, A. İlçim, and G. Kökdil. 2014. "Total phenolic and flavonoid content, mineral composition and antioxidant potential of Trigonella monspeliaca." *Turkish Journal of Pharmaceutical Sciences* 11 (3):255–262.

Variyar, P. S., A. Limaye, and A. Sharma. 2004. "Radiation-induced enhancement of antioxidant contents of soybean (*Glycine max Merrill*)." *Journal of Agricultural and Food Chemistry* 52 (11):3385–3388. doi: 10.1021/jf030793j.

Venkatesan, J., B. Lowe, S. Anil, P. Manivasagan, A. A. A. Kheraif, K. H. Kang, and S. K. Kim. 2015. "Seaweed polysaccharides and their potential biomedical applications." *Starch-Stärke* 67 (5–6):381–390. doi: 10.1002/star.201400127.

Wang, J., S. Hu, S. Nie, Q. Yu, and M. Xie. 2016. "Reviews on mechanisms of in vitro antioxidant activity of polysaccharides." *Oxidative Medicine and Cellular Longevity* 2016:5692852. doi: 10.1155/2016/5692852.

Wani, S. S. A., and P. Kumar. 2015. "Development of cookies using fenugreek seed extract as a functional ingredient." *International Journal of Food, Nutrition and Dietetics* 3 (1):23–28.

Wani, S. S. A., and P. Kumar. 2018. "Fenugreek: A review on its nutraceutical properties and utilization in various food products." *Journal of the Saudi Society of Agricultural Sciences* 17 (2):97–106. doi: 10.1016/j.jssas.2016.01.007.

Wickramarachchi, K. S., M. J. Sissons, and S. P. Cauvain. 2015. "Puff pastry and trends in fat reduction: An update." *International Journal of Food Science & Technology* 50 (5):1065–1075. doi: 10.1111/ijfs.12754.

Xia, Y., J. Song, F. Zhong, J. Halim, and M. ÓMahony. 2020. "The 9-point hedonic scale: Using R-Index Preference Measurement to compute effect size and eliminate artifactual ties." *Food Research International*:109140.

Xiao, H.-W., Z. Pan, L.-Z. Deng, H. M. El-Mashad, X.-H. Yang, A. S. Mujumdar, Z.-J. Gao, and Q. Zhang. 2017. "Recent developments and trends in thermal blanching—a comprehensive review." *Information Processing in Agriculture* 4 (2):101–127. doi: 10.1016/j.inpa.2017.02.001.

Xing, J., and P.-S. Ng. 2016. *Indigenous Culture, Education and Globalization: Critical Perspectives from Asia*. New York: Springer.

Yadav, S. K., and S. Sehgal. 1997. "Effect of home processing and storage on ascorbic acid and β-carotene content of bathua (*Chenopodium album*) and fenugreek (*Trigonella foenum graecum*) leaves." *Plant Foods for Human Nutrition* 50 (3):239–247. doi: 10.1007/BF02436060.

Yadav, S. K., and S. Sehgal. 2003. "Effect of domestic processing and cooking methods on total, HCl extractable iron and in vitro availability of iron in bathua and fenugreek leaves." *Nutrition and Health* 17 (1):61–63. doi: 10.1177/026010600301700107.

Yadav, S. R., D. M. Biyani, and M. J. Umekar. 2019. "*Trigonella foenum-graecum*: A herbal plant review." *World Journal of Pharmaceutical Research* 8 (12):402–419. doi: 10.13040/IJPSR.0975-8232. IJLSR.4(2).15-26.

Yadav, U., R. R. B. Singh, and A. Chatterjee. 2016. "Optimization of physical properties and protein to produce functional extruded snack concocted with composite flour using RSM." *Indian Journal of Dairy Science* 69 (1):24–32.

Yanishlieva, N. V., E. Marinova, and J. Pokorný. 2006. "Natural antioxidants from herbs and spices." *European Journal of Lipid Science and Technology* 108 (9):776–793. doi: 10.1002/ejlt.200600127.

Yashin, A., Y. Yashin, X. Xia, and B. Nemzer. 2017. "Antioxidant activity of spices and their impact on human health: A review." *Antioxidants* 6 (3):1–18. doi: 10.3390/antiox6030070.

Yetim, H., O. Sagdic, M. Dogan, and H. W. Ockerman. 2006. "Sensitivity of three pathogenic bacteria to Turkish Cemen paste and its ingredients." *Meat Science* 74 (2):354–358. doi: 10.1016/j.meatsci.2006.04.001.

Yetneberk, S., H. L. de Kock, L. W. Rooney, and J. R. N. Taylor. 2004. "Effects of sorghum cultivar on injera quality." *Cereal chemistry* 81 (3):314–321. doi: 10.1094/CCHEM.2004.81.3.314.

Yu, H., H. Liang, M. Ren, K. Ji, Q. Yang, X. Ge, B. Xi, and L. Pan. 2019. "Effects of dietary fenugreek seed extracts on growth performance, plasma biochemical parameters, lipid metabolism, Nrf2 antioxidant capacity and immune response of juvenile blunt snout bream (Megalobrama amblycephala)." *Fish & Shellfish Immunology* 94:211–219. doi: 10.1016/j.fsi.2019.09.018.

Żuk-Gołaszewska, K., and J. Wierzbowska. 2017. "Fenugreek: Productivity, nutritional value and uses." *Journal of Elementology* 22 (3):1067–1080. doi: 10.5601/jelem.2017.22.1.1396.

Part VI Regulatory Aspects

22 Fenugreek Based Products in USA, Australia, Canada, and India

Savita Nimse and Sanjeevani Deshkar

CONTENTS

22.1 Introduction 440
22.2 Regulatory Status of Fenugreek-Based Products in the USA.. 440
    22.2.1 Dietary Supplement Health and Education Act, 1994 (DSHEA)......................... 441
    22.2.2 Generally Recognized as Safe (GRAS)... 441
    22.2.3 Online Databases of the Dietary Supplements in the US Marketplace................. 441
    22.2.4 Fenugreek Monographs in Herbal Medicines Compendium................................ 442
22.3 Regulatory Status of Fenugreek-Based Products in Australia 442
    22.3.1 Australian Regulation of Complementary Medicines... 442
    22.3.2 TGA Business Services Ingredients (Australian Approved Names List for
Therapeutic Substances) 443
    22.3.3 TGA Business Services Indications for Listed Medicines 443
    22.3.4 ARTG Register/Database: Fenugreek as Medicinal Ingredient............................. 443
22.4 Regulatory Status of Fenugreek-Based Products in Canada 444
    22.4.1 Natural Health Product Regulation, 2004 444
    22.4.2 NNHPD Monograph.. 445
    22.4.3 Product Monographs.. 445
    22.4.4 Licenses Natural Health Product Database... 445
22.5 Regulatory Status of Fenugreek-Based Products in India... 446
    22.5.1 Drug and Cosmetic Act and Rules (DCA&R) for Ayurvedic, Siddha, Unani
(ASU) drugs... 446
        22.5.1.1 Ayurvedic, Siddha, or Unani Drug 446
        22.5.1.2 The *Ayurvedic Pharmacopoeia of India* and Fenugreek 447
        22.5.1.3 Fenugreek in Online Database in India 447
    22.5.2 Food Safety and Standards Act, 2006 447
    22.5.3 Food Safety and Standard and Regulations, 2016 447
22.6 Conclusions.. 448
References 448

ABBREVIATIONS

API: Ayurvedic Pharmacopoeia of India; **ARGCM:** Australian Regulatory Guidelines for Complementary Medicines; **ARTG:** Australian Register of Therapeutic Goods; **ASU:** Ayurvedic, Siddha, Unani; **DCA&R:** Drug and Cosmetic Act and Rules; **DIN-HM:** Homeopathic Medicine Number; **DSHEA:** Dietary Supplement Health and Education Act;

DSLD: Dietary Supplement Label Database; FDA: Food and Drug Administration; FSSAI: Food Safety and Standards Authority of India; GMP: Good Manufacturing Practice; GRAS: Generally Recognized As Safe; HMC: Herbal Medicines Compendium; NHP: Natural Health Products; NHPID: Natural Health Products Ingredients Database; NIH: National Institutes of Health; NNHPD: Natural and Non-Prescription Health Products Directorate; NPN: Natural Product Number; PCI: Pre-Cleared Information; PLA: Product License Application; TGA: Therapeutic Good Administration; TKDL: Traditional Knowledge Digital Library; USFDA: The United States Food and Drug Administration; USP: United States Pharmacopoeia

22.1 INTRODUCTION

Herbal medicines and dietary supplements are extensively utilized in maintaining healthy lifestyle and wellbeing. They are also the preferred option in the management of chronic diseases including cancer, cardiovascular diseases, diabetes, etc. (Costa et al. 2017; National Research Council Committee 1989). While the use of dietary supplements has increased substantially from the last two decades, the challenges towards their research and regulations are growing. The major challenges in dietary supplement research include concerns related to their quality, safety, and efficacy (Dwyer, Coates, and Smith 2018). The adulteration in dietary supplements or presence of contaminants may lead to serious adverse effects or health hazards. Serious interactions of dietary supplements with medications like antihypertensives or antidiabetics may lead to safety concerns (Sood et al. 2008).

As most of the dietary supplement products may contain multiple ingredients, establishing the efficacy of a product may be challenging. Clinical trials may be often inconclusive, irreproducible, or conflicting and the evidence from the literature may not be sufficient (Dwyer, Coates, and Smith 2018; Lim et al. 2019). The basic regulatory challenge in the case of dietary supplements is the lack of global consensus about the terminology and categorization of dietary supplements, natural health products, food supplements, or contemporary medicines in different countries. While in some countries the products are regarded as dietary supplements, some may categorize them as health products or medicines (Dwyer, Coates, and Smith 2018; Thakkar et al. 2020).

The regulatory framework for use of fenugreek as dietary supplements or herbal medicines remains inconsistent across jurisdictions internationally. Furthermore, the continuously changing regulatory framework over the globe is another challenge. Although fenugreek is a major constituent of many dietary supplements and herbal products in the market, its use as a dietary supplement or herbal medicine or natural health product is under the different regulatory frameworks of different countries in the world.

This chapter describes the regulatory framework and status of fenugreek-based products in the USA, Australia, Canada, and India. We have selected these four countries considering their acceptance and wide use of natural products as medicines and presence with established and regulatory frameworks. Although specific terminologies and documentations differ, the approach of regulatory bodies of other countries of the world are similar to either one or more of the four selected countries.

22.2 REGULATORY STATUS OF FENUGREEK-BASED PRODUCTS IN THE USA

Fenugreek is an acceptable ingredient for use as per the definition of dietary supplement due to its vast history of use as food, in dietary supplements (US FDA 2016a), and its GRAS status in the USA (US FDA 2016b, 2016c).

22.2.1 Dietary Supplement Health and Education Act, 1994 (DSHEA)

In the USA, the FDA regulates the dietary supplements under the Dietary Supplement Health and Education Act of 1994 (DSHEA) (US FDA 2016d). The FDA does not approve dietary supplements based on their safety and efficacy before they are marketed. Under DSHEA, only new dietary ingredients are required to be reviewed by the FDA for safety and other information. The manufacturers and distributors of the dietary supplements are responsible to ensure the compliance to DSHEA. The FDA is responsible for the post-marketing monitoring and takes actions if the product is found to be non-compliant to the regulation (US FDA 2016a). Under DSHEA the safety of any new product in the market starts with the safety of the ingredients used. The ingredients that have been recognized as a food substance and are present in the food supply or used in dietary supplements before October 15, 1994 can be used as a dietary ingredient in the dietary supplements without any approval.

22.2.2 Generally Recognized as Safe (GRAS)

Fenugreek is used in many dietary supplements in the market. Fenugreek (*Trigonella foenum graecum*) is generally recognized as safe (GRAS) for their intended use [Sec. 182.10 Spices and other natural seasonings and flavorings and Sec. 182.20 Essential oils, oleoresins (solvent-free), and natural extractives (including distillates)] by the USFDA (US FDA 2016b, 2016c). Magazines such as *NutraIngredients USA* and *Nutraceutical World* publish news articles regarding independent GRAS affirmation of the standardized fenugreek extract by companies for their fenugreek-related products and others. Self-determination of a substance's GRAS status is referred to as an "independent GRAS determination". There are two ways an ingredient can achieve GRAS status: (1) by submitting a GRAS notice to the FDA and receiving a letter with no questions from the FDA or (2) self-affirmation of GRAS status. A self-affirmed GRAS is done when a person or company has convened a panel of experts to assess the safety of a particular ingredient. For GRAS substances, there is no legal requirement for the FDA's review and approval of the scientific evidence for safety under the conditions of its intended use. However, some companies prefer to ensure that the FDA has reviewed its notice of a GRAS determination, without raising safety or legal issues, before marketing. There is no list or database for the ingredients for which independent GRAS affirmation is done. However, the companies who opt for independent GRAS affirmation do publicly disclose the self-affirmation on newsletters like *Nutraceutical World* or *NutraIngredients USA*.

22.2.3 Online Databases of the Dietary Supplements in the US Marketplace

The information on the dietary supplement products available in the US marketplace can be accessed on online databases such as the Dietary Supplement Label Database (DSLD) from the National Institutes of Health (NIH) (The Office of Dietary Supplements 2013) and the Supplement OWL, a self-regulatory initiative of the dietary supplement industry, spearheaded by the Council for Responsible Nutrition (The Supplement OWL® 2017). The USFDA maintains a list of ingredients that are GRAS for use in foods and a list of ingredients and products of concern but does not maintain a list of ingredients to be used in dietary supplements. The DSLD database can be searched by ingredients, by products, and by manufacturers.

The DSLD database is searched by ingredients with names containing fenugreek, and different categories (i.e. blend, botanical, and chemical) can be retrieved. The category botanical includes a plant, part of a plant, or extract of a plant/part. It also includes algae, fungi, molds, as well as ingredients that are not botanicals but are eaten as food (e.g. mushroom, seaweed, spirulina, chlorella, etc.). As per recent search (June 2021), the ingredient group ID 253, with group name "Fenugreek" and category "botanical" contains 920 products with Fenugreek/Methi/*Trigonella foenum-graecum* ingredients as follows:

- Certified organic extract/powder (65 products)
- Seed powder (82 products)
- Extract: total alcohol free extract (1 product), 4:1 extract (16 products), 6.5:1 extract (2 products), 25:1 extract (1 product), 50% extract (7 products)
- Concentrate/galactomannans/fiber (11 products), 4-hydroxyisoleucine (2 products)
- Extract/liposomal extract with trademarks/brands (e.g. Hyperox, Ferosap, FenuLife (346 products))
- Products with trademarks (e.g. Testofen® (33 products) or Testosurge® or Torabolic® (6 products), Fenupro/Fenusterols/promilin (14 products))
- No specific details for Fenugreek/Methi/*Trigonella foenum-graecum* (335 products)

22.2.4 FENUGREEK MONOGRAPHS IN HERBAL MEDICINES COMPENDIUM

Public quality standards help to ensure the quality of all medicines, including herbal medicines. One such standards is the Herbal Medicines Compendium (HMC), published by the US Pharmacopeial Convention (USP). HMC is a freely available online resource for quality standards for herbal ingredients used in herbal medicines.

Standards are expressed primarily in monographs, containing the general information such as the definition of the herbal ingredient, title, and specification. The specification contains tests for critical quality attributes of the ingredient with analytical test procedures and acceptance criteria of tests. However, the safety and efficacy of herbal articles are outside the scope of the USP's decision-making. The HMC contains the monographs for herbal articles whereas the United States Pharmacopeia and the Dietary Supplements Compendium contain information on botanical (herbal) dietary ingredients. These standards have considerable overlap but differ based on their intended use.

The following monographs of fenugreek have been approved by the USP Expert Committee to be included in the Herbal Medicines Compendium as Final Authorized Standards.

- *Trigonella foenum-graecum* seeds (Final authorized Version 1; posted date Jan. 31, 2014): The article consists of dried ripened seeds of *Trigonella foenum-graecum* L. (family Fabaceae). It contains not less than 0.2% of 4 hydroxy isoleucine, calculated on dried basis.
- *Trigonella foenum-graecum* seed dry extract (Final authorized Version 1; posted date Jan. 10, 2014): The article consists of dried ripened seeds of *Trigonella foenum-graecum* L. (family Fabaceae) by extraction with hydroalcoholic mixtures. The ratio of starting crude plant material to dry extract is between 5:1 and 4:1. It contains not less than 1% 4-hydroxy isoleucine, calculated on anhydrous basis.
- *Trigonella foenum-graecum* seed powder (Final authorized Version 1; posted date Jan. 10, 2014): The article consists of dried ripened seeds of *Trigonella foenum-graecum* L. (family: Fabaceae) reduced to a powder or very fine powder. It contains not less than 0.2% of 4-hydroxyisoleucine, calculated on dried basis.

22.3 REGULATORY STATUS OF FENUGREEK-BASED PRODUCTS IN AUSTRALIA

22.3.1 AUSTRALIAN REGULATION OF COMPLEMENTARY MEDICINES

In Australia, Regulatory Guidelines for Complementary Medicines (ARGCM) is regulated as medicines under the Therapeutic Goods Act 1989 (TGA). The Therapeutic Goods Act 1989 (the Act) and the Therapeutic Goods Regulations 1990 (the Regulations) are the legislative basis for a uniform framework of controls for the import, export, manufacture, and supply of complementary medicines in Australia.

Fenugreek is preapproved for use as an active ingredient in listed complementary medicine and considered to be in the low risk category. The complementary medicines (also known as "traditional" or "alternative" medicines) include vitamin, mineral, herbal, aromatherapy, and homoeopathic products. The TGA assess the risks and benefits during the evaluation, licensing of Australian manufacturers and verifying of overseas manufacturers' compliance with the same standards as their Australian counterparts, and post-market monitoring and enforcement of standards. Complementary medicines must be entered as either "registered" or "listed" or "listed assessed" medicines and must be "included" on the Australian Register of Therapeutic Goods (ARTG) before they may be supplied in or exported from Australia, unless exempted (Therapeutic Goods Administration 2021).

Products such as herbal complementary medicines receive a lesser degree of initial assessment because of lower risk. Detailed evaluations of higher risk (registered) medicines and risk-benefit assessment balance need to be carried out before approval for use in Australia (regardless of registered or listed category in the ARTG). Additionally, it must be made in a licensed facility that conforms to good manufacturing practices (GMP). TGA is authorized to initiate action in case of any problem with a medicine, device, or manufacturer, which can result in regulatory actions such as continued detailed monitoring to withdrawing the product from the market (Therapeutic Goods Administration 2021).

22.3.2 TGA Business Services Ingredients (Australian Approved Names List for Therapeutic Substances)

The TGA provides access to information about ingredients that are permissible to be used in the products. The TGA Business Services website has a searchable database of approved terminology for chemical, biological, and herbal ingredients, including the active ingredients, excipients, and components and equivalents of ingredients that may be used in TGA-regulated therapeutic goods. This database for ingredients is called the "Ingredients table". There are two types of ingredients, active and excipients. An active ingredient is the therapeutically active component in a medicine's final formulation that is responsible for its physiological action. An excipient ingredient is any component of a finished dosage form, other than an active ingredient. The ingredient searched in the ingredient table gives the "Ingredient summary" which provides the approved role of the ingredient (active or excipient). Fenugreek (the ingredient ID 59791) is summarized for use as "an active ingredient" in categories such as Export Only, Listed Medicines, Over the Counter, or Prescription Medicines. It is also available for use as a "Homoeopathic Ingredient" in "Listed Medicines" and for use as an "Excipient Ingredient" in categories of Export Only, Listed Medicines, Over the Counter, or Prescription Medicines.

22.3.3 TGA Business Services Indications for Listed Medicines

The TGA also provides access to information on the list of indications that are permissible to be used in the listed medicines. The TGA Business Services website has a searchable database for permissible indications known as "the Indications for Listed Medicines Table". This table provides a list of indications that can be used in listed medicine applications. These preapproved indications can be used in the listed medicine application.

22.3.4 ARTG Register/Database: Fenugreek as Medicinal Ingredient

When fenugreek is searched on the Australian Register of Therapeutic Goods (ARTG) with search criteria as products containing fenugreek as medicinal ingredient, a total of 81 products are found in the database. These 81 products contain either fenugreek alone as the active ingredient or multiple active ingredients containing fenugreek as one of the active ingredients. There are four products in the ARTG with fenugreek as the only active ingredient. As fenugreek is a preapproved ingredient in the TGA ingredient database, all products with fenugreek as an "active ingredient" in the ARTG are low risk listed medicines (Therapeutic Goods Administration 2020).

The medicines with fenugreek in capsule dosage forms are listed in the ARTG database as follows:

- Microgenics Fenugreek 2000: Indications based on traditional use in Chinese medicine (to dissipate coldness, to clear cold-dampness, to warm and nourish/enrich kidneys, to decrease abdominal pain/discomfort); traditional use in Ayurvedic medicine (to promote healthy appetite, to relieve loss of appetite); and traditional use in Western herbal medicine (to improve healthy appetite, to relieve symptoms of indigestion/dyspepsia, and to be used as a galactagogue/lactogogue to enhance breast milk production in breastfeeding women).
- Nature's Sunshine Fenugreek: Indications based on use in Western herbal medicine (to demulcent/soothe irritated tissues, to reduce occurrence of symptoms of indigestion/dyspepsia, to use as a galactagogue/lactogogue to enhance/improve/promote/increase breast milk production, and to expectorate/clear respiratory tract mucous).
- Nature's Own Fenugreek 1000 mg: Indications based on traditional use in Western herbal medicine (to maintain/support healthy appetite, to maintain/support healthy digestion, to be used as a galactagogue/lactogogue in order to enhance/improve/promote/increase breast milk production, to decrease/reduce excess mucous); traditional use in Chinese medicine (to decrease/reduce/relieve abdominal pain/discomfort); and traditional use in Ayurvedic medicine (to soothe gastro-intestinal tract mucous membranes, to soothe respiratory tract mucous membranes/mucous tissue).
- Andropique®: Indications based on modern scientific evidence (to maintain healthy body fat/muscle composition in male athletes, to maintain/support testosterone formation/synthesis in healthy males, and maintain/support testosterone level in healthy males).

22.4 REGULATORY STATUS OF FENUGREEK-BASED PRODUCTS IN CANADA

22.4.1 Natural Health Product Regulation, 2004

Products such as probiotics, herbal remedies, vitamins and minerals, homeopathic medicines, traditional medicines (e.g. traditional Chinese medicines), amino acids, and essential fatty acids are defined as natural health products (NHP). All NHPs sold in Canada need to be compliant with Natural Health Products Regulations, which came into force on January 1, 2004. Fenugreek is listed as a NHP ingredient in the Natural Health Products Ingredients Database (NHPID) of Health Canada as a medicinal ingredient.

NHPs are over-the-counter products, do not need a prescription to be sold, and need to be safe. In addition, NHP products must have a product license, and the licensed Canadian site to manufacture, package, label, and import. To obtain product and site licenses, compliance to GMP and specific labeling and packaging requirements and proper safety and efficacy evidence for the product must be provided (Health Canada 2021a). The safety and efficacy of NHP and their health claims need to be supported by scientific evidence, which may include clinical trial data or references to published studies, journals, pharmacopoeias, and traditional resources. On assessment of safety, efficacy, and quality, Health Canada can issue a product license along with an eight-digit Natural Product Number (NPN) or Homeopathic Medicine Number (DIN-HM), which must appear on the label.

The Natural Health Products Ingredients Database (NHPID) is a source of medicinal and non-medicinal ingredients for use in NHP. It is a key component of the NHP's online listing of monographs, pre-cleared information (PCI), single ingredient monographs, and product monographs. The term PCI under Health Canada is defined as any form of information that supports the safety, efficacy, or quality of a medicinal ingredient as determined to be acceptable as NHP by NHPID.

PCI is a useful provision to speed up the evaluation of the NHP, and serves as a reliable source of product information for consumers. The Compendium of Monographs of ingredients is made available through the NHPID. The NHPID includes information related to ingredients that are not allowed or are restricted for use in NHP.

22.4.2 NNHPD Monograph

Natural and Non-Prescription Health Products Directorate (NNHPD) has developed and published a Compendium of Monographs that allows applicants to support the safety, efficacy, and quality of an NHP as part of their Product License Application (PLA). A monograph as recorded is a written description of particular elements on an identified ingredient or product.

There are three classes of applications, by use of NNHPD monographs. Class I applications are those that must comply with all of the parameters of an individual NNHPD monograph (exactly as worded in the monograph). Class II applications are general and traditional applications supported entirely by a combination of two or more NNHPD monographs. Class III applications are general, traditional, and homeopathic applications requiring full assessment (not captured in Class I or II) and include a few other scenarios (Health Canada 2019).

Fenugreek is included in single ingredient monographs "Fenugreek Oral" (Health Canada 2018a) and "Fenugreek Topical" (Health Canada 2018b). The products containing fenugreek as a medicinal ingredient can comply with the requirements of Class I applications of NNHPD monographs. Fenugreek Oral (Health Canada 2018a) and Fenugreek Topical (Health Canada 2018b) are single ingredient monographs in NHPID. The uses of fenugreek as an ingredient in herbal medicine as per NHPID for oral route of administration are to help stimulate the appetite, to aid digestion as a digestive tonic, to help as a mild laxative to relieve dyspepsia and gastritis, to act as an expectorant to help relieve excess mucous of the upper respiratory passages (anticatarrhal), to serve as a nutritive tonic, to act as a galactagogue/lactogogue to help promote milk production/secretion, to be supportive therapy for the promotion of healthy glucose levels, to help reduce elevated blood lipid levels (hyperlipidemia). At the same time, the uses of fenugreek as an ingredient of herbal medicine for topical application are to help heal minor skin wounds, burns, irritations, and local inflammations.

22.4.3 Product Monographs

A monograph is a written description of particular elements on an identified topic. Single ingredient monographs apply to formulations containing only one medicinal ingredient. The term "Product monograph" is used for products with more than one medicinal ingredient or to outline the conditions of use based on a product category. For example, fenugreek as an ingredient has a product monograph under the category antioxidants, where use of fenugreek is mentioned as a source of antioxidant(s)/provides antioxidant(s) that help(s) fight/protect (cells) against/reduce (the oxidative effect of/the oxidative damage caused by/cell damage caused by) free radicals. (Health Canada 2021b).

22.4.4 Licenses Natural Health Product Database

The Licensed Natural Health Products Database is the source and listing of all the NPHs that are licensed by Health Canada. The search in this database in June 2021 yielded 848 products licensed since the year 2005, containing "Fenugreek as medicinal ingredient" used alone or with other herbs in multiherbal products. Out of the total 848 products, 156 products contain fenugreek as the sole active medicinal ingredient. The dosage form of these products is either oral (capsule, tablet, tincture, herbal tea, liquid extract, powder for suspension, drops, granules) or topical (cream).

This Health Canada's Licensed Natural Health Products Database gives one a glance of the broad use of fenugreek as a NHP. The indications for these products are mainly based on traditional use: to increase appetite and to be taken as an orexigenic, to stimulate the intestine; to relieve respiratory irritation; to help as a mucilage in catarrh of the upper respiratory passages; to assist as an expectorant, as a mucilaginous demulcent, as an antipyretic to reduce fevers; to treat chronic affections of the stomach and bowels; to be used as a laxative, as an anti-pellagra, as treatment for antiphlogistic action; to treat dyspepsia and gastritis; to help promote milk production/secretion as a galactagogue/lactogogue.

Some of the fenugreek-based products have the following indications: helps support testosterone production in men, helps support normal levels of free testosterone in men, helps support libido (sexual desire/drive/urge) in men, helps support sexual arousal in healthy women, helps improve sexual health including sexual cognition and behavior in women, helps reduce symptoms of age-related androgen decline, helps increase total and free testosterone levels (within the normal range), and helps increase libido, desire, and arousal for adult males suffering from age-related androgen decline.

Some other indications for fenugreek-based products for use in herbal medicine are to: help reduce elevated blood lipid levels/hyperlipidemia, act as supportive therapy for the promotion of healthy glucose levels, help to increase women's sexual desire, help to support women's healthy sex drive and libido, and help to reduce menopausal symptoms such as hot flashes and night sweats.

22.5 REGULATORY STATUS OF FENUGREEK-BASED PRODUCTS IN INDIA

22.5.1 Drug and Cosmetic Act and Rules (DCA&R) for Ayurvedic, Siddha, Unani (ASU) drugs

In India, Drug and Cosmetic Act (D and C) 1940 and Rules 1945 are the regulatory provisions for Ayurveda, Unani, Siddha medicines (Ministry of Health and Family Welfare- Government of India 1940, 1945). At the same time, the Ministry of Ayurveda, Yoga, Naturopathy, Unani, Siddha, Sowa-Rigpa and Homoeopathy (AYUSH) is the regulatory authority with the mandate to provide manufacturing licenses to market the herbal drugs.

The D and C Act extends the control over licensing, manufacture, labeling, packing, and quality of Ayurvedic products. Schedule "T" of the act describes the GMP requirements for the manufacturer of herbal medicines. The official pharmacopoeias and formularies are available for the quality standards of the medicines.

The first schedule of the D and C Act has listed authorized texts, which have to be followed for licensing any herbal product under the two categories (i.e. ASU drugs and Patent or proprietary medicines). The Indian system of registration of herbal medicines has provision in each state via state drug licensing authority of Ayurvedic, Siddha and Unani. The safety and acceptability of new products in the marketplace starts first with the ingredients.

Fenugreek is acceptable as a safe ingredient based on its history of use and listing in authoritative books and regulations for use in Ayurvedic medicines in India.

There are many Ayurvedic products containing fenugreek as the active medicinal ingredient and available on online shopping websites and platforms. For example, SugaHeal (to maintain healthy blood sugar level and glycated hemoglobin (HbA1c), reduces external insulin dosage), Zandu Methi Capsules (to help flush out toxins from the body and maintain healthy sugar levels), Bixa Botanical Methi Powder (to aid as blood sugar controller, nutritive, to increase hemoglobin and to help with kidney stones), Heera Ayurvedic Research Foundation Fenugreek Extract (to promote lactation, reduce cholesterol levels, help with appetite control), Satvayush Khadi Fenugreek (to be used as massage oil for hair and body), and many more.

22.5.1.1 Ayurvedic, Siddha, or Unani Drug

Ayurvedic, Siddha, or Unani drugs include all medicines manufactured exclusively in accordance with the formulae described in the authoritative books of Ayurvedic, Siddha, and Unani Tibb systems of medicine, specified in the First Schedule of Drug and Cosmetic Act (D and C) 1940 and Rules 1945.

22.5.1.2 The *Ayurvedic Pharmacopoeia of India* and Fenugreek

The *Ayurvedic Pharmacopoeia of India* (API) is the authoritative book listed in Schedule 1 of the Drug and Cosmetic Act (D and C) 1940 and Rules 1945 with a collection of standards and quality requirements of referred drugs that are manufactured, distributed, and sold by Ayurvedic license. The monographs on fenugreek are included in the API as Methi, Methi-hydroalcoholic extract, and Methi-water extracts with therapeutic uses of fenugreek for tastelessness, malabsorption syndrome, fever, increased frequency, and turbidity of urine (API 2016).

22.5.1.3 Fenugreek in Online Database in India

The Traditional Knowledge Digital Library (TKDL) is a database containing codified literature from Indian Systems of Medicine (viz., Ayurveda, Unani, Siddha, Sowa Rigpa as well as Yoga) available in the public domain. The inclusion of medicine in the TKDL is based on traditional knowledge from the existing literature that has been converted into a digitized format from the existing Indian (Sanskrit, Urdu, Arabic, Persian, and Tamil) and five international (English, German, Spanish, French, and Japanese) languages. Although TKDL is not the primary source or prior art, any evidence that is already known, in itself, it is a useful tool to understand the codified knowledge in the Indian systems of medicine (Ayurveda, Siddha, Unani, and Yoga) as prior art.

The online search of TKDL for mentions of fenugreek yielded 12 traditional formulations as treatment options for a range of diseases like acute diarrhea, indigestion, dyspepsia, loss of appetite, gastroenteritis, cholera, malabsorption syndrome, and treatment of scar marks after chickenpox and smallpox.

22.5.2 FOOD SAFETY AND STANDARDS ACT, 2006

In 2006, the Central Government established a body, Food Safety and Standards Authority of India (FSSAI), and enacted the Food safety and standards act, 2006 for laying down science-based standards for food articles and to regulate their manufacturing, storage, distribution, sale and import, for safe food for human consumption in India (FSSAI 2006). Section 22 of the Food safety and standard act 2006, defined the foods for special dietary uses or functional foods or nutraceuticals or health supplements.

22.5.3 FOOD SAFETY AND STANDARD AND REGULATIONS, 2016

In 2016, the Food Safety and Standard and Regulations included the definitions for "health supplements" and "nutraceuticals," and allowed the listing of the ingredients that may be used in such products (FSSAI 2016). These standards were intended to regulate Health Supplements, Nutraceuticals, Food for Special Dietary Use, Food for Special Medical Purpose, Functional Food and Novel Food (FSSAI 2016). Under this regulation, food supplements can be categorized under "Health supplements" or "Nutraceuticals" categories. The "Health supplements" are intended to supplement the diet with nutrients, whereas "nutraceuticals" provide the physiological benefits and help in maintaining optimum health (FSSAI 2016). The premarket approval for such listed product categories may not required. On the other hand, the ingredients that are not listed in the Food Safety and Standard and Regulations, 2016, need premarket approval.

Fenugreek is listed as the ingredient for the nutraceutical products under the Schedule VI (List of ingredients as Nutraceuticals) part B s.r. no. 178 (Fenugreek seeds extract). Thus, permission to manufacture/market/export nutraceuticals containing fenugreek seed extract as the ingredient can be obtained without any product approval. Therefore, many nutraceutical products containing fenugreek are the active ingredient available on online platforms, including Biotrex Nutraceuticals

Fenugreek (improving cholesterol levels), Andropique® (male testosterone booster, promotes fat loss), HealthyHey Nutrition Fenugreek Seed Extract (supplement for healthy glucose levels and beneficial for skin aging).

22.6 CONCLUSIONS

The regulatory approaches towards botanical food supplements differ significantly across jurisdictions of different countries. However, based on the history of the safe use of fenugreek in food and medicine, it has been classified as a low-risk/preapproved ingredient and can be used in dietary supplements/complementary medicines/herbal medicines/nutraceuticals. The regulatory status of fenugreek-based products under regulations such as DSHEA (USA), Complementary Medicine (Australia), Natural Health Product (Canada), and the D&C act and Food safety standards Act 2006 (India) is tabulated next:

Country	Notification/Registration	Ingredient Acceptability System	Fenugreek Status
USA	Notification/None	Generally Recognized as Safe, new/old dietary ingredients	GRAS and old dietary ingredient
Australia	Listing (notification)/registration	Preapproved list of ingredients	Preapproved for use as an active ingredient in listed medicine
Canada	Product license (registration)	Monograph	Fenugreek monograph (oral and topical)
India	Manufacturing license (registration)	Ayurvedic monographs/ingredients allowed under the nutraceutical regulation	Ingredient monograph in the authoritative book *Ayurvedic Pharmacopoeia of India* and an allowed ingredient under the nutraceutical regulation.

REFERENCES

API. 2016. "Ayurvedic pharmacopoeia of India, Volume IX- Hydro-alcoholic & water extracts." Pharmacopoeia commission for Indian Medicine (Ayurveda, Unani and Siddha) and Homeopathy, Government of India https://pcimh.gov.in/WriteReadData/RTF1984/1536816382.pdf (Accessed: June 15, 2021).

Costa, C., A. Tsatsakis, C. Mamoulakis, M. Teodoro, G. Briguglio, E. Caruso, D. Tsoukalas, D. Margina, E. Dardiotis, and D. Kouretas. 2017. "Current evidence on the effect of dietary polyphenols intake on chronic diseases." *Food and Chemical Toxicology* 110:286–299. doi: 10.1016/j.fct.2017.10.023.

Dwyer, J. T., P. M. Coates, and M. J. Smith. 2018. "Dietary supplements: Regulatory challenges and research resources." *Nutrients* 10 (1):41. doi: 10.3390/nu10010041.

FSSAI. 2006. "Food safety and standards act—2006." Government of India www.indiacode.nic.in/bitstream/123456789/7800/1/200634_food_safety_and_standards_act%2C_2006.pdf (Accessed: June 10, 2021).

FSSAI. 2016. "Food safety and standards (health supplements, nutraceuticals, food for special dietary use, food for special medical purpose, functional food and novel food) regulations." Ministry of Health and Family Welfare, Government of India https://fssai.gov.in/upload/uploadfiles/files/Nutraceuticals_Regulations.pdf (Accessed: June 10, 2021).

Health Canada. 2018a. "A monograph: FENUGREEK—TRIGONELLA FOENUM-GRAECUM—oral." Natural Health Products Ingredients Database http://webprod.hc-sc.gc.ca/nhpid-bdipsn/atReq.do?atid=fenugreek.oral&lang=eng (Accessed: June 8, 2021).

Health Canada. 2018b. "A monograph: FENUGREEK—TRIGONELLA FOENUM-GRAECUM—topical." Natural Health Products Ingredients Database http://webprod.hc-sc.gc.ca/nhpid-bdipsn/atReq.do?atid=fenugreek.topical&lang=eng (Accessed: June 8, 2021).

Health Canada. 2019. "Natural health products management of applications policy." Government of Canada www.canada.ca/en/health-canada/services/drugs-health-products/natural-health-products/legislation-guidelines/guidance-documents/management-product-licence-applications-attestations.html (Accessed: June 8, 2021).

Health Canada. 2021a. "Natural health products regulations." Government of Canada http://laws-lois.justice.gc.ca (Accessed: June 8, 2021).

Health Canada. 2021b. "Natural health products ingredients database." Health Canada http://webprod.hc-sc.gc.ca/nhpid-bdipsn/monosReq.do?lang=eng&monotype=single (Accessed: June 8, 2021).

Lim, Y. Z., S. M. Hussain, F. M. Cicuttini, and Y. Wang. 2019. "Nutrients and dietary supplements for osteoarthritis." *Bioactive Food as Dietary Interventions for Arthritis and Related Inflammatory Diseases*:97–137.

Ministry of Health and Family Welfare- Government of India. 1940. "Drug and cosmetic act and rules (D&C) 1940." (Accessed: June 15, 2021).

Ministry of Health and Family Welfare- Government of India. 1945. "Drug and Cosmetic Act and Rules (D&C) 1945." (Accessed: June 15, 2021).

National Research Council Committee. 1989. *Diet and Health: Implications for Reducing Chronic Disease Risk*. Washington, DC: National Academies Press.

The Office of Dietary Supplements. 2013. "The dietary supplement label database (DSLD)." National Institutes of Health, https://dsld.od.nih.gov/dsld/ (Accessed: May 31, 2021).

Sood, A., R. Sood, F. J. Brinker, R. Mann, L. L. Loehrer, and D. L. Wahner-Roedler. 2008. "Potential for interactions between dietary supplements and prescription medications." *The American Journal of Medicine* 121 (3):207–211.

The Supplement OWL®. 2017. "Council for responsible nutrition (CRN)." https://supplementowl.org/about-the-owl (Accessed: May 31, 2021).

Thakkar, S., E. Anklam, A. Xu, F. Ulberth, J. Li, B. Li, M. Hugas, N. Sarma, S. Crerar, and S. Swift. 2020. "Regulatory landscape of dietary supplements and herbal medicines from a global perspective." *Regulatory Toxicology and Pharmacology* 114:104647. doi: 10.1016/j.yrtph.2020.104647.

Therapeutic Goods Administration. 2020. "Overview of the regulation of listed medicines and registered complementary medicines, Australian regulatory guidelines." Australian Government www.tga.gov.au/sites/default/files/overview_of_the_regulation_of_listed_medicines_and_registered_complementary_medicines_0.pdf (Accessed: May 26, 2021).

Therapeutic Goods Administration. 2021. "Reference database of the TGA. Australian register of therapeutic goods (ARTG)." www.tga.gov.au/node/4050 (Accessed: May 26, 2021).

TKDL. "Representative database of Ayurvedic, Unani, Siddha and Sowarigpa formulations." Council of Scientific & Industrial Research (CSIR) and Ministry of Ayurveda, Yoga & Naturopathy, Unani, Siddha and Homeopathy (AYUSH) www.tkdl.res.in/tkdl/langdefault/common/Home.asp?GL=Eng (Accessed: June 15, 2021).

US FDA. 2016a. "Questions and answers on dietary supplements." www.fda.gov/food/information-consumers-using-dietary-supplements/questions-and-answers-dietary-supplements (Accessed: June 15, 2021).

US FDA. 2016b. "Generally recognized as safe (GRAS)." www.fda.gov/food/food-ingredients-packaging/generally-recognized-safe-gras (Accessed: June 15, 2021).

US FDA. 2016c. "CFR-code of federal regulations title 21. Chapter IB-food for human consumption, Part 182—Substances generally recognized as safe. Section 182.10 Spices and other natural seasonings and flavorings." U.S. Food and Drug Administration www.accessdata.fda.gov/scripts/cdrh/cfdocs/cfcfr/cfrsearch.cfm?fr=182.10 (Accessed: June 15, 2021).

US FDA. 2016d. "Dietary supplements." www.fda.gov/Food/DietarySupplements/ (Accessed: June 15, 2021).

Index

Note: Page numbers in *italics* indicate a figure and page numbers in **bold** indicate a table on the corresponding page.

4-hydroxyisoleucine
    antidepressant-like effects, 242
    antidiabetic effects, 62–63, *62*, 131
    antihyperglycemic effects, 287
    anti-inflammatory effects, 211, 212
    antioxidant effects, 211
    as chemical constituent, 4, 34, 54, *55*, 71, 94, 96, 97, 105, 130, 150, 188, 270
    cytotoxic effects, 341
    dyslipidemic effects, 63
    immunological, infectious, and malignancy-related effects, 343
    insulinotropic effects, 174–175, **175**, 198, 404
    as listed ingredient, 442
    nephroprotective effects, 286, 288
    neuroprotective efficacy, 247
    physicochemical properties, 364–365
    renal and pulmonary effects, 298

A

abdominal cramps, 7
acid precipitation extraction, 69
active compounds, 94–95, 97
acute lung disorder, 295–296
adjuvant-induced arthritis (AIA), 192
aerogels, 383–384
alanine, 58
albumins, 52, 53–54, 56, 57, 64, 68–69
alcoholism, 198
alkaline extraction/isoelectric precipitation, 68, 69, 71
alkaloids, 52, 94–95, 130, 150, 159, **161**, 188, 190, 191, 210, 212, 216, 218, 284, 306
allergenicity, 60–61, 70
allergic disorders, 324–325
aluminum toxicity, 230–231
Alzheimer's disease (AD), 209, 217, 230–231, 242–243, 245
amino acids, 4, 34, 51, 52–59, *55*, 62, 70, 95, 105, 130, 131, 150, 159, **161**, 212, 242, 341
anemia, 422–423
animal feed, 4, 7, 12, 103, 187, 419–420
animal milk production, 262, 419–420
animal studies, 131–133, **134–136**, 177–178
anti-aging properties, 199
antiapoptotic properties, 297
anti-arthritic properties, 189–193, 200, 326
antibacterial properties, 36, 198–199, 327, **328–329**, 373
anti-cancer properties, 4, 34, 105, 191, 197–198, 199, 200, 267, 333, 334, **335–339**, 342, 386
anti-cataract properties, 200
anti-diabetic properties, 8, 9, 13, 61, 62–63, 71, 78, 83, 96–97, 99, 110, 130, 133, 137, 177, 195–197, 210, 285, 286, 296, 312
anti-fibrotic properties, 297, 312
antifungal medications, 372–374
anti-hypercholesterolemic activities, 63, 296
anti-hypertriglyceridemic activities, 63
anti-inflammation properties, 4, 7, 9, 34, 105, 137, 174, 180–181, 189–193, 199, 211–212, 213, 266, 284, 286, 293, 295, 296, 297, 298, 305–306
antimicrobial properties, 198–199, 373
antinociceptive properties, 212–213, **235–236**
antioxidant properties, 7, 9, 35, 36, 37, 61–62, 78, 80–83, 87, 94–96, 99, 105, 130, 131, 133, 137, 174, 176, 178–179, 192, 194, 195, 199–200, 210, 214–217, 218, 220, 262, 266, 284, 286, 287, 290, 291, 296, 307, 312, 313, 322, 341, 371, 374, 385–386, 404–408, 415–416, 418, 420
anxiety disorders, 240–241, 245
aphrodisiac, 95, 99
apigenin, 4, 212, 270, 308, 312, 313–314
aqueous enzymatic extraction, 69
arachidic acid, 404
arginine, 4, 34, 57, 58, 94, 404
arthritis, 199–200
asparagine, 58, 404
aspartic acid, 58, 406
asthma, 296–298
atherosclerosis, 8, 34, 61, 63, 150, 159, 173, 194–195, 199, 210, 230
Australian Register of Therapeutic Goods (ARTG), 443–444
Australian regulation
    Australian Register of Therapeutic Goods, 443–444
    Australia Regulatory Guidelines for Complementary Medicines (ARGCM), 442–443, 448
    TGA Business Services Indications for Listed Medicines, 443
    TGA Business Services Ingredients, 443
    Therapeutic Goods Act 1989, 442–443
Australia Regulatory Guidelines for Complementary Medicines (ARGCM), 442–443, 448
autoimmune disorders, 325–327
Ayurveda
    Ayurvedic Pharmacopoeia of India, 447
    experiential uses of fenugreek per, 7, 22
    pharmacodynamics attributes of fenugreek per, 3, 5–6, 12–13
    regulatory provisions, 446–447
    therapeutics, 95, 104–105, 190, 210
    types of fenugreek per, 7, 12
Ayurvedic Pharmacopoeia of India (API), 447

B

bacterial infections, 327
bakery products, 3, 16, 57, 69–70, 97, 215, 407–412
barbari flatbread, 408
beverages, 57, 70
bhajji, 413
bhakri, 417
binding agent, 375–376
bioactive ingredient formulations

clinical studies, **367–371**
hair care formulations, 371–372
nano-formulations as antifungal medications, 372–373
nano-formulations for complex wound dressings, 373–374
skin care formulations, 371–372
topical formulations, 371–374
types, 367–374
bio-adhesive film/cream, 383
bioavailability enhancer, 377
biodegradable graft copolymer, 384
biodiversity, 27–28
biology, 4, 93
biscuits, 410, **411**, 422–423
blanching process effects, 407
blood cancer, 340
body composition, 106–107, 113–115
boiling process effects, 57–58, 405–406
bread, 407–408, **409**
breast cancer, 8, 12, 117, 188, 197, 259, 267–268, 341–342
breastfeeding, 4, 34, 95, 260–262, 422
burgers, 415

C

cake, 408–410
Canadian regulation
Licensed Natural Health Products Database, 445–446
Natural Health Product Regulation of 2004, 444–445, 448
NNHPD monographs, 445
product monographs, 445
cancer
breast cancer, 8, 12, 117, 188, 197, 259, 267–268, 341–342
cervical cancer, 4, 259, 267–268
colon cancer, 188, 342–343, 379
Ehrlich ascites carcinoma, 334
gastrointestinal cancer, 331
hepatocellular carcinoma, 340–341
human stomach cancer, 197
inflammation, 189
multiple cancerous cell lines, 333–334
oxidative stress, 178, 214
prostate cancer, 197
skin carcinogenesis, 334–340
T-cell lymphoma, 341
treatment, 22, 35, 61, 83, 95, 96, 176, 197–198, 210, 271, 320–321, 323, 332–343, 385–386, 440
cardiac fibrosis, 312–313
cardiometabolic risk factors, 180–181
cardiovascular diseases (CVDs), 63, 173, 178, 181
cardiovascular system, 34, 105, 130, 173–181
carotenoids, 96
carpaine, 4
cataracts, 200
cell culture studies, 131
centrally mediated pain, 234
central nervous system (CNS) disorders, 7, 12, 229–230, 238–239, 247
cerebral ischemia, 237
cervical cancer, 4, 259, 267–268

chapattis, 416
chemical-induced nephrotoxicity, 291–293
chemical-induced neurotoxicity, 230–231
chemotherapy-induced neuropathy, 217
childbirth, 4, 5, 95, 188
cholesterol regulation, 4, 9, 22, 34–36, 37, 38, 48, 63, 71, 97, 105, 150–152, 156–159, 162, 165, 174, 176, 193–195, 198, 230, 243, 262, 285, 404, 414, 419, 421–422, 446
chromosome numbers, 26–27
chronic kidney disease (CKD), 283–284
cisplatin, 291
clinical studies, 7–11, 130, 133–137, **138–140**, 162–165, 177–178, 190, 210–211, 284, 322–323, **367–371**
cognitive (learning and memory) disorders, 242–245
cognitive (learning and memory) disorders management, 266–267
colon cancer, 188, 342–343, 379
common names, 23, 26, 92, 103, 130, 188
complementary and alternative medicines (CAMs), 130
complex wound dressings, 373–374
controlled release beads, 381
controlled release tablets, 378–379
convulsive disorders, 234
cookies, 410
cooking process effects
blanching process, 407
boiling process, 57–58, 405–406
drying methods, 58, 64, 66
gamma irradiation treatment, 406–407
germination process, 404–405
microwave drying process, 406
refrigeration process, 406
roasting process, 405
soaking process, 405
storage conditions effects, 407
coumarins, 130, 188
c-peptide, 107, 181
creatine, 106, 113, 119–120
cultivation, 3, 4, 21, 23–24, **25**, 93, 99, 103
cumarins, 96
curcumins, 96
cyclooxygenase, 180, 190, 191, 192, 195, 198, 212, 231, 295, 306, 326, 372
cysteine, 58, 404

D

defatting, 58
depression, 241–242, 245
diabetes-induced CNS complications, 246–248
diabetes mellitus
animal studies, 131–133, **134–136**
anti-diabetic properties, 9, 62–63, 71, 96–97, 110, 130
antihyperglycaemic effects, 131–133
clinical studies, 7–11, 130, 133–137, **138–140**
controlling glucose, 4, 7–8, 12, 34, 95, 96–97, 99, 107, 129–137, **138–140**, 176–178, 195–197
food preparations, 420–422
inflammation, 209
types, 7–8, 62, 107, 129–137, **138–140**, 209

Index

use of complementary medicine, 130
diabetic nephropathy, 284–288, **289**, 312
diabetic neuropathy (DN), 211, 215–217, 220, 237
diarrhoea, 7
Dietary Supplement Health and Education Act of 1994 (DSHEA), 441, 448
Dietary Supplement Label Database (DSLD), 441
dietary supplements, 441–442
digestive system, 4, 7, 12, 59–60, 95, 130
dioscin, 343
diosgenin
    anti-cancer effects, 197, **335–340**, 343
    anti-diabetic effects, 196
    antihyperglycaemic effects, 132
    antihyperlipidaemic effect, 157–159
    anti-inflammation effects, 180, 191, 212, 213, 216–218, 220
    as chemical constituent, 34, 105, 131, 150, 188
    convulsive disorders effect, 234
    dysmenorrhea effects, 266
    extraction, 39, 47, 48
    glycemic status effects, 177
    hypoglycaemic and hypocholesterolaemic effects, 97
    lipid profile effects, 176
    management of female-specific cancers, 267
    management of menopause-induced osteoporosis, 267
    metabolic syndrome effects, 180
    neuroprotective effects, 213, 231, 232
    pharmacokinetic properties, 365
    physicochemical properties, 364
    testosterone and anabolic effects, 97
    use in milk production, 262
    vasoprotective effects, 178, 194
disease-modifying anti-rheumatic drugs (DMARDs), 190
distribution, 21, 23–24, 51
doli ki roti, 417
Doshas (humours), 6
Dravya, 6
Drug and Cosmetic Act and Rules (DCA &R), 446, 448
drying methods, 58, 64, 66, 406
dyslipidemia, 34
dysmenorrhea, 7, 265–266

E

Ehrlich ascites carcinoma (EAC), 334
emulsifying properties, 65–66
emulsion stabilizer, 376–377
endurance, 115–116
epigenin, 94
epilepsy, 234
ethnobotany, 103
ethnomedicinal uses, 4–5, **5**
excipients formulations
    aerogels, 383–384
    bio-adhesive film/cream, 383
    biodegradable graft copolymer, 384
    fiber, 377
    galactomannan, 374, 378–379, 381, 382, 383, 384
    gum, 375–377, 379–381, 382
    mucilage, 381, 382
    nanocomposite, 381–382
    nanoparticles, 384–386
    nanoparticulate, 382–383
    types, 367–374
extruded snacks, 413–415

F

fatty acids, 191, 287, 404
female-specific health conditions
    breast cancer, 8, 12, 117, 188, 197, 259, 267–268, 341–342
    breastfeeding, 4, 34, 95, 260–262, 422
    cancer, 267–268
    cervical cancer, 267–268
    cognitive decline, 266–267
    dysmenorrhea, 265–266
    menopausal stress, 239–240
    menopause-induced osteoporosis, 267
    polycystic ovarian syndrome, 10–11, 12, 105, 108, 110, 259, 262–264
    post-menopausal conditions, 9–10, 12, 264–267
    safety concerns, 269–271
    sexual function, 264
females sexual dysfunction (FSD), 264
Fenfuro™, 8, 107–108, 157
fenugreek glycoside supplementation (Fenu-FG), 9, 117, 271
fenugreek gum
    as binding agent, 375–376
    clinical studies, 211
    for colon-targeted drug delivery, 379–380
    as emulsion stabilizer, 376–377
    formulations, 374–382
    for liver-targeted oral delivery, 379
    for microencapsulation of probiotics, 380
    as mucoadhesive nasal gel, 382
    as natural super-disintegrant, 375
    obtaining, 103–104
    physicochemical properties, 362–364
    silica lipid drug delivery system, 380–381
    as a suspending agent, 377
fenugreek husk, 10
fenugreek leaves
    chemical composition, 52, **52**
    ethnomedicinal uses, 4
    external uses, 7
    internal uses, 214
    traditional use, 21
fenugreek oil, 179, 211, 239, 285–287, 333, 403
fenugreek proteins
    allergenicity, 60–61
    anti-nutritive compounds, 59
    application in food systems, 69–70
    biological effects, 61–63
    challenges and trends using, 70–71
    chemistry, 52–58, 188
    digestibility, 59–60
    functional properties in food systems, 63–67
    methods of extraction, 68–69
    nutritional value, 58–59
    structure, 52–56
    thermal properties, 67–68
fenugreek roots, 21
fenugreek seed extraction

beneficial effects, 99, 106–107, 115, 212, 213
novel extracts, 10–11, 107–109
phenolic compounds, **84–86**
protein extraction methods, 68–69
saponins, **84–86**
supercritical carbon dioxide extraction of phytochemicals, 35–48
fenugreek seed powder
  anti-cancer properties, 197
  anti-diabetic properties, 9
  anti-fibrotic properties, 313
  antihyperglycaemic properties, 132, 137
  anti-inflammation properties, 8–9
  antioxidant properties, 199–200, 214–217
  breastfeeding, 261
  central nervous system disorders, 247
  cognitive disorders management, 243
  colon cancer management, 343
  efficacy against acute lung disorder, 296
  efficacy against diabetes-induced renal dysfunction, 286
  efficacy against drug- or chemical-induced nephrotoxicity, 292
  efficacy against urolithiasis, 290
  female disorder management, 270
  food products, 410, 414, 415, 417, 421–422
  fungal infections use, 330
  growth-promoting activities, 321–322
  hepatoprotective properties, 310
  immunocompromised status use, 324
  immunomodulatory properties, 199–200
  management of hyperlipidaemic conditions, 151–152, **153**
  management of primary hyperlipidaemic conditions, **164–165**
  menopausal stress management, 239–240
  neuroprotective properties, 214–216, 218, 230–231, 248
  obtaining, 103
  polycystic ovarian syndrome management, 263
  structure, 54
fenugreek seeds
  cancer treatments, 4
  chemical composition, 52, **52**, 105, 130, 188
  digestibility, 60
  effect of processing conditions, 56–58
  ethnomedicinal uses, 4
  external uses, 7, 17
  hyperlipidaemia management, 151–165
  internal uses, 7–8, 17
  management of T2DM, 7–8
  phytochemical analysis, 214–215
  structure, 52–56, 130
  traditional use, 21
fenugreek supplementation, 98, 266
fenugreek (*Trigonella foenum-graecum* Linn)
  animal studies, 131–133, **134–136**
  anti-aging properties, 199
  antiapoptotic properties, 297
  anti-arthritic properties, 189–193, 200, 326
  antibacterial properties, **328–329**
  anti-cancer properties, 34, 105, 191, 199, 200, 267, 333, 334, **335–339**, 342, 386
  anti-cataract properties, 200

anti-diabetic properties, 8, 9, 13, 61, 62–63, 71, 78, 83, 96–97, 99, 110, 130, 133, 137, 176–178, 195–197, 285, 286, 296, 312
anti-fibrotic properties, 297, 312
antihyperglycaemic properties, 131–133
anti-inflammation properties, 4, 7, 9, 34, 105, 137, 174, 180–181, 189–193, 199, 211–212, 213, 266, 284, 286, 293, 295, 296, 297, 298, 305–306
antimicrobial properties, 198–199, 373
antinociceptive properties, 212–213
antioxidant properties, 7, 9, 35, 36, 37, 61–62, 78, 80–83, 87, 94–96, 99, 105, 130, 131, 133, 137, 174, 176, 178–179, 192, 194, 195, 199–200, 210, 214–216, 218, 220, 262, 266, 284, 286, 287, 290, 291, 292–293, 296, 307, 312, 313, 322, 341, 371, 373, 374, 385–386, 404–408, 415–416, 418, 420
autoimmune disorders, 325–327
in Ayurveda, 3–4, 5–7, 12–13, 104–105, 190, 210
bioactive ingredient formulations, 270, 367–374
biodiversity, 27–28
biology, 4, 93, 174–176
cardiometabolic risk factors, 180–181
cell culture studies, 131
cholesterol regulation, 4, 9, 22, 35–36, 37, 38, 48, 63, 71, 97, 105, 150–152, 156–159, 162, 165, 174, 176, 193–195, 198, 230, 243, 262, 285, 404, 414, 419, 421–422, 446
clinical studies, 7–11, 130, 133–137, **138–140**, 190–200, 210–211, 322–323, **367–371**
common names, 23, 26, 92, 103, 130, 188
controlling glucose in diabetes mellitus, 4, 7–8, 12, 34, 95, 96–97, 99, 107, 129–137, **138–140**, 176–178, 195–197
cooking process effects, 404–407
cultivation, 3, 4, 21, 23–24, **25**, 93, 99, 103, 187
ethnomedicinal uses, 4–5
excipients formulations, 374–386
as galactagogue, 7, 99, 105, 260–262
glycemic status, 176–178
growth-promoting activities., 321–323
health benefits, 105–107
hepato-protective effects, 198
history, 92
hyperlipidaemia management, 151–165
hypertension, 178
hypolipidemic effects, 193–195
immunomodulatory properties, 199–200
neuroprotective properties, 214–216, 218, 230–231, 248, 266
novel delivery technologies, 362–387
nutritional constituents, 94–95
nutritional properties, 403–404
organoleptic properties, 403
oxidative stress, 178–179
peripheral neuropathy management, 211–215
pharmacokinetic properties, 365–367
pharmacological attributes, 4, 7–11, **11**, 34, 105–107, 174, 189, 230
physicochemical properties, 362–365
phytoconstituents, 188, **210**, 211, 230, **238**, **245**, **261**, **268–269**, 270, 284, 288, **311**, 313–314, **313**, 362–365
plant chemistry, 34, 93–94, 105, 130, 188, 190

regulation in Australia, 442–444
regulation in Canada, 444–446
regulation in India, 446–448
regulation in USA, 440–442
safety concerns, 269–271, 284
sensory properties, 403
sports nutrition benefits, 105–107, 113–121
supercritical extraction of phytochemicals, 33–48
taxonomy, 26, 92
therapeutics, 95–97, 188–189
traditional use, 3, **5**, 15–18, 34, 95, 210, 230, 260, 284, 285
use in allergic disorders, 324–325
use in cancer, 4, 8, 12, 22, 33, 61, 83, 95, 96, 117, 176, 178, 188, 189, 197–198, 210, 214, 259, 267–268, 271, 320–321, 323, 331, 332–343, 379, 385–386, 440
use in chemotherapy-induced neuropathy, 217
use in diabetes-induced CNS complications, 246–248
use in diabetic nephropathy, **289**
use in diabetic neuropathy, 215–217
use in female-specific health conditions, 4, 34, 95, 259–271, *272*
use in fibrotic disorders, 305–314
use in food, 3, 7, 12, 16, 34, 57, 69–70, 95–97, **98**, 188, 260, 407–412
use in immunological disorders, 323–327
use in infectious diseases, 327–332
use in neurodegenerative disorders, 199–200, 209–220, 284–288
use in neurological disorders, 230–237, **238**
use in psychological disorders, 238–245, **245–246**
use in pulmonary diseases, 295–298
use in renal diseases, 283–295
FenuSMART™, 10
fibrotic disorders, 305–314
fish feed, 420
flatbreads, 408
flavonoids, 4, 94, 95, 96, 97, 174, 180, 188, 190, 191, 210, 242, 284, 306
flour, 57, 60, 69–70, 97, 215, 410–412, 413, 416–417
foaming properties, 65
food
  animal feed, 4, 7, 12, 103, 187, 419–420
  application in products, 95–97, **98**, 260
  bakery products, 3, 16, 57, 69–70, 97, 215, 407–412
  beverages, 57, 70
  cooking process effects, 404–407
  extruded snack preparations, 413–415
  functional food preparations, 420–423
  main course preparations, 416–417
  pickles, 417–419
  snack preparations, 412–413
  spices, 12, 22, 23, 34, 417–419, **418–419**
Food Safety and Standard and Regulations of 2016, 447–448
Food Safety and Standards Act of 2006, 447
free fatty acids (FFAs), 150
functional food preparations, 420–423
fungal infections, 330–331
Furocyst, 10, 108
Furosap™, 11, 108–109, 117
furostenol glycoside, 284, 366–367

G

galactagogue, 7, 99, 105, 260–262
galactomannan
  analgesic effects, 191
  anti-arthritic effects, 191
  anti-cancer effects, 342, 343
  anti-diabetic effects, 133
  anti-diabetic properties, 422
  anti-inflammation effects, 191
  antioxidant effects, 96, 192, 371
  body composition effects, 113–115
  breast cancer effects, 8
  as chemical constituent, 22, 52, 94, 130, 150, 174, 188, 260, 270, 403
  diabetes mellitus effects, 8
  enzymatic breakdown, 404
  as excipient, 374, 378–379, 381, 382, 383, 384
  in extruded snacks, 414
  fat accumulation prevention, 96
  glycaemic control activities, 137
  gut microbiota regulation, 132, 165
  hypolipidaemic effects, 156–157, **158**, 194
  immunostimulatory effects, 323
  nephroprotective effect, 286
  physicochemical properties, 362–364
  in spent hen patties, 415
  strength and endurance effects, 115–116
  use in food, 105
gamma irradiation treatment effects, 406–407
gastrointestinal cancer, 331
gelation, 66–67
gelicin, 94
generally recognized as safe (GRAS), 440, 441
genetic diversity, 27–28
gentamicin, 291–293
gentianine, 4, 188, 234
germination process effects, 56–57, 404–405
glibenclamide, 107
globulins, 53, 54, 56–57, 64, 68, 69
glucagon-like peptide-1 (GLP-1), 97, 181
glutamic acid, 58, 406
glutamine, 58, 404
glutelins, 52, 53, 56, 57
glycemic status, 176–178
glycine, 57, 58, 406
glycosides
  androgenic and anabolic effects, 107
  anti-cancer effects, 333, 334
  anti-inflammation effects, 211
  antinociceptive effects, 213
  antioxidant effects, 211
  bitter taste, 403
  bone tissue effects, 267
  as chemical constituent, 4, 83, 94, 188, 210, 284
  extraction, 36
  as female libido enhancer, 264
  fibrotic disorders effects, 306, 307, 310, 312, 313, 314
  hyperlipidaemia effects, 164
  nephrotoxicity effects, 291, 293, 295
  neuroprotective effects, 237, 242
  oxidative stress effects, 179
  physicochemical properties, 366–367
  pulmonary disorders effects, 99, 297, 298

resistance training effects, 9
safety concerns, 271
strength and endurance effects, 115
testosterone effects, 97, 117–119
growth-promoting activities., 321–323, 420
Guna (properties), 5
gut microbiota, 98

H

hair care formulations, 371–372
hay fever, 99
heart disease prevention, 99
heating process effects, 57–58
helminth infections, 331–332
hepatic fibrosis, 309–312
hepatocellular carcinoma, 340–341
hepato-protective effects, 198
herbalists, 15–16
herbal medicines, 130
Herbal Medicines Compendium (HMC), 442
herbal tea, 422
hexane, 58
histidine, 4, 34, 57, 58, 94, 404, 406
hormone imbalance, 5, 34, 110, 263
human immunodeficiency virus (HIV), 244
human stomach cancer, 197
hypercholesterolemia, 34, 35, 37, 95, 96–97, 105
hyperglycemia, 35, 96–97, 98, 130
hyperlipidaemia
    characteristics of, 150–151
    clinical studies, 8–9, 12, 37, 98, 162–165, 174, 284
    preclinical (*in vivo*) evidence, 151–162
hypertension, 35, 37, 178
hypolipidemic effects, 193–195

I

IBHB supplementation, 99, 233–234
idiopathic pulmonary fibrosis (IPF), 307
idlis flour, 413
immunity enhancement, 323
immunity-induced nephrotoxicity, 293–295
immunological disorders, 323–327
immunomodulatory properties, 199–200, 213–214
immunostimulants, 321, 323–324
Indian flatbread preparations, 416–417
Indian regulation
    Drug and Cosmetic Act and Rules, 446, 448
    Food Safety and Standard and Regulations of 2016, 447–448
    Food Safety and Standards Act of 2006, 447
    online databases, 447
    provisions for Ayurveda, Unani, Siddha medicines, 446–447
Indian snack preparations, 412–413
infectious diseases, 327–332
inflammation
    in adipose tissues, 132
    allergic disorders, 325
    atherosclerosis, 8, 194–195
    autoimmune disorders, 325–326
    in cancer, 321, 333, 341
    characteristics of, 180, 189–193

complex wounds, 374
diabetic neuropathy, 216–217
infectious diseases, 331
neurodegenerative disorders, 199–200, 209–220
neurological disorders, 231, 237
oxidative stress, 178–179
post-exercise, 119
psychological disorders, 243, 245
pulmonary diseases, 295
renal diseases, 286, 291
topical formulations, 4, 7, 372, 445
injera, 416
insulin-dependent diabetes mellitus, *see* type 1 diabetes
interfacial properties, 64–65
Iranian medicine, 15–18
isoleucine, 57, 58, 404, 406
isoorientin, 52, 94, 105
isovitexin, 4

K

kachori, 413
kaempferol, 188, 239
Kapha, 104
kidney issues, 34

L

lactation, 99
lavash flatbread, 408
L-Dopa therapy, 99
least gelation concentration (LGC), 66–67
leucine, 58, 404
leukemia, 340
Licensed Natural Health Products Database, 445–446
lignin, 96
linoleic acid, 41, 92, 98, 192, 212, 266, 287, 288, 404
linolenic acid, 41, 92, 212, 266, 287, 288
lipid profile, 173–176
lipoxygenase, 180, 190, 192, 195, 198, 212, 306, 326, 372
liver cancer, 340–341
liver issues, 34
liver-targeted oral delivery, 379
low molecular weight galactomannan (LMWGAL-TF), 216, 271
lung delivery, 381–382
lung fibrosis, 307
lutein, 366
luteolin, 4, 230, 234, 270
lysine, 4, 34, 58, 94, 404

M

main course preparations, 416–417
malignancy disorders, *see* cancer
masala parwal, 413
matthi, 412–413
meat patties, 415
menopausal stress, 239–240, 245, 263
menopause-induced osteoporosis, 267
metabolic function, 34, 98, 130
metabolic syndrome, 98, 108, 151, 180, 283, 288, 421
methi aloo, 416

Index

Methika, see fenugreek (*Trigonella foenum-graecum* Linn)
methionine, 58, 94, 404, 406
methi thepla, 416
microencapsulation of probiotics, 380
missi roti, 417
mood disorders, 241–242
mucoadhesive buccal patch, 382
mucoadhesive nasal gel, 382
muffins, 408
multiple cancerous cell lines, 333–334
multiple sclerosis (MS), 219, 237
myeloperoxidase (MPO), 192, 212, 218, 296, 307

N

nano-formulations, 372–374
nanoparticles, 384–386
National Institutes of Health (NIH), 441
Natural and Non-Prescription Health Products Directorate (NNHPD), 445
natural health product regulation (NHP), 444–445
Natural Health Product Regulation of 2004, 444–445, 448
natural super-disintegrant, 375
nephropathy, 220, 237, 284–288, **289**
neurodegenerative disorders, 199–200, 209–220, 284–288
neuroinflammation, 209, 217–219
neurological disorders, 6, 230–237, **238**
neuropathic pain, 237
neuroprotective properties, 214–216, 218, 230–231, 248, 266
nitric oxide synthase, 180, 190, 192, 232, 372
non-insulin dependent diabetes mellitus, *see* type 2 diabetes mellitus
noodles, 414
novel delivery technologies, 362–387
nutraceutical use, 34
nutritional properties, 94–95, 403–404

O

obesity, 34, 131, 132, 150, 154, 157, 165, 176, 180, 263, 265, 288, 309, 413, 421
occupational stress, 239, 245
ocular delivery, 382–383
oleanolic acid, 39, 47, 48
oleic acid, 92, 287, 288, 404
online databases
    Australian regulation, 443–444
    Canadian regulation, 445
    USA regulation, 441–442
organoleptic properties, 403
orientin, 4, 94, 140
osteoporosis, 35, 37, 267
oxidative stress, 8, 12, 97, 133, 178–179, **179**, 199, 213, 214, 215–216, 217, 218, 231–232, 237, 239, 243–248, 263, 287–288, 291–293, 306, 307, 310, 312, 322, 333, 334, 341, 342

P

pakoras, 413
palmitic acid, 41, 404
paratha, 417
Parkinson's disease (PD), 99, 209, 218–219, 231–234
pasta, 414
pastirma, 415
peptides, 61, 63, 70, 71, 94, 107
peripheral neuropathy (PN), 211–215
pharmacodynamics attributes, 5–6
pharmacokinetic properties, 365–367
pharmacological attributes, 4, 7–11, **11**, 34, 105–107, 174, 189, 230
phenolic compounds, 78, 80, 81, **84–86**, 94, 95
phenylalanine, 58, 406
physicochemical properties, 362–365
phytochemicals
    presence of, 22, 96, 188, 214–215, 234
    supercritical extraction of, 35–48
phytoconstituents, 188, **210**, 211, 230, **238**, **245**, **261**, **268–269**, 270, 284, 288, **311**, 313–314, **313**, 362–365
phytoestrogens, 239–240, 260, 262, 263, 264, 266–267, 341–342, 422
phytosterols, 34–35, 37, 38, 242, 334
pigs feed, 420
Pitta, 104
pizza, 410
plant chemistry, 34, 93–94, 105, 130
polycystic ovarian syndrome (PCOS), 10–11, 12, 105, 108, 110, 259, 262–264
polyphenol stilbenes, 131
polysaccharides, 97, 130
post-exercise muscle recovery, 118–119
post-menopausal cognitive decline, 266–267
post-menopausal conditions, 9–10, 12, 264–267
poultry feed, 420
product monographs, 445
prolamins, 52, 53, 56, 57, 60
proline, 57, 58, 406
prostate cancer, 197
protein efficiency ratio (PER), 59
protoalkaloids, 34
protodioscin, 34, 39, 47, 48, 106, 108–109, 268, 343
protozoal infections, 331
psychological disorders, 238–245, **245–246**
puff pastry, 412
pulmonary diseases, 283–284, 295–298, **298**
pulmonary fibrosis, 99, 307–308, *309*

Q

quercetin, 4, 52, 94, 188, 239

R

rabbit feed, 420
Rasa (taste), 5
refrigeration process effects, 406
regulation
    status of fenugreek-based products in Australia, 442–444
    status of fenugreek-based products in Canada, 446–448
    status of fenugreek-based products in USA, 440–442
renal diseases, 283–295
reproductive system, 7, 264, 290

respiratory system, 4, 95, 189, 283–284, 295–298
response surface analysis, 43–44
rhaponticin, 188
rheumatoid arthritis (RA), 189–190, 325–327
rheumatoid arthritis synovial fibroblasts (RASF), 189
roasting process effects, 405
RSM-CCRD analysis, 43, 45–46, 48
rutin, 52

S

safety concerns, 269–271, 284, 313–314
salt extraction (micellization), 34, 68, 69, 71
sapogenins, 4, 34, 35, 36–37, 38–40, 43–46, *44*, **47**, 48, 130, 267, 366
saponins
    analgesic effects, 234
    androgenic and anabolic effects, 117
    anti-cancer effects, 197, 333, 334
    antidepressant effects, 242
    anti-diabetic effects, 196
    antihyperglycaemic effects, 131, 137
    antihyperlipidaemic effects, 154, 157–159, **160**, 165
    anti-inflammation effects, 190, 191, 211, 212, 216, 217, 266, 306, 372
    antimicrobial effects, 373
    antioxidant effects, 211, 214
    anxiety disorders effects, 242
    application in growth and development, 321
    as chemical constituent, 4, 34–35, 52, 93–95, 105, 130, 174, 188, 210, 260, 284, 362, 403
    cognitive disorder effects, 243, 244
    extraction, 36–37, 78, 80, 81, 83–87, **84–86**, 107
    as female libido enhancer, 264
    female-specific cancer effects, 268
    hypoglycaemic and hypocholesterolaemic effects, 97
    hypolipidemic effects, 193
    immune function effects, 214
    immunostimulatory effects, 324
    infectious disease effects, 327
    lung disorder effects, 295
    metabolic syndrome effects, 180
    neuroprotective effects, 213
    pharmacokinetic properties, 366
    physicochemical properties, 364
    safety concerns, 270, 271
    strength and endurance effects, 115
    testosterone and anabolic effects, 97
sarsapogenin, 34, 39, 43, 47, 48
sausage, 415
sensory properties, 403
serine, 58, 406
seviyan, 412–413
Siddha drugs, 446–447
silica lipid drug delivery system, 380–381
sinusitis, 99
skin carcinogenesis, 334–340
skin care formulations, 371–372
skin diseases, 4
snack preparations, 412–413
soaking process effects, 405
soft drinks, 57, 70
solubility, 63–64
soluble fiber, 175, 180, 270, 342, 404, 420
species, 22–23, 93
spent hen patties, 414
sperm profiles, 11, 12, 110
spices, 12, 22, 23, 34, 321, 417–419, **418–419**
sports nutrition
    applications, 113–121
    benefits, 105–107
    body composition, 106–107
    creatine delivery, 119–120
    for endurance, 106–107, 115–116
    post-exercise muscle recovery, 118–119
    for strength increase, 106–107, 115–116
sprouts, 422–423
standardized fenugreek seed extract (SFSE-G), 307
standardized hydroalcoholic extract of fenugreek seeds (SFSE-T), 232–233
stearic acid, 41, 404
steroidal drugs, 34
steroidal sapogenins, 4, 34, 35, 36–37, 38–39, 40, 43, *44*, **47**, 48, 130, 150, 211
steroids, 4
sterols, 37–38, 40, *40*, **41**, 96, 287, 404
storage conditions effects, 407
strength increase, 115–116
supercritical extraction (SCE)
    analysis of phytochemicals, 38–39
    obtaining steroidal sapogenins rich extracts, 36–37
    of phytochemicals, 35–48
    of phytosterols, 37–38
    pre-treatment of plant seeds, 36–37
    response surface analysis, 43–44
    RSM-CCRD analysis, 43, 45–46, 48
    of sapogenins, 35, 36–37, 38–40, 43–46, **47**, 48
    sterols, 37–38, 40
    of unsaturated fatty acids, 38, 40–41, *42*, 48
    of vitamin D, 37–38, 48
    of vitamin E, 37–38, 48
Supplement OWL, 441
surface hydrophobicity ($H0$), 64
suspending agent, 377
sustained release tablets, 378–379
systemic sclerosis, 313

T

taxonomy, 26, 92
T-cell lymphoma, 341
terpenoids, 96
testosterone, 11, 12, 97, 105, 106, 107, 110, 116–118
TGA Business Services Indications for Listed Medicines, 443
TGA Business Services Ingredients, 443
therapeutic claims, 95–97
Therapeutic Goods Act 1989 (TGA), 442–443
thermostable extract of fenugreek seeds (TEFS), 308
threonine, 57, 58, 94, 404, 406
tigogenin, 34
tocopherols, 37
topical formulations, 371–374
toxicity, 270–271
toxicology, 11
Traditional Knowledge Digital Library (TKDL), 447

Index

traditional use, 3, **5**, 15–18, 34, 95, 210, 260, 284, 285
triacylglycerol, **41**, 194
Trigonella
    history of, 3, 103
    species of, 23, 93
Trigonella foenum-graecum Linn, *see* fenugreek
    (*Trigonella foenum-graecum* Linn)
trigonelline, 22, 95, 97, 105, 132, 133, 150, 159, 211, 216, 218, 232, 244–245, 267, 284, 286, 288, 343, 364, 365
triterpene saponins, 37, 38
triterpenoid sapogenins, 38, 39, 43, 48
tryptophan, 58, 94, 406
type 1 diabetes, 129, 209
type 2 diabetes mellitus, 7–8, 62, 107, 129–137, **138–140**, 209
tyrosine, 57, 58, 406

U

ultrafiltration extraction, 69
Unani drugs, 446–447
unsaturated fatty acids (UFA), 38, 40–41, **41**, *42*, 48, 94
upma, 413
urolithiasis, 288–291, **294**
ursolic acid, 39, 47, 48
USA regulation
    Dietary Supplement Health and Education Act of 1994, 441, 448
    generally recognized as safe (GRAS), 440, 441
    Herbal Medicines Compendium (HMC), 442
    online databases of dietary supplements, 441–442

V

vaginal applications, 383
valine, 58, 94, 406
Vanamethika (*Meliotus parviflora* Desf.), 7, 12
vasoprotective properties, 178, 194
Vata, 104
vector-borne diseases, 332
Veerya (potency), 5, 6
vegetables, 416
Vicenin-1, 366
Vipaka (biotransformation), 5, 6
vitamin A, 4, 94
vitamin B1, 4, 94
vitamin C, 4, 94, 199–200
vitamin D, 35, 37, *40*, 48, 94, 105
vitamin E, 35, 37, *40*, 48
vitamins, 96, 105
vitexin, 4, 52, 94, 105, 179, 188

W

waffles, 412
water (aqueous) extraction, 69

Y

yamogenin, 34

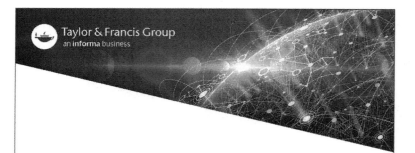

Made in the USA
Middletown, DE
07 September 2023

38146274R00263